D0081461

A Biologist's Guide to Mathematical Modeling in Ecology and Evolution

A Biologist's Guide to Mathematical Modeling in Ecology and Evolution

SARAH P. OTTO

and TROY DAY

Princeton University Press
Princeton and Oxford

Published by Princeton University Press, 41 William Street,
Princeton, New Jersey 08540
In the United Kingdom: Princeton University Press, 3 Market Place,
Woodstock, Oxfordshire
OX20 1SY

Library of Congress Cataloging-in-Publication Data
Otto, Sarah P., 1967-
A biologist's guide to mathematical modeling in ecology and evolution / Sarah P. Otto and Troy Day.
p. cm.
Includes bibliographical references (p.).
ISBN-13: 978-0-691-12344-8 (cl : alk. paper)
ISBN-10: 0-691-12344-6 (cl : alk. paper)
1. Ecology—Mathematical models. 2. Evolution (Biology)—Mathematical models.
I. Day, Troy, 1968- II. Title.

QH541.15.M3O88 2007
577.01′5118—dc22 2006044774

British Library Cataloging-in-Publication Data is available

This book has been composed in ITC Stone Serif

Printed on acid-free paper. ∞

press.princeton.edu

Printed in the United States of America

10 9 8 7 6 5 4 3 2 1

Contents

Preface

Why and How to Read This Book

Mathematics is a subject that most biologists take only in high school and in their first year of college or university. Yet many biologists later come to realize that greater familiarity with mathematics would serve them well: for understanding the papers that they read, for developing new theory to explore their ideas, and for generating appropriate models to interpret data. At that point, it can seem hard to reenter the world of mathematics. This book aims to be the guide to mathematics that biologists need.

Our goal is to teach why mathematics is so useful in biology, how to read and interpret equations, and how to construct and analyze new models. The book is fairly comprehensive, providing a lot of the background material needed to analyze a wide variety of models. Our focus is on developing ecological and evolutionary models that describe how biological systems change over time. That said, most of the techniques presented in the book can be readily applied to model many other phenomena, in biology as well as in other disciplines.

Constructing and analyzing models has been made easier for the average biologist by the advent of computer-aided mathematical tools, including *Mathematica*, Maple, and Matlab. Such packages are called symbolic software packages and are like fancy calculators that can work with symbols (e.g., x, e^x, $\sin(x)$) in addition to numbers. Best of all, these packages know more mathematical rules than any of us ever learned. Symbolic software packages allow biologists to focus on the lessons that a mathematical model can teach, by making many of the calculations involved in modeling much easier. Using these packages nevertheless requires an understanding of the mathematical concepts involved, and it is this conceptual understanding that we hope to foster through this book.

The first six chapters and Primer 1 introduce the art of constructing and analyzing a model. This material assumes that you have had elementary algebra and calculus, although we review techniques as needed. The next section (Chapters 7–12) covers more advanced models describing multiple classes of individuals and interactions among them. For this material, an understanding of how to handle matrices and vectors is necessary. Because most biologists have not had a course on matrices, Primer 2 introduces important concepts and methods of matrix algebra. We end the book with a section (Chapters 13–15) on the importance of chance events in biology and how to incorporate such probabilistic events into biological models. Again, because most biologists have not had a course on probability, Primer 3 introduces important concepts in probability theory.

Each chapter begins with a list of chapter goals, describing the major topics to be learned, as well as a list of key concepts, whose descriptions are highlighted in marginal boxes throughout the chapter. Important definitions are set apart from the text in highlighted boxes, as are key recipes and rules that can be used in the construction and analysis of models. Supplementary material for many of the chapters is available on the book website (http://press.princeton.edu/titles/8458.html), and provides greater depth of information for selected topics. At the end of each chapter we also provide a list of references that go beyond the material presented.

Mastering the mathematical techniques of this book requires practice, and therefore we have developed several problems at the end of each chapter. In addition, we have developed labs (available at the book website) that model important biological problems using the techniques described in this book, and we provide on-line files showing how to generate the figures of the text using the mathematical software package *Mathematica*.

For courses in mathematical modeling, we expect it would take a full year to cover all of the material of the text. Nevertheless, an excellent introduction to biomathematics can be gained from the first six chapters, perhaps with a subset of the material presented from the second half (e.g., Primer 2, Chapters 7, 8, and 10).

We are grateful for the advice, corrections, and words of wisdom received from a large number of students and colleagues. In particular, we thank Peter Abrams, Aneil Agrawal, Dilara Ally, Carl Bergstrom, Erin Cameron, Karen Choo, Andy Gardner, Aleeza Gerstein, Dick Gomulkiewicz, Jessica Hill, Ben Kerr, Simon Levin, Andy Peters, Daniel Promislow, Steve Proulx, Ben Roberts, Locke Rowe, Risa Sargent, Tom Sherratt, Jennifer Slater, Chrissy Spencer, Peter Taylor, Franjo Weissing, Geoff Wild, and a number of anonymous reviewers. We are especially grateful to Michael Whitlock for reading and providing detailed comments on the entire book. The Natural Sciences and Engineering Research Council of Canada provided financial support through both their Discovery Research Grant (to SPO and TD) and Steacie Fellowship (to SPO) programs. Special thanks are due to Patrick Phillips, who provided cheerleading when it was most needed, Mark Blows who provided TD with an ideal Australian habitat in the final stages of this book, and to Sam Elworthy, who steered us clear of several pitfalls. Finally, we are indebted to Laura Nagel and Michael Whitlock, whose unflagging support and encouragement allowed this book to be.

A Biologist's Guide to Mathematical Modeling in Ecology and Evolution

CHAPTER 1

Mathematical Modeling in Biology

Chapter Goals:

- To develop an appreciation for the importance of mathematics in biology

Chapter Concepts:

- Variables
- Dynamics
- Parameters
- Principle of parsimony

1.1 Introduction

Mathematics permeates biology. Unfortunately, this is far from obvious to most students of biology. While many biology courses cover results and insights from mathematical models, they rarely describe how these results were obtained. Typically, it is only when biologists start reading research articles that they come to appreciate just how common mathematical modeling is in biology. For many students, this realization comes long after they have chosen the majority of their courses, making it difficult to build the mathematical background needed to appreciate and feel comfortable with the mathematics that they encounter. This book is a guide to help any student develop this appreciation and comfort. To motivate learning more mathematics, we devote this first chapter to emphasizing just how common mathematical models are in biology and to highlighting some of the important ways in which mathematics has shaped our understanding of biology.

Let's begin with some numbers. According to BIOSIS, 886,101 articles published in biological journals contain the keyword "math" (including math, mathematical, mathematics, etc.) as of April 2006. Some of these articles are in specialized journals in mathematical biology, such as the *Bulletin of Mathematical Biology*, the *Journal of Mathematical Biology*, *Mathematical Biosciences*, and *Theoretical Population Biology*. Many others, however, are published in the most prestigious journals in science, including *Nature* and *Science*. Such a coarse survey, however, misses a large fraction of articles describing theoretical models without using "math" as a keyword.

We performed a more in-depth survey of all of the articles published in one year within some popular ecology and evolution journals (Table 1.1). Given that virtually every statistical analysis is based on an underlying mathematical model, nearly all articles relied on mathematics to some extent. With a stricter definition that excludes papers whose only use of mathematics is through statistical analyses, 35% of *Evolution* and *Ecology* articles and nearly 60% of *American Naturalist* articles reported predictions or results obtained using mathematical models. The extent of mathematical analysis varied greatly, but mathematical equations appeared in almost all of these articles. Furthermore, many of the articles used computer simulations to describe changes that occur over time in the populations under study. Such simulations can be incredibly helpful, allowing the reader to "see" what the equations predict and allowing authors to obtain results from even the most complicated models.

TABLE 1.1
Use of mathematical models in full-length journal articles

Journal (in 2001)	Number of articles	General use of models[a]	Specific use of models[b]	Equations presented[c]
American Naturalist	105	96%	59%	58%
Ecology	274	100%	35%	38%
Evolution	231	100%	35%	33%

[a]General use: Used a mathematical model in the broadest sense, including statistical or phylogenetic analyses with a mathematical basis (e.g., ANOVA, regression, etc.).
[b]Specific use: Used a mathematical model to obtain results (excluding cases that involve only statistical or phylogenetic analyses); the model may or may not be derived in the paper.
[c]Equations presented: Excluding standard statistical equations.

An important motivation for learning mathematical biology is that mathematical equations typically "say" more than the surrounding text. Given the space constraints of many journals, authors often leave out intermediate steps or fail to state every assumption that they have made. Being able to read and interpret mathematical equations is therefore extremely important, both to verify the conclusions of an author and to evaluate the limitations of unstated assumptions.

To describe all of the biological insights that have come from mathematical models would be an impossible task. Therefore, we focus the rest of this chapter on the insights obtained from mathematical models in one tiny, but critically important, area of biology: the ecology and epidemiology of the human immunodeficiency virus (HIV). As we shall see, mathematical models have allowed biologists to understand otherwise hidden aspects of HIV, they have produced testable predictions about how HIV replicates and spreads, and they have generated forecasts that improve the efficacy of prevention and health care programs.

1.2 HIV

On June 5, 1981, the Morbidity and Mortality Weekly Report of the Centers for Disease Control reported the deaths of five males in Los Angeles, all of whom had died from pneumocystis, a form of pneumonia that rarely causes death in individuals with healthy immune systems. Since this first report, acquired immunodeficiency syndrome (AIDS), as the disease has come to be known, has reached epidemic proportions, having caused more than 20 million deaths worldwide (Joint United Nations Programme on HIV/AIDS 2004b). AIDS results from the deterioration of the immune system, which then fails to ward off various cancers (e.g., Karposi's sarcoma) and infectious agents (e.g., the protozoa that cause pneumocystis, the viruses that cause retinitis, and the bacteria that cause tuberculosis). The collapse of the immune system is caused by infection with the human immunodeficiency virus (Figure 1.1). HIV is transmitted

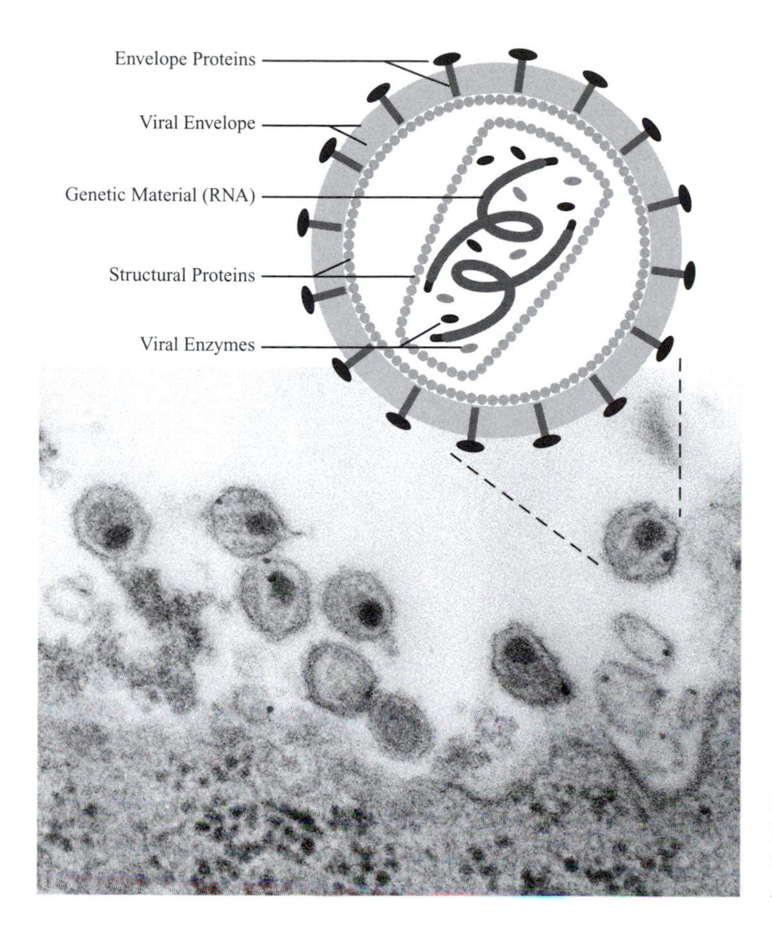

Figure 1.1: The human immunodeficiency virus. Electron micrograph shows HIV co-cultivated with human lymphocytes (courtesy of CDC; A. Harrison, P. Feorino, and E. L. Palmer).

from infected to susceptible individuals by the exchange of bodily fluids, primarily through sexual intercourse without condoms, sharing of unsterilized needles, or transfusion with infected blood supplies (although routine testing for HIV in donated blood has reduced the risk of infection through blood transfusion from 1 in 2500 to 1 in 250,000 [Revelle 1995]).

Once inside the body, HIV particles infect white blood cells by attaching to the CD4 protein embedded in the cell membranes of helper T cells, macrophages, and dendritic cells. The genome of the virus, which is made up of RNA, then enters these cells and is reverse transcribed into DNA, which is subsequently incorporated into the genome of the host. (The fact that normal transcription from DNA to RNA is reversed is why HIV is called a retrovirus.) The virus may then remain latent within the genome of the host cell or become activated, in which case it is transcribed to produce both the proteins necessary to replicate and daughter RNA particles (Figure 1.2). When actively replicating, HIV can produce hundreds of daughter viruses per day per host cell (Dimitrov et al. 1993), often killing the host cell in the process. These virus particles (or virions) then go on to infect other CD4-bearing cells, repeating the process. Eventually, without treatment, the population of CD4+ helper T cells declines dramatically from about 1000 cells per cubic millimeter of blood to about 200 cells, signaling the onset of AIDS (Figure 1.3).

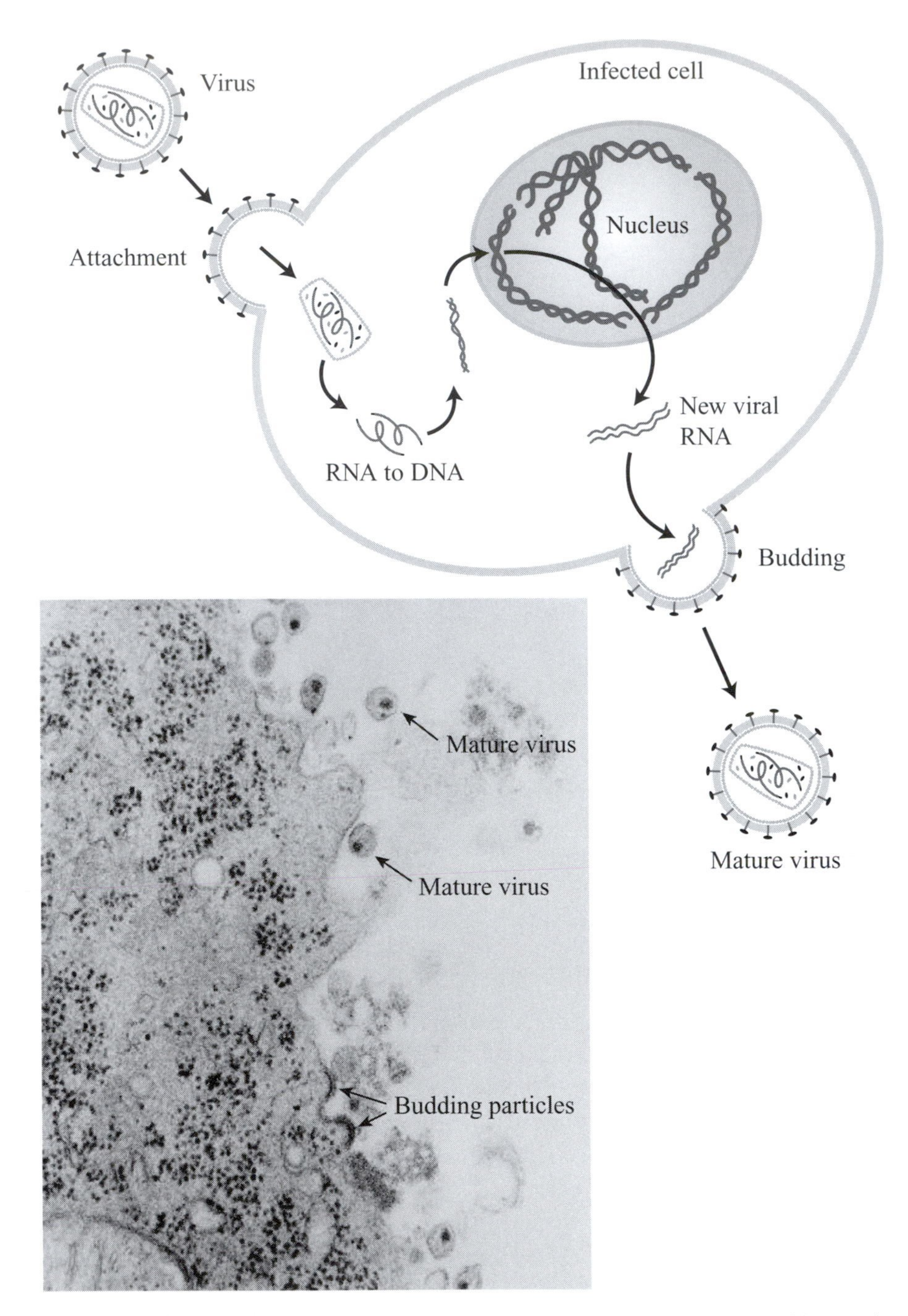

Figure 1.2: The life cycle of HIV within a host cell. Electron micrograph shows budding and mature HIV (courtesy of CDC; A. Harrison, E. L. Palmer, and P. Feorino).

Normally, CD4+ helper T cells function in the cellular immune response by binding to fragments of viruses and other foreign proteins presented on the surface of other immune cells. This binding activates the helper T cells to release chemicals (cytokines), which stimulate both killer T cells to attack the infected cells and B cells to manufacture antibodies against the foreign particles. What makes HIV particularly harmful to the immune system is that the

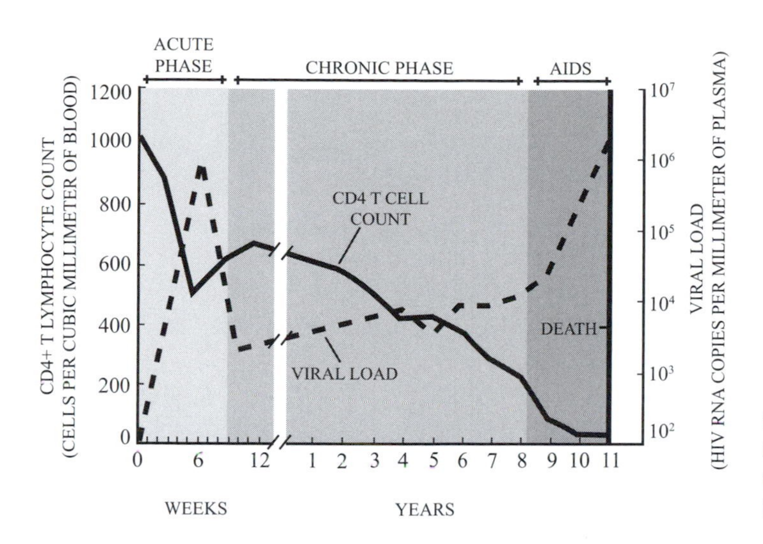

Figure 1.3: The time course of HIV infection within an individual. Viral loads and CD4+ T cell counts are plotted over time since infection. Based on data in Fauci et al. (1996).

virus preferentially attacks activated helper T cells; by destroying such cells, HIV can eliminate the very cells that recognize and fight other infections.

Early on in the epidemic, the median period between infection with HIV-1 (the strain most common in North America) and the onset of AIDS was about ten years (Bacchetti and Moss 1989). The median survival time following the onset of an AIDS-associated condition (e.g., Karposi's sarcoma or pneumocystis) was just under one year (Bacchetti et al. 1988). Survival statistics have improved dramatically with the development of effective antiretroviral therapies, such as protease inhibitors, which first became available in 1995, and with the advent of combination drug therapy, which uses multiple drugs to target different steps in the replication cycle of HIV. In San Francisco, the median survival after diagnosis with an AIDS-related opportunistic infection rose from 17 months between 1990 and 1994 to 59 months between 1995 and 1998 (San Francisco Department of Public Health 2000). Unfortunately, modern drug therapies are extremely expensive (typically over US$10,000 per patient per year) and cannot be afforded by the majority of individuals infected with HIV worldwide. Until effective therapy or vaccines become freely available, HIV will continue to take a devastating toll (Figure 1.4; Joint United Nations Programme on HIV/AIDS 2004a).

1.3 Models of HIV/AIDS

Mathematical modeling has been a very important tool in HIV/AIDS research. Every aspect of the natural history, treatment, and prevention of HIV has been the subject of mathematical models, from the thermodynamic characteristics of HIV (e.g., Hansson and Aqvist 1995; Kroeger Smith et al. 1995; Markgren et al. 2001) to its replication rate both within and among individuals (e.g., Funk et al. 2001; Jacquez et al. 1994; Koopman et al. 1997; Levin et al. 1996; Lloyd 2001; Phillips 1996). In the following sections, we describe four of these models in more detail. These models were chosen because of their implications

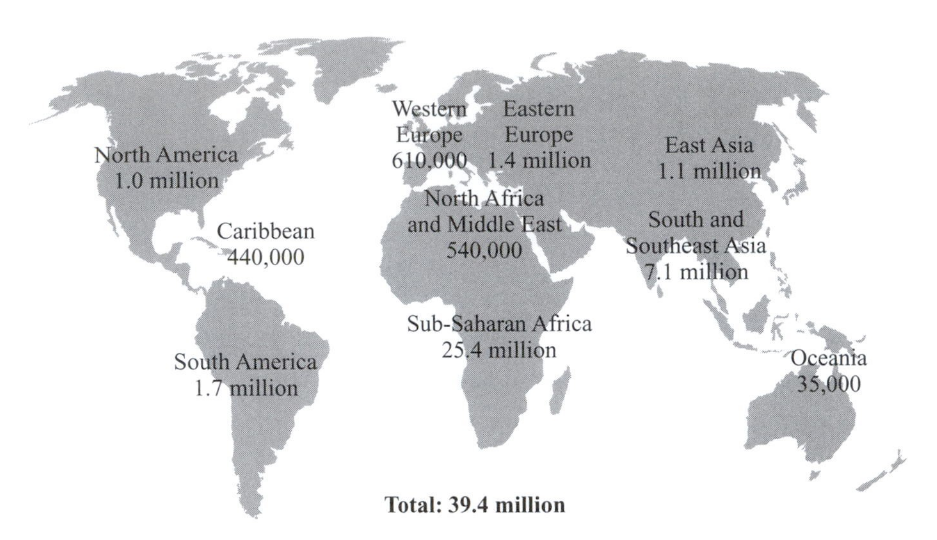

Figure 1.4: Number of individuals living with HIV. The number of adults and children estimated to be living with HIV is shown (Joint United Nations Programme on HIV/AIDS, 2004a).

for our understanding of HIV, but they also illustrate the sorts of techniques that are described in the rest of this book.

1.3.1 Dynamics of HIV after Initial Infection

After an individual is infected by HIV, the number of virions within the bloodstream skyrockets and then plummets again (Figure 1.3). This period of primary HIV infection is known as the acute phase; it lasts approximately 100 days and often leads to the onset of flu-like symptoms (Perrin and Yerly 1997; Schacker et al. 1996). The rapid rise in virus particles reflects the infection of CD4+ cells and the replication of HIV within actively infected host cells. But what causes the decline in virus particles? The most obvious answer is that the immune system acts to recognize and suppress the viral infection (Koup et al. 1994). Phillips (1996), however, suggested an alternative explanation: the number of virions might decline because most of the susceptible CD4+ cells have already been infected and thus there are fewer host cells to infect. Phillips developed a model to assess whether this alternative explanation could mimic the observed rise and fall of virions in the blood stream over the right time frame. In his model, there are four *variables* (i.e., four quantities that change over time): R, L, E, and V. R represents the number of activated but uninfected CD4+ cells, L represents the number of latently infected cells, E represents the number of actively infected cells, and V represents the number of virions in the blood stream. The *dynamics* of each variable (i.e., how the variable changes over time) depend on the values of the remaining variables. For example, the number of viruses changes over time in a manner that depends on the number of cells infected with actively replicating HIV. In the next chapter, we describe the steps involved in building models such as this one (see Chapter 2, Box 2.4).

A *variable* of a model is a quantity that changes over time.

The *dynamics* of a system is the pattern of changes that occur over time.

Phillips' model contains several *parameters*, which are quantities that are constant over time (see Chapter 2, Box 2.4). In particular, the death rate of

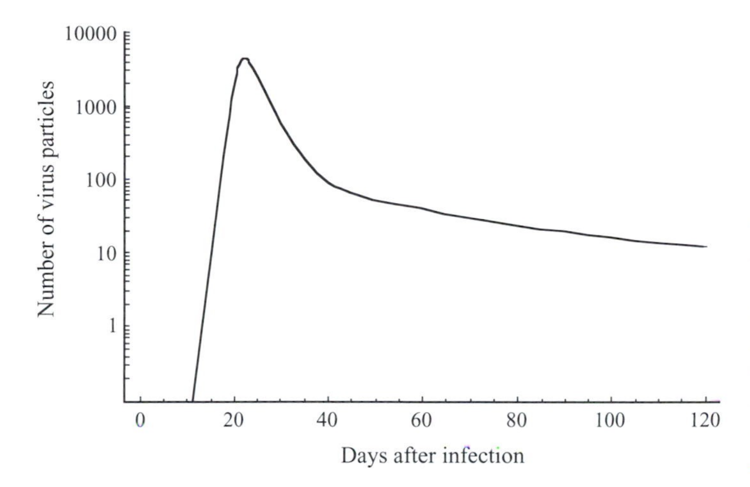

Figure 1.5: Number of virus particles in the blood stream. Based on the model and parameter values of Phillips (1996), the number of virions per mm^3 blood (V) is shown as a function of the number of days since primary infection (y-axis is plotted on a log-scale). Around a month after infection, the number of virus particles declines by about 100-fold even without any specific response by the immune system. See Box 2.4 in Chapter 2 for more details.

actively infected cells (δ) and the death rate of viruses (σ) are parameters in the model and are not allowed to change over time. (δ and σ are the lower-case Greek letters "delta" and "sigma." Greek letters are often used in models, especially for terms that remain constant ("parameters"). See Table 2.1 for a complete list of Greek letters.) Thus, Phillips built into his model the crucial assumption that the body does not get better at eliminating infected cells or virus particles over time, under the null hypothesis that the immune system does not mount a defense against HIV during the acute phase. To model the progression of HIV within the body, Phillips then needed values for each of the parameters in the model. Unfortunately, few data existed at the time for many of them. To proceed, Phillips chose plausible values for each parameter and numerically ran the model (a technique that we will describe in Chapter 4). The numerical solution for the number of virus particles, V, predicted from Phillips' model is plotted in Figure 1.5 (compare to Figure 1.3). Phillips then showed that similar patterns are observed under a variety of different parameter values. In particular, he observed that the number of virus particles typically rose and then fell by several orders of magnitude over a period of a few days to weeks. (An order of magnitude refers to a factor of ten. The number 100 is two orders of magnitude larger than one.)

A *parameter* of a model is a quantity that remains constant over time.

Phillips thus came to the counterintuitive conclusion that "the reduction in virus concentration during acute infection may not reflect the ability of the HIV-specific immune response to control the virus replication" (p. 497, Phillips 1996). The wording of this conclusion is critical and insightful. Phillips did not use his model to prove that the immune system plays no role in viral dynamics during primary infection. In fact, his model cannot say one way or the other whether there is a relevant HIV-specific immune response during this time period. What Phillips *can* say is that an immune response is not necessary to explain the observed data. This result illustrates an important principle in modeling: the *principle of parsimony*. The principle of parsimony states that one should prefer models containing as few variables and parameters as possible to describe the essential attributes of a system. Paraphrasing Albert Einstein, a model should be as simple as possible, but no simpler. In Phillips' case, he

According to the *principle of parsimony*, a simple explanation (or model) should be preferred over a complex explanation if both are equally compatible with the data.

could have added more variables describing an immune response during acute infection, but his results showed that adding such complexity was unnecessary. A simpler hypothesis can explain the rise and fall of HIV in the bloodstream: as infection proceeds, a decline in susceptible host cells reduces the rate at which virus is produced. Without having a good reason to invoke a more complex model, the principle of parsimony encourages us to stick with simple hypotheses.

Phillips' model accomplished a number of important things. First, it changed our view of what was possible. Without such a model, it would seem unlikely that a dramatic viral peak and decline could be caused by the dynamics of a CD4+ cell population without an immune response. Second, it produced testable predictions. One prediction noted by Phillips is that the viral peak and decline should be observed even in individuals that do not mount an immune response (i.e., do not produce anti-HIV antibodies) over this time period. Indeed, this prediction has been confirmed in several patients (Koup et al. 1994; Phillips 1996). Employing a more quantitative test, Stafford et al. (2000) fitted a version of Phillips' model to data on the viral load in ten patients from several time points during primary HIV infection; they found a good fit to the data within the first 100 days following infection. Third, Phillips' model generated a useful null hypothesis: viral dynamics do not reflect an immune response. This null hypothesis might be wrong, but at least it can be tested.

Phillips acknowledged that this null hypothesis can be rejected as a description of the longer-term dynamics of HIV. His model predicts that the viral load should reach an equilibrium (as described in Chapter 8), but observations indicate that the viral load slowly increases over the long term as the immune system weakens (the chronic phase in Figure 1.3). Furthermore, Schmitz et al. (1999) directly tested Phillips' hypothesis by examining the role of the immune system in rhesus monkeys infected with the simian immunodeficiency virus (SIV), the equivalent of HIV in monkeys. By injecting a particular antibody, Schmitz et al. were able to eliminate most CD8+ lymphocytes, which are the killer T cells thought to prevent the replication of HIV and SIV. Compared to control monkeys, the experimentally treated monkeys showed a much more shallow decline in virus load following the peak. This proves that, at least in monkeys, an immune response does play some role in the viral dynamics observed during primary infection. Nevertheless, the peak viral load was observed at similar levels in antibody-treated and untreated monkeys. Thus, an immune response was not responsible for stalling viral growth during the acute phase, which is best explained, instead, by a decline in the number of uninfected CD4+ cells (the targets of HIV and SIV).

1.3.2 Replication Rate of HIV

After the initial acute phase of infection, HIV circulates within the body at low levels until the onset of AIDS (Figure 1.3). These low levels suggest that virus particles might be produced at a low rate per day. This suggestion was, however, shown to be false using mathematical models in conjunction with experimental data (Ho et al. 1995; Nowak et al. 1995; Wei et al. 1995). According to the mathematical models, low numbers of virus particles can result from a low rate of viral production, P, or from a high rate of clearing virus from the body, c (Ho et al. 1995). Determining which of these possibilities is correct is not possible using only the observed number of virus particles in untreated patients. These landmark papers pointed out, however, that you can tease apart these possibilities using mathematical models that predict viral dynamics following the application of antiretroviral drugs (Ho et al. 1995; Nowak et al. 1995; Wei et al. 1995). For example, Ho et al. (1995) treated HIV-infected patients with ABT-538, an antiviral drug that effectively prevents HIV replication (at least in the short term). Thus, the experimental treatment reduced viral production P to zero, causing the viral load to plummet within the bloodstream. The rate at which the viruses decreased in frequency was consistent with a simple mathematical equation that we will encounter in Chapter 6 (equation (6.10b)). Fitting the mathematical model to the data allowed the authors to obtain an important and surprising result: virus particles were rapidly cleared from the body, with the half-life of HIV in plasma being only a couple of days. The authors thus inferred that the production rate of viruses must normally be enormous, on the order of a billion new viruses produced per day, in order to maintain HIV in the face of high clearance rates. Later work, using more precise experimental data and more detailed modeling, demonstrated that the turnover of HIV is even more rapid, with the half-life of HIV being less than a day and with over 10 billion viruses produced per day. This is a remarkable insight, as it was once thought that relatively little was happening during the chronic phase of HIV infection (Perelson et al. 1996).

These papers had an enormous impact on our understanding of HIV. One of the most important conclusions to follow from this work was that we must expect genetic diversity to be rapidly generated in HIV as a result of the high rate of viral production. If resistance to an antiviral drug requires a particular mutation, it is virtually guaranteed that this mutation will arise rapidly. Only combination drug therapies, requiring multiple mutations for resistance, have a long-term chance of success given the enormous evolutionary potential of HIV.

1.3.3 The Effects of Antiretroviral Therapy on the Spread of HIV

The specter of AIDS has softened following the development of effective antiretroviral therapies (ART), involving various drug combinations that have allowed people to live longer with HIV. Public health officials are concerned, however, that this respite will be short lived for two reasons: (a) people may be more inclined to engage in risky behavior knowing that ART exists and (b) HIV might evolve resistance to these drugs, causing the drugs to become ineffective. With these possibilities in mind, Blower et al. (2000) constructed a mathematical model to predict how drug therapy might affect the number of new cases of HIV and the number of deaths due to AIDS. Their model was tailored to data from the San Francisco gay community, where approximately 30% of men were infected with HIV (HIV+) and approximately 50% of these were taking combination ART.

In using their model to predict the future course of HIV, the authors pointed out that uncertainty exists in most model parameters, such as the effect of ART on survival and infectivity, as well as changes in the rate of risky behavior. The authors thus allowed the parameters to be drawn from a range of plausible values rather than assigning one specific number to each. They then drew each parameter from its range and used 1000 different combinations of parameters to predict the spread of HIV. We derive their model in Chapter 2 (Box 2.5) but for the sake of presentation, we simplify it slightly (the model presented by Blower et al. is no more difficult conceptually, it just takes longer to describe). Specifically, we ignore the evolution of resistance by HIV, and we set most parameters to the median value of the range used by Blower et al. (2000). Here, we focus our attention on the parameter describing changes in risky behavior within the gay community following the introduction of ART, and we allow this parameter to take on a range of possible values. Specifically, the average number of sexual partners with whom an uninfected individual (HIV–) has unprotected sex per year is allowed to increase from c before ART by a factor $1 + i$. We examine the impact of ART on risk-taking behavior, from causing no change ($i = 0$) to doubling the risk ($i = 1$). The variables that are tracked over time are the number of uninfected (X), infected but untreated (Y_U), and infected and treated (Y_T) gay men that are sexually active within San Francisco (see Box 2.5).

This model can be used to address the following questions (Blower et al. 2000): What percentage of AIDS deaths (relative to pre-ART levels) are averted by ART? What percentage of new HIV+ cases (relative to pre-ART levels) are averted by ART? How sensitive are these predictions to different choices of parameters, none of which is known with certainty?

The risks that people are willing to take generally depend on the perceived consequences; because antiretroviral therapies have been so effective at countering the progression of HIV and at lowering the infectiousness of HIV, gay men who are sexually active in San Francisco might be more willing to engage in sex without condoms (increasing i). Unfortunately, it is hard to predict exactly how sexual behavior will change. But it is possible, through a model, to investigate a range of possible changes in behavior. Figure 1.6 indicates that the effectiveness of ART depends strongly on these changes in behavior. 43% of the AIDS deaths that would have occurred within the gay population of San Francisco within the next ten years will be avoided by the use of ART if risky behavior does not increase, but only 24% of the AIDS deaths will be averted if risky behavior doubles. In fact, if the amount of risky behavior more than triples, then the model predicts *more* AIDS deaths with drug therapy than without!

A gloomier picture emerges if one looks at the number of new cases of HIV infection (Figure 1.6b). When antiretroviral therapy is first offered to a community, not all HIV+ individuals will immediately begin drug therapy. If half of HIV+ men start taking the drugs but if the amount of unprotected sex doubles ($i = 1$), then the number of new cases of HIV is predicted to *double* within the community following the beginning of antiretroviral therapy. This massive influx of new cases declines over time, however, as more HIV+ men enter treatment. This decline results from the fact that ART lowers the viral load within

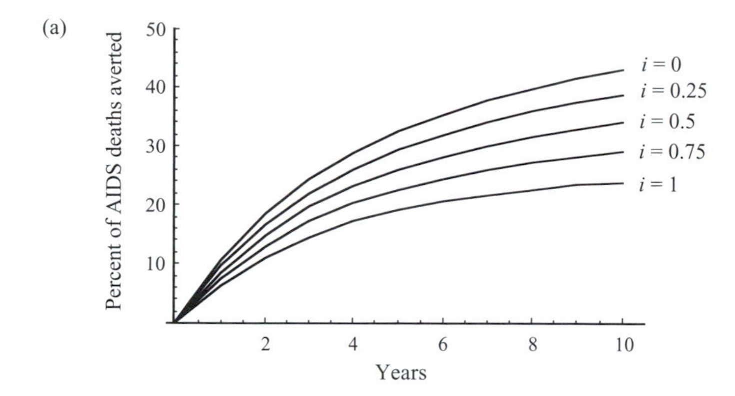

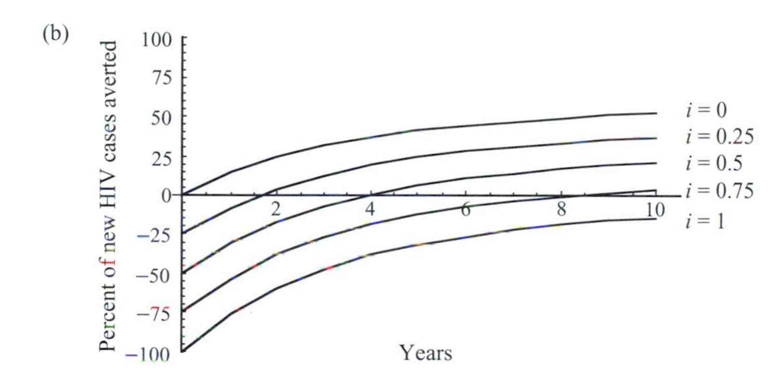

Figure 1.6: HIV, AIDS, and antiretroviral therapy. Percent reduction in the cumulative number of (a) AIDS deaths and (b) new HIV cases in the years following the onset of antiretroviral therapy (ART; modified from Blower et al. 2000). Each figure is drawn for various values of *i*, which measures the increase in risky behavior (unprotected sex) within the gay male community once ART becomes available, based on the equations developed in Box 2.5. A negative number on the *y*-axis implies that the cumulative number of cases increases not decreases.

the body, which decreases the probability that HIV will be transmitted through an exchange of bodily fluids. Over a ten-year period, the cumulative number of new cases of HIV is expected to be lower with ART than without as long as risky behavior increases by less than 80%.

Blower *et al.* also modeled the evolution of drug resistance by including two more variables to describe individuals who are infected with resistant strains of HIV and are either untreated (R_U) or treated (R_T). They allowed drug resistance to evolve at a certain rate in HIV+ individuals under ART and to affect most of the parameters in their model. The evolution of drug resistance led to a more pessimistic outlook, in terms of both the number of AIDS cases and the number of new infections, but the effect was relatively slight over the ten-year period that was considered. Over this time period, the rate at which HIV+ individuals enter into the treatment program and changes in risky behavior were much more important determinants of the forecasted number of HIV+ cases and deaths due to AIDS.

By exploring the future of HIV and AIDS within the San Francisco gay community, the model of Blower et al. (2000) identified a major health care problem: if gay men are willing to take greater risks following the introduction of ART, many more new cases of HIV and many more deaths from AIDS will result than if safe sex practices remain in place. Unfortunately, recent trends in San Francisco have indicated that the level of risky behavior has increased dramatically in recent years (San Francisco Department of Public Health 2000). The percentage of gay men reporting sex with multiple partners and unprotected intercourse rose from 23% in 1994 to 47% in 2000 according to the STOP AIDS project. During the same period of time, the percentage of gay men reporting using condoms all of the time fell from 70% to 50%. With such large increases in risky behavior, there has been an overall increase in HIV incidence, as predicted by the model of Blower et al. Among blood samples drawn at anonymous testing sites, the incidence of HIV increased from 1.3% in 1997 to 4.2% in 1999 (San Francisco Department of Public Health 2000). Similar increases in incidence have been observed in Ontario, Canada (Calzavara et al. 2002).

1.3.4 Estimating the Number of New Infections

Preventing the spread of HIV requires the coordination of social, educational, and medical services. In many countries these services are in very limited supply and must be targeted to the populations most at risk of infection. The level of risk of an individual cannot, however, be directly measured and must be inferred. In contrast, it is relatively straightforward to determine who is already infected with HIV, as long as blood samples from representative individuals can be tested. But these infected individuals may have harbored HIV for many years and may not represent the demographic group most at risk of becoming newly infected. Long-term longitudinal studies, where thousands of uninfected individuals are regularly tested over many years, are the best way to determine risk factors. These studies are costly, however, and take many years before the data become available for use in prevention programs.

Alternatively, mathematical models can be used to estimate the number of new infections from data on the number of currently infected individuals. To determine the age group most at risk of infection, Williams et al. (2001) modeled changes in the HIV status of women aged 14 to 49 in the rural district of Hlabisa in South Africa. We analyze this model in the on-line supplementary material to Chapter 10 (see Sup. Mat. 10.2). HIV has been spreading in South Africa at an exponential rate since the early 1980s, with the number of HIV-infected people doubling approximately every two years. In Hlabisa, a broad survey of HIV status was possible because almost all pregnant women (95%) in this region attend hospitals or clinics for prenatal care. During 1998, 3163 of these women gave blood samples that were tested for HIV. The *prevalence* of HIV among these women (i.e., the fraction of women that were HIV+) is shown in Figure 1.7a.

Williams et al. (2001) developed a model to estimate the risk of contracting HIV (see the on-line Supplementary Material to Chapter 10, Sup. Mat. 10.2). The more specific aim of the study was to estimate the probability per year that an uninfected woman in Hlabisa becomes newly infected with HIV, for women of various ages (this probability is known as the *incidence* of HIV). To estimate this unknown risk, their model related the probability of contracting HIV and the death rate from AIDS to the changes that should occur from year to year in the fraction of infected women. This study uses a model in a fundamentally different way from the previous examples. In all three previous examples, models were used to project into the future, describing how a population (of cells or of individuals) was expected to change over time given various parameters. In this

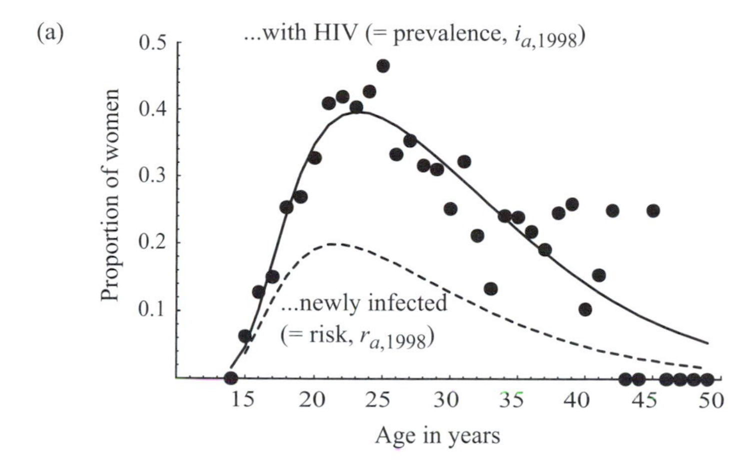

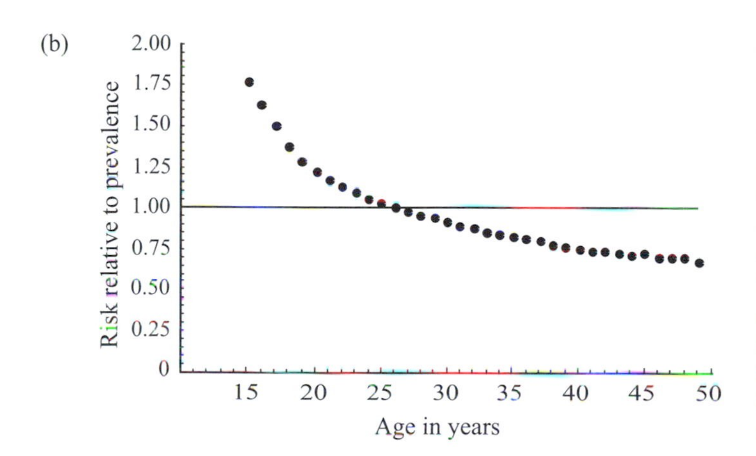

Figure 1.7: The impact of HIV on the population of pregnant women from Hlabisa, South Africa. (a) Shown are the proportion of women who were HIV+ in 1998 (prevalence, given by dots) and the fraction of HIV– women who contract HIV annually (risk, given by the dashed curve, from equation S10.2.8 in Sup. Mat. 10.2). The solid curve is fit to the prevalence data using polynomial smoothing (see Williams et al., 2001). (b) The age distribution of risk is compared to the age distribution of HIV+ prevalence. Each dot represents the probability that a woman who newly acquires HIV is of age *a* (from equation S10.2.8) relative to the probability that a woman who has HIV is of age *a*. This figure tells us which age classes are more at risk of acquiring HIV than expected based on who has HIV (points above one; solid line).

case, however, Williams et al. knew the state of the population in 1998, but they wanted to infer something about the parameters that might have led to that pattern. Thus, models can be used both to generate predictions about the future and to make inferences about the past.

Figure 1.7a shows data from 1998 on the number of women of age a infected with HIV (prevalence $P_{a,1998}$) and compares this to the inferred risk of infection from HIV as a function of age (risk $r_{a,1998}$) based on the model. The peak age of risk (21–22 years) is very nearly equal to the peak prevalence (23 years). This indicates that most HIV+ women attending hospitals and clinics in Hlibasi became infected within the last couple of years. Williams et al. (2001) noted that the shortness of the period since infection most likely reflects the fact that HIV is spreading rapidly through the region, so that many infections are newer than would be expected if HIV were well established within the population (endemic). Another important result is that the current prevalence data do not give a very good picture of who is most at risk. This is illustrated in Figure 1.7b, which gives the risk of contracting HIV for women at each age relative to that expected from the prevalence data. This figure illustrates that teenagers are up to 75% more likely to contract HIV than would be expected from the low prevalence of HIV among younger women. In the absence of a long-term longitudinal study, the estimates provided by Williams et al. (2001) help to assess the risk of HIV infection as a function of age and can be used to better target programs designed to halt the spread of HIV in South Africa.

1.4 Concluding Message

Through reading this book, we hope that you will come to appreciate just how useful mathematical modeling can be in biology. Models can help to guide a scientist's intuition about how various processes interact; they can point out logical flaws in an argument; they can identify testable hypotheses, generate key predictions, and suggest appropriate experiments; and they can reshape fields by providing new ways of thinking about a problem.

But mathematical models also have their limitations. The results of an analysis are only as interesting as the biological questions motivating a model. And even if a scientist has identified an interesting question, it may turn out that a model addressing the question is hopelessly complicated and can be solved only by making a series of assumptions, some of which are dubious. Finally, models, by themselves, can only tell us what is possible. Models can tell us, for example, how HIV levels within the body or HIV incidence within a population might change over time. But without data, collected in the field or the lab, mathematical models can never tell us what has happened or what is happening. Thus, it would be foolish to promote mathematical biology above other areas in biology. Equally, it would be foolish to avoid mathematics altogether. Science will progress faster and further by a marriage of mathematical and empirical biology. This marriage will be even more successful if more biologists can use math, when needed, to advance their own research goals. It is toward this end that we devote this book.

Further Reading

For general information about the immune system and the evolution and ecology of infectious diseases, see

- Frank, S. A. 2002. *Immunology and Evolution of Infectious Diseases*. Princeton University Press, Princeton, N.J.

Further information concerning the life cycle, health impact, and societal implications of HIV is available through the links on the book website (http://press.princeton.edu/titles/8458.html) and at www.zoology.ubc.ca/biomath. Also, see

- Moore, R. D. and Bartlett, J. G. (1998) Improving HIV therapy. *Scientific American* 279: 84–93.

References

Bacchetti, P., and A. R. Moss. 1989. Incubation period of AIDS in San Francisco. *Nature* 338:251–253.

Bacchetti, P., D. Osmond, R. E. Chaisson, S. Dritz, G. W. Rutherford, L. Swig, and A. R. Moss. 1988. Survival patterns of the first 500 patients with AIDS in San Francisco. *J. Infect. Dis* 157:1044–1047.

Blower, S. M., H. B. Gershengorn, and R. M. Grant. 2000. A tale of two futures: HIV and antiretroviral therapy in San Francisco. *Science* 287:650–654.

Calzavera, L., et al. 2002. Increases in HIV incidence among men who have sex with men undergoing repeat diagnostic HIV testing in Ontario, Canada. *AIDS* 16:1655–1661.

Dimitrov, D. S., R. L. Willey, H. Sato, L. J. Chang, R. Blumenthal, and M. A. Martin. 1993. Quantitation of human immunodeficiency virus type 1 infection kinetics. *J. Virol.* 67:2182–2190.

Fauci, A. S., G. Pantaleo, S. Stanley, and D. Weissman. 1996. Immunopathogenic mechanisms of HIV infection. *Ann. Intern. Med.* 124:654–663.

Funk, G. A., M. Fischer, B. Joos, M. Opravil, H. F. Gunthard, B. Ledergerber, and S. Bonhoeffer. 2001. Quantification of in vivo replicative capacity of HIV-1 in different compartments of infected cells. *J. Acquired Immune Defic. Syndr.* 26:397–404.

Hansson, T., and J. Aqvist. 1995. Estimation of binding free energies for HIV proteinase inhibitors by molecular dynamics simulations. *Protein Eng.* 8:1137–1144.

Ho, D. D., A. U. Neumann, A. S. Perelson, W. Chen, J. M. Leonard, and M. Markowitz. 1995. Rapid turnover of plasma virions and CD4 lymphocytes in HIV-1 infection. *Nature* 373:123–126.

Jacquez, J. A., J. S. Koopman, C. P. Simon, and I. M. Longini, Jr. 1994. Role of the primary infection in epidemics of HIV infection in gay cohorts. *J. Acquired Immune Defic. Syndr.* 7:1169–1184.

Joint United Nations Programme on HIV/AIDS. 2004a. AIDS epidemic update, December 2004. UNAIDS, Geneva, Switzerland.

Joint United Nations Programme on HIV/AIDS. 2004b. UNAIDS 2004 Report on the global AIDS epidemic. UNAIDS, Geneva, Switzerland.

Koopman, J. S., J. A. Jacquez, G. W. Welch, C. P. Simon, B. Foxman, S. M. Pollock, D. Barth-Jones, A. L. Adams, and K. Lange. 1997. The role of early HIV infection in the spread of HIV through populations. *J. Acquired Immune Defic. Syndr. Hum. Retrovirol.* 14:249–258.

Koup, R. A., J. T. Safrit, Y. Cao, C. A. Andrews, G. McLeod, W. Borkowsky, C. Farthing, and D. D. Ho. 1994. Temporal association of cellular immune responses with the initial control of viremia in primary human immunodeficiency virus type 1 syndrome. *J. Virol.* 68:4650–4655.

Kroeger Smith, M. B. et al. 1995. Molecular modeling studies of HIV-1 reverse transcriptase nonnucleoside inhibitors: Total energy of complexation as a predictor of drug placement and activity. *Protein Sci.* 4:2203–2222.

Levin, B. R., J. J. Bull, and F. M. Stewart. 1996. The intrinsic rate of increase of HIV/AIDS: epidemiological and evolutionary implications. *Math. Biosci.* 132:69–96.

Lloyd, A. L. 2001. The dependence of viral parameter estimates on the assumed viral life cycle: Limitations of studies of viral load data. *Proc. R. Soc., Seri. B* 268: 847–854.

Markgren, P. O., M. T. Lindgren, K. Gertow, R. Karlsson, M. Hamalainen, and U. H. Danielson. 2001. Determination of interaction kinetic constants for HIV-1 protease inhibitors using optical biosensor technology. *Anal. Biochem.* 291:207–218.

Nowak, M. A., S. Bonhoeffer, C. Loveday, P. Balfe, M. Semple, S. Kaye, M. Tenant-Flowers, and R. Tedder. 1995. HIV results in the frame. Results confirmed. *Nature* 375:193.

Perelson, A. S., A. U. Neumann, M. Markowitz, J. M. Leonard, and D. D. Ho. 1996. HIV-1 dynamics in vivo: virion clearance rate, infected cell life-span, and viral generation time. *Science* 271:1582–1586.

Perrin, L., and S. Yerly. 1997. Clinical presentation, virological features, and treatment perspectives in primary HIV-1 infection. *AIDS Clin. Rev.*: 67–92.

Phillips, A. N. 1996. Reduction of HIV concentration during acute infection: Independence from a specific immune response. *Science* 271:497–499.

Revelle, M. 1995. Progress in blood supply safety. *FDA Consum.* 29:21–124.

San Francisco Department of Public Health. 2000. HIV/AIDS Epidemiology Annual Report 2000 (http://www.dph.sf.ca.us/Reports/HlthAssess.htm). San Francisco, CA.

Schacker, T., A. C. Collier, J. Hughes, T. Shea, and L. Corey. 1996. Clinical and epidemiologic features of primary HIV infection. *Ann. Intern Med.* 125:257–264.

Schmitz, J. E., M. J. Kuroda, S. Santra, V. G. Sasseville, M. A. Simon, M. A. Lifton, P. Racz, K. Tenner-Racz, M. Dalesandro, B. J. Scallon, J. Ghrayeb, M. A. Forman, D. C. Montefiori, E. P. Rieber, N. L. Letvin, and K. A. Reimann. 1999. Control of viremia in simian immunodeficiency virus infection by CD8+ lymphocytes. *Science* 283:857–860.

Stafford, M. A., L. Corey, Y. Cao, E. S. Daar, D. D. Ho, and A. S. Perelson. 2000. Modeling plasma virus concentration during primary HIV infection. *J. Theor. Biol.* 203: 285–301.

Wei, X. *et al.* 1995. Viral dynamics in human immunodeficiency virus type 1 infection. *Nature* 373:117–122.

Williams, B., E. Gouws, D. Wilkinson, and S. A. Karim. 2001. Estimating HIV incidence rates from age prevalence data in epidemic situations. *Stat. Med.* 20:2003–2016.

CHAPTER 2
How to Construct a Model

Chapter Goals:

- To describe the steps involved in developing a model
- To derive equations that describe the dynamics of a biological phenomenon

Chapter Concepts:

- Discrete-time model
- Continuous-time model
- Recursion equations
- Differential equations
- Life-cycle diagrams
- Flow diagrams
- Mass action

2.1 Introduction

If you have seen mathematical models but never constructed one, it may seem like an overwhelming task. Where do you start? What is the goal? How do you know whether the model makes sense? This chapter outlines the typical process of modeling and gives helpful hints and suggestions to break down the overwhelming task into manageable bits. The most important piece of advice is to *start*. Start thinking about problems that puzzle you. Grab a piece of paper and start drawing a flow diagram illustrating the various processes at work. The biggest hurdle preventing most biologists from modeling is the paralysis one feels in the face of mathematics; most of the technical problems that pop up along the way can be surmounted or sidestepped (at the very least by simulation). You will certainly make mistakes (we all do), but there are telltale signs of mistakes, and they can be corrected. Over time, you will learn more tools and techniques that will allow you to avoid pitfalls and to get further with the problems that interest you. Your intuition will develop to help you "see" when something is wrong with your model and to help you interpret your results.

Models can describe any biological phenomenon. In the core of this book, we focus on *dynamical* models, which describe how a system changes over time. Dynamical models are very common in biology as they provide insight into how various forces act to change a cell, an organism, a population, or an assemblage of species. Within dynamical models, two broad classes are distinguished: deterministic and stochastic. "Deterministic" is shorthand for the assumption that the future is entirely predicted (determined) by the model. "Stochastic" is shorthand for the assumption that random (stochastic) events affect the biological system, in which case a model can only predict the probability of various outcomes in the future. In the remainder of this chapter, as well as in Chapters 3–12, we focus on deterministic models. The steps for constructing stochastic models are similar, but we postpone further consideration of stochastic models until Chapters 13–15.

Box 2.1 describes, in seven steps, how to construct a dynamical model. This is like describing how to ride a bike in a series of steps; obviously we can only give an idea about how the process works. Mastering the steps requires practice, and the remainder of this chapter contains a series of seven sections, each corresponding to one of the seven steps in Box 2.1.

Box 2.1: Seven Steps to Modeling a Biological Problem

Step 1: Formulate the question

What do you want to know?

Describe the model in the form of a question.

Boil the question down!

Start with the simplest, biologically reasonable description of the problem.

Step 2: Determine the basic ingredients

Define the variables in the model.

Describe any constraints on the variables.

Describe any interactions between variables.

Decide whether you will treat time as discrete or continuous.

Choose a time scale (i.e., decide what a time step equals in discrete time and specify whether rates will be measured per second, minute, day, year, generation, etc.).

Define the parameters in the model.

Describe any constraints on the parameters.

Step 3: Qualitatively describe the biological system

Draw a life-cycle diagram (see Figure 2.2) for discrete-time models involving multiple events per time unit.

Draw a flow diagram to describe changes to the variables over time.

For models with many possible events, construct a table listing the outcome of every event.

Step 4: Quantitatively describe the biological system

Using the diagrams and tables as a guide, write down the equations.

Perform checks. Are the constraints on the variables still met as time passes? Make sure that the units of the right-hand side equal those on the left-hand side.

Think about whether results from the model can address the question.

Step 5: Analyze the equations

Start by using the equations to simulate and graph the changes to the system over time.

Choose and perform appropriate analyses.

Make sure that the analyses can address the problem.

Step 6: Checks and balances

Check the results against data or any known special cases.

Determine how general the results are.

Consider alternatives to the simplest model.

Extend or simplify the model, as appropriate, and repeat steps 2–5.

Step 7: Relate the results back to the question

Do the results answer the biological question?

Are the results counterintuitive? Why?

Interpret the results verbally, and describe conceptually any new insights into the biological process.

Describe potential experiments.

2.2 Formulate the Question

The first step, coming up with a question, can be more difficult than it sounds. In most biology classes, students are told what the questions are and what answers have been found. Rarely are students asked to formulate scientific questions for themselves. This is very unfortunate because, in any scientific enterprise (modeling or otherwise), the process begins with a question. One hint is to keep an eye out for things that do not make sense or that seem to conflict—there very well might be an interesting and nonintuitive resolution. For now, start simple and don't worry about how profound your question is. Look around you, find a living object, and think up one question about how it might change over time. We did this and came up with the following three questions, which we will use in this chapter to illustrate model construction. (i) How does the number of branches of a tree change over time? (ii) How does a cat change the number of mice in a yard? (iii) How does the number of people with the flu change over the flu season?

The above three questions are "toy" examples that will make it easier to show the steps of modeling. Nevertheless, these simple examples also embody many of the key elements that come together in various combinations when constructing more complicated and realistic models. As we will see, the tree branching model is a special case of a model describing population growth. The mouse model incorporates an important component of immigration that is commonly used in ecology. For example, Blower et al. (2000) used a similar model of immigration to describe individuals moving into the gay male community of San Francisco. Finally, the flu model highlights some important concepts related to interactions among variables. For example, the way that we will model flu transmission is fundamentally similar to the way that Phillips (1996) modeled the infection of cells by HIV. Thus, these toy models provide an excellent background for tackling more complex models.

2.3 Determine the Basic Ingredients

Once you have a question in mind, proceed to Step 2 in Box 2.1. First, think about what entities might change over time; these entities are the *variables* in your model. The number of variables will depend on the question of interest. In our toy examples, we might choose to follow (i) the number of branches on a tree, (ii) the number of mice in a yard, and (iii) the number of people with the flu and the number without the flu. In choosing variables to track, we must always simplify reality. For example, in keeping track of the number of branches, we lose information about their size and age. As a general principle, start simple, adding more variables only when the model fails to address the question.

Next, we assign a letter to represent each variable—it is easier to write "x" than "the number of branches on a tree." The letters n, p, x, and y are commonly used to represent variables, but the choice is arbitrary. A good idea is to choose letters that help you remember what the variable represents, e.g., "n" for number or "p" for proportion. If a model contains multiple variables that are similar in nature, placing subscripts on the variables can help to emphasize their similarity, e.g., n_1 and n_2 for the numbers of two different species. For our models, we will use (i) $n(t)$ for the number of branches on a tree, (ii) $n(t)$ for the number of mice in a yard, and (iii) $n(t)$ for the number of people with the flu and $s(t)$ for the number of *s*usceptible people.

To remind ourselves that a variable, say n, varies over time, we can write it as $n(t)$ where t represents time and there is no space between the n and the (t). The parentheses tell us that our variable is a function of something else (time), and we read $n(t)$ as "n at time t." This notation helps to avoid math errors. For example, without this notation, we might forget that n takes on different values at different times and mistakenly treat it as a constant. Be aware, however, that not all authors use the same notation; they might write n_t instead or might simply state that n is a variable and not write it explicitly as a function of time. The important thing is to be consistent and to remember that, if we write a variable as $n(t)$, we mean "n at time t" not "n times t."

Another way to avoid math errors is to keep a list (at least a mental list) of any constraints that must remain true about the variables. For example, the number of branches on a tree should never become negative. The number of people with the flu and the number without the flu should never be negative and should sum up to the total population size. If a variable describes a frequency, a probability, or a fraction of a whole (e.g., the fraction of the total population with the flu), it should always lie between zero and one ($0 \leq p(t) \leq 1$). Ensuring that your equations and results obey the list of constraints is a good way to check that no errors have crept in.

Once you have a preliminary list of variables, the next step is to choose a type of dynamical model to describe changes in these variables. There are two main types of dynamical models, *discrete time* and *continuous time*, depending on whether time is represented in discrete steps or along a continuous axis. Discrete-time models describe how the variables change from one time unit (e.g., day, year, or generation) to the next. Continuous-time models track the variables over any period of time. Both discrete-time and continuous-time models are idealizations of reality, and they make somewhat different assumptions.

> A *discrete-time model* tracks changes to variables in discrete time steps.

> A *continuous-time model* allows variables to change at any point in time (i.e., time is treated as continuous).

Discrete-time models assume that changes cannot compound within a time unit. For example, in a discrete-time model for the number of branches on a tree, branches that arise during a time unit cannot give rise to new branches within the same time unit. As long as the time unit is short enough (e.g., a day), this assumption is often reasonable. If the time unit were long (e.g., a year), however, then some new branches might very well branch again within the year. These branching events would not be counted in a discrete-time model if the new branches were not present at the beginning of the year.

Continuous-time models assume that variables can change at any point in time, with increments or decrements occurring even within tiny intervals of time. As a consequence, it is possible for a change to occur in one small interval of time followed by the same type of change in the next small interval of time. But this may not be biologically realistic. For example, a continuous-time model might allow a newly formed branch to immediately produce its own new branch. In reality, the new branch must undergo enough cell divisions to produce a new bud, which takes time. If the rate of branching is small, then this won't be much of a problem because the average time between branching events will be large. But if the rate of branching is high, then a continuous-time model will generate incorrect predictions unless it takes into account the time lag between the formation of a branch and the formation of buds on this new branch.

Because discrete- and continuous-time models treat the timing of events in different ways, they display different temporal dynamics. In discrete-time models the variables "jump" from one value to another from one time unit to the next, and the size of these jumps can be small or large depending upon the parameters of the model. In continuous-time models, on the other hand, the variables change smoothly over time. This means that, as a variable goes from one value to another, it passes through all intervening values along the way (Figure 2.1).

In either case, we must also choose a time scale over which changes to the variables are measured. We use a "day" as the basic unit of time for the toy models considered in this chapter. Specifically, we assume that each time step in discrete-time models reflects the passage of 24 hours and that all processes in continuous-time models occur at a rate measured per day.

Just as time can be modeled discretely or continuously, so too can the variables themselves. For example, the number of branches on a tree, the number of mice in a yard, and the number of people with the flu are all discrete, integer-valued quantities (i.e., they are integers such as 0,1,2, . . . , etc.). On the other hand, an organism's metabolic rate or an organism's weight can take on any of a continuum of possible values. Regardless of the true nature of the variables, the majority of models in ecology and evolution treat variables as being continuous, an approach that we follow throughout most of the book (except in Chapters 13–15, which incorporate random events and explicitly track the numbers of each type). There are three main justifications for treating variables as continuous. First, for many questions, the variables of interest take on large enough values that treating them as continuous will introduce very little error in the results (e.g., the number of HIV particles in the blood). Second, a reinterpretation of the variable (e.g., as the total biomass of mice rather than the number of mice) can sometimes justify the use of a continuous variable. Third,

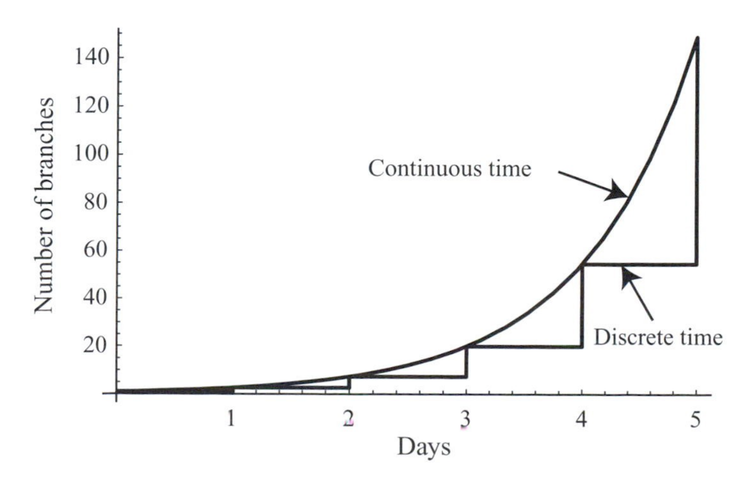

Figure 2.1: Tree branching. A plot of the number of branches on a tree over time using a discrete-time model and a continuous-time model.

it is typically easier mathematically to treat variables as being continuous rather than discrete. Remember, all models are abstractions of biological reality, and treating variables as continuous is often a reasonable abstraction.

A *recursion equation* describes the value of a variable in the next time step.

In discrete-time models, we track changes to a variable using a *recursion* equation, which describes the value of a variable (say, n) in the next time unit as a function of the variable in the current time unit:

$$n(t+1) = \text{“some function of } n(t)\text{.”} \tag{2.1a}$$

Such equations are called recursions, because one can apply them recursively to find out how the variable changes across a number of time units (from t to $t+1$, then from $t+1$ to $t+2$, etc.). An equivalent way to track a variable is to use a *difference* equation. A difference equation specifies how much a variable changes from one time unit to the next, and it is just the difference between the recursion equation for $n(t+1)$ and the current value of the variable $n(t)$:

$$\Delta n = n(t+1) - n(t) = \text{“some function of } n(t)\text{,”} \tag{2.1b}$$

where the capital Greek letter Δ ("Delta," see Table 2.1) denotes "change," and we read Δn as "the change in the variable n." Recursion equations are more commonly used to describe the value of a variable in discrete-time models, but we will occasionally use difference equations when we want to understand how much a variable changes across a time step.

In continuous-time models, equations specify the rate of change of the variables over time:

$$\frac{\mathrm{d}(n(t))}{\mathrm{d}t} = \text{“some function of } n(t)\text{.”} \tag{2.1c}$$

A *differential equation* describes the rate at which a variable changes over time.

Such equations are called *differential* equations. Differential equations are distinct from the more familiar derivatives taught in introductory calculus courses (see Box 2.2). You can think of a differential equation as a description of the ebb and flow in a variable over time. To get a better feel for a differential equation, imagine plotting the value of the variable $n(t)$ as a function of time (see Chapter 4). The slope of the curve would be $\mathrm{d}(n(t))/\mathrm{d}t$ because the derivative of a function at a point gives the slope of the function at that point. If the variable is increasing over time, the slope and thus $\mathrm{d}(n(t))/\mathrm{d}t$ are positive. If the variable is decreasing over time, the slope and thus $\mathrm{d}(n(t))/\mathrm{d}t$ are negative. When the magnitude of $\mathrm{d}(n(t))/\mathrm{d}t$ is small, the variable changes slowly over time, whereas when the magnitude of $\mathrm{d}(n(t))/\mathrm{d}t$ is large, the variable changes rapidly. As we will see, this mental picture is the reverse of how we typically construct models. We usually start by describing how various biological forces change the value of the variable (i.e., contribute to $\mathrm{d}(n(t))/\mathrm{d}t$), and we then try to infer the value of the variable itself (i.e., $n(t)$).

Which type of model should be used? Sometimes, there is a natural choice. If you want to model the number of annual plants on an island, a discrete-time model using a year as the time unit is appropriate because the life cycle of annual

TABLE 2.1
Greek letters. Here, we list the Greek letters commonly encountered in biological models (with alternative characters in parentheses).

Lower case	Upper case	Name
α	A	alpha
β	B	beta
χ	X	chi
δ	Δ	delta
ε	E	epsilon
$\phi(\varphi)$	Φ	phi
γ	Γ	gamma
η	H	eta
ι	I	iota
κ	K	kappa
λ	Λ	lambda
μ	M	mu
ν	N	nu
o	O	omicron
$\pi(\varpi)$	Π	pi
$\theta(\vartheta)$	Θ	theta
ρ	P	rho
$\sigma(\varsigma)$	Σ	sigma
τ	T	tau
υ	Υ	upsilon
ω	Ω	omega
ξ	Ξ	xi
ψ	Ψ	psi
ζ	Z	zeta

plants is itself discrete; that is, the seeds produced during one year will not germinate until the following year. By contrast, if you want to model your blood sugar levels after a meal, a continuous-time model would be more natural because there are no clear demarcations in time. Conceptually, it is sometimes easier to think in terms of discrete-time models where changes describe what happens over an interval of time rather than continuous-time models where changes are described by instantaneous rates. Mathematically, however, continuous-time models can be easier to analyze because one can utilize the various rules of calculus summarized in Appendix 2 (see Chapter 6). As we discuss later (Box 2.6

Box 2.2: Derivatives and Differential Equations

Calculus is the mathematical study of rates of change. The most important concepts and rules of calculus are summarized in Appendix 2, including formulas for differentiating and integrating a variety of functions. For example, the derivative of the polynomial $y = ax^2 + bx + c$ with respect to x is $\mathrm{d}y/\mathrm{d}x = 2ax + b$. Here, the rate of change of the *dependent* variable y is a function only of the *independent* variable x. In many biological problems, however, the rate of change of the dependent variable is a function of the dependent variable itself, e.g., $\mathrm{d}y/\mathrm{d}x = \alpha y + \beta$. Notice that the variable on the right-hand side is y not x. An equation relating the derivative of a variable to a function of the variable itself is called a *differential equation*. Equations (2.8)–(2.10) are differential equations. For example, in equation (2.8), the derivative of the dependent variable describing the number of tree branches, $n(t)$, with respect to the independent variable (time t) is a function of $n(t)$, not t. Differential equations naturally arise in continuous-time biological models because we often expect the rate of change of a variable to be a function of its current value. For example, large trees can have more new branches, a cat can eat more mice if there are more mice available, and more people can catch the flu if there are more susceptible people within the population.

A derivative or differential equation describes how a variable changes. But what we usually want to know is the *value* of the dependent variable (e.g., $n(t)$) as a function of the independent variable (e.g., t). In a typical calculus course, we are taught how to solve for y by taking the antiderivative or integral of both sides. In other words, we could solve the equation $\mathrm{d}y/\mathrm{d}x = 2ax + b$ for $y(x)$ by integrating both sides with respect to x to obtain its *solution*, $y = ax^2 + bx + c$ (see Appendix 2), which gives us the value of y for any value of x. A common error that students make when they first encounter differential equations is to integrate the left-hand side of an equation like $\mathrm{d}n(t)/\mathrm{d}t = bn(t)$ with respect to t but the right-hand side with respect to $n(t)$. This would give $n(t) = bn(t)^2/2$. To see that this is incorrect, take the derivative of both sides with respect to t (see Appendix 2). This would give $\mathrm{d}n(t)/\mathrm{d}t = bn(t)\mathrm{d}n(t)/\mathrm{d}t$, which incorrectly has $\mathrm{d}n(t)/\mathrm{d}t$ on the right-hand side. The error in this procedure crept in when we took the antiderivative of the left-hand side with respect to t, but the antiderivative of the right-hand side with respect to a different variable, $n(t)$. To solve for $n(t)$ we would have to take the antiderivative of *both sides* with respect to t, i.e.,

$$\int \frac{\mathrm{d}n(t)}{\mathrm{d}t}\,\mathrm{d}t = \int bn(t)\,\mathrm{d}t. \tag{2.2.1}$$

The left-hand integral is $n(t)$, as before, but we cannot evaluate the right-hand integral because doing so requires $n(t)$, which is what we are trying to find. In Chapter 6, we will see how to obtain solutions to certain types of differential equations, like the ones presented in this chapter. For now, it is enough to recognize the distinction between derivatives and differential equations and to remember that care must be taken when integrating differential equations.

Before leaving the subject, it is worth mentioning that the term "differential equation" encompasses several types of equations, all of which arise in biology. Differential equations can be written as functions of more than one dependent variable. For example, in our flu model, the

(continued)

Box 2.2 *(continued)*

differential equation (2.10a) for the number of people with the flu, $dn(t)/dt$, will depend on both the number of people with the flu, $n(t)$, and the number of susceptible individuals in the population, $s(t)$. Differential equations can also be written as functions of *both* the dependent variable $n(t)$ and the independent variable t. Such differential equations arise whenever we expect a variable to change as a function both of its current value and of time. For example, in a seasonal environment, the budding rate of a tree should depend on the time of year as well as on the number of branches on a tree. We can model this by treating b as some function of time, $b(t)$, rather than a constant. In addition, differential equations might depend on the past state of a variable as well as (or instead of) its current state. For example, in the tree branching example, the production of new branches at time t might depend on the total number of branches τ days ago, or $n(t-\tau)$, as these branches are now large enough to branch again. Revising equation (2.8) gives $dn(t)/dt = bn(t-\tau)$. Such equations, known as "delay differential equations," arise naturally when describing biological processes involving time lags.

All of the above examples have only one independent variable (time). These fall into the category known as "ordinary differential equations" (ODE). Many biological problems involve more than one independent variable (e.g., space as well as time), and such differential equations are known as "partial differential equations" (PDE).

and Chapter 4), discrete-time and continuous-time models can sometimes exhibit similar behavior over time, and it is possible to predict when they should behave similarly. Thus, in many cases, one is free to choose between the two.

The next step is to describe the *parameters* of the model; these are the various quantities that influence the dynamics of the model, but that remain fixed over time as the variables change. As with variables, each parameter is given its own symbol, which you are free to choose. Commonly used symbols for parameters are italicized roman letters (e.g., a, b, c, d, m, and r) and lower-case greek letters (e.g., α, β, Table 2.1).

A chief difference between discrete-time and continuous-time models is that parameters representing events per unit time are described as the *number of events* (or fraction of the population undergoing the event) per time step in discrete-time models but as the instantaneous *rate of events* per unit time in continuous time. In contrast, parameters that do not represent events per unit time (e.g., the probability that an event is one type or another) retain the same definition in the two types of models. We will discuss the difference in parameter units between discrete- and continuous-time models at greater length in Box 2.6, once we have described how their dynamical equations are derived.

Potential parameters for our discrete-time models include (i) the number of new branches that bud off each old branch per day, b; (ii) the fraction of mice in the yard eaten by the cat per day, d, and the number of mice born per mouse per day, b; (iii) the fraction of healthy people that are exposed to a flu carrier per day, c, and the probability of transmission of the flu between a healthy person and a flu carrier upon exposure, a. The analogous parameters in a continuous-time model would be (i) the rate of budding for each old branch, b; (ii) the rate of consumption of mice, d, and the rate of births per mouse, b; and (iii) the rate of contact between a flu carrier and a susceptible person, c, and the probability of transmission of the flu between a carrier and a healthy person per contact, a. These parameters represent events per unit time and so have slightly different definitions for the discrete-time and continuous-time models except a, which always represents the probability of contracting the flu per contact.

As with variables, one should also keep track of any constraints imposed on each parameter. For example, can a parameter be negative? Does a parameter represent a fraction, proportion, or probability, in which case it must fall between zero and one? These constraints might well depend on the type of model. For example, the parameter d in the cat-mouse model is restricted to lie between zero and one in discrete-time models (because it represents the *fraction* of mice eaten by the cat), whereas the analogous parameter d in the continuous-time model can have any positive value (because it represents the *rate* of consumption of mice per unit time). This is another common difference in the parameters between discrete- and continuous-time models (described more fully in Box 2.6).

In addition to the absolute constraints on each parameter, it is worth keeping track of the range of parameter values that are biologically reasonable. For example, it is reasonable to assume that the number of new branches that bud off each old branch per day is small for most trees ($b << 1$). Similarly, the number of mice born per mouse per day (b) will be much less than one ($b << 1$). We write $b << 1$ to imply that b is much smaller than one. How much smaller depends on the context, but typically this statement implies that b is 0.1 or less. Having a list of constraints and reasonable ranges for parameters can help in two important ways. First, reasonable parameter values must be chosen to carry out realistic simulations and to plot relevant graphs. Second, results from a model often depend on the values of the parameters, e.g., whether a parameter is positive or negative, large or small, so that making accurate predictions from a model depends on choosing appropriate parameter values.

Before proceeding to the next step, it is a good idea to construct a table of all the variables and parameters in your model, as well as any constraints on these terms. You can later revisit this table to ensure that it includes the variables and parameters needed to capture the essence of the biological process and to address the question of interest. It is very common that the first version of a model includes too many variables and parameters, causing the model to be unnecessarily complex, or too few variables and parameters, causing a model to behave in unintended ways (e.g., populations grow to infinite size, or nobody ever recovers from the flu). If a model displays unintended behavior, then think about whether the biological system being modeled includes other processes that should also be incorporated into the model (e.g., competition, recovery).

2.4 Qualitatively Describe the Biological System

Before writing equations down, it is a very good idea to organize your model conceptually with the aid of a diagram or table. Diagrams and tables make it easier to see whether the necessary variables and parameters are included and make it easier to write down dynamical equations (recursion equations or differential equations). We describe three organizational techniques: a life-cycle diagram, a flow diagram, and a table of events.

2.4.1 Life-Cycle Diagrams

A graphical technique, which we call a *life-cycle diagram*, keeps track of the various events occurring during a single time step, along with their order of occurrence. Such diagrams are useful only for discrete-time models, where there is a discrete time period during which various events can occur. As a simple example, consider the tree branching model. Each time step represents a single day, and only one type of event can happen during any given day: the growth of more branches. As result, the life-cycle diagram is extremely simple (Figure 2.2a).

A *life-cycle diagram* illustrates the order of events that occur within each time step (for discrete-time models).

The tree branching model is so simple that a life-cycle diagram is not really required to organize things. Life-cycle diagrams become indispensable when multiple events occur during a single time step. Consider the model of mice being eaten by a cat. Now there are three events that occur each day: mice give birth, mice move in from neighboring areas, and the cat eats mice. In a discrete-time model, one must choose an order for these events, as well as a point in time when the population is censused (e.g., when we count the number of mice, $n(t)$). For example, Figure 2.2b illustrates the case where events occur in the following order: a census, followed by predation by the cat, mouse births, mouse migration, and finally the next census. These events cause changes to the number of mice, which we describe as $n(t)$ at the census point, $n'(t)$ after predation, $n''(t)$ after births, and $n'''(t)$ after migration. Because migration is assumed to occur last in the daily life cycle, the number of mice at the next census, $n(t + 1)$, will equal $n'''(t)$. Alternatively, we might instead assume that births happen first, then migration, and then predation, yielding the life cycle in Figure 2.3. As we shall see, the order of events in a life cycle can affect the results of a model, sometimes substantially.

Finally, consider constructing a life-cycle diagram for the model of flu transmission. The time step is again one day, and as with the tree branching model, there is only a single event that can happen during each day: transmission of the flu. There is an additional wrinkle with this model, however, in that there are now two variables that we are tracking (healthy individuals and people with the flu). As a result, we could construct a life-cycle diagram for each of the variables (Figure 2.2c). But because there is only one event per cycle, these life-cycle diagrams are again not very useful (as was the case with the tree branching model).

2.4.2 Flow Diagrams

A second method for organizing a model, which is often more useful for models containing multiple variables, is a *flow diagram*. A flow diagram illustrates the interconnections among the variables and provides a schematic picture of how each variable affects its own dynamics as well as the dynamics of the other variables. In a typical flow diagram, each circle represents one variable within the model. Returning arrows that exit and come back to the same circle represent a

A *flow diagram* illustrates how each variable affects its own dynamics and those of other variables.

(a)

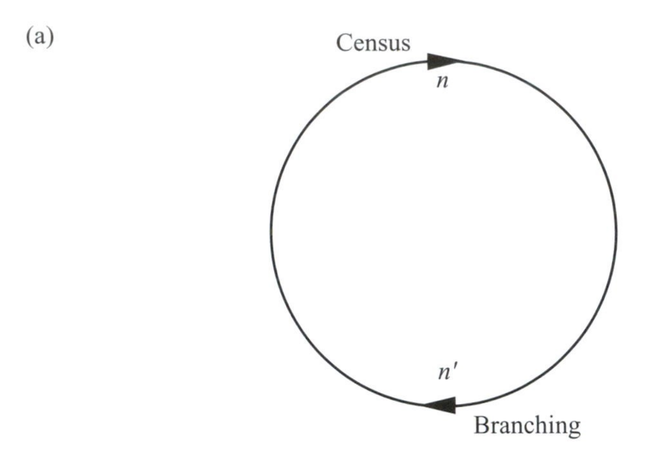

(b)

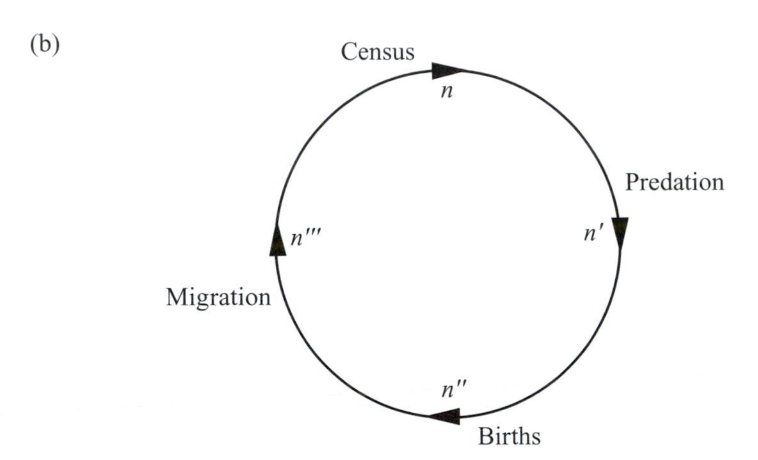

(c)

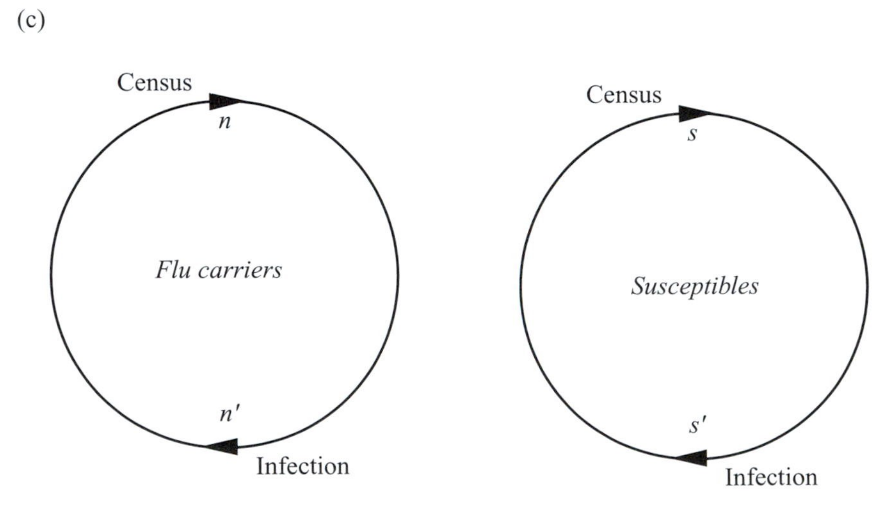

Figure 2.2: Life cycle diagrams. Life cycle diagrams for the three toy models explored in this chapter: (a) the number of tree branches, (b) the number of mice, (c) the number of people with and without the flu.

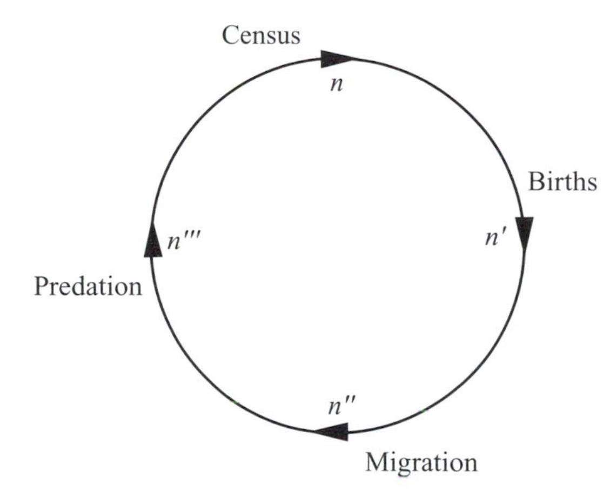

Figure 2.3: Alternate life cycle diagram. A life cycle diagram for the number of mice with a different ordering of events than occurs in Figure 2.2b.

variable that can generate more of itself. As a very simple example, a flow diagram for the tree branching model has a single returning arrow, representing the budding of a tree branch into new tree branches (Figure 2.4a).

Arrows leading into a circle represent the different ways in which a variable can go up over time, while arrows exiting a circle represent the different ways in which the variable can go down over time. A flow diagram for the mouse population has one of each of these, representing immigration and deaths, respectively, along with a returning arrow representing births (Figure 2.4b). Unfortunately, flow diagrams for discrete-time models are cumbersome when multiple events can occur within a single time step (as in this mouse model) because it is difficult to depict the ordering of the events. To be consistent with the ordering of events in Figure 2.2b, we need to consider flow across the death arrow first, then the birth arrow, and finally the migration arrow, updating the variable after each event (e.g., from $n(t)$ to $n'(t)$ after the first event).

Flow diagrams become really useful when there are multiple variables. In our flu model, we are tracking the number of susceptible and infected individuals. There must be an interaction (contact) between an infected and a susceptible person for transmission to occur, and this can be represented on a flow diagram in a variety of ways. In Box 2.3, we describe a convention for building flow diagrams, which is designed to facilitate the process of converting flow diagrams into mathematical equations. According to this convention, an interaction between two variables is represented by the merging of arrows emanating from two circles (Figure 2.4c). Different people use different conventions, but sticking to the same convention is important to avoid mistakes along the way.

Flow diagrams are constructed in the same way for continuous-time and discrete-time models (Box 2.3). For continuous-time models, however, the arrows represent events occurring continuously over time at certain rates, and we do not have to worry about the order in which events take place (e.g., we do not have to update the variables from $n(t)$ to $n'(t)$ after the first event).

On a flow diagram, it is very useful to specify (mathematically) the flow represented by each arrow directly on the diagram, including how this flow depends on the variable(s) themselves (step 8 of Box 2.3). This convention allows us to distinguish between a constant *number* exiting a circle (e.g., D, if

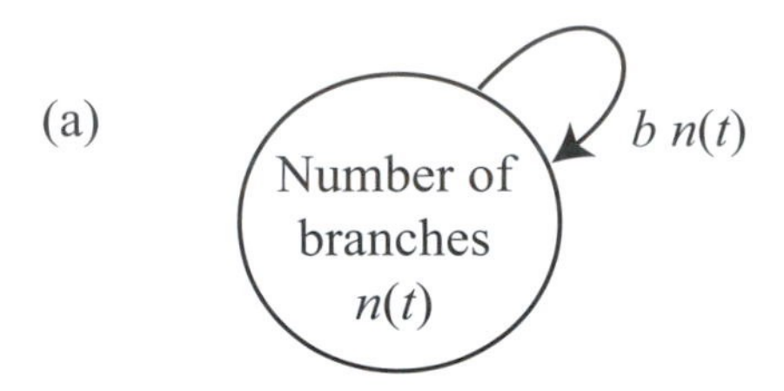

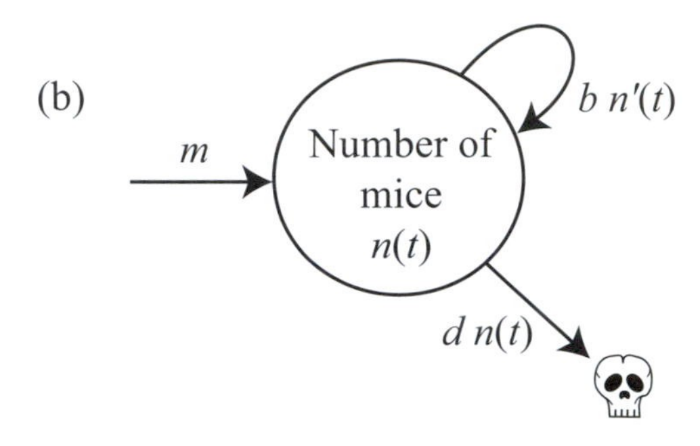

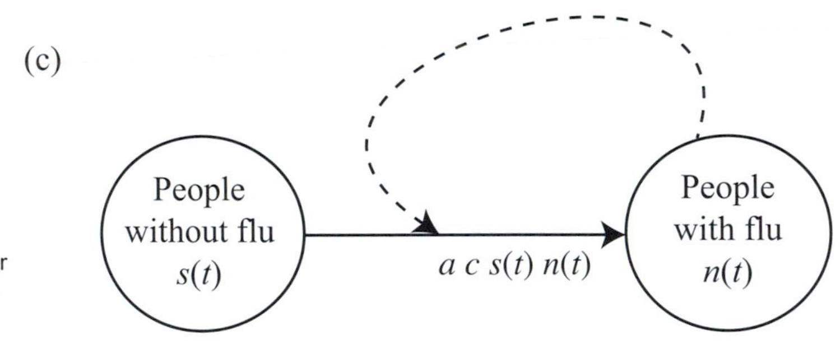

Figure 2.4: Flow diagrams. Flow diagrams for the examples explored in this chapter: (a) the number of tree branches, (b) the number of mice, (c) the number of people with and without the flu.

your cat eats a constant number D of mice per day) and a constant *fraction* exiting a circle (e.g., $d\, n(t)$, if your cat eats a constant fraction d of available mice per day). Specifying the flows is the hardest step in constructing a flow diagram, because it forces us to be very specific about the biological processes that we are modeling.

A *mass-action* interaction assumes that the rate of interaction between two variables is proportional to the values of each.

Specifying the flow for each arrow also forces us to describe if and how the variables interact. Interactions come in many different forms depending on the type and complexity of the interaction. For the flu model in Figure 2.4c, we have used the simplest form of interaction, known as a *mass-action* interaction. Mass action refers to the assumption that two types interact at a rate that is proportional to the number (or density) of the first type times the number (or density) of the second type, just as if the two types were moving about and bumping into each other at random. In the flu model, we assume that individuals with

Box 2.3: Drawing Flow Diagrams

(1) Draw a separate circle to represent each variable in your model.

(2) Use a solid arrow to indicate when a process removes an amount of the variable (arrow exits circle) or contributes an amount to the variable (arrow enters circle).

(3) Use an arrow that comes from nowhere but that enters a circle to indicate when there is an external source for one of the variables (e.g., mice from another field).

(4) Use an arrow that comes from a circle but goes to nowhere (or to a skull) to indicate when a variable exits the system (e.g., by death or emigration).

(5) Use an arrow that starts at one circle and goes to another circle to indicate when one type can become converted into another type (e.g., a susceptible individual catches the flu).

(6) Use a dashed arrow to indicate when a variable influences the flow into another circle but does not represent a decline in the variable from which the arrow begins (e.g., in Figure 2.4c, a carrier of the flu does not lose the flu by passing it on).

(7) Include an arrow that exits and returns to the same circle ("a returning arrow") whenever a variable can generate more of itself (e.g., by new births). A returning arrow can represent changes due to births only, or can describe the net change following both births and deaths.

(8) Write down the total flow along each arrow, specifying how this flow depends on the variable from which the arrow comes and on any interacting variables. If the flow across an arrow represents a conversion from one type to another (e.g., from number of prey to number of predators), there may be a conversion factor (e.g., one prey might represent only $\varepsilon = 1/100$ of the resources needed to produce one predator). Write this factor as "times ε" at the end of the arrow.

(9) For discrete-time models, decide on an ordering for the various events that occur during each time step and put a prime after the variable to indicate its state "after the first event," a double prime after the variable to indicate its state "after the second event," etc.

(10) Check to make certain that your variables are linked together in the way that you want.

(11) Check to make sure that each arrow has a flow rate written by it.

(12) Check to see if there are any variables that are completely unconnected to the rest of the diagram.

(13) Check to see if there are any parameters in your model that do not appear on the flow diagram.

(14) For a discrete-time model, check that there is never more than 100% of a variable leaving a circle.

the flu interact with susceptible individuals at a rate proportional to $n(t)\ s(t)$. Specifically, an infected individual has a probability c of contacting any given one of the $s(t)$ susceptible individuals per day, giving a total number of contacts per day of $c\ s(t)$ per infected individual. With $n(t)$ infected individuals in the population, we expect a total of $c\ n(t)\ s(t)$ contacts per day across the whole population. The probability that any one contact results in the transmission of the influenza virus is a, and therefore we expect a total of $a\ c\ n(t)\ s(t)$ new cases of flu per day. This is written below the arrow in Figure 2.4c.

The use of mass action in the flu model makes qualitative sense; if there are no people with the flu ($n(t) = 0$) or if there are no individuals susceptible to the flu ($s(t) = 0$), then there will be no new cases of the flu, because $a\ c\ n(t)\ s(t)$ is zero. There may, however, be times when you are not yet ready to specify the details of such interactions. If so, you can write the flow in general terms as $g(s(t),n(t))$, indicating that the flow rate is some function $g(\ )$ that depends on the variables $s(t)$ and $n(t)$. This lets you put off the decision of how the interactions depend on the variables until later.

Once a flow diagram has been labeled, steps (11)–(15) of Box 2.3 describe various checks to ensure that the flow diagram accurately reflects your model. If any problems arise, return to section 2.3 and revise your list of variables, parameters, and constraints, adding and subtracting as necessary. It is critical to repeat this process until you are happy that the flow diagram captures the essence of the biological process that you wish to model.

2.4.3 Tables of Events

For discrete-time models involving multiple events within a time step and multiple variables, neither life-cycle diagrams nor flow diagrams easily encapsulate all of the relevant information. In such cases, a *table of events* can be a useful organizational tool. We illustrate the construction of such a table using the flu model (Table 2.2). This model has only one event per time step (infection), so that a table is not really necessary. In more complex models, however, such tables of events are invaluable. For example, Table 8.1 organizes the events for an evolutionary model involving two genes, where we must consider

TABLE 2.2
Interaction table for the flu model. The first column lists every possible pair of individuals that could come into contact. The second column lists the number of each type of contact. The remaining columns list the change in number of infected and susceptible individuals resulting from such a contact. (Alternatively, we could have listed the number of each type after the contact, but here it is easier to list the changes.)

		Result of contact	
Interaction	Number of contacts	Infected	Susceptible
Infected × infected	$c\ n(t)\ n(t)$	No change	No change
Infected × susceptible	$c\ n(t)\ s(t)$	$+a$	$-a$
Susceptible × susceptible	$c\ s(t)\ s(t)$	No change	No change

several events (fertilization, selection, meiosis, and recombination) acting on every possible type within the population.

2.4.4 Rules of Thumb for Qualitatively Describing a Model

The main purpose of these qualitative descriptions is to clarify and organize the biological processes that you want to include in a model. How you decide to do this (using a life-cycle diagram, a flow diagram, a table, or some other approach) is partly a matter of taste, but we suggest the following rules of thumb. Life-cycle diagrams are useful for discrete-time models in which more than one event can occur during a single time step. Flow diagrams are most useful when there are multiple variables in either discrete or continuous time, although care must be taken to specify the order in which arrows should be considered in a discrete-time model. Alternatively, for discrete-time models with multiple events and multiple variables, a table of events is often the clearest way to describe a model.

2.5 Quantitatively Describe the Biological System

At this point, we are ready to derive dynamical equations for the model. Conceptually, dynamical equations track all of the factors that cause a variable to increase or decrease over time and have the form

$$n(t+1) = n(t) + \text{increase} - \text{decrease} \qquad \text{(recursion equations)}, \qquad (2.2a)$$

$$\Delta n = \text{increase} - \text{decrease} \qquad \text{(difference equations)}, \qquad (2.2b)$$

$$\frac{\mathrm{d}(n(t))}{\mathrm{d}t} = \text{rate of increase} - \text{rate of decrease} \qquad \text{(differential equations)}. \qquad (2.2c)$$

For discrete-time models, we describe the value of the variable in the next time step using (2.2a) or the change in the variable across the time step using (2.2b). If the model involves multiple events per time step, we must specify an order to these events (e.g., with a life-cycle diagram) and apply (2.2a) or (2.2b) after each event, updating the value of the variable before the next event. In practice, it is often easiest to first derive the recursion equation (i.e., 2.2a) and then, from this, construct the difference equation (i.e., 2.2b) if desired. For continuous-time models, the procedure is simpler. We sum all of the factors causing the variable to increase or decrease, regardless of how many events occur in the model. In a continuous-time model, we do not have to worry about the order of events within a time step, because the time step is so small (infinitesimally small) that no two events occur at exactly the same point in time.

To make this process more concrete, let us derive dynamical equations for our toy models (Table 2.3). These equations can be derived directly from an understanding of the models or with the aid of the life-cycle diagrams, flow diagrams, or tables of events. Because the type of qualitative description of the model usually depends on whether it is a discrete- or continuous-time model, we will consider these two cases separately.

TABLE 2.3
Dynamic equations derived in this chapter. Type refers to (1) recursion equation in discrete time, (2) difference equation in discrete time, and (3) differential equation in continuous time.

Model	Type	Equation	
Tree-branch model	1	$n(t+1) = n(t) + b\,n(t)$	(2.3a)
	2	$\Delta n = b\,n(t)$	(2.3b)
	3	$\frac{d(n(t))}{dt} = b\,n(t).$	(2.8)
Mouse population	1	$n(t+1) = (1+b)\,(1-d)\,n(t) + m$	(2.4)
	2	$\Delta n = -d\,n(t) + b\,(1-d)\,n(t) + m$	(2.5)
	3	$\frac{d(n(t))}{dt} = b\,n(t) - d\,n(t) + m$	(2.9)
Flu dynamics	1	$n(t+1) = n(t) + a\,c\,n(t)\,s(t)$	(2.7)
		$s(t+1) = s(t) - a\,c\,n(t)\,s(t)$	
	2	$\Delta n = a\,c\,n(t)\,s(t)$	(from 2.7)
		$\Delta s = -a\,c\,n(t)\,s(t)$	
	3	$\frac{d(n(t))}{dt} = a\,c\,n(t)\,s(t)$	(2.10)
		$\frac{d(s(t))}{dt} = -a\,c\,n(t)\,s(t)$	

2.5.1 Discrete-Time Models

Let us start with the simplest of our discrete-time models, the branching model. This model has a single variable (the number of branches) and only a single event can happen during a time step. Given the life-cycle diagram in Figure 2.2a, we can derive the recursion equation by specifying the number of branches existing after the first event (the branching event) occurs.

Recalling that b is the number of new branches that bud off each old branch per day, the total number of branches after the first (and only) event is the number of old branches plus the number of new branches, or $n'(t) = n(t) + n(t)\,b$. The next event on the life-cycle diagram is the census at the time $t+1$, so that $n(t+1)$ equals $n'(t)$. This gives us the recursion equation

$$n(t+1) = n(t) + b\,n(t). \tag{2.3a}$$

From equation (2.3a) we can readily construct the difference equation by subtracting off the current number of branches, $n(t)$:

$$\begin{aligned}\Delta n &= n(t+1) - n(t)\\ &= b\,n(t).\end{aligned} \tag{2.3b}$$

Equation (2.3a) tells us the total number of branches at time $t + 1$, while equation (2.3b) describes how many more branches there are at time $t + 1$ than at time t.

Let us now consider the mouse model (Figure 2.2b). To derive the recursion equation for this model we work our way around the life cycle, updating the value of the variable after each event. Following the logic used in the tree branching example, we have

$$n'(t) = n(t) - d\,n(t) \quad \text{after predation by the cat,}$$

$$n''(t) = n'(t) + b\,n'(t) \quad \text{after births,}$$

$$n'''(t) = n''(t) + m \quad \text{after migration.}$$

After migration, the next event on the life-cycle diagram is the census at the time $t + 1$, so that $n(t + 1)$ equals $n'''(t)$. Plugging the first equation for $n'(t)$ into the second equation, we get $n''(t) = (n(t) - d\,n(t)) + b\,(n(t) - d\,n(t))$, which factors to give $n''(t) = (1 + b)\,(1 - d)\,n(t)$. Plugging this result into the third equation gives the complete recursion

$$\begin{aligned} n(t + 1) &= n'''(t) \\ &= n''(t) + m \\ &= (1 + b)\,(1 - d)\,n(t) + m, \end{aligned} \tag{2.4}$$

which describes the number of surviving mice in the yard on the next day.

Equation (2.4) can be given a relatively simple explanation. A fraction d of mice are eaten by the cat and the remainder $1 - d$ survive, leaving $(1 - d)n(t)$ mice. Next, each surviving mouse gives rise to themselves plus, on average, b babies, resulting in $(1 + b)\,(1 - d)n(t)$ mice. Finally, m new mice arrive, giving equation (2.4).

This general process is summarized in Recipe 2.1:

Recipe 2.1
Writing Recursion Equations from Life-Cycle Diagrams (Discrete-Time Models)

Step 1: Use $n'(t)$, $n''(t)$, $n'''(t)$, etc. to denote the value of the variable after the first, second, third, etc., event in the life cycle and obtain recursions for these according to equation (2.2a).

Step 2: Set $n(t + 1)$ to the value of n after the final event in the life cycle.

Step 3: Substitute the recursion for $n'(t)$ into the recursion for $n''(t)$ and simplify. Then substitute the recursion for $n''(t)$ into the recursion for $n'''(t)$ and simplify, etc., until the resulting expression gives a recursion for $n(t + 1)$ solely in terms of $n(t)$.

If you wish to know the amount of change over the time step, the difference equation can be derived using Recipe 2.2:

Recipe 2.2
Deriving a Difference Equation from a Recursion Equation
Step 1: Calculate $n(t + 1)$ using Recipe 2.1.
Step 2: Subtract $n(t)$ from $n(t + 1)$ and simplify to get the difference equation, $\Delta n = n(t + 1) - n(t)$, describing the change in the variable per time step.

For the cat and mouse model, we get

$$\begin{aligned} \Delta n &= n(t + 1) - n(t) \\ &= (1 + b)(1 - d)\, n(t) + m - n(t) \\ &= -d\, n(t) + b\,(1 - d)\, n(t) + m. \end{aligned} \tag{2.5}$$

Taken together, these terms describe all of the changes in the mouse population per day.

The order of events can have a large impact on the predictions of a model. Consider a rather extreme case in the mouse model, where the cat catches 100% of the mice ($d = 1$), 10% of surviving mice give birth each day ($b = 0.1$), and mice arrive in droves ($m = 100$). If we start with one mouse ($n(0) = 1$) and plug these numbers into equation (2.4), we predict 100 mice after one day. If, however, predation is the last event rather than the first event (Figure 2.3), we have

$$\begin{aligned} n'(t) &= n(t) + b\, n(t) && \text{after births,} \\ n''(t) &= n'(t) + m && \text{after migration,} \\ n'''(t) &= n''(t) - d\, n''(t) && \text{after predation by the cat,} \end{aligned}$$

and the recursion equation will be

$$n(t + 1) = (1 - d)\,(n(t) + b\, n(t) + m). \tag{2.6}$$

Plugging in the same parameters, we now predict 0 mice rather than 100 mice after a day.

Which is the right answer? It depends on when we count the mice, when we let out the cat, and when mice tend to move about and give birth. Consider counting the mice at noon. Mice tend to be nocturnal, and it might be reasonable to assume that those that migrate do not immediately give birth that same night. If the cat is out only in the afternoon, equation (2.4) is a reasonable approximation to the system (afternoon: cat eats; night: mice give birth and then move in; noon: mice get counted). If the cat is out only in the morning, however, equation (2.6) is more appropriate (night: mice give birth and then move in; morning: cat eats; noon: mice get counted). Indeed, the difference in the predicted number of mice makes sense even without a model—if you don't want mice around at your luncheon, then you'd better let the cat out in the morning, not after lunch.

The above example is extreme, but it emphasizes that ordering matters in discrete-time models, and it cannot be ignored. Lest you become too anxious about getting the order of events in a model perfectly right, however, the order typically does not have a large effect as long as little happens during any given time unit (specifically, when each term in the difference equation, Δn, is small relative to $n(t)$). In this case, the results depend less on what just happened within a time unit (which will be relatively little) and more on the value of the variable at the beginning of the time step, $n(t)$. Indeed, many discrete-time models are built by assuming that every change to the variables depends only on their values at the last census. To see this point, try comparing equations (2.4) and (2.6) with more moderate values of the parameters: $d = 0.1$, $b = 0.1$, $m = 1$, and $n(0) = 10$.

Our flu model has two variables. In Figure 2.2c, we drew a life-cycle diagram for each variable. You should try using Recipe 2.1 and Figure 2.2c to construct a recursion equation for each of the variables. Here, we will follow a different approach and use the flow diagram in Figure 2.4c.

Recipe 2.3

Writing Recursion Equations from Flow Diagrams (Discrete-Time Models)

Step 1: Considering each solid arrow in turn, update the value of each variable by taking its previous value

- plus the flow if the arrow enters the circle
- plus the flow if the arrow leaves and returns to the circle
- minus the flow if the arrows leaves the circle.

Step 2: Set $n(t + 1)$ to the value of n after the final arrow has been considered.

There is only one solid arrow in Figure 2.4c. Thus, we need only consider how it affects the number of people with the flu (plus $a\ c\ n(t)\ s(t)$ because the arrow enters the circle representing the number of flu carriers)

$$n(t + 1) = n(t) + a\ c\ n(t)\ s(t) \tag{2.7a}$$

and the number of susceptible individuals (minus $a\ c\ n(t)\ s(t)$ because the arrow leaves the circle representing the number of healthy individuals)

$$s(t + 1) = s(t) - a\ c\ n(t)\ s(t). \tag{2.7b}$$

Alternatively, these equations can be derived using a table of events (Table 2.2), by multiplying the number of contacts by the change caused to the number of infected and susceptible individuals.

2.5.2 Continuous-Time Models

For continuous-time models, differential equations are derived by summing the rates of all changes that occur to a variable, as described by a flow diagram. In fact, we can use the same flow diagrams (Figure 2.4) as before, remembering that the flows across the arrows are now described as rates and that we don't have to worry about the order of events (because continuous-time models consider infinitesimally small time intervals, during which two events are unlikely to occur simultaneously: Box 2.6).

Recipe 2.4:
Writing Differential Equations from Flow Diagrams (Continuous-Time Models)

$\frac{\mathrm{d}(n(t))}{\mathrm{d}t}$ = the flow rates along arrows entering the circle

+ the flow rates along arrows leaving and returning to the circle

− the flow rates along arrows exiting the circle.

For the branching model (Figure 2.4a), there is only one way that the number of branches changes (by the budding off of new branches), and the differential equation is

$$\frac{\mathrm{d}(n(t))}{\mathrm{d}t} = b\, n(t). \tag{2.8}$$

The right-hand side is the same as the difference equation (2.3b) for Δn in the discrete-time model. This makes sense because both difference equations and differential equations describe changes to the variables. In contrast, the recursion equation (2.1) also has $n(t)$ on the right-hand side because it describes the value of the variable rather than how it changes.

As mentioned above, the order of events within a time interval is irrelevant in continuous-time models because the change per time interval considered is infinitesimally small. Thus, in our mouse example (Figure 2.4b), we do not have to update the variable after each event, and we can drop the prime notation ($n'(t)$, etc.). Applying Recipe 2.4 to Figure 2.4b, the differential equation describing the number of mice is then

$$\frac{\mathrm{d}(n(t))}{\mathrm{d}t} = b\, n(t) - d\, n(t) + m, \tag{2.9}$$

whose terms take into account changes due to births, predation, and immigration, respectively. Now, however, the right-hand side does not look the same as the difference equation Δn. As you may have surmised, the reason is that the difference equation (2.5) allows only one bout of deaths, followed by births, followed by migration, whereas the differential equation (2.9) allows these events to occur continuously throughout the day.

Finally, for the flu model (Figure 2.4c), we can apply Recipe 2.4 to translate the flow diagram into a pair of differential equations modeling the number of people with the flu and those that are susceptible:

$$\frac{\mathrm{d}(n(t))}{\mathrm{d}t} = a\,c\,n(t)\,s(t), \tag{2.10a}$$

$$\frac{\mathrm{d}(s(t))}{\mathrm{d}t} = -a\,c\,n(t)\,s(t). \tag{2.10b}$$

The above three toy examples illustrate how flow diagrams can be used to derive the equations of simple models. But the organizational techniques that we have described really become indispensable when constructing more complex models. To illustrate this, Boxes 2.4 and 2.5 derive the differential equations used in the HIV models introduced in Chapter 1. Box 2.4 develops Phillips' (1996) model for the dynamics of HIV within an individual. Phillips used these equations to predict how the numbers of virus particles within the bloodstream might change following infection by HIV. Box 2.5 develops the model of Blower et al. (2000) for the dynamics of HIV spread among individuals within the San Francisco gay male community. Blower *et al.* used these equations to predict the effects of antiretroviral therapies on the spread of HIV and on the total rate of death from AIDS. While these models are more complex and address more important biological questions, the steps involved in deriving the models are identical (Box 2.1).

Although we have derived the above differential equations with the aid of flow diagrams, they can also be derived directly from the discrete-time models by letting the time step shrink, as shown in Box 2.6. Box 2.6 also sheds light on several key differences between discrete- and continuous-time models. In particular, Box 2.6 clarifies the meaning of rate parameters and why constraints on these parameters differ in the two types of models. Box 2.6 also provides insight into when discrete- and continuous-time models will exhibit similar behavior and why they need not.

2.6 Analyze the Equations

At this point, we say that our model has been *fully specified*. We know the variables, the type of model, the parameters, and the equations describing changes in the variables (Table 2.3). The next step is to analyze the model. There are many different ways of analyzing equations, several of which we will discuss in this book. These include (in order of increasing difficulty)

- Graphical analyses (Chapter 4)
- Simulations (Chapter 4)
- Equilibrium and stability analyses (Chapters 5, 7, and 8)
- Deriving general solutions (Chapters 6 and 9)
- Determining long-term or asymptotic behavior (Chapter 10)
- Analyzing the model for periodic behavior (Chapter 11)

Box 2.4: Deriving the Equations in Phillips (1996)

We illustrate Phillips' (1996) model in the form of a *flow diagram* (Figure 2.4.1). The circles represent the number of susceptible CD4+ cells, $R(t)$, the number of latently infected cells, $L(t)$, the number of actively infected cells, $E(t)$, and the number of virions in the blood stream, $V(t)$. The arrows connecting these circles represent the rate per day at which one category leads to another, where the total flow rate is written beside each arrow. When two arrows meet, this represents an interaction that must occur between two categories to give rise to another category (e.g., an uninfected cell must encounter a virus to become infected).

Let us walk through this flow diagram from left to right. By doing so, we are essentially describing all of the assumptions made by Phillips (1996). At a rate of Γ per day, the immune system produces new uninfected CD4+ cells, of which a fraction τ become susceptible to attack by HIV. Even without HIV infection, CD4+ cells die or are eliminated from the body at a rate μ per susceptible cell per day, leading to a total flow out of the circle of $\mu\, R(t)$ per day. In addition, susceptible CD4+ cells become infected if they encounter a virus. New infections are assumed to occur at a rate $\beta\, V(t)$ per susceptible cell per day, leading to a total flow of $\beta\, V(t)\, R(t)$ per day. This is the simplest equation that captures the fact that cells should become infected at a faster rate if there are more cells to be infected ($R(t)$) or more viruses to do the infecting ($V(t)$). β is a constant that determines whether infections occur slowly (low β) or rapidly (high β); it is analogous to the product of the contact rate (c) and the probability of infection (a) in the flu model. The rate of new infections, $\beta\, V(t)\, R(t)$, employs the "mass-action" assumption.

Once infected, a CD4+ cell may harbor HIV in a latent, nonreplicating state or in its actively replicating state; Phillips lets p describe the probability that HIV becomes latent within a newly infected cell so that $1 - p$ is the probability that HIV becomes actively replicating. Because HIV

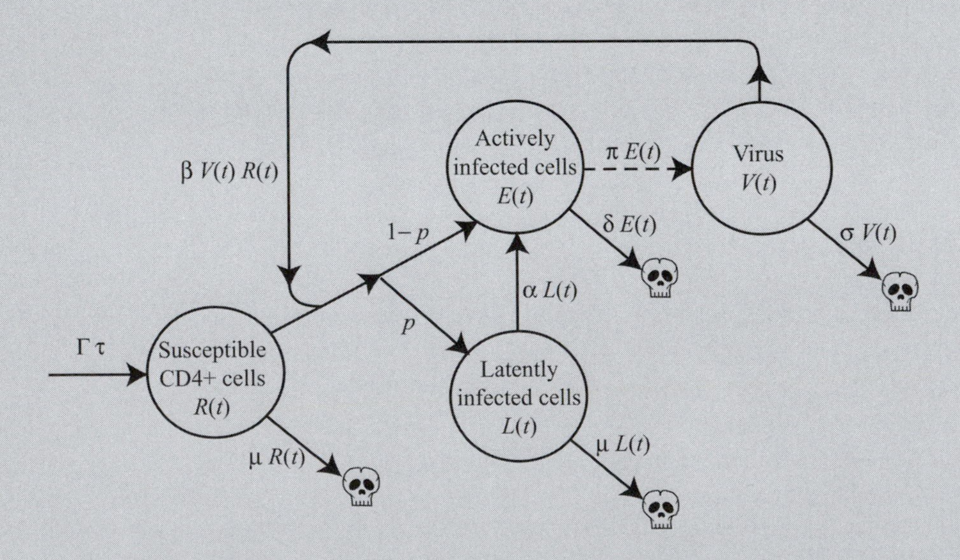

Figure 2.4.1: Flow diagram for viral load. The model describes the number of viruses in the blood stream after HIV infection (Phillips 1996).

(continued)

Box 2.4 *(continued)*

is hidden within the genome of latently infected CD4+ cells, it is assumed that these cells die at the same rate as uninfected cells (μ per cell per day). Latently infected cells may also be activated, however, which occurs at rate α per cell per day. Actively infected CD4+ cells are thus produced by two means: by the immediate conversion of an uninfected cell at rate $(1 - p)\, \beta\, V(t)\, R(t)$ or by the conversion of a latently infected cell at a rate $\alpha\, L(t)$. Actively infected cells die at a much faster rate δ per cell per day, due to the continual budding of virus particles at a rate π per infected cell per day. (We use a dashed arrow between actively infected cells and viruses because viral production by budding does not directly eliminate an infected cell.) Finally, virus particles degrade or are eliminated from the body at a rate σ per virion per day.

From the flow diagram illustrated in Figure 2.4.1, we can write down differential equations, describing the rate of change of each variable over time (e.g., $\mathrm{d}V(t)/\mathrm{d}t$ for the rate of change of virus particles). Each variable represented by a circle in Figure 2.4.1 changes at a rate equal to the sum of all of the arrows entering the circle minus all of the arrows exiting the circle:

$$
\begin{aligned}
\frac{\mathrm{d}R(t)}{\mathrm{d}t} &= \Gamma\, \tau - \mu\, R(t) - \beta\, V(t)\, R(t), \\
\frac{\mathrm{d}L(t)}{\mathrm{d}t} &= p\, \beta\, V(t)\, R(t) - \mu\, L(t) - \alpha\, L(t), \\
\frac{\mathrm{d}E(t)}{\mathrm{d}t} &= (1 - p)\, \beta\, V(t)\, R(t) + \alpha\, L(t) - \delta\, E(t) \\
\frac{\mathrm{d}V(t)}{\mathrm{d}t} &= \pi\, E(t) - \sigma\, V(t).
\end{aligned}
\tag{2.4.1}
$$

(Technically, the rate at which virus particles infect susceptible cells should also be subtracted off from $\mathrm{d}V(t)/\mathrm{d}t$, but this rate is assumed small relative to the large number of virus particles in the bloodstream.) These equations were used by Phillips' (1996) to predict how the number of viral particles varied over time after initial infection with HIV (see Chapter 1 and Figure 1.5).

Box 2.5: Deriving the Equations in Blower et al. (2000)

Blower et al. (2000) developed a model to predict changes in HIV incidence in the San Francisco community of gay males. The authors were particularly concerned that effective antiretroviral therapies (ART) might cause people to be less cautious when engaging in behavior posing a risk for HIV transmission. Here we present a slightly simplified version of their model that ignores the evolution of HIV resistance. Their model assumes that ART has an influence on survival rates, sexual behavior, and the spread of HIV among gay men that are sexually active within San Francisco. In particular, it assumes that the average number of sexual partners with whom an HIV− individual has unprotected sex per year increases from c before ART to $c(1 + i)$.

(continued)

Box 2.5 ***(continued)***

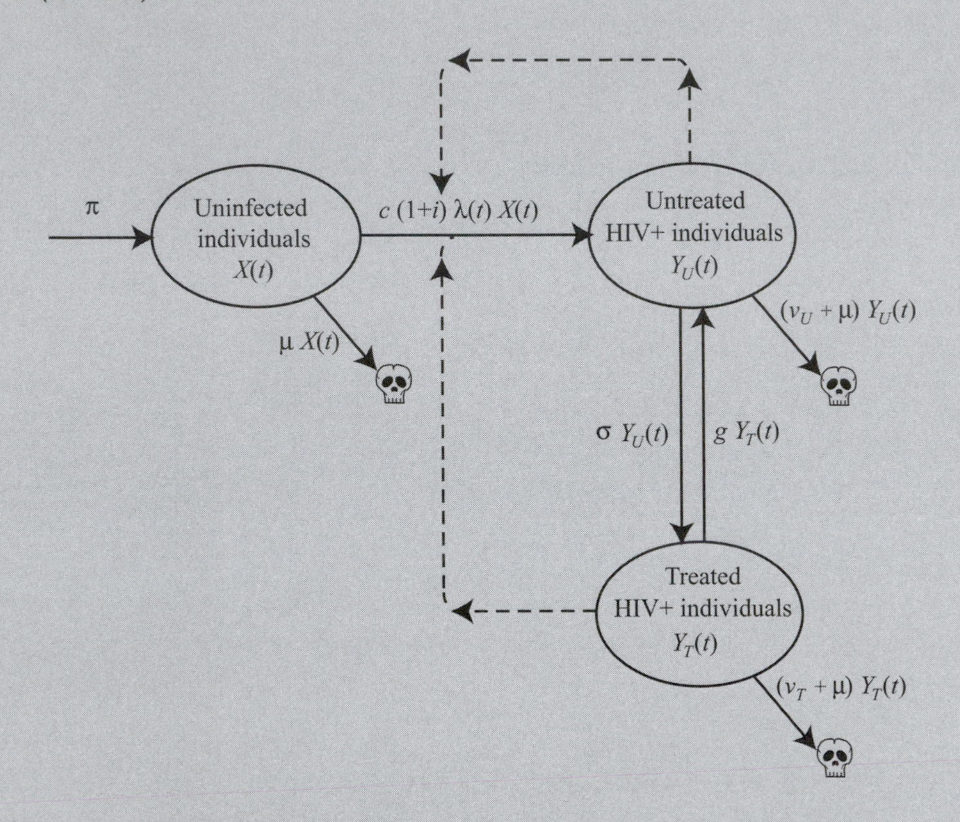

Figure 2.5.1: Flow diagram for HIV and AIDS cases. The model describes the number of cases of HIV and AIDS in the gay male community of San Francisco (Blower et al. 2000). See Table 2.5.1 for further description of the parameters.

A flow diagram for the model of Blower et al. (2000) is illustrated in Figure 2.5.1. The circles represent the number of uninfected individuals, $X(t)$, the number of infected individuals taking drug therapy, $Y_T(t)$, and the number of infected individuals not taking drug therapy, $Y_U(t)$. The arrows represent the rate per year at which one category leads to another category, where the total flow rate is written beside each arrow. When two arrows meet, this represents an interaction that must occur between two categories to give rise to another category (e.g., between infected and uninfected individuals). The parameters describing the flow rates in Figure 2.5.1 are defined in Table 2.5.1.

From the flow diagram, we can determine the rate of change of each variable using Recipe 2.4:

$$\begin{aligned}
\frac{\mathrm{d}X(t)}{\mathrm{d}t} &= \pi - c(1+i)\,\lambda(t)\,X(t) - \mu\,X(t) \\
\frac{\mathrm{d}Y_U(t)}{\mathrm{d}t} &= c(1+i)\,\lambda(t)\,X(t) + g\,Y_T(t) - \sigma\,Y_U(t) - \mu\,Y_U(t) - \nu_U\,Y_U(t), \qquad (2.5.1) \\
\frac{\mathrm{d}Y_T(t)}{\mathrm{d}t} &= \sigma\,Y_U(t) - g\,Y_T(t) - \mu\,Y_T(t) - \nu_T\,Y_T(t).
\end{aligned}$$

(continued)

Box 2.5 *(continued)*

TABLE 2.5.1

Parameters in HIV/AIDS model. Parameters in the model predicting HIV incidence following antiretroviral therapy (Blower et al. 2000). All rates are per year, and the model assumes that changes to the community are occurring continuously.

Parameter	Description	Value
π	The rate at which HIV– men join the gay community in SF	2133
μ	The rate at which gay men leave the sexually active community[a]	1/30
c	The number of partners per year with whom risky sex occurs (before ART)	1.7
$c\,(1+i)$	The number of partners per year with whom risky sex occurs (after ART)	$1.7\,(1+i)$
β_U	Chance of infection per HIV+ partner (untreated) with whom risky sex occurs	0.1
β_T	Chance of infection per HIV+ partner (treated) with whom risky sex occurs	0.025
$\lambda(t)$	Force of infection per partner; $\lambda(t) = \dfrac{\beta_U Y_U(t) + \beta_T Y_T(t)}{X(t) + Y_U(t) + Y_T(t)}$	Varies
σ	The rate at which untreated HIV+ men enter treatment	0.5[b]
g	The rate at which treated HIV+ men abandon treatment	0.05
v_U	The death rate of untreated HIV+ men from AIDS (expected survival of 12 years following infection)	1/12
v_T	The death rate of treated HIV+ men from AIDS (expected survival of 27 years following infection)	1/27

[a]By moving away, becoming sexually inactive, or dying for reasons unrelated to HIV/AIDS.
[b]The current rate in San Francisco.

In this model, it is assumed that uninfected individuals engage in behavior that puts them at risk of contracting HIV at a rate of $c(1 + i)$ per uninfected individual per year, where i equals zero before ART but rises to some unknown value after ART. Unlike the flu model (2.7), this rate is assumed to be a personal decision that does not depend on the number or density of possible sexual partners. That is, individuals don't just bump into each other randomly as assumed in a mass-action model; instead they actively seek out sexual partners at a particular rate $c(1 + i)$. For each sexual contact, the composition of the population determines the probability that the contact results in an infection. At time t, this probability is given by $\lambda(t)$ (the per capita "force of infection"), which incorporates the probability that a sexual partner is HIV+ times the probability of acquiring HIV from this partner during sex, β. Specifically, if $N(t)$ is the total number of potential partners, $N(t) = X(t) + Y_U(t) + Y_T(t)$, then the probability that a sexual partner is HIV+ but not undergoing treatment is $Y_U(t)/N(t)$; such partners tend to have a higher transmission

(continued)

Box 2.5 ***(continued)***

probability β_U. Similarly, the probability that a sexual partner is HIV+ and undergoing treatment is $Y_T(t)/N(t)$; such partners tend to have a lower transmission probability β_T. Accounting for the possibility of contracting HIV from either type of individuals, the force of infection is

$$\lambda(t) = \beta_T \frac{Y_T(t)}{N(t)} + \beta_U \frac{Y_U(t)}{N(t)}. \tag{2.5.2}$$

Using the parameter values in Table 2.5.1, equations (2.5.1) were solved numerically to generate Figure 1.6 (as described in Chapter 4).

Box 2.6: The Relationship between Discrete-Time and Continuous-Time Models

Although discrete- and continuous-time models are different, they share several fundamental similarities. In fact, one can derive a continuous-time model directly from a discrete-time model by shrinking the length of the time unit down to zero. By describing this procedure, we gain a much clearer understanding of the relationship between discrete- and continuous-time models.

Consider the mouse model, which was derived using a day as the unit of time. What would happen in a shorter unit of time, Δt? In order for this procedure to work, we have to assume that the same set of events could occur in the same order in successively smaller time units. (If this does not make biological sense, e.g., if migration only happens at night, then we should not use a continuous-time model to describe the process.) In the mouse model, the first event that happened was that the cat ate a fraction d of the mouse population. Now, in half a day, we would expect the cat to eat half this amount, $d/2$. In general, in a shorter amount of time Δt, we would expect the cat to eat a fraction $d\,\Delta t$ of the mouse population. In this procedure, we assume that d retains the same value. Nevertheless, as the time interval shrinks, it is possible for d to take on larger and larger values without depleting the entire population of mice. For instance, in cutting the day in half, $d/2$ is the fraction of mice eaten in half a day, and this must still lie between zero and one. Now, however, d can lie anywhere between zero and two. Similarly, for smaller time increments we have the restriction that $d\Delta t$ must lie between zero and one and therefore that d must lie between zero and $1/\Delta t$. As Δt gets smaller and smaller, the maximum allowable value of d gets larger and larger. This reveals why parameters that describe flow are restricted to be less than one in discrete-time models but can have no upper limit in continuous-time models. The same argument applies to both births and migration events, of which we now expect $b\,\Delta t$ and $m\,\Delta t$ to occur in the time unit Δt. Thus, for a discrete-time model with time unit Δt, we would replace d, b, and m in Figure 2.2b with $d\,\Delta t$, $b\,\Delta t$, and $m\,\Delta t$.

(continued)

Box 2.6 *(continued)*

We then proceed, event by event, through the life cycle, using Recipe 2.1 to generate a recursion equation:

$$n'(t) = n(t) - d\,\Delta t\, n(t) \qquad \text{after predation by the cat,}$$

$$n''(t) = n'(t) + b\,\Delta t\, n'(t) \qquad \text{after births,}$$

$$n'''(t) = n''(t) + m\,\Delta t \qquad \text{after migration.}$$

Now $n'''(t)$ is $n(t+\Delta t)$, the number of mice after the time unit, Δt has passed. By plugging the first equation for n' into the second equation and the resulting equation for $n''(t)$ into the third equation, we get

$$n(t + \Delta t) = (1 + b\,\Delta t)\,(1 - d\,\Delta t)\,n(t) + m\,\Delta t. \tag{2.6.1}$$

Next, we can use the definition of a derivative to convert recursion (2.6.1) into a differential equation (Box 2.2). According to the definition of a derivative (Appendix 2),

$$\frac{\mathrm{d}n(t)}{\mathrm{d}t} \equiv \lim_{\Delta t \to 0}\left[\frac{n(t + \Delta t) - n(t)}{\Delta t}\right] \tag{2.6.2}$$

Here's how to read equation (2.6.2) in words: the derivative of n with respect to t is defined ("$\equiv$") as the change in n over a time interval (that is, $n(t + \Delta t) - n(t)$) divided by the length of the time interval (Δt), in the limit as the time interval shrinks to zero ("$\lim_{\Delta t \to 0}$").

We begin the conversion process by plugging (2.6.1) into the term in square brackets in (2.6.2):

$$\begin{aligned}\frac{n(t + \Delta t) - n(t)}{\Delta t} &= \frac{(1 + b\Delta t)(1 - d\,\Delta t)\,n(t) + m\Delta t - n(t)}{\Delta t} \\ &= b\,n(t) - d\,n(t) + m - b\,d\,n(t)\Delta t.\end{aligned} \tag{2.6.3}$$

Next, we let the time interval Δt go to zero, which causes the last term to drop out. We are left with the same differential equation (2.9) that we derived directly from the flow diagram for the continuous-time model. This procedure works for any discrete-time model as long as it is reasonable to allow each event to occur in successively smaller periods of time Δt.

As another example, consider the discrete-time flu model. In a short amount of time Δt, we expect that an infected person contacts any given susceptible person with probability $c\,\Delta t$. Therefore, the total number of contacts with susceptible individuals is $c\,\Delta t\, s(t)$ per infected individual in this short period of time. Again, while the parameter c must lie between zero and one in the discrete-time model, the maximum allowable value of this parameter now increases without bound as we shrink the time interval Δt to zero. Every time a contact occurs, whether in discrete or continuous time, the probability that the flu is transmitted to the healthy person is a. Given that there are $n(t)$ such infected individuals, we expect a total number of new flu cases in

(continued)

Box 2.6 *(continued)*

the time interval Δt to be $a\, c\, \Delta t\, n(t)\, s(t)$. Using Recipe 2.1, the number of flu cases after an interval of time Δt would be

$$n(t + \Delta t) = n(t) + a\, c\, \Delta t\, n(t)\, s(t). \tag{2.6.4}$$

Plugging (2.6.4) into (2.6.2) and taking the limit

$$\frac{dn(t)}{dt} \equiv \lim_{\Delta t \to 0}\left[\frac{a\, c\, \Delta t\, n(t)\, s(t)}{\Delta t}\right] = a\, c\, n(t)\, s(t), \tag{2.6.5}$$

we regain the differential equation (2.10a).

The above procedure illustrates that the way in which we scale down the flow rates as the time interval Δt decreases determines the restrictions on the parameters in the continuous-time models. For example, in deriving the continuous-time flu model from equation (2.6.4), we saw that the maximum allowable value of c increased to infinity. But the parameter a, which represents the probability of transmission per contact, retains the restriction of having to lie between zero and one from the discrete-time model because we did not scale this parameter at all when shrinking the time interval Δt to zero.

Because it is a considerable source of confusion for new (and experienced) modelers, it is also worth clarifying the difference between discrete- and continuous-time models in terms of the units of the parameters. In our discrete-time mouse model, d was a *fraction* (and therefore was constrained to lie between zero and one) whereas such parameters are referred to as *rates* in continuous-time models. But how and why did a fraction become a rate when moving to continuous time, and what is the relationship between the two? One way to understand this is to notice that the fraction of mice eaten in any time interval can always be written as a rate of consumption per unit time, d, multiplied by the length of the time interval in question, Δt. Of course, as we have seen, we must ensure that d and Δt are chosen so that $d\, \Delta t$ lies between zero and one because this represents the *fraction* eaten. In a discrete-time model, the unit of time is arbitrarily assigned the value of $\Delta t = 1$, so that we can view d as really being $d\, \Delta t$, where d remains a rate and $d\, \Delta t$ remains a fraction. This would, however, make the discrete-time equations harder to read, so that it is much clearer to refer to $d\, \Delta t$ as simply d. In deriving the continuous-time model, we put Δt explicitly back into the discrete-time model. But when we applied equation (2.6.2) to obtain a differential equation, we divided our expression for the change in $n(t)$ over the time interval by the length of the interval Δt. As a result, the continuous-time equations involve d alone on the right-hand side of the equation (rather than $d\, \Delta t$), which is a consumption *rate* (i.e., consumption per unit time) rather than a fraction.

Finally, just because we can derive continuous-time equations from discrete-time equations does not guarantee that they will behave in the same way. In fact, we will see some spectacular differences between these two types of models in Chapter 4. By deriving continuous-time equations from discrete-time equations, however, we gain some insight into when and why the dynamics of these models should differ. In the above, we shrunk the time interval, from one to 1/2 to Δt, which we then allowed to shrink to zero. This procedure changes how often the variables in

(continued)

Box 2.6 *(continued)*

the model are updated, from once to twice to $1/\Delta t$ times per original time unit. Every time we update the variables, we allow the changes that occur within one time interval to impact the changes that occur within the next time interval. If none of the variables change by much within one unit of time, then updating the variables will make little difference. If, however, the variables undergo large changes within a time unit (i.e., changes by more than just a few percent), then it will matter whether we fix the value of the variables to their initial values within a time unit or update these variables after every small interval of time Δt. Continuous-time models represent the extreme case where the variables are continuously updated over time. We will return to this issue in Chapters 4 and 6, but for now we conclude with the following important point: *The behavior of discrete-time and continuous-time models will be similar if each variable changes little over the time unit considered in the discrete-time model.*

When changes over a time unit are not small, all bets are off, and the discrete-time and continuous-time models can behave quite differently. In this case, discrete-time models are also quite sensitive to the ordering of events within a time unit (see the luncheon discussion in section 2.5.1). Unless there is a good biological reason to believe in one type of model (discrete-time versus continuous-time) and one type of ordering, then you should be careful before placing too much stock in any predictions from a model in which large changes can occur over a time unit. When such large changes are possible, it might be worthwhile deriving both a discrete-time and a continuous-time version of the model to see how sensitive the results are to the way in which the problem is modeled.

If you are just starting to model, the number of different mathematical techniques that are available is daunting. Keep in mind that even the best mathematicians do not know them all. Any modeler knows only a subset of possible techniques. It helps to remember this—not only because it's easier to manage learning math when you don't feel that you have to learn everything, but also because it is important to recognize that you should always keep an eye out for useful new techniques to add to your mathematical toolbox. We can always learn (and develop!) more techniques. A good idea is to read papers in the area that interests you to decide which mathematical techniques to learn first. While no one person can master all mathematical techniques, knowing the basic steps of modeling can allow you to collaborate effectively with modelers who do know the techniques that you need.

2.7 Checks and Balances

The process of mathematical modeling is rarely smooth. Rarely does the first set of equations that you write down end up being the final set. Generally, modeling is an iterative procedure. First of all, everybody makes mistakes. This means that it is critical to check your equations and analyses thoroughly and to start over again whenever you discover a mistake. The most obvious way to check for mistakes is to rederive everything. Oddly, many mistakes are not caught this way, probably because our minds are likely to make the same error twice. Therefore, it is a good idea to get into the habit of checking your results using other pieces of information.

If there are any constraints on the variables, make sure that your results obey these constraints. For example, if you are modeling the proportion of females within a population and the proportion of males, then the sum of these proportions should equal one. If you are modeling the number of mice within a population, you should stop the model as soon as the number becomes negative, which means that the mice have gone extinct. If the biological processes considered only add to a variable, that variable should never decrease over time.

Similarly, make sure that each equation has the right units or "dimensionality"—if you are modeling the number of individuals in a population, your answer should have the dimensions of a number, not a number squared. Plus, the units of the right-hand side of an equation should equal the units of the left-hand side. For example, in the cat-mouse model, equation (2.4) has units of number of mice on the left, $n(t+1)$. On the right, we have $(1+b)\,(1-d)\,n(t)+m$. The term $(1+b)$ has units "number of mice per mouse." The term $(1-d)$ measures the fraction of surviving mice and therefore has no units. Thus, $(1+b)\,(1-d)\,n(t)$ correctly has units of "number of mice," as does m, the number of (migrant) individuals. In the flu model, equation (2.7a) has units of number of infected individuals on the left, $n(t+1)$. On the right, we have $n(t)+a\,c\,n(t)\,s(t)$. The first term, $n(t)$, has the units "number of infected individuals." At first, it might seem as if the second term $a\,c\,n(t)\,s(t)$, has the dimensions of "(number of infected individuals) × (number of susceptible individuals)" because it involves the product of $n(t)$ and $s(t)$. But c is the probability of any given infected individual contacting a susceptible individual per number of susceptible individuals in the population (see section 2.4.2) and therefore has units of "1/(number of susceptible individuals)," while a is a probability and has no units. Thus, $a\,c\,n(t)\,s(t)$ also has the units "number of infected individuals."

Another way to check your results is to look at special cases where you know what should happen. For example, in the mice model, if predation and immigration are absent ($d = 0$ and $m = 0$), the equations should describe the same growth process as the tree-branching model (e.g., the recursion equation (2.4) becomes $n(t+1) = (1+b)\,n(t)$, which is identical to equation (2.3)). Therefore, any results that you obtain for the mice model with $d = 0$ and $m = 0$ should be the same as those for the branching model. Another good idea is to check results against simulations, which represent a special case where all parameters and starting conditions are specified.

Conversely, you can save a lot of effort if you notice that your model is a special case of another model or can be written in the same form as another model. This is helpful because you can then apply the known results of the other model to your own problem. For example, the equations for the branching model have the same form as the equations describing exponential growth (Chapter 3). The exponential growth model has been well studied, and we know a lot about its behavior. Realizing this, we can apply our knowledge about exponential growth (e.g., Figure 4.1 and equation (4.1)) to our tree branching model.

Finally, but most importantly, you should check your results by seeing if they make sense. What did you expect to happen? Do the results match your expectations? If the results match, they are more likely to be correct. If they don't match, then either the results are wrong or your intuition is wrong. If, after extensive checking, you cannot find any errors in the math, then try to figure out why your intuition was wrong. This is often an extremely valuable exercise, allowing you to correct and refine your understanding of the biological system. For example, the results of the model studied by Phillips (1996) led the research community to reevaluate what forces were driving HIV dynamics after infection (see Box 2.4 and section 1.3.1). Phillips' model showed that it was possible for viral loads to rise rapidly and then decline without the immune system kicking in. In hindsight, this result makes sense, even if it was difficult to foresee before modeling the problem.

Another reason why your final model might be different from the initial model is that you can get through the entire process and realize that your initial model was too simple or too complex. Sometimes, your results will indicate that your model was not exactly what you intended. For example, because the tree-branching model has the same solution as the model of exponential growth, the number of branches will grow exponentially (see Figure 4.1). This makes sense, because we did not include anything in the model to slow growth once the tree gets large. Realizing this, we might want to redo our model and include the possibility that, as the tree grows, it experiences more competition and shading from nearby branches. One way to do this would be to let the growth rate b decline as the number of branches grows. We shall talk about an extension to the exponential growth model that does exactly this in Chapter 3 (the logistic growth model). Similarly, as we have modeled the flu, susceptible individuals get infected until everybody has the flu. But in the real world there is never a time when everybody has the flu, because people recover from the flu and can also become resistant (see the SIR model developed in Chapter 3).

It is also easy to make your initial model too complex by including too many variables and parameters. With an overly complex model, you are much more likely to run into a brick wall in the analysis. Always consider whether every variable and parameter is necessary for you to address the biological question. Sometimes the answer will be "yes," in which case explore the model as best you can. At other times, certain details that, upon reflection, are less important can be dropped. For example, in the cat and mouse model, you might initially keep track of both the number of male mice and the number of female mice. But after running some simulations of the model, you might realize that it is only the number of female mice that matters to the dynamics, except when there are not enough males to fertilize the females. At this point, you might then decide to reduce the complexity of the model (following the principle of parsimony) and focus only on the total number of female mice.

Part of the art of mathematical modeling is learning how models can be simplified. Sometimes, parameters can be grouped together to reduce the total number of parameters in a model. For example, in the flu model (2.7), the contact rate c and the probability of infection per contact, a, always enter into the equations as the product $a\,c$. Thus, we can define a new variable $\beta = a\,c$, which measures the infectivity of the flu. Replacing a and c with β allows us to reduce the number of parameters by one. Doing so also allows us to see that increasing the contact rate or the probability of transmission per contact should have equivalent effects on the spread of the flu.

It is also sometimes possible to reduce the number of variables in a model. For example, in the flu model, the total number of individuals, $N = n(t) + s(t)$, remains constant. (You can show this for the discrete model by adding together $n(t + 1) + s(t + 1)$ and showing that the sum is the same as in the previous generation.) Therefore, you can rewrite equation (2.7a) by substituting $s(t) = N - n(t)$ to get an equation that involves only one variable ($n(t)$):

$$\begin{aligned} n(t+1) &= n(t) + a\,c\,n(t)\,(N - n(t)) \\ &= n(t) + \beta\,n(t)\,(N - n(t)). \end{aligned} \tag{2.11}$$

Reducing the number of parameters and variables makes a model more elegant, but more importantly it can make the model easier to analyze, as we shall see throughout the book. We will describe some methods to reduce the number of parameters and variables in a model in Chapter 9. But the best advice is to keep an eye out for features of a model (e.g., that the number of individuals remains constant) that might help to describe a model in the simplest and most elegant terms.

2.8 Relate the Results Back to the Question

You might be tempted to think that you are done once you have analyzed your model. Modeling biological processes, however, is worthwhile only if the mathematical results are related back to biological problems. At its best, theory is closely tied to empirical observations and tests. For example, empirical observations suggest a theoretical model, which generates an empirical test, which suggests that the model needs to be refined in particular ways. This interplay is extremely fruitful.

Even in the absence of such a tight interplay between modeling and empirical research, it is always desirable to go beyond the mathematical analysis and determine the broader biological insights that can be gained. How do the results alter the way scientists should think about a problem? What predictions can be made based on the model? What experiments could test these predictions? Are there any data that can be explained or better understood in light of the model? For example, the models that we described in the last chapter on HIV dynamics were interesting, not because of their mathematical equations, but because they helped us better understand how HIV might replicate and spread. They changed the way we thought about HIV, made counterintuitive predictions, and suggested empirical tests. Essentially, the difference between an important and widely read model and an irrelevant and obscure model lies, not in the modeling steps described above, but in this final step: describing how the model helps us better understand and interpret the biological world around us.

2.9 Concluding Message

In this chapter we have introduced the process of model construction by decomposing the task into a series of seven steps (Box 2.1). We have illustrated these steps with a series of toy examples as well as with models from the literature on HIV described in Chapter 1 (Boxes 2.4 and 2.5). In the next chapter we apply these same steps to derive some classic models from ecology, evolution, and epidemiology.

Problems

Problem 2.1: In Phillips' model of HIV dynamics within the body (Box 2.4), what parameter would you alter to incorporate an immune response to HIV particles? In three or four sentences, say how you might expand the model to incorporate this immune response.

Problem 2.2: Ground squirrels engage in alarm calls to alert their fellow squirrels that a predator may be present. Upon hearing a call, silent squirrels may start calling. Over time, calling squirrels may stop calling if the danger has not materialized. Draw a flow diagram with two circles representing the number of silent and calling ground squirrels over time. Place flow rates above each arrow in your diagram and describe in words what the flow represents.

Problem 2.3: The genome of any organism consists of a number of purine nucleotides (adenine and guanine) and pyrimidine nucleotides (cytosine and thymine). During DNA replication, however, mutations occasionally occur, causing a purine to be incorrectly replaced by a pyrimidine or vice versa. Figure 2.5 illustrates a flow diagram for this mutation process.

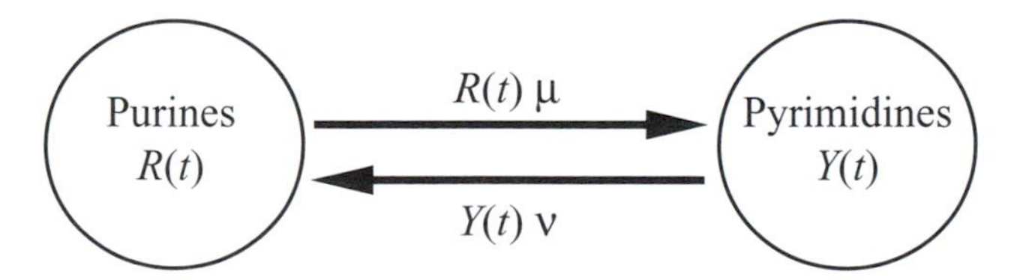

Figure 2.5: Two-state mutation model

(a) Based on the flow diagram, write down discrete-time equations for the number of purines, $R(t)$, and pyrimidines, $Y(t)$. You may choose to write either recursion or difference equations, but you should specify which type of equation you have chosen. (b) Write down continuous-time equations for the number of purines, $R(t)$, and pyrimidines, $Y(t)$.

Problem 2.4: Yeast and bacterial cells can be grown so that they divide continually using a "chemostat." Chemostats are tanks carrying a complete medium with all of the sugars and essential elements necessary for microbial growth, as illustrated in Figure 2.6. New medium is added to the tank via a constant drip (inflow), while

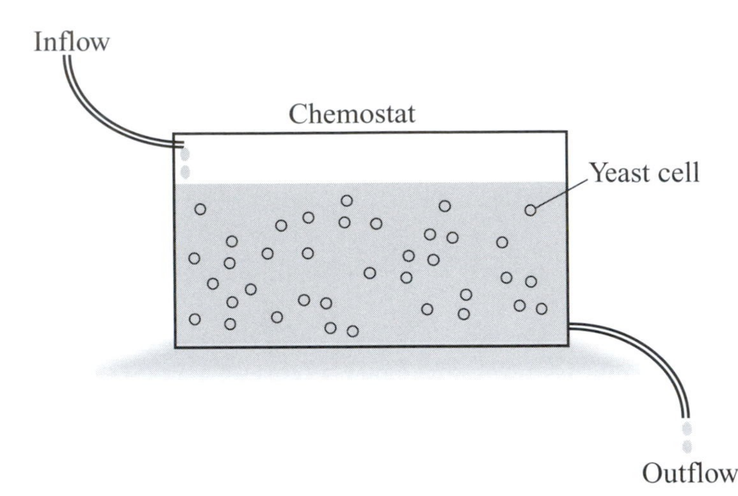

Figure 2.6: Yeast in a chemostat

used medium and cells exit via an effluent tube (outflow). To model the dynamics of a yeast population grown in a chemostat, (a) list all of the variables that you would want to include, (b) list all of the parameters that you think might be relevant, (c) describe the type of model that you are considering (discrete or continuous), and (d) specify any restrictions on the variables and parameters (e.g., $x(t)$ must be positive). Don't forget to describe the units for the variables that you choose (e.g., "number of individuals" or "density per milliliter") and the parameters (e.g., "rate of loss per cell per minute").

Problem 2.5: In Chapter 8 we will analyze a model for disease transmission based on the following equations: $dS/dt = \theta - d\,S - \beta\,S\,I + \gamma\,I$ and $dI/dt = \beta\,S\,I - (d + \nu + \gamma)I$. The variables S and I denote the number of susceptible and infected individuals. (a) Draw and label a flow diagram for these two variables. (b) Suggest a plausible biological interpretation of the parameters γ and ν.

Problem 2.6: Suppose that after contracting the flu, people are initially resistant to reinfection, but this immunity eventually wanes. (a) Alter the flow diagram for the flu model in Figure 2.4c to include a "recovered and immune" class with these properties. (b) Suppose that immune individuals have a constant per capita rate of losing immunity. What are the continuous-time equations (2.10) for this modified flu model?

Problem 2.7: There are six different possible orderings of events in the mouse model of the text. There are, however, only four different recursion equations, because some equations are compatible with more than one ordering of events. Match the orders of events (a)–(f) to their corresponding recursion equations (i)–(iv):

(a) Census, births, predation, migration
(b) Census, births, migration, predation
(c) Census, predation, births, migration
(d) Census, predation, migration, births
(e) Census, migration, births, predation
(f) Census, migration, predation, births

(i) $n(t + 1) = (1 + b)(n(t)\,(1 - d) + m)$
(ii) $n(t + 1) = (1 - d)(n(t)\,(1 + b) + m)$
(iii) $n(t + 1) = (1 - d)\,(1 + b)\,(n(t) + m)$
(iv) $n(t + 1) = (1 - d)\,(1 + b)\,n(t) + m$

Problem 2.8: The flow diagram in Figure 2.7 might describe the dynamics of colonial animals (e.g., naked mole rats) with reproductive individuals, nonreproductive "workers," and a specialized group of workers ("soldiers") that defend the colony and recruit new soldiers from among the worker class. (a) Infer which variables

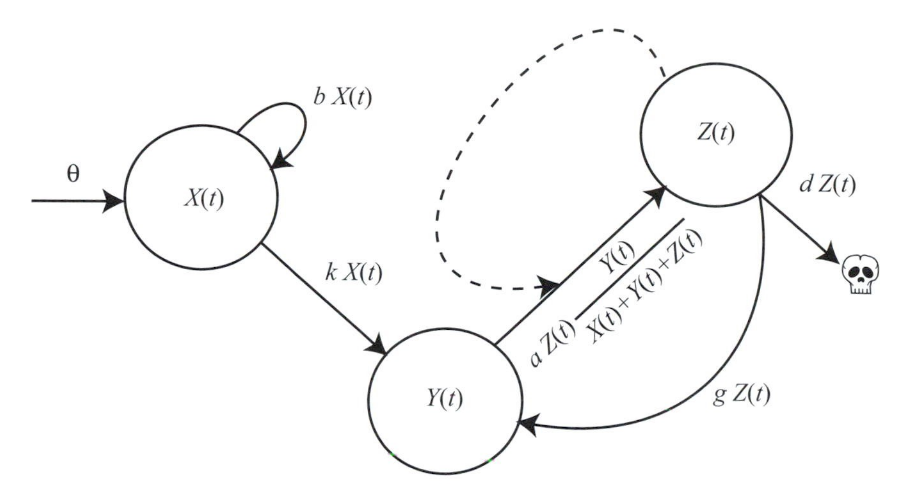

Figure 2.7: Class structure in a naked mole rate colony

correspond to the reproductive class, the soldier class, and the worker class. Specify what parts of the flow diagram you used to draw your inferences. (b) Derive continuous-time differential equations for the three variables $X(t)$, $Y(t)$, and $Z(t)$.

References

Blower, S. M., H. B. Gershengorn, and R. M. Grant. 2000. A tale of two futures: HIV and antiretroviral therapy in San Francisco. *Science* 287:650–654.

Phillips, A. N. 1996. Reduction of HIV concentration during acute infection: Independence from a specific immune response. *Science* 271:497–499.

CHAPTER 3

Deriving Classic Models in Ecology and Evolutionary Biology

Chapter Goals:

- To introduce some classic models in ecology, evolution, and epidemiology
- To derive dynamical equations for analysis in subsequent chapters

Chapter Concepts:

- Exponential growth
- Logistic growth
- Linear and nonlinear models
- Population-genetic models
- Symmetric models
- Normalization
- Competition models
- Consumer-resource models
- SIR epidemiological models

3.1 Introduction

In this chapter, we describe a variety of models whose behavior we will explore in subsequent chapters. The models we have chosen are classics in ecology, evolution, and epidemiology. These classic models incorporate many of the elements commonly encountered when developing new models, and they illustrate many of the dynamical patterns that you are likely to encounter, from fairly predictable to weird and wonderful.

In section 3.2 we introduce the exponential and logistic models describing population growth. In section 3.3, the haploid and diploid single-gene models of evolution by natural selection are introduced. Section 3.4 then introduces the Lotka-Volterra competition and predator-prey models, as well as consumer-resource models. Section 3.5 introduces SIR epidemiological models for the spread of infectious diseases. We will describe the basic assumptions underlying all of these models and derive their dynamical equations using the steps described in Chapter 2 (see Table 3.1 for quick reference). Finally, in section 3.6 we discuss how one can work backward, starting with published equations for a model and deducing its underlying assumptions.

3.2 Exponential and Logistic Models of Population Growth

In any species, the number of individuals changes over time in response to resource availability, competition, predation, disease, weather, and chance events. The simplest models describing changes in population size are exponential and logistic growth. Both assume that the environment is constant, and both ignore any interactions with other species (no competing species, predators, parasites, etc.). The two models of growth differ in what they assume about the availability of resources (e.g., food, water, nesting sites, etc.). The exponential growth model assumes that the amount of resources available to each individual is constant, regardless of the population size, whereas the logistic growth model assumes that fewer resources are available to each individual as the population size increases.

3.2.1 Exponential Population Growth

The discrete-time exponential growth model assumes that each reproducing parent is replaced by a constant number of individuals, R, in the next time unit. Technically, this assumes that all individuals in the population are capable of

TABLE 3.1
Models derived in this chapter. Type refers to (1) recursion equation in discrete time, (2) difference equation in discrete time, and (3) differential equation in continuous time.

Model	Type	Equation	
Exponential growth	1	$n(t+1) = R\, n(t)$	(3.1b)
	2	$\Delta n = (R-1)\, n(t)$	(3.2)
	3	$\frac{dn}{dt} = r\, n(t)$	(3.3)
Logistic growth	1	$n(t+1) = n(t) + r\, n(t)\left(1 - \frac{n(t)}{K}\right)$	(3.5a)
	2	$\Delta n = r\, n(t)\left(1 - \frac{n(t)}{K}\right)$	(3.5b)
	3	$\frac{dn}{dt} = r\, n(t)\left(1 - \frac{n(t)}{K}\right)$	(3.5c)
Haploid selection	1	$p(t+1) = \frac{W_A\, p(t)}{W_A\, p(t) + W_a\, q(t)}$	(3.8c)
	2	$\Delta p = \frac{(W_A - W_a)\, p(t)\, q(t)}{W_A\, p(t) + W_a\, q(t)}$	(3.9)
	3	$\frac{dp}{dt} = s\, p(t)\, q(t)$	(3.11b)
Diploid selection	1	$p(t+1) = p(t)^2 \frac{W_{AA}}{\bar{W}} + p(t)\, q(t) \frac{W_{Aa}}{\bar{W}}$	(3.13a)
Competition equations	1	$n_1(t+1) = n_1(t) + r_1 n_1(t)\left(1 - \frac{n_1(t) + \alpha_{12}\, n_2(t)}{K_1}\right)$	(3.14)
		$n_2(t+1) = n_2(t) + r_2 n_2(t)\left(1 - \frac{n_2(t) + \alpha_{21}\, n_1(t)}{K_2}\right)$	
	3	$\frac{dn_1}{dt} = r_1\, n_1(t)\left(1 - \frac{n_1(t) + \alpha_{12}\, n_2(t)}{K_1}\right)$	(3.15)
		$\frac{dn_2}{dt} = r_2\, n_2(t)\left(1 - \frac{n_2(t) + \alpha_{21}\, n_1(t)}{K_2}\right)$	
Consumer-resource equations	3	$\frac{dn_1}{dt} = f(n_1) - g(n_1, n_2)$	(3.16)
		$\frac{dn_2}{dt} = \varepsilon\, g(n_1, n_2) - h(n_2)$	
SIR equations	3	$\frac{dS}{dt} = b - d\, S(t) - a\, c\, S(t)\, I(t) + \sigma\, R(t)$	(3.19)
		$\frac{dI}{dt} = a\, c\, S(t)\, I(t) - \delta\, I(t) - \rho\, I(t)$	
		$\frac{dR}{dt} = \rho\, I(t) - \sigma\, R(t) - d\, R(t)$	

reproduction (as in a hermaphroditic or asexual species). The model can also be applied to species with separate male and female sexes, however, by assuming that the number of offspring is limited by the number of females and then counting females only.

To derive a general discrete-time exponential model, we allow births and deaths to affect the number of individuals in the population, $n(t)$. With two processes, we must specify an order to these events, and we shall assume that births are followed by deaths. Using b to denote the per capita number of births and d to denote the fraction of the population that dies, the life cycle and flow diagrams are then given by Figure 3.1.

This simple exponential model does not track the age of individuals (see Chapter 10 for models that do). In other words, individuals are treated equally regardless of whether they are one or several time units old. Consequently, the same exponential growth model can describe populations in which all parents die (nonoverlapping generations) or only some of them die (overlapping generations)—in either case, d measures the total fraction of the population that dies after the round of births.

We can now derive the recursion equation by applying Recipe 2.1 to Figure 3.1:

$$n'(t) = n(t) + b\,n(t) = (1+b)\,n(t) \qquad \text{(after births)},$$

$$n''(t) = n'(t) - d\,n'(t) = (1-d)\,n'(t) \qquad \text{(after deaths)}.$$

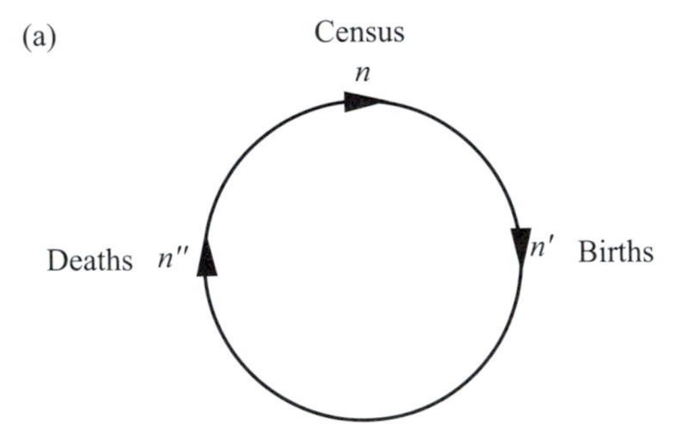

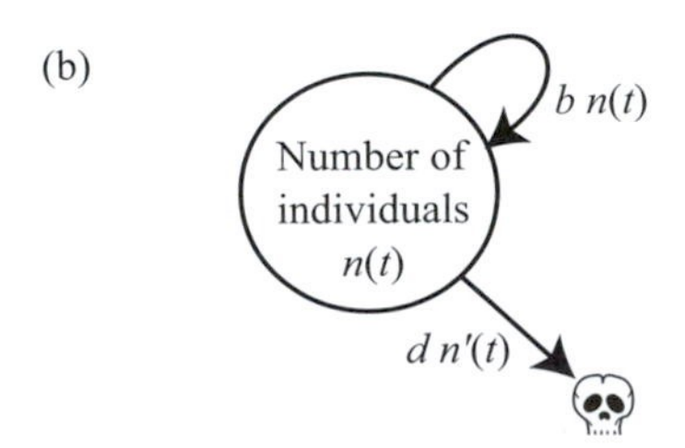

Figure 3.1: Exponential growth model. The (a) life cycle and (b) flow diagram for the exponential growth model.

As deaths represent the final event in the life cycle, we have $n(t + 1) = n''(t)$. Plugging the first equation into the second, we get

$$n(t + 1) = n''(t) = (1 - d)\,(1 + b)\,n(t). \tag{3.1a}$$

This recursion can be simplified by defining the compound parameter $R = (1 - d)(1 + b)$, so that

$$n(t + 1) = R\,n(t). \tag{3.1b}$$

R equals the number of surviving individuals per parent and is known as the *reproductive factor*. The key assumption of the exponential growth model is that R is constant, regardless of the population size.

We have already encountered a special case of the exponential growth model in Chapter 2. In the tree-branching example, each branch gave rise to b new branches in the next time step, and branches were never lost ($d = 0$). Thus, $R = 1 + b$, where R represents the parent branch plus b new branches (Figure 2.2a).

We can also describe the dynamics of the discrete-time exponential model using a difference equation, where $\Delta n = n(t + 1) - n(t)$:

$$\Delta n = (R - 1)\,n(t). \tag{3.2}$$

The quantity $(R - 1)$ is denoted by r in the biological literature and is the per capita change in the number of individuals from one generation to the next (known as the *growth rate*). Here we denote the growth rate by r_d, where the subscript d refers to the fact that this is the per capita change in the discrete-time model. From equation (3.1a) we have $r_d = R - 1 = b - d - b\,d$. If R equals one (each reproducing individual exactly replaces itself) then r_d equals zero (no change in the number of individuals), and the population size remains constant. Either of equations (3.1) and (3.2) can be used to predict the growth of a population over time (Chapter 4).

If births and deaths can occur at any point in time rather than during specific seasons, a continuous-time model of exponential growth is more appropriate. In the continuous-time exponential growth model, there is a per capita birth rate b and a per capita death rate d. When multiplied by a very small interval of time Δt, these give the number of births per individual and the fraction of the population that dies within that time interval (see Box 2.6). The flow diagram remains the same (Figure 3.1), except that, in any instant in time, only one event can happen. Therefore, we need not order the events, and we can drop the prime notation ($n'(t) = n(t)$). Applying Recipe 2.4 to the flow diagram, the differential equation for exponential population growth is

The *exponential growth* model assumes that the variable changes at a rate that is proportional to its current value.

$$\frac{dn}{dt} = b\,n(t) - d\,n(t) = r_c\,n(t) \tag{3.3}$$

where $r_c = b - d$ is the per capita rate of change in the number of individuals in the continuous-time model.

We can also derive the differential equation for exponential growth from the discrete-time recursion by shrinking the time interval over which events occur (Box 2.6). In a shorter unit of time Δt within the discrete-time model, we would expect the per capita number of births to be proportionately smaller (i.e., $b\,\Delta t$). The same is true of the fraction of the population that dies (i.e., $d\,\Delta t$). Consequently, after a short interval of time, the population size is

$$n(t + \Delta t) = (1 - d\,\Delta t)\,(1 + b\,\Delta t)\,n(t). \tag{3.4a}$$

Plugging $n(t + \Delta t)$ into the definition of a derivative (Appendix 2) and simplifying gives

$$\begin{aligned} \frac{\mathrm{d}n(t)}{\mathrm{d}t} &= \lim_{\Delta t \to 0}\left[\frac{n(t + \Delta t) - n(t)}{\Delta t}\right] \\ &= \lim_{\Delta t \to 0}\left[\frac{(1 - d\,\Delta t)(1 + b\,\Delta t)\,n(t) - n(t)}{\Delta t}\right] \\ &= \lim_{\Delta t \to 0}[(-d + b - d\,b\,\Delta\,t)\,n(t)]. \end{aligned} \tag{3.4b}$$

The term $-d\,b\,\Delta t$ disappears in the limit as Δt goes to zero, leaving us with $(-d + b)\,n(t)$, which is equivalent to the differential equation (3.3) derived from the flow diagram.

The derivation of equations (3.4) also helps to reveal why the discrete- and continuous-time parameters r_d and r_c differ by $-b\,d$. In the discrete-time model, deaths follow births within a single time step, and therefore newborns can potentially die before the next census. As we shrink down the time unit Δt, however, this possibility becomes negligible because very few births and deaths happen within a time unit.

As an example of exponential growth, in 1937, eight pheasants were introduced onto Protection island off the coast of Washington State, United States of America. Over the next five years the population grew exponentially (Lack 1954), nearly tripling in size every year ($R = 3$; Figure 3.2). If this trend had continued, we would be overrun by pheasants. Starting with eight pheasants, we would have $8 \times 3 = 24$ pheasants after one year, $8 \times 3 \times 3 = 72$ pheasants after two years,

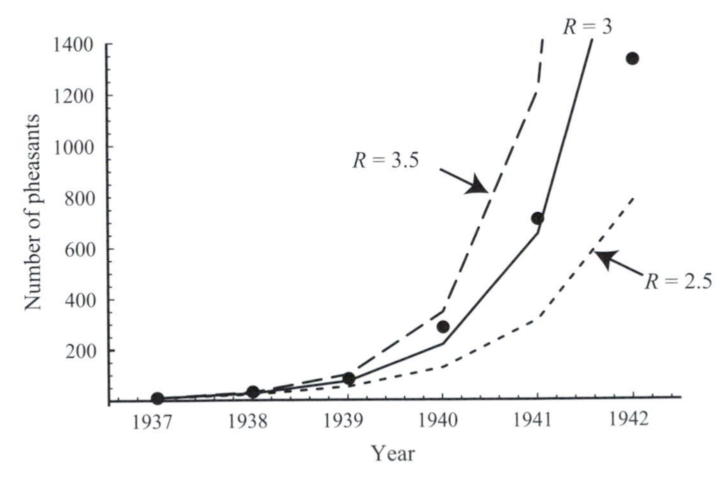

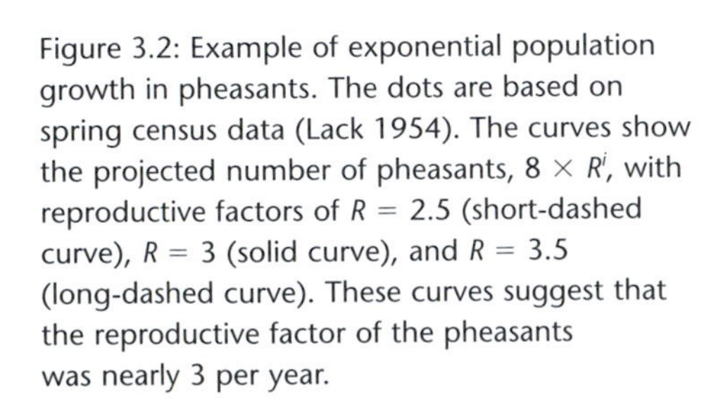
Figure 3.2: Example of exponential population growth in pheasants. The dots are based on spring census data (Lack 1954). The curves show the projected number of pheasants, $8 \times R^i$, with reproductive factors of $R = 2.5$ (short-dashed curve), $R = 3$ (solid curve), and $R = 3.5$ (long-dashed curve). These curves suggest that the reproductive factor of the pheasants was nearly 3 per year.

$8 \times 3 \times 3 \times 3 = 216$ pheasants after 3 years, etc. Generalizing from this trend, there would be 8×3^i pheasants after i years of growth. With the average pheasant weighing about 1.5 kilograms, it would take only 50 years for the total mass of pheasants (i.e., $8 \times 3^{50} \times 1.5 = 8.6\ 10^{24}$ kilograms) to exceed the mass of the earth (5.98×10^{24} kilograms). Obviously exponential growth cannot continue unabated. In fact, Lack observed that "the increase was slowing down and was about to cease (the number in 1942 falls short of a threefold increase), but at this point the island was occupied by the military and many of the birds shot."

3.2.2 Logistic Population Growth

Many factors slow the rate of growth of a population, including declining resource availability, increased predation pressure, and a higher incidence of disease. The logistic model in discrete time describes these processes indirectly by assuming that R declines with increasing population size. The exact form of this decline has been modeled in many ways (see Problem 3.4). The standard logistic model assumes that the number of surviving individuals per parent declines linearly with population size (Figure 3.3).

Writing the reproductive factor as $R(n)$ to emphasize its dependence on the population size, $R(n)$ starts at $(1 + r_d)$ when the population size is near zero and there is no competition for resources. The parameter r_d is known as the *intrinsic* rate of growth because it measures whether the population tends to grow ($r_d > 0$) or shrink ($r_d < 0$) when there is no competition for resources. As the population size increases, the reproductive factor decreases and eventually reaches a point where each individual exactly replaces itself; the population size at this point is called the *carrying capacity* K, because it is the maximum population size at which the population can sustain itself. From Figure 3.3, the equation for the line describing the reproductive factor as a function of population size is

$$R(n) = \underbrace{(1 + r_d)}_{\text{intercept}} + \underbrace{\left(-\frac{r_d}{K}\right)}_{\text{slope}} n(t).$$

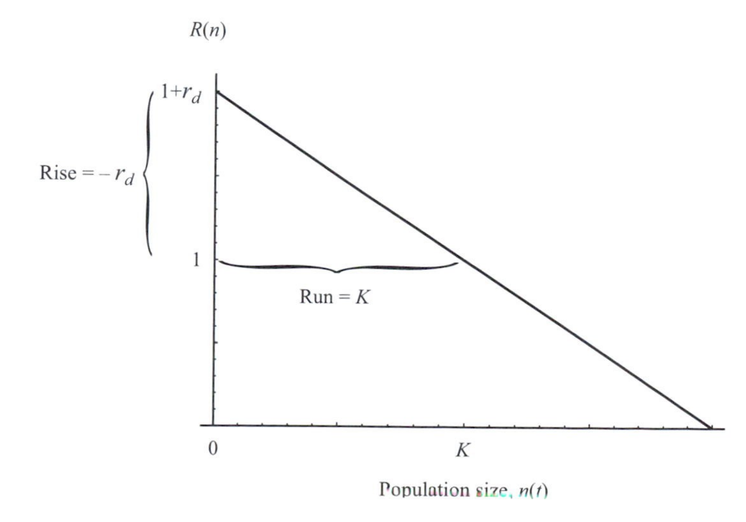

Figure 3.3: The number of surviving individuals per parent. The logistic model assumes that the number of surviving individuals declines linearly from a maximum of $1 + r_d$ when $n(t)$ is zero (the intercept) to one when $n(t)$ is K. The slope of the line is given by the rise ($-r_d$) over the run (K) between these two points.

The *logistic growth* model assumes that the rate of change of a variable decreases linearly as the variable increases in value, with no change occurring if the variable is at the carrying capacity K.

Replacing R with $R(n)$ and gathering terms that depend on r_d, the recursion equation (3.1b) for population growth becomes

$$n(t+1) = n(t) + r_d\, n(t)\left(1 - \frac{n(t)}{K}\right). \tag{3.5a}$$

Similarly, the difference equation (3.2) becomes

$$\Delta n = r_d\, n(t)\left(1 - \frac{n(t)}{K}\right). \tag{3.5b}$$

The logistic model in continuous time may be derived by assuming that the per capita growth rate declines linearly with population size. Denoting this growth rate by $r(n)$, try to derive an expression for this function assuming that $r(n)$ declines linearly with $n(t)$, from a maximum equal to the intrinsic growth rate r_c when there are no competitors (i.e., $r(0) = r_c$) to zero when the population is at its carrying capacity (i.e., $r(K) = 0$). Doing so, and replacing the constant growth rate r_c in equation (3.3) with this function, gives the continuous-time logistic equation

$$\frac{dn}{dt} = r_c\, n(t)\left(1 - \frac{n(t)}{K}\right). \tag{3.5c}$$

As an example, Mable and Otto (2001) cultured haploid and diploid populations of the yeast *Saccharomyces cerevisiae* in separate flasks containing nutrients. "Haploid" means that there is one copy of every gene within the genome, while "diploid" means that there are two copies. Bacteria are haploid; animals and plants are typically diploid; and many single-celled organisms, fungi, and algae can persist in either state. By counting cell densities over time, population sizes were estimated for the two ploidy levels (Figure 3.4 based on Figure 2 of Mable and Otto 2001). Although the populations grew nearly exponentially at first, growth rate decreased as the population size increased. The observed carrying capacity (K) is larger for the haploid cells, mainly because haploid cells are smaller than diploid cells and presumably require fewer resources per cell. Technically, yeast cells grow until the resources are depleted at which point the culture is said to be in "stationary phase" rather than at carrying capacity. But do the haploid and diploid cells have different intrinsic rates of increase, r_c? By fitting the logistic equation (3.5c) to the data, Mable and Otto (2001) estimated the parameters of the logistic equation as $K_{\text{haploid}} = 3.7 \times 10^8$, $r_{\text{haploid}} = 0.55$, $K_{\text{diploid}} = 2.3 \times 10^8$, and $r_{\text{diploid}} = 0.55$. (To find the best fitting parameters, a transformation was performed to turn the logistic equation into a line, and then fitting a line to the transformed data by a standard statistical technique known as linear regression. Details are available online in the *Mathematica* notebook for the figures in Chapter 3.) Despite the large difference in carrying capacities, the difference between the haploid and the diploid growth rate was negligible under these conditions.

Before moving on to the next section, we should introduce some terminology. The exponential growth equations (3.1), (3.2), and (3.3) represent

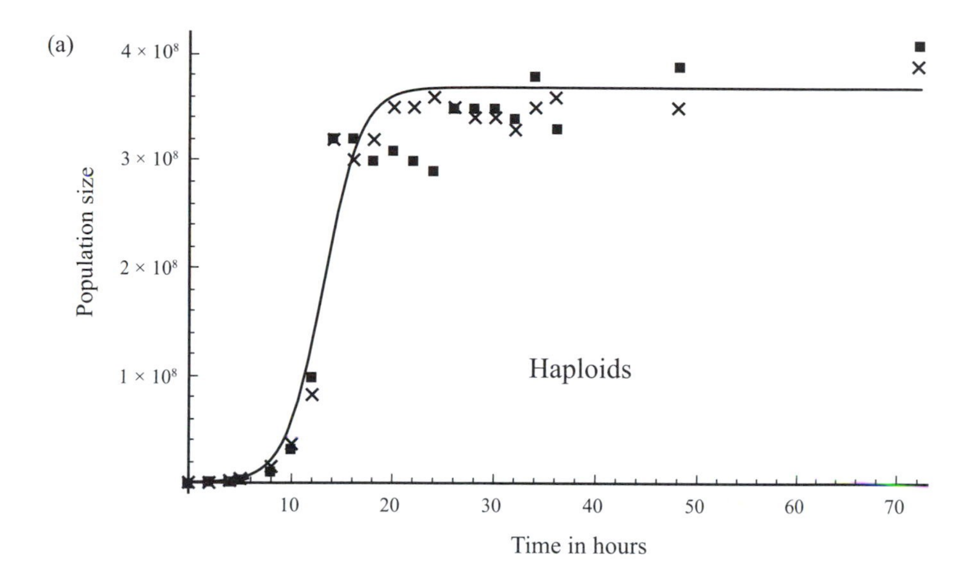

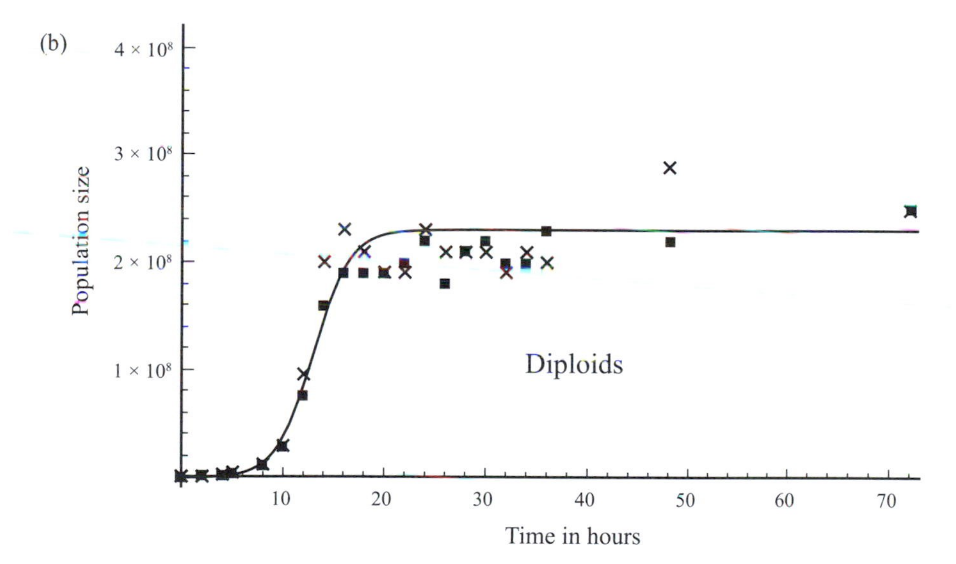

Figure 3.4: Example of logistic population growth in yeast. Population sizes from Mable and Otto (2001) are plotted over time based on data from (a) haploids and (b) diploids. Crosses and squares represent data from two different replicates. The solid curves are plots of the population size over time based on equations (3.5c) and the estimated parameter values (see Chapter 4).

linear models because the change that occurs per unit time depends on the variable only through a term that is proportional to $n(t)$. That is, these equations do not involve more complicated functions like $n(t)^2$, $1/n(t)$, $\exp(n(t))$, etc. In contrast, the logistic growth equations (3.5) represent *nonlinear* models because they are not linear functions of $n(t)$. This is more obvious if we multiply out the terms in the logistic equation; for example, equation (3.5b) can be rewritten as $\Delta n = r_d\, n(t) - r_d\, n(t)^2/K$, which involves the square of the variable, $n(t)^2$. As we shall see in Chapter 6, it is straightforward to obtain the *general solution* of linear models, meaning that there is an explicit equation that

In a *linear model*, the change per unit time is a linear function of the variables; models that involve more complicated functions of the variables are called *nonlinear models*.

predicts exactly what the value of the variable will be at any future point in time. But it can be difficult or even impossible to find general solutions for nonlinear equations. Importantly, most models in biology are nonlinear, because we typically want to understand how cells/organisms/populations interact, and interactions generate nonlinear equations.

It is confusing that equations (3.1), (3.2), and (3.3) are sometimes referred to as linear equations and, at other times, referred to as equations for exponential growth. These different terms reflect different ways of viewing a dynamical model: we can focus either on *changes* to the system as a function of the variables themselves, or on the *value* of the variable as a function of time. Consider equation (3.3). This equation for dn/dt tells us that the rate at which the variable changes is a linear function of the variable itself, $n(t)$. We therefore call this a linear differential equation. But just because a differential equation is linear does not mean that the value of the variable is a linear function of time. We must analyze the model further to determine the exact value of the variable as a function of time. For the exponential growth model, the value of the variable as a function of time is given by the general solution $n(t) = n(0)e^{r_c t}$ (see Chapters 4 and 6). Thus, the value of the variable $n(t)$ is an exponential function of the independent variable t.

3.3 Haploid and Diploid Models of Natural Selection

Population-genetic models describe how variants of a gene (alleles) change in frequency over time.

In the above section, survival and reproduction were assumed to be the same for every member of a population. What if genetic variation in these characteristics exists within a population? This question is addressed by population-genetic models that track the frequency of different variants within a population over time. In this section, we explore two models of evolution by natural selection, one that applies to a haploid population where each individual carries one *allele* (one variant) of a gene, and one that applies to a diploid population where each individual carries two alleles. Figure 3.5 illustrates the life cycle for the two models, specifying where selection acts relative to reproduction.

3.3.1 Haploid Models of Natural Selection

Consider a population of two types of individuals (A and a), which both breed true (i.e., type A produces only A offspring, and type a produces only a offspring). This scenario describes the case of a haploid population with two alleles (A and a) at one genetic locus. The model also describes two types of individuals (A and a) within an asexual population. First, consider the number of each type of individual: n_A and n_a. To be concrete, we assume that the population is censused immediately after juveniles are born, although equivalent equations can be derived using other census points (e.g., by counting haploid adults). Each type survives and reproduces according to Figure 3.1, but we now allow R to differ between A and a individuals. It is traditional in evolutionary

(a)

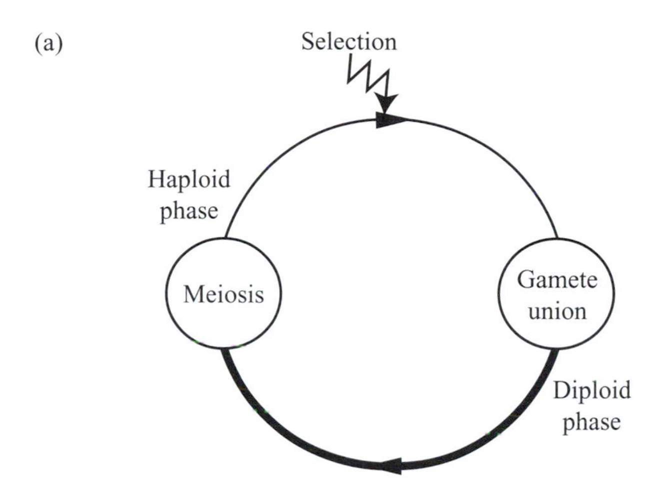

(b)

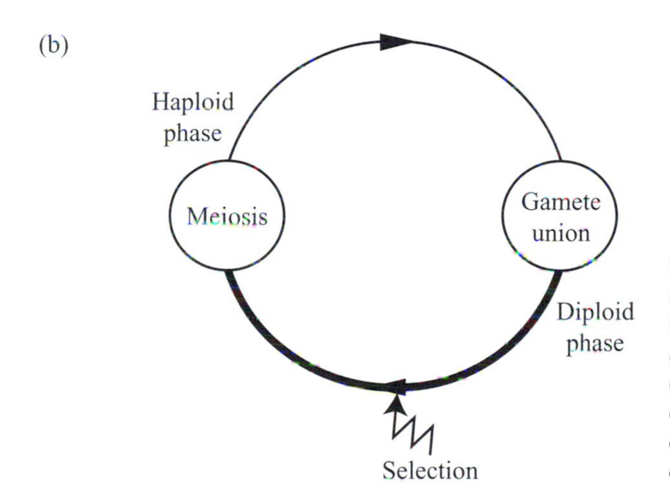

Figure 3.5: Life-cycle diagram for models of selection. All sexual life cycles can be conceptualized as having a haploid and a diploid phase, although there might be no growth in one phase (for example, the haploid phase of the human life cycle consists only of the single cells, sperm and eggs). Selection can act in the (a) haploid phase or (b) diploid phase.

biology to denote the reproductive factor by W, and therefore, using equation, (3.1) we have

$$n_A(t+1) = W_A\, n_A(t), \tag{3.6a}$$

$$n_a(t+1) = W_a\, n_a(t). \tag{3.6b}$$

In terms of the birth and death parameters, we have $W_A = (1 - d_A)\,(1 + b_A)$ and $W_a = (1 - d_a)\,(1 + b_a)$ both of which must be positive (or zero).

In model (3.6), it is arbitrary which allele we label A. If we exchange the label A for a (and vice versa) in equations (3.6), we end up with the same two equations. Such models are called *symmetric*. Observing that a model is symmetric is helpful because it provides an excellent way to check a set of equations; exchanging the parameters and variables for the symmetric terms should not affect the equations or their predictions.

In a *symmetric* model, the labels of the variables are arbitrary; interchanging the variable and parameter names does not alter the form of the equations.

Most evolutionary models focus on the frequency of each type rather than their absolute numbers. Typically, p and q are used to denote the proportions of A and a alleles. By definition, a proportion is equal to the number of a type divided by the total number of individuals ($n_A + n_a$):

$$p = \frac{n_A}{n_A + n_a}, \tag{3.7a}$$

$$q = \frac{n_a}{n_A + n_a}. \tag{3.7b}$$

Because they are proportions, p and q must lie between zero and one, and their sum must always equal one. Thus, if we know the value of one of these variables, the other can be obtained from the fact that $p + q = 1$. Consequently, the dynamics of the allele frequencies can be described using only a single variable, e.g., p.

By plugging equation (3.6) into (3.7a), the frequency of allele A in the next time unit $p(t + 1)$ can be found:

$$p(t+1) = \frac{n_A(t+1)}{n_A(t+1) + n_a(t+1)} \tag{3.8a}$$

$$= \frac{W_A\, n_A(t)}{W_A\, n_A(t) + W_a\, n_a(t)}. \tag{3.8b}$$

Equation (3.8b) gives the allele frequency at time $t + 1$, but it does so in terms of the number of each type at time t. To obtain a recursion equation that can predict the dynamics of the allele frequency over the course of several generations, we need to write the right-hand side of (3.8b) in terms of $p(t)$. There are many ways to accomplish this transformation using the relationship (3.7) between allele frequencies and allele numbers. The easiest is to use a trick: if we divide the top and bottom by ($n_A + n_a$), then each term involves only p or q:

$$p(t+1) = \frac{W_A \dfrac{n_A(t)}{n_A(t) + n_a(t)}}{W_A \dfrac{n_A(t)}{n_A(t) + n_a(t)} + W_a \dfrac{n_a(t)}{n_A(t) + n_a(t)}}$$

$$= \frac{W_A\, p(t)}{W_A\, p(t) + W_a\, q(t)} \tag{3.8c}$$

$$= \frac{W_A\, p(t)}{W_A\, p(t) + W_a(1 - p(t))}$$

where, in the last line, we use the fact that the allele frequencies sum to one to rewrite $q(t)$ as $1 - p(t)$.

While most experienced modelers would use this trick (because they can "see" that it would simplify the equation), such insight only comes with experience. An alternative and more general method toward the same end is to solve equation (3.7) for the old variable n_A in terms of the new variable p. Multiplying both sides of equation (3.7a) by $n_A + n_a$, we get $p\,(n_A + n_a) = n_A$.

Gathering the terms involving the old variable n_A on the left, we get $p\, n_A - n_A = -p\, n_a$. Finally, dividing both sides by the factor $(p - 1)$, we get

$$n_A = \frac{-p}{p-1} n_a = \frac{p}{1-p} n_a.$$

Plugging this term at time t into the right-hand side of equation (3.8b) gives

$$p(t+1) = \frac{W_A \dfrac{p(t)}{1-p(t)} n_a(t)}{W_A \dfrac{p(t)}{1-p(t)} n_a(t) + W_a\, n_a(t)}.$$

By factoring out $n_a(t)$ and multiplying the top and bottom by $1 - p(t)$, we obtain equation (3.8c).

In this model, the numbers of surviving individuals per parent for the two alleles are W_A and W_a. These numbers are known as the *absolute fitnesses* of the two types. Although equation (3.8c) contains both of these parameters, we can use a technique similar to that used when deriving (3.8c) to reduce the number of parameters in this model. Again, it might not be obvious how to do this, but with experience you will be able to see that dividing the top and bottom of (3.8c) by W_a produces an equation with only one parameter:

$$p(t+1) = \frac{V_A\, p(t)}{V_A p(t) + (1 - p(t))} \tag{3.8d}$$

where $V_A = W_A/W_a$ is the *relative fitness* of allele A (i.e., the fitness of allele A relative to allele a).

The fact that the dynamics of allele frequency depend only on the relative fitnesses of the two alleles reveals important information about the causes of evolutionary change. For example, suppose that resource availability fluctuates over time in a way that causes the number of surviving offspring per parent to equal $\sigma(t)\, W_A$ and $\sigma(t)\, W_a$ for alleles A and a, where $\sigma(t)$ is a factor that varies over time and modulates the otherwise constant reproductive factors W_A and W_a. Despite this added environmental complexity, the allele frequency dynamics continue to be described by equation (3.8d), because $\sigma(t)$ cancels out when calculating the relative fitness V_A. Similarly, if the density of the population $(n_A + n_a)$ affects the number of surviving offspring per parent carrying allele A by the same factor as those parents carrying allele a, the allele frequency dynamics will remain the same. This is a key insight: evolutionary change can be studied without reference to the ecological context of a population as long as the ecological context affects the reproductive output of each allele by the same factor. This result does not hold, however, if the ecological context differs between the alleles, e.g., if they differ in their sensitivity to density-dependent competition (see Problems 3.10 and 3.17).

Given the recursion equation for the discrete-time haploid model, we can derive a continuous-time differential equation following the method of

Box 2.6. To determine the allele frequency change, we calculate the difference in allele frequency over one generation by subtracting the current allele frequency $p(t)$ from both sides of equation (3.8c):

$$\begin{aligned}\Delta p = p(t+1) - p(t) &= \frac{W_A\, p(t)}{W_A\, p(t) + W_a\,(1 - p(t))} - p(t) \\ &= \frac{(W_A - W_a)\, p(t)\,(1 - p(t))}{W_A\, p(t) + W_a\,(1 - p(t))},\end{aligned} \tag{3.9}$$

which is factored by placing each term over the common denominator $W_A\, p(t) + W_a\,(1 - p(t))$. It is useful to define the *selection coefficient* for the discrete-time model, $s_d = (W_A - W_a)/W_a$, which is the proportional difference in fitness between allele A and allele a. We can use this definition to replace W_A in equation (3.9) with $W_A = W_a s_d + W_a$, giving us

$$\begin{aligned}\Delta p &= \frac{W_a s_d\, p(t)(1 - p(t))}{(W_a s_d + W_a)\, p(t) + W_a\,(1 - p(t))} \\ &= \frac{s_d\, p(t)(1 - p(t))}{1 + s_d\, p(t)}.\end{aligned} \tag{3.10}$$

Equation (3.10) describes the amount of allele frequency change over one generation. We can derive the differential equation describing the rate of allele frequency change in continuous time by supposing that the number of births and the fraction of the population that dies in a time interval, Δt, are $b\,\Delta t$ and $d\,\Delta t$, respectively, just as we did in the exponential growth model (3.4a). The reproductive outputs of A and a individuals over this interval become $W_A(\Delta t) = (1 - d_A\Delta t)(1 + b_A\Delta t)$ and $W_a(\Delta t) = (1 - d_a\Delta t)(1 + b_a\Delta t)$, and therefore the selection coefficient will be a function of the length of the time interval Δt. The selection coefficient $s_d(\Delta t)$ can be written explicitly as

$$\begin{aligned}s_d(\Delta t) &= \frac{W_A(\Delta t) - W_a(\Delta t)}{W_a(\Delta t)} \\ &= \frac{((b_A - d_A) - (b_a - d_a))\Delta t - (b_A\, d_A - b_a\, d_a)\Delta t^2}{1 + (b_a - d_a)\Delta t - b_a\, d_a\, \Delta t^2}.\end{aligned}$$

We can derive a differential equation for the allele frequency using equation (3.10) and the definition of the derivative (Appendix 2):

$$\frac{\mathrm{d}p}{\mathrm{d}t} = \lim_{\Delta t \to 0} \frac{\Delta p}{\Delta t} = \lim_{\Delta t \to 0}\left(\frac{s_d(\Delta t)}{\Delta t}\,\frac{p(t)(1 - p(t))}{1 + s_d(\Delta t)\, p(t)}\right). \tag{3.11a}$$

The term $s_d(\Delta t)$ enters equation (3.11a) in two places. It first enters as $s_d(\Delta t)/\Delta t$. Dividing $s_d(\Delta t)$ by Δt and letting Δt go to zero, we get $(b_A - d_A) - (b_a - d_a)$, which is the selection coefficient for the continuous-time model, s_c. The second place it enters is $1 + s_d(\Delta t)\, p(t)$, which goes to one as Δt goes to zero.

Consequently, as we shrink the time interval in equation (3.11a), we are left with

$$\frac{dp}{dt} = s_c p(t)(1 - p(t)). \tag{3.11b}$$

The same differential equation can be derived from the flow diagram for each of the alleles (Figure 3.1) using the quotient rule of calculus (Box 3.1).

The differential equation (3.11b) describes how the allele frequency changes over time and provides insight into the action of natural selection (Figure 3.6). First, when allele *A* or *a* is absent from the population (i.e., when p or q equals zero), dp/dt will be zero, and p remains constant. In other words, if there is no genetic variation, there will be no change in allele frequency due to selection. Second, the rate of evolutionary change is maximized when the alleles are equally frequent (Figure 3.6; Problem 3.19), i.e., when genetic variation is most abundant.

At first sight, the differential equation (3.11b) describing the change in allele frequency in continuous time looks substantially different from the difference equation (3.10) describing the change in allele frequency over a discrete generation. Yet these equations are not as different as they seem. When selection is weak (i.e., when s_d is near zero), the denominator in equation (3.10), $1 + s_d\, p(t)$, will be very nearly one. Making this approximation, the difference and differential equations become identical (see Primer 1 for an overview of approximations). Consequently, when selection is weak the evolution of allele frequencies should be similar, regardless of whether they are modeled in discrete or continuous time (Figure 3.6).

3.3.2 Diploid Models of Natural Selection

As diploid organisms, it is natural for us to wonder if evolutionary change in diploids differs from evolutionary change in haploids. For example, does the efficiency with which natural selection changes allele frequencies differ between haploids and diploids? To answer this question, we need to develop a model of selection for diploids. Even with only two alleles present (again, *A* and *a* at frequencies p and q), we must keep track of three genotypic combinations within a diploid population: *AA* homozygotes, *Aa* heterozygotes, and *aa* homozygotes. You might wish to repeat the derivations used above, starting with the number of each diploid genotype and converting these equations into the frequencies of each allele as we did in equations (3.6)–(3.11) (see Supplementary Material 3.1). In the following, we take a different approach, which allows us to introduce several shortcuts that are often used in deriving evolutionary models.

The first shortcut involves censusing the population at the stage in the life cycle that is simplest to describe. In a sexual population, diploid organisms are produced by the union of haploid gametes. Because there are only two types of gametes (*A* and *a*) rather than three diploid genotypes, it is simpler to census at the gamete stage, letting the frequency of *A* and *a* gametes be $p(t)$ and $q(t)$, respectively. As long as every diploid individual is formed by gametes uniting

Box 3.1: Haploid Selection and the Quotient Rule

In the text we considered how a discrete-time haploid model of selection can be derived from the exponential model of population growth (Figure 3.1) by considering two types of individuals, those that carry allele A and those that carry allele a. The same method can be used to derive a continuous-time haploid model of selection. Again, we consider the number of each type of individual, n_A and n_a, and we let each type reproduce according to Figure 3.1 (except that only one event happens at any instant in time and $n' = n$). Let the growth rate $r = (b - d)$ depend on the allele carried by an individual, r_A or r_a. We can write down differential equations for the dynamics of the two types (see equation 3.3):

$$\frac{\mathrm{d}n_A}{\mathrm{d}t} = r_A\, n_A(t), \tag{3.1.1a}$$

$$\frac{\mathrm{d}n_a}{\mathrm{d}t} = r_a\, n_a(t). \tag{3.1.1b}$$

Equations (3.1.1) describe how the numbers of each type change, but what if we want to know how the allele frequency changes over time? The first step is to rewrite $\mathrm{d}p/\mathrm{d}t$ using equation (3.7) as $\mathrm{d}(n_A/(n_A + n_a))/\mathrm{d}t$. (Here, we simplify the notation by dropping the (t) and calling the variables p, n_A, and n_a.) We can use the quotient rule (A2.13) to evaluate this derivative:

$$\frac{\mathrm{d}p}{\mathrm{d}t} = \frac{\mathrm{d}\left(\dfrac{n_A}{n_A + n_a}\right)}{\mathrm{d}t}$$

$$= \frac{\dfrac{\mathrm{d}n_A}{\mathrm{d}t}(n_A + n_a) - n_A\dfrac{\mathrm{d}(n_A + n_a)}{\mathrm{d}t}}{(n_A + n_a)^2}. \tag{3.1.2}$$

Because the derivative of a sum, $\mathrm{d}(n_A + n_a)/\mathrm{d}t$, is the sum of the derivatives, $\mathrm{d}n_A/\mathrm{d}t + \mathrm{d}n_a/\mathrm{d}t$ (Rule A2.3), equation (3.1.2) equals

$$\frac{\mathrm{d}p}{\mathrm{d}t} = \frac{\dfrac{\mathrm{d}n_A}{\mathrm{d}t}(n_A + n_a) - n_A\dfrac{\mathrm{d}n_A}{\mathrm{d}t} - n_A\dfrac{\mathrm{d}n_a}{\mathrm{d}t}}{(n_A + n_a)^2}.$$

Canceling out terms in the numerator leaves

$$\frac{\mathrm{d}p}{\mathrm{d}t} = \frac{n_a\dfrac{\mathrm{d}n_A}{\mathrm{d}t} - n_A\dfrac{\mathrm{d}n_a}{\mathrm{d}t}}{(n_A + n_a)^2}.$$

(continued)

Box 3.1 *(continued)*

Now, we can use equations (3.1.1) to replace the derivatives on the right-hand side:

$$\frac{dp}{dt} = \frac{n_a(r_A n_A) - n_A(r_a n_a)}{(n_A + n_a)^2}$$

$$= (r_A - r_a)\frac{n_A n_a}{(n_A + n_a)^2}.$$

Finally, we can use (3.7) to rewrite the product $n_A\, n_a/(n_A + n_a)^2$ as pq, leaving us with

$$\frac{dp}{dt} = (r_A - r_a)\, pq. \tag{3.1.3}$$

If we set $q = 1 - p$ and define the selection coefficient s_c as the difference in growth rates, $s_c = r_A - r_a$, equation (3.1.3) becomes equation (3.11b).

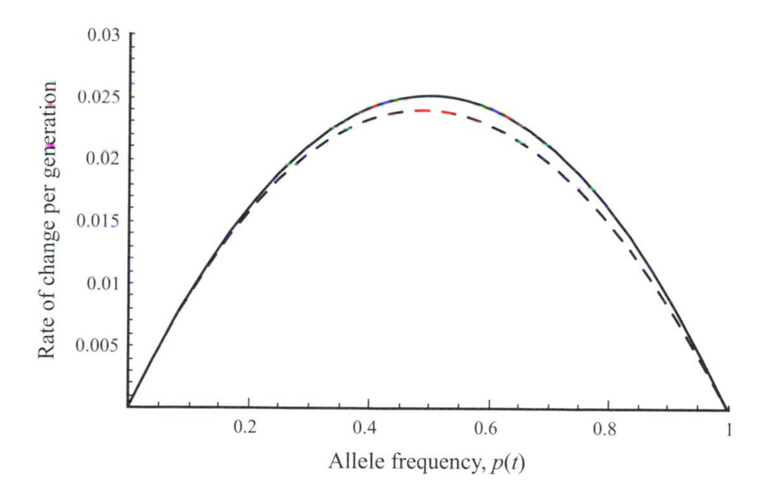

Figure 3.6: Allele frequency change with selection. The thick curve illustrates the instantaneous rate of change in the frequency of allele A, dp/dt, for the continuous-time model of haploid selection, given by the differential equation (3.11b). The dashed curve illustrates the per generation amount of change Δp in the discrete-time model, given by the difference equation (3.10). The two equations predict similar rates of change as long as selection is not too strong ($s = 0.1$ shown here).

at random with one another (i.e., as long as there is no selfing, assortative mating, sexual selection, spatial structure, surviving individuals from the previous generation, etc.), then the frequency of each diploid genotype can be calculated from the gamete frequencies alone. (Fortunately, the same holds when individuals, not gametes, unite at random; see Problem 3.7.) When two gametes unite, the chance that the first gamete (the egg, say) carries allele A is $p(t)$. The chance that the second gamete (the sperm, say) carries allele A is also $p(t)$. Thus, of the $p(t)$ offspring formed from an A egg, a fraction $p(t)$ will involve an A sperm. But a fraction $p(t)$ of $p(t)$ is just $p(t)^2$ (e.g., 1/6 of 1/6 is 1/36). Thus, $p(t)^2$ of the offspring will be AA. This logic may be repeated for all of the possible combinations of gametes as illustrated in Figure 3.7, demonstrating that the

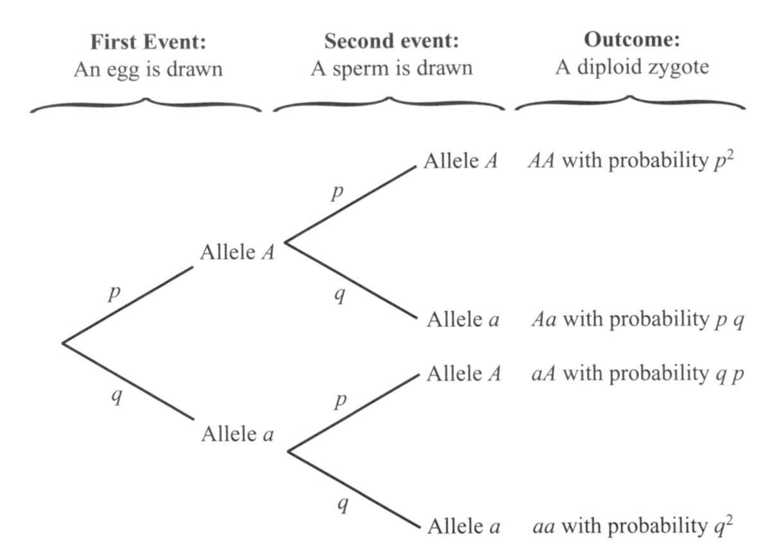

Figure 3.7: A "decision" tree for the union of two gametes. The outcome of this decision tree is a zygote (far right), which is formed by drawing an egg (first event) followed by a sperm (second event) from a large gamete pool. At each branching point in a decision tree, all of the possible alternatives are illustrated as lines, with the chance of observing each alternative written above the line. For each gamete, the chance of drawing allele *A* is *p*, and the chance of drawing allele *a* is *q*. As long as the events are independent of one another, then the probability of observing a particular outcome is equal to the product of the probabilities over all lines leading to that outcome. Because we assume that the gametes unite at random, the types of eggs and sperm are, by definition, independent of one another. If there are alternative ways of achieving the same outcome, the total probability of observing the outcome is the sum of the probabilities for each alternative. For example, an *Aa* heterozygote may be produced by the combination of an *A* egg and an *a* sperm, or vice versa, each of which occurs with probability *pq*. Thus, the total probability of observing a heterozygote is 2*pq*.

frequencies of the diploid genotypes at the beginning of a generation (among the "zygotes") are $p(t)^2$ for *AA*, $2\,p(t)q(t)$ for *Aa*, and $q(t)^2$ for *aa*. These are the so-called Hardy-Weinberg proportions. As proportions, they should sum to one, which indeed they do: $p(t)^2 + 2\,p(t)q(t) + q(t)^2 = (p(t) + q(t))^2 = (1)^2 = 1$.

The second short-cut is to recognize that selection alters the frequency of each genotype by an amount proportional to its fitness. This is clearly true when dealing with the numbers of each genotype (see equations (3.6)), but it also holds if we convert to genotype frequencies by dividing by the total population size. This short-cut allows us to skip directly to a description of the diploid genotype frequencies after selection, which are proportional to $p(t)^2 W_{AA}$, $2p(t)q(t)W_{Aa}$, and $q(t)^2 W_{aa}$. These quantities no longer sum to one, however, but instead sum to what is known as the mean fitness of the population:

$$\overline{W} = p(t)^2\, W_{AA} + 2\,p(t)q(t)\, W_{Aa} + q(t)^2 W_{aa}. \tag{3.12}$$

To ensure that the frequencies sum to one and that the frequency of each genotype after selection is proportional to its fitness, we divide by the mean fitness to get $p(t)^2W_{AA}/\overline{W}$, $2p(t)q(t)W_{Aa}/\overline{W}$, and $q(t)^2W_{aa}/\overline{W}$. This process of dividing a set of non-negative numbers by their sum is called *normalization*; it converts the numbers into frequencies. These calculations are summarized in Table 3.2. Such a table of events is very handy, especially when the number of genotypes becomes large, in which case flow diagrams become hopelessly complex.

To convert a set of numbers into the frequency of each type, we *normalize* the numbers by dividing by their sum; this ensures that the frequencies sum to one.

Finally, the diploid adults that have survived selection undergo meiosis to produce gametes. With Mendelian segregation, the percentage of gametes that carry the A allele is 100% from AA individuals, 50% from Aa individuals, and 0% from aa individuals (ignoring mutation). This information is provided in the second to last column of Table 3.2. Multiplying this column by the frequency of each adult, gives us the total frequency of the A gamete in the next generation:

$$p(t+1) = p(t)^2 \frac{W_{AA}}{\overline{W}} + \frac{1}{2}\left(2\,p(t)\,q(t)\,\frac{W_{Aa}}{\overline{W}}\right). \tag{3.13a}$$

Similarly, the total frequency of the a gamete is:

$$q(t+1) = \frac{1}{2}\left(2\,p(t)\,q(t)\,\frac{W_{Aa}}{\overline{W}}\right) + q(t)^2\,\frac{W_{aa}}{\overline{W}}. \tag{3.13b}$$

Because this model is also symmetric, we can check equation (3.13a) by replacing A with a and p with q, which correctly has the form of equation (3.13b). Another check is that the gamete frequencies in the next generation should sum to one, which they do: $p(t+1) + q(t+1) = (p(t)^2W_{AA} + p(t)\,q(t)W_{Aa} + p(t)\,q(t)W_{Aa} + q(t)^2\,W_{aa})/\overline{W}$, which equals one by the definition (3.12) of the mean fitness.

In the above derivation, the fitnesses (W) act as "weights" that increase or decrease the representation of each genotype relative to the others. In biological terms, W could represent differences in survival ability ("viability") or differences in an individual's ability to make offspring ("fertility") or both. Furthermore, we could multiply each of the W's by any common factor σ without

TABLE 3.2
Genotype frequencies in the diploid model of natural selection

Gametes uniting to form a zygote	Frequency before selection	Frequency weighted by fitness	Frequency after selection	Gametes produced	
				A	a
$A \times A$	p^2	p^2W_{AA}	$p^2\frac{W_{AA}}{\overline{W}}$	1	0
$A \times a$	$p\,q$	$p\,q\,W_{Aa}$	$p\,q\frac{W_{Aa}}{\overline{W}}$	$\frac{1}{2}$	$\frac{1}{2}$
$a \times A$	$q\,p$	$q\,p\,W_{Aa}$	$q\,p\frac{W_{Aa}}{\overline{W}}$	$\frac{1}{2}$	$\frac{1}{2}$
$a \times a$	q^2	$q^2\,W_{aa}$	$q^2\frac{W_{aa}}{\overline{W}}$	0	1

changing the behavior of the model, because the factor would cancel out when dividing by the mean fitness. Thus, as in the haploid model of natural selection, changes in allele frequencies among diploid organisms depend only on the relative success of each genotype under the assumptions we have made (e.g., no mutation, random mating, large population size).

Although the haploid and diploid models of natural selection are similar in form, the efficiency of natural selection differs substantially. As described in Problem 3.9, the diploid model requires twice the amount of selection to accomplish the same change in allele frequency as the haploid model. At an intuitive level, diploid selection is less effective because the benefits of the most fit allele are partially masked in heterozygotes by the expression of the least fit allele.

3.4 Models of Interactions among Species

Species do not exist in isolation from one another. The simple models of exponential and logistic growth fail to capture the fact that the growth of a species must depend on its interactions with other species. Do the species compete with or aid each other in obtaining resources? Do they fight for territories or provide a habitat for one another? Does a species consume or provide fodder for another species? A complete model describing every species interaction would be prohibitively complex except in the simplest of biological communities. Yet some insight can be obtained by focusing on specific types of interactions and how these interactions affect population growth. Below we focus on competition models (section 3.4.1) and consumer-resource models (section 3.4.2).

3.4.1 The Lotka-Volterra Model of Competition

Let us begin by introducing the Lotka-Volterra model of competition, which builds upon the one-species logistic model. In the logistic equation, intraspecific competition becomes more severe as the population size increases, resulting in a lower per capita growth rate (Figure 3.3). The Lotka-Volterra model extends the logistic model by allowing multiple species to compete for resources, resulting in both inter- and intraspecific competition.

To keep things simple, we consider only two competing species, whose numbers $n_1(t)$ and $n_2(t)$ change over time. The two species can have different intrinsic growth rates r_1 and r_2, and different carrying capacities K_1 and K_2. To account for competition between the two species, the Lotka-Volterra model assumes that each individual of species i experiences competition as if its own species had a population size of $n_i(t) + \alpha_{ij}\, n_j(t)$. (Here we use i to refer to either species 1 or 2, depending on which species is the current focus of our attention, and use j to refer to the competing species. This saves us from having to write down equations for both species each time.) The parameter α_{ij} is known as the *competition coefficient*, which converts the strength of competition exerted by an individual of species j on an individual of species i into the equivalent amount of competition that would be exerted if both individuals were of

> *Competition models* describe how the number of individuals of each species changes when more than one species uses the same resource.

species i. Assuming that the reproductive factor of species i declines linearly with the amount of competition, we now have

$$R_i = \underbrace{(1 + r_i)}_{\text{intercept}} + \underbrace{\left(-\frac{r_i}{K_i}\right)}_{\text{slope}} (n_i(t) + \alpha_{ij}\, n_j(t))$$

Incorporating this reproductive factor into the discrete-time recursion equations $n_i(t + 1) = R_i\, n_i\,(t)$ and gathering together terms involving r_i, the Lotka-Volterra model becomes

$$n_1(t + 1) = n_1(t) + r_1\, n_1(t)\left(1 - \frac{n_1(t) + \alpha_{12}\, n_2(t)}{K_1}\right), \tag{3.14a}$$

$$n_2(t + 1) = n_2(t) + r_2\, n_2(t)\left(1 - \frac{n_2(t) + \alpha_{21}\, n_1(t)}{K_2}\right). \tag{3.14b}$$

Similarly, the continuous-time differential equations for the Lotka-Volterra model are

$$\frac{dn_1}{dt} = r_1\, n_1(t)\left(1 - \frac{n_1(t) + \alpha_{12}\, n_2(t)}{K_1}\right), \tag{3.15a}$$

$$\frac{dn_2}{dt} = r_2\, n_2(t)\left(1 - \frac{n_2(t) + \alpha_{21}\, n_1(t)}{K_2}\right). \tag{3.15b}$$

As a check, if the species do not interact in any way (i.e., $\alpha_{12} = \alpha_{21} = 0$), then equations (3.14) and (3.15) should reduce to the logistic equations (3.5a) and (3.5c), which indeed they do. Another check is provided by the fact that this model is also symmetric: which species is labeled 1 and which is labeled 2 is arbitrary. Thus, we can interchange the indices referring to species 1 and 2 in the recursion equation (3.14a) and regain equation (3.14b), and vice versa. The same holds for the differential equations (3.15a) and (3.15b).

Although we have been discussing competition, species j could facilitate the growth of species i, in which case α_{ij} would be negative. For example, species 1 could excavate holes that are used as nesting sites by species 2. The signs of α_{12} and α_{21} therefore reflect the relationship between the two species:

α_{12}	α_{21}	Relationship
–	–	Mutualistic
–	0	Commensal
0	–	Commensal
+	–	Parasitic
–	+	Parasitic
+	+	Competitive

Equations (3.14) and (3.15) can thus be used to describe the dynamics of populations experiencing many different types of species interactions.

3.4.2 Consumer-Resource Models

Consumer-resource models describe the dynamics of a resource and the dynamics of a population that consumes this resource.

Throughout the above, we have treated resources as constant. While this might be reasonable for some physical resources (e.g., light striking a patch of land or nutrients flowing down a river past a stationary aquatic community), in many cases the level of resources will be impacted by the growth of the species that we are modeling. For example, nutrients such as nitrogen might decline as a plant population grows, and the number of plankton might decline as a fish population grows. To account for such phenomena, we must construct models that explicitly track the dynamics of a resource as well as the population consuming the resource. Such models are referred to as *consumer-resource* models or as *predator-prey* models if the resource is itself an organism (a prey).

The general structure of a consumer-resource model is illustrated in Figure 3.8 for the case with one resource $n_1(t)$ and one consumer $n_2(t)$. From this flow diagram, we can derive the dynamical equations for this model in continuous time by summing the rate of change over all arrows for each circle (Recipe 2.4):

$$\frac{dn_1}{dt} = f(n_1(t)) - g(n_1(t),n_2(t)),$$
$$\frac{dn_2}{dt} = \varepsilon\, g(n_1(t),n_2(t)) - h(n_2(t)). \tag{3.16}$$

Similar equations can be derived in discrete time, once we have specified the order in which events occur.

We have chosen to illustrate the flow diagram and derive these differential equations in very general terms by not specifying the functions $f(n_1)$, $g(n_1,n_2)$, and $h(n_2)$. The advantage of developing such a general model is that it makes it possible to derive many different consumer-resource models by appropriate choices for the functions $f(n_1)$, $g(n_1,n_2)$, and $h(n_2)$, as described next.

The term $f(n_1)$ represents the rate of change of the resource through means other than consumption (i.e., assuming $n_2(t) = 0$). This resource-renewal function

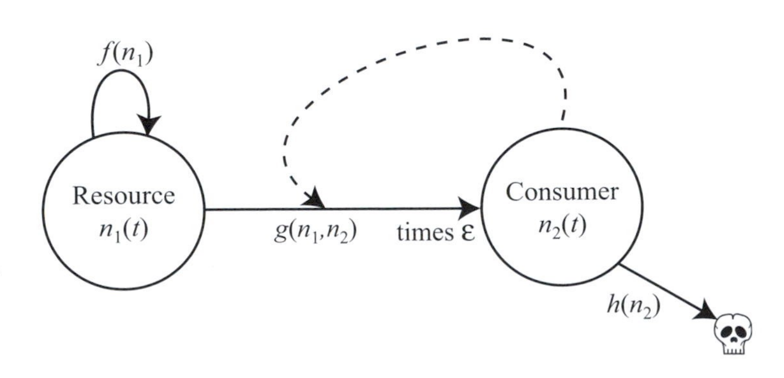

Figure 3.8: Flow diagram for consumer-resource model. A general flow diagram describing the levels of a resource n_1 and a consumer n_2, where the rate of renewal of resources, $f(n_1)$, the rate of resource consumption, $g(n_1,n_2)$, and the rate of loss of consumers, $h(n_2)$, are left unspecified.

might involve inflow and outflow of abiotic resources (as in a chemostat; see Problem 2.4) or immigration, emigration, births, and deaths of biotic resources. Furthermore, the rate of change of the resource might depend on the current density of resources, or it might not. Table 3.3 lists some possible choices for $f(n_1)$. To simplify the notation, we drop the (t) notation, but we must remember that n_1 and n_2 are functions of time.

The term $g(n_1,n_2)$ represents the rate of consumption of the resource by the consumer. In the simplest case, a "mass-action" rate of consumption is assumed, as in the flu model of Chapter 2. That is, the total rate of contact between consumers and resources within the community is assumed to equal $c\, n_1\, n_2$. At each contact, the probability that the consumer successfully uses the resource is a. Hence, $g(n_1,n_2) = a\, c\, n_1\, n_2$, which is known as a "linear" or "type-I" consumption rate (sometimes referred to as a type-I functional response). Table 3.3 lists other common choices for $g(n_1,n_2)$. After having gathered a resource, the consumer converts it into the biomass needed for offspring production. In the flow diagram and in equation (3.16), the conversion factor by which resource units are turned into consumers is given by ε. For example, one prey might represent only a fraction ε of the resources needed to produce one predator.

Finally, $h(n_2)$ represents the rate at which the number of consumers changes in the absence of resources (i.e., assuming $n_1 = 0$). Typically, it is assumed that

TABLE 3.3
Consumer-resource models. Examples of functions that can be used in the consumer-resource model (3.16), where n_1 refers to the level of resources (e.g., number of prey) and n_2 refers to the level of consumers (e.g., number of predators).

Function	Description
$f(n_1) = \theta$	Inflow of resources at a constant rate
$f(n_1) = -\psi$	Outflow of resources at a constant rate
$f(n_1) = r\, n_1$	Constant per capita growth of resource species
$f(n_1) = r n_1\left(1 - \frac{n_1}{K}\right)$	Per capita growth of resource species declines linearly with resource level (logistic)
$f(n_1) = r\, n_1\, e^{-\alpha n_1}$	Per capita growth of resource species declines exponentially with resource level
$g(n_1, n_2) = a\, c\, n_1\, n_2$	Linear (type I) rate of resource consumption
$g(n_1,n_2) = \frac{a\, c\, n_1}{b + n_1} n_2$	Saturating (type II) rate of resource consumption
$g(n_1,n_2) = \frac{a\, c\, n_1^k}{b + n_1^k} n_2$	Generalized (type III) rate of resource consumption
$h(n_2) = \delta\, n_2$	Constant per capita death rate of consumer
$h(n_2) = (\delta\, n_2)\, n_2$	Per capita death rate of consumer increases linearly with consumer population size

consumers die off at a constant per capita rate δ, so that $h(n_2) = \delta\, n_2$. But again, the rate of change of consumers might involve immigration, emigration, and even births if the consumer can utilize alternative resources. Furthermore, this rate might or might not depend on the current density of consumers. Table 3.3 lists two possible alternatives for $h(n_2)$.

By combining the functions $f(n_1)$, $g(n_1,n_2)$, and $h(n_2)$ in different ways, a large number of consumer-resource models are possible. Next, we present equations for two of these models, which we will return to in later chapters.

We start with the model that incorporates the simplest choices for each of these functions: a constant inflow (immigration) of resources at rate $f(n_1) = \theta$, a type-I rate of resource consumption $g(n_1,n_2) = a\, c\, n_1\, n_2$, and a constant death rate of consumers $h(n_2) = \delta\, n_2$. Equations (3.16) then become

$$\begin{aligned} \frac{dn_1}{dt} &= \theta - a\, c\, n_1(t)\, n_2(t), \\ \frac{dn_2}{dt} &= \varepsilon\, a\, c\, n_1(t)\, n_2(t) - \delta\, n_2(t). \end{aligned} \tag{3.17}$$

Equation (3.17) can be used, for example, to model the inflow into a lake of a nutrient (e.g., nitrogen) that is required for and limits the growth of an algal species, where $n_1(t)$ and $n_2(t)$ represent the levels of nutrients and algae, respectively. By setting $\theta = 0$, equation (3.17) can be used to model the growth of a community when resources begin at some level $n_1(0)$ and are not replenished. Interestingly, as we shall see in Chapters 4 and 9, the logistic equation and equation (3.17) without immigration and death ($\theta = \delta = 0$) generate identical predictions about the level of consumers over time.

The second consumer-resource model that we consider is a classic one in ecology, known as the Lotka-Volterra predator-prey model. The only difference from the previous model (3.17) is that the resources are prey that undergo exponential growth in the absence of predators, $f(n_1) = r\, n_1$. Equations (3.16) then become

$$\begin{aligned} \frac{dn_1}{dt} &= r\, n_1(t) - a\, c\, n_1(t)\, n_2(t), \\ \frac{dn_2}{dt} &= \varepsilon\, a\, c\, n_1(t)\, n_2(t) - \delta\, n_2(t). \end{aligned} \tag{3.18}$$

As we shall see in Chapter 4, the Lotka-Volterra predator-prey model predicts that the number of prey and number of predators cycle over time, demonstrating that species interactions can lead to interesting dynamical behavior that might help explain the cyclic dynamics of several species (e.g., Fussmann et al. 2000; Krebs et al. 1995).

Although we have focused on continuous-time consumer-resource and predator-prey models, discrete-time models are straightforward to develop given the tools that you have now learned. As an example, we develop the discrete-time version of the Lotka-Volterra predator-prey model in Supplementary Material 3.2.

3.5 Epidemiological Models of Disease Spread

SIR epidemiological models track the dynamics of infectious diseases by modeling the number of susceptible, infected, and recovered individuals within a population.

The final model that we derive describes the spread of a disease within a population. This is an extension of the flu model that we explored in the previous chapter, but we now add more realism by accounting for births, deaths, recovery, and the gain and loss of immunity. In the previous chapter we considered only two types of individuals, those susceptible to the flu and those infected with the flu. We now allow for a third class of individuals: those who have recovered from the disease and are currently resistant. This model is known as the SIR model in epidemiology, which stands for Susceptible-Infected-Recovered. Consequently, we use the variables $S(t)$, $I(t)$, and $R(t)$ to denote these three types of individuals at time t. We focus on the continuous-time model; the discrete-time version is similar, but with the variables updated after each event.

We begin by using the "mass-action" assumption for the rate at which susceptible individuals contract the disease as in Chapter 2: $a\ c\ S(t)\ I(t)$, where an infected individual contacts susceptible individuals at a rate c per susceptible individual, and a is the probability of disease transmission upon contact (see Figure 2.4c). As illustrated in the flow diagram (Figure 3.9), infected individuals can now recover from the disease at a per capita rate ρ. We assume that, while infected, these individuals developed antibodies to the pathogen that enable them to resist reinfection. Resistance need not be permanent, however, and individuals that have recovered from the disease become susceptible again at a per capita rate σ.

We also incorporate the possibility that individuals die and that infected individuals have a different death rate δ than healthy individuals, d. Finally, we assume that new susceptible individuals enter the population at a rate θ through immigration. This assumption is reasonable if immigrants arrive from a location that has not yet been exposed to the disease. The model can also be generalized to account for reproduction from within the population, by replacing θ with the desired growth rate (e.g., by $b(S(t) + I(t) + R(t))$ assuming that infection does not alter the birth rate of individuals and that all offspring are born susceptible).

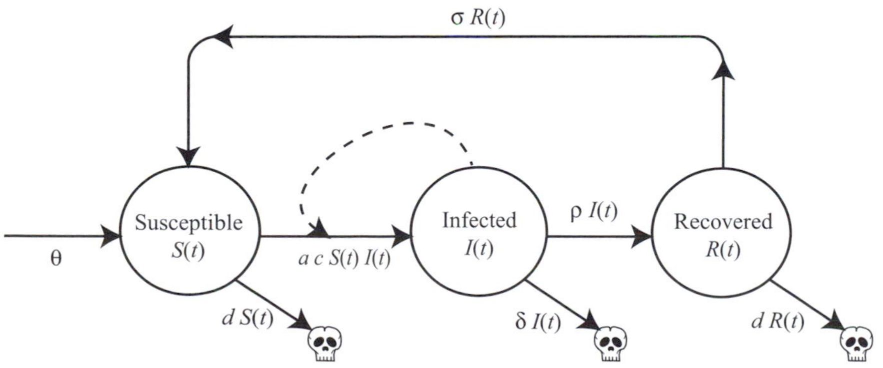

Figure 3.9: A flow diagram for the susceptible-infected-recovered (SIR) model

Using Figure 3.9, the differential equations for the SIR model are

$$\frac{dS}{dt} = \theta - d\,S(t) - a\,c\,S(t)\,I(t) + \sigma\,R(t), \tag{3.19a}$$

$$\frac{dI}{dt} = a\,c\,S(t)\,I(t) - \delta\,I(t) - \rho\,I(t), \tag{3.19b}$$

$$\frac{dR}{dt} = \rho\,I(t) - \sigma\,R(t) - d\,R(t) \tag{3.19c}$$

(Recipe 2.4). As a check, we should be able to regain the differential equations for our previous flu model if we remove the recovered category and let the parameters that describe the added processes of immigration, death, and recovery equal zero ($\theta = \delta = d = \rho = \sigma = 0$). Equations (3.19) then become

$$\frac{dS}{dt} = -a\,c\,S(t)\,I(t),$$

$$\frac{dI}{dt} = a\,c\,S(t)\,I(t).$$

Translating to the variable names used in Chapter 2 (where $S(t)$ was called $s(t)$ and $I(t)$ was called $n(t)$), we do indeed regain equations (2.10).

It is not always reasonable to assume that the transmission of disease between infected and susceptible individuals follows a law of mass action (McCallum et al. 2001). We can generalize the transmission term in epidemiological models by decomposing the rate of disease transmission from an infected individual into the product of three factors: (i) the rate of contact with other individuals in the population, (ii) the probability that, if a contact occurs, the individual contacted is susceptible to the disease, and (iii) the probability that, if a susceptible individual is contacted, the disease is transmitted.

Let us first show how the mass-action assumption is just a special case of this general approach. Under mass action, the rate at which an infected individual contacts other members of the population (regardless of their disease status) is $cN(t)$ per infected individual, where $N(t) = S(t) + I(t) + R(t)$ is the total population size. The probability that a contacted individual is susceptible to the disease equals $S(t)/N(t)$ (the fraction of the population that is susceptible). These two factors, corresponding to factors (i) and (ii) above, can be multiplied together into a single factor $c\,S(t)$, representing the rate of contact with susceptible individuals per infected individual. Multiplying by the total number of infected individuals and by the probability of transmission, a, gives $a\,c\,S(t)\,I(t)$, which is the total transmission rate that we used for mass action.

For sexually transmitted diseases, it is often more realistic to assume that the rate of contact between an infected individual and other individuals does not depend strongly on the population size, because people seek out sexual relationships regardless of the population size. In this case, the rate of contact with other individuals in the population will be nearly constant, c. Given that susceptible individuals represent a fraction $S(t)/N(t)$ of the population, the rate of contact with susceptible individuals is then $c\,S(t)/N(t)$ per infected individual.

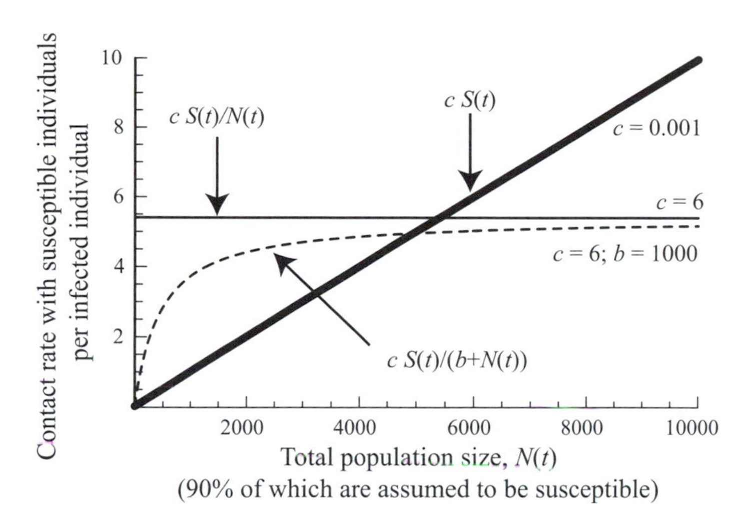

Figure 3.10: Contact rate between infected and susceptible individuals. When modeling disease transmission, the rate at which a particular infected individual contacts susceptible individuals depends on the assumptions of the model. Here, we compare the assumption of mass action (thick line, $c\,S(t)$), the assumption of a constant contact rate regardless of population size (thin line, $cS(t)/N(t)$), and the assumption of a diminishing-returns contact rate as the population size increases (dashed curve, $c\,S(t)/(b + N(t))$. In each case, we assume that 90% of the population is susceptible ($S(t) = 0.9\,N(t)$).

Accounting for all of the infected individuals within the population and for the probability of transmission per contact, a, the total rate of transmission becomes $a\,c\,S(t)\,I(t)/N(t)$. This is sometimes termed "proportional transmission", "frequency-dependent transmission," or "standard incidence." Proportional transmission was assumed by Blower et al. (2000) in their model of HIV (see Box 2.5).

Presumably, in very small populations, the contact rate for sexually transmitted diseases should depend on the population size, because it becomes difficult to find compatible partners. Thus, we might want to modify the above transmission model to allow the contact rate to be proportional to the total population size for low population sizes but to reach a roughly constant value as the population size increases. As described in Primer 1, one potential function with this property is $c\,N(t)/(b + N\,(t))$ where b is a positive parameter. Multiplying this contact rate by the fraction of individuals susceptible to the disease, $S(t)/N(t)$, the rate of contact with susceptible individuals becomes $c\,S(t)/(b + N(t))$ per infected individual. Multiplying by the number of infected individuals and by the probability of transmission per contact, the total rate of transmission becomes $a\,c\,S(t)\,I(t)/(b + N(t))$. These three different types of interaction represent the most commonly used transmission functions and are compared in Figure 3.10.

3.6 Working Backward—Interpreting Equations in Terms of the Biology

Most of this book is aimed at teaching the process of model construction and analysis. This process involves starting with an idea, constructing mathematical equations to describe the idea, and then analyzing these equations in order to extract meaningful biological information. As biologists, however, we are often required to work through this process in reverse. For example, many biological papers that contain models have very terse and sometimes incomplete descriptions of the assumptions and methods used. As a result, it is crucial to be able to "read" equations and decipher the assumptions that are implicit in their derivation. This not only is useful for gaining a better understanding of such papers but is also a crucial skill for ensuring that a model accurately describes the processes that it is meant to. Additionally, models can often be applied to a much broader set of questions than those stated in the papers in which they appear, and therefore it is also important to be able to understand what the equations mean so that you can determine when it is safe to port them over to an entirely different question. As we will see, there are often multiple biological interpretations that are consistent with a single model.

Suppose you are reading a paper that models the population dynamics of a single species subject to exploitation, such as hunting or fishing. The paper claims to model a population in which there is logistic growth in the absence of exploitation, along with a constant hunting pressure. Here are three different possible differential equations for such a model, each of which makes slightly different assumptions about these processes:

$$\frac{dn}{dt} = r\,n(t)\left(1 - \frac{n(t)}{K}\right) - \theta \tag{3.20a}$$

where r is the intrinsic growth rate parameter, K is the carrying capacity parameter, and θ is the constant harvesting rate parameter;

$$\frac{dn}{dt} = r\,n(t)\left(1 - \frac{n(t)}{K}\right) - H\,n(t) \tag{3.20b}$$

where H is the constant harvesting rate parameter;

$$\frac{dn}{dt} = r\,n(t)\left(1 - \frac{n(t)}{K}\right) - H\,(n(t) - P) \tag{3.20c}$$

where P and H are constant harvesting rate parameters.

These three equations have two terms; the first term represents within-population reproduction and the second term represents the loss of individuals through exploitation. According to the first term in all three models, the growth of the population in the absence of exploitation is described by the logistic equation (3.5c). Therefore, if there is no exploitation, we would expect the population to increase in size and plateau at a value of K (Figure 3.4).

All three equations also purport to model a constant harvesting rate, but precisely what is meant by a constant harvest rate is quite different for each model. There are many different ways in which the harvesting rate might be set, but we consider two possibilities: (i) the government regulates the number of organisms that can be taken from the population per time unit (e.g., year), or (ii) the government regulates the amount of "hunter-hours" that the population must endure per time unit. Let us consider each equation in turn to decipher the assumptions that were made about these processes.

We can see from equation (3.20a) that the *total* loss rate of individuals through harvesting is constant and independent of population size. In other words, the rate of loss of individuals through harvesting is the same regardless of the current population size. This means that, of the two different possibilities for setting the harvest rate, this model more closely represents the assumption that the government sets the number of organisms that can be harvested from the population. Indeed, equation (3.20a) can be used to model any process in which a population grows logistically and is subject to a constant total rate of loss or gain of individuals. For example, it can be used to model density-independent emigration or immigration instead of harvesting.

What is meant by a constant harvest rate in equation (3.20b)? We can see from this equation that the loss of individuals through harvesting occurs at a constant per capita rate. Consequently, each individual in the population suffers a constant probability of being harvested, regardless of the population size. This contrasts with equation (3.20a) in which the *total* harvest rate is constant (i.e., each individual is less likely to be harvested as the population size increases). This model more closely represents the assumption that the government sets the total number of hunter-hours allowed, because we would expect the total loss rate to increase as the population size increases under this scheme. More generally, equation (3.20b) can be used to model any process in which a population is subject to a constant per capita rate of loss or gain of individuals.

The expression for the harvest rate in equation (3.20c) is more complex. Mathematically, we can see that the total loss rate of individuals through harvesting is proportional to $n(t) - P$. What assumptions about harvesting are consistent with this formulation? To answer this question we need to consider how the total loss rate is affected by the parameters and by the population size $n(t)$. The term $n(t) - P$ is positive as long as the population size is larger than P, meaning that there is a loss of individuals in this case through exploitation. If the population size is smaller than P, however, then the "exploitation" term actually results in an influx of individuals into the population. Consequently, this model represents a situation in which there is exploitation of the population provided that the population size is large enough, but there is stocking of the population if the population size decreases below the critical value P. As with most models, however, there are also other possible interpretations or applications. For example, if we multiply out the exploitation term in (3.20c), we obtain

$$\frac{dn}{dt} = r\,n(t)\left(1 - \frac{n(t)}{K}\right) - H\,n(t) + H\,P. \tag{3.21}$$

This result suggests an alternative interpretation of equation (3.20c): the population is subject to a constant per capita harvesting rate of H and is being stocked at a constant total rate of $H\,P$, where the amount of harvesting depends on the population size, but the stocking does not.

As one final example, consider a model of a sexually transmitted disease that tracks the number of susceptible, S, and infected, I, individuals within a sexually active adult population. The rate at which an infected individual transmits the disease is equal to the rate of unprotected sexual contact with all other individuals, c, times the probability that the partner is susceptible, $S/(S + I)$, times the probability that sex between an infected and susceptible individual results in disease transmission, a (see section 3.5). In addition, new individuals are added to the population at a constant total rate b (via immigration or maturation) and are removed from the population at a constant per capita rate d (via emigration or death). The following equations might be used:

$$\frac{dS}{dt} = -a\,c\,I\frac{S}{S+I} + b - d\,S, \tag{3.22a}$$

$$\frac{dI}{dt} = a\,c\,I\frac{S}{S+I} - d\,I. \tag{3.22b}$$

In our verbal description of this model, we did not specify whether the new arrivals are disease-free or infected (or a mixture of the two). The differential equations do, however, indicate what assumption was made about the disease status of immigrants. By being able to read and interpret the equations, we can identify this assumption. The rate at which new individuals are added to the population was said to equal b, and we can see from equations (3.22) that this term affects only the differential equation for S. As a result, we can conclude that all new arrivals are assumed to be susceptible. Different assumptions about the pool of immigrants would generate different equations (Problem 3.18).

3.7 Concluding Message

In this chapter we introduced classic models in ecology, evolution, and epidemiology (Table 3.1). As we have witnessed, models often involve generalizations of, or alterations to, previous models. We hope that you are gathering a sense of the interconnectedness of many dynamical models. This interconnectedness makes it easier to develop new models because you can use a previous model as a foundation upon which to build. As we shall see in later chapters (especially Chapters 6 and 9), this interconnectedness also makes it easier to analyze new models, by drawing upon known results from previous models. We turn next to methods for analyzing the equations that we have developed in this chapter.

Problems

Problem 3.1: In equation (3.1), it was assumed that parents first give birth and then all individuals (parents and offspring alike) have a probability d of dying. (a) How will the number of surviving individuals per parent, R, differ if deaths happen first and then births? (b) What will the number of surviving individuals per parent, R, equal if births happen first but then all parents (but no offspring) die, so that the model describes a population with nonoverlapping generations?

Problem 3.2: In the logistic model, we assumed that the number of surviving offspring per parent, R, declines linearly with population size n. Show that growth is still described by equation (3.5a) if we assume instead that (a) only the birth rate

b depends linearly on n or (b) only the death rate d depends linearly on n. (c) Show, however, that if both b and d depend linearly on n, then $R(n)$ is not a linear function of n. [Recall the definition $R = (1 - d)(1 + b)$ and assume that $R(0) = 1 + r$ and $R(K) = 1$.]

Problem 3.3: According to the recursion equation (3.5a) for the logistic model,

$$n(t + 1) = n(t) + r\,n(t)\left(1 - \frac{n(t)}{K}\right),$$

it is possible for $n(t + 1)$ to be negative even if $n(t)$, r, and K are all positive. (a) Solve for $n(1)$ by hand using $r = 1$ and $K = 100$ starting from the population sizes $n(0) = 50$, 100, 200, and 500. (b) By rearranging the recursion equation, determine the population size $n(t)$ above which $n(t + 1)$ becomes negative and the population goes extinct. That is, find n^* in terms of r and K such that $n(t + 1) < 0$ whenever $n(t) > n^*$. Check that your answer to part (b) is consistent with your answer to part (a).

Problem 3.4: Many alternatives to the logistic equation have been described, each of which incorporates different assumptions about how density affects the per capita growth term $R(n)$. (a) Write the recursion and difference equations for n under the assumption that $R(n)$ decreases exponentially from $1 + r$ as the population size increases, so that $R(n)$ can be written as $(1 + r)\,e^{-\alpha n(t)}$. (b) According to the recursion equation that you derive, is it possible for $n(t + 1)$ to be negative if $n(t)$ is positive (justify your answer)? (This model is known as the Ricker model.)

Problem 3.5: In the equations for logistic growth, (3.5), if a population has a high intrinsic growth rate r and grows rapidly when the population is very small, then it must also decline rapidly when the population is very large and above the carrying capacity. (a) Use equation (3.5) to prove this assertion for specific choices of r, K, and n. (b) Describe how you might generalize the logistic model so that a species that grows rapidly does not necessarily decline rapidly as well. (c) Illustrate your argument with an appropriate differential equation.

Problem 3.6: In the derivation of the haploid model of selection, we let the number of surviving individuals in the next generation equal W_A for allele A and W_a for allele a. Consider, instead, the case where these alleles alter the *growth rate* r by a factor, W_A for allele A and W_a for allele a. That is, the number of each allele changes over time according to

$$n_A(t + 1) = (1 + W_A\,r)\,n_A(t),$$
$$n_a(t + 1) = (1 + W_a\,r)\,n_a(t).$$

(a) Derive the recursion equation for the allele frequency $p(t + 1)$ as a function of $p(t)$ for this model. (b) Can r be factored out of this recursion equation? (c) Show that if we measure selection by the constant $s = (W_A r - W_a r)/(1 + W_a r)$, i.e., as the difference in growth rates divided by the absolute fitness of the a allele, we regain the recursion equation $p(t + 1) = p(t) + \Delta p$, where Δp is given by equation (3.10).

Problem 3.7: In the diploid model of selection in the text, we assumed that gametes unite at random to produce zygotes (Table 3.2). A more realistic model for many animal populations is that diploid individuals, not their gametes, mate at random. In a discrete-time model, let the frequency of diploid zygotes at time t be $d(t)$, $h(t)$,

TABLE 3.4
Genotype frequencies in the diploid model of natural selection with random mating among adults

Mating pair	Frequency of mating pair (after selection)	Offspring produced		
		AA	Aa	aa
$AA \times AA$	$\left(d(t)\frac{W_{AA}}{\overline{W}}\right)\left(d(t)\frac{W_{AA}}{\overline{W}}\right)$			
$AA \times Aa$	$2\left(d(t)\frac{W_{AA}}{\overline{W}}\right)\left(h(t)\frac{W_{Aa}}{\overline{W}}\right)$			
$AA \times aa$	$2\left(d(t)\frac{W_{AA}}{\overline{W}}\right)\left(r(t)\frac{W_{aa}}{\overline{W}}\right)$			
$Aa \times Aa$	$\left(h(t)\frac{W_{Aa}}{\overline{W}}\right)\left(h(t)\frac{W_{Aa}}{\overline{W}}\right)$			
$Aa \times aa$	$2\left(h(t)\frac{W_{Aa}}{\overline{W}}\right)\left(r(t)\frac{W_{aa}}{\overline{W}}\right)$			
$aa \times aa$	$\left(r(t)\frac{W_{aa}}{\overline{W}}\right)\left(r(t)\frac{W_{aa}}{\overline{W}}\right)$			

and $r(t)$ for AA, Aa, and aa individuals, respectively. Similarly, let W_{AA}, W_{Aa}, and W_{aa} equal the relative fitnesses of the three genotypes. Accounting for such fitness differences, the mating table with random mating among diploid adults is given by Table 3.4. Here, we have combined all matings involving the same genotypes into the same row; for example, the second row combines matings involving an AA female $\times$ Aa male with matings involving an Aa female $\times$ AA male. (a) Complete the mating table by filling in the proportion of each offspring genotype produced by each mating pair. (b) Calculate recursion equations for $d(t)$, $h(t)$, and $r(t)$ by taking the product of the "Frequency of mating pair" column with each "Offspring" column, in turn, and summing across rows. (c) Use these recursions to prove that Hardy-Weinberg proportions are achieved in the next generation. Specifically, show that $d(t+1) = p'(t)^2$, $h(t+1) = 2\,p'(t)\,q'(t)$, and $r(t+1) = q'(t)^2$ where $p'(t) = (d(t)W_{AA} + \frac{1}{2}h(t)W_{Aa})/\overline{W}$ and $q'(t) = (\frac{1}{2}h(t)W_{Aa} + r(t)W_{aa})/\overline{W}$ are the frequencies of the A and a alleles among the surviving parents (after selection).

Problem 3.8: (a) Show that you can rewrite $\overline{W}$ using the composite fitness terms $W_{A^\bullet} = p(t)W_{AA} + q(t)W_{Aa}$ and $W_{a^\bullet} = p(t)W_{Aa} + q(t)W_{aa}$. (b) Show that the diploid recursion equation (3.13a) can be written in a form equivalent to the haploid recursion equation (3.8c) with $W_{A^\bullet}$ and $W_{a^\bullet}$ taking the places of W_A and W_a. (c) What do $W_{A^\bullet}$ and $W_{a^\bullet}$ represent?

Problem 3.9: Show that if the fitness of a diploid individual is the product of the fitness effects of each of its alleles (i.e., $W_{AA} = W_A W_A$, $W_{Aa} = W_A W_a$, and $W_{aa} = W_a W_a$) then the recursion equation (3.13) for natural selection in diploids reduces to that observed in the haploid model (3.8c). To accomplish the same change in allele frequency per generation as the haploid model, the diploid model requires that fitness be equivalent to two rounds of haploid selection (e.g., $W_A W_A$). Thus, selection is half as effective in diploid organisms as in haploid organisms (see Crow and Kimura 1970 for the more general case with arbitrary heterozygous fitness).

Problem 3.10: [Challenging] We described the Lotka-Volterra model of competition, equation (3.14), assuming that the two competing entities were species. But the same model can be applied to the case where $n_1(t)$ and $n_2(t)$ are the numbers of two different genotypes within a single species (e.g., the numbers of haploids carrying alleles A and a, respectively). (a) Using equation (3.14), obtain recursions for the frequency of allele A, $p(t)$, and the total population size $n(t)$. (Write these recursions in terms of the variables $p(t)$ and $n(t)$, only.) (b) By equating your recursion for $p(t+1)$ to equation (3.8c),

$$p(t+1) = \frac{W_A\, p(t)}{W_A\, p(t) + W_a\,(1 - p(t))},$$

specify what the relative fitnesses W_A and W_a equal in this model. (c) Show that, even after simplification, these relative fitnesses depend on both the allele frequencies and the population size. (d) For what special values of the parameters will the relative fitnesses be independent of the allele frequency?

Problem 3.11: The consumer-resource model described by equations (3.16) is very general. Show that it also describes the Lotka-Volterra model of competition by determining the forms of the functions $f(n_1)$, $g(n_1,n_2)$, and $h(n_2)$ that cause equations (3.16) to reduce to equations (3.15).

Problem 3.12: The Lotka-Volterra model of predator-prey dynamics discussed in the text assumes a single prey species and a single predator species, but it can be readily generalized to include more species. Figure 3.11 illustrates a slightly more complex

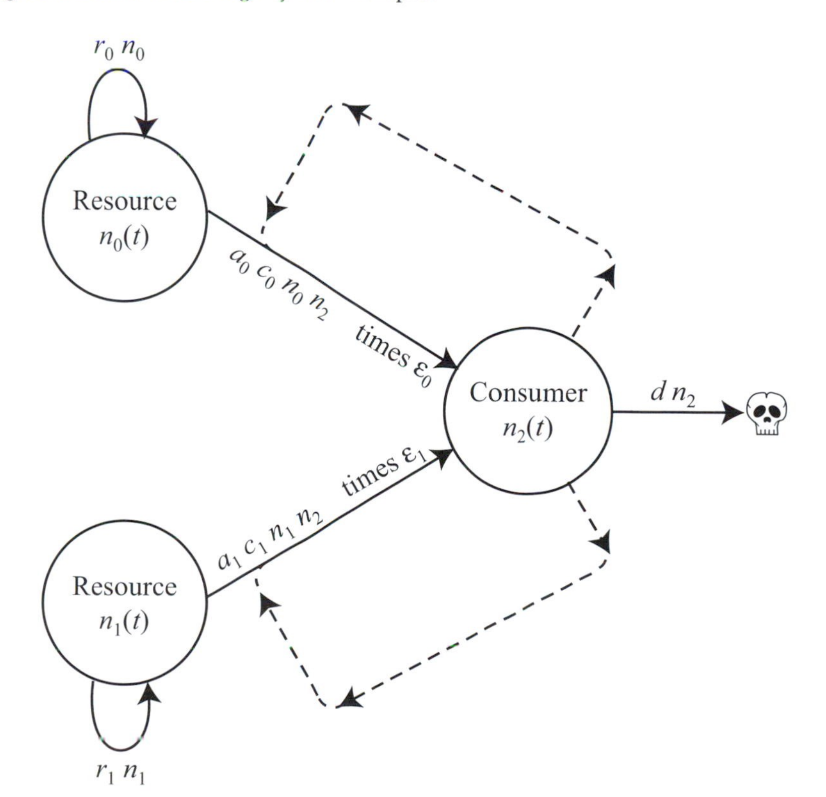

Figure 3.11: A flow diagram for the Lotka-Volterra predator-prey model with two prey species.

food web, with two species of prey, n_0 and n_1. Derive the differential equations for this two-prey, single-predator Lotka-Volterra model.

Problem 3.13: Head lice are readily spread among children, especially in day care and school settings. Individuals do not develop resistance to head lice even after infection; instead, following treatment, they become susceptible again. (a) Modify the flow diagram for the SIR model (Figure 3.9) to describe the spread of lice in a population of susceptible, $S(t)$, and infected, $I(t)$, children. (b) Write down differential equations for $S(t)$ and $I(t)$.

Problem 3.14: Mating within a population is often not random. If a population is spatially structured or if kin prefer to mate with one another, then gametes carrying the same genes unite more often than one would expect under random mating. This is because the parents are related to one another and therefore are likely to carry the same genes, a phenomenon known as *inbreeding*. The simplest way to account for inbreeding in population-genetics models is to say that each egg has a chance $1 - f$ of being fertilized by a sperm drawn randomly from the gamete pool and a chance f of being fertilized by a sperm carrying the same allele. Convince yourself that the probability of producing an AA zygote will then equal $(1 - f)\, p(t)^2 + f\, p(t)$. (a) Calculate the probability of producing an Aa zygote and an aa zygote. (b) How much less common are heterozygous zygotes than expected at Hardy-Weinberg? Using these results, modify the second through fourth columns of Table 3.2, making sure that the second column sums to one and that the fourth column sums to one. (c) Based on this modified table, derive the recursion equation for $p(t + 1)$, generalizing (3.13a) to the case of nonrandom mating. (d) Using this recursion, prove that the allele frequency does not change over time if the genotypes are equally fit ($W_{AA} = W_{Aa} = W_{aa} = 1$).

Problem 3.15: In Chapter 2, we developed a model of the spread of the flu described by the recursion equation (2.11):

$$n(t + 1) = n(t) + a\, c\, n(t)\, (N - n(t)),$$

where a, c, and N are constants (infection probability, per capita contact number per time step, and total population size, respectively). This equation is mathematically equivalent to the logistic equation (3.5a). That is, one can write r and K as functions of a, c, and N that when plugged into the logistic equation (3.5a) give (2.11). What must these functions for r and K be?

Problem 3.16: Defining $W_{AA} = 1 + s$, $W_{Aa} = 1 + hs$, and $W_{aa} = 1$, show that the recursion equation (3.13a) can be used to obtain the following continuous-time version of the diploid model of selection:

$$\frac{dp}{dt} = s\, p(1 - p)\, (p + h(1 - 2p)),$$

by shrinking the time step over which changes are observed (see Box 2.6 and the derivation of equation (3.11)). Check that this differential equation is equivalent to

$$\frac{dp}{dt} = p(1 - p)\, (p(W_{AA} - W_{Aa}) + (1 - p)(W_{Aa} - W_{aa})).$$

Problem 3.17: Suppose the population size at time t is $N(t)$, and that the absolute fitnesses $W_{ij}(N(t))$ are arbitrary functions of the population size at time t. Typically we expect absolute fitness to decrease as the population size gets large enough, and it might do so in different ways for different genotypes. (a) Starting with the numbers of each genotype, show that the dynamics of the population size and the frequency of allele A obey the following pair of recursion equations:

$$N(t+1) = \overline{W}(N(t), p(t))\, N(t),$$

$$p(t+1) = \frac{p(t)\, W_{AA}(N(t)) + (1 - p(t))\, W_{Aa}(N(t))}{\overline{W}(N(t), p(t))} p(t).$$

Assume that gametes unite at random, so that the population of newly formed zygotes is in Hardy-Weinberg proportions at time t. (b) What must be true about the form of the fitnesses $W_{ij}(N(t))$ so that the population size drops out of the recursion equation for the allele frequency? [Hint: If you get stuck, read through Sup. Mat. 3.1.]

Problem 3.18: (a) Modify equations (3.22) to reflect an assumption that a fraction f of individuals immigrating into the population already carry the sexually transmitted disease. (b) Describe one way to check your differential equations using equations (3.22).

Problem 3.19: Use the fact that the derivative of a function equals zero at a maximum (Appendix 2) to determine the allele frequency p that maximizes the rate of evolutionary change $dp/dt = s\, p\, (1 - p)$ in the haploid model of selection (see Figure 3.6).

Further Reading

For more detailed information about classic models in ecology and evolution, consult

- Case, T. J. 2000. *An Illustrated Guide to Theoretical Ecology*. Oxford University Press, Oxford.
- Roughgarden, J. 1996. *Theory of Population Genetics and Evolutionary Ecology: An Introduction*. Prentice-Hall, Upper Saddle River, N.J.

References

Blower, S. M., H. B. Gershengorn, and R.M. Grant. 2000. A tale of two futures: HIV and antiretroviral therapy in San Francisco. *Science* 287:650–654.

Crow, J. F., and M. Kimura. 1970. *An Introduction to Population Genetics Theory*. Harper & Row, New York.

Fussmann, G. F., S. P. Ellner, K. W. Shertzer, and N. G. J. Hairston. 2000. Crossing the Hopf bifurcation in a live predator-prey system. *Science* 290:1358–1360

Krebs, C. J., S. Boutin, R. Boonstra, A.R.E. Sinclair, J.N.M. Smith, M.R.T. Dale, K. Nartin, and R. Turkington. 1995. Impact of food and predation on the snowshoe hare cycle. *Science* 269:1112–1115.

Lack, D. L. 1954. *The Natural Regulation of Animal Numbers*. Clarendon Press, Oxford.

Mable, B. K., and S. P. Otto. 2001. Masking and purging mutations following EMS treatment in haploid, diploid and tetraploid yeast (*Saccharomyces cerevisiae*). *Genet. Res.* 77:9–26.

McCallum, H., N. Barlow, and J. Hone. 2001. How should pathogen transmission be modeled? *Trends Ecol. Evol.* 16:295–300.

PRIMER 1
Functions and Approximations

Rarely is it possible to construct and analyze models in biology without making some sort of approximation or simplifying assumption. Typically, we must make compromises, and part of the art of modeling is being able to choose those compromises that are most appropriate. The reasons for this are twofold.

First, it is not always possible, or desirable, to include everything that we know about the details of a biological process in a model. The purpose of modeling is to abstract only the important parts. A crude analogy can be drawn with choosing a scale when constructing a map. If we want to use the map only for planning a driving route across the country, then it is unnecessary, and undesirable, to choose a scale that reveals all of the traffic lights, intersections, and curves in the roads in a precise way. We might not have enough information to do so anyhow. Rather, a coarser-scale map that ignores these details and that depicts the roads in a stylized fashion might be all that is possible or desirable.

Second, when it comes time to analyze a model, we are often forced to make further approximations or simplifications. Sometimes the analysis becomes too messy or difficult for extracting relevant information, and therefore we must approximate the process as best we can. Returning to our analogy, if we wanted to find the shortest route on a map, we might approximate the distance along a road that doesn't curve too much using a straight line. This would make it much easier to compare the driving distances of different alternative routes.

This primer provides a mathematical foundation for making appropriate choices for simplifying and approximating models. In the construction of models, we are often required to describe biological processes in a stylized way, by choosing functions that relate one quantity to another. Section P1.1 presents guidelines for these choices. In the analysis of models we often use straight-line approximations much as in the map analogy above, and section P1.2 presents a straightforward technique for doing so. Section P1.3 is more advanced and presents the Taylor series, a powerful mathematical concept. Taylor series are generalizations of the linear approximations of section P1.2, and they allow us to derive more accurate approximations to a function than is possible using straight lines.

P1.1 Functions and Their Forms

When modeling a biological process, we must necessarily limit ourselves to a certain frame of reference. For example, when we model the growth of a population, we typically ignore biological phenomena at a lower level (e.g., cell growth and development) and at a higher level (e.g., extinction and speciation). Nevertheless, we often link a model to biological phenomena occurring at other levels without explicitly modeling the mechanistic details of these. Rather, we incorporate these other phenomena by describing them with a function that behaves in a way that is consistent with our understanding of these processes or with data. Typically, the function is required to display certain desired qualitative characteristics but is otherwise arbitrary.

The modeling approach where a function is used to describe underlying biological processes is called *phenomenological*. In contrast, the modeling approach where the details at another level are explicitly tracked is called *mechanistic*. In Chapter 3, the logistic model (3.5a) was phenomenological, because we described competition using a linear function, whereas the consumer-resource models (3.16)–(3.18) were more mechanistic in that they explicitly described how predators compete in the consumption of prey. The term "mechanistic" is clearly relative, however, because we still had to choose how the consumers and prey interact in a consumer-resource model (Table 3.3 presents several potential choices).The advantage of a mechanistic model is that it can be used to incorporate all of the details known to influence a biological system, which is especially appropriate if data exist that allow us to choose parameter values describing these details. The advantage of a phenomenological model is that it generally contains fewer variables and parameters and is often easier to analyze and understand, because it is less mired down by details. For a phenomenological model to describe a biological process adequately, however, we must choose a function that behaves in a way that is consistent with our understanding of the processes or data. But how do we choose an appropriate function? In this section, we describe the shape of some of the most commonly encountered functions used in modeling biological processes, providing examples of how and why they might be chosen.

To motivate this section, let us return to the logistic model of intraspecific competition introduced in Chapter 3. Rather than explicitly modeling the behavioral decisions made by individuals during the consumption of resources, we used a phenomenological description of competition by assuming that the number of surviving offspring per parent can be described by a function $R(n)$ that decreases as the population size increases. Specifically, we used a function that decreases linearly as population size increases. This choice was somewhat arbitrary, however, and Figure P1.1 illustrates other possible functions that still have a maximum reproductive factor of $1 + r$ and a carrying capacity of K (defined as the population size at which the parental population exactly replaces itself, $R(K) = 1$). If we wish to model a species whose reproductive factor is fairly constant until the population size approaches the carrying capacity, after which point the number of surviving individuals per parent plummets, then we might wish $R(n)$ to have a shape like the thick curve in Figure P1.1. Alternatively, we might be modeling a species in which the severity of competition tapers off as the population size increases, as in the dashed curve. But how do we choose functions that behave in these ways? One of the best guiding principles is to choose a function that is as simple as possible while still having the desired shape. In the following, we present several functions that are

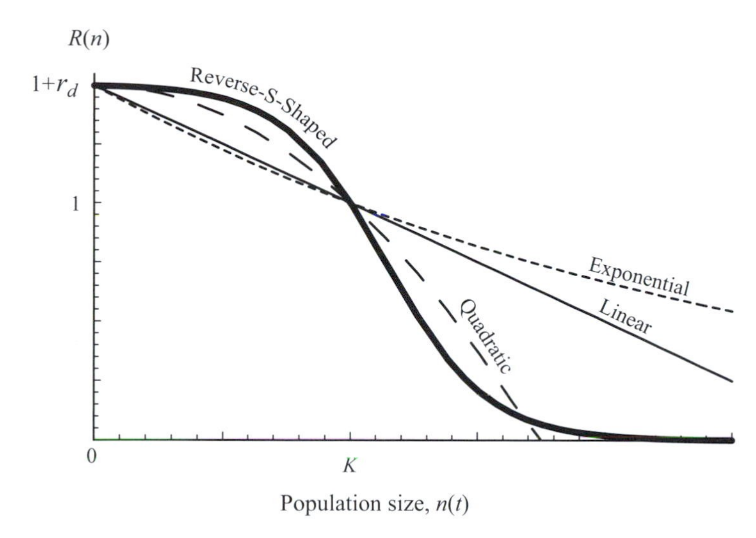

Figure P1.1: Functions describing competition. The number of surviving individuals per parent, $R(n)$, is plotted as a function of population size, $n(t)$. This function can, potentially, take on many different shapes, four of which are shown.

commonly encountered in biological models and that take on a wide variety of different shapes.

If the desired function is a line, the function will have the form

> **Definition P1.1:**
>
> $$f(x) = \underbrace{b}_{\text{slope}}\, x + \underbrace{c}_{\text{intercept}} \qquad \text{(linear function)}$$

The slope b determines how much $f(x)$ changes (the "rise") for a given change in x (the "run"; Figure P1.2). The intercept c determines the value of $f(x)$ when $x = 0$; altering c alters the height of the line.

If the desired function curves up or down in the shape of a parabola, it will have the form

> **Definition P1.2:**
>
> $$f(x) = a\, x^2 + b\, x + c \qquad \text{(quadratic function)}$$

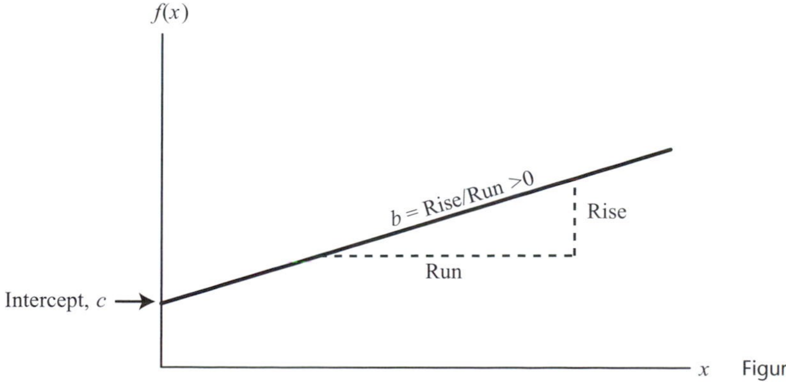

Figure P1.2: Linear functions

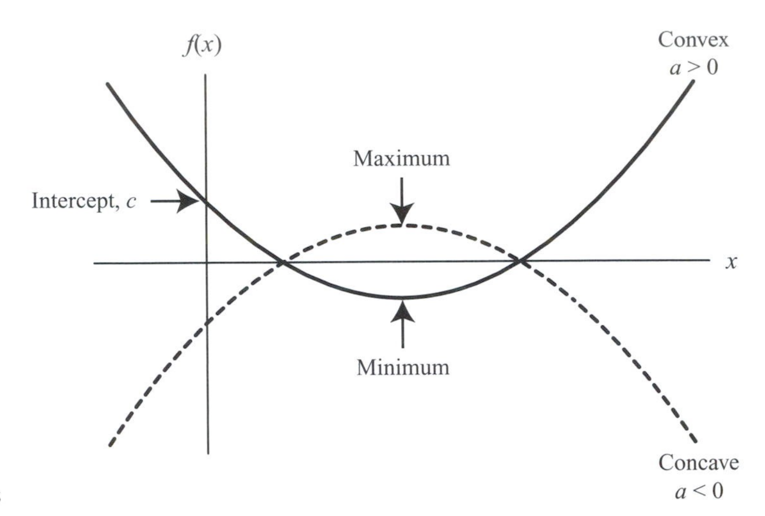

Figure P1.3: Quadratic functions

The shape of the quadratic function is largely determined by a. If $a > 0$, the parabola is *convex* (points up; Figure P1.3), if $a < 0$, the parabola is *concave* (points down; Figure P1.3), and if $a = 0$, then the quadratic function reduces to a linear function. The larger is the magnitude of a, the narrower the parabola. The minimum ($a > 0$) or maximum ($a < 0$) of the quadratic occurs at $x = -b/(2a)$, so that the placement of the maximum/minimum can be chosen by setting b. (Show this by finding the value of x that causes $dy/dx = 0$; Appendixes 2 and 4). Again, c is the intercept. Linear and quadratic functions are special cases of polynomial functions, where an nth-degree polynomial is a function of the form

Definition P1.3:

$$f(x) = \sum_{i=0}^{n} a_i x^i \qquad \text{(polynomial function)}$$

Polynomials typically have $n - 1$ maxima and/or minima, and their behavior as x goes to negative or positive infinity is determined by the sign of the term with the highest power $a_n x^n$.

If the desired function rises or falls exponentially, it will have the form

Definition P1.4:

$$f(x) = c\, e^{ax} \qquad \text{(exponential function)}$$

The shape of the function is largely determined by a. If $a > 0$, the function rises exponentially (Figure P1.4), while if $a < 0$, the function declines exponentially to zero (Figure P1.4). Again, c is the intercept.

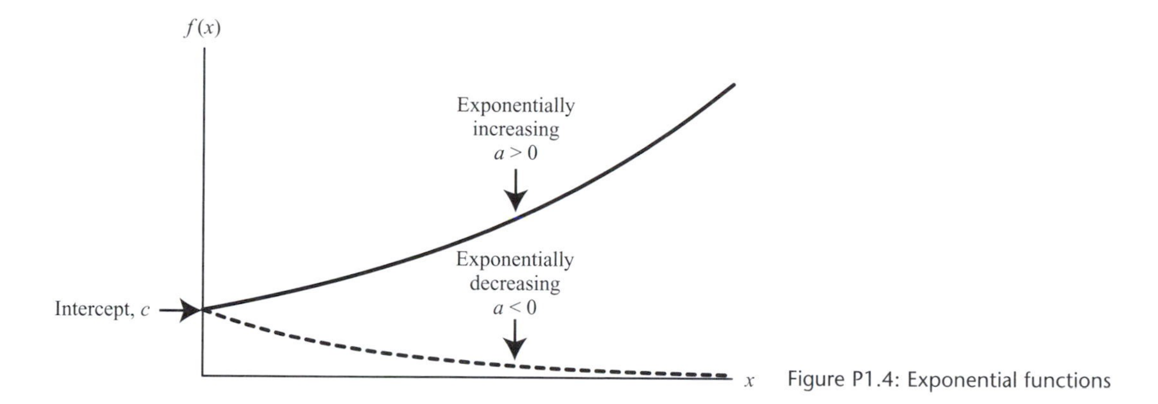

Figure P1.4: Exponential functions

If the desired function is nearly linear for small values of x but reaches a constant value as x becomes large (an "asymptote"), one possible choice is

> **Definition P1.5:**
>
> $$f(x) = \frac{a\,x + c}{b\,x + d} \qquad \text{(rational function)}$$

This function starts at a height c/d (the intercept) when $x = 0$. It rises or falls linearly, at first, with slope $(a\,d - b\,c)/d^2$ (the value of df/dx when x is zero). As x becomes very large, the term $a\,x$ dominates c in the numerator, and the term $b\,x$ dominates d in the denominator. Consequently, the function eventually reaches an asymptote at $(a\;x)/(b\;x) = a/b$ (Figure P1.5). More generally, a rational function is any polynomial divided by a polynomial.

If the desired function is bell shaped, one possible choice is

> **Definition P1.6:**
>
> $$f(x) = \max\, e^{-(x-b)^2/a} \qquad \text{(bell-shaped function)}$$

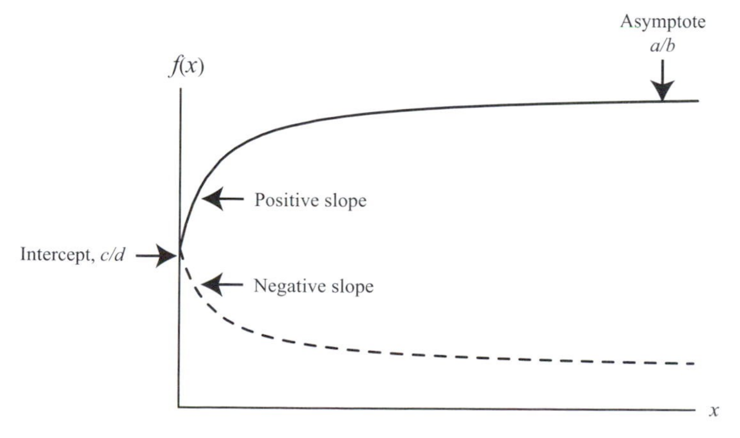

Figure P1.5: Rational functions

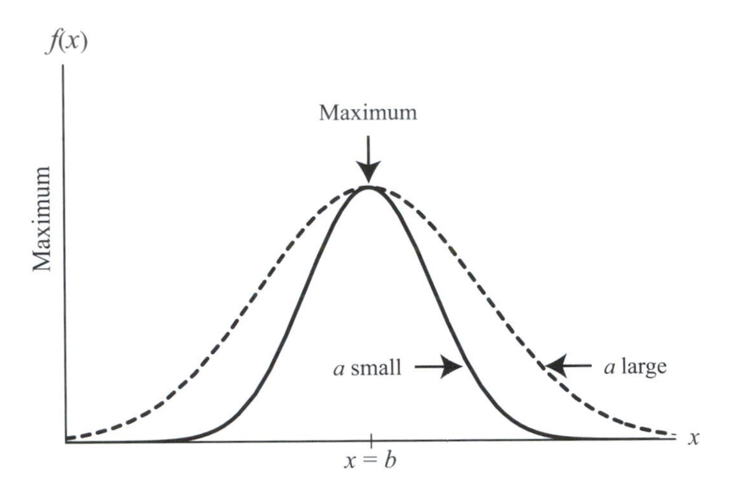

Figure P1.6: Bell-shaped functions

This function has a maximum of $f(x) = \text{max}$ when $x = b$. The shape is determined by a, with larger values of a causing the bell shape to be broader (Figure P1.6). This function is sometimes called "normal," because of its close relationship to the normal distribution in statistics (where $a = 2\sigma^2$ and $\text{max} = 1/\sqrt{2\pi\sigma^2}$; see Definition P3.14a in Primer 3).

If the desired function is S shaped (sigmoidal), one possible choice is

Definition P1.7:

$$f(x) = \frac{c\, e^{a x}}{c\, e^{a x} + (1 - c)} \qquad \text{(S-shaped function)}$$

Here c is a fraction ($0 < c < 1$) that can be thought of as the proportion of the way up the "S" that the function is at $x = 0$. If $a > 0$, the function rises to one (Figure P1.7), while if $a < 0$, it falls to zero (reverse S shape, Figure P1.7). The larger is the magnitude of a, the sharper the S-shaped function. This function is sometimes called "logistic," because of its close relationship to the rise in population size observed in the logistic model of population growth (see Figure 3.4 and equation (6.14b)).

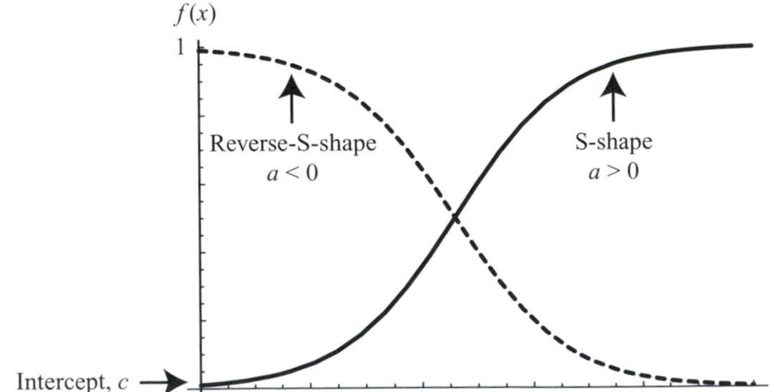

Figure P1.7: S-shaped functions

The above functions are commonly encountered in models of biological phenomena. They can be combined or manipulated to take on a shape that best matches what is understood about the biological process. They can be shifted, stretched, or flipped using Rule P1.1.

Rule P1.1: Changing the Shape of a Function

The following rules can be used to alter a function $f(x)$ to match a desired shape:

- A function can be shifted to the right by an amount d by replacing x with $(x - d)$.
- A function can be shifted to the left by an amount d by replacing x with $(x + d)$.
- The height of a function can be increased by an amount d by replacing $f(x)$ with $f(x) + d$.
- A function can be stretched along the horizontal axis by a factor d by replacing x with x/d.
- A function can be stretched along the vertical axis by a factor d by replacing $f(x)$ with $f(x) \times d$.
- A function can be reflected across the horizontal axis by replacing $f(x)$ with $-f(x)$.

For example, imagine modeling the effect of an inhibitor on the level of phosphorus within a cell. The level of phosphorus might decline exponentially at rate α from an initial level P_{max} as more of the inhibitor is applied, but it might asymptote at some minimum level P_{min} among the cells that remain alive. We could model such a process by modifying the exponential function (Definition P1.4), setting $a = -\alpha$ to reflect the fact that the function is declining at rate α, choosing the intercept such the function has the appropriate range of y values, $c = P_{max} - P_{min}$, and then shifting the height of the function up by P_{min}. Using Rule P1.1, this is accomplished by the function

$$f(x) = (P_{max} - P_{min})e^{-\alpha x} + P_{min}. \tag{P1.1}$$

Returning to the case of disease transmission, let us identify a contact rate function that rises proportionally with the total population size N but eventually reaches a constant value in very large populations. The rational function (Definition P1.5) has the right shape, where its intercept should be zero ($c = 0$), as no contacts occur when an infected individual is isolated ($N = 0$). This gives us the function

$$f(N) = \frac{aN}{bN + d}. \tag{P1.2}$$

Equation (P1.2) has the same form as the contact rate function used in Chapter 3 (there we divided the numerator and the denominator by b and redefined a/b and d/b using the parameters c and h, respectively).

Exercise P1.1: Determine functions $R(N)$ for the reproductive factor whose shapes are consistent with Figure P1.1. In each case, choose the parameters such that the intercept is $R(0) = 1 + r$ and the value when $N = K$ is $R(K) = 1$.

(a) A function that declines exponentially to zero.
(b) A quadratic function with a maximum at $N = 0$.
(c) A reverse-S-shaped function that declines from $1 + r$ to zero. Keep a arbitrary (allowing the steepness to vary), and use Rule P1.1 to increase the intercept to $1 + r$.

[Answers to the exercises are provided at the end of the primer.]

When summarizing complicated biological processes at other levels by a function, the hope is that a model and its results will not be very sensitive to the exact form of the function. There are, however, often multiple functions that are compatible with a set of assumptions. We saw this in Figure P1.1, where we described several possible functions to describe competition phenomenologically. Nevertheless, for low intrinsic growth rates (Figure P1.8a), these very distinct functions for $R(n)$ predict very similar population sizes over time (obtained by iterating the recursion equation $n(t + 1) = R(n)\, n(t)$ repeatedly, as we will describe in Chapter 4). It would be difficult to discern from these curves which underlying functions were used to build the model. In this case, we would be justified in not worrying too much about the exact shape of the function describing competition among individuals. This is not always true, however. Indeed, for high intrinsic growth rates, the extent of population growth depends strongly on the underlying function used to describe competition (Figure P1.8b). In this case, we must be very careful in our choice of competition function—basing the function on either a mechanistic model of competition or data. As a general rule, the less sure you are about the function used to represent a biological phenomenon, the more effort you should expend to explore alternative functions to determine the sensitivity of the results to the shape of the function. It is also a good rule of thumb to keep a model as general as possible for as long as possible, e.g., by writing $f(N)$ for the number of offspring per parent. While some analyses will require specifying the form of the function, others might not (see Chapter 12).

P1.2 Linear Approximations

The previous section focused on building functions whose shape matches what is known (or assumed) about a biological process. When analyzing models we are often confronted with a related question: Is it possible to choose a simpler function that approximates a complicated function? To be useful, the approximation must match the shape of the complex function, at least within a region of interest. Sometimes, we will want to approximate a dynamical equation when a variable lies near a particular point (e.g., when the population size

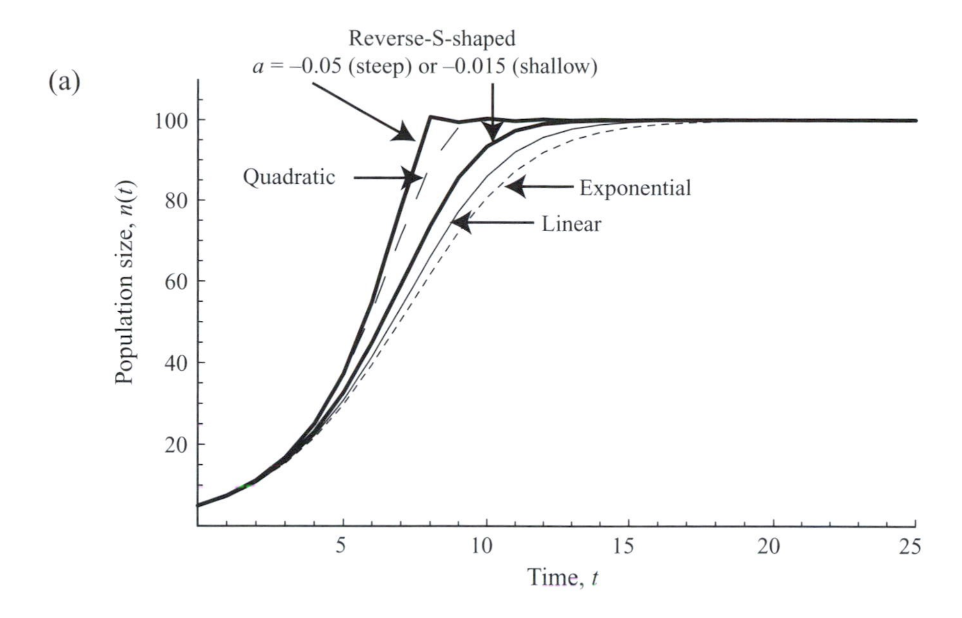

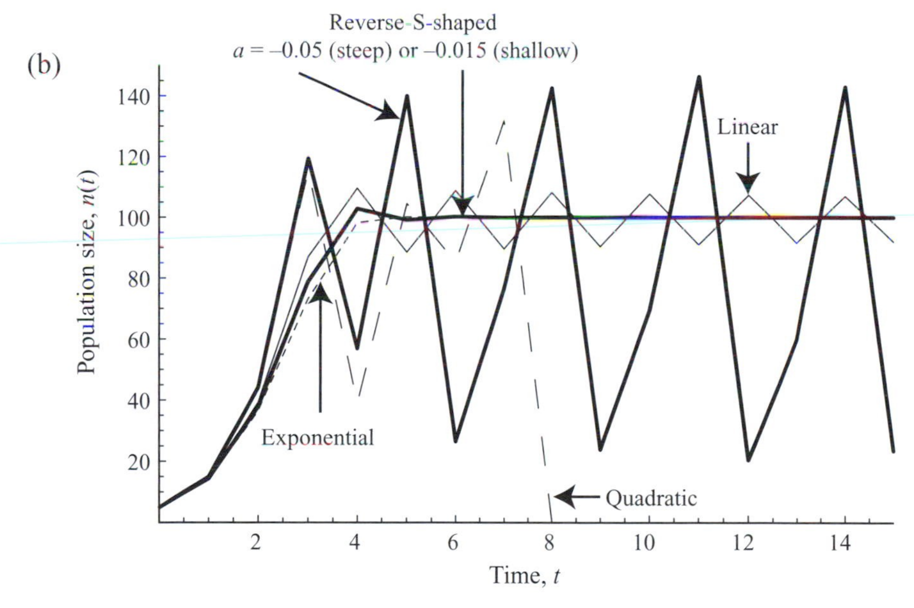

Figure P1.8: Population size over time using different functions for the reproductive factor. The population size was tracked over time (see Chapter 4) using the functions for $R(n)$ illustrated in Figure P1.1 with $K = 100$. We considered two versions of the reverse-S-shaped function: $a = -0.05$ (as in Figure P1.1) and $a = -0.015$ (a more shallow decline). (a) Low intrinsic growth rate, $r = 0.5$. (b) High intrinsic growth rate, $r = 2$. We shall discuss the odd behavior of the discrete-time logistic model in Chapter 4.

is near 0). At other times, we might wish to assume that a parameter is near a particular value (e.g., the mutation rate is near zero). Knowing how to approximate an equation accurately is arguably the most important technique in modeling and one that we will encounter throughout this book.

The most straightforward approximation is based on the idea that any curve looks roughly like a line if we look at it from a close enough distance. This is exactly the idea behind measuring distances on a map using straight lines (at least over stretches of road that aren't too curved). In particular, a function $f(x)$ can be approximated around a particular point a by a line tangent to the function at $x = a$ (see Figure A2.1 in Appendix 2). This assumes that the function is well behaved around a and does not go off to infinity or exhibit a "kink" at this point, assumptions that we make throughout this section and that we formalize in the next section.

We can use Definition P1.1 to find a formula for this tangent line, which we call $\tilde{f}(x)$ to emphasize that it is an approximation ($\sim$) to $f(x)$. The slope of the line is given by $\mathrm{d}f/\mathrm{d}x$ evaluated at the point a (Appendix 2), which we write as $\mathrm{d}f/\mathrm{d}x|_{x=a}$. To match the height of the function, we assume that $\tilde{f}(x)$ equals $f(x)$ at $x = a$ and plug these values into Definition P1.1:

$$f(a) = \underbrace{\left(\left.\frac{\mathrm{d}f}{\mathrm{d}x}\right|_{x=a}\right)}_{\text{slope}} a + \underbrace{c}_{\text{intercept}} .$$

Rearranging, this equation can be solved for the intercept:

$$c = f(a) - \left(\left.\frac{\mathrm{d}f}{\mathrm{d}x}\right|_{x=a}\right) a.$$

The result is a linear function that provides a good approximation to $f(x)$ at a:

$$\tilde{f}(x) = \underbrace{\left(\left.\frac{\mathrm{d}f}{\mathrm{d}x}\right|_{x=a}\right)}_{\text{slope}} x + \underbrace{f(a) - \left(\left.\frac{\mathrm{d}f}{\mathrm{d}x}\right|_{x=a}\right) a}_{\text{intercept}} .$$

Rearranging this formula further, we obtain the useful recipe:

Recipe P1.1
Approximating a Function $f(x)$ by a Line at $x = a$.
For points x near a, the function $f(x)$ can be approximated by the line

$$\tilde{f}(x) = f(a) + \left(\left.\frac{\mathrm{d}f}{\mathrm{d}x}\right|_{x=a}\right)(x - a). \tag{P1.3}$$

Equation (P1.3) is also called a *linear Taylor series approximation* (see next section).

The linear approximation (P1.3) is extremely handy for simplifying complicated equations. For example, consider equation (P1.1) describing the amount of phosphorus, $f(x)$, within a cell as a function of the amount of inhibitor applied, x. Imagine an experiment was performed using only small amounts of inhibitor, and a roughly linear decrease in phosphorus was observed. To relate these experimental data to equation (P1.1), we can use Recipe P1.1 to find a line that approximates equation (P1.1) when the amount of inhibitor is small (i.e., when x is near $a = 0$). To use Recipe P1.1, we need $f(a)$ and $df/dx|_{x=a}$ evaluated at $a = 0$. From equation (P1.1), $f(0) = P_{max}$ and $df/dx = -\alpha (P_{max} - P_{min})e^{-\alpha x}$ (Appendix 2 reviews the calculus needed to evaluate such derivatives). Evaluating the derivative at $x = a = 0$, we get $df/dx|_{x=0} = -\alpha(P_{max} - P_{min})e^{-\alpha 0} = -\alpha (P_{max} - P_{min})$. This gives us a linear approximation to equation (P1.1):

$$\tilde{f}(x) = P_{max} - \alpha(P_{max} - P_{min})\, x. \tag{P1.4}$$

Equation (P1.4) can then be fitted to the data using a linear regression analysis (described in most introductory statistics books). Using the fitted line, we can determine the intercept (giving us P_{max}) and the slope (giving us $-\alpha (P_{max} - P_{min})$).

The most subtle part about using Recipe P1.1 is to identify the term x that we wish to vary and the point a near which we assume x lies. The following example illustrates this process. In Chapter 3, we showed that the change in allele frequency per generation is given by the difference equation (3.10):

$$\Delta p = \frac{s_d p(t)\,(1 - p(t))}{1 + s_d p(t)}. \tag{P1.5}$$

What is a good linear approximation to this equation assuming that selection is weak? This question implies that we are interested in varying selection and finding a line that is a function of s_d. Thus, s_d plays the role of x. The question also implies that we are interested in values of s_d near 0 (weak selection). Defining the difference equation (P1.5) as the function of interest, $f(s_d)$, we are now ready to use Recipe P1.1. We first find the constant term by setting $s_d = a = 0$ in $f(s_d)$ to get $f(0) = 0$. We next find the derivative term by taking the derivative of $f(s_d)$ with respect to s_d, giving us

$$\frac{df}{ds_d} = \frac{d\left(\dfrac{s_d p(t)\,(1 - p(t))}{1 + s_d p(t)}\right)}{ds_d} = \frac{p(t)\,(1 - p(t))}{(1 + s_d\, p(t))^2}. \tag{P1.6}$$

(Remember, we are not varying $p(t)$ in this approximation, so it is treated as a constant.) Evaluating this derivative at $s_d = 0$ gives $p(t)\ (1 - p(t))$. Plugging these terms into the linear approximation (P1.3) leaves us with

$$\begin{aligned} \tilde{f}(s_d) &= 0 + p(t)\,(1 - p(t))\,(s_d - 0) \\ &= s_d\, p(t)\,(1 - p(t)). \end{aligned} \tag{P1.7}$$

This confirms the claim made in Chapter 3 that the change in allele frequency per generation in the discrete-time model is similar to the differential equation (3.11b), $dp/dt = s_c p(t)(1 - p(t))$, when selection is weak.

Exercise P1.2: Use Recipe P1.1 to find the following linear approximations. (These functions were chosen as they are often encountered and approximated in the biological literature.)

(a) $e^r \approx 1 + r$ assuming r is small. (P1.8a)

(b) $\frac{1}{1+s} \simeq 1 - s$ assuming s is small. (P1.8b)

(c) $\ln(t) \approx t - 1$ assuming t is near one. (P1.8c)

(d) $\frac{1}{x} \simeq \frac{1}{a} - \frac{1}{a^2}(x - a)$ assuming x is near a. (P1.8d)

Using a calculator and setting $a = 1$, explore the accuracy of these linear approximations over a range of values of the variables, some close to and others far from the assumed values (e.g., try $s = 0.01$, 0.1, and 1).

P1.3 The Taylor Series

The above linear approximations are a special case of a much more general and powerful technique known as the Taylor series. The Taylor series allows us to understand, more rigorously, when a function can be approximated. It also provides a method to obtain approximations to any order of accuracy desired. No other mathematical approximation is more often used in biological models than the Taylor series approximation.

Before we describe the Taylor series to approximate a function, we must first define what is meant by a sequence and a series:

Definition P1.8:

A *sequence* is a list of mathematical terms, typically labeled by some index (e.g., $a_0, a_1, a_2, \ldots, a_k$). Examples include (i) $0, 1, 4, 9, \ldots, k^2$ and (ii) $1, 0.5, 0.5^2, 0.5^3, \ldots, 0.5^k$.

Definition P1.9:

A *series* is a sum of the terms of a sequence (e.g., $a_0 + a_1 + a_2 + \cdots + a_k$). Examples include (i) $0 + 1 + 4 + 9 + \cdots + k^2$ and (ii) $1 + 0.5 + 0.5^2 + 0.5^3 + \cdots + 0.5^k$. If there are an infinite number of terms in the series, we say that it is an "infinite series" and represent it in one of two equivalent ways:

$$a_0 + a_1 + a_2 + \cdots \tag{P1.9a}$$

or

$$\sum_{i=0}^{\infty} a_i \tag{P1.9b}$$

where the notation in (P1.9) is really shorthand for writing $\lim_{n\to\infty} \sum_{i=0}^{n} a_i$

As the number of terms that we include in a series increases, the summation in (P1.9) sometimes approaches a particular value. In this case, the infinite series is said to *converge* upon this value. Not all series converge, however. For example, the series $1 + 2 + 3 + 4 + \cdots + k$ representing the sum of all integers always gets larger as more terms are included. Such infinite series are said to *diverge* (or, equivalently, to be *undefined*).

Consider our two example series. The infinite series (i) $0 + 1 + 2 + \cdots = \Sigma_{i=0}^{\infty} i^2$ diverges; as more terms are included, the summation gets larger and larger without bound. In contrast, the infinite series (ii) $1 + 0.5 + 0.5^2 + \cdots = \Sigma_{i=0}^{\infty} 0.5^i$ is nicely behaved and converges upon 2. (To verify this answer, use Rule A1.19 of Appendix 1 to rewrite this sum and then see what happens in the limit as, the number of terms summed, gets large.)

With these definitions in hand, we will now describe a remarkable mathematical property: most functions can be rewritten as the series

$$f(x) = b_0 + b_1 x + b_2 x^2 + \cdots, \tag{P1.10}$$

which has an infinite number of terms, and is referred to as a "power series," because it contains terms that are integer powers of x. More generally, we can focus on the behavior around a particular point a and write the function as

$$f(x) = b_0 + b_1(x - a) + b_2(x - a)^2 + \cdots. \tag{P1.11}$$

Here, we have shifted the power series by an amount a by replacing x in equation (P1.10) with $x - a$ (Rule P1.1). Expression (P1.11) is referred to as a power series in x, around the point $x = a$.

To be of any use, however, we must be able to find the coefficients b_i in the power series. The Taylor series does just that.

Definition P1.10: The Taylor Series of a Function, *f(x)*.
Most functions $f(x)$ can be represented as a power series around the point $x = a$ given by:

$$f(x) = b_0 + b_1(x - a) + b_2(x - a)^2 + \cdots, \tag{P1.12a}$$

whose coefficients equal

$$b_i = \frac{1}{i!} \left.\frac{\mathrm{d}^i f}{\mathrm{d}x^i}\right|_{x=a} \tag{P1.12b}$$

where $\mathrm{d}^i f/\mathrm{d}x^i$ is the ith derivative of $f(x)$ with respect to x and where $i! = 1 \times 2 \times \cdots \times i$ ("i factorial"). The coefficients b_i are evaluated at $x = a$ and do not depend on x. This is called the Taylor series of the function.

For the function $f(x)$ to equal the power series (P1.12a), all derivatives in (P1.12b) must be finite. This restriction determines which functions can and cannot be represented by a Taylor series.

Because $0! = 1$ (by definition) and the 0th derivative of a function is the function itself, the first three coefficients in (P1.12b) are

$$b_0 = \frac{1}{0!}\left.\frac{\mathrm{d}^0 f}{\mathrm{d}x^0}\right|_{x=a} = f(a), \tag{P1.13a}$$

$$b_1 = \frac{1}{1!}\left.\frac{\mathrm{d}^1 f}{\mathrm{d}x^1}\right|_{x=a} = \left.\frac{\mathrm{d}f}{\mathrm{d}x}\right|_{x=a}, \tag{P1.13b}$$

$$b_2 = \frac{1}{2!}\left.\frac{\mathrm{d}^2 f}{\mathrm{d}x^2}\right|_{x=a} = \frac{1}{2}\left.\frac{\mathrm{d}^2 f}{\mathrm{d}x^2}\right|_{x=a}, \tag{P1.13c}$$

To become comfortable with the Taylor series, let us work through two examples.

Example

Find the Taylor series of $f(x) = e^x$ around the point $a = 0$.

- The coefficient b_0 equals $f(a)$. Plugging $a = 0$ in for x in $f(x)$ gives $b_0 = 1$.
- The coefficient b_1 equals $\mathrm{d}f/\mathrm{d}x|_{x=a}$. Differentiating $f(x)$ once with respect to x (giving e^x) and evaluating the result at $x = a = 0$ (giving 1) gives $b_1 = 1$.
- The coefficient b_2 equals $(\mathrm{d}^2 f/\mathrm{d}x^2|_{x=a})/2$. Differentiating $f(x)$ twice with respect to x (giving e^x), evaluating the result at $x = a = 0$ (giving 1), and dividing by 2 gives $b_2 = 1/2$.
- The coefficient b_i equals $(\mathrm{d}^i f/\mathrm{d}x^i|_{x=a})/i!$. Differentiating $f(x)$ i times with respect to x (giving e^x), evaluating the result at $x = a = 0$ (giving 1), and dividing by $i!$ gives $b_i = 1/i!$.

Consequently, placing $a = 0$ and $b_i = 1/i!$ in equation (P1.12a), the function e^x can be rewritten as the Taylor series

$$e^x = \frac{1}{1} + \frac{x}{1} + \frac{x^2}{2} + \cdots. \tag{P1.14}$$

Focusing only on the first two terms, $e^x = 1 + x \ldots$ is identical to the linear approximation (P1.8a), based on the tangent to the exponential function (Recipe P1.1).

Before proceeding, it is important that you have a clear understanding of what is meant by equation (P1.14). The left-hand side is a *function* of x, and the right-hand side is an infinite series of terms, each of which is a function of x. The equality means that, if we calculated the right-hand side with only one term, and then only two terms, and then only three terms, etc, and for each of these we plotted the resulting function of x, then the plots would get closer to e^x in the limit as we include more terms (Figure P1.9).

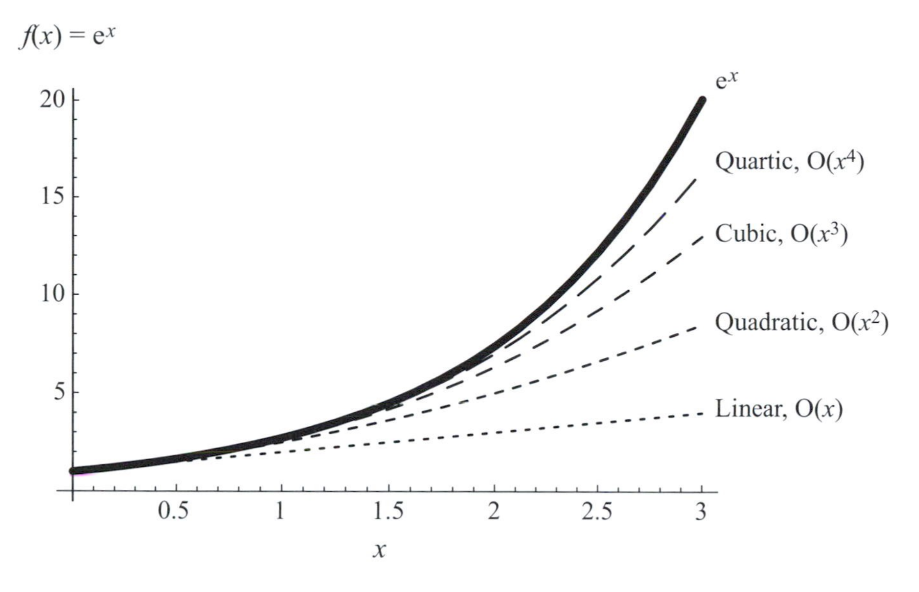

Figure P1.9: Taylor series of the exponential function. The shape of a series as a function of x depends on the number of terms included in the series. The solid curve is the function e^x, and dashed lines illustrate the Taylor series (P1.14) as an increasing number of terms are included in the series (up to and including the order given).

Example

Find the Taylor series of $f(x) = \ln(x)$ around the point $a = 1$.

- The coefficient b_0 equals $f(a)$. Plugging $a = 1$ in for x in $f(x)$ gives $b_0 = 0$.
- The coefficient b_1 equals $df/dx|_{x=a}$. Differentiating $f(x)$ once with respect to x (giving $1/x$) and evaluating the result at $x = a = 1$ (giving 1) gives $b_1 = 1$.
- The coefficient b_2 equals $(d^2 f/dx^2|_{x=a})/2$ Differentiating $f(x)$ twice with respect to x (giving $-1/x^2$), evaluating the result at $x = a = 1$ (giving -1), and dividing by 2 gives $b_2 = -1/2$.
- The coefficient b_i equals $(d^i f/dx^i|_{x=a})/i!$. Differentiating $f(x)$ i times with respect to x (giving $(-1)^{i-1}(i-1)!/x^i$), evaluating the result at $x = a = 1$ (giving $(-1)^{i-1}(i-1)!$), and dividing by $i!$ gives $b_i = (-1)^{i-1}/i$.

Placing $a = 1$ and $b_i = (-1)^{i-1}/i$ in equation (P1.12a), the function $\ln(x)$ can be rewritten as the Taylor series

$$\ln(x) = 0 + (x - 1) - \frac{1}{2}(x - 1)^2 + \frac{1}{3}(x - 1)^3 - \frac{1}{4}(x - 1)^4 + \cdots \quad \text{(P1.15)}$$

(Figure P1.10). Again, focusing on the first two terms, $\ln(x) = (x - 1) \cdots$ is identical to the linear approximation (P1.8c).

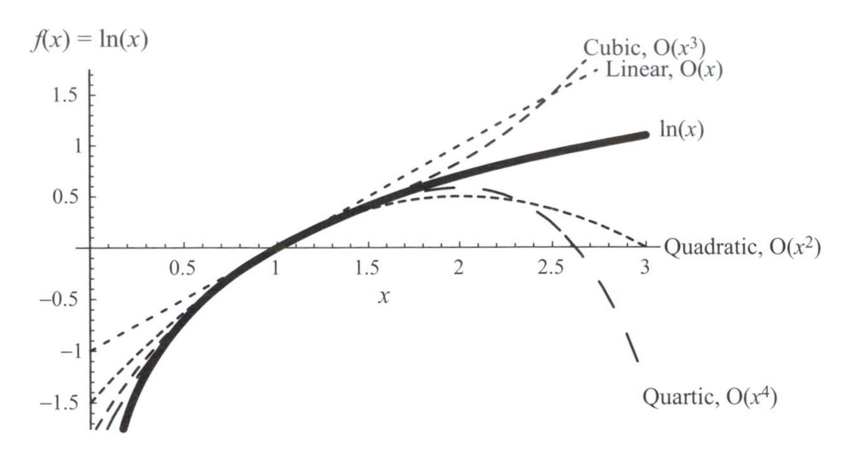

Figure P1.10: Taylor series of the logarithmic function. The solid curve is the function ln(x), and dashed lines illustrate the Taylor series as an increasing number of terms are included.

Exercise P1.3: Use Definition P1.10 to write the Taylor series for the following functions. Include terms up to cubic order, $i = 3$, and write the remaining terms as " . . . ".

(a) $f(x) = e^x$ assuming x is near one ($a = 1$)
(b) $f(x) = x\,e^x$ assuming x is near zero ($a = 0$)
(c) $f(x) = x^2$ assuming x is near one ($a = 1$)

As already mentioned, a series need not converge. Even if a series does converge for some values of x, it might not converge for other values. In practical terms, when a Taylor series does not converge, it will be of little use because we cannot match the function of interest with a series no matter how many terms we include. The Taylor series (P1.14) for e^x appears to converge for all values of x (Figure P1.9). In contrast, the Taylor series (P1.15) for ln(x) appears to converge for values of x between 0 and 2 but not for $x > 2$ (Figure P1.10); that is, when $x > 2$, adding more terms to the series does not improve the fit.

To get a better feeling for whether a series will converge, consider the terms in the Taylor series (Definition P1.10). As i increases, the terms $(x-a)^i/i!$ eventually go to zero (because the factorial function $i!$ rises faster than the power function $(x-a)^i$), and they should do so more rapidly for values of x near the point, a. Consequently, as long as the derivative terms $\mathrm{d}^i f/\mathrm{d}x^i|_{x=a}$ do not increase too dramatically as i increases, the series will converge (at least for values of x near the point a). In the Taylor series for e^x around $x = a = 0$, the derivative terms were always equal to one, so that convergence is guaranteed. In the Taylor series for ln(x) around $x = a = 1$, however, the magnitude of the derivative terms was $(i-1)!$, which increases with i. Overall, the magnitude of the ith-order term in the Taylor series for ln(x) is $(x-1)^i/i$, which converges only if $(x-1)$ is less than one in magnitude, leading to the requirement that $0 < x < 2$ for convergence.

Even if the series converges only for some interval of x, this interval always contains the value $x = a$. Consequently, for values of x near a, the terms $b_i (x - a)^i$ in the Taylor series (P1.12) get smaller and smaller as i increases (unless the series diverges for all values of x). This fact is enormously useful because it means that we can almost always approximate a function $f(x)$ near the point $x = a$ using the first few terms of the Taylor series (Recipe P1.2).

Recipe P1.2:
The Taylor Series Approximation of a Function $f(x)$

Most functions $f(x)$ can be approximated around the point $x = a$ by truncating the Taylor series (P1.12) to the degree of accuracy desired and ignoring the remaining terms. If the terms that we ignore are on the order of $(x - a)^i$, we write these terms as $O(x - a)^i$. Doing so clarifies the degree of accuracy of the approximation.

The crudest approximation is the constant:

$$f(x) = f(a) + O(x - a) \qquad \text{(constant approximation).}$$

A constant approximation is correct at $x = a$, but it provides no information about what happens as we move x away from the point a. A more accurate approximation is the line:

$$f(x) = f(a) + \left(\left.\frac{df}{dx}\right|_{x=a}\right)(x - a) + O(x - a)^2 \qquad \text{(linear approximation).}$$

This equation is identical to equation (P1.3) in Recipe P1.1. The term $df/dx|_{x=a}$ is the slope of the tangent line, which describes whether the function increases or decreases as we move x away from the point a. Including the next order gives

$$f(x) = f(a) + \left(\left.\frac{df}{dx}\right|_{x=a}\right)(x - a) + \left(\left.\frac{d^2f}{dx^2}\right|_{x=a}\right)\frac{(x - a)^2}{2} + O(x - a)^3 \qquad \text{(quadratic approximation).}$$

According to the quadratic approximation, the function curves upward around the point $x = a$ when $d^2f/dx^2|_{x=a}$ is positive (convex) but curves downward if it is negative (concave).

Example

To illustrate the Taylor series approximation, let us approximate the function $\sin(x)$ around $x = 1$ (Figure P1.11). The constant approximation, $\sin(1)$, provides a poor approximation except at $x = 1$ (dotted line). The linear approximation is $\sin(1) + \cos(1)\,(x - 1)$, which is fairly accurate for x values between

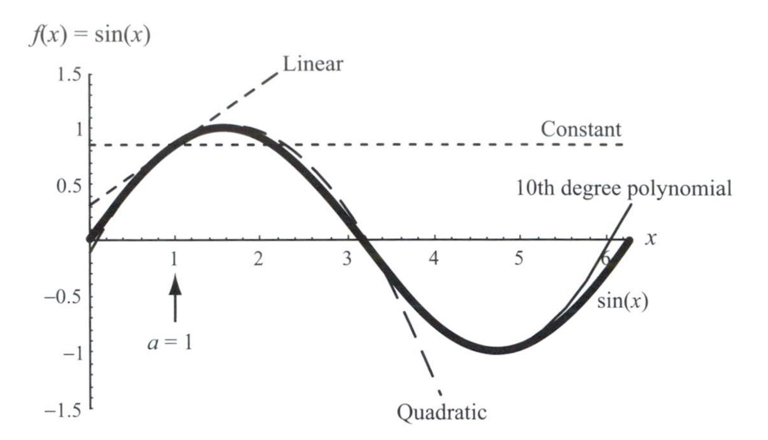

Figure P1.11: Taylor series of the sine function. The sinusoidal function, $f(x) = \sin(x)$, is shown in bold and is compared to Taylor series approximations around the point $a = 1$ using as an increasing number of terms in equation (P1.12). Every approximation is accurate at $a = 1$, but more terms are needed to adequately describe $\sin(x)$ for values of x further away from 1.

about 0.5 and 1.5 (short-dashed curve). The quadratic approximation is $\sin(1) + \cos(1)\,(x - 1) - \sin(1)\,(x - 1)^2/2$ (try deriving this), which works well between about 0 and 3.5 (long-dashed curve). Including terms in the Taylor series from $i = 0$ to 10 gives an approximation that is fairly accurate over the whole cycle (thin curve).

The more terms that are included in a Taylor series approximation, the more accurate the approximation tends to be. How accurate the approximation should be depends on the question. For a wide variety of problems, we might know that x is very close to some value a; in this case, a linear approximation will almost always suit our needs (Recipe P1.1). For example, in Chapter 5, we shall examine whether a biological system approaches an equilibrium value (defined as a point where the system remains constant over time). In this case, a linear approximation to the dynamical equations assuming that the variable(s) is near its equilibrium value is perfectly adequate.

Before leaving the topic, we must point out that the Taylor series is not a complete panacea. We only expect a Taylor series approximation to be accurate near the point a around which it is taken, as is evident from Figure P1.11. For those functions in which convergence is an issue, the Taylor series approximation can give a completely wrong answer for values of x outside the radius of convergence (e.g., see Exercise P1.7). Finally, the Taylor series can fail completely for functions whose derivatives don't exist at certain points and/or whose higher-order derivatives are too large to ignore. Such functions tend to be jagged, with discontinuities ("kinks"), or have points where the function goes off to plus or minus infinity. A good rule of thumb is to be cautious when using a Taylor series on functions that are not smooth. And, in general, it is a good idea to ensure that any approximation is accurate by plotting the original function and its approximation (see Chapter 4).

Exercise P1.4: Perform a Taylor series of the difference equation for logistic growth, $\Delta n = r\,n - r\,n^2/K$, around the point $n = 0$.

(a) Find all terms in the Taylor series up to and including the third-order term in n.

(b) Compare the resulting Taylor series to the original difference equation and explain why the n^3 term is zero in the Taylor series.

(c) Determine what condition must hold for the second-order term (n^2) to be small relative to the linear-order term (n) in the Taylor series.

Exercise P1.5: An alternative to the logistic model of density dependence is the Beverton-Holt model: $n(t+1) = \lambda\, n(t)/(1 + \alpha\, n(t))$.

(a) Find a linear approximation to the Beverton-Holt equation assuming that α is small (α near the point $a = 0$).

(b) Evaluate the next term in the Taylor series approximation, and show that this term can be ignored only when the population size is small.

Exercise P1.6: Find a linear approximation to the difference equation in the haploid model of selection, $\Delta p = s\, p\, (1 - p)/(1 + s\, p)$, with respect to the allele frequency p around the point:

(a) $p = 0$

(b) $p = 1$

Exercise P1.7: This exercise illustrates what it means to have a function whose Taylor series does not converge, using $f(x) = \ln(x+1)$. (a) Find the first five terms of the Taylor series of $f(x)$ around the point $x = 0$. (b) Infer the value of the ith-order term in the Taylor series. (c) The radius of convergence is one for $f(x)$, meaning that x must be less than one for the terms in the Taylor series to converge to zero as i gets larger. Setting $x = 0.9$, find the values of the ith-order terms of the Taylor series for $i = 5$, 10, and 100 and show that these converge upon zero. Repeat for $x = 1.1$ and show that these do not converge.

Answers to Exercises

Exercise P1.1

(a)
$$R(N) = (1+r)\, e^{-\ln(1+r)\, N/K} = (1+r)^{1-N/K}.$$

The intercept tells us that $R(0) = c = 1 + r$. Because the value at $N = K$ must be one, $R(K) = c\, e^{a\,K} = 1$, and $c = 1 + r$, a must equal $-\ln(1+r)/K$ (see rules for simplifying logarithmic and exponential functions in Appendix 2).

(b)
$$R(N) = (-r\, N^2/K^2) + 1 + r.$$

Because the maximum occurs at $N = 0$, we know that $-b/(2a) = 0$, so that $b = 0$. The intercept tells us that $R(0) = c = 1 + r$. Because the value at $N = K$ must be one, $R(K) = a\, K^2 + b\, K + c = 1$, $b = 0$, and $c = 1 + r$, a must equal $-r/K^2$.

(c)

$$R(N) = (1+r)\frac{e^{a\,N}}{c\,e^{a\,N} + (1-c)} = (1+r)\frac{e^{a\,N}}{\left(1 - \dfrac{r\,e^{a\,K}}{1-e^{aK}}\right)e^{a\,N} + \left(\dfrac{r\,e^{aK}}{1-e^{aK}}\right)}.$$

First, we must stretch the vertical axis by multiplying by $(1+r)/c$ (Rule P1.1), so that the intercept lies above one (recall that c must be less than one), giving $R(N) = (1+r)\,e^{a\,N}/[c\,e^{a\,N} + (1-c)]$. Because the value at $N = K$ must be one, $R(K) = (1+r)\,e^{a\,K}/[c\,e^{a\,K} + (1-c)] = 1$. Solving for c gives $1 - (r\,e^{a\,K})/(1 - e^{a\,K})$. The value of a can be chosen to alter the shape; more negative values of a give reproductive factors that remain near $1+r$ for longer and then plummet down more rapidly as the population size reaches its carrying capacity.

Exercise P1.3

(a)
$$e^x = e + e(x-1) + \frac{e(x-1)^2}{2} + \frac{e(x-1)^3}{6} + \cdots.$$

(b)
$$x\,e^x = x + x^2 + \frac{x^3}{2} + \cdots.$$

(c) $x^2 = 1 + 2(x-1) + (x-1)^2 + 0 + \cdots$.
In this last example, the first three terms on the right-hand side factor to give x^2, and all higher derivatives equal zero. More generally, an nth-degree polynomial is exactly given by the first $n+1$ terms in a Taylor series, with all remaining terms equal to zero.

Exercise P1.4

(a) Taking derivatives with respect to n and using $a = 0$,

$$\Delta n = 0 + r(n-0) - 2r/K\frac{(n-0)^2}{2!} + 0\,\frac{(n-0)^3}{3!} + \cdots.$$

(b) The sum of the first three terms in the Taylor series (up to and including the n^2 term) equals the original logistic equation. Because the original equation represents a quadratic equation in n, taking the derivative of the equation with respect to n more than twice results in zero.
(c) To ignore the second-order term, $r\,n^2/K$ must be small relative to $r\,n$, which requires that $n/K << 1$ (i.e., the population size is far below the carrying capacity). When this condition is met, the difference equation for the logistic model can be approximated by the difference equation for the exponential model, $\Delta n = r\,n$.

Exercise P1.5

(a) Taking derivatives with respect to α and using $a = 0$, $n(t+1) = \lambda\, n(t) - \lambda\, n(t)^2\, (\alpha - 0) + \cdots$. While this equation is linear in α it is quadratic in $n(t)$. In fact, this approximation is equivalent to the logistic equation (3.5a) if we set $\lambda = 1 + r$ and $\alpha = r/(K\,(1 + r))$.

(b) The next-order term is $2\,\lambda\, n(t)^3\, (\alpha - 0)^2/2!$. This term might not be small, even though α is small, if the population size is large. Comparing this term to the magnitude of the lower-order term $\lambda\, n(t)^2\, \alpha$ indicates that the higher-order term is smaller and can be neglected only if $n(t)\, \alpha << 1$.

Exercise P1.6

(a) Taking derivatives with respect to p and using $a = 0$, $\Delta p = 0 + s\,(p - 0) + \cdots$. Consequently, when p is very small, the difference equation for the model of selection is approximately equal to the exponential model, $\Delta p = s\, p$.

(b) Taking derivatives with respect to p and using $a = 1$,

$$\Delta p = 0 - \frac{s}{1 + s}(p - 1) + \cdots.$$

Exercise P1.7

(a) Taking derivatives with respect to x and using $a = 0$,

$$f(x) = 0 + (x - 0) - \frac{(x - 0)^2}{2!} + 2\frac{(x - 0)^3}{3!} - 6\frac{(x - 0)^4}{4!}\cdots.$$

(b) Simplifying, each of these terms has the form $(-1)^{i+1}\, x^i/i$.

(c) When $x = 0.9$, $(-1)^{i+1}\, x^i/i$ equals 0.118 for $i = 5$, -0.0349 for $i = 10$, and -2.66×10^{-7} for $i = 100$, which is declining. When $x = 1.1$, $(-1)^{i+1}\, x^i/i$ equals 0.322 for $i = 5$, -0.259 for $i = 10$, and -137.8 for $i = 100$, which is increasing. In general, when $x > 1$, the numerator x^i grows exponentially while the denominator, i, grows only linearly, so that the coefficients of the Taylor series get larger in magnitude.

CHAPTER 4

Numerical and Graphical Techniques—Developing a Feeling for Your Model

Chapter Goals:

- To describe numerical and graphical techniques for exploring a model's behavior

Chapter Concepts:

- Initial conditions
- Variable over time plots
- General solution
- Variable versus variable plots
- Phase-plane diagrams
- Vector-field plots
- Null clines

4.1 Introduction

In the next few chapters we explore various mathematical techniques commonly used to analyze models. Before launching into these techniques, it is worthwhile having a sense of what to expect from a model. This chapter describes how to get a feel for a model's behavior through graphs. The basic approach is to specify numerical values for all of the parameters, and for the initial values of each variable, and then to use the model's equations to predict what happens over time.

If, for your organism(s), you knew the exact values of each parameter and the initial values of each variable, then a graph might be all that you need to answer your questions. Most of the time, however, biologists don't have such precise knowledge. Furthermore, we often want to use a model to understand general processes occurring in nature rather than a specific process in a specific species. The downside of the numerical techniques discussed in this chapter is that they are not well suited for understanding the behavior of a model in general terms. For example, they cannot answer questions like: Do the variables always grow under all conditions? When do the variables tend toward one value versus another? What are the conditions under which the variables change over time without ever settling down?

Even when your goal is to answer general questions, however, exploring numerical examples is worthwhile. Comparing mathematical results to numerical examples is a great way to check your calculations. If you find that a variable goes to zero in a graph under a particular set of conditions but your mathematical analysis suggests that this cannot happen, then clearly an error has crept into either the code generating the graph or the mathematical analysis. Graphs illustrating the dynamics of the model in particular cases can also guide which mathematical analyses to perform. Humans are great at seeing patterns, and it is often easier to prove that a certain pattern exists than to analyze equations blindly without any sense of how they might behave.

Although some of the graphs we discuss can be drawn by hand, computers are tremendously helpful in the numerical analysis of models. In particular, software packages like *Mathematica* (Wolfram Research), Maple (Waterloo Maple Inc.), and Matlab (MathWorks, Inc.) allow us to enter equations and generate graphs in a matter of a few lines of computer code. These "mathematical software packages" are essentially fancy calculators that manipulate symbols as well as numbers. In the on-line material, we provide introductory notebooks for using mathematical software packages ("Labs"). If you have access to the relevant software, you should familiarize yourself with some of the more commonly used commands by working through the first Lab. We also provide the code used to generate each type of graph in the on-line supplementary material for Chapter 4.

For dynamical equations, there are three main types of graphs. In section 4.2, we discuss graphs that plot a variable as a function of time (i.e., as a function of the independent variable). Such graphs illustrate the dynamics of a variable for a given set of parameters. In section 4.3, we discuss graphs that plot the change in a variable as a function of the variable itself (i.e., as a function of the dependent variable). Such graphs clarify the conditions under which a variable grows or shrinks. In section 4.4, we discuss graphs that plot one variable as a function of another variable (i.e., as a function of another dependent variable). Such graphs help us see how interactions among the variables affect their dynamics.

4.2 Plots of Variables Over Time

Variables in discrete-time models are straightforward to graph as functions of time. Because they can be written as recursion equations (e.g., giving $n(t+1)$ as a function of $n(t)$), discrete-time models can always be "iterated." Given the starting values for each variable at time 0 (the *initial conditions*) and values for each of the parameters, the recursion equations determine the values of the variables at time 1. We can then plug these new values into the recursions again to find the values of the variables at time 2, etc. Such iteration of a discrete-time model can be used to obtain numerical values for the variables at any future point in time, given specific parameters and initial conditions. Having done so, a variable of interest can be plotted on the vertical axis against time on the horizontal axis.

> The *initial conditions* (or *starting conditions*) are the numerical values of every variable at the initial time point.

4.2.1 The Dynamics of the Exponential Population Growth Model

Consider the recursion equation for exponential growth, $n(t+1) = R\,n(t)$, and the initial condition $n(0) = 1000$ (i.e., a population of size 1000 at time $t = 0$). If the number of surviving offspring per parent, R, is 1.01, the recursion tells us that $n(1)$ will be $1.01 \times 1000 = 1010$. With $n(1)$ in hand, the recursion then tells us that $n(2) = 1.01 \times 1010 = 1020.1$. Continuing in this fashion, we get the series of population sizes graphed in Figure 4.1 (bottom curve in panel a). We can repeat this process using different values of the parameter, R, to see how this parameter affects the shape of the graph (Figure 4.1). A key point emerges from the figure: exponential growth causes dramatic increases in population size for large values of R, but the population size declines for those values of R that are less than one. This makes intuitive sense: if each individual leaves fewer than one surviving individual in the next time step, then the population should decrease over time until it goes extinct.

> *Variable over time plots* illustrate the behavior of the variable of interest (vertical axis) versus time (horizontal axis).

The recursion equation for exponential growth is simple enough that we can even use the above iterative procedure without specifying parameter values.

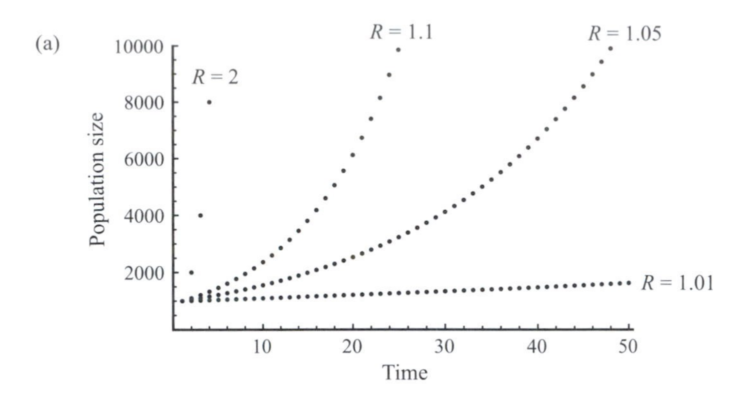

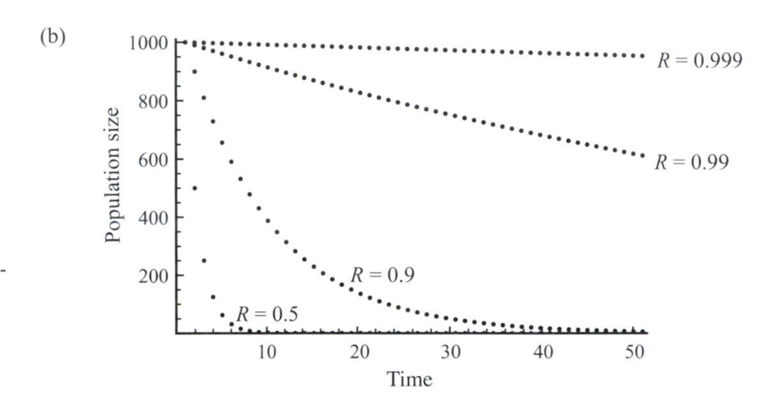

Figure 4.1: The dynamics of the exponential model in discrete time. The recursion equation (3.1b) is iterated for 50 time steps, giving the population size, $n(t)$, on the vertical axis as a function of time on the horizontal axis for various values of R (the number of surviving individuals at $t + 1$ per individual at t). The initial population size is $n(0) = 1000$.

Starting with a population of size $n(0)$ at time 0, there will be $R\,n(0)$ individuals after one time unit. After the second time unit, $n(2) = R\,n(1)$, but we already know that $n(1) = R\,n(0)$, and therefore $n(2) = R^2\,n(0)$. Repeating this procedure, after three time units there will be $R^3\,n(0)$ individuals. Every time we iterate this recursion, a 1 is added to the exponent of R. Therefore, after t time units, the number of individuals is

$$n(t) = R^t\,n(0). \qquad (4.1)$$

A *general solution* to a dynamic model is an equation giving the value of a variable at any future point in time, as a function of the parameters, the initial conditions, and the amount of time that has elapsed.

This is the *general solution* to the exponential growth model, and it gives the population size at any future point in time for any value of the parameter R without having to calculate intermediate population sizes along the way.

The general solution allows us to determine exactly when the population will grow or shrink without having to explore all possible values of R numerically. $R^t\,n(0)$ always equals $n(0)$ when R equals one, it declines as time passes whenever R is less than one, and it increases as time passes whenever R is greater than one.

Does modeling population growth in continuous-time generate different predictions? To address this question graphically requires that we predict future population sizes using the differential equation $dn/dt = r\,n(t)$. That is, we must "solve" the differential equation so that we know $n(t)$ itself at time t, and not just the rate at which $n(t)$ changes over time. Just as with discrete-time recursion equations, differential equations can be solved analytically or numerically. We will describe analytical methods for solving differential equations in Chapter 6. For now, we solve the exponential growth model in continuous time using an educated guess. While this sounds like a crazy idea, many differential equations have been solved in this way.

In the continuous-time model for exponential growth, we might guess that the solution is similar to the discrete-time solution (4.1): $n(t) = R^t\, n(0)$. If we take the derivative of this function for $n(t)$ with respect to t (using Rule A2.7 in Appendix 2), we get $\mathrm{d}n/\mathrm{d}t = \ln(R)\, R^t\, n(0)$, which is equivalent to $\mathrm{d}n/\mathrm{d}t = \ln(R)\, n(t)$ according to our purported solution, $n(t) = R^t\, n(0)$. This differential equation is not exactly the same as the one that we are trying to solve, $\mathrm{d}n/\mathrm{d}t = r\, n(t)$. But if we set $\ln(R)$ to r (equivalently, R to e^r), these two differential equations match up. Therefore, we can correct our initial guess by replacing R with e^r:

$$n(t) = e^{r\,t}\, n(0). \tag{4.2}$$

If we take the derivative of (4.2) with respect to time, we get $\mathrm{d}n/\mathrm{d}t = r\, e^{r\,t}\, n(0)$. We can then substitute in solution (4.2) to regain $\mathrm{d}n/\mathrm{d}t = r\, n(t)$, confirming that (4.2) is indeed the general solution. Equation (4.2) clarifies why this is called the "exponential growth" model; the population size changes exponentially with time. Equation (4.2) can then be plotted over time, resulting in curves that pass through the dots shown in Figure 4.1 (as long as we replace R with e^r; see Lab exercise on plotting functions).

4.2.2 The Dynamics of the Logistic Population Growth Model

For most models, it is not so simple to obtain a general solution (see Chapter 6). For example, if you try to iterate the logistic equation (3.5a),

$$n(t+1) = n(t) + rn(t)\left(1 - \frac{n(t)}{K}\right),$$

it quickly becomes an enormous headache (we have simplified the notation from that of eqn. 3.5a by using r rather than r_d for the intrinsic growth rate). If you specify the parameters and initial values of the variables, however, the recursion equation can be easily iterated using a calculator, a spreadsheet program, or a mathematical software package (Figure 4.2; see Lab exercise on the logistic model). This model has two parameters, r and K. The qualitative behavior of the model does not change much for different values of the carrying capacity K (we used $K = 1000$; see Problem 4.3), but it appears to be extremely sensitive to the value of the intrinsic growth rate r.

When the population starts below carrying capacity, it increases toward K, growing more rapidly for larger values of r. When r is small, the population size smoothly approaches the carrying capacity, following an S-shaped curve. When r is large enough, however, growth is so rapid that the population *overshoots* K. When r is slightly above two, the population size appears to cycle between two points. For example, when $r = 2.1$ (Figure 4.2a), the low point in the cycle is near 823. If we plug $K = 1000$, $r = 2.1$, and $n(t) = 823$ into the logistic

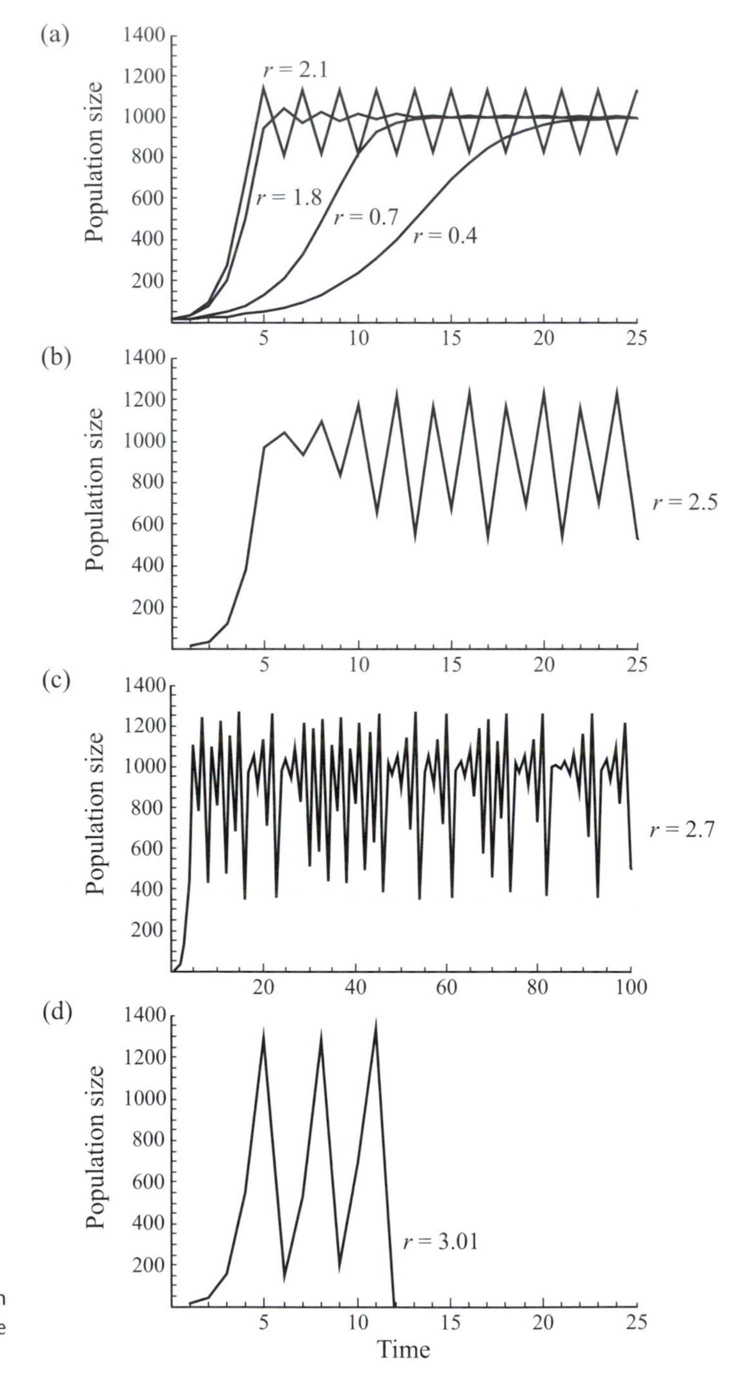

Figure 4.2: The dynamics of the logistic model in discrete time. The recursion equation (3.5a) is iterated for various values of the intrinsic growth rate, r. In each case, the initial population size is $n(0) = 10$ and the carrying capacity is $K = 1000$. More time steps are illustrated in Figure 4.2c to emphasize that the population dynamics do not follow a simple cycle when $r = 2.7$. Because time is discrete, the model only predicts the population sizes at integer values of time, but we connect the values of $n(t)$ with lines to clarify the behavior.

recursion, we find that the population size in the next generation is 1129, far above the carrying capacity. If we then plug $K = 1000$, $r = 2.1$, and $n(t) = 1129$ into the logistic recursion, we find that the population size plummets back to 823, and the two-point cycle begins again. As we increase the growth rate further, the number of points visited on the cycle increases from two (e.g., when $r = 2.1$; Figure 4.2a), to four (e.g., when $r = 2.5$; Figure 4.2b), to eight (e.g., when $r = 2.56$), to sixteen (e.g., when $r = 2.565$), etc. But for slightly larger values of

r (e.g., when $r = 2.7$; Figure 4.2c), something bizarre happens: the population size fluctuates all over the place, with no detectable pattern. The dynamics of the model have now entered the world of *chaos* (Box 4.1). Superficially, chaotic dynamics appear to be entirely erratic and random, but the remarkable thing is that they are, in fact, entirely predictable and deterministic. If we know the exact values of the parameters and variables, we can predict the population size at any future time by simply iterating the recursion equation. On the other hand, chaotic dynamics imply that the future state of the system will be extremely sensitive to imprecision in the calculations, e.g., due to rounding errors (Box 4.1).

Even when the variables are changing chaotically, they eventually settle down and remain within a specific range of values. As r increases, this range increases until, at $r = 3$, the range includes negative values of the population size (Problem 4.12). At this point the population fluctuates so wildly that extinction ensues (e.g., $r = 3.01$; Figure 4.2d). Thus, counterintuitively, populations with higher growth rates are not necessarily better protected from extinction (see Figure 4.1.2 in Box 4.1).

What values of r are typical of population growth? This question was addressed by Hassell et al. (1976) in a survey of population growth in 28 insect species. Of these species, 26 exhibited intrinsic growth rates for which the logistic recursion predicts that the population size will eventually settle down at the carrying capacity, one led to the prediction of a periodic cycle (the Colorado potato beetle), and only one led to the prediction of chaotic behavior (blowflies under laboratory conditions). This suggests that periodic cycles and chaotic fluctuations might not be particularly relevant to the population growth of many species, although biologists continue to debate the importance of chaos in natural populations (Dennis et al. 2001; Hastings et al. 1993; Turchin and Ellner 2000). Nevertheless, chaos theory has fundamentally changed the way scientists view the world, forcing us to recognize that seemingly disordered behavior can be generated from very orderly rules, such as that embodied by the logistic equation.

Do we also see cycling and chaos in the continuous-time model of logistic population growth? While it is possible to obtain the general solution for the differential equation (3.5c), which we could then plot (see Chapter 6), let us instead consider how to analyze the differential equation numerically. A straightforward numerical approach is to use the definition of a derivative to relate a differential equation to a discrete-time difference equation:

$$\frac{dn}{dt} \equiv \lim_{\Delta t \to 0}\left[\frac{n(t + \Delta t) - n(t)}{\Delta t}\right].$$

Although it would be impossible to simulate an infinitely small time step, we can choose some very small time interval Δt to obtain an approximation for the differential equation:

$$\frac{dn}{dt} \approx \left[\frac{n(t + \Delta t) - n(t)}{\Delta t}\right]_{\text{for } \Delta t \text{ very small}}$$

Box 4.1: Chaos

The fact that deterministic equations can lead to seemingly unpredictable dynamics was first discovered by Edward Lorenz in 1961 (see Lorenz 1963). As a meteorologist working at MIT, Lorenz was developing computer simulations to predict weather patterns using a complex model involving twelve differential equations. One day, he took the predicted weather from a printout of a previous simulation and started a new set of simulations from these values, holding everything else constant. Soon thereafter, however, he noticed that the predicted weather was completely different from his previous results. He later realized that he started the second simulation at a very slightly different position because he had rounded a variable when printing it out. While the initial position was off by just 0.000127, the long-term weather predictions were entirely different. A defining characteristic of chaos is this *sensitivity to initial conditions*, such that two trajectories that start near one another grow apart over time until they are no nearer than two trajectories that started far apart. This sensitivity to initial conditions has become known as the "butterfly effect:"

> The flapping of a single butterfly's wing today produces a tiny change in the state of the atmosphere. Over a period of time, what the atmosphere actually does diverges from what it would have done. So, in a month's time, a tornado that would have devastated the Indonesian coast doesn't happen. Or maybe one that wasn't going to happen, does. Stewart (1997), p. 141.

The emergence of chaos from entirely deterministic equations led Lorenz to conclude that there was little hope of predicting long-term weather patterns.

In 1974, the biologist Robert May (1974) published the simplest equation known to exhibit chaos: the logistic equation in discrete time. The logistic model involves a single variable and describes population growth as a quadratic function of the current population size (equation 3.5a). Because the population size can overshoot the equilibrium, chaotic fluctuations around the equilibrium are observed when the growth rate is large (e.g., $r = 2.7$ in Figure 4.2). Because the dynamics are chaotic, the dynamics are sensitive to initial conditions. For example, two populations whose initial sizes are very similar (e.g., 10,000 and 10,001) eventually become as different in population size as a population whose initial size is dramatically different (see Figure 4.1.1).

In the continuous-time logistic model, however, chaos is not observed (see Figure 4.3). Indeed, continuous-time models with only one or two variables never exhibit chaos. This does not mean that continuous-time models are always nicely behaved. In fact, continuous-time models with three or more variables can exhibit chaos. Indeed, chaotic dynamics are a common feature of food webs involving more than two species (Hastings and Powell 1991; Klebanoff and Hastings 1994; McCann and Hastings 1997). Interestingly, these models have been used to show that some forms of species interactions (e.g., linear food webs) are more prone to exhibit chaos than other forms of species interactions (e.g., omnivory).

We have mentioned that a defining feature of chaos is sensitivity to initial conditions. There are other clues that a model might exhibit chaos. One of them is that the system tends to oscillate between a number of states that doubles ("bifurcates") repeatedly as a parameter in the model is altered. This sort of behavior is observed in the logistic model as we increase the intrinsic

(continued)

Box 4.1 *(continued)*

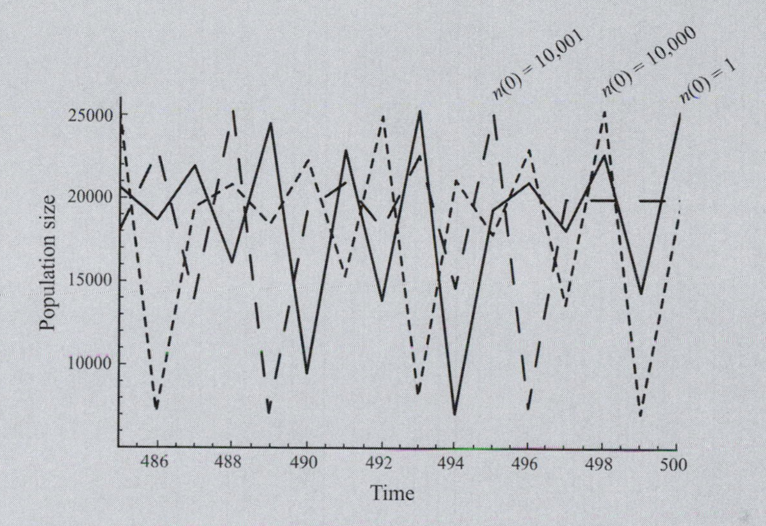

Figure 4.1.1: Sensitivity to initial conditions. A diagram of the population size versus time for the logistic equation (3.5a), starting from three different population sizes. After a few hundred generations, the population size dynamics of the two trajectories whose initial sizes were $n(0) = 10{,}000$ (short dashed lines) and $n(0) = 10{,}001$ (long dashed lines) are no closer together than they are to the population size dynamics starting from $n(0) = 1$ (solid lines). The parameters used were $r = 2.7$ and $K = 20{,}000$. Only time points 485 to 500 are shown to help see the differences between the trajectories.

growth rate, r (Figure 4.2). For low values of r, the population size approaches a single value K (e.g., $r = 0.7$ in Figure 4.2a). For higher values of r, the population size approaches an oscillation between two values, one above and one below K (e.g., $r = 2.1$ in Figure 4.2a). As r is increased further, the system settles down to a cycle involving four population sizes (e.g., $r = 2.5$ in Figure 4.2b). And as r is increased even further, the number of points through which the cycle passes (the *period*) doubles again and again. This period-doubling behavior is easiest to visualize using a *bifurcation diagram*.

Bifurcation diagrams illustrate the eventual states of a system on the vertical axis as a function of a parameter of interest on the horizontal axis. In the logistic model, the dynamics of the logistic model are quite sensitive to the intrinsic growth rate r, but not to other parameters such as the carrying capacity K (Problem 4.3). Thus, we use r as the parameter of interest in our bifurcation diagram (Figure 4.1.2). To produce this diagram, we iterated the recursion equation (3.5a) for a large number of generations (200) until the dynamics approached an equilibrium, a cycle, or showed no tendency to settle down. We then took the last 20 time points and plotted their values on the vertical axis. For small growth rates ($r < 2$), these last 20 time points were always very near the carrying capacity ($K = 1000$). The first period doubling occurred at $r = 2$, above which the population size cycled between two values (e.g., with $r = 2.1$, the population size

(continued)

Box 4.1 *(continued)*

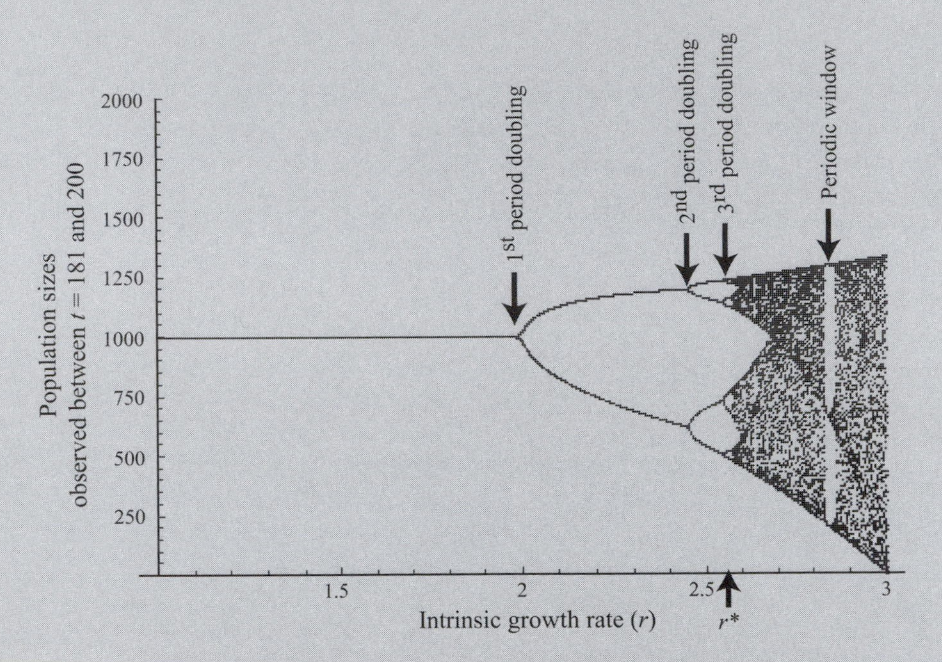

Figure 4.1.2: A bifurcation diagram for the logistic model. The recursion equation (3.5a) was iterated for 200 time steps, the last 20 of which are plotted on the vertical axis over a range of r values along the horizontal axis (r was increased from 1 to 3 in steps of 0.001). In every case, the initial population size was $n(0) = 10$, and the carrying capacity was $K = 1000$.

cycled between $n = 823$ and $n = 1129$, which give the vertical positions of the two points on the bifurcation diagram for $r = 2.1$). These two values grew further apart from one another until about $r = 2.45$, where the next period-doubling event occurred. Each subsequent period-doubling event occurred after shorter and shorter intervals in r. Eventually, the period doublings occurred so rapidly as r increased that the dynamics passed through a point at which an infinite number of period doublings had occurred. It is at this point that the dynamics become chaotic (at about $r^* = 2.569944$).

Period doubling is one route to chaos, and bifurcation diagrams allow us to visualize this process. If the bifurcation diagram for a model exhibits a forklike shape with tongs that divide faster and faster as the parameter of interest increases (as in Figure 4.1.2), expect to see chaos in the model.

A bifurcation diagram also illustrates that chaos does not imply a complete lack of order. In fact, for r values slightly above r^*, the population size remains within a fairly narrow region around the carrying capacity. This region expands as r increases and eventually includes zero at $r = 3$. Between the onset of chaos at r^* and extinction at $r = 3$, "periodic windows" appear, where the population size no longer fluctuates chaotically but cycles once again between a limited set of values. For example, a cycle among three points is observed at $r = 2.84$. Forklike bifurcations also occur within these periodic windows as r is increased further until, once again, the dynamics become chaotic.

Rearranging, we then get a recursion equation

$$n(t + \Delta t) = n(t) + \frac{dn}{dt}\Delta t,$$

which we can iterate, just as we did in Figure 4.2. As long as we choose a sufficiently small time interval Δt, the numerical solution for $n(t)$ should be consistent with the differential equation. By "consistent," we mean that the slope of the curve of $n(t)$ versus t should equal equation (3.5c) for dn/dt at every point in time.

The above method is known as Euler's method. Unfortunately, it is not very efficient because very small values of Δt are needed to obtain accurate predictions when n changes rapidly. Much more powerful and efficient algorithms have been developed to solve differential equations numerically (e.g., see Press 2002). Fortunately for us, mathematical software packages, including Maple, Matlab, and *Mathematica*, incorporate these algorithms "behind the scenes," making it fairly painless to solve differential equations numerically. For example, *Mathematica* contains code that is over 500 pages long to obtain numerical solutions for differential equations that are both accurate and efficient; all we have to do is remember the command name: NDSolve for "Numerical Differential-equations Solve" (see Lab exercise on solving differential equations).

Figure 4.3 illustrates the numerical solution to the logistic differential equation for various values of the intrinsic growth rate r. Regardless of how large r gets, there is no sign of overshooting, cycles, chaos, or extinction! The continuous- and discrete-time formulations agree when the intrinsic growth rate is small (Figure 4.4), but they lead to completely different predictions when r is large. For example, when r is greater than 3, the discrete-time model predicts that the population will go extinct, while the continuous-time model predicts that the population rises quickly and steadily to carrying capacity, where it remains forever after.

What is going on? Conceptually, the reason for the discrepancy between discrete- and continuous-time models is that the population growth that accrues in the discrete-time model over one time step is based on the population

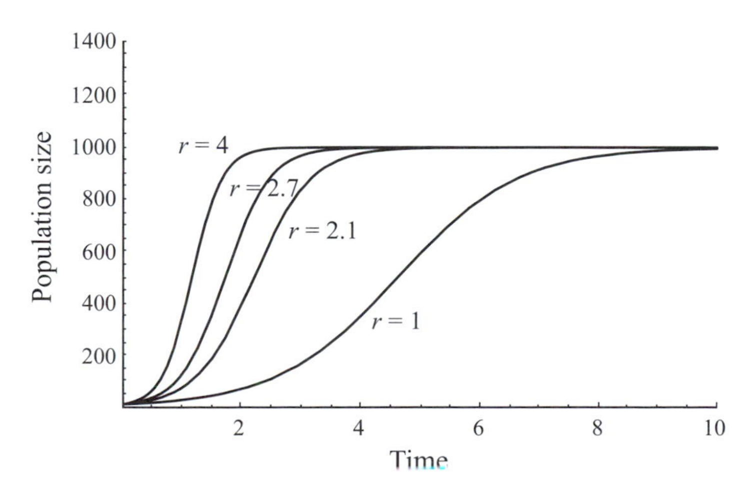

Figure 4.3: The dynamics of the logistic model in continuous time. The differential equation (3.5c) is solved numerically for various values of the intrinsic growth rate r using the command NDSolve in *Mathematica*. In each case, the initial population size is $n(0) = 10$ and the carrying capacity is $K = 1000$.

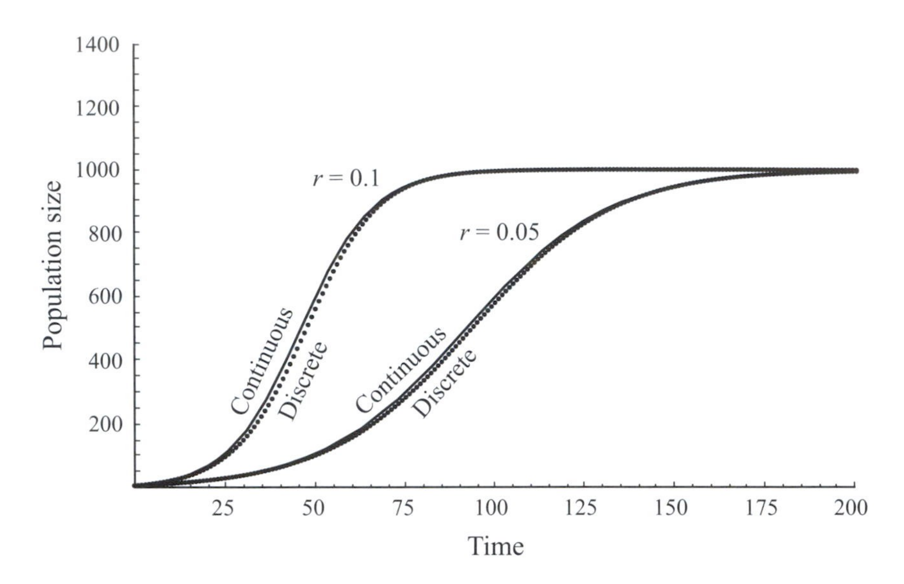

Figure 4.4: A comparison of the discrete-time and continuous-time logistic models. For low intrinsic growth rates, the two models predict similar trajectories for the population size over time. See Figures 4.2 and 4.3 for further details.

size at the beginning of the time step. There is nothing to prevent the population from growing past the carrying capacity by the end of the time step. In contrast, in the continuous-time model, the effect of population size feeds back continuously on the growth rate as the number of individuals changes. This instantaneous feedback prevents the population from overshooting and oscillating around its carrying capacity.

This sensitivity to the parameters and to the way that we model population growth must make us cautious in applying the logistic model uncritically to any natural population. To use the discrete-time model, we must be sure that the population satisfies the assumption that each event (e.g., births and deaths) happens only once per time unit and depends only on the preceding values of the variables. To use the continuous-time model, we must be sure that the population satisfies the assumption that each event (e.g., births and deaths) can happen in any order in any small interval of time, and that all individuals are equivalent, regardless of their age.

Other aspects of the logistic model are suspect and deserve to be critically evaluated. For example, we have assumed that the number of individuals can take on any real value, but in reality the number of individuals must be an integer. Furthermore, the number of offspring per parent will vary due to chance events (see Chapter 14). Making the logistic model more realistic in these ways causes dramatic changes to the predictions and eliminates the possibility of chaos (Henson et al. 2001).

At this point, you might be worried about the value of modeling if the results are so sensitive to the parameters and assumptions. Even when some predictions are sensitive, however, other predictions might not be. Results that do not depend very much on the parameters and underlying assumptions are

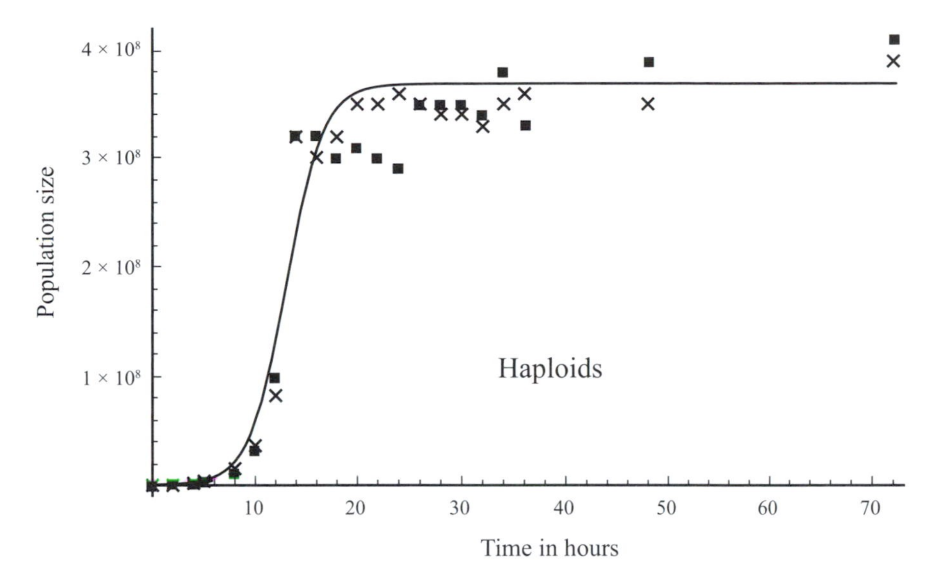

Figure 4.5: Dynamics of population growth in haploid yeast. The number of yeast in a flask is plotted over time using the consumer-resource model (3.17) where the resources are non-renewable ($\theta = 0$) and cell death is ignored ($d = 0$). The dynamics predicted by the continuous-time logistic model and the consumer-resource model lie exactly on top of each other, providing equally good fits to the data from Mable and Otto (2001) (crosses and squares, see Figure 3.4a). Parameter a was set to 1.47×10^{-9} ($= r_{\text{haploid}}/K_{\text{haploid}}$) and the initial resource level measured in terms of the number of cells that could be produced, $\varepsilon\, n_1(0)$, was set to 3.70×10^8 ($= K_{\text{haploid}} - n_2(0)$) (see Chapter 9). The fact that these models are mathematically equivalent explains why yeast in a vial display logistic growth, even though the resources are not constant over time.

called *robust*, and we tend to place more stock in such results. For example, as long as growth rates are not too large, populations grow according to an S-shaped curve in both discrete and continuous time (Figure 4.4). Furthermore, identical behavior is observed in a consumer-resource model in which we explicitly track the level of resources as well as the number of consumers (Figure 4.5; proven formally in Chapter 9). Thus, growth according to an S-shaped curve is a fairly robust prediction of the logistic model, breaking down only with high growth rates in a discrete-time model.

Second, even when a model is sensitive to assumptions, it can suggest new and interesting avenues of research. Perhaps the differences in model behavior that occur as a result of different assumptions provide an explanation for different population dynamics in nature. This might be addressed, for example, by determining if those species that persist despite having a high growth rate ($r > 3$) tend to replicate continuously over time.

4.2.3 The Dynamics of Models of Natural Selection

Not all models are as sensitive to the underlying assumptions as the logistic model. For example, the haploid model of natural selection displays very similar dynamics in both discrete and continuous time (Figure 4.6). The same is

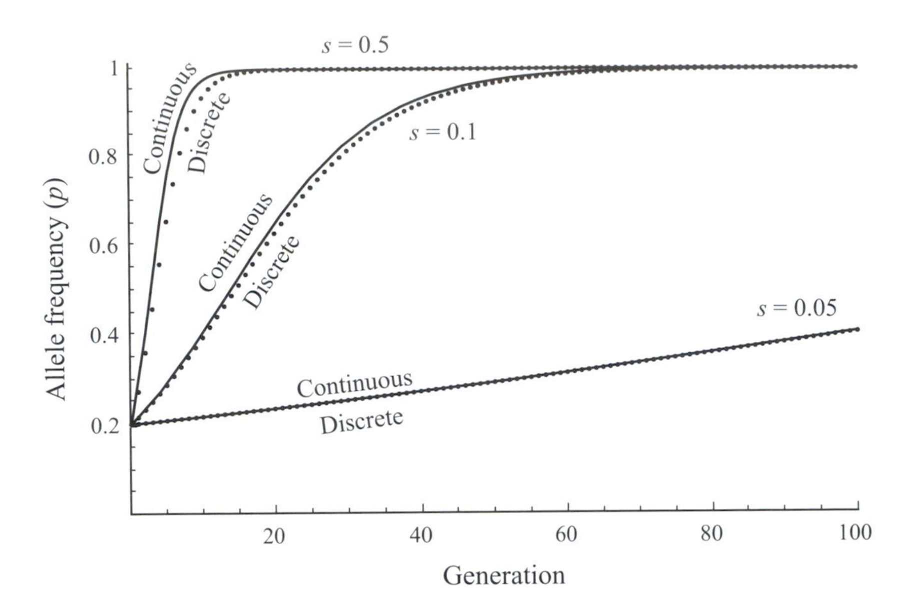

Figure 4.6: A comparison of the discrete-time and continuous-time models of haploid selection. For various values of the selection coefficient, s, the recursion equation (3.9) is iterated, and the differential equation (3.11b) is solved numerically using NDSolve in *Mathematica*. The predictions from the two models are in qualitative agreement even when selection is strong. The initial frequency of allele A is $p(0) = 0.2$.

true of the diploid model (not shown). As a result, both give a similar picture of the dynamics of natural selection.

Nevertheless, the behavior of models of selection is sensitive to the ordering of the fitnesses. In Figure 4.7, we plot the allele frequency as a function of time using the discrete-time, diploid model to gain some understanding of the effects of natural selection. These graphs appear to reveal three main types of behavior. If the heterozygotes have the highest fitness (e.g., $W_{AA} = 0.8$, $W_{Aa} = 1$, and $W_{aa} = 0.95$), natural selection tends to preserve both alleles (Figure 4.7a). Such selection is called "heterozygote advantage" or "overdominance." In contrast, when heterozygotes have intermediate fitness (e.g., $W_{AA} = 1$, $W_{Aa} = 0.95$, and $W_{aa} = 0.8$), natural selection increases the frequency of one allele over the other (Figure 4.7b). This form of selection is called "directional selection." Finally, when heterozygotes have the lowest fitness (e.g., $W_{AA} = 0.95$, $W_{Aa} = 0.8$, and $W_{aa} = 1$), natural selection favors different alleles depending on the initial allele frequency (Figure 4.7c). Such selection is called "heterozygote disadvantage" or "underdominance." These results are restricted to the parameter values chosen, but the methods introduced in Chapter 5 will confirm that the behavior of the model always falls into one of the above three classes, depending on the ordering of the fitnesses.

4.2.4 The Dynamics of Competing Species

The above examples involve only a single variable. Similar plots can be drawn for models involving more than one variable, with different curves illustrating the dynamics of different variables. In discrete-time models, such plots

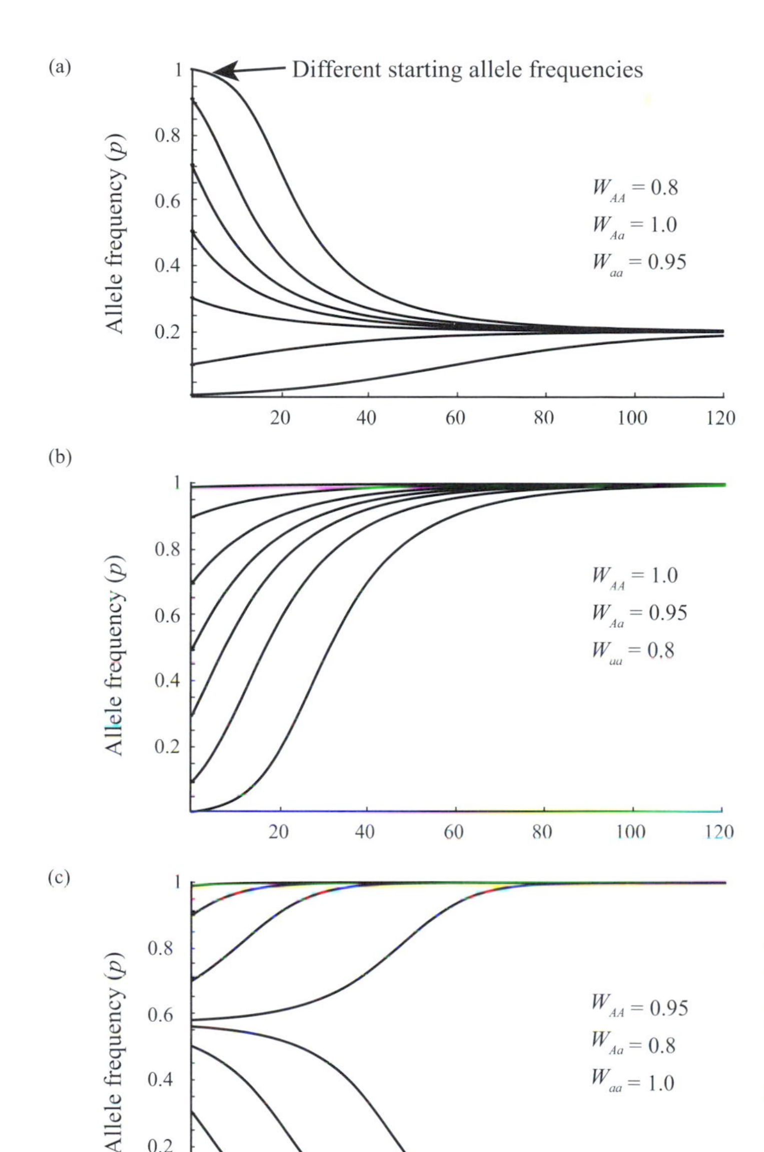

Figure 4.7: The discrete-time model of diploid selection. The recursion equation (3.13a) is iterated for various values of the initial frequency of allele A, $p(0) = 0.01, 0.1, 0.3, 0.5, 0.7, 0.9, 0.99$ (as indicated by the intercept with the y-axis). (a) Heterozygote advantage ($W_{AA} = 0.8$, $W_{Aa} = 1$, and $W_{aa} = 0.95$). (b) Directional selection favoring allele A ($W_{AA} = 1$, $W_{Aa} = 0.95$, and $W_{aa} = 0.8$). (c) Heterozygote disadvantage ($W_{AA} = 0.95$, $W_{Aa} = 0.8$, and $W_{aa} = 1$). Because time is discrete, the model only predicts the frequencies at integer values of time, but we connect the values of $p(t)$ with lines to clarify the behavior.

are again constructed by iteration of the recursion equations, where now each variable must be calculated in the next time step, as a function of all of the variables in the current time step.

For example, Figure 4.8 illustrates the dynamics of two species following the Lotka-Volterra equations with $r_1 = r_2 = 0.5$, $\alpha_{12} = 0$, $\alpha_{21} = 0.5$, and $K_1 = K_2 = 1000$. Under this (arbitrary) choice of parameters, species 1 competes for the resources of species 2 ($\alpha_{21} > 0$) while species 2 does not compete for the resources of species 1 ($\alpha_{12} = 0$). Starting from a population size of ten for each

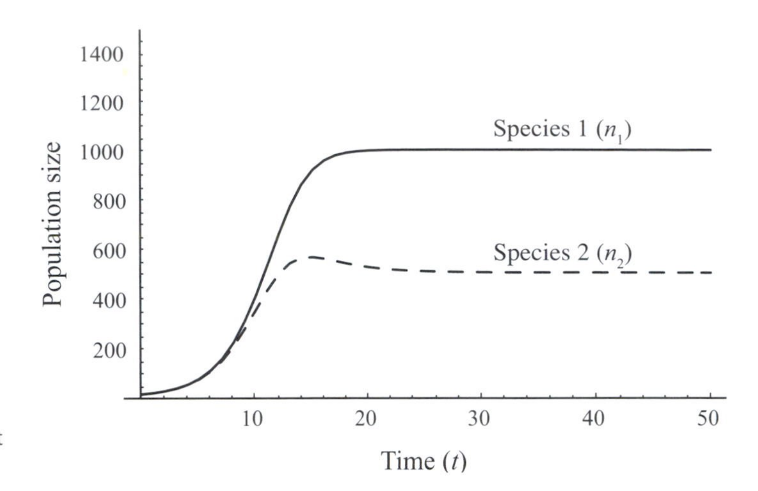

Figure 4.8: The Lotka-Volterra model of competition in discrete time. The recursion equations (3.14a) are iterated, beginning from a population size of ten for each species ($n_1(0) = 10$, $n_2(0) = 10$). The parameters in the model were set to $r_1 = r_2 = 0.5$, $\alpha_{12} = 0$, $\alpha_{21} = 0.5$, and $K_1 = K_2 = 1000$. Because time is discrete, the model only predicts population sizes at integer values of time, but we connect these values with lines to clarify the behavior.

species, $n_1(0) = 10$, $n_2(0) = 10$, we can plug these parameters into equation (3.14) to predict the population size in the next time step:

$$\begin{aligned} n_1(1) &= n_1(0) + r_1 n_1(0)\left(1 - \frac{n_1(0) + \alpha_{12} n_2(0)}{K_1}\right) \\ &= 10 + 0.5 \times 10\left(1 - \frac{10 + 0 \times 10}{1000}\right) \\ &= 14.95, \end{aligned}$$

$$\begin{aligned} n_2(1) &= n_2(0) + r_2 n_2(0)\left(1 - \frac{n_2(0) + \alpha_{21} n_1(0)}{K_2}\right) \\ &= 10 + 0.5 \times 10\left(1 - \frac{10 + 0.5 \times 10}{1000}\right) \\ &= 14.925. \end{aligned}$$

After only one time step, the expected number of offspring is lower for species 2, not because it has a lower intrinsic growth rate or carrying capacity, but because it experiences more competition from the other species. This difference in growth continues, until eventually the system reaches a point where species 1 is near its carrying capacity, while species 2 sustains only half the number of individuals expected on the basis of its carrying capacity. For these parameters, a very similar graph is produced if we numerically solve the differential equations (3.15) for the continuous-time model of competition (Problem 4.10).

Variable versus variable plots illustrate the future state (or change) of a variable (vertical axis) versus the current state of the variable (horizontal axis).

4.3 Plots of Variables as a Function of the Variables Themselves

In the above graphs, we plotted the variable of interest over time (the independent variable). The second type of graph that we discuss is easier to generate; we simply plot the equation (recursion, difference, or differential) as a function of the variable itself. Such plots are particularly helpful in understanding the dynamics of one-variable models because they clarify how the direction of change depends on the current state of the system.

4.3.1 Plotting the Value of a Variable at Time $t + 1$ versus at Time t

Consider the haploid model of natural selection with the recursion equation (3.9),

$$p(t+1) = \frac{W_A p(t)}{W_A p(t) + W_a(1 - p(t))}.$$

Figure 4.9 plots $p(t + 1)$ as a function of $p(t)$ when W_A is greater than W_a (solid curve in Figure 4.9a) and when W_A is less than W_a (solid curve in Figure 4.9b). In both cases, we have also drawn a dashed diagonal 1:1 line. The diagonal line represents those special cases where $p(t + 1) = p(t)$. Wherever the recursion curve crosses the diagonal line, the allele frequency in the next time step $p(t + 1)$ will equal the allele frequency in the previous time step $p(t)$. Such a point is called an "equilibrium" because it remains unchanged over time (see Chapter 5). Thus, if the system starts at an equilibrium value for the allele frequency, it will remain there forevermore.

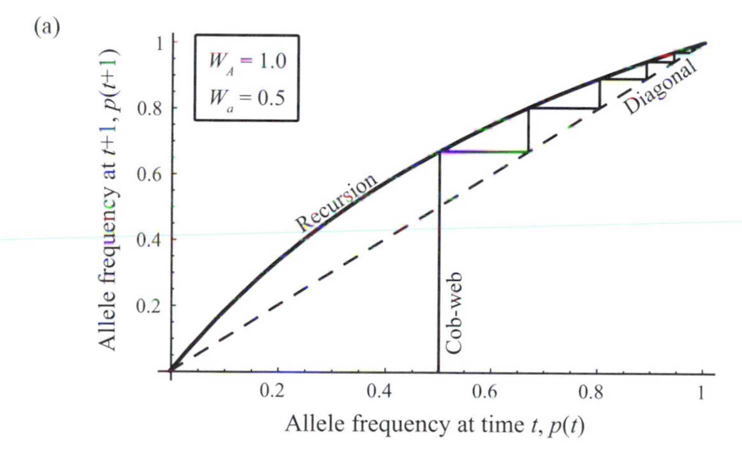

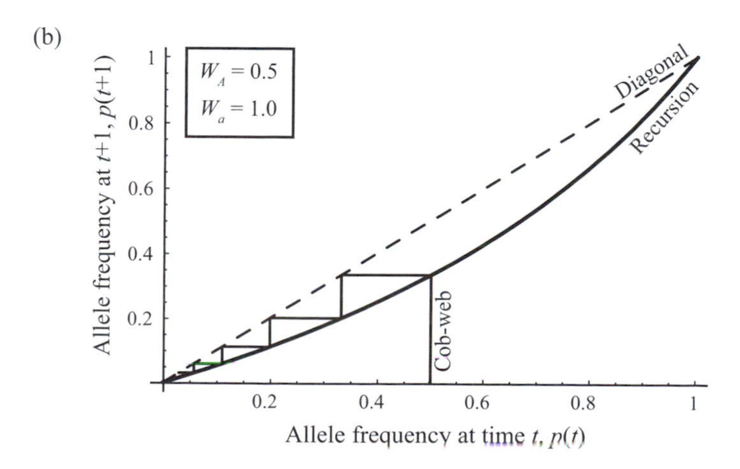

Figure 4.9: Allele frequency recursion in the haploid model of selection. The frequency of allele A at $t + 1$ is plotted against the frequency of allele A at t, using the recursion equation (3.9). The diagonal line (dashed) represents the case where $p(t + 1) = p(t)$. At any point where the recursion curve falls above the diagonal line, the allele frequency increases over time, as in (a) where $W_A = 1$ and $W_a = 0.5$. At any point where the recursion curve falls below the diagonal line, the allele frequency decreases, as in (b) where $W_A = 0.5$ and $W_a = 1$. The vertical and horizontal lines starting at $p(0) = 0.5$ illustrate "cobwebbing," a procedure that can be used to determine changes to the variable over time (Box 4.2).

In the haploid model, there are two places where the recursion function crosses the diagonal: when the frequency of A is zero and when it is one (Figure 4.9). Again, we know that this is true only for the particular fitnesses used in the figure, but we will see in Chapter 5 that these two points represent the only two equilibria of the haploid model with constant fitnesses.

Recipe 4.1
Identifying Equilibria Graphically

A system at "*equilibrium*" does not change over time.

- In a plot of $n(t+1)$ versus $n(t)$, any point where the recursion equation crosses the diagonal line represents an equilibrium, because $n(t+1) = n(t)$.
- In a plot of Δn versus $n(t)$, any point where the difference equation crosses the horizontal axis represents an equilibrium, because $\Delta n = 0$.
- In a plot of $\mathrm{d}n/\mathrm{d}t$ versus $n(t)$, any point where the differential equation crosses the horizontal axis represents an equilibrium, because $\mathrm{d}n/\mathrm{d}t = 0$.

Now, focus your attention on whether the recursion lies above or below the diagonal. Before reading on, think about what this means. Whenever the recursion lies above the diagonal, $p(t+1)$ is greater than $p(t)$, and allele A increases in frequency. For the parameter values in Figure 4.9a (where $W_A > W_a$), the recursion is above the diagonal for all values of the allele frequency ($0 < p(t) < 1$). As a result, allele A rises in frequency every generation until it approaches one ("fixation"). Conversely, whenever the recursion falls below the diagonal (as in Figure 4.9b, where $W_A < W_a$), $p(t+1)$ is always lower than $p(t)$. As a result, allele A decreases in frequency until it is lost from the population. You can even use these figures to run your own iterations by hand using a technique known as "cobwebbing" (Figure 4.9; see Box 4.2).

Recipe 4.2
Identifying Direction of Change Graphically

- In a plot of $n(t+1)$ versus $n(t)$, if the recursion equation lies above the diagonal line, the variable increases over time ($n(t+1) > n(t)$) and vice versa.
- In a plot of Δn versus $n(t)$, if the difference equation lies above the horizontal line, the variable increases over time ($\Delta n > 0$) and vice versa.
- In a plot of $\mathrm{d}n/\mathrm{d}t$ versus $n(t)$, if the differential equation lies above the horizontal line, the variable increases over time ($\mathrm{d}n/\mathrm{d}t > 0$) and vice versa.

There are two general types of equilibria: *attracting* and *repelling*. As these categories suggest, variables tend to move toward attracting equilibria over time, whereas they tend to move away from repelling equilibria over time. Attracting states are also known as *stable* equilibria, while repelling states are known as *unstable* equilibria. As an example, imagine trying to balance on a bicycle whose pedals are locked in place. If you could manage to start perfectly upright and remain completely motionless, then in the absence of any external forces (e.g., a gust of wind), you would remain upright indefinitely. Starting slightly off vertical, however, you would fall either to the left or to the right. Measuring the angle of the bike from the horizontal position, the bike has three equilibrium states: lying flat on its left-hand side (0°), lying flat on its right-hand side (180°), and standing perfectly upright (90°). The first two equilibria (0° and 180°) are both attracting, or stable, equilibria because the bicycle will fall to the ground once it has started to tip over. The third equilibrium (90°) is repelling, or unstable, because the bicycle will fall over once it starts to tip away from this point.

Stable equilibria are subdivided further into *locally stable* equilibria, which are attracting from nearby states, and *globally stable* equilibria, which are attracting from all states. There are two locally stable equilibria in the bicycle example: lying on the left or lying on the right. The bicycle can end up in either state depending on the direction in which it starts to fall. If there were a person on the right pushing the bike back if it ever started to fall, then the bicycle would only have two equilibria, an unstable equilibrium in the upright position (i.e., 90°) and a stable equilibrium on its left-hand side (0°). The latter equilibrium is globally stable because it is attracting for all possible starting conditions (except for the perfectly balanced upright position).

Returning to the haploid model of selection illustrated in Figure 4.9a (where $W_A > W_a$), the frequency of allele A always rises toward $p = 1$ because the recursion lies above the diagonal. Therefore, $p = 1$ is a globally stable equilibrium. On the other hand, $p = 0$ is unstable, and the population moves further and further away from this point. The opposite is true in Figure 4.9b (confirm to yourself that $p = 1$ is unstable and $p = 0$ is globally stable).

If an equilibrium is globally stable, it must also be locally stable, but the opposite is not true. Consider the diploid model of natural selection when the heterozygote is least fit. Figure 4.10 shows $p(t+1) = p(t)^2 W_{AA}/\overline{W} + p(t)\,q(t) W_{Aa}/\overline{W}$ plotted against $p(t)$ for the case $W_{AA} = 0.6$, $W_{Aa} = 0.2$, and $W_{aa} = 1$. Judging from where the recursion curve crosses the diagonal line, there are now three equilibria ($p = 0$, $p = 1$, and $p = 0.66$). For small initial allele frequencies ($0 < p < 0.66$), the recursion lies below the diagonal, implying that the allele frequency decreases in each generation and approaches $p = 0$. Therefore $p = 0$ is a locally stable equilibrium. For high initial allele frequencies ($0.66 < p < 1$), the recursion lies above the diagonal, implying that the allele frequency increases in each generation and approaches $p = 1$. Therefore, $p = 1$ is also a locally stable equilibrium. Thus, there is a region (known as a *basin of attraction*) within which p moves toward 0, and another region within which p moves toward 1. As a result, neither of these equilibria is globally stable (see Chapter 5).

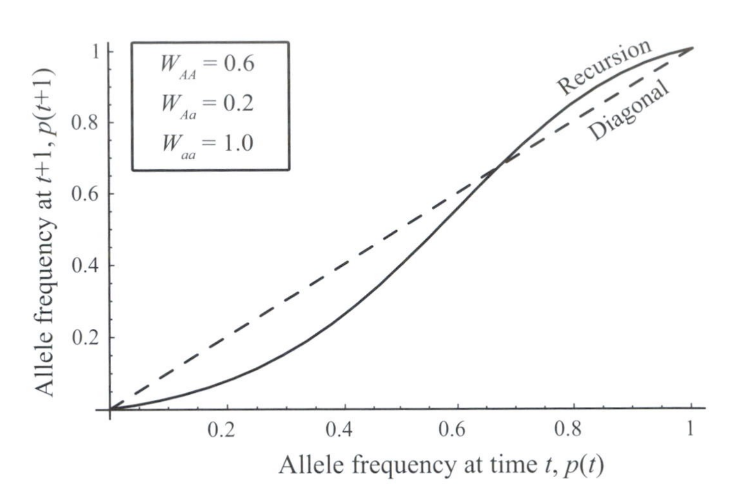

Figure 4.10: Allele frequency recursion in the diploid model of selection. The frequency of allele A at $t + 1$ is plotted against the frequency of allele A at t, using the recursion equation (3.13a). The diagonal line (dashed) represents the case where $p(t + 1) = p(t)$. The allele frequency decreases over time when $p(t)$ is less than 0.66 (recursion lies below the diagonal) but increases over time when $p(t)$ is greater than 0.66 (recursion lies above the diagonal). Here, $W_{AA} = 0.6$, $W_{Aa} = 0.2$, and $W_{aa} = 1$.

We will return to the issue of stability in greater depth in Chapter 5. For the time being, however, we can foreshadow a key result by thinking about the feature of these graphs that determines whether an equilibrium is locally stable or unstable. For an equilibrium to be locally stable, the variable must rise in frequency when it begins below the equilibrium (i.e., the recursion equation must lie above the diagonal), and the variable must decline in frequency if it begins above the equilibrium (i.e., the recursion equation must fall below the diagonal). Considering that the diagonal line has a slope of one (it is the 1:1 line), this implies that the slope of the recursion at the equilibrium must therefore be less than one for the equilibrium to be locally stable. Conversely, if the slope of the recursion is greater than one at an equilibrium, then the equilibrium will be unstable. Check to see that this claim is valid for the equilibria in Figures 4.9 and 4.10. The haploid and diploid models of selection are good examples to begin to think about stability, because the slope of the recursion with respect to the variable p is always positive (Problems 4.5 and 4.6). This is not true of all models, however, and more complicated dynamics arise when the slope is negative (Box 4.2).

4.3.2 Plotting the Change in a Variable versus the Variable at time t

Let us now consider a different type of plot, where instead of plotting the value of the variable in the next generation on the vertical axis, we plot the *change* in the variable on the vertical axis. Again we use the current value of the variable on the horizontal axis. The change in a variable is given by a difference equation in a discrete-time model or by a differential equation in a continuous-time model.

We first revisit the haploid model of selection. With weak selection, the change in the frequency of allele A is approximately $sp(t)(1-p(t))$ according to either the difference equation (3.10) for Δp or the differential equation (3.11b) for dp/dt (Chapter 3). In Figure 4.11, the rate of change, $sp(t)(1-p(t))$, is plotted as a function of $p(t)$. The population is at equilibrium if there is no change in

Box 4.2: Cobweb Plots of the Logistic Equation

Cobweb plots provide a graphical way of iterating a recursion equation involving a single variable. In a cobweb plot, the value of the variable at time $t + 1$ is plotted on the vertical axis as a function of the variable at time t along the horizontal axis (as in section 4.3). Starting from any initial position along the horizontal axis, the value of the variable in the next time step is determined by drawing a vertical line up to the curve for the recursion equation. The predicted value of the variable equals the height of the curve at this point. This value can then be used as the initial state of the system in the next time step. This procedure may be repeated for as many time steps as desired.

The iteration procedure is made easier by adding a diagonal line to the plot (i.e., the line where $n(t + 1) = n(t)$) and employing the method illustrated in Figure 4.2.1 for the logistic equation (3.5a). At step (a), the variable changes from an initial value of $n(0) = 400$ to $n(1) = 568$, which is determined by drawing a vertical line starting at $n(0)$ up to the recursion curve. (b) The starting value of the variable in the next time step is given by the height of this recursion curve, $n(1) = 568$. To use this height as the starting value in the next generation, we draw a horizontal line from the recursion curve to the diagonal line, which effectively sets the starting position in the next generation to 568, the final position in the previous generation (see Figure 4.2.1). (c) At this point, a vertical line is drawn to the recursion equation again, giving the variable in the next time step, $n(2) = 740$. From this point on, each time step corresponds to drawing a horizontal line from the recursion curve to the diagonal to obtain the starting value (d), then drawing a vertical line from the diagonal to the recursion curve (e) to determine the value of the variable in the next time step. This cobwebbing procedure can be used to predict the state of the variables over multiple time steps (Figure 4.2.1).

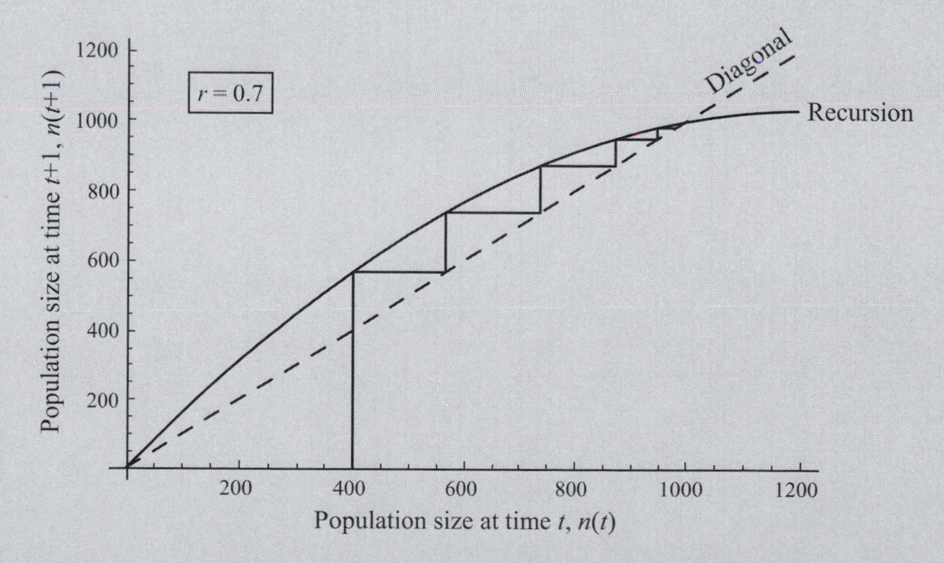

Figure 4.2.1: A cobweb diagram for the logistic model (low r). The starting condition and parameters of the logistic equation were $n(0) = 400$, $r = 0.7$, and $K = 1000$.

(continued)

Box 4.2 ***(continued)***

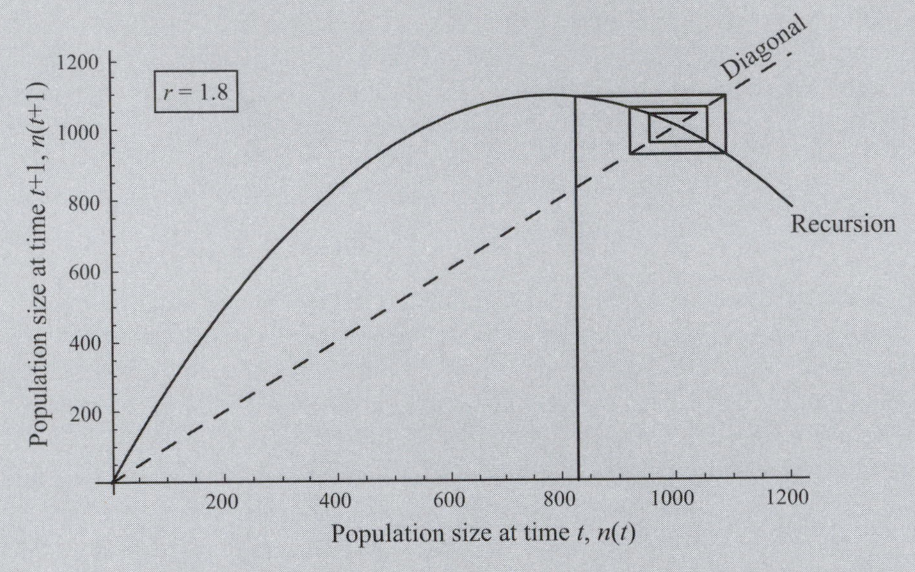

Figure 4.2.2: A cobweb diagram for the logistic model (intermediate r). The starting condition and parameters of the logistic equation were $n(0) = 825$, $r = 1.8$, and $K = 1000$.

In Figure 4.2.1, the intrinsic growth rate is low ($r = 0.7$), and the cobwebbing procedure suggests a smooth approach to the carrying capacity at $K = 1000$ (see Figure 4.2a). Overshooting the equilibrium does not occur in this case. As we increase r, however, the height of the recursion rises until it eventually lies above K when the system starts near to, but below, K (Figure 4.2.2 for $r = 1.8$). For example, when starting with an initial population size of $n(0) = 825$ (far left vertical line), the population grows to 1085 individuals, above the carrying capacity at $K = 1000$. Foreshadowing Chapter 5, the key feature of figures like Figure 4.2.2, where overshooting an equilibrium is possible, is that the slope of the recursion equation at the equilibrium of interest is negative (here, the slope at K is -0.8). This is mathematically equivalent to the statement that it is possible to rise above K starting at points slightly below K.

Even in Figure 4.2.2, the system approaches the carrying capacity (i.e., the cobwebs are moving toward the equilibrium). If r is increased even further, however, there comes a point where the height of the recursion equation rises so much that populations starting slightly below K grow so much that they end up further away from K on the other side (Figure 4.2.3 for $r = 2.5$). For example, when starting with an initial population size of $n(0) = 905$ (below the carrying capacity by 95 individuals), the population grows to 1186 individuals (above the carrying capacity by 186 individuals). The key feature of figures like Figure 4.2.3, where a system started near an equilibrium overshoots the equilibrium by so much that it moves further from the equilibrium, is that the slope of the recursion equation at the equilibrium of interest is less than -1 (here, the slope at K is -1.5). In Chapter 5, we will see that the insights provided by these cobweb plots are consistent with analytical methods that determine when an equilibrium is stable or unstable.

(continued)

Box 4.2 *(continued)*

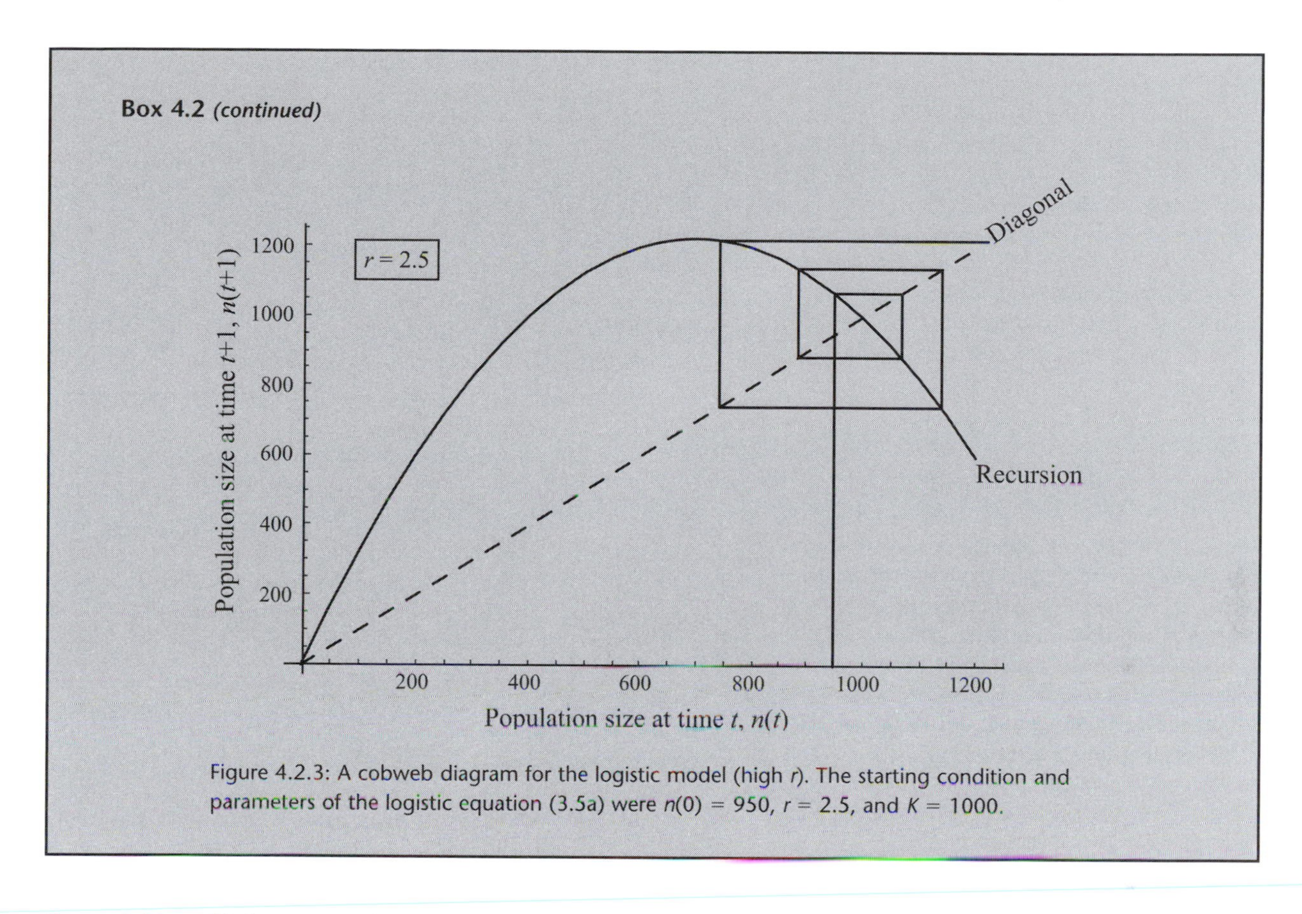

Figure 4.2.3: A cobweb diagram for the logistic model (high r). The starting condition and parameters of the logistic equation (3.5a) were $n(0) = 950$, $r = 2.5$, and $K = 1000$.

the variable, which occurs wherever the curve of the difference or differential equation crosses the horizontal axis (Recipe 4.1). Once again, we find equilibria at $p = 0$ and $p = 1$. As mentioned in Recipe 4.2, whenever the change in p is positive (i.e., lies above the horizontal axis), allele A increases in frequency (as in Figure 4.11a). This is consistent with the rise in allele frequency observed in Figure 4.6, where we plotted the allele frequency as a function of time. Similarly, whenever the change in p is negative, allele A decreases in frequency (as in Figure 4.11b).

Again, we can use these figures to gain some insight into the conditions that must be met for an equilibrium to be locally stable. Look at Figure 4.11 and think about how the stability of an equilibrium depends on the slope of the difference or differential equation. If the slope is positive at an equilibrium, then a variable that begins slightly below the equilibrium will decrease away from it (change is negative), while a variable that begins slightly above the equilibrium will increase away from it (change is positive). Thus, when the slope is positive, an equilibrium is unstable.

Conversely, if the slope is negative at an equilibrium, then a variable that begins slightly below the equilibrium will increase (change is positive), and a variable that begins slightly above the equilibrium will decrease (change is negative), so the equilibrium might be stable. We've waffled a bit, here, by saying "might be," because, for discrete-time models, it is possible for the variable to move toward the equilibrium, but overshoot it and end up even further away

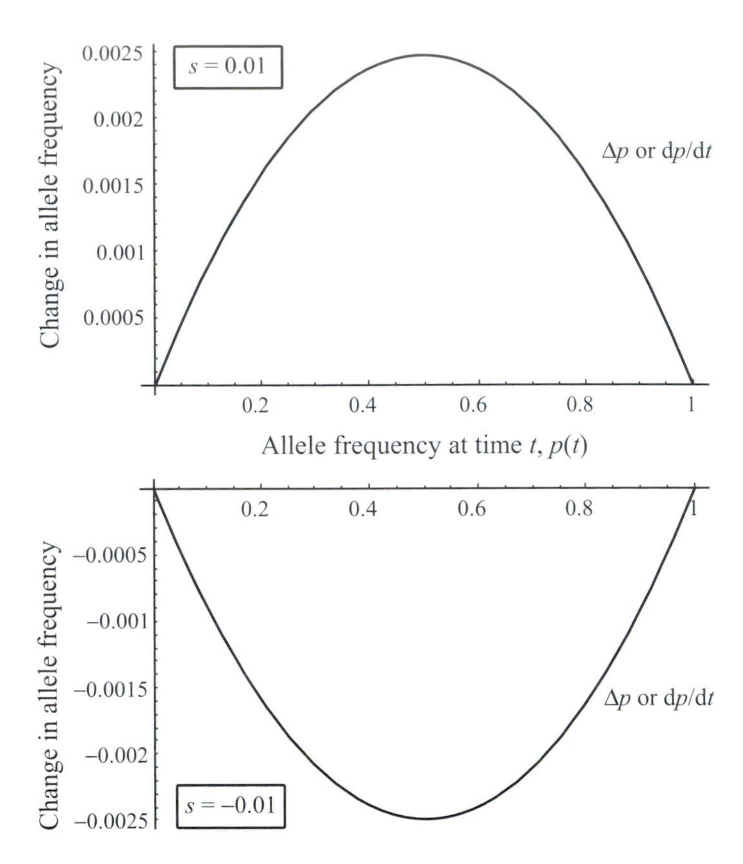

Figure 4.11: Change in allele frequency as a function of the current allele frequency in the haploid model of selection. The vertical axis describes the change in allele frequency, $\Delta p \approx dp/dt = s\, p(t)\, (1-p(t))$. The horizontal axis gives the current allele frequency $p(t)$. At any point where the curve falls above the horizontal axis, the allele frequency increases over time, as occurs in (a) for $s = 0.01$. At any point where the curve falls below the horizontal axis, the allele frequency decreases over time, as occurs in (b) for $s = -0.01$.

on the other side. We will address this issue in much greater detail in the next chapter, but for now it is enough to recognize that the slope of a difference or differential equation when plotted against the variable of interest provides important information about the stability of an equilibrium.

As another example, consider the logistic model of population growth. The change in population size is $rn(t)(1 - n(t)/K)$ according to either the difference equation (3.5b) for Δn or the differential equation (3.5c) for dn/dt (Chapter 3). Figure 4.12 plots this change as a function of $n(t)$ for $r = 1$ and $r = 2.5$, assuming a carrying capacity of 1000. In both cases, the function crosses the horizontal axis at $n = 0$ and $n = K$, which correspond to the two equilibria of the model. For both $r = 1$ and $r = 2.5$, if the population size begins below the carrying capacity it grows, while if it begins above the carrying capacity it shrinks. This suggests that the equilibrium $n = 0$ is unstable and that the $n = K$ equilibrium is stable. But we must be careful! In a discrete-time model, overshooting the equilibrium is possible (Figure 4.2). Thus, even though the population size moves in the direction of an equilibrium this does not mean that it gets closer to the equilibrium. For example, when the population begins at $n(0) = 900$, the change in population size is $+225$ for $r = 2.5$ (check by using the difference equation or by reading off the value from the graph). The population at the next time is thus $n(1) = 900 + 225 = 1125$, which is even further away from the equilibrium on the other side. Indeed, as we shall see in Chapter 5,

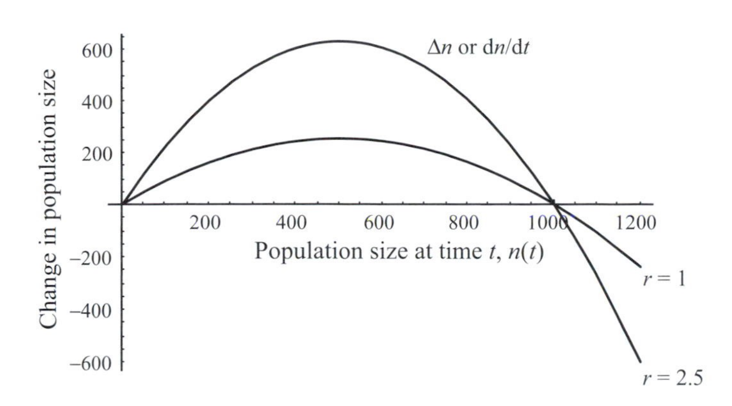

Figure 4.12: Changes in population size as a function of the current population size in the logistic model. The vertical axis describes the change in population size $\Delta n = dn/dt = r\, n(t)\, (1 - n(t)/K)$. The horizontal axis gives the current population size $n(t)$. At any point where the curve lies above the horizontal axis, the population size increases over time, as occurs below the carrying capacity of $K = 1000$. At any point where the curve falls below the horizontal axis, the population size decreases over time, as occurs above the carrying capacity of $K = 1000$. Two curves are shown, assuming $r = 1$ and $r = 2.5$.

the carrying capacity is not a stable equilibrium of the discrete-time model when the intrinsic growth rate is too high.

The problem of overshooting the equilibrium does not arise in the continuous-time logistic model. In a continuous-time model, a variable rises smoothly over time, passing through all possible points between where it begins and where it ends. For a variable to overshoot an equilibrium (e.g., at $K = 1000$), it would have to pass through the equilibrium. For example, for the population size to change smoothly from $n(0) = 900$ to $n(1) = 1125$ in a continuous-time model would require that the population size equal 1000 at some intermediate point in time. But when the variable hits the equilibrium, the change in the variable will equal zero (by the definition of an equilibrium), and the variable will not move any further. This provides another perspective on why the behavior of the logistic model in continuous time is much more tame than in discrete time. The logistic model, like any other continuous-time model with only a single variable, cannot exhibit overshooting or chaos because the variable cannot pass through an equilibrium without getting stuck.

The graphs considered in this section are less natural for models with more than one variable. The reason is that one cannot predict the state of the system in the next time step as a function of only one variable. Rather, one must know the current state of all variables to predict their future states. For models with two variables, you could draw a three-dimensional graph where the two axes in the plane (x and y) correspond to the current states of the two variables, and the vertical axis (z) corresponds to the state of the first variable in the next time step. But such a graph is not informative on its own. The first variable might reach a state where it does not change (i.e., the recursion equation may fall on the diagonal where $z = x$), but this is based on the current value of the second variable, which might still be changing. A much clearer picture emerges from phase-plane diagrams, introduced in the next section.

4.4 Multiple Variables and Phase-Plane Diagrams

In this third type of graph, each axis represents a different dependent variable. Time is represented as a series of dots, one for each time step, or as arrows pointing in the direction of change. Such graphs are called "phase-plane diagrams." Phase-plane diagrams are particularly helpful in understanding the dynamics of two-variable models because they indicate how each variable changes as a function of the other variable.

4.4.1 Phase-Line Diagrams

Before describing phase-plane diagrams for two-variable models, let us start with the simpler case of phase-line diagrams in single-variable models. The idea is to draw a line or axis spanning the range of possible values of the variable. At several points along the line, arrows are drawn that indicate the direction in which the variable moves. Figure 4.13 illustrates phase-line diagrams for the haploid and diploid models considered earlier. These phase-line diagrams

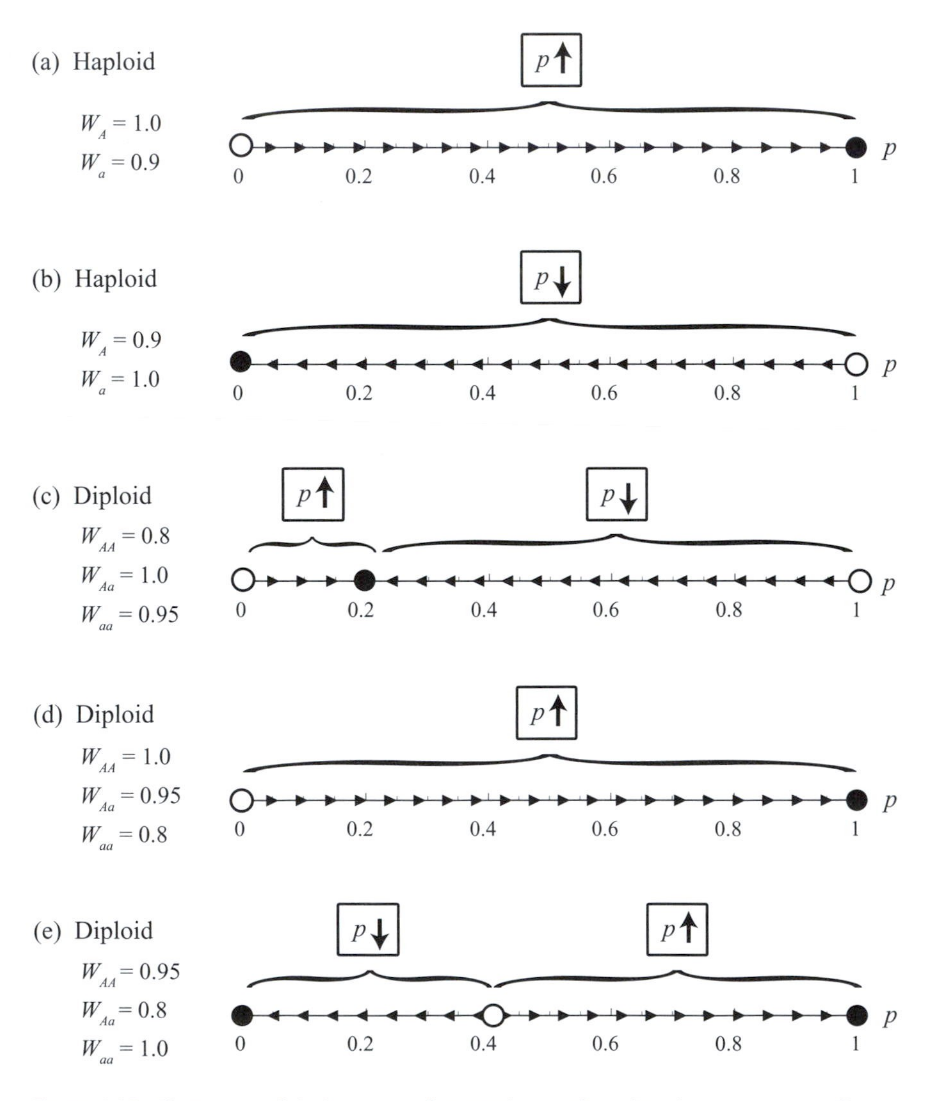

Figure 4.13: Changes in allele frequency illustrated on a phase-line diagram. Arrows illustrate the direction in which allele frequencies evolve. Arrows point toward solid circles and away from hollow circles.

provide a nice visual summary of the behavior of a model. At a glance, it is possible to see the direction in which a variable changes for a particular set of parameters.

In the haploid model of selection, the direction of the arrows is given by the sign of the difference equation (3.10) in discrete time:

$$\Delta p = \frac{s_d p(t)(1 - p(t))}{1 + s_d p(t)},$$

or the sign of the differential equation (3.11b) in continuous time:

$$\frac{dp}{dt} = s_c p(t)(1 - p(t)).$$

In either case, the direction of the arrows depends only on the sign of the selection coefficient because $p(t)(1 - p(t))$ is always positive (we ignore the case where $p(t)(1 - p(t))$ equals zero, which implies that one allele is absent), as must be the mean fitness in the denominator of Δp. Thus, when $s > 0$, the arrows always point to the right, and the allele frequency increases. We can conclude that $p = 1$ is globally stable whenever $s > 0$, regardless of the exact values of the selection coefficient or the initial conditions. Similarly, when $s < 0$, the arrows always point to the left, and $p = 0$ is globally stable.

In continuous-time models, phase-line diagrams tell us immediately which equilibria are locally stable (arrows point in) and which are unstable (arrows point out). In discrete-time models, however, we have to be careful in interpreting phase-line diagrams whenever it is possible for the system to overshoot the equilibrium. In the haploid model of selection, overshooting is not a concern, because it would require one of the allele frequencies to become negative, which is not possible (because the fitnesses cannot be negative, equation 3.8c never becomes negative). But, in general, it is difficult to tell from phase-line diagrams in discrete time whether movement along the arrows might lead a variable further away from an equilibrium on the other side.

4.4.2 Phase-Plane Diagrams

We can extend phase-line diagrams to models with two variables using phase-plane diagrams, with the horizontal axis representing the range of one variable and the vertical axis representing the range of the other variable. On a phase-plane diagram, we can trace how the variables change over time, starting from a particular value along the horizontal and vertical axes. This is best illustrated by example.

A *phase-plane diagram* illustrates the state of one variable (vertical axis) versus the state of a second variable (horizontal axis).

In Figure 4.14, the dynamics of two competing species in the Lotka-Volterra model is illustrated by a phase-plane diagram. Starting from $n_1(0) = 10$, $n_2(0) = 10$, the recursion equations (3.14) were iterated for 50 time steps. At time step t, the population sizes were used to plot a point whose horizontal coordinate is the number of species 1 and vertical coordinate is the number of species 2. Four simulations were conducted using different competitive effects of species 2 on

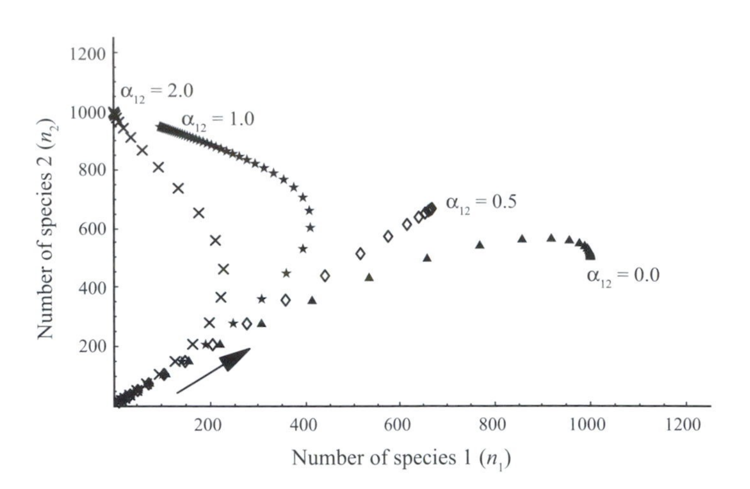

Figure 4.14: Dynamics of the two-species Lotka-Volterra model of competition. The recursion equations (3.14) were iterated for 50 time steps starting with ten individuals of each species, $n_1(0) = n_2(0) = 10$. This was repeated four times using different competitive effects of species 2 on species 1: $\alpha_{12} = 0$ (triangles), $\alpha_{12} = 0.5$ (diamonds), $\alpha_{12} = 1$ (stars), and $\alpha_{12} = 2$ (×). The remaining parameters were: $r_1 = r_2 = 0.5$, $\alpha_{21} = 0.5$, and $K_1 = K_2 = 1000$.

species 1: $\alpha_{12} = 0$ (triangles), $\alpha_{12} = 0.5$ (diamonds), $\alpha_{12} = 1$ (stars), and $\alpha_{12} = 2$ (×), holding all other parameters constant ($r_1 = r_2 = 0.5$, $\alpha_{21} = 0.5$, $K_1 = K_2 = 1000$, $n_1(0) = n_2(0) = 10$). An arrow is added to the graph to show the direction in which time flows. For example, we saw above that, after one time step, the population sizes become $n_1(1) = 14.95$ and $n_2(1) = 14.925$ with $\alpha_{12} = 0$. Thus, in Figure 4.14, the $\{x, y\}$ coordinates of the first triangle corresponding to $t = 0$ are placed at $\{10,10\}$, and the $\{x, y\}$ coordinates of the second triangle corresponding to $t = 1$ are placed at $\{14.95,14.925\}$. These values of the variables are then plugged back into the recursion to get the coordinates at $t = 2, 3, 4$, etc. up until 50 time steps. This process is then repeated for the next value of α_{12} starting back at the initial point $\{10,10\}$.

The population sizes grow in the same direction at first regardless of the value of α_{12} (along the arrow on Figure 4.14) because there are few competitors around. As the population sizes get larger, competition begins to play a more decisive role, causing the trajectories to diverge for different values of α_{12}. When species 2 has no effect on species 1 ($\alpha_{12} = 0$), the trajectory approaches the point $\{1000,500\}$ after 50 times steps (filled triangles), as seen in Figure 4.8. For larger values of α_{12}, competition limits the number of species 1, shifting the trajectories to the left in Figure 4.14. Indeed, for $\alpha_{12} = 2$, species 2 eventually outcompetes species 1, which is driven to extinction (×'s). In Figure 4.14, the symbols start piling on top of one another toward the end of the 50 time steps, implying that smaller and smaller changes occur per time step. This is a hallmark feature of variables as they near an equilibrium.

As a good check that the iterations have been coded correctly, consider the case with $\alpha_{12} = 0.5$ (diamonds). In this case, the parameters for species 1 and 2 are identical: $r_1 = r_2$, $\alpha_{12} = \alpha_{21}$, $K_1 = K_2$, and $n_1(0) = n_2(0)$. With such symmetry, we expect the number of each species to be the same every generation. Figure 4.14 shows this to be true, as the diamonds fall along the diagonal line.

4.4.3 Vector-Field Plots

A drawback of the above phase-plane plots is that they illustrate how the variables change only along certain trajectories (e.g., starting at $n_1(0) = n_2(0) = 10$). A more complete visual representation of the dynamics is possible by

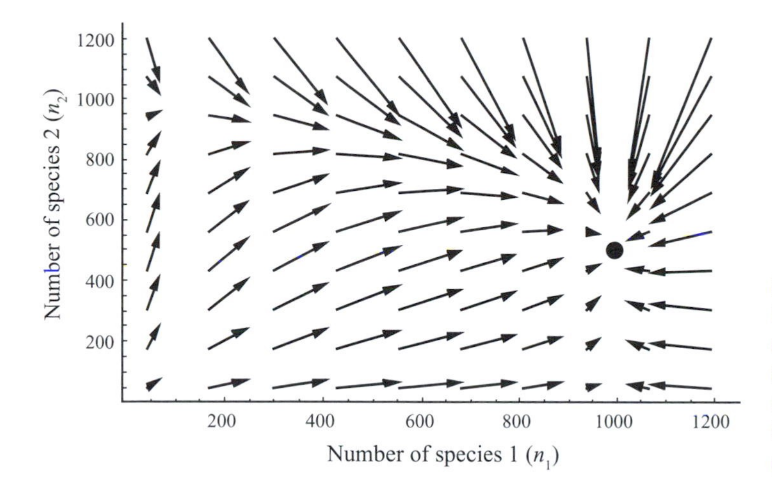

Figure 4.15: Vector-field plot for the two-species Lotka-Volterra model of competition. The change over one generation is shown from a series of different starting positions. All of the parameters are the same as in Figure 4.14 with α_{12} set to 0. The filled circle represents an equilibrium of this model ($n_1 = 1000$; $n_2 = 500$).

drawing arrows specifying the direction of change from a variety of different positions. Such a phase-plane diagram is known as a *vector-field plot*. Figure 4.15 is an example of a vector-field plot for the discrete-time model of competition, where each arrow represents the change over a time step in the numbers of species 1 (along the horizontal axis) and species 2 (along the vertical axis). To see how the arrows are drawn, consider the top left arrow. The base of this arrow is at {50,1200}, which represents the starting population sizes of the two species. Plugging in these initial conditions as well as the parameters ($r_1 = r_2 = 0.5$, $\alpha_{12} = 0$, $\alpha_{21} = 0.5$, and $K_1 = K_2 = 1000$) into equation (3.14) predicts the numbers of species in the next time step:

A *vector-field plot* places arrows on a phase-plane diagram to indicate the direction of change of the system.

$$
\begin{aligned}
n_1(1) &= n_1(0) + r_1 n_1(0)\left(1 - \frac{n_1(0) + \alpha_{12} n_2(0)}{K_1}\right) \\
&= 50 + 0.5 \times 50\left(1 - \frac{50 + 0 \times 1200}{1000}\right) \\
&= 73.75, \\
n_2(1) &= n_2(0) + r_2 n_2(0)\left(1 - \frac{n_2(0) + \alpha_{21} n_1(0)}{K_2}\right) \\
&= 1200 + 0.5 \times 1200\left(1 - \frac{1200 + 0.5 \times 50}{1000}\right) \\
&= 1065.
\end{aligned}
$$

Thus, the tip of the arrow is placed at {73.75,1065}. The same procedure is followed for as many arrows as desired (100 in this plot). It should be noted that sometimes the change per time step is too short or too long to plot with an arrow in this way. Thus, the arrows are often rescaled by a factor to make their lengths more visually appealing.

Vector-field plots are a great way to visualize interactions between two variables. In this example, the arrows help us to see that the variables tend to converge to the equilibrium (arrows point toward the filled circle), suggesting that the equilibrium is stable. Also, the variables do not spiral around the equilibrium, so we would not expect cycling. It is also possible to see that the number of species 2 (position along the vertical axis) can rise above its equilibrium value of 500 before falling back down as the number of species 1 rises (position along the horizontal axis), as observed in Figure 4.8.

4.4.4 Null Clines

Another important feature that is often added to phase-plane diagrams are curves indicating when each variable remains unchanged from one time point to the next. These are known as *null clines*. Along the null cline corresponding to the variable on the horizontal axis, the system can move only up or down, without any horizontal movement. Similarly, along the null cline corresponding to the variable on the vertical axis, the system can move only left or right, without any vertical movement.

> A *null cline* is a curve on a phase-plane diagram that indicates when one variable remains constant. Different variables typically have different null clines.

Using null clines, we can distinguish regions of the phase plane in which each variable grows or shrinks. For example, from the arrows in the vector-field plot (Figure 4.15), we can tell that the number of species 1 increases whenever its current value is low (left side of plot) and decreases whenever its current value is high (right side of plot). The null cline for species 1 will tell us exactly where the cutoff lies between these two regions. Similarly, the number of species 2 increases whenever its current value is low (bottom of plot) and decreases whenever its current value is high (top of plot), and the null cline for species 2 will tell us exactly where these two regions meet.

Let us see how to identify the null clines in the Lotka-Volterra model. From equation (3.14), the change in species 1 is

$$\Delta n_1 = n_1(t+1) - n_1(t) = r_1 n_1(t)\left(1 - \frac{n_1(t) + \alpha_{12} n_2(t)}{K_1}\right). \tag{4.3a}$$

When Δn_1 equals zero, species 1 remains at a constant population size. There are three ways for (4.3a) to equal zero: if $r_1 = 0$, if $n_1 = 0$, or if $1 - (n_1(t) + \alpha_{12} n_2(t))/K_1 = 0$. The first condition is known as a *special case of the parameters*; it will not usually be of biological interest, unless that special case happens to pertain to a species. The second condition occurs when species 1 is absent, indicating that the vertical axis at $n_1 = 0$ is a null cline for species 1. The third condition gives us a more interesting null cline for species 1. Using the parameters $\alpha_{12} = 0$ and $K_1 = 1000$, the third condition reduces to $1 - n_1(t)/1000 = 0$, which is satisfied when $n_1(t) = 1000$ (see dashed vertical line in Figure 4.16).

Similarly, species 2 remains at a constant population size when Δn_2 equals zero, where

$$\Delta n_2 = n_2(t+1) - n_2(t) = r_2 n_2(t)\left(1 - \frac{n_2(t) + \alpha_{21} n_1(t)}{K_2}\right). \tag{4.3b}$$

There are three ways for (4.3b) to equal zero: $r_2 = 0$, $n_2(t) = 0$, or $1 - (n_2(t) + \alpha_{21} n_1(t))/K_2 = 0$. Again, the first corresponds to a special case of the parameters,

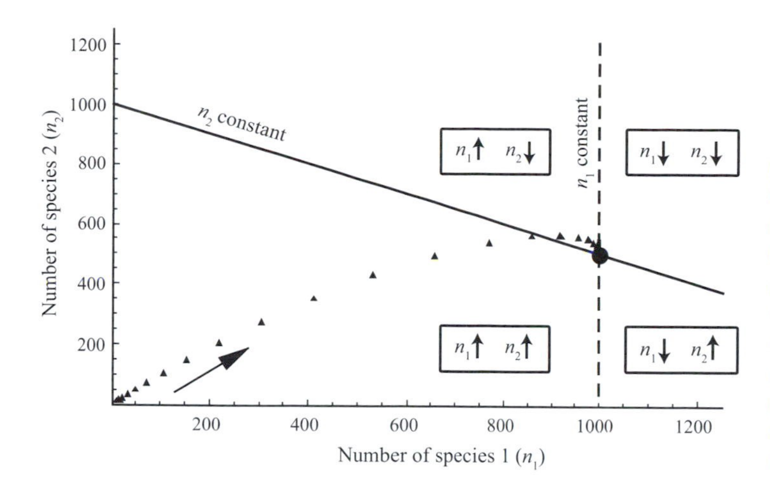

Figure 4.16: Null clines for the two-species Lotka-Volterra model of competition. The dashed line indicates when the number of species 1 remains the same in the next time step. The solid line indicates when the number of species 2 remains the same in the next time step. Only when both species remain constant over time is the system at equilibrium (filled circle). All of the parameters are the same as in Figure 4.14 with α_{12} set to 0, and the triangles again represent 50 iterations starting from $n_1(0) = n_2(0) = 10$.

and the second corresponds to extinction of species 2. The third condition is more interesting. Using the parameters $\alpha_{21} = 0.5$ and $K_2 = 1000$, the third condition tells us that the number of species 2 remains constant when $1 - (n_2(t) + 0.5\, n_1(t))/1000 = 0$. We need to rearrange this equation to get an equation for $n_2(t)$ as a function of $n_1(t)$, so we can plot the null cline on Figure 4.16. Doing so, we get $n_2(t) = 1000 - 0.5\, n_1(t)$, which is a line that decreases from 1000 at $n_1(t) = 0$ to 0 at $n_1(t) = 2000$ (see solid line in Figure 4.16).

Each null cline tells us when one of the two species remains constant in number. Both species remain constant in number, however, only where two null clines cross. This occurs at the equilibrium that we previously identified ($n_1 = 1000$; $n_2 = 500$). At any other point along the null cline for species 1 (dashed line), species 1 remains constant in size, but species 2 changes in number. As a result, the system gets knocked off the null cline.

It is sometimes possible to use phase-plane plots to draw more general conclusions about a model that go beyond a specific set of parameters. As one moves up in a phase-plane plot, the predicted change in a variable always has the same sign until one of its null clines is reached. Similarly, as one moves to the right along the horizontal axis, the predicted change in a variable always has the same sign until hitting one of its null clines. Only at a null cline do we see flips in the direction of change of the variable. This follows from the fact that, for a variable to change from increasing to decreasing (or vice versa), its rate of change per unit time must pass through zero. When it equals zero exactly, the variable will lie on a null cline. If we can determine the direction of change on either side of each null cline, then we can infer the behavior of the system across the entire phase plane, often for arbitrary values of the parameters.

We again consider the Lotka-Volterra model. At the null cline of species 1, $1 - (n_1(t) + \alpha_{12}\, n_2(t))/K_1$ equals zero. If we decrease the number of species 1 by a little bit, $1 - (n_1(t) + \alpha_{12}\, n_2(t))/K_1$ becomes positive, because we are now subtracting off a smaller number, $-n_1(t)/K_1$, from 1. Conversely, if we increase the number of species 1, $1 - (n_1(t) + \alpha_{12}\, n_2(t))/K_1$ becomes negative. The sign of this term, along with the sign of the growth rate r_1, determines the direction of change of species 1 (see equation (4.3a)). Assuming that the growth rates are positive, the above logic indicates that the number of species 1 will increase to the left of its null cline and decrease to the right of its null cline (dashed line). This allows us to write $n_1\uparrow$ in the two boxes on the left of Figure 4.16 and $n_1\downarrow$ in the two boxes on the right. The same procedure can be applied to species 2 to infer that it increases below its null cline ($n_2\uparrow$) and decreases above its null cline ($n_2\downarrow$). From these boxes, we can infer the direction in which the system will move. For example, in Figure 4.16, we can infer that the system will move toward the equilibrium. (Movement away from an equilibrium is also possible, depending on how the null clines cross.)

There are two caveats that must be added to the above discussion. First, even if we know the direction of change in each quadrant, the changes can be so large in magnitude in a discrete-time model that the variables overshoot their null clines. Because of this possibility, we cannot use a phase-plane analysis to determine whether a system approaches an equilibrium or oscillates around it in a discrete-time model. We will return to this issue in Chapter 8.

Second, we have implicitly assumed that the difference or differential equations are *continuous* in the variables and do not show any "discontinuities." Discontinuities, i.e., jumps or kinks in the trajectory of a variable, are not typical in biological models but can arise when change is described by different rules for different values of the variables. For example, you could build a model with the rule that growth of species 1 is exponential at rate $+r$ when the number of species 2 is less than or equal to some threshold τ, and is exponential at rate $-r$ when there are more than τ members of species 2. This model is perfectly acceptable, but we must account for the possibility that a variable could change from increasing to decreasing without passing through zero on either side of the discontinuity at τ. By drawing discontinuities as well as null clines on a phase plane, one can infer the regions in which each variable increases or decreases over time.

4.4.5 Phase-Plane Diagrams in Continuous-Time Models

The phase-plane diagrams that we have considered so far illustrate dynamics in discrete-time models, but the methods are similar for drawing phase-plane diagrams in continuous time. Rather than iterating recursion equations, differential equations must be solved numerically, and the results used to plot changes over time in the two variables. Alternatively, we can generate a vector-field plot directly from the differential equations by drawing arrows throughout the phase plane. Each arrow begins at some point $\{n_1, n_2\}$ and ends at the point $\{n_1 + \mathrm{d}n_1/\mathrm{d}t, n_2 + \mathrm{d}n_2/\mathrm{d}t\}$ (stretched or shrunk, if necessary, to make the arrows visible). Finally, we can add null clines to a phase-plane diagram by determining the curves along which $\mathrm{d}n_1/\mathrm{d}t$ and $\mathrm{d}n_2/\mathrm{d}t$ are zero.

In the on-line Lab exercises on the Lotka-Volterra model of competition, you can compare the dynamics of the discrete-time and continuous-time versions. As you will see, these figures are qualitatively similar, as long as the intrinsic growth rates of both species are low.

As another example, we consider phase-plane diagrams for the continuous-time predator-prey model described by equations (3.18). In Figure 4.17, the

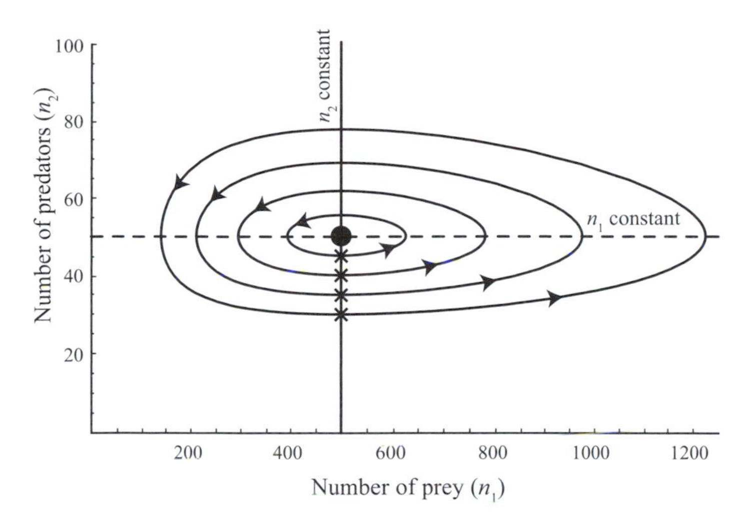

Figure 4.17: Dynamics of the predator-prey model in continuous time. The differential equations (3.18) were numerically solved using NDSolve in *Mathematica*. The resulting number of predators was plotted against the number of prey over time. The parameters used were: $r = 0.5$, $a = 1$, $c = 0.01$, $\varepsilon = 0.02$, $\delta = 0.1$, with an initial number of prey equal to $n_1(0) = 500$ and an initial number of predators $n_2(0)$ set to 30, 35, 40, or 45 (×'s). The direction in which time flows is indicated by arrows. Null clines are also shown indicating where the number of prey is constant ($dn_1/dt = 0$; dashed line) or the number of predators is constant ($dn_2/dt = 0$; solid line).

horizontal axis represents the number of prey, and the vertical axis represents the number of predators. Each curve represents a numerical solution to the differential equations, using the same parameter values but starting from different initial conditions (×'s). In this case, the curves form rings around a point with 500 prey and 50 predators, and the system retraces the same cycle over time. The vector-field plot suggests that this type of dynamic holds in general (at least for these parameter values) (Figure 4.18).

Figures 4.14–4.18 imply that different forms of species interactions can lead to vastly different dynamics and suggest that interactions at the same trophic level (i.e., competition) more readily generate stable equilibria than interactions across trophic levels (i.e., predation). This speculation is based, however, on only a small set of parameter values. We might see the opposite behavior with different parameter values, and it would be impossible to make graphs with every possible combination of parameters to verify this claim. To move beyond speculation, we need to learn methods to demonstrate when an equilibrium

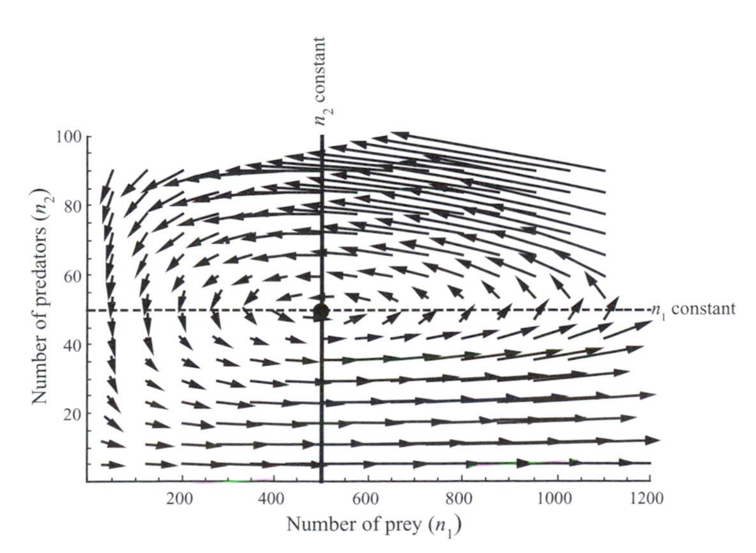

Figure 4.18: Vector-field plot for the predator-prey model in continuous time. The rate of change in the number of prey and the number of predators is shown from a series of different starting positions. Each arrow starts at a point $\{n_1, n_2\}$ and ends at $\{n_1 + dn_1/dt, n_2 + dn_2/dt\}$. All parameters and null clines are the same as in Figure 4.17.

is stable (Chapter 5) and to determine when cycling is possible (Chapters 8 and 11).

Interestingly, we can use phase-plane diagrams to understand why chaotic dynamics cannot be observed in continuous-time models with only two variables (Box 4.1). Imagine starting two replicate trajectories initially close to one another somewhere within a phase-plane diagram. These trajectories can potentially take on many different shapes, but it is not possible for the two trajectories to cross in a continuous-time model. If they were to cross, then the values of the variables would be exactly the same for both trajectories at the point in time when the trajectories cross. But from this point on, the two replicates would have identical dynamics in any deterministic model (i.e., as long as chance plays no role in the model). This fact ensures that there will always be order to trajectories started nearby one another in a two-variable continuous-time model. The trajectories might form closed orbits (as in Figure 4.17), they might spiral in together, or they might spiral out together. It is not possible, however, for two replicate trajectories to crisscross each other. Adding a third variable to a continuous-time model, however, releases the dynamics from this constraint. At an intuitive level, two trajectories can now "cross" in a phase-plane diagram of two of the variables, as long as they have different values of the third variable. Indeed, continuous-time models that exhibit chaos with more than two variables have been described in the biological literature, e.g., in food-web models involving three species (Hastings and Powell 1991; Klebanoff and Hastings 1994).

Example: Dynamics of HIV after Initial Infection (Revisited)

Now that we have a suite of useful graphical techniques for analyzing models, let us return to the model of Phillips (1996) for the dynamics of HIV within the human body. Recall that the concentration of HIV in the blood during the first 100–150 days of an infection rapidly increases, reaches a maximum, and then tapers off (Figure 1.3). Phillips' model explored the possibility that this decline in HIV concentration is due to the depletion of uninfected lymphocytes.

We begin by numerically solving the differential equations in Box 2.4, using a mathematical software package such as *Mathematica*. Figure 1.5 presented a plot of the number of virions (virus particles) in the bloodstream over time using the parameter estimates of Table 1 in Phillips (1996). Figure 4.19a presents the same plot, along with the corresponding plots for the number of latently infected and actively infected cells (see Figure 1 of Phillips (1996)).

One of the most useful applications of the graphical techniques outlined in this chapter is to see how the model's predictions depend on parameter values. Phillips (1996) obtained independent estimates for most of the parameter values used in the model, but there were some parameters for which little information was available. For example, the rate at which virions infect susceptible CD4+ cells per day, β, was unknown. Given the importance of this parameter in HIV transmission among cells, we might expect errors in the estimate of β to have a large impact on the model's predictions.

The sensitivity of the results can be checked using the graphical techniques of section 4.2. First, suppose that the value of β used in Figure 4.19a is an order

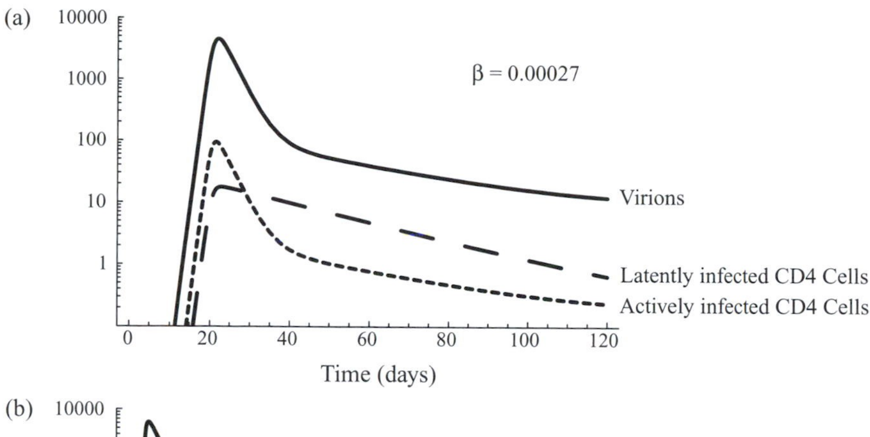

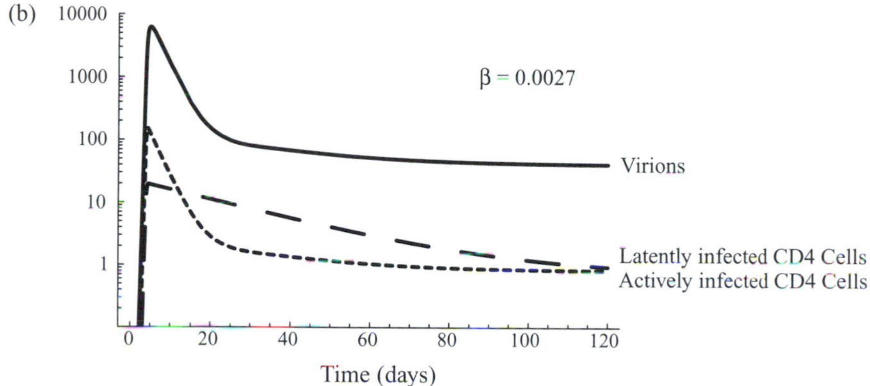

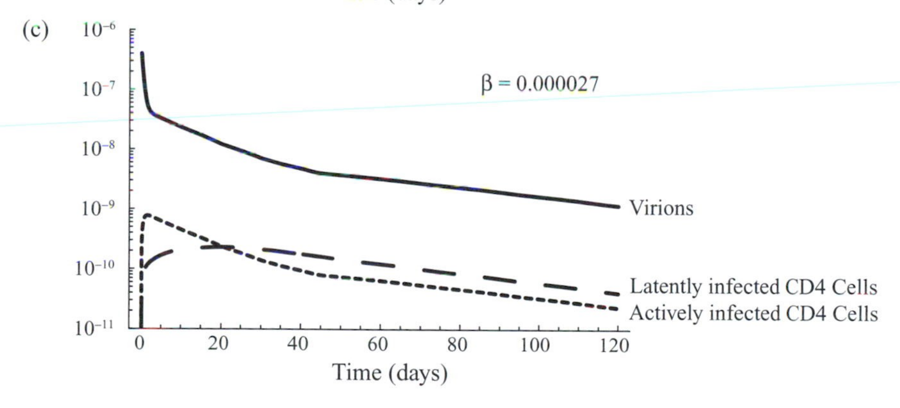

Figure 4.19: HIV dynamics within the bloodstream. The number of virions, latently infected CD4+ cells, and actively infected CD4+ cells are plotted over the first 120 days of an HIV infection. Results are based on simulation of the equations in Box 2.4 with the following parameter values from Phillips (1996): $\Gamma = 1.36$, $\mu = 1.36 \times 10^{-3}$, $\tau = 0.2$, $p = 0.1$, $\alpha = 3.6 \times 10^{-2}$, $\sigma = 2$, $\delta = 0.33$, $\pi = 100$, along with (a) $\beta = 0.00027$, (b) $\beta = 0.0027$, or (c) $\beta = 0.000027$ (note change in vertical axis). The initial number of virions was 4×10^{-7} per mm^3 of blood and the initial number of susceptible cells was 1000 τ per mm^3 of blood (no infected cells were initially present).

of magnitude too low (i.e., it is 10 times too small). This amount of uncertainty is not uncommon in such parameters. If we change the value of β from $\beta = 0.00027$ to $\beta = 0.0027$, the predictions of Figure 4.19b result. The model still does a reasonable job of explaining the qualitative pattern that is observed in the virus concentration in the blood (i.e., the rise and fall). The model also makes similar predictions regarding the maximum viral load that is reached. What has changed significantly, however, is the amount of time that it takes to reach the maximum viral load. The increased transmission coefficient β has resulted in the viral concentration reaching its peak much earlier, around 5 days post infection rather than around 21–22 days post infection.

In the same way we can determine how the predictions are altered if β is decreased by an order of magnitude, to $\beta = 0.000027$. Interestingly, doing so causes the number of virions in the blood to immediately decay to zero (Figure 4.19c). This reveals that the qualitative predictions of the model change dramatically as β is decreased below some critical value. In this case the infection is immediately eliminated from the body. This makes sense—if the virions are not effective enough at infecting susceptible CD4+ cells, then the body will be able to clear HIV out of the bloodstream. This suggests that we try to analyze the model mathematically in order to characterize when we expect the model to exhibit one behavior (establishment of the virus) versus another (elimination of the virus). In Chapter 8 we will develop mathematical techniques, known as stability analyses, for doing exactly that. For now, however, we can numerically simulate the dynamics using a range of β to find that there is a critical value of β near $\beta = 0.000034$. Above this value the virus rises and falls in concentration as is observed in a typical infection (as in Figures 4.19a and 4.19b), but below this value the virus is cleared from the body (as in Figure 4.19c).

Graphical techniques can also suggest which mathematical analyses might help answer particular questions of interest. For example, Figure 4.19 suggests that the curves for the concentration of virions in the blood and for the concentration of actively infected CD4+ cells are almost exactly proportional to one another. Put another way, the concentration of actively infected CD4+ cells is approximately some constant fraction k of the concentration of virions in the blood. Mathematically, this means that, although the concentration of virions, V, and the concentration of actively infected CD4+ cells, E, both change through time, the approximate relationship $V \approx k\,E$ appears to hold. Why might this be so? The reason is closely tied to an important mathematical technique known as *separation of time scales*. This technique will be developed in greater detail in Chapter 9, but let us get a flavor of it here.

To understand why the concentration of virions and actively infected CD4+ cells might always differ by a constant factor we need to consider the underlying differential equations of the model in more detail (Box 2.4). Although the four differential equations of the model allow all four variables to change simultaneously through time, some variables might change faster than others depending upon the parameter values. Examining the parameter values for the HIV dynamics (see caption to Figure 4.19), the parameters π and σ in the differential equation for the viral load (i.e., dV/dt) are much larger than the other parameters, by about two orders of magnitude. Therefore, we might expect the dynamics of the viral load to operate much faster (i.e., on a shorter time scale) than those of the other variables. This would cause the viral population size to reach an approximate steady state given the current values of the other variables. As the other variables change slowly, the viral load will then reequilibrate quickly and reach a new steady state corresponding to the new combination of the other variables. Consequently, we might expect the viral population size to rapidly track changes in the other variables.

If the viral load reequilibrates rapidly, we would expect V to be approximately constant at any point in time. That is, we would expect $dV/dt = \pi E - \sigma V$ to be nearly zero at all times. Solving $\pi E - \sigma V = 0$ for V gives us a steady state solution for the number of viruses in the bloodstream: $V = \pi E/\sigma$. This steady state solution depends only on the number of actively infected CD4+ cells, E, explaining why these two variables change in parallel. Furthermore, this relationship accurately predicts that the number of virions should be $\pi/\sigma = 50$ times the number of actively infected CD4+ cells, which is a perfect match to the graphs (Figure 4.19). This example illustrates how patterns apparent in graphs can be used to guide us toward appropriate mathematical analyses.

4.5 Concluding Message

In this chapter we have discussed three graphical techniques that allow us to obtain a good qualitative understanding of a model's behavior without conducting any formal mathematical analyses. Plotting a variable over time illustrates the temporal dynamics of a model, giving a good sense of how a model behaves (section 4.2). Plotting a variable in the next time step (or the change in the variable) versus the current state of the variable indicates the direction in which the system tends to move (section 4.3). Such plots also allow us to identify the equilibria of a model, defined as those points where the system remains unchanged. Phase-line and phase-plane diagrams also illustrate the direction of change as a function of the current state of the system (section 4.4). When a model contains two variables, phase-plane diagrams are particularly useful, as these diagrams indicate how interactions between the variables influence the behavior of the system.

The above graphical techniques allow us to build our intuition about a model, and they provide a way of communicating this understanding to others. There is wisdom in the old adage: a picture is worth a thousand words. Nevertheless, while graphical approaches are useful, they do have the drawback of requiring us to specify parameter values, and they have a limited capacity for providing very general results. For this we must learn mathematical techniques for analyzing models.

Problems

Problem 4.1: Plot the population size $n(t)$ with respect to time using the recursion equation for logistic growth, $n(t + 1) = n(t) + r\, n(t)\, (1 - n(t)/K)$, for the populations of yeast studied by Mable and Otto (2001). Use the parameters estimated from the growth of (a) haploid yeast ($K_{\text{haploid}} = 3.7 \times 10^8$, $r_{\text{haploid}} = 0.55$, $n(0) = 3.0 \times 10^5$) and (b) diploid yeast ($K_{\text{diploid}} = 2.3 \times 10^8$, $r_{\text{diploid}} = 0.55$, $n(0) = 1.9 \times 10^5$). Compare your figures to Figure 3.4.

Problem 4.2: Plot the population size $n(t)$ with respect to time using the differential equation for logistic growth, $dn/dt = r\, n(t)\, (1 - n(t)/K)$, for the populations of yeast studied by Mable and Otto (2001). Use the parameters estimated from the growth of (a) haploid yeast ($K_{\text{haploid}} = 3.7 \times 10^8$, $r_{\text{haploid}} = 0.55$, $n(0) = 3.0 \times 10^5$) and (b) diploid yeast ($K_{\text{diploid}} = 2.3 \times 10^8$, $r_{\text{diploid}} = 0.55$, $n(0) = 1.9 \times 10^5$). Compare your figures to Figure 3.4.

Problem 4.3: Plot the population size $n(t)$ with respect to time using the recursion equation for logistic growth, $n(t + 1) = n(t) + r\, n(t)\, (1 - n(t)/K)$. Starting with a population of size 100, show that the population grows in a qualitatively similar manner regardless of the carrying capacity (use $K = 500$, $K = 1000$, and $K = 5000$) but with drastically different behavior depending on the growth rate r. That is, show that (a) the population smoothly approaches the carrying capacity without overshooting it when $r = 1.0$, (b) the population settles upon a two-point cycle when $r = 2.1$, (c) the population settles upon a four-point cycle when $r = 2.5$, and (d) the population varies erratically and is chaotic when $r = 2.7$.

Problem 4.4: Figure 4.20 illustrates the diploid recursion equation (3.13a) as a function of the allele frequency $p(t)$ for a different set of fitnesses than Figure 4.10. (a) Place circles on the diagonal line indicating where the three equilibria are. (b) Starting from an allele frequency of 0.1, use the cobwebbing procedure of Box 4.2 to predict the allele frequency after three generations of selection. (c) Draw a phase-line plot with arrows showing the direction of allele frequency change that is consistent with figure 4.20. (d) For each of the three equilibria, specify whether it is locally stable or unstable.

Problem 4.5: In a plot of the allele frequency in one generation, $p(t + 1)$, versus the allele frequency in the previous generation, $p(t)$, under haploid selection, prove that the slope is never negative. That is, use the recursion equation

$$p(t + 1) = \frac{W_A p(t)}{W_A p(t) + W_a(1 - p(t))}$$

to calculate and factor the derivative $\mathrm{d}p(t + 1)/\mathrm{d}p(t)$ which describes the slope in this plot. By examining the sign of each factor, show that this derivative must always be positive.

Problem 4.6: In a plot of the allele frequency in one generation, $p(t + 1)$, versus the allele frequency in the previous generation, $p(t)$, under diploid selection, prove that the slope is never negative. That is, use the recursion equation

$$p(t + 1) = \frac{W_{AA}\, p(t)^2 + W_{Aa}\, p(t)(1 - p(t))}{W_{AA}\, p(t)^2 + W_{Aa}\, 2p(t)(1 - p(t)) + W_{aa}(1 - p(t))^2}$$

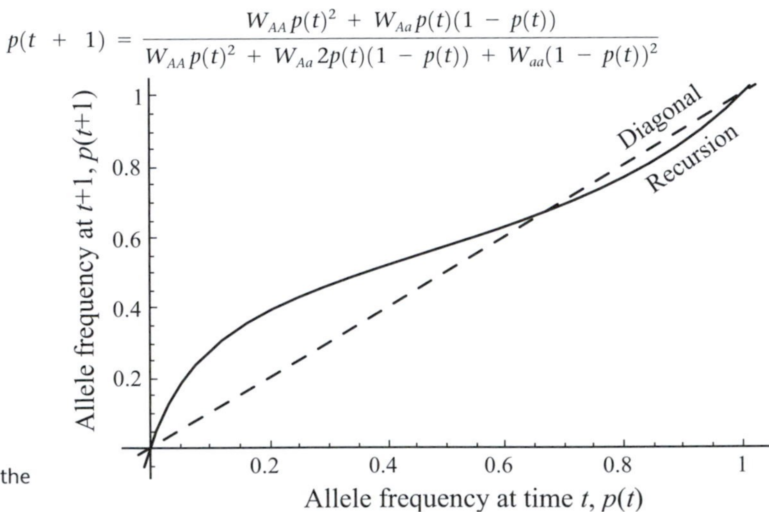

Figure 4.20: Allele frequency recursion in the diploid model of selection

to calculate and factor the derivative $dp(t+1)/dp(t)$, which describes the slope in this plot. By examining the sign of each factor, show that this derivative must always be positive.

Problem 4.7: Find the equations for the null clines of the Lotka-Volterra predator-prey model described by equation (3.18):

$$\frac{dn_1}{dt} = r n_1 - a\, c\, n_1 n_2,$$

$$\frac{dn_2}{dt} = \varepsilon\, a\, c\, n_1 n_2 - \delta\, n_2.$$

Specify which equations must be satisfied for the number of prey (n_1) to remain constant and which equations must be satisfied for the number of predators (n_2) to remain constant.

Problem 4.8: In the consumer-resource model described by equations (3.17), resources (n_1) enter the system at a constant rate and are continuously eaten by a consumer species (n_2):

$$\frac{dn_1}{dt} = \theta - a\, c\, n_1 n_2,$$

$$\frac{dn_2}{dt} = \varepsilon\, a\, c\, n_1 n_2 - \delta\, n_2.$$

For example this model might describe the flow of insect larvae into a pond containing minnow predators. Assume that 1000 new larvae arrive per minute (θ = 1000), that the per capita contact rate is $c = 0.01$ per minute, that the probability that a minnow will consume a larva once detected is $a = 1$, that one larva is the energetic equivalent of $\varepsilon = 0.0005$ minnows, and that the per capita death rate of minnows is $\delta = 0.001$. (a) Draw a vector-field plot for this model. (b) Determine the null clines for this model and add them to the plot. (c) Describe what happens to the number of larvae and minnows over time, and discuss any limitations of the vector-field plot in this case.

Problem 4.9: The logistic equation (3.5a) differs from the recursion equation often considered in mathematical discussions of chaos, which is given by

$$x(t+1) = \mu\, x(t)\, (1 - x(t)).$$

Show that if we define $x(t) = (\mu - 1)\, n(t)/(\mu\, K)$ and $\mu = 1 + r$, we can rearrange the above equation to get (3.5a). This "transformation" explains why the transition to chaos occurs at $r^* = 2.569944$ in this book but at $\mu^* = 3.569944$ in other treatments.

Problem 4.10: Use a mathematical software package to plot the differential equations for the Lotka-Volterra model of competition between two species given by (3.15):

$$\frac{dn_1}{dt} = r_1 n_1(t)\left(1 - \frac{n_1(t) + \alpha_{12} n_2(t)}{K_1}\right),$$

$$\frac{dn_2}{dt} = r_2 n_2(t)\left(1 - \frac{n_2(t) + \alpha_{21} n_1(t)}{K_2}\right).$$

Assume that $r_1 = r_2 = 0.5$, $\alpha_{12} = 0$, $\alpha_{21} = 0.5$, and $K_1 = K_2 = 1000$ and that the initial population size ten for each species ($n_1(0) = 10$, $n_2(0) = 10$). Compare the resulting graphs to Figure 4.8. Do the discrete-time and continuous-time graphs differ substantially for this example?

Problem 4.11: After recovering from a cold, we are susceptible to catching another cold, because many different viruses can cause colds. For such diseases, the Susceptible-Infected-Recovered (SIR) model can be simplified to a Susceptible-Infected (SI) model, where individuals that recover from a cold return to the susceptible class. We can modify the differential equations for the SIR model (3.19) to describe the number of individuals susceptible to colds, S, and infected with colds, I:

$$\frac{dS}{dt} = \theta - d\,S(t) - a\,c\,S(t)\,I(t) + \rho\,I(t),$$

$$\frac{dI}{dt} = a\,c\,S(t)\,I(t) - \delta\,I(t) - \rho\,I(t).$$

Let $\theta = 4$ be the rate at which individuals enter the population, $c = 0.01$ be the per capita rate of contact between a susceptible and an infected individual, $a = 0.1$ be the probability of transmission of the disease per contact, $\rho = 0.2$ be the rate at which infected individuals recover from a cold, and $d = 0.01$ and $\delta = 0.02$ be the death (or emigration) rates of susceptible and infected individuals. Figure 4.21 illustrates a vector-field plot for this model.

Each arrow starts at an (x,y) coordinate equal to some combination of (S, I) and points in the direction of $(S + dS/dt, I + dI/dt)$. To make the arrows easier to see, each arrow was drawn with a constant length of five.

(a) Determine the null-cline along which the number of susceptible individuals, S, remains constant. Write this null cline in terms of what its height must equal $(I = \cdots)$. (b) Determine the null cline along which the number of infected individuals, I, remains constant. Write this null cline in terms of what its height must equal $(I = \cdots)$. (c) Add curves for these null clines to the figure, specifying what remains

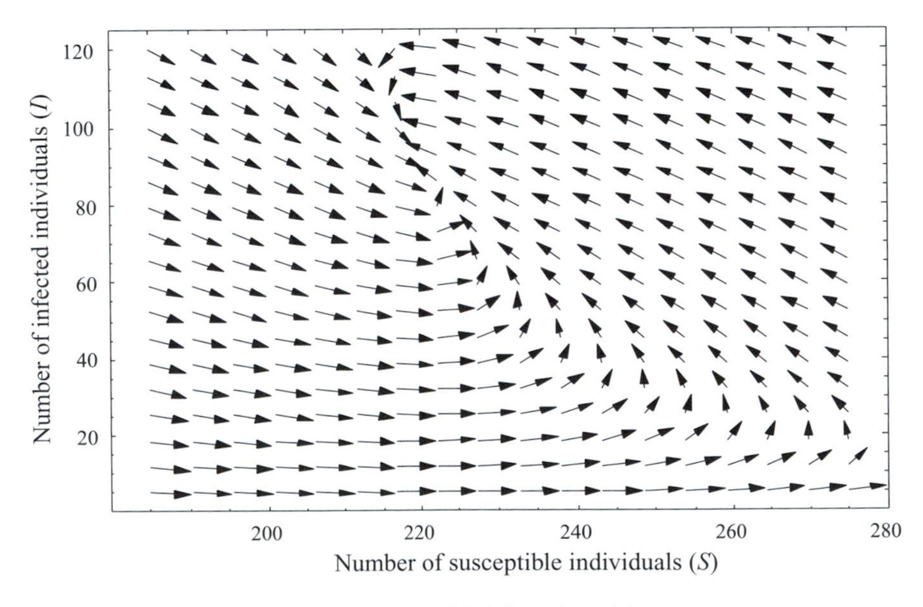

Figure 4.21: Vector-field plot for the susceptible-infected model

constant along each curve. (d) Add a dot to the figure where both variables remain constant (the equilibrium). Convince yourself that if the null clines have been drawn correctly, any arrow starting on a null cline where S is constant should point straight up or down and any arrow starting on a null cline where I is constant should point either left or right.

Problem 4.12: When simulating the logistic model

$$n(t+1) = n(t) + rn(t)\left(1 - \frac{n(t)}{K}\right)$$

we found that extinction ensues for values of r greater than 3 (Figure 4.2d) because the population can overshoot the carrying capacity to such a large degree that the population size declines below zero in the next generation. (a) Find the population size $n(t)$ that leads to the maximum population size in the next generation, $n(t+1)$. [Hint: Use the derivative of $n(t+1)$ with respect to $n(t)$ to find where the maximum occurs.] (b) Find the maximum population size that can be reached by determining $n(t+1)$ starting from the $n(t)$ found in (a). (c) Find $n(t+2)$ starting from the maximum population size $n(t+1)$. (d) Factor the result from (c) and show that $n(t+2)$ will be less than zero and the population will go extinct if r is greater than 3.

Problem 4.13: As suggested by Figure 4.19, we can approximate Phillips' HIV model by assuming that the number of virions equilibrates rapidly ($dV/dt = 0$), while the other variables change more slowly over time. This assumption allows us to reduce the number of differential equations in the model from four to three, giving

$$\frac{dR}{dt} = \Gamma\tau - \mu R - \beta R \frac{\pi}{\sigma} E,$$

$$\frac{dL}{dt} = p\beta R \frac{\pi}{\sigma} E - \mu L - \alpha L,$$

$$\frac{dE}{dt} = (1-p)\beta R \frac{\pi}{\sigma} E + \alpha L - \delta E.$$

(Here, we have used the fact that dV/dt will be 0 if $V = \pi E/\sigma$ to replace V in the remaining three equations of Box 2.4.) Plot the dynamics of these differential equations over time, using the same parameters as in Figure 4.19. Describe any differences between your figure and Figure 4.19 (they should be similar).

Problem 4.14: Here, we ask whether the inclusion of the latent class of CD4+ cells is critical to the results of Phillips' model of HIV. Suppose that there is no latent class of CD4+ cells (set $L = 0$ and $p = 0$). This reduces the three-variable version of Phillips' HIV model described in Problem 4.13 to a model with two equations:

$$\frac{dR}{dt} = \Gamma\tau - \mu R - \beta R \frac{\pi}{\sigma} E,$$

$$\frac{dE}{dt} = \beta R \frac{\pi}{\sigma} E - \delta E.$$

Simulate this two-variable model using the same parameter values as in Figure 4.19. Based on your results, is the latent class of cells crucial for the conclusions drawn by Phillips about the initial dynamics of the virion population?

Problem 4.15: (a) Generate a vector-field plot for the HIV model described by the two equations in Problem 4.14. (b) Find the null clines of this model and draw them on the vector-field plot.

Further Reading

For more information on chaos, consult

- Stewart, I. 1997. *Does God Play Dice? The New Mathematics of chaos*. Penguin Books, London.
- Baker, G. L., and J. P. Gollub. 1996. *Chaotic Dynamics: An Introduction*. Cambridge University Press, Cambridge.

References

Dennis, B., R. A. Desharnais, J. M. Cushing, S. M. Henson, and R. F. Costantino. 2001. Estimating chaos and complex dynamics in an insect population. *Ecol. Monogr.* 71:277–303.

Hassell, M. P., J. H. Lawton, and R. M. May. 1976. Patterns of dynamical behaviour in single-species populations. *J. Anim. Ecol.* 48:471–486

Hastings, A., C. L. Hom, S. Ellner, P. Turchin, and H. C. J. Godfray. 1993. Chaos in ecology—Is mother-nature a strange attractor? *Annu. Rev. Ecol. Syst.* 24:1–33.

Hastings, A., and T. Powell. 1991. Chaos in a three-species food chain. *Ecology* 72:896–903.

Henson, S. M., R. F. Costantino, J. M. Cushing, R. A. Desharnais, B. Dennis, and A. A. King. 2001. Lattice effects observed in chaotic dynamics of experimental populations. *Science* 294:602–605.

Klebanoff, A., and A. Hastings. 1994. Chaos in three species food chains. *J. Math. Biol.* 32:427–451.

Lorenz, E. N. 1963. Deterministic nonperiodic flow. *J. Atmos. Sci.* 20:130–141.

May, R. M. 1974. Biological populations with nonoverlapping generations: stable points, stable cycles, and chaos. *Science* 186:645–647.

McCann, K., and A. Hastings. 1997. Re-evaluating the omnivory-stability relationship in foodwebs. *Proc. R. Soc. London, Ser. B* 264:1249–1254.

Phillips, A. N. 1996. Reduction of HIV concentration during acute infection: Independence from a specific immune response. *Science* 271:497–499.

Press, W. H. 2002. *Numerical Recipes in C++: the Art of Scientific Computing*. Cambridge University Press, Cambridge.

Stewart, I. 1997. *Does God Play Dice? The New Mathematics of Chaos*. Penguin Books, London.

Turchin, P., and S. P. Ellner. 2000. Living on the edge of chaos: Population dynamics of Fennoscandian voles. *Ecology* 81:3099–3116.

Wolfram Research, 2003. *Mathematica*, Version 5.0, Champaign, IL.

CHAPTER 5

Equilibria and Stability Analyses—One-Variable Models

Chapter Goals:

- To determine the equilibria of a model with one variable
- To determine the stability of equilibria
- To find approximate values of an equilibrium when needed

Chapter Concepts:

- Equilibrium condition
- Implicit solution/Explicit solution
- Local stability analysis
- Perturbation analysis

5.1 Introduction

Determining the equilibria of a system is one of the most important steps in analyzing a dynamic model. An equilibrium is a point at which a variable (or variables) remains unchanged over time:

> **Definition 5.1: Equilibrium**
>
> A system at *equilibrium* does not change over time (plural: *equilibria*). A particular value of a variable is called an equilibrium value if, when the variable is started at this value, the variable never changes.
>
> - At an equilibrium in a discrete-time model, $n(t+1)$ must equal $n(t)$ for each variable. Equivalently, the amount by which each variable changes, $\Delta n = n(t+1) - n(t)$, must equal zero.
> - At an equilibrium in a continuous-time model, dn/dt must equal zero for each variable.

Knowing the equilibria of a model is very useful because these states are good candidates for where a system might eventually end up. We can winnow down these candidates even further by determining which equilibria are stable (attracting) or unstable (repelling).

> **Definition 5.2: Stability**
>
> - An equilibrium is *locally stable* if a system near the equilibrium approaches it (locally attracting).
> - An equilibrium is *globally stable* if a system approaches the equilibrium regardless of its initial position.
> - An equilibrium is *unstable* if a system near the equilibrium moves away from it (repelling).

In sections 5.2 and 5.3 we describe how to determine the equilibria of a model, and how to assess whether an equilibrium is stable or unstable. Section 5.4 then presents a method for approximating equilibria and stability conditions in models where exact solutions are not possible. We focus on models with a single variable in this chapter (e.g., one species or one allele frequency). The corresponding analyses for finding equilibria and determining stability in models with more than one variable will be described in Chapters 7 and 8.

5.2 Finding an Equilibrium

The method for finding equilibria is straightforward. To clarify the notation, we will place a caret over a variable, e.g., $\hat{n}$, to represent an equilibrium value of that variable. In a discrete-time model, if $n(t)$ is at an equilibrium value $\hat{n}$, then the value of n in the next time step, $n(t+1)$, will equal $\hat{n}$ as well. Thus, to find an equilibrium for a model, we replace every instance of $n(t)$ and $n(t+1)$ with $\hat{n}$ in a recursion equation. Alternatively, in a difference equation, there must be no change in the variable at an equilibrium, so we set Δn to zero and $n(t)$ to $\hat{n}$. Throughout this chapter, we assume that the dynamics of a model are entirely determined by the current state of the system, $n(t)$, and do not also depend on the past history of states (as in a model with time lags) or on the actual time t (as in a model where the environment changes over time). Similarly, in a continuous-time model, there must be no change in the variable at an equilibrium, so we set dn/dt to zero and $n(t)$ to $\hat{n}$. This procedure results in one equation in one unknown (i.e., the equilibrium value $\hat{n}$); we call this equation the *equilibrium condition*. Any value of $\hat{n}$ that satisfies this condition is an equilibrium of the model. The number of equilibria in a model is determined by how many values of $\hat{n}$ satisfy the equilibrium condition.

An *equilibrium condition* is an equation that is satisfied by the equilibria of a model.

Recipe 5.1
Finding an Equilibrium, $\hat{n}$

Step 1: Obtain the equilibrium condition from the dynamical equation:

- In a discrete-time recursion equation, replace $n(t+1)$ and $n(t)$ with $\hat{n}$. For example, the exponential growth model $n(t+1) = R\,n(t)$ becomes $\hat{n} = R\hat{n}$.
- In a discrete-time difference equation, replace $n(t)$ with $\hat{n}$ and set Δn to zero. For example, the exponential growth model $\Delta n = (R-1)\,n(t)$ becomes $0 = (R-1)\hat{n}$.
- In a continuous-time differential equation, replace $n(t)$ with $\hat{n}$ and set dn/dt to zero. For example, the exponential growth model $dn/dt = rn(t)$ becomes $0 = r\hat{n}$.

Step 2: Solve the equilibrium condition for $\hat{n}$. When canceling a term from both sides of an equilibrium condition, check if that term could equal zero for some value of $\hat{n}$. If so, that value of $\hat{n}$ is an equilibrium of the model.

(continued)

Recipe 5.1 *(continued)*

Step 3: Check each equilibrium by plugging it back into the original dynamical equation and confirming that the system remains constant. Also check that each equilibrium is biologically valid (e.g., is non-negative if the variable represents the number of individuals).

CAUTION: Remember that in step 2, we are solving for values of the *variables* that satisfy the equilibrium condition, not values of the *parameters*.

While the concept behind finding equilibria is straightforward, calculating the equilibria might not be. In fact, depending on your model, it can be an easy, difficult, or even impossible task. To illustrate, we start with cases in which it is easy to determine the equilibria and work up to a case where the equilibria cannot be found explicitly.

5.2.1 Exponential and Logistic Models of Population Growth

To identify equilibria of the exponential growth model in discrete time, we replace $n(t+1)$ and $n(t)$ with $\hat{n}$ in the recursion equation, $n(t+1) = Rn(t)$, to get the equilibrium condition, $\hat{n} = R\hat{n}$ (Step 1). We then solve for $\hat{n}$ (Step 2). You might be tempted to divide both sides by $\hat{n}$ to simplify this equation, leaving $1 = R$. At this point you might conclude that there is no equilibrium of the model unless R happens to equal one. This is known as a *special case of the parameters*, because it is extremely unlikely that R will equal exactly one for any real population. In general, it is a great idea to simplify an equilibrium condition by canceling out a term that multiplies both sides of the condition, but when we do so, we must check whether the canceled term could itself equal zero. For the exponential model, when we divide both sides of the equilibrium condition $\hat{n} = R\hat{n}$ by the term $\hat{n}$, the canceled term could be zero if $\hat{n} = 0$. Therefore, $\hat{n} = 0$ satisfies the equilibrium condition and is a valid equilibrium of the exponential model.

We can check to make sure that we have correctly identified an equilibrium by setting $n(t)$ to the potential equilibrium value (zero in the present case) and confirming that $n(t+1)$ remains at this value (Step 3). Indeed, plugging $n(t) = 0$ into the recursion equation for the exponential model gives $n(t+1) = R \times 0 = 0$. Thus, an equilibrium of the exponential model occurs when there are no individuals within the population.

An alternative way to solve the equilibrium condition $\hat{n} = R\hat{n}$ is to subtract $\hat{n}$ from both sides, giving $0 = R\hat{n} - \hat{n}$, which can be factored into $0 = (R - 1)\hat{n}$. Solving for $\hat{n}$, we can again see that $\hat{n} = 0$ is an equilibrium. Indeed, it is the only equilibrium of the model. The only other way to satisfy this equilibrium condition is to set R to a specific value, $R = 1$. Remember, however, that we are seeking values of the *variable* n at which no change is predicted, and therefore it does not make sense to say that $R = 1$ is an equilibrium. Rather, when $R = 1$, the equilibrium condition is always satisfied, and all initial values of n represent equilibria, because the population exactly replaces itself regardless of the population size.

In the logistic model, applying Recipe 5.1 to the recursion equation,

$$n(t+1) = n(t) + r\,n(t)\left(1 - \frac{n(t)}{K}\right) \qquad \text{(recursion equation from 3.5a)}$$

gives the equilibrium condition

$$\hat{n} = \hat{n} + r\,\hat{n}\left(1 - \frac{\hat{n}}{K}\right) \qquad \text{(equilibrium condition for 3.5a).}$$

Subtracting $\hat{n}$ from both sides demonstrates that $r\hat{n}(1 - \hat{n}/K)$ must equal zero at an equilibrium. Besides the special case of zero growth ($r = 0$, in which case all populations remain constant in size), the equilibrium condition has two possible solutions; either $\hat{n} = 0$ or $(1 - \hat{n}/K) = 0$. The second solution is satisfied when $\hat{n} = K$. Thus, we conclude that there are two states in the logistic model at which the population size will remain constant: when there are no individuals present ($\hat{n} = 0$) or when the population is at the carrying capacity ($\hat{n} = K$).

If you are well practiced in solving equations, it might be obvious that the solution to $(1 - \hat{n}/K) = 0$ is $\hat{n} = K$. Until solving such equations becomes second nature, however, there is a general recipe to find values of $\hat{n}$ that satisfy any linear equation.

Recipe 5.2
Solving Linear Equations

A linear equation in a variable n is one that can be written in the form $a + bn = c + dn$, where a, b, c, and d are parameters that do not depend on n. For example, $1 - n/K = 0$ has the form of a linear equation with $a = 1$, $b = -1/K$, and $c = d = 0$.

Step 1: Move all terms involving n to one side of the equation and all terms not involving n to the other side, by adding or subtracting appropriate terms to both sides. The resulting equation is $bn - dn = c - a$.

Step 2: Factor the variable n out of every term on the side of the equation containing n to get $n\,(b - d) = c - a$.

Step 3: Divide both sides of the equation by the factor multiplying n to get the solution $n = (c - a)/(b - d)$.

For both the exponential and logistic models, we get equivalent equilibrium conditions (and thus the same equilibria) whether we use the recursion equation in discrete time, the difference equation in discrete time, or the differential equation in continuous time. For the exponential model, we can write the equilibrium condition as $0 = r\hat{n}$ using equation (3.1b), (3.2), or (3.3), if we define $r = (R - 1)$. Similarly, for the logistic model, the equilibrium condition is $0 = r\hat{n}(1 - \hat{n}/K)$ using either (3.5a), (3.5b), or (3.5c). In general, recursion and difference equations always give equivalent equilibrium conditions, because the difference equation equals the recursion equation with $n(t)$ subtracted from both sides (Recipe 2.2). Consequently, we can use either when solving for the equilibria of a discrete-time model.

It is not generally true, however, that the equilibria of a differential equation are the same as the equilibria of a recursion or difference equation. As described in Chapter 2, if there are different types of events that occur over time, the dynamical equations for continuous-time and discrete-time models might differ because they make different assumptions about the possibility of concurrent events. In continuous time, it is assumed that two things do not happen in the same instant in time. In discrete time, however, it is possible for two events to occur before the next time step (e.g., an individual could be born and die within the same year).

Consider the toy model that we built in Chapter 2 in which we tracked the number of mice in a yard, given that births, deaths, and migration occur. We obtained the dynamical equations

$$n(t+1) = (1+b)\,(1-d)\,n(t) + m \qquad \text{(recursion equation from 2.4)},$$

$$\Delta n = -d\,n(t) + b\,(1-d)\,n(t) + m \qquad \text{(difference equation from 2.5)},$$

$$\frac{dn}{dt} = b\,n(t) - d\,n(t) + m \qquad \text{(differential equation from 2.9)}.$$

These give the equilibrium conditions

$$\hat{n} = (1+b)(1-d)\,\hat{n} + m \qquad \text{(equilibrium condition for 2.4)},$$

$$0 = -d\,\hat{n} + b\,(1-d)\,\hat{n} + m \qquad \text{(equilibrium condition for 2.5)},$$

$$0 = b\,\hat{n} - d\,\hat{n} + m \qquad \text{(equilibrium condition for 2.9)}.$$

These equations are "implicit solutions" for $\hat{n}$, meaning that they are equations that $\hat{n}$ must satisfy. Really, we want an "explicit solution," where we can write $\hat{n} = stuff$, where *stuff* is a function of the parameters only and not $\hat{n}$. Using Recipe 5.2, we can obtain explicit solutions for $\hat{n}$ from these equations:

An *implicit solution* is an equation that $\hat{n}$ must satisfy. An explicit solution for $\hat{n}$ gives $\hat{n}$ as some function of the parameters.

$$\hat{n} = \frac{m}{1-(1+b)(1-d)} \qquad \text{(equilibrium of 2.4)},$$

$$\hat{n} = \frac{m}{d-b\,(1-d)} \qquad \text{(equilibrium of 2.5)},$$

$$\hat{n} = \frac{m}{d-b} \qquad \text{(equilibrium of 2.9)}.$$

The first two equilibria can both be written as $\hat{n} = m/(d - b + bd)$ and are equivalent, but the equilibrium for the continuous-time model is different. At any of these equilibria, we expect more mice at equilibrium if the migration rate of mice into the yard is higher (higher m), if the mice give birth at a higher rate (higher b), or if they die at a lower rate (lower d). In the discrete-time models, however, births occur before deaths so that mice can die in the same time step in which they are born. This is not allowed in the continuous-time model. Consequently, there is an additional term bd in the denominator of the discrete-time models. This additional term is always positive, and hence the discrete-time model predicts a slightly lower equilibrium level of mice in the yard.

5.2.2 Haploid and Diploid Models of Natural Selection

Next, let us turn to the discrete-time model of haploid selection, described by

$$p(t+1) = \frac{W_A p(t)}{W_A p(t) + W_a(1 - p(t))} \qquad \text{(recursion equation from 3.8c).}$$

This gives the equilibrium condition

$$\hat{p} = \frac{W_A \hat{p}}{W_A \hat{p} + W_a(1 - \hat{p})} \qquad \text{(equilibrium condition from 3.8c).}$$

CAUTION: In Chapter 3, equation (3.8c) was written in an alternative form, with $q(t)$ instead of $(1 - p(t))$ in the denominator. When solving for the equilibrium value of an equation involving both $p(t)$ and $q(t)$, we have to account for the fact that these variables are interrelated. We cannot treat $q(t)$ as a constant when solving for the equilibrium values of $p(t)$, because $q(t) = 1 - p(t)$. The safest method is to replace $q(t)$ with $1 - p(t)$ when calculating equilibria.

There are many ways to solve this equilibrium condition. For example, we could start by canceling out $\hat{p}$ from both sides of the equation, making a mental note that this implies that $\hat{p} = 0$ is one equilibrium:

$$1 = \frac{W_A}{W_A \hat{p} + W_a(1 - \hat{p})} \qquad \text{(note: } \hat{p} = 0 \text{ is an equilibrium).}$$

Then we could multiply both sides by the denominator $W_A \hat{p} + W_a(1 - \hat{p})$, leaving

$$W_A \hat{p} + W_a(1 - \hat{p}) = W_A \qquad \text{(note: } \hat{p} = 0 \text{ is an equilibrium).}$$

This is a linear equation in $\hat{p}$, which can be solved using Recipe 5.2. Bringing all terms involving $\hat{p}$ to the left-hand side and factoring

$$\hat{p}(W_A - W_a) = W_A - W_a \qquad \text{(note: } \hat{p} = 0 \text{ is an equilibrium).}$$

Then we could divide both sides by $(W_A - W_a)$; the equilibrium condition would be satisfied if this factor were equal to zero, $(W_A - W_a) = 0$, which again represents a special case of the parameters (with no fitness difference between the alleles). We are left with

$$\hat{p} = 1 \qquad \text{(note: } \hat{p} = 0 \text{ is an equilibrium).}$$

Thus, the haploid model of selection has two equilibria, $\hat{p} = 0$ and $\hat{p} = 1$. Furthermore, in the special case of the parameters where $W_a = W_A$, all values of p are equilibria.

Alternatively, we could derive an equilibrium condition for the discrete-time model of haploid selection from the difference equation

$$\Delta p = \frac{(W_A - W_a)p(t)(1 - p(t))}{W_A p(t) + W_a(1 - p(t))} \qquad \text{(difference equation from 3.9)}$$

giving the equilibrium condition

$$0 = \frac{(W_A - W_a)\hat{p}(1 - \hat{p})}{W_A \hat{p} + W_a(1 - \hat{p})} \qquad \text{(equilibrium condition from 3.9).}$$

While this equilibrium condition is equivalent to the one that we just analyzed, it is much easier to solve. The right-hand side is zero only if $\hat{p} = 0$ (an equilibrium), $\hat{p} = 1$ (another equilibrium), or $W_A = W_a$ (a special case of the parameter values). In general, if the difference equation has already been factored, use it to solve for the equilibria because setting each factor to zero immediately provides a list of all possible equilibria.

Regardless of how you choose to solve for the equilibrium of the haploid model, you should always get the same answer (if not, check your math). There are only two equilibria, $\hat{p} = 0$ and $\hat{p} = 1$, which are called "boundary" equilibria because they represent the most extreme values that $\hat{p}$ (a frequency) can take. It makes sense that $\hat{p} = 0$ and $\hat{p} = 1$ are equilibria for the haploid model of selection. If allele A (or allele a) is absent at any point in time, then there is nothing in the model that will regenerate the allele. Biologically, mutations can regenerate lost alleles, but we have yet to incorporate mutations into the model (Problem 5.4).

Now consider the diploid model of selection (equation (3.13a)). Before analyzing this model, let us discuss what equilibria we might find, so that we can hone our biological intuition. As in the haploid model of selection, if allele A (or allele a) is absent, then there is nothing in the model to regenerate it, and we therefore again expect $\hat{p} = 0$ and $\hat{p} = 1$ to be equilibria. Now, however, we have reason to suspect that another equilibrium might exist, because we saw that the allele frequency could remain constant at an intermediate value of $\hat{p}$ in Figure 4.10 (roughly at 0.66 for the parameters considered). The advantage of using the previous graphical results to anticipate the number of equilibria in the model is that it provides a good way to check our math. If we fail to find an equilibrium that we think should exist, then we've either made an error in our analysis, or we were wrong to think that the equilibrium should exist in the first place (in which case we should reevaluate our rationale for expecting the equilibrium).

Now let us identify the equilibria mathematically. By plugging in equation (3.12) for the mean fitness, we can write the recursion equation (3.13a) as

$$p(t + 1) = \frac{p(t)^2 W_{AA} + p(t)q(t)W_{Aa}}{p(t)^2 W_{AA} + 2p(t)q(t)W_{Aa} + q(t)^2 W_{aa}}$$

(recursion equation from 3.13a).

To avoid errors, replace $q(t)$ with $1 - p(t)$. The equilibrium condition then becomes

$$\hat{p} = \frac{\hat{p}^2 W_{AA} + \hat{p}(1 - \hat{p})W_{Aa}}{\hat{p}^2 W_{AA} + 2\hat{p}(1 - \hat{p})W_{Aa} + (1 - \hat{p})^2 W_{aa}}$$

(equilibrium condition from 3.13a).

It is easiest at this stage to cancel a $\hat{p}$ from both sides of the equation, noting that $\hat{p} = 0$ must be one equilibrium. For variety's sake, however, we proceed without canceling terms. A good first step is to multiply both sides by the denominator of the right, so that we don't have any fractions in the equation, leaving:

$$\hat{p}^3 W_{AA} + 2\hat{p}^2(1 - \hat{p})W_{Aa} + \hat{p}(1 - \hat{p})^2 W_{aa} = \hat{p}^2 W_{AA} + \hat{p}(1 - \hat{p})W_{Aa}. \quad (5.1)$$

Equation (5.1) looks messy, and it can be hard to see what to do next. A trick that often helps when factoring is to focus on one parameter at a time, simplifying all terms involving that parameter. Here we have three parameters W_{AA}, W_{Aa}, and W_{aa}, so let us start by bringing all terms involving W_{AA} to the left-hand side of the above equation. These terms are $\hat{p}^3 W_{AA} - \hat{p}^2 W_{AA}$, which can be factored into $-\hat{p}^2(1 - \hat{p})W_{AA}$. Next, we move all terms involving W_{Aa} to the left-hand side. These terms are $2\hat{p}^2(1 - \hat{p})W_{Aa} - \hat{p}(1 - \hat{p})W_{Aa}$, which can be factored into $(2\hat{p} - 1)\hat{p}(1 - \hat{p})W_{Aa}$. Finally, there is only one term involving W_{aa}, $\hat{p}(1 - \hat{p})^2 W_{aa}$, which is already on the left-hand side and is already factored. Putting all of these terms together, we can rewrite equation (5.1) as

$$-\hat{p}^2(1 - \hat{p})W_{AA} + (2\hat{p} - 1)\hat{p}(1 - \hat{p})W_{Aa} + \hat{p}(1 - \hat{p})^2 W_{aa} = 0. \quad (5.2)$$

Because each term contains $\hat{p}(1 - \hat{p})$, we can simplify equation (5.2) substantially by factoring:

$$\hat{p}(1 - \hat{p})(-\hat{p}W_{AA} + (2\hat{p} - 1)W_{Aa} + (1 - \hat{p})W_{aa}) = 0. \quad (5.3)$$

The equilibrium condition will be satisfied if any one of the three factors in equation (5.3) equals zero. The first two factors confirm that, indeed, $\hat{p} = 0$ and $\hat{p} = 1$ are equilibria of the diploid model of selection. Setting the third factor to zero,

$$-\hat{p}W_{AA} + (2\hat{p} - 1)W_{Aa} + (1 - \hat{p})W_{aa} = 0,$$

gives us another potential equilibrium. The third equation is linear in $\hat{p}$ and can be solved using Recipe 5.2. First, move all terms involving $\hat{p}$ to one side:

$$-\hat{p}W_{AA} + 2\hat{p}W_{Aa} - \hat{p}W_{aa} = W_{Aa} - W_{aa}.$$

Second, factor out the $\hat{p}$:

$$\hat{p}(-W_{AA} + 2W_{Aa} - W_{aa}) = W_{Aa} - W_{aa}.$$

Third, divide both sides by the factor multiplying $\hat{p}$:

$$\hat{p} = \frac{W_{Aa} - W_{aa}}{-W_{AA} + 2W_{Aa} - W_{aa}},$$

which is typically written with the positive terms first:

$$\hat{p} = \frac{W_{Aa} - W_{aa}}{2W_{Aa} - W_{AA} - W_{aa}}. \tag{5.4}$$

Let us now perform a numerical check of equation (5.4). In Figure 4.10, we saw an equilibrium at about $\hat{p} = 0.66$ using the parameters $W_{AA} = 0.6$, $W_{Aa} = 0.2$, and $W_{aa} = 1$. Plugging these parameter values into equation (5.4) gives a consistent answer of $\hat{p} = (-0.8)/(-1.2) = 2/3$.

In Figure 4.7b, we tracked the allele frequency dynamics with fitnesses equals to $W_{AA} = 1$, $W_{Aa} = 0.95$, and $W_{aa} = 0.8$. In this case, we saw that the frequency of allele A rose over time toward one, regardless of the initial allele frequency. What happened to the equilibrium (5.4)? Plugging these parameters into (5.4) gives $(0.15)/(0.1) = 1.5$. This is a perfectly fine mathematical solution to the equilibrium condition, but it isn't a biologically valid solution, because p is a frequency and must lie between 0 and 1. This emphasizes an important step when identifying equilibria: *You should check that each equilibrium is biologically valid* (Step 3 in Recipe 5.1).

In the diploid model of selection, when is the third equilibrium (5.4) biologically valid? That is, when will $0 \leq \hat{p} \leq 1$? $\hat{p}$ will equal zero if $W_{Aa} = W_{aa}$ and will equal one if $W_{Aa} = W_{AA}$ (you should prove this). In these special cases, the third equilibrium falls on one of the two "boundary" equilibria that we have already identified ($\hat{p} = 0$ and $\hat{p} = 1$). More importantly, when will there be an "internal" equilibrium at which both alleles A and a are present (i.e., when will there be a polymorphism with $0 < \hat{p} < 1$)? From equation (5.4), the equilibrium frequency of allele A will be positive when

$$0 < \frac{W_{Aa} - W_{aa}}{2W_{Aa} - W_{AA} - W_{aa}} \tag{5.5}$$

It is tempting to multiply both sides of this condition by the denominator, but we must be careful:

Rule 5.1: Simplifying inequalities

CAUTION: When multiplying both sides of an inequality by a negative factor, one must reverse the inequality. Thus, $a > b/c$ is equivalent to $a\,c > b$ when c is positive but is equivalent to $a\,c < b$ when c is negative. If the sign of c is not known, it is safest to avoid multiplication by c.

(continued)

Rule 5.1 *(continued)*

Subtracting terms from both sides of an inequality, however, never reverses the condition. Thus, $a > b/c$ is equivalent to $a - b/c > 0$, regardless of the signs of these terms. Similarly, one can always place terms over a common denominator without altering the condition. Thus, $a - b/c > 0$ can be written as $(a\,c - b)/c > 0$. (If this is unfamiliar, try a numerical example with $a = 4$, $b = -4$, $c = -2$.)

In the case of equation (5.5), we do not know whether the denominator is positive or negative, and so it is safer not to touch it. As written, there are two ways to satisfy condition (5.5): either the numerator and denominator are both positive (case A) or they are both negative (case B).

Next, we ask when the frequency of allele A will be less than one ($\hat{p} < 1$). This is equivalent to asking when $0 < 1 - \hat{p}$, which is the same as asking when the frequency of allele a will be positive ($0 < \hat{q}$). Using equation (5.4), we can rewrite $0 < 1 - \hat{p}$ as:

$$0 < 1 - \frac{W_{Aa} - W_{aa}}{2W_{Aa} - W_{AA} - W_{aa}}. \tag{5.6a}$$

Placing the terms on the left-hand side over the same denominator ($2W_{Aa} - W_{AA} - W_{aa}$) and simplifying gives:

$$0 < \frac{W_{Aa} - W_{AA}}{2W_{Aa} - W_{AA} - W_{aa}} \tag{5.6b}$$

Inequality (5.6b) will be satisfied either when the numerator and denominator are both positive or when they are both negative. Because the denominator is the same in (5.6b) and (5.4), this implies that there are only two cases that ensure that both $0 < \hat{p}$ and $\hat{p} < 1$.

Case A: The numerators and denominators of both (5.4) and (5.6b) are positive. For the numerators to be positive, we must have $W_{Aa} > W_{aa}$ and $W_{Aa} > W_{AA}$. For the denominators to be positive, we must have $2W_{Aa} - W_{AA} - W_{aa} > 0$. Because these conditions involve the same terms, it is worth checking that they are mutually compatible. The easiest way to do this is to rewrite $2W_{Aa} - W_{AA} - W_{aa}$ as $(W_{Aa} - W_{AA}) + (W_{Aa} - W_{aa})$, which compares the heterozygote fitness to each of the homozygote fitnesses and allows us to see that the denominator must be positive when the numerators are positive.

Case B: The numerators and denominators of both (5.4) and (5.6b) are negative. For the numerators to be negative, we must have $W_{Aa} < W_{aa}$ and $W_{Aa} < W_{AA}$. For the denominators to be negative, we must have $2W_{Aa} - W_{AA} - W_{aa} < 0$. Convince yourself that this last condition will always hold when $W_{Aa} < W_{aa}$ and $W_{Aa} < W_{AA}$.

We conclude that a biologically valid polymorphic equilibrium exists only when the heterozygote has the highest fitness (Case A) or the lowest fitness

(Case B). If the heterozygote has intermediate fitness, only the boundary equilibria are biologically valid, indicating that the allele frequency will not remain constant over time except if allele A is absent ($\hat{p} = 0$) or fixed ($\hat{p} = 1$).

5.2.3 An Example where the Equilibria Cannot be Found Explicitly

Although it is possible to determine at least some equilibria in many models, it is often the case that the equilibrium condition cannot be fully solved for all equilibria. To illustrate this problem, we add migration to a variant of the logistic equation of population growth known as the Ricker model. In the Ricker model without migration (see Problem 3.4 of Chapter 3), the number of offspring per parent, $R(n)$, is assumed to decrease exponentially as a function of the population size. This yields the recursion equation

$$n(t+1) = (1+r)\, e^{-\alpha n(t)}\, n(t). \tag{5.7}$$

To find the equilibrium without migration, we solve the equilibrium condition

$$\hat{n} = (1+r)\, e^{-\alpha \hat{n}}\, \hat{n}. \tag{5.8}$$

We can cancel $\hat{n}$ from both sides of equation (5.8), indicating that $\hat{n} = 0$ is one equilibrium. This leaves us with $1 = (1+r)e^{-\alpha\hat{n}}$. We need to isolate $\hat{n}$, so divide both sides by $(1+r)$ to get $1/(1+r) = e^{-\alpha\hat{n}}$. This is not yet an explicit function for $\hat{n}$, but we can solve for $\hat{n}$ by taking the natural logarithm of both sides, obtaining $\ln(1/(1+r)) = -\alpha\hat{n}$ and then dividing by $-\alpha$ to get the equilibrium solution $\hat{n} = \ln(1/(1+r))/(-\alpha)$, which depends only on the parameters. Because $\ln(1/x)$ equals $-\ln(x)$ (Rule A1.13), we can simplify this equilibrium further to get $\hat{n} = \ln(1+r)/\alpha$.

So far, so good, but now let us extend this model to include migration. We suppose that m individuals arrive from outside the population of interest in every generation. If the migrants arrive as adults immediately before counting individuals at time $t+1$, the recursion equation becomes

$$n(t+1) = (1+r)\, e^{-\alpha n(t)}\, n(t) + m, \tag{5.9}$$

whose equilibrium condition is

$$\hat{n} = (1+r)e^{-\alpha\hat{n}}\,\hat{n} + m \tag{5.10}$$

Now $\hat{n}$ is no longer present in all of the terms, and it cannot be canceled. If we rewrite equation (5.10) to bring all of the terms involving $\hat{n}$ to the same side, we get $\hat{n}(1 - (1+r)e^{-\alpha\hat{n}}) = m$. We could try taking the logarithm of both sides again, to obtain $\ln(\hat{n}(1 - (1+r)e^{-\alpha\hat{n}})) = \ln(m)$, which can be written as $\ln(\hat{n}) + \ln(1-(1+r)e^{-\alpha\hat{n}}) = \ln(m)$. This does not help much, though, because we still cannot write $\hat{n}$ as an explicit function of the parameters. You can try other ways of simplifying the equilibrium condition, but there is no way to isolate $\hat{n}$.

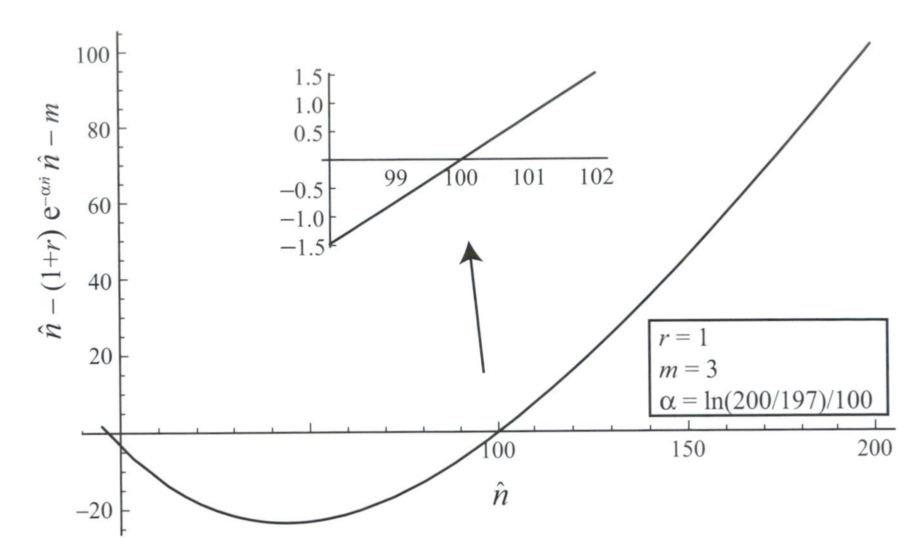

Figure 5.1: A graphical method for finding the solution to an equation. Placing all terms of equation (5.10) on the same side, $\hat{n} - (1 + r)e^{-\alpha\hat{n}}\,\hat{n} - m = 0$, which is satisfied at the equilibrium points, $\hat{n}$. These points can be found graphically for specific values of the parameters (here, $r = 1$, $m = 3$, and $\alpha = \ln(200/97)/100$) by plotting the equation on the vertical axis as a function of $\hat{n}$ on the horizontal axis. Where the curve crosses the horizontal axis, the term $\hat{n} - (1 + r)e^{-\alpha\hat{n}}\,\hat{n} - m$ equals zero, and a solution for $\hat{n}$ has been identified (inset expands the region around $\hat{n} = 100$). A curve can cross the horizontal axis at multiple points, as in this example where it crosses at both $\hat{n} = 100$ and $\hat{n} = -2.8$. Only the former equilibrium is biologically valid.

This does not imply that there is no equilibrium to the recursion equation (5.9). For example, $\hat{n} = 100$ satisfies the equilibrium condition when $r = 1$, $m = 3$, and $\alpha = \ln(200/97)/100$ (try plugging in the numbers). Rather, it means that the implicit solution for $\hat{n}$ given in equation (5.10) cannot be manipulated into an explicit solution. Nevertheless, we can use numerical methods to solve explicitly for an equilibrium of interest, as illustrated in Figure 5.1.

It is worth mentioning how we developed a model whose equilibrium condition could not be solved explicitly. Equations that are mixtures of functions like "exp," "log," and "sine" (i.e., not polynomials) plus polynomial functions like $\hat{n}$ or $\hat{n}^2$ belong to a category known as "transcendental equations." Even innocuous-looking transcendental equations like $e^{\hat{n}} + \hat{n} = 0$, cannot be solved explicitly for $\hat{n}$. So to find a model whose equilibria could not be solved explicitly, we built a model whose equilibrium condition would involve both exponential and polynomial terms.

Even equilibrium conditions that are straightforward polynomial functions of $\hat{n}$ can be impossible to solve if the order of the polynomial is high. If the order of the polynomial is two (i.e., the largest power of $\hat{n}$ is $\hat{n}^2$), then we can use the quadratic equation (Rule A1.10) to solve for $\hat{n}$. In theory, the polynomial could also be solved if its highest power is three or four, but in practice, the solutions to such a cubic or quartic equation are so complicated that they are rarely useful. If the highest power of the equilibrium condition is greater than four and if this equation cannot be factored (e.g., $\hat{n}^5 - 3\hat{n}^3 + 1 = 0$), however, no explicit solution for $\hat{n}$ exists.

Problems solving for equilibria are further exacerbated in models with more than one variable. Indeed, you are guaranteed to find an equilibrium solution only when the recursion equations are linear in all of the variables (see Chapter 7). For example, in a model of natural selection where fitness depends on two loci rather than just one, not all of the equilibria are known explicitly (see Chapter 8).

So what should you do when you end up with a complicated equilibrium condition? First, try bringing all terms to one side of the equation and factoring it (something that software packages such as *Mathematica* are particularly adept at), because the equilibrium condition may turn out to be the product of simpler terms. Alternatively, one can try to find approximate values for the equilibria, a method that we return to at the end of this chapter (Section 5.4).

5.3 Determining Stability

Once the equilibria are identified, the next step is to determine when each equilibrium is stable versus unstable. If we plug an equilibrium back into a dynamical equation, the system will remain at that equilibrium (by definition). But what if a population starts near, but not exactly at, an equilibrium? Will it approach or move away from that equilibrium? Whenever you see the word "near," it implies that some aspect of a model is small. In this case, we assume that the distance to an equilibrium is small. Whenever a term is small, the Taylor series (Primer 1) provides an extremely useful way to approximate a model. In this section, we describe the theory used to determine whether an equilibrium is stable or not. We summarize the method in Recipe 5.3, which provides a step-by-step recipe for carrying out a stability analysis. We then turn to a series of examples to solidify your understanding of the method.

Consider a population that starts at a point $n(t)$ where $n(t)$ is very close to an equilibrium point, $\hat{n}$. We can focus on the distance to the equilibrium by writing $n(t)$ as $n(t) = \hat{n} + \varepsilon(t)$, where ε ("epsilon") measures the *displacement* of the system from the equilibrium. The value of $\varepsilon(t) = n(t) - \hat{n}$ tells us how close the system is to the equilibrium at time t, as well as whether it is above ($\varepsilon(t) > 0$) or below ($\varepsilon(t) < 0$) the equilibrium. The key to determining stability of an equilibrium is to determine whether the distance to the equilibrium grows or shrinks over time. In the next few paragraphs, we will see how to write recursion equations for this displacement. This procedure is simplified enormously by assuming that the population is near the equilibrium, in which case the displacement, $\varepsilon(t)$, is small, and we can safely ignore higher order terms ($\varepsilon(t)^2$, $\varepsilon(t)^3$, etc.; see Primer 1).

First let us derive a recursion equation for the displacement $\varepsilon(t)$ from a general discrete-time model given by $n(t + 1) = f(n(t))$. If at time t, the system is displaced by an amount, $\varepsilon(t) = n(t) - \hat{n}$ from an equilibrium point $\hat{n}$, then in the next generation, the displacement will equal $\varepsilon(t + 1) = n(t + 1) - \hat{n}$. Using the original recursion equation $n(t + 1) = f(n(t))$, this can be written

$$\varepsilon(t + 1) = f(n(t)) - \hat{n}, \tag{5.11a}$$

Equation (5.11a) is not yet in the form of a recursion equation for the displacement, however, because the right-hand side depends on the original variable $n(t)$, not the displacement $\varepsilon(t)$. To rewrite the right-hand side in terms of $\varepsilon(t)$, we use the fact that $n(t) = \hat{n} + \varepsilon(t)$ (by the definition of the displacement). Replacing $n(t)$ with $\hat{n} + \varepsilon(t)$ produces a recursion equation solely in terms of the displacement $\varepsilon(t)$:

$$\varepsilon(t+1) = f(\hat{n} + \varepsilon(t)) - \hat{n}. \tag{5.11b}$$

Equation (5.11b) is our desired recursion, but because the system is near the equilibrium, we can assume $\varepsilon(t)$ is small and use a Taylor series approximation (Recipe P1.4) for $f(\hat{n} + \varepsilon(t))$ around the equilibrium at which $\varepsilon(t) = 0$:

$$f(\hat{n} + \varepsilon(t)) = f(\hat{n}) + \left(\left.\frac{\mathrm{d}f}{\mathrm{d}n}\right|_{n=\hat{n}}\right)\varepsilon(t) + O(\varepsilon(t)^2). \tag{5.12}$$

Here, we have used a linear approximation, because we assume that the displacement $\varepsilon(t)$ is very small, so that terms on the order of $\varepsilon(t)^2$ are negligible; we now drop these terms (represented by $O(\varepsilon(t)^2)$). The term $(\mathrm{d}f/\mathrm{d}n)|_{n=\hat{n}}$ represents the slope of the recursion equation with respect to the variable $n(t)$ evaluated at the equilibrium point $\hat{n}$. To simplify the notation, we define $\lambda = (\mathrm{d}f/\mathrm{d}n)|_{n=\hat{n}}$. Plugging equation (5.12) and this definition into equation (5.11b), the recursion for the displacement becomes

$$\varepsilon(t+1) \approx f(\hat{n}) + \lambda\varepsilon(t) - \hat{n}. \tag{5.13}$$

The term $f(\hat{n})$ represents the state of the system one time step after it starts at the equilibrium $\hat{n}$. But the system will not change if it starts at an equilibrium, so $f(\hat{n})$ must equal $\hat{n}$. Thus, we can simplify this recursion to

$$\varepsilon(t+1) \approx \lambda\varepsilon(t). \tag{5.14}$$

Let us now step back and think about what equation (5.14) says. A population displaced from an equilibrium by a small amount $\varepsilon(t)$ will, at the next time step, be displaced from that equilibrium by $(\mathrm{d}f/\mathrm{d}n)|_{n=\hat{n}}$ times the original displacement. Because $(\mathrm{d}f/\mathrm{d}n)|_{n=\hat{n}}$ is evaluated at $\hat{n}$, it is a function of the parameters only—it does not depend on the variable $n(t)$ or the displacement $\varepsilon(t)$.

Equation (5.14) is an incredibly useful result. No matter how complicated the recursion equation, we can obtain a simple approximate equation (5.14) that describes the dynamics near an equilibrium point. Even better, equation (5.14) is a linear equation in $\varepsilon(t)$ and has the same form as the model of exponential growth. In particular, we can iterate equation (5.14) for any number of generations as we did in equation (4.1); each generation, the displacement changes by a factor λ. Starting from an initial displacement $\varepsilon(0)$, the size of the displacement that we expect after t generations is

$$\varepsilon(t) = \lambda^t\varepsilon(0). \tag{5.15}$$

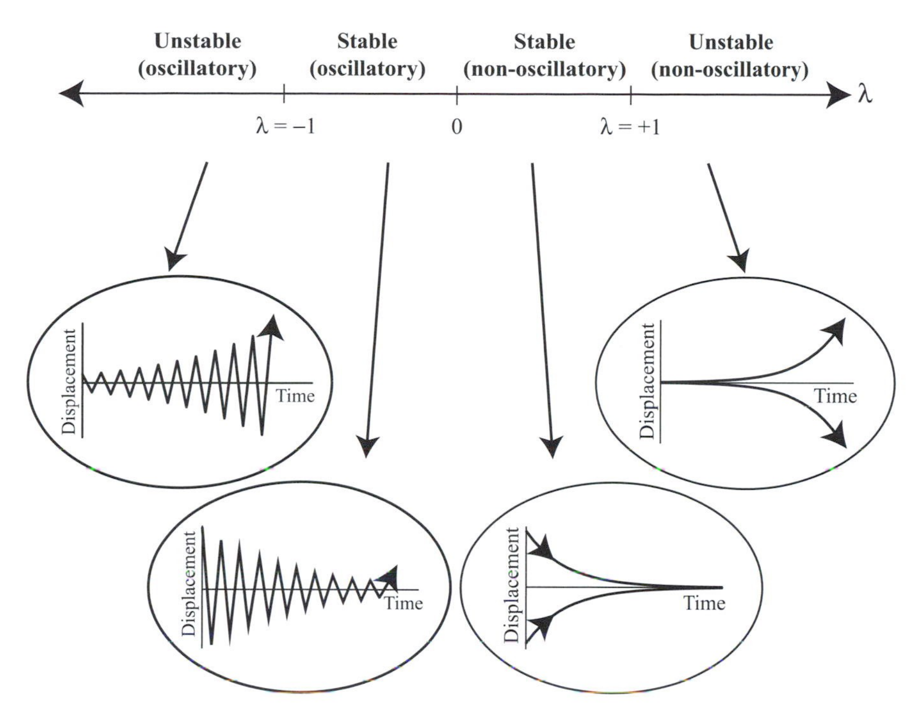

Figure 5.2: Analyzing the stability of an equilibrium in a discrete-time model. For a model whose recursion equation is $n(t+1) = f(n(t))$, the value of $\lambda = (df/dn)|_{n=\hat{n}}$ determines the dynamics near the equilibrium point $\hat{n}$. The inset ovals illustrate the dynamics of the displacement from the equilibrium $\varepsilon(t)$ over time for λ values from each region.

From equation (5.15), we can infer that the behavior of a dynamical system near an equilibrium depends on the magnitude of $\lambda = (df/dn)|_{n=\hat{n}}$ (Figure 5.2). The displacement will shrink over time (indicating that the equilibrium is *stable*) whenever λ is smaller than one in magnitude ($-1 < \lambda < 1$). Conversely, the displacement will grow over time (indicating that the equilibrium is *unstable*) whenever λ is greater than one in magnitude (either $\lambda > 1$ or $\lambda < -1$). A positive value of λ implies that the displacement remains on the same side of the equilibrium over time. A negative value of λ implies that the displacement flips sign and oscillates from one side of the equilibrium to the other at each time step.

In Chapter 4, we plotted the recursion equation $f(n(t))$ against the variable itself, $n(t)$, and discussed how such plots can be used to determine whether a system approaches an equilibrium. The graphical approach and the analytical approach described above are closely related to one another. In the graphical approach, we noted that an equilibrium $\hat{n}$ must fall along the diagonal line (along which $f(n(t)) = n(t)$) and that the slope of the recursion at $\hat{n}$ provides information about the stability of the equilibrium. The term $\lambda = (df/dn)|_{n=\hat{n}}$ is just the mathematical way of writing this slope. As illustrated in Figure 5.3, if the slope is larger than one ($\lambda > 1$), then a population starting slightly above or slightly below an equilibrium will move away from the equilibrium. If the slope is positive but less than one ($0 < \lambda < 1$), the population will move toward the equilibrium. If the slope is negative ($\lambda < 0$), the population will actually

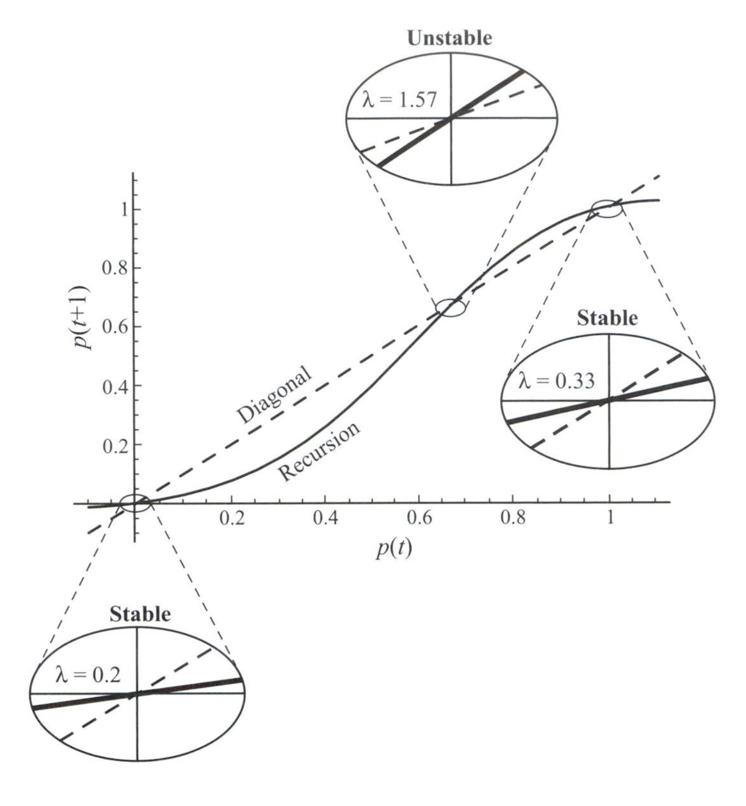

Figure 5.3: A graphical interpretation of the stability condition for the diploid model of selection in discrete time. The recursion equation (3.13a), $p(t+1) = f(p(t+1))$, is plotted against the allele frequency (as in Figure 4.10). The inset ovals show the slope of the recursion equations at each of the three equilibria. These slopes equal $\lambda = (df/dp)|_{p=\hat{p}}$ and are calculated in section 5.3.2. For the equilibrium to be stable, λ must lie between -1 and $+1$. Parameters: $W_{AA} = 0.6$, $W_{Aa} = 0.2$, $W_{aa} = 1$.

shoot past the equilibrium to the other side. If the slope is very negative (in particular, $\lambda < -1$), the population shoots so far past the equilibrium that it ends up even further from the equilibrium than where it started, as observed in the logistic model (Box 4.2; Figure 4.2.3).

It is important to emphasize that, as in Figure 5.3, each equilibrium is associated with its own value of $\lambda = (df/dn)|_{n=\hat{n}}$. Thus, we must calculate λ for each equilibrium of a discrete-time model and determine when that particular equilibrium is stable ($-1 < \lambda < 1$) or unstable ($\lambda > 1$ or $\lambda < -1$).

At this stage we must add two words of caution. CAUTION 1: if the displacement grows at each time step by a factor λ greater than one in magnitude, then the displacement will eventually become large, no matter how small it was initially. As the displacement gets larger, the linear approximation (5.14) for the recursion equation will get worse and worse, until eventually it is completely off. Once this happens, we must return to using the full recursion equation (in all its gory detail) to predict further changes to the system. CAUTION 2: for some equilibria, λ might be equal to $+1$ (or -1). In this case, you might think that the displacement would never grow or shrink, but this isn't necessarily true. Remember, we ignored the higher-order terms in the Taylor series when we derived equation (5.14). If we were to keep the next term in the Taylor series (involving the second derivative $(d^2f/dn^2)|_{n=\hat{n}}$), perhaps the balance would tip toward the equilibrium or away from it. As a result, when the magnitude of λ is one, higher-order terms in the Taylor series must be examined to determine whether an equilibrium is stable or unstable.

One can also analyze the stability of an equilibrium in a discrete-time model using a difference equation. The method is only slightly different from the one described above and is summarized in Recipe 5.3. To avoid confusion, we recommend using a recursion equation rather than a difference equation when determining stability.

Next let us consider a continuous-time model and derive a differential equation for the displacement $\varepsilon = n - \hat{n}$ from the differential equation for the original variable, $dn/dt = f(n)$. (For ease of reading, we now drop the (t) notation, but remember that both n and ε are functions of time.) By the definition of the displacement, $d\varepsilon/dt$ equals $d(n - \hat{n})/dt$. But because the equilibrium $\hat{n}$ is a constant, $d(n - \hat{n})/dt$ is just dn/dt (using Rules A2.3 and A2.1). As a result, we have $d\varepsilon/dt = f(n)$. Again, the right-hand side involves the original variable, not the displacement, but it can be rewritten as $d\varepsilon/dt = f(\hat{n} + \varepsilon)$ using the fact that $n = \hat{n} + \varepsilon$. Taking the Taylor series of the function $f(\hat{n} + \varepsilon)$ around the equilibrium at which $\varepsilon = 0$ then gives the linear approximation

$$\frac{d\varepsilon}{dt} \approx f(\hat{n}) + \left.\frac{df}{dn}\right|_{n=\hat{n}} \varepsilon(t). \tag{5.16}$$

The quantity $f(\hat{n})$ represents the rate of change of the system at an equilibrium, which is zero by definition. The term $(df/dn)|_{n=\hat{n}}$ represents the slope of the differential equation with respect to the variable n evaluated at the equilibrium point $\hat{n}$. Again, we simplify the notation by defining $r = (df/dn)|_{n=\hat{n}}$. (We use r rather than λ to emphasize the fact that it is based on the differential equation not the recursion equation.) Equation (5.16) then becomes

$$\frac{d\varepsilon}{dt} \approx r\varepsilon(t). \tag{5.17}$$

Equation (5.17) has the form of the exponential growth model in continuous time, whose general solution is given by equation (4.2). Thus, the displacement grows or shrinks over time according to the exponential function

$$\varepsilon(t) \approx e^{rt}\,\varepsilon(0). \tag{5.18}$$

When r is negative, the displacement has a negative growth rate and shrinks until the equilibrium is approached (a *stable* equilibrium). When r is positive, the displacement has a positive growth rate and expands over time, causing the system to move away from the equilibrium (an *unstable* equilibrium). In this case, there will come a point when the displacement has grown so much that the linear approximation (5.17) no longer accurately describes the dynamics. At this point, the full nonlinear differential equation must be used. The displacement never oscillates from side to side of the equilibrium in a continuous-time model involving one variable because the displacement follows a continuous path and cannot cross the equilibrium without remaining at the equilibrium forevermore. These conclusions are summarized in Figure 5.4.

Again, there is a close relationship between the stability analysis described above and the change versus variable plots used in the previous chapter. When

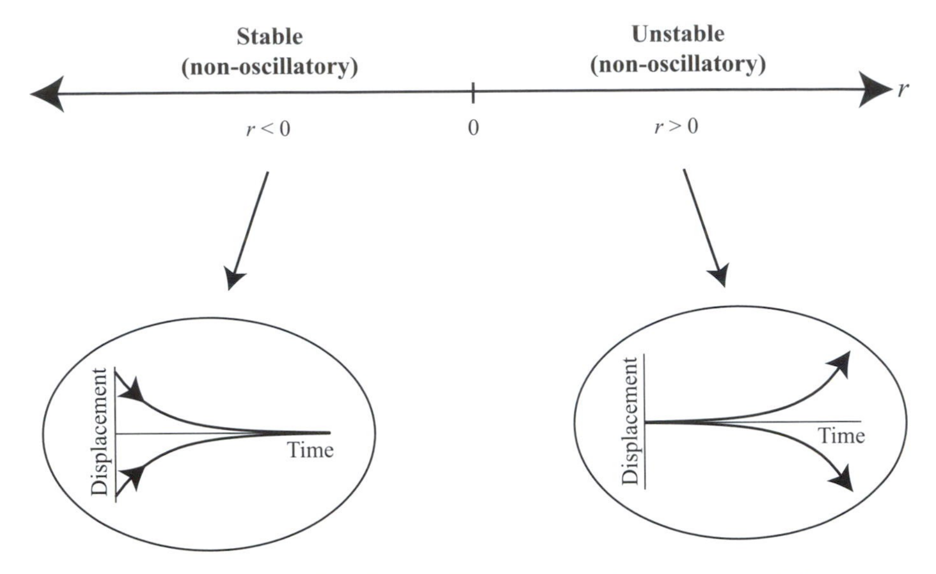

Figure 5.4: Analyzing the stability of an equilibrium in a continuous-time model. For a model whose differential equation is $dn/dt = f(n)$, the value of $r = (df/dn)|_{n=\hat{n}}$ determines the dynamics near the equilibrium point $\hat{n}$. The inset ovals illustrate the dynamics of the displacement from the equilibrium, $\varepsilon(t)$, over time for r values from each region.

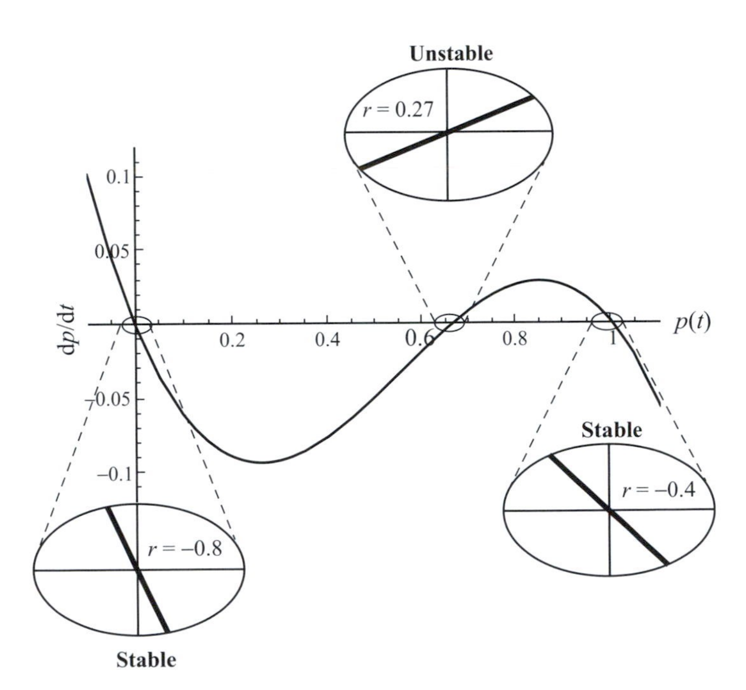

Figure 5.5: A graphical interpretation of the stability condition for the diploid model of selection in continuous time. The differential equation, $f(p) = dp/dt = p(1-p)(p(W_{AA}-W_{Aa}) + (1-p)(W_{Aa}-W_{aa}))$ (from Problem 3.16), is plotted against the allele frequency for the same parameters as in Figure 5.3. The inset ovals show the slope of the differential equations at each of the three equilibria. These slopes equal $r = (df/dp)|_{p=\hat{p}}$, which must be negative for the equilibrium to be stable.

the differential equation df/dn is plotted against the variable n, an equilibrium value is stable if the system grows when started below the equilibrium but shrinks when started above it. This requires that df/dn crosses the horizontal axis at $\hat{n}$ with a negative slope (Figure 5.5). This is exactly equivalent to the condition for stability, $r < 0$, because $r = (df/dn)|_{n=\hat{n}}$ is the mathematical way of describing this slope. Conversely, the equilibrium is unstable if the variable decreases when started below the equilibrium and grows when started above it, which occurs when df/dn crosses the horizontal axis at $\hat{n}$ with a positive slope ($r > 0$). Graphs thus help us visualize why the stability of an equilibrium is determined by the sign of $r = (df/dn)|_{n=\hat{n}}$.

We have just outlined the theory behind an important modeling technique, known as a local stability analysis, used to determine whether an equilibrium is stable. A local stability analysis approximates the dynamics of a model near an equilibrium with a linear equation that can be used to predict whether the system moves toward the equilibrium (the equilibrium is then stable) or away from the equilibrium (the equilibrium is then unstable).

A *local stability anlaysis* (or *linear stability analysis*) determines whether an equilibrium is stable or unstable by approximating the dynamics of the system displaced slightly from the equilibrium.

Recipe 5.3
Performing a Local Stability Analysis

Case A: Recursion equations in a discrete-time model

Consider a recursion equation $n(t + 1) = f(n)$, where $f(n)$ is some function of the variable at time t. The local stability properties of any of its equilibria $\hat{n}$ are determined as follows:

Step 1: Differentiate $f(n)$ with respect to n to obtain df/dn.

Step 2: Replace every instance of n in this derivative with the equilibrium value $\hat{n}$ to obtain $(df/dn)|_{n=\hat{n}}$.

Step 3: Define $\lambda \equiv (df/dn)|_{n=\hat{n}}$.

Step 4: Determine the sign and magnitude of λ.

Step 5: Evaluate the stability of the equilibrium $\hat{n}$ according to Figure 5.2 or the following table:

Case A	$\lambda < 0$	$\lambda > 0$
$\lvert\lambda\rvert < 1$	$\hat{n}$ *is stable (oscillatory)*	$\hat{n}$ *is stable (nonoscillatory)*
$\lvert\lambda\rvert > 1$	$\hat{n}$ *is unstable (oscillatory)*	$\hat{n}$ *is unstable (nonoscillatory)*

"Nonoscillatory" implies that the population remains on the same side of the equilibrium over time. "Oscillatory" implies that the population alternates from side to side of the equilibrium.

Step 6: Repeat Steps 2 through 5 for each equilibrium of interest.

Case B: Difference equations in a discrete-time model

Consider a difference equation $\Delta n = f(n)$. The local stability properties of any of its equilibria $\hat{n}$ are determined as in Case A, with the exception of Step 3:

Step 3: Define $\lambda \equiv (df/dn)|_{n=\hat{n}} - 1$

Case C: Differential equations in a continuous-time model

Consider a differential equation $dn/dt = f(n)$. The local stability properties of any of its equilibria $\hat{n}$ are determined using Steps 1 and 2 above followed by:

(continued)

Recipe 5.3 ***(continued)***

Step 3: Define $r \equiv (\mathrm{d}f/\mathrm{d}n)|_{n=\hat{n}}$.
Step 4: Determine the sign of r.
Step 5: Evaluate the stability of the equilibrium $\hat{n}$ according to Figure 5.4 or the following table:

Case C:	$r < 0$	$r > 0$
	$\hat{n}$ is stable	$\hat{n}$ is unstable

In one-variable, continuous-time models of the form $\mathrm{d}n/\mathrm{d}t = f(n)$, where $f(n)$ is a function of n but not t, the population never oscillates.
Step 6: Repeat Steps 2 through 5 for each equilibrium of interest.

A local stability analysis is an extremely powerful method that can be used to obtain insight into a variety of aspects of biological systems. In the rest of this section, we give examples of the application of Recipe 5.3. After a few examples, consider returning to this section to become truly comfortable with how and why stability analyses work.

5.3.1 Exponential and Logistic Models of Population Growth

Although we already know how the discrete-time model of exponential growth behaves from the general solution (4.1) it is instructive to return to this model for our first local stability analysis. In this case, the recursion equation is $n(t+1) = f(n) = R\,n(t)$, and there is only one equilibrium: $\hat{n} = 0$. We now walk through Recipe 5.3 (Case A). *Step 1:* Take the derivative of $f(n)$ with respect to $n(t)$ to get $\mathrm{d}f/\mathrm{d}n = R$. *Step 2:* Because R does not depend on n, $(\mathrm{d}f/\mathrm{d}n)|_{n=0}$ remains R. *Step 3:* The key determinant of stability is thus $\lambda = R$. *Step 4:* Because R represents the number of surviving individuals per parent, λ must be positive. *Step 5:* We conclude that $\hat{n} = 0$ is unstable if R is greater than one; this makes sense because parents can then produce more than enough offspring to replace themselves. Conversely, the equilibrium $\hat{n} = 0$ is stable if $0 < R < 1$. These conclusions are entirely consistent with the discussion of the general solution in Chapter 4 (see also Figure 4.1).

Next, let us turn to the more interesting logistic model. In this case, the recursion equation is $n(t+1) = f(n) = n(t) + rn(t)(1 - n(t)/K)$, and there are now two equilibria to consider: $\hat{n} = 0$ and $\hat{n} = K$. We first perform a stability analysis of the $\hat{n} = 0$ equilibrium using Recipe 5.3 (Case A). *Step 1:* Take the derivative of $f(n)$ with respect to n, giving

$$\frac{\mathrm{d}f}{\mathrm{d}n} = 1 + r - 2\frac{r\,n(t)}{K}. \tag{5.19}$$

Step 2: Plugging $\hat{n} = 0$ into (5.19) gives $(\mathrm{d}f/\mathrm{d}n)|_{n=0} = 1 + r$, which is the number of surviving offspring per parent when the population size is near zero, $R(0)$. *Step 3:* We thus set λ to $R(0)$. *Step 4:* Because $R(0)$ represents the number of surviving individuals per parent when the population size is low, λ must be positive. *Step 5:* As in the exponential model, $\hat{n} = 0$ is unstable if $R(0)$ is greater than one and stable if $R(0) < 1$. In short, the population size moves away from $\hat{n} = 0$ if parents have more surviving offspring than needed to replace themselves when the population size is low.

Now let us evaluate the stability of the second equilibrium, $\hat{n} = K$. We can skip Step 1, which is the same for all equilibria. *Step 2:* Plugging $\hat{n} = K$ into (5.19) gives $(\mathrm{d}f/\mathrm{d}n)|_{n=K} = 1 - r$, and so, *Step 3:* $\lambda = 1 - r$. *Step 4:* λ can be positive or negative depending on the size of r. *Step 5:* The equilibrium will be stable if $-1 < \lambda < 1$. This looks harmless but the implications of this stability analysis hint at the bizarre behavior of the logistic model.

Because $R(0) = 1 + r$ must be positive (it represents the number of surviving offspring per parent when the population is very small), the smallest that r can be is -1. For r between -1 and 0, $\lambda = 1 - r$ is greater than one, and the equilibrium is unstable. In other words, if the population size is below carrying capacity and the intrinsic growth rate is negative, then the population size will decline toward zero. What if the population size is initially above carrying capacity? The stability analysis also indicates that the population size will grow away from the carrying capacity and become larger over time! Did we make a mistake in the stability analysis? Figure 5.6a indicates not: populations with a negative r that start above K do indeed expand in size toward infinity.

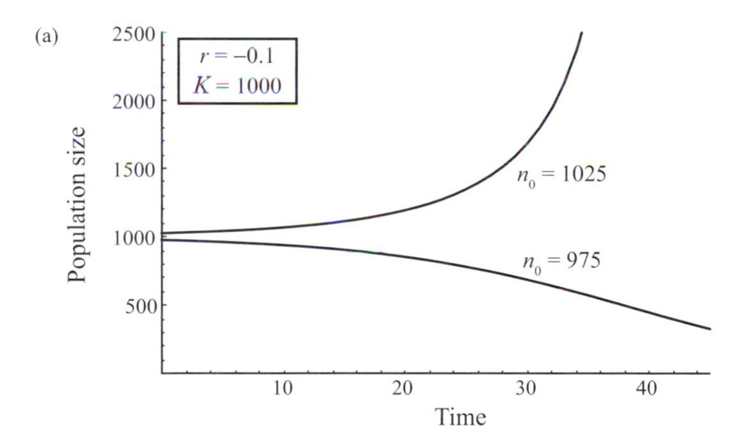

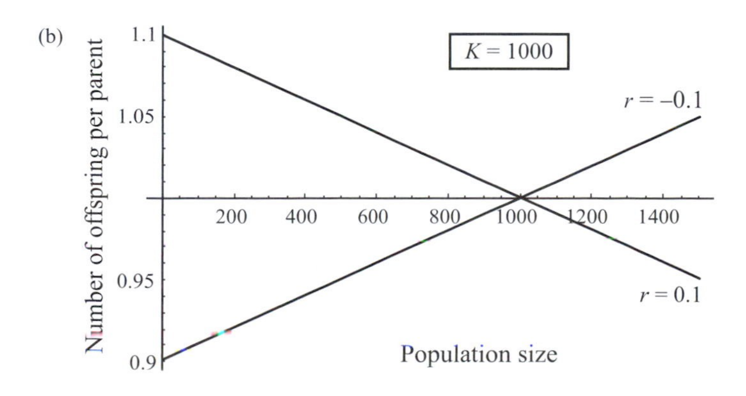

Figure 5.6: Pathological behavior of the logistic model. (a) When the growth rate is negative, a population started below the carrying capacity declines over time, but a population started above the carrying capacity increases in size ($r = -0.1$). (b) The reason for this odd behavior is that the influence of density dependence is reversed when the growth rate is negative, such that each individual leaves more surviving offspring in larger populations than in small populations (compare the line with $r = -0.1$ to the expected decline observed with $r = 0.1$). In both figures, $K = 1000$.

Why does this happen? The logistic model is based on the assumption that density dependence decreases the number of surviving offspring per parent as a linear function of the current population size: $R(n) = 1 + r\,(1 - n(t)/K)$. If the intrinsic growth rate r is negative, however, the density-dependent term in parentheses causes the number of surviving offspring per parent to *rise* as the population size increases (Figure 5.6b). If the population size is above the carrying capacity, then both r and $1 - n(t)/K$ are negative, implying that there is more than one surviving offspring per parent and explaining why the population grows. Furthermore, the problem does not go away if you assume that density dependence causes an exponential rather than linear decline in growth, leading to the Ricker model (5.7) (see Problem 5.7).

One reaction to the above results is to say that the logistic equation just cannot be used when r is negative, but this is an ad hoc restriction. There is no reason, a priori, to believe that the intrinsic growth rate of a species must always be positive; for example, r might very well decrease and become negative as a result of habitat destruction. The best way to resolve this sort of issue is to develop an explicit model of resource use as we did in Chapter 3, where resources (e.g., prey) never become negative, but the growth rate of a population can be.

For the sake of argument, let us continue and assume that r is positive. When r lies between 0 and 1, so too will $\lambda = 1 - r$, implying that $\hat{n} = K$ is stable and that populations near this equilibrium will approach the equilibrium without oscillating around it (see Figure 5.2). When r lies between 1 and 2, $\lambda = 1 - r$ will range from -1 to 0, implying that $\hat{n} = K$ is stable but that populations near this equilibrium will approach it via a series of oscillations. Finally, for r greater than 2, $\lambda = 1 - r$ becomes less than -1, indicating that the population will oscillate around the equilibrium and move further away from it over time. These conclusions are entirely consistent with Figure 4.2 and Box 4.1. Now, however, we know the exact cutoff between nonoscillatory and oscillatory behavior (i.e., $r = 1$) and between stability and instability (i.e., $r = 2$).

Before leaving the discrete-time logistic equation, it is worth discussing its behavior a bit more. The reason that oscillations are observed for high r but not low r is that, by assumption, density dependence has an effect that is proportional to r (think about the equation for $R(n) = 1 + r\,(1 - n(t)/K)$). Thus, populations that grow faster when population size is low will also crash faster when population sizes are large. This may or may not be true in nature; it is certainly possible for a population with a fast intrinsic growth rate to modulate its behavior as the population size grows such that individuals invest less in reproduction and/or more in survival, causing only small changes in population size when density is high. If so, the logistic model will not describe the dynamics of such a species.

The above analysis reveals the possibility of fascinating behavior in the logistic model (e.g., oscillations), but it also reveals that the assumptions underlying the logistic equation (3.5a) constrain population growth in ways that we did not necessarily intend. We found, for example, that density dependence has a positive effect on population growth when the intrinsic growth rate is negative. Furthermore, populations with higher intrinsic growth rates necessarily crash at faster rates when the population size starts above carrying capacity. Always be on the lookout for unintended behavior in a model. If the unintended behavior results from biologically unreasonable assumptions, then you must revise the model if you can or scrap it you can't. Several ecologists have argued that the unintended behavior of the logistic is unacceptable and have instead promoted models that explicitly keep track of the resources that individuals must consume to survive and reproduce (see section 3.4).

Next, we consider the continuous-time logistic model. Using the differential equation (3.5c),

$$\frac{dn}{dt} = r\,n(t)\left(1 - \frac{n(t)}{K}\right),$$

we can define $f(n) = r\,n(t)(1-n(t)/K)$ and then go through the steps outlined in Recipe 5.3 (Case C).

Again, there are two equilibria of the logistic model in continuous time: $\hat{n} = 0$ and $\hat{n} = K$. Let us start with $\hat{n} = 0$. *Step 1:* Taking the derivative of $f(n)$ with respect to n gives

$$\frac{df}{dn} = r - 2\frac{r\,n(t)}{K} \tag{5.20}$$

Step 2: Plugging $\hat{n} = 0$ into (5.20) demonstrates that $(df/dn)|_{n=0} = r$. *Steps 3–5:* The equilibrium $\hat{n} = 0$ is stable when $r < 0$ (extinction predicted) and will be unstable when $r > 0$ (the population grows away from $\hat{n} = 0$).

We now repeat this procedure for the second equilibrium, $\hat{n} = K$. Again, Step 1 is the same. *Step 2:* Plugging $\hat{n} = K$ into equation (5.20) gives $(df/dn)|_{n=K} = -r$. *Steps 3–5:* As a result, the stability conditions for $\hat{n} = K$ are exactly the reverse of the stability conditions for $\hat{n} = 0$. If the population has a positive growth rate ($r > 0$), then $\hat{n} = K$ will be stable, and populations near carrying capacity will approach K over time. If the population has a negative growth rate ($r < 0$), $\hat{n} = K$ will be unstable. Instability, however, implies that populations move away from the equilibrium whether they are slightly above or slightly below the equilibrium. Once again, we are forced to conclude that the logistic model in continuous time behaves oddly, because a population that starts above carrying capacity will increase in size and move away from K if it has a negative intrinsic growth rate.

5.3.2 Haploid and diploid models of natural selection

Determining the stability of the two equilibria, $\hat{p} = 0$ and $\hat{p} = 1$, in the haploid model of selection is a very good problem to work through for your first stability analysis (Problem 5.6), and so here we focus on the diploid model of selection in discrete time. In the diploid model, the recursion equation is

$$p(t+1) = f(p) = \frac{p(t)^2 W_{AA} + p(t)q(t)W_{Aa}}{p(t)^2 W_{AA} + 2p(t)q(t)W_{Aa} + q(t)^2 W_{aa}}, \tag{5.21}$$

which has three equilibria $\hat{p} = 0$ and $\hat{p} = 1$, and $\hat{p}$ given by equation (5.4).

First we analyze the stability of the equilibrium at which the A allele is absent ($\hat{p} = 0$) using Recipe 5.3 (Case A). *Step 1:* We must take the derivative of $f(p)$ with respect to the variable p. Remember, however, that $q(t)$ is a function of $p(t)$, and we must substitute $q(t) = 1 - p(t)$ before taking this derivative. Doing this, applying the quotient rule (A2.13), and rearranging, we get

$$\frac{\mathrm{d}f}{\mathrm{d}p} = \frac{p^2 W_{AA} W_{Aa} + 2p(1-p) W_{AA} W_{aa} + (1-p)^2 W_{Aa} W_{aa}}{(p^2 W_{AA} + 2p(1-p) W_{Aa} + (1-p)^2 W_{aa})^2}. \tag{5.22}$$

To accomplish this simplification, bring together and factor all terms that are multiplied by the same function of the parameters, e.g., by $W_{AA}W_{Aa}$, and repeat this for each combination of parameters.

Steps 2–3: Plugging $\hat{p} = 0$ into equation (5.22), we find that $\lambda = W_{Aa}/W_{aa}$. *Step 4:* The fitness of an individual cannot be negative (it is a measure of the number of surviving offspring per parent genotype), and therefore λ must be positive. As a result, we need only concern ourselves with determining if λ is greater or less than one. Step 5: If $W_{Aa} > W_{aa}$ then λ will be greater than one, and $\hat{p} = 0$ will be an unstable equilibrium. Consequently, the A allele spreads when rare if the fitness of heterozygotes is greater than the fitness of aa homozygotes.

Why doesn't the fitness of AA individuals enter into this stability condition? The stability analysis assumes that we are very near the equilibrium so that p must be very small. In this case, the frequency of the AA genotype (p^2) must be extremely small. Therefore, most A alleles within the population will be present in heterozygous individuals, and so the spread of the A allele is governed by the fitness of heterozygotes relative to the resident aa individuals. Mathematically, the fitness of AA homozygotes is second order with respect to the deviation from the equilibrium (i.e., on the order of $\varepsilon(t)^2$) and so is ignored in the linear approximation used in the derivation of equation (5.14).

Analyzing the second equilibrium ($\hat{p} = 1$) reveals a similar picture. Plugging $\hat{p} = 1$ into (5.22), λ becomes W_{Aa}/W_{AA}. Thus, if the heterozygote is more fit than the resident AA homozygote ($W_{Aa} > W_{AA}$), the equilibrium will be unstable, and the a allele will spread. Conversely, if the heterozygote is less fit ($W_{Aa} < W_{AA}$), the equilibrium will be locally stable to the invasion of a.

We are left with one more equilibrium, given by equation (5.4). Plugging (5.4) into equation (5.22) and simplifying gives

$$\lambda = \left.\frac{\mathrm{d}f}{\mathrm{d}n}\right|_{p=\hat{p}} = \frac{W_{AA}W_{Aa} + W_{aa}W_{Aa} - 2W_{AA}W_{aa}.}{W_{Aa}^2 - W_{AA}W_{aa}} \tag{5.23a}$$

Time for a confession: we didn't derive equation (5.23a) by hand. It is straightforward to do, but it is tedious. This is exactly the sort of ugly algebra that symbolic mathematical software such as *Mathematica* and *Maple* can do in a flash. Even so, to interpret the answer that a computer spits out often involves some educated rearrangements.

To determine where λ lies (positive or negative, below or above one), it is easiest to rearrange equation (5.23a) to make the answer more apparent. We are particularly interested in interpreting (5.23a) in cases where the heterozygote is either more fit or less fit than the two homozygotes, because this is when the polymorphic equilibrium is valid. So let's measure all of the fitnesses relative to the heterozygous fitness. By dividing the top and bottom of equation (5.23a) by W_{Aa}^2, we get

$$\lambda = \frac{\dfrac{W_{AA}}{W_{Aa}} + \dfrac{W_{aa}}{W_{Aa}} - 2\dfrac{W_{AA}}{W_{Aa}}\dfrac{W_{aa}}{W_{Aa}}}{\left(1 - \dfrac{W_{AA}}{W_{Aa}}\dfrac{W_{aa}}{W_{Aa}}\right)}. \tag{5.23b}$$

We can then rewrite (5.23b) in terms (in parentheses) that are positive when the heterozygote is most fit and negative when the heterozygote is least fit:

$$\lambda = \frac{\dfrac{W_{AA}}{W_{Aa}}\left(1 - \dfrac{W_{aa}}{W_{Aa}}\right) + \dfrac{W_{aa}}{W_{Aa}}\left(1 - \dfrac{W_{AA}}{W_{Aa}}\right)}{\left(1 - \dfrac{W_{AA}}{W_{Aa}}\dfrac{W_{aa}}{W_{Aa}}\right)}. \tag{5.23c}$$

By rearranging (5.23a) in this way, we can more easily see that λ is positive whenever the heterozygote is most fit (all parenthetical terms are positive) or least fit (all parenthetical terms are negative). Because λ is positive at the polymorphic equilibrium in the diploid model of selection, the allele frequencies will never oscillate around this equilibrium.

The above rearrangements help us to see if λ is positive or negative, but a different rearrangement is needed to determine whether λ is less than or greater than one. This rearrangement is important as it will tell us when the equilibrium will be stable or unstable. The key insight is that λ will lie below one whenever $(\lambda - 1)$ is negative and above one when $(\lambda - 1)$ is positive. Often $(\lambda - 1)$ factors into easily interpretable parts:

$$(\lambda - 1) = -\frac{(W_{Aa} - W_{aa})(W_{Aa} - W_{AA})}{(W_{Aa}^2 - W_{AA}W_{aa})}. \tag{5.24a}$$

Again, this is easier to interpret if we measure fitness relative to the heterozygous fitness by dividing the top and bottom of (5.24a) by W_{Aa}^2:

$$(\lambda - 1) = -\frac{\left(1 - \dfrac{W_{aa}}{W_{Aa}}\right)\left(1 - \dfrac{W_{AA}}{W_{Aa}}\right)}{\left(1 - \dfrac{W_{AA}}{W_{Aa}}\dfrac{W_{aa}}{W_{Aa}}\right)}. \tag{5.24b}$$

If the heterozygote is most fit, then each parenthetical term is positive, and $(\lambda - 1)$ becomes negative. This proves that λ is less than one, and the equilibrium is locally stable when the heterozygote is most fit. Conversely, if the heterozygote is least fit, then each parenthetical term is negative, and $(\lambda - 1)$ becomes positive. This proves that λ is greater than one, and the polymorphic equilibrium is

TABLE 5.1
Stability conditions for the equilibria of the diploid model of selection

Equilibrium:	$\hat{p} = 0$	$\hat{p} = \dfrac{W_{Aa} - W_{aa}}{2W_{Aa} - W_{AA} - W_{aa}}$	$\hat{p} = 1$
Stable	$W_{Aa} < W_{aa}$	$W_{aa} < W_{Aa} > W_{AA}$	$W_{Aa} < W_{AA}$
Unstable	$W_{Aa} > W_{aa}$	$W_{aa} > W_{Aa} < W_{AA}$	$W_{Aa} > W_{AA}$

unstable when the heterozygote is least fit. We don't have to worry about other possible fitness orderings, because the equilibrium allele frequency given by equation (5.4) lies between 0 and 1 only when the heterozygote is most fit or least fit (Table 5.1).

5.4 Approximations

For many models there comes a point at which the mathematical analysis becomes either too messy to be useful or impossible to conduct. When this occurs we are faced with two choices. First, we can resort to numerical and simulation techniques using computers, provided that we are willing to specify values for all of the parameters. Second, we can seek approximations that allow us to proceed further with the mathematical analysis. Often, our interest in a model focuses on cases where some parameter values are either very small (e.g., mutation rates) or very large (e.g., egg production in some fish species). In this section, we describe how to use such restrictions on the magnitude of parameters to find approximate values for equilibria and to perform approximate stability analyses. This material is more advanced and should be read only once you are comfortable with finding equilibria (section 5.2) and determining their stability (section 5.3).

5.4.1 Approximating equilibria

In cases where it is impossible to find explicit solutions for the equilibria, or when such solutions are so complex that they are difficult to interpret, it can be worth trying to obtain an approximation for the equilibria. Here, we describe a method based on the Taylor series (Primer 1) for finding an approximate solution for $\hat{n}$. The method is known as a *perturbation analysis*, and it is an extremely powerful and useful technique (Box 5.1).

A *perturbation analysis* identifies an approximate solution to an equation by assuming that a parameter is small.

The first step is to determine when you want the equilibrium approximation to be accurate. This involves choosing one or more parameters that you believe are small relative to other parameters in the model. We will write such small parameters in terms of a new parameter ζ ("zeta"), which is small. We use ζ to represent a small perturbation of a *parameter* from some special value (e.g., zero), whereas we used ε to represent a small displacement of the *variable* from some special value (e.g., the equilibrium). For example, you might be willing to assume that mutation rates are small or that a population is sufficiently isolated

Box 5.1: Theoretical Basis of a Perturbation Analysis

In this box, we describe a method for finding an approximate solution to an equation. To begin, we identify a solution to the equation under some special condition (e.g., when a particular parameter is zero). We then "perturb" the equation a little bit away from this special case and determine how the solution to the equation is altered. For this reason, the method is known as a "perturbation analysis." Our focus in this chapter is to apply a perturbation analysis to an equilibrium condition in the hopes of finding an approximate solution that is close to the exact equilibrium. This method is more general, however, and can be used to approximate the solution to any equation of interest.

To apply the technique, we must have an equation that we wish to solve for a term of interest, say $\hat{n}$. We must first identify a parameter (or a set of parameters) that we are willing to assume is small. Let's call this small parameter ζ ("zeta"). For example, if immigration is rare, then we could replace m with ζ in the equilibrium condition (5.10). If we have more than one parameter that we want to be small, then we can rewrite each one as some constant times ζ. Thus, by setting ζ to zero, all of the small parameters in the equation are simultaneously set to zero.

We want to find the value of $\hat{n}$ that solves an equation, but we cannot solve this equation explicitly. (To simplify the notation, we bring all of the terms in the equation of interest to one side, so that it has the form stuff = 0.) Suppose that we write $\hat{n}$ as a sum of terms:

$$\hat{n} = \hat{n}_0 + \hat{n}_1\zeta + \hat{n}_2\zeta^2 + \hat{n}_3\zeta^3 + \cdots. \tag{5.1.1}$$

At an intuitive level, we can think of each successive ζ^i term as providing a more refined estimate for the value of $\hat{n}$. At a more technical level, expression (5.1.1) is the Taylor series of the solution, $\hat{n}$, with respect to ζ around $\zeta = 0$, with $\hat{n}_i$ representing $(1/i!)(d^i\hat{n}/d\zeta^i)|_{\zeta=0}$ (see equation (P1.12) in Primer 1). We now plug expression (5.1.1) into the equation that we wish to solve. To emphasize that this equation depends on the small parameter ζ we write it as $f(\zeta) = 0$. Our goal is to solve this equation for the $\hat{n}_i$ needed to obtain a sufficiently accurate approximation for $\hat{n}$ from expression (5.1.1).

For this method to help, we must be able to solve the original equation for the case where the small parameter is set to zero, $\zeta = 0$. That is, we must be able to solve $f(0) = 0$ for $\hat{n}_0$. If we still cannot solve this equation explicitly, even in this special case, then we must go back to the drawing board and identify other parameters that might be small.

How do we find the $\hat{n}_i$ terms needed in expression (5.1.1)? We certainly cannot find $(d^i\hat{n}/d\zeta^i)|_{\zeta=0}$ by taking the derivative of $\hat{n}$ with respect to the parameter ζ, because we do not know $\hat{n}$. The crux of the method is that we take the Taylor series of the equation we wish to solve, $f(\zeta)$:

$$f(\zeta) = f(0) + \left.\frac{df(\zeta)}{d\zeta}\right|_{\zeta=0}\zeta + \frac{1}{2!}\left.\frac{d^2f(\zeta)}{d\zeta^2}\right|_{\zeta=0}\zeta^2 + \frac{1}{3!}\left.\frac{d^3f(\zeta)}{d\zeta^3}\right|_{\zeta=0}\zeta^3 + \cdots. \tag{5.1.2}$$

A perfect approximation for $\hat{n}$ would cause each term in this sum to be zero. If any term in the Taylor series (5.1.2) did not equal zero, the subsequent terms would be too small to ensure that

(continued)

Box 5.1 ***(continued)***

$f(\zeta) = 0$ is satisfied by the approximation for $\hat{n}$. Thus, our goal is to find the values of $\hat{n}_i$ that cause each term in (5.1.2) to equal zero.

To accomplish this, we set $(d^i f(\zeta)/d\zeta^i)|_{\zeta=0}$ to zero for i, starting with $i = 0$ up to the order that is needed for the desired level of accuracy. We already know that $\hat{n}_0$ causes the first term to equal zero, $f(0) = 0$, because this is the solution when the small parameter ζ is absent. We then move to the next term, plugging our result for $\hat{n}_0$ into $(df(\zeta)/d\zeta)|_{\zeta=0}$, setting this to zero, and solving for $\hat{n}_1$ (or any other remaining $\hat{n}_i$). This can be repeated ad nauseum, for an ever more precise approximation for $\hat{n}$. In practice, however, we often focus on the first two terms in expression (5.1.1), which provide an approximation that is linear in terms of the small parameters. This technique is summarized in Recipe 5.4.

As an example, let us find an approximate solution to the transcendental equation $e^{\alpha\hat{n}} + \hat{n} = 5$ using Recipe 5.4. *Step 1:* In this example, there is only one parameter α, which we assume is small, $\alpha = \zeta$. *Step 2:* Plugging in (5.1.1) for $\hat{n}$ and moving all terms to the right gives

$$f(\zeta) = e^{\zeta(\hat{n}_0 + \hat{n}_1\zeta + \hat{n}_2\zeta^2 + \hat{n}_3\zeta^3 + \cdots)} + (\hat{n}_0 + \hat{n}_1\zeta + \hat{n}_2\zeta^2 + \hat{n}_3\zeta^3 + \cdots) - 5 = 0. \quad (5.1.3)$$

Step 3: The first three derivatives in the Taylor series (5.1.2) are $f(0) = \hat{n}_0 - 4$, $(df(\zeta)/d\zeta)|_{\zeta=0} = \hat{n}_0 + \hat{n}_1$, and $(d^2f(\zeta)/d\zeta^2)|_{\zeta=0} = \hat{n}_0^2 + 2\hat{n}_1 + 2\hat{n}_2$. *Step 4:* Next, we determine the values of $\hat{n}_i$ that cause each of these terms to equal zero. Starting with the first term, we get $\hat{n}_0 = 4$. Using this result in the second term, $(d^2f(\zeta)/d\zeta^2)|_{\zeta=0}$ equals zero when $\hat{n}_1 = -4$. Moving to the third term, $(d^2f(\zeta)/d\zeta^2)|_{\zeta=0} = 0$ is satisfied when $\hat{n}_2 = -4$. We could continue evaluating higher-order terms in the Taylor series, but let us stop and evaluate how we are doing. *Step 5:* To the level of accuracy obtained so far, $\hat{n} \approx 4 - 4\zeta - 4\zeta^2$, which is $\hat{n} \approx 4 - 4\alpha - 4\alpha^2$ in terms of the original parameter α.

If we assume that $\alpha = 0.01$ and we solve $e^{0.01\hat{n}} + \hat{n} = 5$ numerically (e.g., using *Mathematica*), we get $\hat{n} = 3.95961$. If we instead use our approximation, we get $\hat{n} \approx 4 - 4(0.01) - 4(0.01)^2 = 3.9596$. Not a bad match! Even if we only use the first two terms in the approximation, $\hat{n} \approx 4 - 4\alpha$, we get a decent approximation ($\hat{n} \approx 3.96$).

from other populations that immigration is rare relative to the number of individuals born within a population. In other cases, you might think that a parameter is very near a particular value (say, c). For example, fitnesses in a population-genetic model might be expected to be near one. To make it easier to apply the method, we will rewrite any such parameter as c plus a term that we believe to be near zero. For example, if we believe that the fitness of an allele (say, W_A) is near one, then we would write $W_A = 1 + s$, where s is assumed to be small and is replaced by ζ. Similarly, you can also consider cases where a parameter is very large, say K, by rewriting it as $K = 1/b$, where b is small and replaced by ζ.

To make the procedure described in Box 5.1 more concrete, let us find an approximation for $\hat{n}$ in the Ricker model with exponential density dependence and migration. The equilibrium condition (5.10) cannot be solved explicitly, but it can be approximated assuming that migration is rare. Specifically, we assume that m is small and replace it with ζ. As described in Box 5.1, we then replace $\hat{n}$ in the equilibrium condition (5.10) with $\hat{n}_0 + \hat{n}_1\zeta + \hat{n}_2\zeta^2 + \hat{n}_3\zeta^3 + \cdots$, which represents the equilibrium as a main term ($\hat{n}_0$), a small term of the same order as the migration rate ($\hat{n}_1\zeta$), and even smaller terms ($\hat{n}_2\zeta^2 + \hat{n}_3\zeta^3 + \cdots$). Making these substitutions in equation (5.10) gives

$$f(\zeta) = (\hat{n}_0 + \hat{n}_1\zeta + \cdots) - (1 + r)e^{-\alpha(\hat{n}_0 + \hat{n}_1\zeta + \cdots)}(\hat{n}_0 + \hat{n}_1\zeta + \cdots) - \zeta = 0. \tag{5.25}$$

Taking the Taylor series of (5.25) with respect to ζ near the point $\zeta = 0$ gives

$$f(\zeta) = f(0) + \left(\left.\frac{\mathrm{d}f}{\mathrm{d}\zeta}\right|_{\zeta=0}\right)\zeta + \cdots = 0, \tag{5.26}$$

where $f(0) = \hat{n}_0 - (1 + r)e^{-\alpha\hat{n}_0}\hat{n}_0$ and $(\mathrm{d}f/\mathrm{d}\zeta)|_{\zeta=0} = \hat{n}_1 - (1 + r)e^{-\alpha\hat{n}_0}(1 - \alpha\hat{n}_0)\hat{n}_1 - 1$. [Use Rules A2.6 and A2.11 to take the derivative of (5.25) with respect to ζ.]

To ensure that equation (5.26) equals zero, it must be the case that each term in the Taylor series equals zero; if one term did not equal zero, the higher-order terms would be too small to cancel it out. Therefore, we seek the values of $\hat{n}_i$ that ensure that each term in the Taylor series is zero; these values provide us with the approximation for $\hat{n}$ that we desire.

The zeroth-order term $f(0)$ is zero when $\hat{n}_0 - (1 + r)e^{-\alpha\hat{n}_0}\hat{n}_0 = 0$, which is the same as the equilibrium condition (5.8) for the model without migration. In other words, $\hat{n}_0$ represents the term in the approximation for $\hat{n}$ that does not involve the small parameter measuring migration (ζ). As discussed after equation (5.8), the equilibrium condition without migration has two solutions, $\hat{n}_0 = 0$ and $\hat{n}_0 = \ln(1 + r)/\alpha$. Assuming that the population is able to maintain itself in the absence of migration, we focus on the second solution. If, instead, we wished to assume that the population would go extinct in the absence of migration, then we would focus on $\hat{n}_0 = 0$ (Problem 5.14).

We turn next to the linear-order term, $(\mathrm{d}f/\mathrm{d}\zeta)|_{\zeta=0} = \hat{n}_1 - (1 + r)e^{-\alpha\hat{n}_0}(1 - \alpha\hat{n}_0)\hat{n}_1 - 1$, which must equal zero. Plugging in the solution for $\hat{n}_0 = (\ln(1 + r)/\alpha)$, we get

$$\hat{n}_1 - (1 + r)e^{-\ln(1+r)}\left(1 - \alpha\frac{\ln(1 + r)}{\alpha}\right)\hat{n}_1 - 1 = 0.$$

Because $e^{-\ln(1 + r)}$ equals $1/(1 + r)$ (Rules A1.13 and A1.16), we can simplify the above to

$$\ln(1 + r)\hat{n}_1 - 1 = 0. \tag{5.27}$$

This equation is satisfied when

$$\hat{n}_1 = \frac{1}{\ln(1 + r)}. \tag{5.28}$$

Thus, to linear order in ζ (i.e., ignoring terms that are ζ^2 and smaller), an equilibrium of the Ricker model with migration is

$$\begin{aligned}\hat{n} &= \hat{n}_0 + \hat{n}_1\zeta + O(\zeta^2)\\ &= \frac{\ln(1+r)}{\alpha} + \frac{\zeta}{\ln(1+r)} + O(\zeta^2).\end{aligned} \tag{5.29}$$

It is more meaningful to write this approximation in terms of the original parameter $m = \zeta$ describing the migration rate:

$$\hat{n} \approx \frac{\ln(1+r)}{\alpha} + \frac{m}{\ln(1+r)}. \tag{5.30}$$

Earlier, we concocted an example ($r = 1$, $m = 3$, and $\alpha = \ln(200/97)/100$), where α was chosen so that the equilibrium condition (5.10) was satisfied at the equilibrium point $\hat{n} = 100$. Using a calculator to plug these parameter values into (5.30), gives $\hat{n} \approx 100.12$, which is not too far off the exact equilibrium of 100. The advantage of (5.30) is that it is an explicit approximation for $\hat{n}$, which is easy to evaluate for other parameter values, provided that m is small. Furthermore, equation (5.30) clarifies the nature of the equilibrium: it rises with the migration rate with a slope that equals $1/\ln(1 + r)$. Even greater accuracy can be obtained by taking higher order terms in the Taylor series (Problem 5.15). The procedure for using perturbation analysis to approximate an equilibrium is summarized in Recipe 5.4:

Recipe 5.4
Finding an Approximate Equilibrium

Suppose we have an equilibrium condition for $\hat{n}$, and we wish to find an approximate solution for $\hat{n}$ under the assumption that some parameter is small.

Step 1: Identify a parameter(s) that you can assume is small in your model. Write this parameter as ζ. If there is more than one small parameter, write each as some constant times ζ.

Step 2: Plug $\hat{n} = \hat{n}_0 + \hat{n}_1\zeta + \hat{n}_2\zeta^2 + \hat{n}_3\zeta^3 + \cdots$ into the equation that you wish to solve, and write this equation with all terms on the same side as $f(\zeta) = 0$.

Step 3: Calculate $\mathrm{d}^i f(\zeta)/\mathrm{d}\zeta^i|_{\zeta=0}$ for $i = 0$ up to the order desired. (The zeroth derivative is just the function evaluated at zero, $f(0)$.)

Step 4: Starting with $i = 0$, solve $\mathrm{d}^i f(\zeta)/\mathrm{d}\zeta^i|_{\zeta=0} = 0$ for any $\hat{n}_i$ that it contains. Repeat for higher values of i, plugging in any $\hat{n}_i$ that have already been determined, until you have obtained a sufficiently accurate estimate for $\hat{n}$.

(continued)

Recipe 5.4 ***(continued)***

Step 5: Once you have determined $\hat{n}_i$ for $i = 0$ to some order k, plug these values into expression (5.1.1) to obtain an approximation for $\hat{n}$ that is accurate to order ζ^k. To interpret your result, rewrite ζ in terms of the original parameters, by reversing Step 1.

Let us work through another example where we approximate an equilibrium using a perturbation analysis. In the models of natural selection considered so far, mutation was ignored; we basically assumed that each allele is faithfully reproduced every generation. In reality, DNA replication is not error-free, and there is some chance μ that the DNA encoding an A allele will be incorrectly replicated as an alternative allele a. Similarly, there is some chance ν that allele a will mutate into allele A. We can fix this omission by incorporating mutation into the diploid model of natural selection. First, we must specify the life cycle. We census the population at the gamete stage, after which gametes unite at random to produce diploids that experience selection, and finally mutation occurs during the production of gametes. In reality, mutations can occur at other times during the life cycle, but alternative life cycles give similar dynamics.

As described in equation (3.13a), the allele frequency in the population after selection is

$$p' = \frac{p(t)^2 W_{AA} + p(t)(1 - p(t))W_{Aa}}{p(t)^2 W_{AA} + 2p(t)(1 - p(t))W_{Aa} + (1 - p(t))^2 W_{aa}}; \tag{5.31a}$$

this would be the frequency of allele A in the gametes if mutation were absent. With mutation, however, a fraction μ of A alleles is converted to a and vice versa. Thus, among the next generation of gametes,

$$p(t + 1) = (1 - \mu)p' + \nu(1-p'). \tag{5.31b}$$

We obtain the equilibrium condition for this model by plugging (5.31a) into (5.31b) and setting $p(t + 1) = p(t) = \hat{p}$. The result is messy: a cubic polynomial in $\hat{p}$. As mentioned earlier, cubic polynomials can be solved, but it would take several lines of text to write down the answer.

Mutation rates are typically small, on the order of 10^{-6} per gene per generation, so let us set the mutation rate to the small parameter ζ (Step 1, Recipe 5.4). There is a new twist, however, because we have two mutation rates. To allow both mutation rates to be small, we set $\mu = \tilde{\mu}\zeta$ and $\nu = \tilde{\nu}\zeta$, which implies that both mutation rates are proportional to the small parameter ζ. (The terms $\tilde{\mu}$ and $\tilde{\nu}$ are not small, they just tell us the exact factor by which the mutation rates differ from ζ.) Following Step 2, Recipe 5.4, we next replace $\hat{p}$ with $\hat{p}_0 + \hat{p}_1\zeta + \hat{p}_2\zeta^2 + \cdots$ in the equilibrium condition. Writing this equilibrium condition as $f(\zeta) = 0$, we take the Taylor series with respect to ζ near the point $\zeta = 0$, under the assumption that mutations are rare (Step 3, Recipe 5.4).

Moving to Step 4, Recipe 5.4, we set the zeroth-order term in the Taylor series to zero, $f(0) = 0$, and solve for $\hat{p}_0$. Doing so, we regain the three equilibria in the model without mutation: $\hat{p}_0 = 0$, $\hat{p}_0 = 1$, and the polymorphic equilibrium (5.4). A small rate of mutation causes a slight perturbation to each of these equilibria. Let us focus on $\hat{p}_0 = 1$ under the assumption that AA is the most fit genotype ($W_{AA} = 1$), followed by Aa ($W_{Aa} = 1 - h\,s$), and finally aa ($W_{aa} = 1 - s$). In this case, we expect allele A to rise toward fixation (Figure 4.7b) with an equilibrium frequency near one. Next, we turn to the linear order term of the Taylor series, which again we set to zero to find the approximate equilibrium. Calculating $\mathrm{d}f/\mathrm{d}\zeta|_{\zeta=0}$ and plugging in the value for $\hat{p}_0$ that interests us ($\hat{p}_0 = 1$) gives $-\tilde{u} + hs\hat{p}_1 = 0$. Consequently, $\hat{p}_1 = -\tilde{\mu}/(hs)$.

Finally, in Step 5, Recipe 5.4, gathering these terms and rewriting $\tilde{u}$ in terms of the mutation rate using the definition $\mu = \tilde{\mu}\zeta$, the equilibrium frequency of alleles A and a are approximately

$$\hat{p} = \hat{p}_0 + \hat{p}_1\zeta + O(\zeta^2) \approx 1 - \frac{\mu}{hs},$$

$$\hat{q} = 1 - \hat{p} \approx \frac{\mu}{hs}. \tag{5.32}$$

This equilibrium is known as the "mutation-selection balance" and is a classic result in evolutionary biology (Haldane 1927). To this order of approximation, the equilibrium does not depend on the mutation rate ν between allele a and A. Intuitively, ν drops out of the equilibrium approximation because very few alleles within the population are a, and the extremely rare occasion that these few alleles mutate has a negligible influence on the equilibrium of the model. (ν would enter into $\hat{p}_2$ if we derived a more accurate approximation for the equilibrium by solving for the next order term in the Taylor series.) Equation (5.32) makes a lot of sense: the less fit allele, a, is less common within a population when mutations are rare or when selection against it is strong.

In this example, we have assumed that the mutation rates μ and ν are small and that both are the same order of magnitude (on the order of the small parameter ζ). You can make even more sophisticated assumptions about the relative magnitudes of parameters in a perturbation analysis. For example, if there are many ways for an allele to lose its function by mutation but very few ways in which a mutant allele can regain function, you might wish to assume that the forward mutation rate μ is much higher than the backward mutation rate ν. You can express this assumption in mathematical terms by setting $\mu = \tilde{\mu}\zeta$ and $\nu = \tilde{\nu}\zeta^2$; if ζ is small, ζ^2 will always be much smaller. Alternatively, in a model with both mutation and migration, you might assume that both processes are rare but that the mutation rate is much smaller than the migration rate (e.g., by setting $m = \tilde{m}\zeta$, $\mu = \tilde{\mu}\zeta^2$, and $\nu = \tilde{\nu}\zeta^2$).

A perturbation analysis can be used to find an approximate solution to any equation, not just an equilibrium condition (Box 5.1). For example, it can be used to approximate eigenvalues, which arise in stability analyses of models with more than one variable (see Chapter 8). It can also be used to determine approximate solutions to a differential equation (Simmonds and Mann 1988).

Perturbation analyses do occasionally fail, however. For example, the equations that you must solve to obtain the terms $\hat{n}_0$, $\hat{n}_1$, etc., might themselves be too complex to solve. But the technique is powerful because it allows you to state from the outset what you assume about the magnitude of the various parameters and to use this additional information to understand a model's behavior. Sup. Mat. 5.1 discusses problems that can arise when performing a perturbation analysis and how to circumvent some of these.

5.4.2 Stability when the Equilibrium Is Known by Approximation

If we only know the approximate value of an equilibrium, can we determine whether it is stable or unstable? Fortunately, the answer is yes. The underlying conceptual approach is identical to the stability analysis of an equilibrium that we know exactly (Recipe 5.3), except that we cannot evaluate $\lambda = (\mathrm{d}f/\mathrm{d}n)|_{n=\hat{n}}$ (for a discrete-time model) or $r = (\mathrm{d}f/\mathrm{d}n)|_{n=\hat{n}}$ (for a continuous-time model) exactly at the equilibrium. We can, however, approximate λ or r using a Taylor series in the small parameter. In this section, we describe the method for a discrete-time model; continuous-time models are handled analogously.

Let's suppose, as in Box 5.1 and Recipe 5.4, that the exact equilibrium can be written as a series of terms involving some small parameter ζ, as $\hat{n} = \hat{n}_0 + \hat{n}_1\zeta + \hat{n}_2\zeta^2 + \hat{n}_3\zeta^3 + \cdots$. If the system begins slightly away from the equilibrium of interest (i.e., the variable is displaced slightly from $\hat{n}$), we would normally predict changes in the displacement over time using equation (5.14). We continue to do so, but now we substitute the series expression for the equilibrium in $\lambda = (\mathrm{d}f/\mathrm{d}n)|_{n=\hat{n}_0+\hat{n}_1\zeta+\hat{n}_2\zeta^2+\cdots}$ rather than the exact (and unknown) value $\hat{n}$. Because λ now depends on the small parameter ζ, we write it as $\lambda(\zeta)$ to emphasize this fact.

This expression for $\lambda(\zeta)$ is exact provided that we include an infinite number of terms in the series expression for the equilibrium; of course, in practice, we cannot do this. Therefore, the key to the approach is to approximate $\lambda(\zeta)$ using a Taylor series in the small parameter ζ around the point $\zeta = 0$. This gives

$$\lambda(\zeta) = \lambda(0) + \left(\left.\frac{\mathrm{d}\lambda}{\mathrm{d}\zeta}\right|_{\zeta=0}\right)\zeta + \frac{1}{2!}\left(\left.\frac{\mathrm{d}^2\lambda}{\mathrm{d}\zeta^2}\right|_{\zeta=0}\right)\zeta^2 + \frac{1}{3!}\left(\left.\frac{\mathrm{d}^3\lambda}{\mathrm{d}\zeta^3}\right|_{\zeta=0}\right)\zeta^3 + \cdots. \quad (5.33)$$

Equation (5.33) depends on the terms in our approximation, $\hat{n}_i$, but we need only know the values of $\hat{n}_i$ up to the order of the approximation desired in equation (5.33). For example, if we want a first-order approximation (i.e., terms up to and including ζ in equation (5.33)) then we typically need only $\hat{n}_0$ and $\hat{n}_1$. On the other hand, if we want a second-order approximation (i.e., terms up to and including ζ^2 in equation (5.33)) then we typically will need $\hat{n}_2$ as well.

Equation (5.33) provides an important qualitative insight. If the magnitude of $\lambda(0)$ is not near one, then the stability properties of an equilibrium will not change when the parameter that is considered small is added to a model. The only time that the higher-order terms in (5.33) can cause a change in the stability of an equilibrium is when the magnitude of $\lambda(0)$ is close enough to one that a small term proportional to ζ can tip the balance between stability and instability. (Technically, this assumes that the derivatives of $\lambda(\zeta)$ are never infinite and that the Taylor series converges, at least for values of ζ near 0 (see Primer 1), although these issues rarely arise.)

As an example, let us return to the diploid model of natural selection. In the absence of mutation, we found that the stability of the equilibrium, $\hat{p} = 1$, is determined by $\lambda = W_{Aa}/W_{AA}$. Let us assume that directional selection favors the A allele and write the fitnesses as $W_{AA} = 1$, $W_{Aa} = (1 - hs)$, and $W_{aa} = (1 - s)$. In this case, $\lambda = (1 - hs)$, which is less than one by assumption, so the equilibrium is stable. We also showed that if we add a small rate of mutation to the model, the equilibrium allele frequency becomes approximately $\hat{p} \approx 1 - \mu/(hs)$ (equation (5.32)). If we perform a stability analysis of this equilibrium using equation (5.33) with $\mu = \tilde{\mu}\,\zeta$ and $\nu = \tilde{\nu}\,\zeta$, we get $\lambda(\zeta) = \lambda(0) + d\lambda(\zeta)/d\zeta|_{\zeta=0} + \cdots$, where the terms involving the mutation rate are so small that stability is still governed by $\lambda(0) = (1 - hs)$ unless hs is also very small. In short, the presence of a small rate of mutation does not alter the fact that an equilibrium near fixation on allele A is stable when selection favors this allele.

This is an extremely useful fact: if you know the stability of an equilibrium from one model, then you will know the stability of nearby equilibria from a variety of similar models whose recursions represent slight perturbations to the original recursions. Only when the original equilibrium lies on the boundary between stability and instability will smaller-order terms in the perturbations affect stability (for more details, see Karlin and McGregor 1972a, b).

5.5 Concluding Message

In this chapter we presented techniques that allow you to identify the equilibria of a model (summarized in Recipe 5.1) and to assess whether these equilibria are stable or unstable to small changes in the variable (summarized in Recipe 5.3). These techniques are immensely powerful and allow you to answer a broad variety of questions. For example, you can use the methods for finding equilibria to determine the long-term impact of harvesting on the size of a population (Problem 5.9). You can perform a stability analysis to determine whether a genetically modified organism is likely to spread within a population (Muir and Howard 1999, 2001). Or you can determine when migration will prevent adaptation of a population to its local environment by swamping the effects of natural selection (Problem 5.10).

We also introduced an extremely powerful technique, known as a perturbation analysis, which allows you to obtain approximate solutions to equations that cannot be solved exactly. This technique can, for example, allow you to identify equilibria approximately, even for complicated models whose equilibrium condition cannot be solved (summarized in Recipe 5.4).

The methods described in this chapter for finding equilibria and determining their stability, while valuable, have two main limitations. The first is that they only describe the behavior of a model at or near equilibria. In the next chapter, we describe methods for finding general solutions to models involving one variable. When successful, these methods can be used to determine the global behavior of a model. The second is that the methods of this chapter have been limited to models involving a single variable. To take full advantage of these techniques, we must describe how the methods for finding equilibria and determining stability can be extended to models with more than one variable. Chapters 7 and 8 are devoted to this task.

Problems

Problem 5.1: The logistic equation (3.5a) assumes that the number of surviving offspring per parent declines linearly with population size. Ecologists have considered many other forms of density dependence (Henle et al. 2004; May et al. 1974). Find the equilibria of the following density-dependent models of population size:

(a) $$n(t+1) = \lambda\, n(t)\left(1 - \left(\frac{n(t)}{K}\right)^{\theta}\right)$$ (θ-logistic model).

(b) $$n(t+1) = \lambda\, n(t)\, e^{-\alpha n(t)}$$ (Ricker model).

(c) $$n(t+1) = \lambda\, n(t)\frac{1}{1+\alpha\, n(t)}$$ (Beverton-Holt model).

(d) $$n(t+1) = \lambda\, n(t)\frac{1}{(1+\alpha\, n(t))^{\theta}}$$ (Hassell model).

(e) $$n(t+1) = \lambda\, n(t)\frac{1}{1+\alpha\, n(t)^{\theta}}$$ (Maynard-Smith and Slatkin model).

Problem 5.2: Use the recursion equation (3.13a) for the diploid model of natural selection,

$$p(t+1) = \frac{p(t)^2 W_{AA} + p(t)q(t)W_{Aa}}{p(t)^2 W_{AA} + 2p(t)q(t)W_{Aa} + q(t)^2 W_{aa}}$$

to prove that if the fitnesses are equal to one another ($W_{AA} = W_{Aa} = W_{aa}$), then the allele frequency remains constant, regardless of its initial value. This represents a special case of the parameters.

Problem 5.3: In the haploid model of selection considered in the text, it was assumed that the fitnesses W_A and W_a were constants. In many examples of biological interest, however, the success of each type might depend on the frequency of the other type, a phenomenon known as "frequency-dependent selection." For example, allele A might produce a toxin that reduces the fitness of individuals carrying allele a, or individuals carrying allele A might produce a compound that can be consumed by a individuals. (a) Find the equilibria of the haploid model if $W_A = 1 + \alpha q$ and $W_a = 1 + \beta p$. (b) Specify when the equilibria in (a) will be biologically valid. (c) Show that you cannot obtain explicit solutions for all of the equilibria if the fitness of A decreases exponentially with the frequency of a using $W_A = e^{-\alpha q}$ and $W_a = 1 + \beta p$.

Problem 5.4: DNA replication is not entirely error-free, causing mutations to occur within a population. Here, we incorporate mutations into the haploid model of selection. Assuming that individuals carrying allele A are most fit, define fitnesses of A and a as $W_A = 1$ and $W_a = 1 - s$. Mutation causes a proportion μ of the haploid offspring produced by an A individual to become a. Mutations from a to A are ignored (such mutations will be extremely infrequent whenever the less fit allele a is rare). Under these assumptions, the frequency of the A allele among offspring will equal $p(t+1) = (1-\mu)p'$, where p' is the frequency of A after selection given by equation (3.8c):

$$p' = \frac{W_A p(t)}{W_A p(t) + W_a(1 - p(t))}.$$

(a) Determine the two equilibria for p. (b) When are these equilibria biologically valid? (c) At the polymorphic equilibrium, show that the mean fitness of the population is $\overline{W} = W_A\hat{p} + W_a(1 - \hat{p}) = 1 - \mu$, which does not depend on the strength of selection. Mutations that are more strongly selected against are less frequent at equilibrium, and these effects exactly balance, causing the selection coefficient to cancel out of $\overline{W}$. (d) Determine the stability of the equilibrium when the A allele is absent. (e) Determine the stability of the equilibrium when the a allele is absent. (f) Discuss the conditions under which each equilibrium will be locally stable. Is it possible for both equilibria to be locally stable simultaneously?

Problem 5.5: When deleterious mutations are perfectly recessive ($W_{AA} = 1$, $W_{Aa} = 1$, $W_{aa} = 1 - s$) and back mutations are absent ($\nu = 0$), the diploid model of natural selection with mutations given by equations (5.31) becomes

$$p(t+1) = (1 - \mu)\frac{p(t)^2 + p(t)(1 - p(t))}{p(t)^2 + 2p(t)(1 - p(t)) + (1 - p(t))^2(1 - s)}.$$

In this example, the equilibrium can be solved exactly. (a) Solve for the three equilibria of this model. (b) Specify which equilibria are biologically valid under the assumption that s is positive.

Problem 5.6: Perform a local stability analysis of the haploid model of natural selection in discrete time, using the recursion equation (3.8c)

$$p(t+1) = \frac{W_A p(t)}{W_A p(t) + W_a(1 - p(t))}.$$

(a) Determine the conditions under which the equilibrium $\hat{p} = 0$ is stable. (b) Determine the conditions under which the equilibrium $\hat{p} = 1$ is stable. Check that your results make sense and are consistent with Figure 4.9.

Problem 5.7: If density dependence causes an exponential decline in the growth rate, as in the Ricker model (5.7), we find two equilibria, $\hat{n} = 0$ and $\hat{n} = \ln(1 + r)/\alpha$. We can use the nonzero equilibrium to rewrite the Ricker model in terms of the carrying capacity of the habitat, K, rather than the parameter α, using the definition $\ln(1 + r)/\alpha = K$. (a) Show that the Ricker equation is then equivalent to

$$n(t+1) = (1 + r)^{1-n(t)/K}\, n(t).$$

(b) Determine the conditions under which the equilibrium $\hat{n} = 0$ is stable. (c) Determine the conditions under which the equilibrium $\hat{n} = K$ is stable. (d) Confirm that if $r < 0$, populations whose initial size is greater than $\hat{n} = K$ will grow over time, which is an undesirable property of density-dependent models that assume a constant carrying capacity regardless of the growth rate.

Problem 5.8: Mating within a population is often not random. In Problem 3.14, recursion equations were developed for a model in which each egg has a chance f of being fertilized by a sperm carrying the same allele and a chance $1 - f$ of being fertilized by a sperm drawn randomly from the gamete pool. To simplify the algebra, assume that the fitnesses are $W_{AA} = 1 - s$, $W_{Aa} = 1$, and $W_{aa} = 1 - t$. After some rearrangement, these recursions become

$$p(t+1) = \frac{p(t) - s(1-f)p(t)^2 - sfp(t)}{1 - s(1-f)p(t)^2 - sfp(t) - t(1-f)(1-p(t))^2 - tf(1-p(t))}.$$

(a) Find the three equilibria of this model. (b) What conditions must hold for the polymorphic equilibrium to be biologically valid? (c) Determine the stability conditions for all three equilibria. (d) Offer a biological explanation for why the stability of each equilibrium now depends on the fitnesses of both the AA and aa homozygotes. Check your answers against those given in the text for the special case when inbreeding is absent ($f = 0$). (You might wish to use a mathematical software package to carry out these calculations.)

Problem 5.9: Harvesting prevents the population size of a species from attaining its natural carrying capacity. We can add harvesting to the logistic model by assuming that the per capita harvest rate is m per day in a population whose intrinsic growth rate is r per day and whose carrying capacity is K in the absence of harvesting. (a) Derive a differential equation describing the dynamics of the population size. (b) Determine the equilibria for this model. (c) Determine the stability of these equilibria. (d) What condition must hold for the population to persist? (e) What is the maximum allowable harvest rate that ensures that the population size will remain stable at a size greater than 1000, which is considered by some to represent a minimum viable population size?

Problem 5.10: Species living at the edge of their natural range can fail to adapt to local conditions because of the constant inflow of migrants from the center of the range. Consider a haploid model of selection where selection in a marginal patch favors allele a: $W_A = 1 - s$ and $W_a = 1$. Adults migrate into the marginal patch from a more favorable area at rate m. We will assume that these migrants all carry allele A, which is favored in the core habitat (Figure 5.7).

After migration, the frequency of the locally unfit allele A becomes $p(t+1) = (1-m)p' + m$, where p' is the frequency of allele A in the local population after selection but before migration:

$$p' = \frac{p(t)(1-s)}{p(t)(1-s) + (1-p(t))}$$

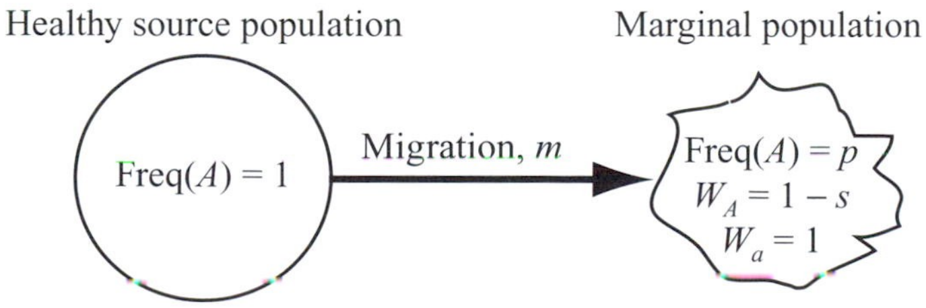

Healthy source population
Marginal population
Freq(A) = 1
Migration, m
Freq(A) = p
$W_A = 1 - s$
$W_a = 1$

Figure 5.7: Evolution in a source and marginal habitat

(assuming random mating). (a) Find the two equilibria of this model. (b) What conditions must hold for polymorphic equilibrium to be biologically valid? (c) Determine when allele a will disappear from the population when rare despite the fact that it is locally favored by examining the stability of the equilibrium at $\hat{p} = 1$.

Problem 5.11: "Memes" are cultural traits, such as inventions (e.g., DVDs) and fads (e.g., wearing hats on backward), whose spread within a population can be modeled. Consider a meme that is adopted slowly, only after an individual sees two other individuals with the meme. A recursion equation for the fraction of the population with the meme f might be

$$f(t + 1) = f(t)(1-\delta) + \alpha f(t)^2(1-f(t)),$$

where δ represents the loss of the meme and the fraction of individuals that adopt the meme is $\alpha f(t)^2$ per time step (the fact that $f(t)^2$ is squared reflects our assumption that the meme spreads only after witnessing two individuals with the meme). (a) Find the three equilibria of this model. (b) Figure 5.8 illustrates the recursion $f(t + 1)$ versus $f(t)$ for $\delta = 0.4$ and $\alpha = 2.0$ along with the diagonal line (dashed) representing where $f(t + 1) = f(t)$. Mark each equilibrium with an X. (c) From the slope of the recursion equation at each equilibrium (on the graph), specify which equilibria are stable (S) and which are unstable (U).

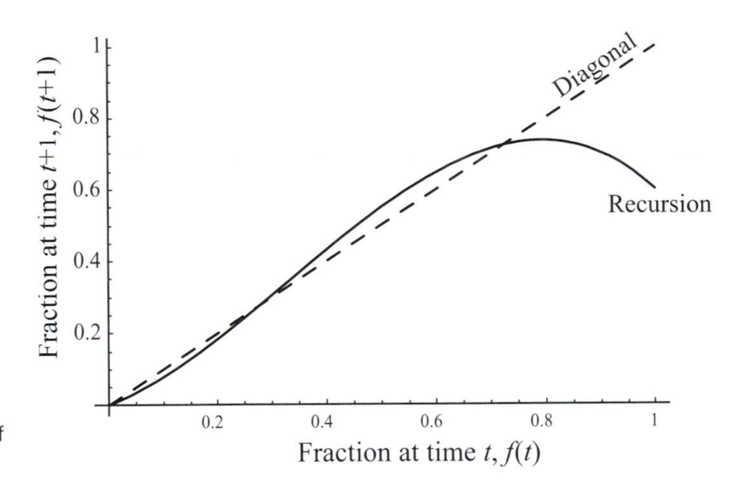

Figure 5.8: Cultural evolution and the dynamics of a "meme."

Problem 5.12: Population size might be regulated by competition for suitable territories. Consider a large number of suitable territories or patches. At time t, a fraction $p(t)$ of these patches are occupied. Of the unoccupied sites, a fraction $mp(t)$ are recolonized from occupied patches. Subsequently, each occupied site suffers a risk of local extinction e through catastrophic events such as fire or disease. These assumptions are consistent with the following discrete-time recursion equation for the fraction of occupied sites:

$$p(t + 1) = (1-e)(p(t) + mp(t)(1-p(t))).$$

(a) Find the two equilibria of this model. (b) Under what conditions is there a biologically valid equilibrium with the species present? (c) Given that the equilibrium in (b) is valid, when is it stable? (d) If we assume that m is less than one so that not all patches can be immediately recolonized, can the fraction of occupied sites overshoot the equilibrium?

Problem 5.13: The extinction-recolonization model described in Problem 5.12 can also be studied in continuous time, using the differential equation

$$\frac{dp}{dt} = m\,p(1 - p) - e\,p.$$

(a) Find the two equilibria of this model. (b) Under what conditions is there a biologically valid equilibrium with the species present? (c) Given that the equilibrium in (b) is valid, when is it stable? (d) Is it possible for the fraction of occupied sites to overshoot the equilibrium?

Problem 5.14: For the variant of the logistic model with immigration given by the recursion equation (5.9), use a perturbation analysis to find a linear approximation for the equilibrium population size $\hat{n}$ assuming that the population would go extinct in the absence of migration (i.e., $\hat{n}_0 = 0$). Use this approximation to show that $\hat{n}$ becomes positive when migration is present as long as the growth rate r is negative.

Problem 5.15: For the variant of the logistic model with immigration given by the recursion equation (5.9), find an approximation for the equilibrium population size $\hat{n}$ to second order in the migration rate. In other words, use the equilibrium condition (5.10) and the method described in Box 5.1 and Recipe 5.4 to extend (5.30) to include a term proportional to m^2. Into this second-order approximation, plug in the parameter values $r = 1$, $m = 3$, and $\alpha = \ln(200/97)/100$, to show that this second-order approximation gives an equilibrium population size of $\hat{n}_0 \approx 99.99$, which is very close to the true solution of 100.

Problem 5.16: Find the approximate mutation-selection balance in the diploid model of selection under the assumption that allele A *decreases* fitness. That is, use a perturbation analysis to repeat the derivation of (5.32) under the assumption that $\hat{p}_0 = 0$. Show that your answer depends on the mutation rate ν from allele a to A but not on μ. [Note: Specify the relative fitnesses that you are assuming. You can continue to measure fitness relative to the AA genotype, using $W_{AA} = 1$, $W_{Aa} = 1 - hs$, and $W_{aa} = 1 - s$, although s must be negative for A to decrease fitness. Alternatively, you can redefine the fitnesses.]

Further Reading

For more information on perturbation methods, consult

- Hinch, E. J. 1991. *Perturbation Methods*. Cambridge University Press, Cambridge.
- Simmonds, J. G., and J. E. Mann. 1988. *A First Look at Perturbation Theory*. Dover Publications, Mineola, N.Y.

References

Haldane, J. B. S. 1927. A mathematical theory of natural and artificial selection, Part V: Selection and Mutation. *Proc. Cambridge Philoso. Soc.* 23:838–844.

Henle, K., S. Sarre, and K. Wiegand. 2004. The role of density regulation in extinction processes and population viability analysis. *Biodiversity Conservation* 13:9–52.

Karlin, S., and J. McGregor. 1972a. Application of method of small parameters to multi-niche population genetic models. *Theor. Popul. Biol.* 3:186–209.

Karlin, S., and J. McGregor. 1972b. Polymorphisms for genetic and ecological systems with weak coupling. *Theor. Popul. Biol.* 3:210–238.

May, R. M., G. R. Conway, M. P. Hassell, and T. R. E. Southwood. 1974. Time delays, density-dependence and single-species oscillations. *J. Anim. Ecolo.* 43:747–770.

Muir, W. M., and R. D. Howard. 1999. Possible ecological risks of transgenic organism release when transgenes affect mating success: sexual selection and the Trojan gene hypothesis. *Proc. Natl. Acad. Sci. U.S.A.* 96:13853–13856.

Muir, W. M., and R. D. Howard. 2001. Fitness components and ecological risk of transgenic release: A model using Japanese medaka (*Oryzias latipes*). *Am. Nat.* 158:1–16.

Simmonds, J. G., and J. E. Mann. 1988. *A First Look at Perturbation Theory*. Dover Publications, Mineola, N.Y.

CHAPTER 6

General Solutions and Transformations—One-Variable Models

Chapter Goals:

- To describe methods for obtaining general solutions for models with one variable
- To describe transformation methods for simplifying models

Chapter Concepts:

- Transformation
- Affine model
- Brute force iteration
- Separation of variables

6.1 Introduction

In Chapter 5 we focused on finding equilibria of models and determining the local stability properties of these equilibria. In a small fraction of models, it is possible to go beyond identifying equilibria and their stability by finding the *general solution* of the equations:

> **Definition 6.1: General Solution**
> A description of the state of the system for all future points in time that depends only on the parameters, the initial state, and the amount of time that has passed.

We have already seen an example of a general solution for a discrete-time recursion equation. The recursion equation $n(t+1) = R\,n(t)$ can be iterated by hand to get the general solution (4.1):

$$n(t) = R^t n(0). \tag{6.1}$$

Any model where the variable is multiplied by the same factor each generation to obtain the value of the variable in the next generation has a general solution of this form. We will describe recipes that can be followed to find the general solution for any linear model with one variable (this chapter) or multiple variables (Chapter 9). Because these recipes exist, there is some hope of finding the general solution for any linear model (although the equations can become so complicated that we might get bogged down along the way). For nonlinear models, however, there may or may not be a general solution, and it is a bit of a black art to be able to massage equations in just the right way to obtain a general solution.

Many of the techniques in this chapter involve *transformations* from the original set of variables to another set of variables. Transformations often simplify the dynamical equations. For example, in some cases, it is possible to "see" how to obtain a general solution only after a transformation is performed. More generally, however, transformations can be used to gain more insight into models and to obtain approximate dynamical equations when an exact general solution cannot be obtained.

> A *transformation* rewrites the dynamical equations of a model in terms of a new set of variables, which is chosen to simplify the equations or to provide more biological insight than the original set of variables.

In section 6.2 we describe transformations in a general context. Sections 6.3 and 6.4 then discuss how to obtain general solutions to linear and nonlinear models in discrete time, typically with the help of a transformation. Finally, sections 6.5 and 6.6 do the same for models in continuous time. Throughout, we focus on models with one variable, leaving the corresponding analyses for models with multiple variables until Chapter 9.

6.2 Transformations

As you get more experience analyzing models, you will come to appreciate techniques that simplify the calculations and that help extract useful biological information. One of the most powerful techniques involves transforming the variables of a model. To perform a transformation, we define a new set of variables in terms of the original variables in the model. We then derive "transformed" dynamical equations that describe how these new variables change over time. For example, if we have a differential equation describing the size of a population, N, over time, $dN/dt = f(N)$, we might define a new variable y as $y = \ln(N)$ or perhaps $y = a + bN$. We could then determine how the new variable changes over time by rewriting the differential equation in terms of the new variable. This transformation changes how we keep track of the system (in terms of y rather than N), but it does not alter the underlying process being modeled.

The utility of transformations is that, when chosen well, the equation describing the dynamics of the new variable can be much simpler than the original equation. Given a choice between analyzing a simple equation in terms of y or a complicated equation in terms of N, it makes sense to go the simple route. Not only does this save effort, but it can also be the only way to "see" how to solve a model. Furthermore, transformations often provide insight into why a model behaves the way that it does. And once we have analyzed the simpler transformed equation, we can then transform back to the original variable to understand the dynamics of the original variable.

In fact, we have already introduced several transformations in this book. In Chapter 3, we described a useful transformation in models of selection, going from the number of individuals carrying allele A and allele a to the proportion p carrying allele A (Box 3.1). And in Chapter 5, we described how the stability of an equilibrium can be assessed by performing a transformation from the original variable to a new variable describing the displacement from an equilibrium, $\varepsilon(t) = n(t) - \hat{n}$. We then transformed the recursion equation involving the original variable $n(t)$ into a recursion equation involving the new variable $\varepsilon(t)$. This transformed recursion equation was particularly helpful because we could approximate it with a linear equation (5.14) in $\varepsilon(t)$ under the assumption that the system is near the equilibrium.

This all sounds great, but you are probably asking yourself: how do I choose an appropriate transformation? This is a good question, which, unfortunately, does not have a simple answer. Different transformations work best in different situations. In this chapter, we discuss some methods and transformations that often work for obtaining general solutions. We also discuss clues that you can use to guide your choice of transformation. Using these clues and being willing to play around trying different transformations increases the likelihood that you will hit upon a useful one.

6.3 Linear Models in Discrete Time

With only one variable in discrete time, there are two types of linear models that we need to solve: those of the form

$$n(t+1) = Rn(t), \tag{6.2a}$$

and those that also involve a constant,

$$n(t+1) = Rn(t) + m. \tag{6.2b}$$

Equations of the first form arise when a process or processes act independently on every individual on a per capita basis. For example, each individual might give birth, die, or emigrate out of a population, leading the total number of births, deaths, or emigrants to be proportional to the total number of individuals within the population, $n(t)$. In contrast, equations of the second form, known as *affine* models, arise when there is a constant input or outflow from the system that does not depend on the current state $n(t)$. For example, there might be a constant number of migrants into a population from a source population, as in the toy model of mice migrating in from surrounding fields (equation (2.4) has the same form as (6.2b) with $R = (1 + b)\,(1 - d)$). Or there might be a constant number of individuals that are harvested from a population per time step, causing m to be negative. Equation (6.2b) could also be used to describe the number of red blood cells in the body on day t, where R is the fraction of red blood cells that are not eliminated by the liver and m is the number of new red blood cells produced by the bone marrow per day. The important point is that in affine models with recursion equations like (6.2b), the variable is replenished or eliminated by a process that is not dependent on the current value of the variable.

An *affine model* depends linearly on the variables and contains a constant term representing any input or outflow to the system.

According to equation (6.2a), every time step that passes causes the previous solution to be multiplied by R, so $n(1) = Rn(0)$, $n(2) = Rn(1) = R^2n(0)$, etc. Consequently, at any future point in time, $n(t) = Rn(t-1) = \cdots = R^tn(0)$, and we get the general solution (6.1). If we try this technique with (6.2b), it also works, but the general solution is not pretty. Equation (6.2b) tells us also that $n(2)$ must be $Rn(1) + m$. Using equation (6.2b) to substitute in for $n(1)$, we get $R\,(R\,n(0) + m) + m = R^2\,n(0) + R\,m + m$. It is not yet clear what is happening as we iterate (6.2b). If we calculate $n(3)$ in the same manner, we get $n(3) = R\,n(2) + m = R(R^2\,n(0) + R\,m + m) + m = R^3n(0) + R^2\,m + R\,m + m$. Thus, it appears that the general solution has the form

$$n(t) = R^t\,n(0) + m\sum_{i=0}^{t-1} R^i. \tag{6.3a}$$

Brute force iteration is a method to solve recursion equations by repeatedly plugging in the recursion equation for the value of the variable in the previous time step.

This method, known as *brute force iteration*, again works in this situation. But this is not always the case; often brute force iteration generates a complicated mess of equations that do not follow any obvious pattern, especially when there is more than one variable.

Life is much simpler if we first transform the affine model (6.2b) into a linear model without constants (6.2a). The transformation that accomplishes this goal defines a new variable $\delta(t)$ as the distance between the original variable $n(t)$ and the equilibrium of the system. Using this transformation, we can solve any affine model:

Recipe 6.1

Solving a Linear Discrete-Time Model with a Constant Term

Affine recursion equations of the form $n(t+1) = \rho n(t) + c$ can be solved as follows:

Step 1: Solve for the equilibrium $\hat{n}$. Here, $\hat{n} = c/(1-\rho)$.

Step 2: Define a new variable $\delta(t)$ as the distance of the system from the equilibrium, $\delta(t) = n(t) - \hat{n}$. Reversing this equation implies that $n(t) = \delta(t) + \hat{n}$.

Step 3: The recursion equation for the transformed variable is $\delta(t+1) = \rho\delta(t)$. To prove this, use the recursion equation for $n(t+1)$ to write a recursion for $\delta(t+1)$:

$$\delta(t+1) = n(t+1) - \hat{n} = \rho n(t) + c - \hat{n}.$$

Replacing $n(t)$ with $\delta(t) + \hat{n}$ gives $\delta(t+1) = \rho\delta(t) + \rho\hat{n} + c - \hat{n}$. Plugging in the equilibrium from Step 1 and factoring causes the last three terms to equal zero (try this).

Step 4: The general solution for the distance to the equilibrium is thus $\delta(t) = \rho^t\,\delta(0)$.

Step 5: The general solution for the original variable is found by using the result from Step 4 in $n(t) = \delta(t) + \hat{n}$, resulting in $n(t) = \rho^t\,\delta(0) + \hat{n}$. Rewriting $\delta(0)$ in terms of $n(0)$ gives the general solution:

$$\begin{aligned} n(t) &= \rho^t\,(n(0) - \hat{n}) + \hat{n} \\ &= \rho^t\, n(0) + (1 - \rho^t)\,\hat{n}. \end{aligned}$$

As this recipe always works, Steps 2–4 can be skipped, and the above equation can be used after having identified $\hat{n}$ and ρ.

The advantage of a good transformation is that it can reveal an underlying simplicity to the dynamics that might be impossible to see in the original variable. For example, in an affine model, the fact that $\delta(t+1) = \rho\,\delta(t)$ tells us that the distance to the equilibrium changes by a factor ρ at every time step. Thus, there is exponential growth away from ($\rho > 1$) or decay toward ($\rho < 1$) the equilibrium. This is not at all apparent when looking at the original equation (6.2b).

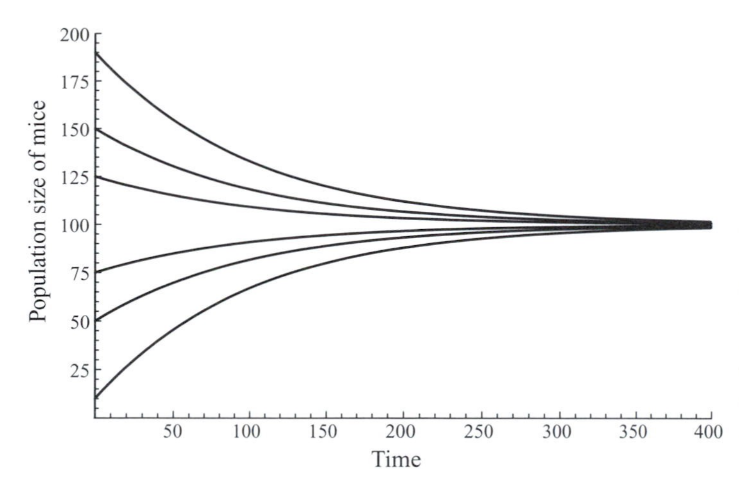

Figure 6.1: The general solution for the model of cats eating mice. The cat-mouse model given by equation (2.4) is equivalent to (6.2b), where $R = (1 + b)(1 - d)$. For $d = 0.1$, $b = 0.1$, and $m = 1$, the predicted equilibrium occurs at $\hat{n} = m/(1 - R) = 100$ and $R = 0.99$. According to the general solution (6.3b), the distance from the equilibrium declines by a factor, $R = 0.99$, at every time step regardless of the initial starting position.

All we have to do to apply Recipe 6.1 is to find $\hat{n}$ and ρ. The equilibrium condition for (6.2b) is $\hat{n} = R\hat{n} + m$, which has one equilibrium at $\hat{n} = m/(1 - R)$. The term ρ represents the factor multiplying the variable, which in this case equals R. Consequently, Recipe 6.1 indicates that the distance from the equilibrium has the general solution $\delta(t) = R^t\delta(0)$, which decays exponentially over time if $R < 1$ and grows exponentially over time if $R > 1$. Plugging $\hat{n}$ and ρ into Step 5 gives the general solution for equation (6.2b):

$$n(t) = R^t\, n(0) + \left(1 - R^t\right)\frac{m}{1 - R}. \tag{6.3b}$$

The general solution (6.3b) is certainly nicer than (6.3a), which involved a summation, but it is a bit disconcerting that the two solutions for the same model look so different. In fact they aren't different. Using Rule A1.19, the sum $\sum_{i=0}^{t-1} R^i$ equals $(1 - R^t)/(1 - R)$, which we can insert into (6.3a) and rearrange to get (6.3b). Thus, brute force iteration gives the same general solution as using a transformation. The advantage of using a transformation is that the steps are straightforward and the answer is simple without having to evaluate a sum. Even more importantly, the transformation gave us insight into how the model behaves—the system moves closer to or further from the equilibrium by the same factor R at each step (Figure 6.1).

6.4 Nonlinear Models in Discrete Time

Unfortunately, there is no general recipe to follow for solving nonlinear models in discrete time, even when there is only one variable. One look at the extraordinarily complex dynamics of the logistic model (Figure 4.2), and it is no surprise that there isn't a general solution. But even the diploid model of natural selection, which is pretty nicely behaved (Figure 4.7), does not have a general solution. If a model exhibits chaotic dynamics, then you can rest assured that there is no general solution. For most nonlinear models, however, it is hard to tell whether or not a general solution exists. Typically, we must try a bunch of different transformations to see if any of them dramatically simplify the recursion equation.

The haploid model of selection is a good example where the dynamics can be simplified by using a transformation. As described in Chapters 3 and 5, the recursion equation for this model is

$$p(t+1) = \frac{W_A\, p(t)}{W_A\, p(t) + W_a\, q(t)}, \tag{6.4}$$

and the equilibria are $\hat{p} = 0$ and 1. Equation (6.4) is a nonlinear function of the allele frequency $p(t)$, and it is not clear that it can be solved. First, we could try brute force iteration, using (6.4) to define $p(t + 2)$ as a function of $p(t + 1)$, plugging in (6.4) for $p(t + 1)$ as a function of $p(t)$, and simplifying to infer how each generation of selection affects the recursion. This method does generate a general solution, but it is cumbersome.

Let us try a transformation instead. You might first consider a transformation that measures the distance to an equilibrium. Well, $p(t)$ already measures the distance to the equilibrium of $\hat{p} = 0$. We could rewrite (6.4) in terms of the distance to the other equilibrium, $\hat{p} = 1$. This distance is measured by $q(t) = 1 - p(t)$, whose recursion equation is

$$q(t+1) = 1 - \frac{W_A\, p(t)}{W_A\, p(t) + W_a\, q(t)} = \frac{W_a\, q(t)}{W_A\, p(t) + W_a\, q(t)}. \tag{6.5}$$

But equation (6.5) is no simpler than (6.4).

What other transformations might you try? Let us step back a bit and think about the model. It is a symmetrical model in the sense that we could reverse which allele is called A and which allele is called a and get the same recursion. It is a good idea when seeking a transformation to preserve symmetry. That is, transformations like $x(t) = 2\,p(t) - 3\,q(t)$ or $x(t) = e^{p(t)} - q(t)$ are unlikely to help because they break the symmetry of the model by weighting the frequency of alleles A and a differently. One possible symmetric transformation is $d(t) = p(t) - q(t)$. Let us see if this transformation simplifies the model:

$$\begin{aligned} d(t+1) &= p(t+1) - q(t+1) \\ &= \frac{W_A\, p(t)}{W_A\, p(t) + W_a\, q(t)} - \frac{W_a\, q(t)}{W_A\, p(t) + W_a\, q(t)} \\ &= \frac{W_A\, p(t) - W_a\, q(t)}{W_A\, p(t) + W_a\, q(t)}. \end{aligned} \tag{6.6a}$$

This is a mixed-up recursion equation, with the variable d on the left and the variables p and q on the right. To fix this, we can plug $q(t) = 1 - p(t)$ into $d(t)$ to get $d(t) = 2p(t) - 1$, which implies that $p(t) = (d(t) - 1)/2$. Substituting $q(t) = 1 - p(t)$, then $p(t) = (d(t) - 1)/2$, and factoring, gives

$$d(t+1) = \frac{W_A - W_a + W_A\, d(t) + W_a\, d(t)}{W_A + W_a + W_A\, d(t) - W_a\, d(t)}. \tag{6.6b}$$

This is even worse than the original recursion equation.

As tempting as it is to give up, let us forge ahead and try one more symmetrical transformation $f(t) = p(t)/q(t)$:

$$\begin{aligned} f(t+1) &= \frac{p(t+1)}{q(t+1)} = \frac{W_A\, p(t)}{W_a\, q(t)} \\ &= \frac{W_A}{W_a} f(t). \end{aligned} \tag{6.7}$$

Wow! We now have a simple recursion that has the same form as the exponential growth model (6.2a), with R equal to W_A/W_a. Therefore, we can apply the general solution for the exponential model to this model: $f(t) = (W_A/W_a)^t f(0)$. Once we have the general solution in terms of the transformed variable, we can "back transform" to obtain the general solution in terms of the original variable. Because $q(t)$ equals $1 - p(t)$, $f(t) = p(t)/(1 - p(t))$, which can be rearranged to describe the old variable $p(t)$ in terms of the new variable $f(t)$ as $p(t) = f(t)/(f(t) + 1)$. Therefore,

$$\begin{aligned} p(t) &= \frac{f(t)}{1 + f(t)} = \frac{\left(\dfrac{W_A}{W_a}\right)^t \dfrac{p(0)}{q(0)}}{1 + \left(\dfrac{W_A}{W_a}\right)^t \dfrac{p(0)}{q(0)}} \\ &= \frac{W_A^t\, p(0)}{W_A^t\, p(0) + W_a^t\, q(0)}, \end{aligned} \tag{6.8}$$

where we have multiplied the top and the bottom by $W_a^t\, q(0)$ to get the last line. Equation (6.8) is the same general solution that we would have obtained by brute force iteration.

There is an additional advantage gained from having performed this transformation. Equation (6.7) provides insight into the nature of selection in the haploid model of selection. It tells us that selection alters the ratio of one allele frequency to the other by a constant factor equal to the ratio of their fitnesses. Consequently, the ratio of the allele frequencies undergoes exponential growth (when $W_A > W_a$) or decline (when $W_A < W_a$) over time.

The fact that equation (6.7) is so simple also allows us to broaden the scope of the results. For example, so far we have assumed that the fitnesses of the two alleles are constant over time, but what if they vary from generation to generation? Suppose that at time t the fitnesses of alleles A and a are $W_{A,t}$ and $W_{a,t}$, respectively. Equation (6.7) indicates that the ratio of the allele frequencies is multiplied by $W_{A,t}/W_{a,t}$ each generation. Thus, we can iterate (6.7) to obtain the general solution

$$f(t) = \left(\prod_{i=0}^{t-1} \frac{W_{A,i}}{W_{a,i}}\right) f(0) \tag{6.9a}$$

(the Π indicates a product, as described in Appendix 1). From Rule A1.26, we can rewrite the product in (6.9) in terms of the geometric mean of the relative fitnesses over the time span from 0 to $t-1$:

$$f(t) = \left(\text{Geometric mean}\left(\frac{W_A}{W_a}\right)\right)^t f(0)$$

or, equivalently,

$$f(t) = \left(\frac{\text{Geometric mean}(W_A)}{\text{Geometric mean}(W_a)}\right)^t f(0). \tag{6.9b}$$

The geometric mean of t numbers averages a set of numbers like the arithmetic mean, except that we multiply the numbers together and take the tth root of the result (geometric mean $\sqrt[t]{x_1\, x_2 \ldots x_t}$) rather than add the numbers together and divide the result by t (arithmetic mean $(x_1 + x_2 + \cdots + x_t)/t$). If the numbers are non-negative (as with fitnesses), the geometric mean is always less than or equal to the arithmetic mean and is much more sensitive to low values. For example, if fitness is zero in any one generation, then the geometric mean fitness is zero, even if the arithmetic mean fitness is high.

Equation (6.9b) tells us that allele A spreads over time and $f(t)$ increases if the geometric mean fitness of individuals carrying A is greater than the geometric mean fitness of individuals carrying a. Cannings (1971) demonstrated this result in more general terms and pointed out that an allele with less fitness variability over time tends to have a higher geometric mean fitness and is thus selectively favored over alleles with the same arithmetic mean fitness but greater temporal variability in fitness (Problem 6.1).

To find a general solution, it is a good idea to try multiple approaches: brute force iteration, transformations, and/or using a mathematical software package (e.g., using the command "RSolve" in *Mathematica*). The truth of the matter, however, is that most nonlinear recursion equations do not have a general solution. We turn next to differential equations in continuous time, which can be solved for a much broader array of models. One way to obtain a rough general solution for a discrete-time model is to approximate the recursion equation by a differential equation and then use the methods described next. This approximation works well as long as the changes that occur over a time step are small (Box 2.6).

6.5 Linear Models in Continuous Time

As with discrete-time equations, linear differential equations have two forms:

$$\frac{dn}{dt} = rn \tag{6.10a}$$

and an affine form:

$$\frac{dn}{dt} = rn + m. \tag{6.10b}$$

Equation (6.10a) implies that changes in the system arise from processes occurring to each individual independently, causing a constant rate of change per capita (due to births, deaths, emigration, etc.). In contrast, equation (6.10b) also allows an inflow or outflow of individuals (or whatever the variable describes) at a rate that does not depend on the value of the variable. For example, equation (6.10b) was used by Ho et al. (1995) to model the rate of HIV turnover within the bloodstream (see section 1.3.2). In their model, the number of viruses within the plasma, V, was the variable (in place of n), and viruses were cleared from the bloodstream at a per capita rate c (in place of r). Regardless of the current number of viruses within the plasma, however, infected CD4+ cells produced new viruses at a constant rate P (in place of m).

Both differential equations (6.10a) and (6.10b) can be solved using the method known as a *separation of variables* (Recipe 6.2).

Recipe 6.2
Solving Differential Equations Using a Separation of Variables

Differential equations that can be written as $dn/dt = f(n)\, g(t)$ can be solved as follows:

Step 1: Rewrite the differential equation as $(1/f(n))\,dn = g(t)\,dt$.

Step 2: Take the indefinite integral of both sides $\int (1/f(n))\,dn = \int g(t)\,dt$, integrating the left-hand side with respect to the dependent variable n and the right-hand side with respect to the independent variable t. Don't forget to add a constant of integration.

Step 3: Attempt to solve the resulting equation for n.

Step 4: Use an initial condition (e.g., at $t = 0$, there are $n(0)$ individuals) to determine the constant of integration.

A *separation of variables* is a technique for solving differential equations that are some function of the dependent variable multiplied by some other function of the independent variable.

Let us first solve the exponential-growth model (6.10a) using a separation of variables. We begin by writing the differential equation in the form $dn/dt = f(n)\, g(t)$. We could choose $f(n) = r\,n$ and $g(t) = 1$ or we could choose $f(n) = n$ and $g(t) = r$; either way we would get the same answer. We arbitrarily make the first choice. The two integrals we have to evaluate are then $\int (1/f(n))\, dn = \int (1/(r\,n))\,dn = (1/r)\ln(n) + c_1$ and $\int g(t)\, dt = \int 1\, dt = t + c_2$. Setting these equal to one another and merging together the constants of integration by defining $c_2 - c_1 = c$, we get $(1/r)\ln(n) = t + c$.

Now we proceed to solve for n by multiplying both sides by r and taking the exponential of both sides, getting $n = e^{r(t+c)} = e^{rt} e^{rc}$. Setting t to zero, indicates that $n(0) = e^{rc}$. Replacing e^{rc} with $n(0)$, we obtain the general solution for the continuous-time model of exponential growth:

$$n = e^{rt}\, n(0). \tag{6.11}$$

This is the same solution as equation (4.2), which we obtained by educated guesswork.

You might feel uncomfortable (reasonably enough) breaking apart a differential equation in Step 1 of Recipe 6.2 as if dn/dt represented a regular fraction. The best way to ensure that a mathematical shortcut works is to check that the solution satisfies the original equation. In this case, we can take the derivative of the solution $n = e^{rt}\, n(0)$ with respect to t to get $dn/dt = d(e^{rt}\, n(0))/dt = r\, e^{rt} n(0)$.

According to our general solution, $e^{rt}n(0)$ is equivalent to n. Making this replacement correctly gives the differential equation that we were trying to solve: $dn/dt = rn$. Thus, a separation of variables was successful in producing a general solution (6.11) that satisfies the differential equation (6.10a). The general solution to the exponential-growth model is widely used in all aspects of life, including in the calculation of interest by banks (Box 6.1).

It is worth comparing the general solutions for the continuous-time model, $n = e^{rt}n(0)$ and the discrete-time model, $n = R^t n(0)$, of exponential growth. If R were equal to e^r, the two solutions would be identical and would predict the same trajectory for the population size. But if we defined r as the per capita per generation change in population size, then we would set $R = 1 + r$ in the discrete-time model (see Chapter 3). According to the Taylor series (P1.14), $e^r = 1 + r + \frac{1}{2}r^2 + \cdots$ is always greater than $(1 + r)$ (Figure 6.2a). Thus, for the same per capita per generation change in population size, populations grow faster in continuous time than in discrete time (Figure 6.2b). Intuitively, this is because every offspring in continuous time can immediately reproduce and add to future growth, whereas offspring must wait until the next time step to reproduce in the discrete-time model.

Equation (6.10b) describing exponential growth with a constant inflow or outflow can be solved directly using the method of separation of variables. We leave this task to Problem 6.4, as it represents the next most complicated problem for you to tackle. Alternatively, we can transform (6.10b) into the form of (6.10a) to obtain a linear differential equation without a constant term, whose solution we have already calculated:

Recipe 6.3

Solving a Linear Continuous-Time Model with a Constant Term

Linear differential equations of the form $dn/dt = \rho\, n + c$ can be solved as follows:

Step 1: Solve for the equilibrium $\hat{n}$. Here, $\hat{n} = -c/\rho$.

Step 2: Define a new variable δ as the distance of the system from the equilibrium, $\delta = n - \hat{n}$. Reversing this equation implies that $n = \delta + \hat{n}$.

Step 3: The differential equation for δ is the same as the differential equation for n, because $dn/dt = d(\delta + \hat{n})/dt = d\delta/dt = \rho n + c$ given that $\hat{n}$ is a constant. Replacing n with $\delta + \hat{n}$ and factoring leaves a differential equation of the form $d\delta/dt = \rho\,\delta$, which does not involve a constant term.

Step 4: From equation (6.11), the general solution for the distance to the equilibrium is $\delta(t) = e^{\rho t}\delta(0)$.

Step 5: The general solution for the original variable is found by replacing δ with $n - \hat{n}$, and simplifying to get $n(t) = e^{\rho t}n(0) + (1 - e^{\rho t})\hat{n}$.

Next, we turn to nonlinear models in continuous time.

Box 6.1: Getting the Most for Your Money

The exponential growth model can help you make wiser financial decisions. If, for example, you inherit \$10,000 and decide to invest it in a five-year fixed-term account, you might be offered 4% interest per year from your bank. But how is this interest calculated? At one of our banks (a major Canadian bank), this interest is calculated on an annual basis, meaning that you would have \$10,000 until the very end of the year, at which point \$400 ($0.04 \times$ \$10,000) would be added to the account. Other banks (and even other types of accounts at the same bank) calculate interest on a different schedule. If you found a bank that offers a 4% annual interest rate compounded monthly, then your \$10,000 would grow at the end of the first month by \$33.33 (($0.04/12$) $\times$ \$10,000).

How much would your money be worth at the end of a five-year fixed term at these two banks? To answer this question, we can use the general solution (6.1), $n(t) = R^t\, n(0)$. Here, the variable $n(t)$ represents the amount of money in your account, the time step represents the period over which interest is calculated (one year for the first bank, one month for the second bank), and R is the factor by which money grows over the time step ($R_{\text{annual}} = 1 + 0.04$ for the first bank, $R_{\text{monthly}} = 1 + (0.04/12)$ for the second bank). After the five-year term, the amount of money you would have in each bank equals

$$(1 + 0.04)^5 \times \$10{,}000 = \$12{,}166.53 \qquad \text{(4\% annual interest rate compounded yearly)},$$

$$\left(1 + \frac{0.04}{12}\right)^{12\times 5} \times \$10{,}000 = \$12{,}209.97 \qquad \text{(4\% annual interest rate compounded monthly)}.$$

Thus, you would earn \$43.44 more over five years from the second bank. You could earn an extra \$3.92 if you found a bank that compounded interest daily:

$$\left(1 + \frac{0.04}{365}\right)^{365\times 5} \times \$10{,}000 = \$12{,}213.89 \qquad \text{(4\% annual interest rate compounded daily)}.$$

As you can see, the more often the interest is compounded, the more the account grows over time. The best that you could do, given a 4% annual interest rate, would be to find a bank that compounds interest continuously. To calculate how much more you would earn, we must use the general solution (6.11) to the continuous-time exponential model, $n = e^{r\,t}\, n_0$. Measuring time in years, r represents the annual growth rate (4%), and after five years the amount of money in your account would be

$$e^{0.04\times 5} \times \$10{,}000 = \$12{,}214.03 \qquad \text{(4\% annual interest rate compounded continuously)}.$$

You would gain an additional 14 cents if your savings were compounded continuously rather than daily.

The continuous-time formula can also be arrived at directly from the discrete-time compounding formula. In particular, if the interest was compounded a total of k times during the year, then after 5 years your account would contain

$$\left(1 + \frac{0.04}{k}\right)^{k\times 5} \times \$10{,}000 \qquad \text{(4\% annual interest rate compounded } k \text{ times/year)}.$$

(continued)

Box 6.1 *(continued)*

As k gets larger and larger (i.e., as the compounding becomes more and more frequent), this quantity gets closer and closer to $e^{0.04\times 5} \times \$10{,}000$. Indeed, in the limit as k goes to infinity, $\lim_{k\to\infty}(1 + x/k)^k = e^x$ (see Box 7.4).

These calculations demonstrate that, if all else is equal, you should choose a bank that compounds your savings as often as possible. The converse argument holds for loans, however. You should choose a bank that compounds your loans as infrequently as possible. The same Canadian bank that calculates interest on an annual basis for investments calculates interest for outstanding Visa bills on a daily basis! By compounding investment accounts on an annual basis, but calculating loans on a daily basis, the bank earns a substantial amount of money.

Often, banks offer different interests rates. These are difficult to compare when the interest rates are compounded differently. The general solutions to the exponential growth model can help choose the bank whose rate is truly in your best interest. For example, if one bank offers a mortgage loan with an interest rate of 4.75% compounded monthly, would this be better or worse than a mortgage loan with an interest rate of 4.8% compounded annually? By compounding monthly, the total amount by which the 4.75% loan would grow over the course of the year would be $R_{\text{annual}} = (1 + 0.0475/12)^{12} = 1.04855$, i.e., a 4.855% increase, which means that you would owe more money over the course of a year than had you taken out the loan at 4.8% compounded annually. The difference seems pretty trivial, but home mortgages typically involve a lot of money paid back over a long period of time. To get the best interest rate possible, ask your banker to translate their offered rate of interest to the "effective interest rate" (EIR), giving the total factor by which your account or mortgage would grow over a year if you did not withdraw or deposit funds.

6.6 Nonlinear Models in Continuous Time

Most differential equations arising in biology are nonlinear functions of the variable of interest. Fortunately, these can still be solved using a separation of variables, as long as they can be written in the form $\mathrm{d}n/\mathrm{d}t = f(n)g(t)$. Many models in biology have this form because changes to the system are assumed to depend only on the current composition of the system and not on the exact time, so that $g(t) = 1$ and $\mathrm{d}n/\mathrm{d}t = f(n)$. For such models, Recipe 6.2 can be applied.

We have already seen two examples of differential equations of the appropriate form, the logistic model

$$\frac{\mathrm{d}n}{\mathrm{d}t} = r\,n\left(1 - \frac{n}{K}\right) \tag{6.12}$$

and the haploid model of selection

$$\frac{\mathrm{d}p}{\mathrm{d}t} = s\,p\,(1 - p). \tag{6.13}$$

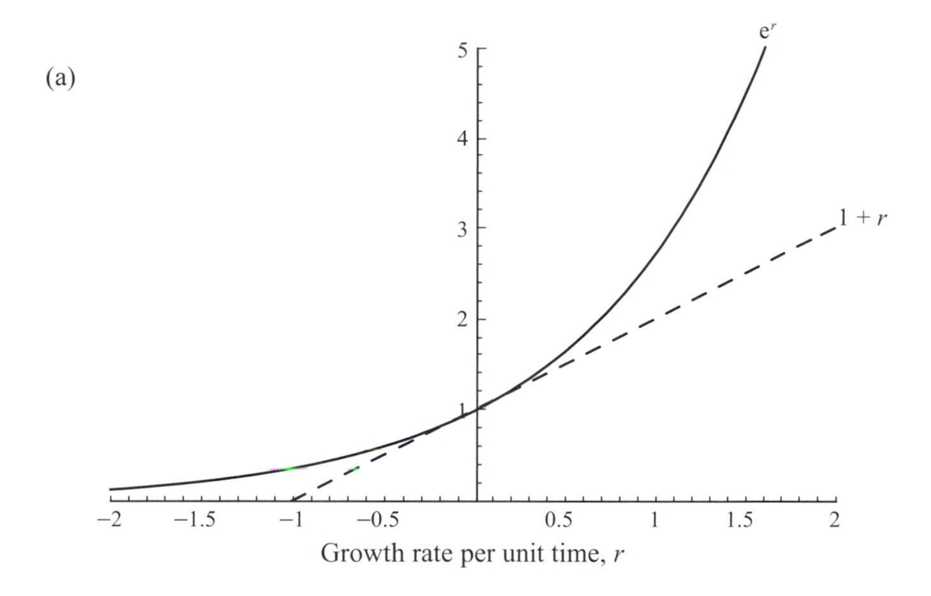

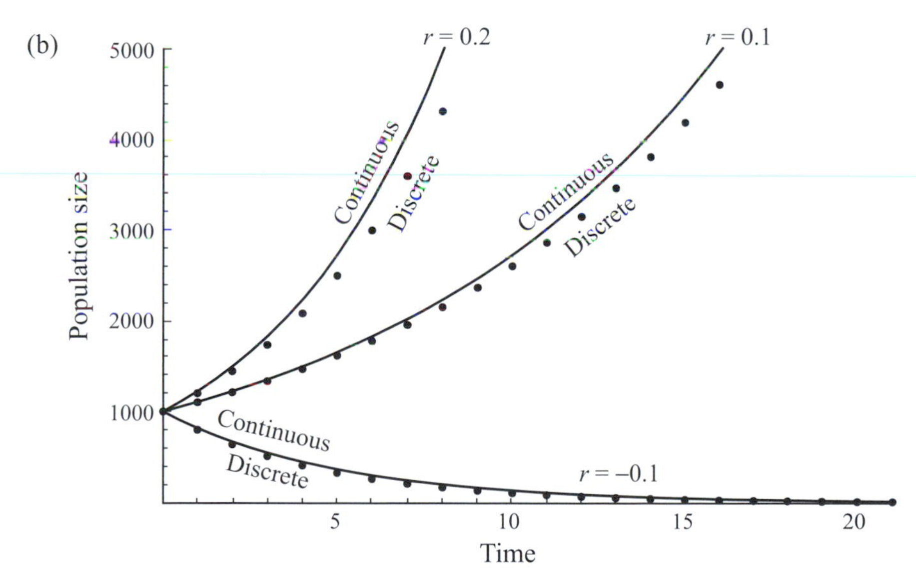

Figure 6.2: A comparison of exponential growth in discrete and continuous time. (a) e^r is larger than $1 + r$ for any value of the growth rate, r. (b) Consequently, the population size is predicted to be larger in the continuous-time model of exponential growth, $n = e^{rt}\, n(0)$ (solid), than in the discrete-time model, $n = (1 + r)^t\, n(0)$ (dots), given the same value of r.

These models seem different, but in fact they are closely related. Take a close look at equations (6.12) and (6.13). The model of selection on haploids represents a special case of logistic growth with $n = p$, $K = 1$, and $r = s$. This realization makes our life easier because if we solve (6.12) then we have solved (6.13) as well.

So let us apply Recipe 6.2 to the logistic model. First, we define $f(n) = rn(1 - n/K)$ and $g(t) = 1$. The integrals we have to evaluate are then

$$\int \frac{1}{f(n)}\,\mathrm{d}n = \int \frac{1}{rn\left(1 - \frac{n}{K}\right)}\,\mathrm{d}n$$

and $\int g(t)\mathrm{d}t = \int 1\ \mathrm{d}t = t + c_3$. The first integral is a bit complicated, but it can be solved in a number of ways, including consulting tables of integrals (e.g., Appendix 1; Rule A2.22) or using software packages like Maple or *Mathematica*.

We will solve this integral, however, using partial fractions. Whenever you have a fraction like

$$\frac{1}{rn\left(1 - \frac{n}{K}\right)},$$

involving the product of two linear functions in the denominator (here, $r\,n$ and $1 - n/K$), we can always break the fraction into two pieces using Rule A1.9. In this example,

$$\frac{1}{rn\left(1 - \frac{n}{K}\right)} = \frac{1}{rn} + \frac{\frac{1}{rK}}{\left(1 - \frac{n}{K}\right)}$$

These two terms can be integrated separately. We've already calculated

$$\int \frac{1}{rn}\,\mathrm{d}n = \frac{1}{r}\ln(n) + c_1$$

for the exponential-growth model. The second term integrates to

$$\frac{1}{rK}\int \frac{1}{\left(1 - \frac{n}{K}\right)}\,\mathrm{d}n = -\frac{1}{r}\ln\left(1 - \frac{n}{K}\right) + c_2.$$

Adding these terms together, setting them equal to $\int g(t)\ \mathrm{d}t = \int 1\ \mathrm{d}t = t + c_3$, and merging the constants of integration leaves us with

$$\frac{1}{r}\ln(n) - \frac{1}{r}\ln\left(1 - \frac{n}{K}\right) = t + c.$$

We next proceed to Step 3: solving this equation for the population size. Multiplying both sides by r and using Rule A1.14 to gather together the logarithmic terms leaves us with $\ln(n/(1 - n/K)) = rt + rc$. Taking the exponential of both sides gives $n/(1 - n/K) = e^{rt + rc} = e^{rt} e^{rc}$. This result is the same as the exponential-growth model except for the term in the denominator on the left, $1 - n/K$. This term measures the proportional distance to the carrying capacity (for example, if $n = K/3$, then the population is 2/3 away from the carrying capacity). Thus, this result suggests that the population size grows exponentially when measured relative to the distance of the population from the carrying capacity. Solving this equation for n (Recipe 5.2) gives us the solution to the continuous-time logistic model:

$$n(t) = \frac{e^{rt}e^{rc}}{1 + e^{rt}e^{rc}/K} \tag{6.14a}$$

To solve for c, we could use either equation (6.14a) or the previous equation $n/(1 - n/K) = e^{rt}e^{rc}$, but the latter is easier because c appears only once. Setting $t = 0$, we find that $n_0/(1 - n_0/K) = e^{rc}$. Making this replacement for e^{rc} in equation (6.14a), we get the general solution for the population size:

$$n(t) = \frac{e^{rt}\left(\dfrac{n_0}{1 - n_0/K}\right)}{1 + \dfrac{e^{rt}}{K}\left(\dfrac{n_0}{1 - n_0/K}\right)} \tag{6.14b}$$

$$= \frac{e^{rt}n_0}{1 - \dfrac{n_0}{K} + \dfrac{e^{rt}n_0}{K}}.$$

To simplify the fractions, we have multiplied the top and bottom by $1 - n_0/K$ in the last line.

Again, it is a good idea to check the general solution for n by taking its derivative and making sure that we can rewrite it as (6.12). We can also compare equation (6.14b) to numerical solutions of the differential equation (see the online Lab exercise on solving differential equations; Figure 6.3). Both checks confirm that (6.14b) satisfies the differential equation for the logistic model.

Now let us take advantage of the fact that the haploid model of selection is a special case of the logistic model with $n = p$, $K = 1$, and $r = s$ to solve this model too:

$$p(t) = \frac{e^{st}p_0}{1 - p_0 + e^{st}p_0}. \tag{6.15a}$$

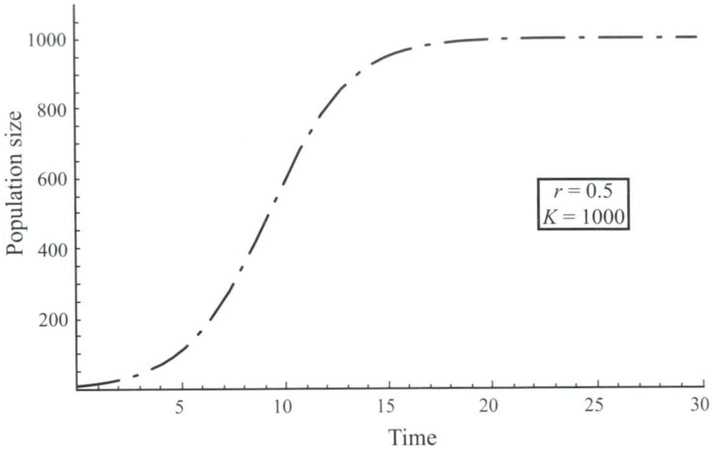

Figure 6.3: The general solution and a numerical solution of the logistic model in continuous-time. NDSolve was used in *Mathematica* to provide a numerical solution to the differential equation (6.12) (long-dashed curve). This is compared to the analytical solution (6.14b) (short-dashed curve). The two curves match precisely.

Recalling that the selection coefficient s describes the difference in growth rates between the two alleles, $r_A - r_a$ (Box 3.1), equation (6.15a) can be written as

$$p(t) = \frac{e^{(r_A - r_a)t}\, p_0}{(1 - p_0) + e^{(r_A - r_a)t}\, p_0}. \tag{6.15b}$$

Multiplying the top and bottom by $e^{r_a t}$,

$$p(t) = \frac{e^{r_A t}\, p_0}{e^{r_a t}(1 - p_0) + e^{r_A t}\, p_0}. \tag{6.15c}$$

Equation (6.15c) is analogous to the discrete-time model (6.8), but with e_j^r replacing W_j.

At this point, it might seem puzzling that the haploid model of selection is a special case of the logistic model in continuous time but that the haploid model and logistic model behave so differently in discrete time. Unlike the continuous-time model, there is no choice of parameters that converts the logistic model in discrete time into the haploid model of selection (compare equations (3.5b) and (3.9)). Furthermore, while there is a general solution (6.8) describing the haploid model at any future point in discrete time, there is no general solution for the logistic model in discrete time. The oscillations and chaos exhibited by the logistic model cannot be described by any simple function of time. The underlying reason why the haploid model and logistic model behave so differently in discrete time is that their nonlinearity arises in different ways. In the haploid model of selection, the nonlinearity arises because the allele frequencies are normalized to sum to one. Because each allele frequency is positive, the normalizing factor (the mean fitness) is positive, and the allele frequencies remain positive after normalization (see equation (6.4)). Because no allele frequency can ever become negative, the allele frequencies cannot overshoot either equilibrium, $\hat{p} = 0$ or $\hat{p} = 1$. In contrast, the logistic recursion equation is nonlinear because the number of surviving offspring per parent is assumed to depend on the population size. There is no normalization procedure in the logistic model, and the population size is free to rise above and fall below the carrying capacity. It is because the equilibrium can be overshot in the logistic model that chaos is possible.

We have described how a separation of variables can be used to solve differential equations of the form $dn/dt = f(n)g(t)$. Even for such differential equations, however, a separation of variables is not guaranteed to yield a solution. The integrals in Step 2 of Recipe (6.2) might be impossible to evaluate. Even when the integrations can be performed, they can yield an equation for the variable n that cannot be explicitly solved (e.g., in the diploid model of natural selection; see Problem 6.7). On the other hand, just because a differential equation cannot be written in the form $dn/dt = f(n)g(t)$ does not mean that it cannot be solved. Indeed, entire books are devoted to methods for solving various types of differential equations (see further reading). In Box 6.2, we provide the solutions to other forms of differential equations that are commonly encountered in biological models (see Problems 6.9–6.12).

6.7 Concluding Message

In this chapter we have discussed the utility of transformations as well as how to obtain general solutions to some models with one variable. A general solution predicts the state of a system at any future point in time, as a function of the initial state of the system, the parameters, and the amount of time that has passed. Thus, a general solution describes the behavior of a model in one formula and can be used to answer any question about the model. For example, the long-term behavior of a system can be determined by allowing time to increase to infinity in the general solution. Also, the form of the general solution often provides insight into the fundamental manner by which a biological system changes. In the haploid model of selection, for instance, we learned that each generation of selection alters the ratio of allele frequencies by a factor equal to the ratio of their fitnesses.

For models involving a single variable, we have discussed several methods for obtaining a general solution. For a linear model, there is always a general solution, whether the model involves a constant input or output term (an *affine* model) or not (Recipe 6.1 and 6.3; equations (6.1) and (6.11)). The picture is not so rosy for nonlinear models. Except for a small subset of models in continuous time (see, for example, Box 6.2), there is no recipe to follow that is guaranteed to yield a general solution. Various techniques, from brute force iteration to transformations, must be tried in the hopes of hitting upon a general solution. As a rough guiding principle, general solutions are most likely to exist for models with only one stable equilibrium, but even such models cannot always be solved (e.g., the logistic model in discrete time). When a general solution is elusive, plotting numerical solutions (Chapter 4) alongside a stability analysis (Chapter 5) can be used to obtain as comprehensive a picture as possible.

Box 6.2: Some Additional Methods for Solving Differential Equations

Many methods exist to solve differential equations (see further reading). In this box, we summarize some useful techniques that help solve differential equations that are commonly encountered in biological models.

Linear differential equations have the form

$$\frac{dn}{dt} = f(t)n + g(t). \tag{6.2.1}$$

The solution to a linear differential equation is

$$n = e^{\mu(t)}\left(\int e^{-\mu(t)}g(t)\,dt + c\right), \tag{6.2.2}$$

Where $\mu(t) = \int f(t)dt$. This equation and its solution apply, for example, to a population that experiences immigration or harvesting at a variable rate over time, $g(t)$ (independent of the current population size) and that undergoes exponential growth or decline at a variable rate $f(t)$ (see Problems 6.9 and 6.13).

Homogeneous differential equations have the form

$$\frac{dn}{dt} = F\left(\frac{n}{t}\right), \tag{6.2.3}$$

where F is any function that depends only on the ratio n/t and not on n or t independently. Homogeneous equations can be solved by defining a new variable $v = n/t$. The differential equation for v will then be

$$\frac{dv}{dt} = \frac{d\left(\frac{n}{t}\right)}{dt} = \frac{1}{t}\frac{dn}{dt} - \frac{n}{t^2}.$$

Plugging in the equation for dn/dt and replacing n with $v\,t$ allows us to simplify this equation to

$$\frac{dv}{dt} = \frac{F(v) - v}{t},$$

which is always separable and can be solved using Recipe 6.2, even if the original differential equation for n was not separable.

Bernoulli differential equations have the form

$$\frac{dn}{dt} = n\,f(t) + n^a g(t). \tag{6.2.4}$$

(continued)

Box 6.2 *(continued)*

Bernoulli equations can be solved by defining a new variable $v = n^{1-a}$. The differential equation for v will then be

$$\frac{dv}{dt} = \frac{d(n^{1-a})}{dt} = (1 - a)n^{-a}\frac{dn}{dt}.$$

Plugging in dn/dt for the Bernoulli equation gives

$$\frac{dv}{dt} = (1 - a)(n^{1-a} f(t) + g(t)).$$

By definition, n^{1-a} is v, so a Bernoulli equation can be rewritten as

$$\frac{dv}{dt} = (1 - a)(v f(t) + g(t)). \tag{6.2.5}$$

This is a linear differential equation and has the solution (6.2.2) once f and g are redefined to include the constant term $(1 - a)$ in (6.2.5). The logistic equation (6.12) is a special case of (6.2.4) with $f(t) = r$, $a = 2$, and $g(t) = -r/K$ (Problem 6.10). The power of this method is that it allows us to obtain solutions to a generalized logistic equation in which the intrinsic growth rate r and/or the carrying capacity K vary over time (Problem 6.12).

Problems

Problem 6.1: Consider the haploid model of selection with two alleles A and a, where the fitness of allele A relative to that of allele a alternates from generation to generation between $W_A = 3/2$ and $W_A = 1/2$. (a) Calculate the arithmetic average fitness of the two alleles over a time period from 0 to $t - 1$, which may be even or odd (Rule A1.21). (b) Calculate the geometric average fitness of the two alleles over the same time span (Rule A1.26). (c) Over the long term, which allele will spread?

Problem 6.2: In the presence of mutations, allele frequencies change over time even when selection is absent. Assume that a fraction μ of A alleles is converted to a each generation, while a fraction ν of a alleles is converted to A each generation. The frequency of A is then described by the recursion equation $p(t + 1) = (1 - \mu)$ $p(t) + \nu q(t)$. (a) Determine the equilibrium for this model of mutation (don't forget to rewrite q in terms of p). (b) Determine the general solution for this model using recipe 6.1. (c) How rapidly does the allele frequency approach the equilibrium, assuming that the mutation rates are very low?

Problem 6.3: Here, we expand the exponential growth model to consider the case where the per capita number of surviving offspring, R, declines as an inverse function of the current population size $R = R_1 n(t)^{-b}$, where R_1 represents the number of surviving offspring when there is only one individual in the population. (a) By brute force iteration, determine the general solution for the resulting recursion equation $n(t + 1) = R_1\, n(t)^{1-b}$. Simplify your result using Rules A1.1 and A1.19. (b) Check the result from (a) when $b = 1$, by noting that $n(t + 1)$ then equals R_1 regardless of the initial population size. (c) Assuming that $0 < b < 1$, let t go to infinity in your general solution and determine the population size toward which the system will head. (You can check this result by finding the equilibrium for the recursion equation.)

Problem 6.4: Messenger RNA levels within the cell reflect the production of new mRNA molecules by transcription of genes and the decay of existing mRNA transcripts. Let m represent the number of new transcripts produced per second and r represent the rate of decay of existing transcripts. The change in the number of mRNA molecules within the cell can then be described by the differential equation (6.10b):

$$\frac{dn}{dt} = -rn + m.$$

(a) Solve for the equilibrium of this differential equation, $\hat{n}$. (b) By blocking the production of new transcripts ($m = 0$), Iyer and Struhl (1997) estimated the half-life of mRNA transcripts for the histidine gene of yeast, *his3*, to be 660 seconds. Use the general solution for the exponential decay model $n(t) = e^{-rt}\, n(0)$ to estimate r given $m = 0$ and $n = n_0/2$ after 660 seconds. (c) Iyer and Struhl (1997) also determined that the normal number of mRNA transcripts within yeast cells was about 7. Assuming that the normal transcript levels are at equilibrium, use your answers to (a) and (b) to estimate the rate of transcriptional initiation m for *his3*. (d) Solve the differential equation by the method of a separation of variables. Use the initial condition, $n = n_0$ at $t = 0$, to replace the constant of integration. [Show your work.] (e) If the mRNA became degraded by heat such that $n_0 = 0$, how long would it take for the cell to regain approximately half of the normal level of mRNA molecules ($n = \hat{n}/2$)? Does this time depend on the rate of mRNA decay r or the rate of transcription initiation m? (f) Using your estimates for r and m from (b) and (c), plot the general solution from (d) giving the expected number of mRNA molecules as a function of time (using *Mathematica* or any other method). Check that your answer to (e) is consistent with this plot. [Note that these methods were the very ones used by Iyer and Struhl (1997) to characterize mRNA decay rates and rates of transcriptional initiation.]

Problem 6.5: Habitat degradation can cause the growth rate of a population to decline over time. This can be modeled by modifying the exponential growth model such that r becomes $r_0 - \delta t$, where δ represents the rate of habitat destruction. The size of a population then follows the differential equation

$$\frac{dn}{dt} = (r_0 - \delta t)\, n.$$

(a) Using the method of a separation of variables, solve for the population size at time t given that the population was at size n_0 at time $t = 0$. (b) Using your answer to (a), determine the predicted extinction time as the time it would take for the population to decline to a single individual ($n_0 = 1$). [The quadratic formula will give you two solutions for t. Choose the appropriate root and explain your choice.]

Problem 6.6: By differentiating $p(t)$ with respect to time, check that the solution to the model of haploid selection, (6.15a), satisfies the differential equation (6.13).

Problem 6.7: Consider the diploid model of selection in continuous time. As shown in Problem 3.16, the differential equation for the A allele frequency can be written as

$$\frac{dp}{dt} = s\,p(1-p)(p + h(1-2p)),$$

where $W_{AA} = 1 + s$, $W_{Aa} = 1 + h\,s$, and $W_{aa} = 1$.

(a) Try solving this differential equation using DSolve in *Mathematica*. (b) The answer to (a) is not pretty and involves an InverseFunction, which is *Mathematica*'s way of saying that it cannot find an explicit solution for p even though it knows a function that p must satisfy. Find this function by performing a separation of variables (use *Mathematica* for help with the integral). (c) Show that this function cannot be simplified even if allele A is recessive ($h = 0$) or dominant ($h = 1$) but that it can be simplified when selection is additive ($h = 1/2$). (d) By comparing the above differential equation to (6.13), infer the general solution for the continuous-time model of diploid selection when selection is additive ($h = 1/2$).

Problem 6.8: Within a population, say of university students, one can model the spread of infectious diseases, like colds, using the flow diagram in Figure 6.4.

The parameter c represents the *per capita* rate at which an infected individual contacts a susceptible individual, a is the probability of transmission of the disease per contact, and σ is the rate at which individuals recover from the disease. In this model, the population is assumed constant over the time frame of interest (e.g., we can treat the number of students as roughly constant over a school year), and individuals who recover from the disease are assumed to be susceptible again. This assumption is reasonable for colds, which are caused by a large number of different viruses. The following differential equations can be used to track the number of susceptible individuals S and infected individuals I over time:

$$\frac{dS}{dt} = -a\,c\,S\,I + \sigma I, \tag{Q6.1a}$$

$$\frac{dI}{dt} = a\,c\,S\,I - \sigma I. \tag{Q6.1b}$$

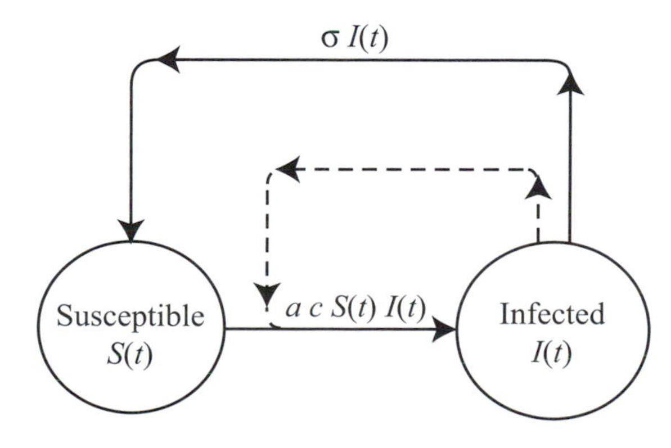

Figure 6.4: Flow diagram for a susceptible-infected model.

(a) Use the quotient rule (see Box 3.1) and the above equations to prove that the proportion of infected individuals, $P = I/(S + I)$, satisfies the differential equation

$$\frac{dP}{dt} = \alpha P(1 - P) - \sigma P \tag{Q6.2}$$

where $\alpha = ac\,(S + I)$ represents the infectivity of the disease, a, times the total rate of contact among individuals in the population. Because we assume that the total population size $S + I$ remains constant, the differential equation (Q6.2) depends only on the fraction of infected individuals, P. (b) Determine the equilibria for (Q6.2) and determine when each equilibrium is valid given that P represents a proportion. (c) Determine the local stability condition for each equilibrium. Describe in words what these conditions imply. (d) Assuming that the force of infection α is greater than the recovery rate σ and using the information obtained from parts (b) and (c), sketch the shape that the differential equation must have in a plot of dP/dt (vertical axis) versus P (horizontal axis). Use this sketch to determine which equilibrium is globally stable over the range $0 \leq P \leq 1$. (e) Determine the general solution for (Q6.2). [There are multiple ways of doing this, including using a separation of variables. The simplest method is to show how (Q6.2) and the logistic equation (6.12) are related.]

Problem 6.9: Sink populations are defined as populations that are maintained by migration from another source population. Here we consider a sink population with a negative intrinsic growth rate that receives immigrants at rate m per year, which is decreasing over time due to habitat deterioration in migration corridors, $m = m_0 - \delta t$. A differential equation describing this situation is

$$\frac{dn}{dt} = r\,n + (m_0 - \delta\,t).$$

(a) Use the solution for a linear model from Box 6.2 to obtain the general solution for this model. (b) Plot the result using the parameters $r = -0.01$, $m_0 = 0.5$, $\delta = 0.02$, and initial population size $n_0 = 40$. (c) Explain why the curve has the shape that it does and specify when you expect the sink population to go extinct.

Problem 6.10: Solve the logistic model in continuous time using the recipe for solving a Bernoulli differential equation, $dn/dt = n\,f(t) + n^a\,g(t)$ where $f(t) = r$, $a = 2$, and $g(t) = -r/K$ (see Box 6.2). Check your answer against the solution (6.14b).

Problem 6.11: Solve the haploid model of selection in continuous time, $dp/dt = s\,p\,(1 - p)$, under the assumption that the selection coefficient varies sinusoidally with time, $s = s_0 + \sigma \sin(\theta t)$, as might be the case in a seasonal environment. Use the fact that the model is a Bernoulli differential equation with $a = 2$ (Box 6.2) and the fact that $\int e^{f(t)}\,(df(t)/dt)\,dt = e^{f(t)} + c$ (see Rule A2.6). Use the initial allele frequency p_0 to solve for the constant of integration, c. Check that your answer is consistent with (6.15a) when $\sigma = 0$. If $s_0 > 0$, what happens as time goes to infinity?

Problem 6.12: By decreasing the density of resources available, habitat degradation could act to reduce the reproductive potential of a species or to increase the amount of territory needed to sustain each individual. Here we alter the logistic model by allowing r or K to decrease over time. (a) Solve the logistic equation in continuous time when $r(t) = r_0(1 - \delta t)$ but $K(t) = K_0$ is constant. (b) Solve the logistic equation in continuous time when $K(t) = K_0/(1 + \delta t)$ but $r(t) = r_0$ is constant. (c) Plot and compare your solutions starting from $n_0 = 500$ individuals with $r_0 = 1$, $K_0 = 1000$, and $\delta = 0.1$. Use the fact that the logistic equation with time varying parameters, written as $dn/dt = n\, r(t) - n^2\, r(t)/K(t)$, corresponds to a Bernoulli differential equation (Box 6.2). [Hint: Rules A2.23 and A2.29 can help with the integrals. Do not forget to include the constant of integration.]

Problem 6.13: The cat-mouse model discussed in Chapter 2 is described by the differential equation $dn/dt = b\, n - d\, n + m$. (a) Solve this differential equation using a separation of variables, assuming that the initial number of mice is n_0. (b) Solve using the solution to a linear differential equation (6.2.2). (c) Check that these two solutions are consistent with each other.

Problem 6.14: Here we generalize the exponential growth model (6.1) to allow the environment to vary over time, causing the number of surviving individuals per parent, R_t, to depend on time. (a) By brute force iteration, solve the model of exponential growth $n(t + 1) = R_t\, n(t)$. (b) Rewrite your solution to (a) in terms of the geometric mean value of R_t over the time span from 0 to $t-1$. (c) Based on these calculations, what would you expect to happen over the long term if the environment fluctuated such that $R_t = 1/3$ in every odd time step and $R_t = 2$ in every even time step?

Further Reading

For further information on solving differential equations, consult

- Arnold, V.I., and R. Cooke. 1994. *Ordinary Differential Equations and Their Applications*, 3rd ed. Springer-Verlag, New York.
- Boyce, W.E., and R. C. Di Prima. 2004. *Elementary Differential Equations and Boundary Value Problems*, 8th ed. Wiley, New York.
- Braun, M. 1983, *Differential Equations and Their Applications*, 3rd ed. Springer-Verlag, New York.
- Bronson. R. 1994. *Schaum's Outline of Differential Equations*, 2nd ed. McGraw-Hill Trade, New York.
- Polking, J. et al. 2002. *Differential Equations with Boundary Value Problems*, 1st edition, Prentice-Hall, Englewood Cliffs, N.J.

References

Cannings, C. 1971. Natural selection at a multiallelic autosomal locus with multiple niches. *J. Genet.* 60:255–259.

Ho, D. D., A. U. Neumann, A. S. Perelson, W. Chen, J. M. Leonard, and M. Markowitz. 1995. Rapid turnover of plasma virions and CD4 lymphocytes in HIV-1 infection. *Nature* 373:123–126.

Iyer, V., and K. Struhl. 1996. Absolute mRNA levels and transcriptional initiation rates in *Saccharomyces cerevisiae*. *Proc. Natl. Acad. Sci. U.S.A.* 93:5208–5212.

PRIMER 2
Linear Algebra

The mathematical field of linear algebra provides many useful techniques for analyzing dynamical models. Linear algebra can be thought of as a series of bookkeeping techniques for linear equations. An advantage of these techniques is that complicated equations involving more than one variable can be written in a pleasantly compact form. An even more important advantage is that we can use theorems from linear algebra to prove certain facts about the behavior of our models.

P2.1 An Introduction to Vectors and Matrices

Linear algebra describes mathematical operations on *lists of information.* Each element of a list may be a number, a parameter, a function, or a variable. A *vector* represents a list of elements arranged in one of two ways. A "column vector" arranges elements from top to bottom:

A ***vector*** is a list of elements.

$$\begin{pmatrix} 12 \\ 19 \end{pmatrix} \quad \begin{pmatrix} 1 \\ 5 \\ 9 \\ 7 \end{pmatrix} \quad \begin{pmatrix} x \\ y \end{pmatrix} \quad \begin{pmatrix} x \\ y \\ z \end{pmatrix} \quad \begin{pmatrix} n_1 \\ n_2 \\ \vdots \\ n_d \end{pmatrix}.$$

A "row vector" arranges elements from left to right:

$$(12, 19) \qquad (1, 5, 9, 7) \qquad (x, y) \qquad (x, y, z) \qquad (n_1, n_2, \ldots n_d).$$

Row vectors can be written with large spaces rather than commas separating each element, e.g., $(12 \quad 19)$ instead of $(12, 19)$. Vectors are sometimes written in curly brackets { } or in square brackets [ ], but this is a matter of preference. The number of elements in a vector indicates its "dimensionality." For example, the vector $\binom{x}{y}$ is a two-dimensional column vector, while (x, y) is a two-dimensional row vector.

There is a simple, but very important, graphical interpretation of vectors. We can visualize a vector with two elements as a line on a plane, starting at the origin and ending at the point whose x and y coordinates are given by the elements of the vector $\binom{x}{y}$ (Figure P2.1). Similarly, a vector with three elements can be visualized as a line in three dimensions (Figure P2.2). This interpretation holds for higher dimensions as well, although visualizing the vector becomes problematic! This graphical interpretation indicates that a d-dimensional vector can be thought of as a line with a certain length and direction in d-dimensional

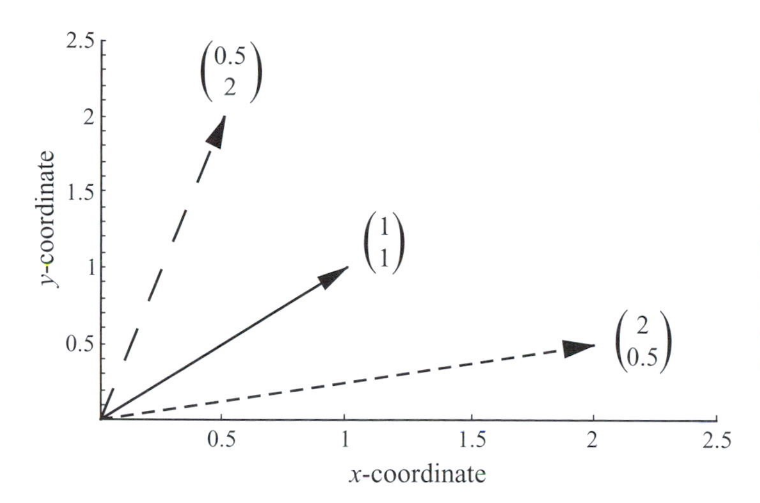

Figure P2.1: Vectors represented as arrows on a two-dimensional plot. A vector is plotted by starting at the origin and ending at the point whose x- and y-coordinates are given by the vector $\binom{x}{y}$. Thus, each vector represents a line pointing in a particular direction with a particular length. For example, the vector $\binom{x}{y} = \binom{2}{0.5}$ points in a direction that is $\arctan(y/x) = 14°$ above the horizontal axis and $\sqrt{x^2 + y^2} = 2.06$ units in length.

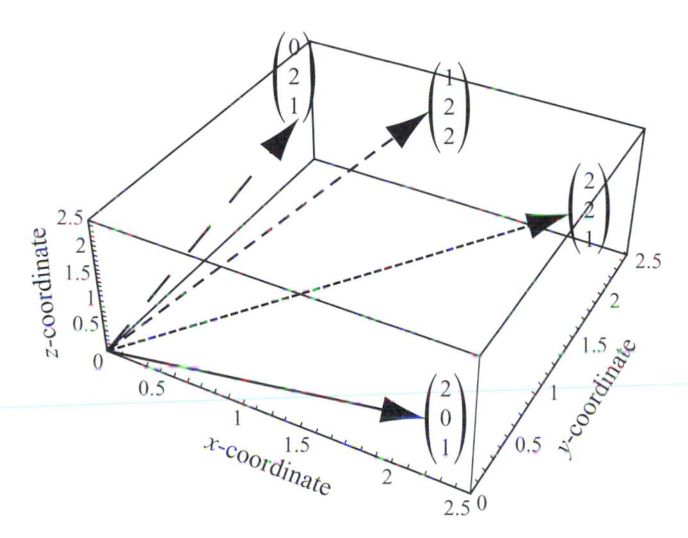

Figure P2.2: Vectors represented as arrows on a three-dimensional plot. A vector is plotted by starting at the origin and ending at the point whose x-, y- and z-coordinates are given by the vector $\begin{pmatrix} x \\ y \\ z \end{pmatrix}$.

space. For example, the length of a two-dimensional vector $\binom{a}{b}$ can be calculated using the Pythagorean theorem for a triangle: $\sqrt{a^2 + b^2}$ (Figure P2.1).

While a column vector has only one column and a row vector has only one row, a *matrix* (plural: matrices) represents a more general list that can contain any number of rows and columns:

A ***matrix*** is a table of elements with r rows and c columns.

$$\begin{pmatrix} 12 & 19 \\ 10 & 23 \end{pmatrix} \quad \begin{pmatrix} 1 & 5 & 9 \\ 7 & 7 & 7 \end{pmatrix} \quad \begin{pmatrix} a & b \\ c & d \\ e & f \end{pmatrix} \quad \begin{pmatrix} m_{11} & m_{12} & \cdots & m_{1c} \\ m_{21} & m_{22} & \cdots & m_{2c} \\ \vdots & \vdots & \vdots & \vdots \\ m_{r1} & m_{r2} & \cdots & m_{rc} \end{pmatrix}.$$

The dimensionality of a matrix is always written as the number of rows by the number of columns (the dimensionality of the above matrices is 2×2, 2×3, 3×2, and $r \times c$). A vector can be thought of as a special case of a matrix; a column vector is a $d \times 1$ matrix and a row vector is a $1 \times d$ matrix.

We will use the convention that m_{ij} describes the element in the ith row and jth column of a matrix. That is, we always refer to an element according to its *row first and column second*. We will identify a matrix by assigning it a label using a boldface symbol (e.g., $\mathbf{M}$) and a vector by assigning it a label with a "half arrow" above it (e.g., $\vec{x}$).

Matrices also have a graphical interpretation. A matrix can be considered as a way of "transforming" (i.e., moving) a vector. The transformation might rotate or stretch the original vector so that the vector points in a different direction and/or has a different length. As illustrated in Figure P2.3, a 2×2 matrix maps the x and y coordinates of a point given by the vector $\binom{x}{y}$ to a new set of x and y coordinates. These new coordinates are determined by matrix multiplication, as described in section P2.4.

Most matrices used in mathematical biology are "square," meaning that the number of rows equals the number of columns (i.e., the matrix has dimensionality $d \times d$). There are several special square matrices that we will encounter often. A *diagonal matrix* contains elements that are all zero except along the diagonal from upper left to lower right:

Definition P2.1:

$$\mathbf{D} = \begin{pmatrix} m_{11} & 0 & \cdots & 0 \\ 0 & m_{22} & \cdots & 0 \\ \vdots & \vdots & \vdots & \vdots \\ 0 & 0 & \cdots & m_{dd} \end{pmatrix} \qquad \text{(diagonal matrix)}$$

An *identity matrix* is a diagonal matrix with ones along the diagonal:

Definition P2.2:

$$\mathbf{I} = \begin{pmatrix} 1 & 0 & \cdots & 0 \\ 0 & 1 & \cdots & 0 \\ \vdots & \vdots & \vdots & \vdots \\ 0 & 0 & \cdots & 1 \end{pmatrix} \qquad \text{(identity matrix)}$$

An *upper triangular matrix* is a matrix whose elements below the diagonal are zero:

Definition P2.3:

$$\begin{pmatrix} m_{11} & m_{12} & \cdots & m_{1c} \\ 0 & m_{22} & \cdots & m_{2c} \\ \vdots & \vdots & \vdots & \vdots \\ 0 & 0 & \cdots & m_{rc} \end{pmatrix} \qquad \text{(upper triangular matrix)}$$

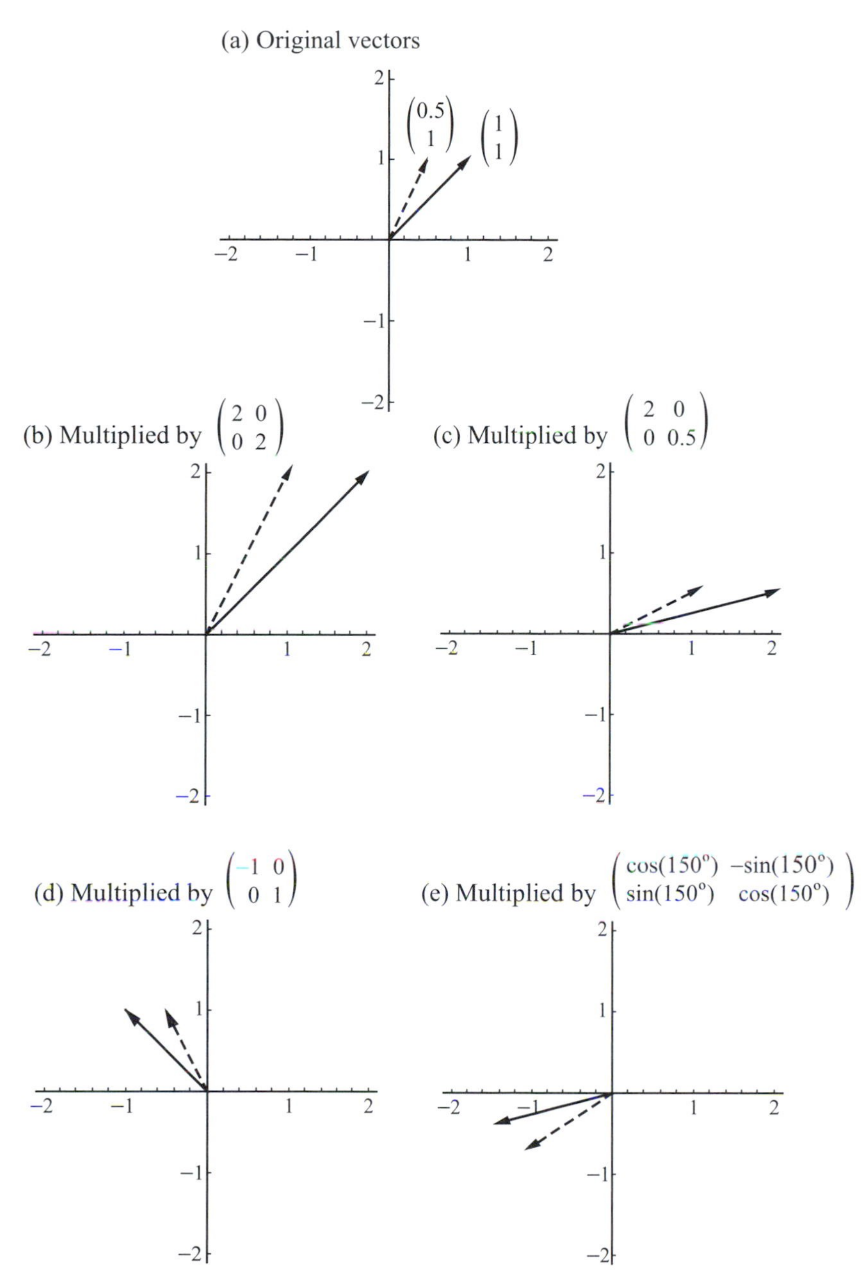

Figure P2.3: Matrices transform vectors into new vectors. When the original vectors in (a) are multiplied on the left by a matrix, the vectors are stretched and rotated. (b) Multiplication by $\begin{pmatrix} \gamma & 0 \\ 0 & \gamma \end{pmatrix}$ stretches the vectors by an amount γ ($\gamma = 2$). (c) Multiplication by $\begin{pmatrix} \gamma_x & 0 \\ 0 & \gamma_y \end{pmatrix}$ stretches the vectors by an amount γ_x long the x-axis and by γ_y along the y-axis ($\gamma_x = 2$, $\gamma_y = 0.5$). (d) Multiplication by $\begin{pmatrix} -1 & 0 \\ 0 & 1 \end{pmatrix}$ reflects the vectors across the y-axis. (Conversely, $\begin{pmatrix} 1 & 0 \\ 0 & -1 \end{pmatrix}$ reflects across the x-axis.) (e) Multiplication by the matrix $\begin{pmatrix} \cos(\theta) & -\sin(\theta) \\ \sin(\theta) & \cos(\theta) \end{pmatrix}$ rotates the vectors counterclockwise by an amount θ ($\theta = 150°$).

A *lower triangular matrix* is a matrix whose elements above the diagonal are zero:

Definition P2.4:

$$\begin{pmatrix} m_{11} & 0 & \cdots & 0 \\ m_{21} & m_{22} & \cdots & 0 \\ \vdots & \vdots & \vdots & \vdots \\ m_{r1} & m_{r2} & \cdots & m_{rc} \end{pmatrix} \qquad \text{(lower triangular matrix)}$$

It is sometimes useful to write a matrix in terms of submatrices, by blocking certain elements together. For example, the 4×4 matrix

$$\begin{pmatrix} m_{11} & m_{12} & m_{13} & m_{14} \\ m_{21} & m_{22} & m_{23} & m_{24} \\ m_{31} & m_{32} & m_{33} & m_{34} \\ m_{41} & m_{42} & m_{43} & m_{44} \end{pmatrix}$$

could be written in block form as $\begin{pmatrix} \mathbf{A} & \mathbf{B} \\ \mathbf{C} & \mathbf{D} \end{pmatrix}$, where $\mathbf{A} = \begin{pmatrix} m_{11} & m_{12} \\ m_{21} & m_{22} \end{pmatrix}$, $\mathbf{B} = \begin{pmatrix} m_{13} & m_{14} \\ m_{23} & m_{24} \end{pmatrix}$, $\mathbf{C} = \begin{pmatrix} m_{31} & m_{32} \\ m_{41} & m_{42} \end{pmatrix}$, and $\mathbf{D} = \begin{pmatrix} m_{33} & m_{34} \\ m_{43} & m_{44} \end{pmatrix}$. But this is not the only possible block form. We could also write it as $\begin{pmatrix} \mathbf{E} & \mathbf{F} \\ \mathbf{G} & \mathbf{H} \end{pmatrix}$, where

$$\mathbf{E} = \begin{pmatrix} m_{11} & m_{12} & m_{13} \\ m_{21} & m_{22} & m_{23} \\ m_{31} & m_{32} & m_{33} \end{pmatrix}, \quad \mathbf{F} = \begin{pmatrix} m_{14} \\ m_{24} \\ m_{34} \end{pmatrix},$$

$\mathbf{G} = (m_{41} \quad m_{42} \quad m_{43})$, and $\mathbf{H} = (m_{44})$. For square matrices, we will confine our attention to block forms where the diagonal submatrices (**A** and **D** or **E** and **H**) are also square. Block form is particularly useful when one or both of the off-diagonal submatrices consist entirely of zeros. If matrix **B** *and* **C** consists of zeros, we say that the matrix has a "block diagonal form." If matrix **B** *or* **C** consists of zeros, we say that the matrix has "block triangular form."

Sometimes, it is useful to "transpose" a matrix, which makes the rows of the original matrix into the columns of a new matrix. The transpose is represented by a nonitalicized T superscript:

Rule P2.1:

$$\mathbf{M}^{\mathrm{T}} = \begin{pmatrix} m_{11} & m_{12} & \cdots & m_{1c} \\ m_{21} & m_{22} & \cdots & m_{2c} \\ \vdots & \vdots & \vdots & \vdots \\ m_{r1} & m_{r2} & \cdots & m_{rc} \end{pmatrix}^{\mathrm{T}} = \begin{pmatrix} m_{11} & m_{21} & \cdots & m_{r1} \\ m_{12} & m_{22} & \cdots & m_{r2} \\ \vdots & \vdots & \vdots & \vdots \\ m_{1c} & m_{2c} & \cdots & m_{rc} \end{pmatrix} \qquad \text{(matrix transpose)}$$

For example, the transpose of $\begin{pmatrix} 1 & 2 \\ 3 & 4 \end{pmatrix}$ is $\begin{pmatrix} 1 & 3 \\ 2 & 4 \end{pmatrix}$ and the transpose of $\begin{pmatrix} a \\ b \end{pmatrix}$ is (a, b). The transpose can be performed on matrices of any dimension. For example, the transpose of a column vector is a row vector, and vice versa.

Exercise P2.1: Determine the dimensionality of the following vectors and matrices and write down the transpose of each.

(a)
$$(2 \quad 4).$$

(b)
$$\begin{pmatrix} 1 \\ 2 \\ 3 \end{pmatrix}.$$

(c)
$$\begin{pmatrix} 6 & -4 \\ 2 & 0 \end{pmatrix}.$$

(d)
$$\begin{pmatrix} 1 + x & 2 + x & 3 + x \\ 5 & 7 & 9 \end{pmatrix}.$$

(e)
$$\begin{pmatrix} a & b \\ c & d \\ e & f \end{pmatrix}.$$

[Answers to the exercises are provided at the end of the primer.]

P2.2 Vector and Matrix Addition

The real power and utility of linear algebra comes from the rules by which we add, subtract, and multiply vectors and matrices. These rules are simply things that we have to commit to memory, just as we committed the rules of addition, subtraction, and multiplication of numbers to memory in grade school. In sections P2.2–P2.6, we describe these basic matrix operations. We then turn our attention to important techniques that build upon these basic operations. When describing matrix operations, we will provide specific examples as well as general formulas using the following arbitrary vectors and matrices in d dimensions:

$$\vec{v}_1 = \begin{pmatrix} x_1 \\ x_2 \\ \vdots \\ x_d \end{pmatrix}, \ \vec{v}_2 = \begin{pmatrix} y_1 \\ y_2 \\ \vdots \\ y_d \end{pmatrix}, \ \mathbf{M} = \begin{pmatrix} m_{11} & m_{12} & \cdots & m_{1d} \\ m_{21} & m_{22} & \cdots & m_{2d} \\ \vdots & \vdots & \vdots & \vdots \\ m_{d1} & m_{d2} & & m_{dd} \end{pmatrix}, \ \mathbf{N} = \begin{pmatrix} n_{11} & n_{12} & \cdots & n_{1d} \\ n_{21} & n_{22} & \cdots & n_{2d} \\ \vdots & \vdots & \vdots & \vdots \\ n_{d1} & n_{d2} & \cdots & n_{dd} \end{pmatrix}.$$

Vector *addition* involves adding each element of two vectors, one position at a time, and placing the answer in the same position of the resulting vector:

Rule P2.2:

$$\vec{v}_1 + \vec{v}_2 = \begin{pmatrix} x_1 \\ x_2 \\ \vdots \\ x_d \end{pmatrix} + \begin{pmatrix} y_1 \\ y_2 \\ \vdots \\ y_d \end{pmatrix} = \begin{pmatrix} x_1 + y_1 \\ x_2 + y_2 \\ \vdots \\ x_d + y_d \end{pmatrix} \quad \text{(vector addition)}$$

For example, adding the row vectors (2, 3) and (8, 20) gives the row vector (10, 23). Vector *subtraction* is similarly straightforward:

Rule P2.3:

$$\vec{v}_1 - \vec{v}_2 = \begin{pmatrix} x_1 \\ x_2 \\ \vdots \\ x_d \end{pmatrix} - \begin{pmatrix} y_1 \\ y_2 \\ \vdots \\ y_d \end{pmatrix} = \begin{pmatrix} x_1 - y_1 \\ x_2 - y_2 \\ \vdots \\ x_d - y_d \end{pmatrix} \quad \text{(vector subtraction)}$$

CAUTION: Vector addition and subtraction can be carried out only on vectors with the same dimensionality because corresponding positions must exist for each vector.

Graphically, adding two-dimensional vectors is akin to starting one vector at the end of the other vector, giving a new total vector that starts at the origin and ends at the tip of the second vector (Figure P2.4).

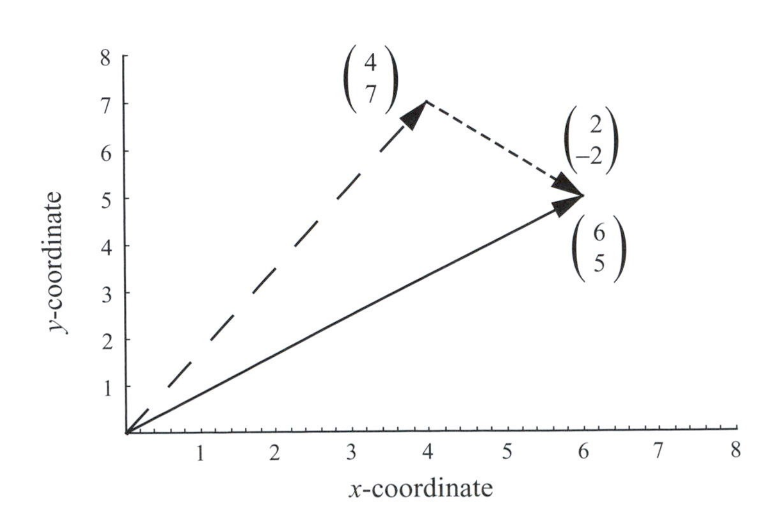

Figure P2.4: Vector addition. Vector addition represents the placement of one vector at the tip of another vector. Here, the short-dashed vector $\binom{2}{-2}$ is added to the long-dashed vector $\binom{4}{7}$ to get the solid vector $\binom{6}{5}$.

Exercise P2.2: Specify which of the following vector sums are valid and perform these summations.

(a)
$$\begin{pmatrix} 5 \\ 7 \end{pmatrix} + (2 \quad 4).$$

(b)
$$\begin{pmatrix} 5 \\ 7 \end{pmatrix} + \begin{pmatrix} 2 \\ 4 \end{pmatrix}.$$

(c)
$$\begin{pmatrix} 1 \\ 2 \\ 3 \end{pmatrix} + \begin{pmatrix} 4 \\ 5 \\ 6 \end{pmatrix}.$$

(d)
$$(1 \quad 2 \quad 3 \quad 4) + (5 \quad 6 \quad 7).$$

(e)
$$(0.5 \quad 1.3 \quad 2) + (a \quad b \quad c).$$

Once you are comfortable with vector addition, matrix addition is straightforward. When adding two matrices ($\mathbf{M} + \mathbf{N}$), one makes a new matrix, with each position equal to the sum of the corresponding positions in the two original matrices:

Rule P2.4:

$$\mathbf{M} + \mathbf{N} = \begin{pmatrix} m_{11} + n_{11} & m_{12} + n_{12} & \cdots & m_{1d} + n_{1d} \\ m_{21} + n_{21} & m_{22} + n_{22} & \cdots & m_{2d} + n_{2d} \\ \vdots & \vdots & \vdots & \vdots \\ m_{d1} + n_{d1} & m_{d2} + n_{d2} & \cdots & m_{dd} + n_{dd} \end{pmatrix} \quad \text{(matrix addition)}$$

Rule P2.5:

$$\mathbf{M} - \mathbf{N} = \begin{pmatrix} m_{11} - n_{11} & m_{12} - n_{12} & \cdots & m_{1d} - n_{1d} \\ m_{21} - n_{21} & m_{22} - n_{22} & \cdots & m_{2d} - n_{2d} \\ \vdots & \vdots & \vdots & \vdots \\ m_{d1} - n_{d1} & m_{d2} - n_{d2} & \cdots & m_{dd} - n_{dd} \end{pmatrix} \quad \text{(matrix subtraction)}$$

As a concrete numerical example, $\begin{pmatrix} 12 & 10 \\ 8 & 6 \end{pmatrix}$ plus $\begin{pmatrix} 8 & 1 \\ 10 & -3 \end{pmatrix}$ equals $\begin{pmatrix} 20 & 11 \\ 18 & 3 \end{pmatrix}$.

CAUTION: Matrix addition and subtraction can only be carried out on matrices with the same dimensionality.

Exercise P2.3: Specify which of the following matrix sums are valid and perform these summations.

(a)
$$\begin{pmatrix} 1 & 2 \\ 3 & 4 \end{pmatrix} + \begin{pmatrix} 6 & 4 \\ 2 & 0 \end{pmatrix}.$$

(b)
$$\begin{pmatrix} 1 & 2 \\ 3 & 4 \end{pmatrix} + \begin{pmatrix} 1 & 2 & 3 \\ 5 & 7 & 9 \end{pmatrix}.$$

(c)
$$\begin{pmatrix} a & b \\ c & d \end{pmatrix} - \begin{pmatrix} 1 & 2 \\ 3 & 4 \end{pmatrix}.$$

(d)
$$\begin{pmatrix} a & b & c \\ d & e & f \end{pmatrix} + \begin{pmatrix} a & a & a \\ d & d & d \end{pmatrix}.$$

(e)
$$\begin{pmatrix} 1 & 2 & 3 \\ 4 & 5 & 6 \\ 7 & 8 & 9 \end{pmatrix} - \begin{pmatrix} 1 & 2 & 3 \\ 4 & 5 & 6 \\ 7 & 8 & 9 \end{pmatrix}.$$

P2.3 Multiplication by a Scalar

A ***scalar*** is a single element.

Numbers, parameters, variables, and functions (basically anything that is not a vector or a matrix) are known as *scalars*. Multiplication of a vector by a scalar k is performed by multiplying each element in the vector by k:

Rule P2.6:

$$k\,\vec{v}_1 = \begin{pmatrix} k\,x_1 \\ k\,x_2 \\ \vdots \\ k\,x_d \end{pmatrix}. \qquad \text{(vector multiplication by a scalar)}$$

Multiplying a vector by a scalar stretches the vector.

For example, if we were to multiply the column vector $\binom{1}{2}$ by the scalar x we would get $\binom{x}{2x}$. One can also undo this procedure and factor out a term from each element in a vector. For example, $\binom{x/W}{y/W}$ is the same as $\frac{1}{W}\binom{x}{y}$. Vector multiplication by a scalar has a very important graphical interpretation: it corresponds to stretching or compressing the vector by a factor k, without altering its direction (Figure P2.5).

Matrix multiplication by a scalar proceeds in the same manner, multiplying every element in the matrix by the scalar:

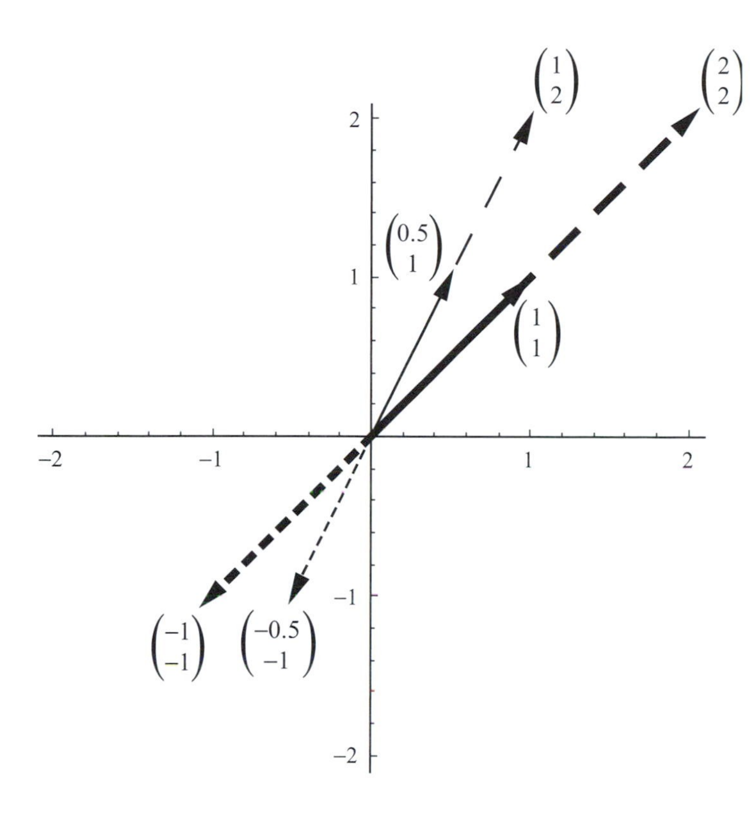

Figure P2.5: Multiplication of a vector by a scalar. A vector stretches or shrinks when multiplied by a scalar γ. The original thin-lined vector $\binom{0.5}{1}$ and thick-lined vector $\binom{1}{1}$ are multiplied by the scalar $\gamma = 2$ (long-dashed vectors) and by the scalar $\gamma = -1$ (short-dashed curves). A vector multiplied by a scalar always remains on the same line, although it can point in the opposite direction. Multiplication of two-dimensional vectors by a scalar, γ, is equivalent to multiplication by the matrix $\begin{pmatrix}\gamma & 0\\ 0 & \gamma\end{pmatrix}$ (see Figure P2.3b).

Rule P2.7:

$$k\,\mathbf{M} = \begin{pmatrix} k\,m_{11} & k\,m_{12} & \cdots & k\,m_{1d} \\ k\,m_{21} & k\,m_{22} & \cdots & k\,m_{2d} \\ \vdots & \vdots & \vdots & \vdots \\ k\,m_{d1} & k\,m_{d2} & \cdots & k\,m_{dd} \end{pmatrix} \quad \text{(matrix multiplication by a scalar)}$$

When it comes to multiplication by a scalar, no warnings are needed. The multiplication can proceed regardless of the dimensionality of the vector or matrix.

Exercise P2.4: Specify which of the following scalar multiplications are valid and determine the result.

(a) $$3\,(2 \quad 4).$$

(b) $$(-1)\begin{pmatrix}4\\5\\6\end{pmatrix}.$$

(c) $$2\begin{pmatrix}1 & 2\\ 3 & 4\end{pmatrix}.$$

(continued)

Exercise P2.4 (*Continued*)

(d)
$$a\begin{pmatrix}1 & 2 & 3\\ 5 & 7 & 9\end{pmatrix}.$$

(e)
$$(2+a)\begin{pmatrix}1 & 2\\ 3 & 4\end{pmatrix}.$$

P2.4 Multiplication of Vectors and Matrices

Vector and matrix *multiplication* is less straightforward and requires practice. The basic procedure is the multiplication of a row vector by a column vector of the same length:

Rule P2.8:

$$(x_1, x_2, \ldots, x_d)\begin{pmatrix}y_1\\ y_2\\ \vdots\\ y_d\end{pmatrix} = (x_1\,y_1 + x_2\,y_2 + \cdots + x_d\,y_d) \quad \text{(vector multiplication)}$$

Multiplication of a $1 \times d$ row vector and a $d \times 1$ column vector results in a 1×1 element, which we can write compactly as $\sum_{i=1}^{d} x_i\,y_i$.

Here's a technique that can help you get used to vector multiplication. Put your left index finger on the first element of the left-hand vector and your right index finger on the first element of the right-hand vector. Multiply these together. Then move your left hand over to the right and your right hand down by one element. Multiply these together and add them to the previous result. Continue until your hands have moved all the way through the vectors. If the vectors have the same length, both hands will reach the last element at the same time. Try this technique using these examples: $(1, 1)\binom{x}{y} = (x + y)$, $(a, b)\binom{1}{2} = (a + 2b)$, and $(3, 5)\binom{1}{2} = (13)$.

To multiply a matrix and a vector together, this basic procedure is repeated, beginning with the first row of the matrix, proceeding to the second row of the matrix, etc. The resulting sums then go in the corresponding rows of a new vector:

Rule P2.9a:

$$\begin{pmatrix}m_{11} & m_{12} & \cdots & m_{1c}\\ m_{21} & m_{22} & \cdots & m_{2c}\\ \vdots & \vdots & \vdots & \vdots\\ m_{r1} & m_{r2} & \cdots & m_{rc}\end{pmatrix}\begin{pmatrix}y_1\\ y_2\\ \vdots\\ y_c\end{pmatrix} = \begin{pmatrix}\sum_{i=1}^{c} m_{1i}\,y_i\\ \sum_{i=1}^{c} m_{2i}\,y_i\\ \vdots\\ \sum_{i=1}^{c} m_{ri}\,y_i\end{pmatrix} \quad \text{(matrix multiplication)}$$

For example, a two-dimensional matrix multiplied by a two-dimensional vector gives

Rule P2.9b:

$$\begin{pmatrix} a & b \\ c & d \end{pmatrix}\begin{pmatrix} x \\ y \end{pmatrix} = \begin{pmatrix} ax + by \\ cx + dy \end{pmatrix} \qquad (2 \times 2 \text{ matrix multiplication})$$

Multiplying a vector by a matrix stretches and rotates the vector.

As illustrated in Figure P2.3, matrix multiplication of a vector *transforms* or moves the vector to a new position. Depending on the form of this matrix, this transformation might stretch, shrink, reflect, or rotate the vector (Figure P2.3). For example, the matrix $\begin{pmatrix} 2 & 0 \\ 0 & 2 \end{pmatrix}$ multiplied by the vector $\begin{pmatrix} 0.5 \\ 1 \end{pmatrix}$ stretches the solid arrow in Figure P2.3a to get the solid arrow in Figure P2.3b. Using Rule P2.9b, this matrix multiplication results in the vector $\begin{pmatrix} 2 \times 0.5 + 0 \times 1 \\ 0 \times 0.5 + 2 \times 1 \end{pmatrix} = \begin{pmatrix} 1 \\ 2 \end{pmatrix}$.

We can extend this procedure to multiply together two matrices. Basically, we treat one column at a time in the second matrix and apply Rule P2.9, repeating the above procedure, beginning with the first column of the second matrix, then the second column, until the ends of the matrices are reached:

Rule P2.10:

$$\begin{pmatrix} m_{11} & m_{12} & \cdots & m_{1c} \\ m_{21} & m_{22} & \cdots & m_{2c} \\ \vdots & \vdots & \vdots & \vdots \\ m_{r1} & m_{r2} & \cdots & m_{rc} \end{pmatrix}\begin{pmatrix} n_{11} & n_{12} & \cdots & n_{1d} \\ n_{21} & n_{22} & \cdots & n_{2d} \\ \vdots & \vdots & \vdots & \vdots \\ n_{c1} & n_{c2} & \cdots & n_{cd} \end{pmatrix} = \begin{pmatrix} \sum_{i=1}^{c} m_{1i}\, n_{i1} & \sum_{i=1}^{c} m_{1i}\, n_{i2} & \cdots & \sum_{i=1}^{c} m_{1i}\, n_{id} \\ \sum_{i=1}^{c} m_{2i}\, n_{i1} & \sum_{i=1}^{c} m_{2i}\, n_{i2} & \cdots & \sum_{i=1}^{c} m_{2i}\, n_{id} \\ \vdots & \vdots & \vdots & \vdots \\ \sum_{i=1}^{c} m_{ri}\, n_{i1} & \sum_{i=1}^{c} m_{ri}\, n_{i2} & \cdots & \sum_{i=1}^{c} m_{ri}\, n_{id} \end{pmatrix}$$

(general matrix multiplication)

This is where the trick of moving your hands can really help. To calculate the *ij*th element of the resulting matrix, place the index finger of your left hand at the start of the *i*th row of the first matrix and the index finger of your right hand at the start of the *j*th column of the second matrix. Multiply together these two terms, then move to the right with your left hand and down with your right hand. Multiply these two terms and add them to the previous result. Continue moving your hands, taking the product of the elements that you are pointing to and adding the product to the running total, until you reach the end of the *i*th row. The sum total of all of these products is then placed in the *i*th row and *j*th column of the new matrix.

CAUTION: Matrix multiplication requires that the left-hand matrix has the same number of columns as the right-hand matrix has rows. That is, an $r \times c$ matrix can only be multiplied on the right by a $c \times d$ matrix (c must be the same). The resulting matrix will have dimension $r \times d$.

Practicing matrix multiplication is critical for avoiding errors, so try these examples:

$$\begin{pmatrix} a & b \\ c & d \end{pmatrix}\begin{pmatrix} e & f \\ g & h \end{pmatrix} = \begin{pmatrix} ae + bg & af + bh \\ ce + dg & cf + dh \end{pmatrix} \quad (2 \times 2 \text{ multiplied by } 2 \times 2 \rightarrow 2 \times 2).$$

$$(a, b)\begin{pmatrix} x \\ y \end{pmatrix} = (ax + by) \quad (1 \times 2 \text{ multiplied by } 2 \times 1 \rightarrow 1 \times 1).$$

$$\begin{pmatrix} a \\ b \end{pmatrix}(x, y) = \begin{pmatrix} ax & ay \\ bx & by \end{pmatrix} \quad (2 \times 1 \text{ multiplied by } 1 \times 2 \rightarrow 2 \times 2).$$

$$\begin{pmatrix} a & b & c \\ d & e & f \end{pmatrix}\begin{pmatrix} x \\ y \\ z \end{pmatrix} = \begin{pmatrix} ax + by + cz \\ dx + ey + fz \end{pmatrix} \quad (2 \times 3 \text{ multiplied by } 3 \times 1 \rightarrow 2 \times 1).$$

$$\begin{pmatrix} a & b \\ c & d \end{pmatrix}(x, y) = \text{Cannot be done} \quad (2 \times 2 \text{ multiplied by } 1 \times 2 \text{ is not allowed}).$$

Now we arrive at the first main difference between linear algebra and regular algebra. We're used to thinking that the order in which we multiply two things together does not matter; 2 times 3 is the same as 3 times 2. This is true of scalars, but it is not generally true of matrices and vectors! For example,

$$\begin{pmatrix} 1 & 2 \\ 3 & 4 \end{pmatrix}\begin{pmatrix} 5 & 6 \\ 7 & 8 \end{pmatrix} = \begin{pmatrix} 19 & 22 \\ 43 & 50 \end{pmatrix} \text{ does not equal } \begin{pmatrix} 5 & 6 \\ 7 & 8 \end{pmatrix}\begin{pmatrix} 1 & 2 \\ 3 & 4 \end{pmatrix} = \begin{pmatrix} 23 & 34 \\ 31 & 46 \end{pmatrix}.$$

In mathematical jargon, matrix multiplication is not *commutative* ($\mathbf{AB} \neq \mathbf{BA}$). Matrix multiplication does, however, satisfy the following rules:

Rule P2.11:

$(\mathbf{AB})\mathbf{C} = \mathbf{A}(\mathbf{BC})$ (associative law)

Rule P2.12:

$(\mathbf{A} + \mathbf{B})\mathbf{C} = \mathbf{AC} + \mathbf{BC}$ (distributive law)

Rule P2.13:

$\mathbf{A}(\mathbf{B} + \mathbf{C}) = \mathbf{AB} + \mathbf{AC}$ (distributive law)

Rule P2.14:

$k(\mathbf{AB}) = (k\mathbf{A})\mathbf{B} = \mathbf{A}(k\mathbf{B}) = (\mathbf{AB})k$ (commutative law for scalars)

There is another important concept that is similar in linear algebra and regular algebra. In both, there is a special object that leaves other objects unchanged upon multiplication. In regular algebra, this is the number one; multiplying anything by one has no effect. In linear algebra, this special object is the identity matrix. Any square matrix (**M**) with dimensionality $d \times d$ can be multiplied by the $d \times d$ identity matrix on either the right or the left with no effect:

Rule P2.15:
MI = **IM** = **M** (multiplication by **I**)

Demonstrate this to yourself using $\mathbf{I} = \begin{pmatrix} 1 & 0 \\ 0 & 1 \end{pmatrix}$ and any 2×2 matrix that pops to mind. Similarly, a vector remains unaltered when multiplied by an identity matrix; the vector retains its original length and direction.

Exercise P2.5: Using the matrices

$$\mathbf{A} = \begin{pmatrix} a & b \\ c & d \end{pmatrix}, \quad \mathbf{B} = \begin{pmatrix} e & f \\ g & h \end{pmatrix}, \quad \mathbf{C} = \begin{pmatrix} i & j \\ k & l \end{pmatrix}, \quad \mathbf{I} = \begin{pmatrix} 1 & 0 \\ 0 & 1 \end{pmatrix},$$

(a) prove that **AB** does not equal **BA** (matrix multiplication is not commutative)
(b) prove that **A**(**B** + **C**) does equal **AB** + **AC** (matrix multiplication is distributive)
(c) prove that (**AB**)**C** does equal **A**(**BC**) (matrix multiplication is associative)
(d) prove that **AI** and **IA** equal **A** (multiplication by the identity matrix leaves the matrix unchanged)

There is one last fact about matrix multiplication that we will need. As long as the dimensions are appropriate, we are free to multiply any equation on both sides by a matrix *as long as we multiply both sides of the equation on the right or both sides of the equation on the left*. For example, if **A**, **B**, **C**, and **D** are square matrices of dimension d and if **AB** = **C**, then **ABD** = **CD**. To prove this, replace **AB** with **C** to get **CD** = **CD**, which is always true. Similarly, we could multiply on the left to get **DAB** = **DC**. But we will not generally preserve the validity of the equation if we multiply by **D** on the left on one side of the equation and on the right on the other side of the equation (that is, **ABD** ≠ **DC** and **DAB** ≠ **CD**). Try checking that **ABD** = **CD** and **DAB** = **DC** but that **ABD** ≠ **DC** using the 2×2 matrices

$$\mathbf{A} = \begin{pmatrix} 1 & 2 \\ 2 & 1 \end{pmatrix}, \quad \mathbf{B} = \begin{pmatrix} 1 & 1 \\ 3 & 1 \end{pmatrix}, \quad \mathbf{C} = \begin{pmatrix} 7 & 3 \\ 5 & 3 \end{pmatrix}, \quad \mathbf{D} = \begin{pmatrix} 2 & 1 \\ 2 & 5 \end{pmatrix}.$$

So far, we have focused on a type of matrix multiplication known as the "*dot product*" or "*inner product*" (Rules P2.8–P2.10). The dot product is the one most commonly encountered in mathematical biology, but a few other products make occasional appearances, and these are described in Sup. Mat. P2.1.

P2.5 The Trace and Determinant of a Square Matrix

The trace of a square matrix is simply the sum of its diagonal elements ($\mathrm{Tr}(\mathbf{M}) = \sum_{i=1}^{d} m_{ii}$). In two dimensions we have

The *trace* sums the diagonal elements of a matrix.

Definition P2.5:

$$\mathrm{Tr}\begin{pmatrix} a & b \\ c & d \end{pmatrix} = a + d \qquad \text{(trace of a } 2 \times 2 \text{ matrix)}$$

The trace of a matrix is a scalar. The trace has no obvious intuitive meaning, but it simplifies other matrix calculations that we will encounter later.

The determinant of a square matrix is another scalar. For a 2×2 matrix, the determinant is calculated as

The *determinant* measures whether the rows of a matrix are independent.

Definition P2.6:

$$\mathrm{Det}\begin{pmatrix} a & b \\ c & d \end{pmatrix} = \begin{vmatrix} a & b \\ c & d \end{vmatrix} = ad - bc \qquad \text{(determinant of a } 2 \times 2 \text{ matrix)}$$

and is denoted either as Det(**M**) or simply as |**M**|. The determinant measures whether the rows of a matrix are *linearly independent* of one another. By linearly independent, we mean that we cannot write any row x as a linear function of the other rows (that is, there is no equation like $\mathrm{row}_x = a_1\ \mathrm{row}_1 + \cdots + a_{x-1}\ \mathrm{row}_{x-1} + a_{x+1}\ \mathrm{row}_{x+1} + \cdots + a_d\ \mathrm{row}_d$ that holds true no matter how we choose the constants a_i). If $\mathrm{Det}(\mathbf{M}) = 0$, then the rows are not independent, and a linear relationship does exist among the rows. For example,

$$\mathrm{Det}\begin{pmatrix} 1 & 2 \\ 4 & 8 \end{pmatrix} = 8 - 8 = 0,$$

This reflects the fact that we can write the second row as four times the first row. But

$$\mathrm{Det}\begin{pmatrix} 1 & 2 \\ 4 & 7 \end{pmatrix} = 7 - 8 = -1,$$

and there is no number by which we can multiply the first row to get the second row. If you are familiar with statistics, the determinant is closely related to the χ^2-statistic, which can be used to test independence between the rows and columns of a data table.

Graphically, when we use a 2×2 matrix whose determinant is zero to multiply a series of two-dimensional vectors, the resulting vectors always lie on the same line. For example, in Figure P2.6a, five vectors of length one were chosen to point in different directions. These vectors "span" the two-dimensional plane, meaning

(a) Original vectors

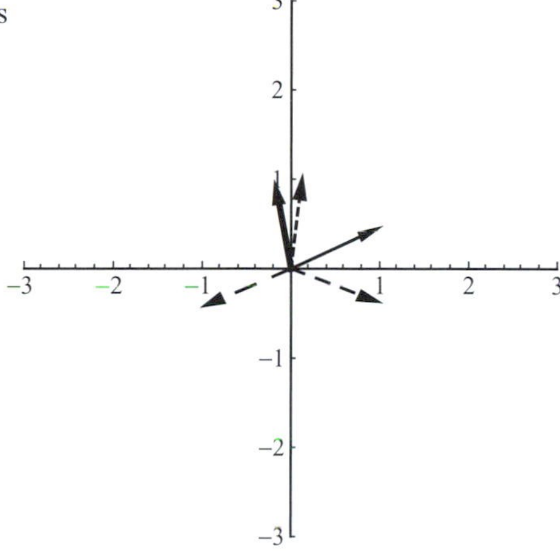

(b) Multiplied by $\begin{pmatrix} 0.5 & 1 \\ -1 & 2 \end{pmatrix}$

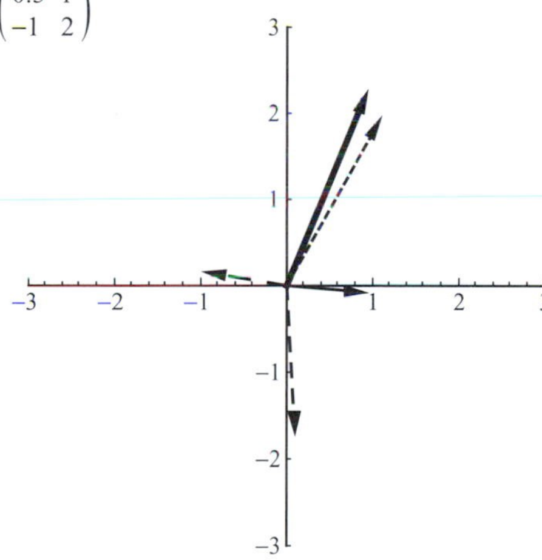

(c) Multiplied by $\begin{pmatrix} 0.5 & 1 \\ 1 & 2 \end{pmatrix}$

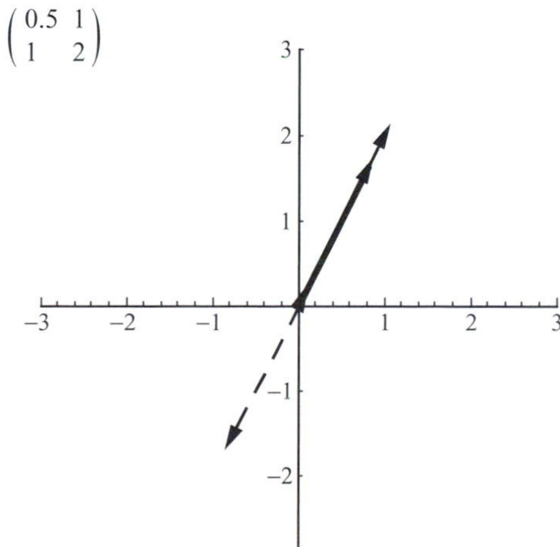

Figure P2.6: Comparing matrices with non-zero and zero determinants. (a) Five vectors of length one were chosen to point in different directions. (b) The five vectors were multiplied on the left by the matrix $\begin{pmatrix} 0.5 & 1 \\ -1 & 2 \end{pmatrix}$, whose determinant is non-zero. (c) The five vectors were multiplied on the left by the matrix $\begin{pmatrix} 0.5 & 1 \\ 1 & 2 \end{pmatrix}$, whose determinant is zero.

that we cannot draw a one-dimensional line that lies on top of all of the vectors. When we multiply these vectors by the matrix $\begin{pmatrix} 0.5 & 1 \\ -1 & 2 \end{pmatrix}$, whose determinant is 2, we get five new vectors, which again span the plane (Figure P2.6b). If, however, we multiply these vectors by $\begin{pmatrix} 0.5 & 1 \\ 1 & 2 \end{pmatrix}$, whose determinant is 0, we get five new vectors, which all fall along the same line (Figure P2.6c). In general, multiplying a vector by a matrix whose determinant is zero causes a loss of information about some dimensions of the vector. While we can undo multiplication by a matrix whose determinant is not zero to get back the original vectors (see section P2.6 on inverses), we cannot undo multiplication by a matrix whose determinant is zero, because we have lost some information about the original vector.

Conceptually, this fact has an analogy in regular (nonmatrix) algebra. Any number x can be multiplied by a number a to get a new number b. We can undo this process by dividing b by a to get back $x = b/a$. This does not work, however, if a equals zero because all information about the original number x is lost once we multiply it by zero.

Exercise P2.6: Calculate the trace and determinant of the following matrices:

$$\mathbf{A} = \begin{pmatrix} 1 & 2 \\ 3 & 4 \end{pmatrix}, \quad \mathbf{B} = \begin{pmatrix} 1 & 3 \\ 2 & 4 \end{pmatrix}, \quad \mathbf{C} = \begin{pmatrix} -1 & -2 \\ -3 & -4 \end{pmatrix}.$$

Exercise P2.7: In which of the following matrices are the rows independent of one another? When the rows are not independent, specify how they are related.

$$\mathbf{D} = \begin{pmatrix} 1 & 2 \\ -3 & -6 \end{pmatrix}, \quad \mathbf{E} = \begin{pmatrix} 1 & 3 \\ 3 & 3 \end{pmatrix}, \quad \mathbf{F} = \begin{pmatrix} 1 & 3 \\ x & 3x \end{pmatrix}.$$

For a 2×2 matrix, it is pretty straightforward to eyeball the matrix to see whether the rows are independent, even without calculating the determinant. But for larger matrices, it can be very hard to see whether any relationship exists among the rows. The determinant is straightforward (if tedious) to calculate for any square matrix, however, and is nonzero if and only if all rows are linearly independent of each other. Again, multiplication of a vector by a $d \times d$ matrix whose determinant is zero causes a loss of information about some of the dimensions of the original vector.

For a 3×3 matrix, the determinant is

Rule P2.16:

(determinant of a 3×3 matrix)

$$\begin{vmatrix} m_{11} & m_{12} & m_{13} \\ m_{21} & m_{22} & m_{23} \\ m_{31} & m_{32} & m_{33} \end{vmatrix} = m_{11}\begin{vmatrix} m_{22} & m_{23} \\ m_{32} & m_{33} \end{vmatrix} - m_{12}\begin{vmatrix} m_{21} & m_{23} \\ m_{31} & m_{33} \end{vmatrix} + m_{13}\begin{vmatrix} m_{21} & m_{22} \\ m_{31} & m_{32} \end{vmatrix}$$

That is, one calculates the determinant by moving across the first row, taking each term m_{1j}, and multiplying it by the determinant of the smaller matrix obtained by deleting the first row and the jth column, alternately adding or subtracting this product to the running total. As an example,

$$\begin{vmatrix} 1 & 2 & 3 \\ 7 & 3 & 5 \\ 11 & 0 & 1 \end{vmatrix} = 1\begin{vmatrix} 3 & 5 \\ 0 & 1 \end{vmatrix} - 2\begin{vmatrix} 7 & 5 \\ 11 & 1 \end{vmatrix} + 3\begin{vmatrix} 7 & 3 \\ 11 & 0 \end{vmatrix} = 3 + 96 - 99 = 0.$$

In this example, it is hard to tell that the rows are related, but the third row is just twice the second row minus three times the first row: $(11 \quad 0 \quad 1) = 2(7 \quad 3 \quad 5) - 3(1 \quad 2 \quad 3)$. The fact that the determinant is zero tells us that some such relationship exists.

For a larger square matrix, the procedure for calculating the determinant is just a generalization of the three-dimensional case:

Rule P2.17:

$$|\mathbf{M}| = \sum_{j=1}^{d} (-1)^{j+1} m_{1j} |\mathbf{M_{1j}}| \qquad \text{(determinant of a } d \times d \text{ matrix)}$$

Again, the determinant is calculated by rewriting it in terms of the elements of the first row m_{1j} times determinants of smaller matrices $\mathbf{M_{1j}}$, each of which is obtained by eliminating the first row and jth column. This procedure seems straightforward until you realize that the determinant of a 10×10 matrix is a function of the determinants of ten 9×9 matrices, which are each in turn functions of determinants of nine 8×8 matrices, etc. In practice, what this means is that we would almost always use a computer to calculate the determinant of a large matrix.

If a matrix contains a row or a column with several zeros in it, it is easier to calculate the determinant by moving across the row with the most zeros (or moving down the column with the most zeros). The only caveat is that we must multiply the determinant by -1 if we use an even-numbered row or an even-numbered column. Thus, we can generalize Rule P2.17 to allow us to use the kth row to find the determinant:

Rule P2.18a:

$$|\mathbf{M}| = (-1)^{k+1} \sum_{j=1}^{d} (-1)^{j+1} m_{kj} |\mathbf{M_{kj}}| \qquad \text{(determinant of a } d \times d \text{ matrix)}$$

or the kth column to find the determinant:

Rule P2.18b:

$$|\mathbf{M}| = (-1)^{k+1} \sum_{j=1}^{d} (-1)^{j+1}\, m_{jk}\, |\mathbf{M_{jk}}| \qquad \text{(determinant of a } d \times d \text{ matrix)}$$

These rules can be used to prove the following other handy rules:

Rule P2.19:
The determinant of a matrix is the same as the determinant of its transpose.

Rule P2.20:
The determinant of a diagonal, upper triangular, or lower triangular matrix is the product of the elements along the diagonal ($|\mathbf{M}| = \prod_{i=1}^{d} m_{ii}$).

Rule P2.21:
The determinant of a block-diagonal matrix or a block-triangular matrix is the product of the determinants of the diagonal submatrices.

Rule P2.22:
If a $d \times d$ matrix $\mathbf{M}$ contains a row (or a column) that is all zeros except for the element along the diagonal (m_{ii}), then the determinant of $\mathbf{M}$ equals m_{ii} times the determinant of a smaller matrix obtained by deleting the ith row and ith column of $\mathbf{M}$.

One way to calculate the determinant of a large matrix is to turn the original matrix into a triangular matrix by combining rows together (or columns together) as described in Sup. Mat. P2.2. Because the determinant of a triangular matrix is easy to calculate (Rule P2.20), these operations can simplify matters substantially, especially if the original matrix is "sparse" (meaning that it already contains several zeros).

Exercise P2.8: Using the definitions for the determinant of a 2×2 and a 3×3 matrix, demonstrate to yourself that Rule P2.20 holds for the following matrices:

$$\mathbf{A} = \begin{pmatrix} 2 & 4 \\ 0 & 3 \end{pmatrix}, \qquad \mathbf{B} = \begin{pmatrix} 2 & 4 & 7 \\ 0 & 3 & 8 \\ 0 & 0 & 1 \end{pmatrix}.$$

P2.6 The Inverse

While we have discussed matrix addition, subtraction, and multiplication, we have not mentioned one fundamental algebraic operation: division. Interestingly, the standard division operator is absent in linear algebra. We cannot just take $\mathbf{AB} = \mathbf{C}$ and rewrite it as $\mathbf{B} = \mathbf{C}/\mathbf{A}$. If standard division were possible, we could undo it by multiplying both sides by the denominator to get $\mathbf{BA} = \mathbf{C}$, but this equation is not the same as $\mathbf{AB} = \mathbf{C}$ because matrix multiplication is not commutative. There is, however, a matrix operation on square matrices whose role is analogous to division, which involves the *inverse* of a matrix. By definition, a matrix $\mathbf{M}$ times its inverse, which we write as $\mathbf{M}^{-1}$, equals the identity matrix $\mathbf{I}$:

The *inverse* matrix, $\mathbf{M}^{-1}$, reverses the stretching and rotating accomplished by the matrix, $\mathbf{M}$.

Rule P2.23a:

$$\mathbf{M}\,\mathbf{M}^{-1} = \mathbf{I} \quad \text{and} \quad \mathbf{M}^{-1}\,\mathbf{M} = \mathbf{I}. \qquad \text{(inverse matrix)}$$

In essence, multiplication by $\mathbf{M}^{-1}$ "reverses" the stretching and rotating accomplished by $\mathbf{M}$ itself. Thus, a vector multiplied by $\mathbf{M}$ and then $\mathbf{M}^{-1}$ has the exact same length and direction as the original vector, $\mathbf{M}^{-1}\mathbf{M}\,\vec{v} = \vec{v}$. The notation used to denote inverses (i.e., $\mathbf{M}^{-1}$) hints at this interpretation. In a sense the inverse is like "one over $\mathbf{M}$" in that it reverses the effect of multiplication by the matrix.

If we write a 2×2 matrix in the general form $\mathbf{M} = \begin{pmatrix} a & b \\ c & d \end{pmatrix}$, its inverse is

Rule P2.23b:

$$\mathbf{M}^{-1} = \begin{pmatrix} a & b \\ c & d \end{pmatrix}^{-1} = \begin{pmatrix} \dfrac{d}{ad - bc} & \dfrac{-b}{ad - bc} \\ \dfrac{-c}{ad - bc} & \dfrac{a}{ad - bc} \end{pmatrix} \qquad \text{(inverse of a 2×2 matrix)}$$

It is worth memorizing the inverse of a 2×2 matrix. This is made easier by noticing that the inverse involves reversing the elements on the main diagonal (a and d), changing the signs of the off-diagonal elements (b and c), and then dividing everything by the determinant Det(**M**) $= ad - bc$. The inverse of a 2×2 matrix should always be a 2×2 matrix, and you should always get the identity matrix if you multiply a matrix by its inverse. We run into problems, however, if the determinant of a matrix is zero, because the inverse involves division by zero and becomes undefined. In general, inverse matrices exist only for square matrices with nonzero determinants. Such matrices are called "nonsingular" or "invertible." In contrast, matrices whose determinant is zero are known as "singular" or "noninvertible."

Exercise P2.9: Determine the inverse of the following matrices (if in doubt, check your answer by multiplying the matrix by its inverse, which should return the identity matrix):

(a) $$\begin{pmatrix} 1 & 0 \\ 0 & 1 \end{pmatrix}.$$

(b) $$\begin{pmatrix} 1 & 0 \\ 0 & 3 \end{pmatrix}.$$

(c) $$\begin{pmatrix} 1 & 3 \\ 1 & 3 \end{pmatrix}.$$

(d) $$\begin{pmatrix} 1 & 2 \\ 1 & 4 \end{pmatrix}.$$

(e) $$\begin{pmatrix} 1 & 2 \\ 3 & 4 \end{pmatrix}.$$

Exercise P2.10: Multiply the general 2×2 matrix $\left(\begin{smallmatrix} a & b \\ c & d \end{smallmatrix}\right)$ by its inverse

$$\begin{pmatrix} \dfrac{d}{ad - bc} & \dfrac{-b}{ad - bc} \\ \dfrac{-c}{ad - bc} & \dfrac{a}{ad - bc} \end{pmatrix},$$

either on the left or the right. Show that you can simplify the result to get the identity matrix $\left(\begin{smallmatrix} 1 & 0 \\ 0 & 1 \end{smallmatrix}\right)$.

The inverse of larger matrices can be found using the method of row reduction (see Sup. Mat. P2.3), although a much less tedious method is to use a mathematical software package, such as *Mathematica* or Maple. If the matrix happens to be diagonal (see Definition P2.1), however, its inverse is found simply by inverting each diagonal element:

Rule P2.24:

$$\mathbf{D}^{-1} = \begin{pmatrix} 1/m_{11} & 0 & \cdots & 0 \\ 0 & 1/m_{22} & \cdots & 0 \\ \vdots & \vdots & \vdots & \vdots \\ 0 & 0 & \cdots & 1/m_{dd} \end{pmatrix} \quad \text{(inverse of a diagonal matrix)}$$

P2.7 Solving Systems of Equations

Knowing the inverse of a matrix is very handy for solving problems involving multiple variables. Consider first a nonmatrix linear equation $m\ n = v$, which we want to solve for n. To find n, we divide both sides by m. This is equivalent to multiplying both sides by m^{-1}, giving the solution $n = m^{-1}v$. This is also the way we will solve matrix equations using inverse matrices.

Suppose you wanted to solve the system of linear equations

$$\begin{aligned} a n_1 + b n_2 &= v_1, \\ c n_1 + d n_2 &= v_2 \end{aligned} \tag{P2.1}$$

for the two unknown variables n_1 and n_2. For example, equation (P2.1) might represent the conditions for a model to be at equilibrium (see Chapter 7). First, let us solve (P2.1) by hand using Recipe P2.1.

Recipe P2.1

Solving a system of d linear equations for the d unknowns $n_1, \ldots, n_d$.

Step 1: Choose the simplest of the equations, and solve for one of the unknowns (say, n_j) in terms of the other unknowns. Record this solution.

Step 2: Eliminate the equation chosen in Step 1 and replace n_j in all remaining equations with the solution found in Step 1. Simplify the remaining equations where possible.

Step 3: Return to Step 1 until one equation remains. Solve this one remaining equation for the one remaining unknown.

Step 4: Work backward through the solutions that you have recorded, at each step updating the solution by replacing all other unknowns by their solutions. When you have finished, you should have d solutions for the d unknowns.

Equations (P2.1) are equally complex, so let us solve the first equation for n_1 (Step 1), giving $n_1 = (v_1 - b\ n_2)/a$, which we record. We then toss out the first equation and plug this solution for n_1 into the second equation (Step 2), giving $c\ (v_1 - b\ n_2)/a + d\ n_2 = v_2$. We can simplify this equation by bringing all terms involving the unknown variable n_2 to one side, $n_2\ (-c\ b/a + d) = v_2 - c\ v_1/a$. As this is our last equation, we solve it for the one remaining unknown variable n_2, and then multiply the numerator and denominator by a to avoid fractions over fractions, giving us $n_2 = (a\ v_2 - c\ v_1)/(a\ d - b\ c)$ (Step 3). This provides a solution for the unknown n_2 in terms of the parameters. Finally, we plug this solution for n_2 into the recorded equation from Step 1, $n_1 = (v_1 - b\ n_2)/a$, which can be factored to give the solution $n_1 = (d\ v_1 - b\ v_2)/(a\ d - b\ c)$ (Step 4).

Alternatively, we can write equations (P2.1) using matrix notation as

$$\mathbf{M}\ \vec{n} = \vec{v} \tag{P2.2}$$

where $\mathbf{M} = \begin{pmatrix} a & b \\ c & d \end{pmatrix}$, $\vec{n} = \begin{pmatrix} n_1 \\ n_2 \end{pmatrix}$, and $\vec{v} = \begin{pmatrix} v_1 \\ v_2 \end{pmatrix}$. Using the definition of an inverse matrix, equation (P2.2) can by solved immediately by multiplying both sides on the left by $\mathbf{M}^{-1}$, giving the solution

$$\vec{n} = \mathbf{M}^{-1}\ \vec{v}. \tag{P2.3}$$

(We multiply on the left by $\mathbf{M}^{-1}$ so that $\mathbf{M}\ \vec{n}$ becomes $\mathbf{M}^{-1}\ \mathbf{M}\ \vec{n} = \mathbf{I}\ \vec{n}$, which equals $\vec{n}$. Multiplying on the right would give us $\mathbf{M}\ \vec{n}\ \mathbf{M}^{-1}$, which does not equal $\vec{n}$.) You should calculate $\mathbf{M}^{-1}\ \vec{v}$ using Rule P2.23b and show that it generates the same solutions as in the last paragraph. There is even a third method for solving linear equations, known as the *method of row reduction*, which is particularly useful for numerical solutions to larger matrix equations (it is described in Supplementary Material P2.3). All three methods are mathematically equivalent, however, so you are free to choose the one that is easiest for any particular application.

Exercise P2.11: tRNA molecules are key to the translation of mRNA into proteins. tRNAs can exist in one of two states, unbound or bound to an amino acid. Let the number of unbound tRNA molecules within a cell equal n_1, and let the number that are bound equal n_2. If α is the rate of tRNA production by transcription, β is the rate of amino acid binding, γ is the rate of amino acid loss (e.g., by translating an mRNA into a protein), and δ is the rate of tRNA degradation, a reasonable model for the change in the number of tRNA molecules over time might be

$$\frac{dn_1}{dt} = \alpha - \beta n_1 + \gamma n_2 - \delta n_1,$$

$$\frac{dn_2}{dt} = \beta n_1 - \gamma n_2 - \delta n_2.$$

These equations can be written in matrix form as $d\vec{n}/dt = \mathbf{M}\vec{n} - \vec{v}$. Determine the matrix $\mathbf{M}$ and the constant vector $\vec{v}$. Any equilibrium to

(continued)

the model must satisfy $\mathbf{M}\vec{n} = \vec{v}$. Find the inverse matrix $\mathbf{M}^{-1}$, and use it to obtain the equilibrium number of unbound and bound tRNAs from $\vec{n} = \mathbf{M}^{-1}\vec{v}$.

P2.8 The Eigenvalues of a Matrix

Eigenvalues can be used to determine whether a matrix stretches or shrinks a system.

In biological models, arguably the most important attributes of a matrix are its eigenvalues, which can be used to predict whether variables grow or shrink over time (Chapters 7–9). By definition, an eigenvalue λ of an invertible square matrix $\mathbf{M}$ satisfies the equation

$$\mathbf{M}\,\vec{u} = \lambda\,\vec{u} \tag{P2.4a}$$

for some nonzero vector $\vec{u}$. ($\vec{u}$ must point in some direction; it cannot be a vector of all zeros.)

But how do we find the eigenvalues λ of a matrix? First, let us try to find $\vec{u}$ by bringing all terms involving $\vec{u}$ to one side: $\mathbf{M}\,\vec{u} - \lambda\vec{u} = \vec{0}$, where $\vec{0}$ is a vector of zeros. At this point, we cannot just factor out $\vec{u}$ to get $(\mathbf{M} - \lambda)\vec{u} = \vec{0}$, because this involves an invalid subtraction (the scalar λ does not have the same number of dimensions as the matrix $\mathbf{M}$). A trick that we will use repeatedly in such cases is to multiply terms that are not multiplied by a matrix by the identity matrix $\mathbf{I}$. Here, we rewrite $\mathbf{M}\,\vec{u} - \lambda\vec{u} = \vec{0}$ as $\mathbf{M}\,\vec{u} - \lambda\mathbf{I}\,\vec{u} = \vec{0}$. In accordance with the distributive law (Rule P2.12), this equation can now be factored to give

$$(\mathbf{M} - \lambda\mathbf{I})\vec{u} = \vec{0} \tag{P2.4b}$$

If the matrix $(\mathbf{M} - \lambda\mathbf{I})$ were invertible, then the vector $\vec{u}$ would be given by $\vec{u} = (\mathbf{M} - \lambda\mathbf{I})^{-1}\vec{0}$. Any time a matrix is multiplied by a vector of zeros, the result is a vector of zeros, $\vec{0}$, indicating that $\vec{u}$ would be a vector of zeros. But we said earlier that $\vec{u}$ must point in some direction and that it cannot be a vector of zeros. The only way out of this apparent contradiction is to conclude that $(\mathbf{M} - \lambda\mathbf{I})$ must not be invertible, which requires that the determinant of $(\mathbf{M} - \lambda\mathbf{I})$ be zero. Indeed, this conclusion allows us to determine the eigenvalues of a matrix without having to calculate the vector $\vec{u}$:

Definition P2.7:
The *characteristic polynomial* of an invertible $d \times d$ matrix $\mathbf{M}$ is defined as $\text{Det}(\mathbf{M} - \lambda\mathbf{I}) = 0$, which is a dth-order polynomial in λ. The d eigenvalues of the matrix $\mathbf{M}$ are the d roots of this characteristic polynomial: $\lambda_1, \lambda_2, \ldots, \lambda_d$.

Eigenvalues are found by determining the roots of a *characteristic polynomial.*

For example, in the $d = 2$ case with $\mathbf{M} = \begin{pmatrix} a & b \\ c & d \end{pmatrix}$,

$$(\mathbf{M} - \lambda\mathbf{I}) = \begin{pmatrix} a - \lambda & b \\ c & d - \lambda \end{pmatrix}, \tag{P2.5}$$

so that

$$\text{Det}(\mathbf{M} - \lambda\mathbf{I}) = (a - \lambda)(d - \lambda) - bc \tag{P2.6}$$
$$= \lambda^2 - \lambda(a + d) + (ad - bc) = 0.$$

Equation (P2.6) is the characteristic polynomial of **M**. For a 2×2 matrix, the characteristic polynomial is a quadratic equation. The two roots of this equation are the two eigenvalues of **M**. Using the quadratic formula (Rule A1.10), these eigenvalues are

$$\lambda_1 = \frac{(a + d) + \sqrt{(a + d)^2 - 4(ad - bc)}}{2}, \tag{P2.7a}$$

$$\lambda_2 = \frac{(a + d) - \sqrt{(a + d)^2 - 4(ad - bc)}}{2}. \tag{P2.7a}$$

Alternatively, we can write the eigenvalues for a 2×2 matrix in terms of its trace and determinant:

$$\lambda_1 = \frac{\text{Tr}(\mathbf{M}) + \sqrt{(\text{Tr}(\mathbf{M}))^2 - 4\,\text{Det}(\mathbf{M})}}{2} \tag{P2.8a}$$

$$\lambda_2 = \frac{\text{Tr}(\mathbf{M}) - \sqrt{(\text{Tr}(\mathbf{M}))^2 - 4\,\text{Det}(\mathbf{M})}}{2}. \tag{P2.8b}$$

A good check that you've done your algebra correctly is that the eigenvalues of a 2×2 matrix **M** should sum to Tr(**M**) (i.e., to the sum of the diagonal elements of **M**). Another good check is that the product of the eigenvalues should equal Det(**M**). (You can demonstrate these facts using equation (P2.8).) These relationships remain true for larger matrices and can be extremely handy. For example

Rule P2.25:
The two eigenvalues of a 2×2 matrix must have the same sign if their product Det(**M**) is positive. When Det(**M**) > 0, the real parts of both eigenvalues will be positive if their sum Tr(**M**) is positive and will be negative if Tr(**M**) is negative. When Det(**M**) < 0, one eigenvalue will be positive and one negative.

To get some practice finding eigenvalues, let us consider some numerical examples:

$$\mathbf{A} = \begin{pmatrix} 2 & 4 \\ 0 & 3 \end{pmatrix}, \quad \mathbf{B} = \begin{pmatrix} 2 & 4 & 7 \\ 0 & 3 & 5 \\ 0 & 0 & 1 \end{pmatrix}, \quad \mathbf{C} = \begin{pmatrix} 1 & 0 & 0 \\ 3 & 4 & -3 \\ 2 & 2 & -1 \end{pmatrix}. \tag{P2.9}$$

Following Definition P2.7, the two eigenvalues of **A** are the two roots of Det $(\mathbf{A} - \lambda I) = 0$. Taking the determinant of

$$\begin{pmatrix} 2 - \lambda & 4 \\ 0 & 3 - \lambda \end{pmatrix},$$

we get the characteristic polynomial of **A**: $(2 - \lambda)(3 - \lambda) = 0$. This equation has two solutions $\lambda = 2$ and 3, which are the two eigenvalues of **A**. Alternatively, because **A** is a 2×2 matrix, we can use equation (P2.8) to write the eigenvalues in terms of the trace and determinant, $\text{Tr}(\mathbf{A}) = 5$ and $\text{Det}(\mathbf{A}) = 6$:

$$\lambda_1 = \frac{5 + \sqrt{(5)^2 - 4(6)}}{2} = 3,$$

$$\lambda_2 = \frac{5 - \sqrt{(5)^2 - 4(6)}}{2} = 2.$$

Similarly, we can use Definition P2.7 to find the eigenvalues of **B**, by first finding the determinant,

$$\text{Det}(\mathbf{B} - \lambda \mathbf{I}) = \begin{vmatrix} 2 - \lambda & 4 & 7 \\ 0 & 3 - \lambda & 5 \\ 0 & 0 & 1 - \lambda \end{vmatrix}.$$

Because this is an upper triangular matrix, we can use Rule P2.20 to identify the determinant quickly as the product of the diagonal elements: $(2 - \lambda)(3 - \lambda)$ $(1 - \lambda)$. Setting this to zero gives a characteristic polynomial whose three roots (the three eigenvalues of **B**) are easy to find: $\lambda = 2$, $\lambda = 3$, and $\lambda = 1$.

These examples suggest a more general rule. Because the determinant of any diagonal, upper triangular, or lower triangular matrix is the product of the diagonal elements (Rule P2.20), the characteristic polynomial of such a matrix **M** always factors into a product of terms $(m_{ii} - \lambda)$ involving only the diagonal elements m_{ii}. As a consequence,

Rule P2.26:
The eigenvalues of a diagonal or triangular matrix are the elements along the diagonal ($\lambda_1 = m_{11}, \lambda_2 = m_{22}, \ldots, \lambda_d = m_{dd}$).

For a broader class of matrices, the following rules can help to identify eigenvalues:

Rule P2.27:
The eigenvalues of a block-diagonal matrix or a block-triangular matrix are the eigenvalues of the submatrices along the diagonal.

Rule P2.28:
If a matrix **M** contains a row (or a column) that is all zeros except for the element along the diagonal (m_{ii}), one of the eigenvalues of the matrix is m_{ii} and the remainder are the eigenvalues of a smaller matrix obtained by deleting the ith row and ith column of **M**.

As a final example, let us determine the eigenvalues of matrix **C**. The characteristic polynomial of **C** is found by calculating the determinant:

$$\text{Det}(\mathbf{C} - \lambda\mathbf{I}) = \begin{vmatrix} 1-\lambda & 0 & 0 \\ 3 & 4-\lambda & -3 \\ 2 & 2 & -1-\lambda \end{vmatrix} = 0.$$

Using Rule P2.16, this determinant is

$$\text{Det}(\mathbf{C} - \lambda\mathbf{I}) = (1-\lambda)\begin{vmatrix} 4-\lambda & -3 \\ 2 & -1-\lambda \end{vmatrix} - (0)\begin{vmatrix} 3 & -3 \\ 2 & -1-\lambda \end{vmatrix} + (0)\begin{vmatrix} 3 & 4-\lambda \\ 2 & 2 \end{vmatrix} = 0.$$

The last two terms are zero, which leaves us with

$$\begin{aligned} \text{Det}(\mathbf{C} - \lambda\mathbf{I}) &= (1-\lambda)\begin{vmatrix} 4-\lambda & -3 \\ 2 & -1-\lambda \end{vmatrix} \\ &= (1-\lambda)((4-\lambda)(-1-\lambda) + 6) \\ &= (1-\lambda)(2 - 3\lambda + \lambda^2) \\ &= (1-\lambda)(2-\lambda)(1-\lambda). \end{aligned}$$

This characteristic polynomial also factors nicely (we chose **C** so that it would), allowing us to identify the three roots of the characteristic polynomial: $\lambda = 1$, $\lambda = 2$, and $\lambda = 1$. Because $\lambda = 1$ appears twice in this list, we say that matrix **C** has a *repeated* eigenvalue.

In this example, we can find the eigenvalues of **C** faster by recognizing that it is a block-triangular matrix and applying Rule P2.27. Matrix **C** can be written in block form as $\begin{pmatrix} \mathbf{E} & \mathbf{F} \\ \mathbf{G} & \mathbf{H} \end{pmatrix}$, where $\mathbf{E} = (1)$, $\mathbf{F} = (0\ \ 0)$, $\mathbf{G} = \begin{pmatrix} 3 \\ 2 \end{pmatrix}$, and $\mathbf{H} = \begin{pmatrix} 4 & -3 \\ 2 & -1 \end{pmatrix}$. Then according to Rule P2.27, the eigenvalues of **C** are the eigenvalues of the diagonal submatrices **E** and **H**. A 1×1 matrix, like **E**, is the simplest example of a diagonal matrix, and the one element in the matrix is its eigenvalue (Rule P2.26). Here, $\lambda = 1$. That leaves us to find the eigenvalues of **H**. Taking the determinant $\text{Det}(\mathbf{H} - \lambda\mathbf{I})$, we get the characteristic polynomial of **H**, $(4 - \lambda)(-1 - \lambda) + 6 = 0$, which can be factored into $(2 - \lambda)(1 - \lambda) = 0$, indicating that the second and third eigenvalues are $\lambda = 2$ and $\lambda = 1$.

This example illustrates why writing matrices in block-diagonal form or block-triangular form can be so helpful: doing so allows you to determine the eigenvalues of a large matrix from the eigenvalues of smaller matrices. Furthermore, there are elementary matrix operations that can be applied to the matrix ($\mathbf{M} - \lambda\mathbf{I}$) without altering its eigenvalues (see Sup. Mat. P2.2). These operations can be used to massage ($\mathbf{M} - \lambda\mathbf{I}$) into block-triangular form so that Rule P2.27 can be applied. Alternatively, these operations can be used to create a row (or a column) whose entries are all zero except on the diagonal, so that Rule P2.28 can be applied. These techniques allow you to reduce the size of a matrix until it becomes manageable.

In the above examples, the eigenvalues have been real numbers. Even when a matrix contains only real elements, its eigenvalues can be complex numbers. We can see this clearly for 2×2 matrices from expression (P2.7), which will yield complex values whenever $(a + d)^2 - 4(ad - bc) < 0$. The matrix $\begin{pmatrix} 1 & -1 \\ 1 & 1 \end{pmatrix}$ is a numerical example, having complex eigenvalues $\lambda = 1 + \sqrt{-1}$ and $\lambda = 1 - \sqrt{-1}$. As we shall see, complex eigenvalues are typical of models in which the variables cycle over time (see Chapters 7–9). For certain matrices, it is possible to tell whether or not the eigenvalues will be real. For example,

Eigenvalues that are complex numbers indicate cycling.

Rule P2.29:
If a matrix $\mathbf{M}$ is symmetric above and below the diagonal (i.e., $m_{ij} = m_{ji}$ for all i and j), all of its eigenvalues are real.

In Appendix 3, we describe the Perron-Frobenius theorem, which can be used to determine the type of eigenvalues for a broader class of matrices. As this material is more advanced, we recommend that it be read later, along with Chapter 10, where it is more extensively used.

In section P2.6, we said that the determinant must not be zero for a matrix to have an inverse. Yet we can still calculate the eigenvalues of a matrix whose determinant is zero. Interestingly, when the determinant is zero, one eigenvalue always equals zero. That is, when $\text{Det}(\mathbf{M}) = 0$, λ factors out of the characteristic polynomial, because the constant terms in the characteristic polynomial (i.e., the terms not involving λ) are given by $\text{Det}(\mathbf{M})$. As mentioned earlier, multiplication by a matrix whose determinant is zero causes a loss of information along some axis (see Figure P2.6). Regardless of where the system starts, it reaches zero along this axis in the next time step. The eigenvalue of zero reflects this loss of information, indicating that there is some direction in which the system immediately shrinks to zero. Eigenvalues of zero commonly arise whenever the variables of a model are constrained to satisfy a particular relationship (for example, when genotype frequencies are constrained to be at Hardy-Weinberg equilibrium).

Finally, as mentioned earlier, eigenvalues are useful for determining when multiplication by a matrix expands or shrinks the variables of a system. For example, we will use eigenvalues to predict things such as whether a mutation is likely to spread (a variable is expanding) or a population is likely to go extinct (a variable is shrinking). As we will see, the long-term fate of a system does not actually depend equally on all of its eigenvalues. By definition, the *leading eigenvalue*

The *leading eigenvalue* dominates the long-term dynamics of a system.

is the eigenvalue that dominates a system's long-term behavior. In discrete-time models, the leading eigenvalue is the eigenvalue with the largest magnitude. In continuous-time models, it is the eigenvalue with the largest real part. (We shall see why these eigenvalues dominate the long-term behaviour in Chapter 9.) For example, the leading eigenvalues of matrices **A**, **B**, and **C** are 3, 3, and 2, respectively. We will delve more deeply into the topic of leading eigenvalues in Chapters 7–9.

Exercise P2.12: Determine the eigenvalues of the following matrices using Rules P2.26–P2.28 as appropriate.

(a)
$$\begin{pmatrix} \alpha & 0 \\ \beta & 3\delta \end{pmatrix}.$$

(b)
$$\begin{pmatrix} 1 & 2 \\ 3 & 2 \end{pmatrix}.$$

(c)
$$\begin{pmatrix} 2 & 7 & 5 \\ 0 & 19 & 0 \\ 4 & 3 & 1 \end{pmatrix}.$$

(d)
$$\begin{pmatrix} 1 & 2 & 0 \\ 3 & 2 & 0 \\ x & y & z \end{pmatrix}.$$

(e)
$$\begin{pmatrix} \alpha & \beta & \alpha\delta & \beta(\delta-\gamma) \\ \beta & \alpha & \beta(\delta-\gamma) & \alpha\delta \\ 0 & 0 & \delta & \delta-\gamma \\ 0 & 0 & \delta-\gamma & \delta \end{pmatrix}.$$

Exercise P2.13: A simple model of ecological succession in a forest assumes that early successional forests (E) give rise at some rate to mid-successional forests (M), which in turn give rise to late successional forests (L), where E, M, and L represent the fraction of all forests in each state. If late successional forests undergo disturbance (e.g., fire) at some rate to regenerate early successional forests, ecological succession can be described by a system of differential equations:

$$\begin{pmatrix} dE/dt \\ dM/dt \\ dL/dt \end{pmatrix} = \begin{pmatrix} -a & 0 & c \\ a & -b & 0 \\ 0 & b & -c \end{pmatrix} \begin{pmatrix} E \\ M \\ L \end{pmatrix}.$$

Using Definition P2.7 and Rule P2.16, what is the characteristic polynomial for the above rate matrix? What are its three eigenvalues?

P2.9 The Eigenvectors of a Matrix

We began the last section by saying that an eigenvalue λ of a matrix $\mathbf{M}$ satisfies the equation $\mathbf{M}\vec{u} = \lambda\vec{u}$ for some nonzero vector $\vec{u}$. But we were able to solve for the eigenvalues using Definition P2.7 without ever specifying what this vector was. The vector $\vec{u}$ is known as an *eigenvector* of matrix $\mathbf{M}$. Eigenvectors play an important role in dynamical models too, as they specify the directions along which a system tends to expand or shrink without being rotated. Each eigenvalue of a matrix, λ_i, has an associated eigenvector, $\vec{u}_i$. Once an eigenvalue is identified, its associated eigenvector can be found by solving

Eigenvectors specify directions in which a system expands or shrinks.

$$\mathbf{M}\vec{u}_i = \lambda_i \vec{u}_i. \tag{P2.10}$$

Equation (P2.10) tells us something interesting and special about an eigenvector: when multiplied by the matrix $\mathbf{M}$, an eigenvector is stretched or shrunk by a factor equal to its associated eigenvalue λ_i, but the direction of the eigenvector remains unchanged (Figure P2.7). This is the defining feature of all eigenvectors.

Equation (P2.10) describes d equations in d unknowns (the elements of the column vector $\vec{u}_i$). These equations can be solved using either Recipe P2.1 or the method of row reduction described in Supplementary Material P2.3. You might be tempted to solve equation (P2.10) by matrix manipulation, but the answer, $\vec{u}_i = (\mathbf{M} - \lambda_i \mathbf{I})^{-1}\vec{0}$, involves an inverse that is undefined because $\text{Det}(\mathbf{M} - \lambda_i \mathbf{I})$ is zero by the definition of an eigenvalue. Therefore, to illustrate how (P2.10) can be solved, let us work through a few examples.

First, we calculate the eigenvectors of $\mathbf{A} = \begin{pmatrix} 2 & 4 \\ 0 & 3 \end{pmatrix}$. We already identified the two eigenvalues as 2 and 3. To begin, we must choose one eigenvalue, say $\lambda = 2$, and find the eigenvector that satisfies equation (P2.10) with $\lambda = 2$:

$$\begin{pmatrix} 2 & 4 \\ 0 & 3 \end{pmatrix}\begin{pmatrix} u_1 \\ u_2 \end{pmatrix} = 2\begin{pmatrix} u_1 \\ u_2 \end{pmatrix}.$$

This provides us with two equations that both must be satisfied by the eigenvector $\begin{pmatrix} u_1 \\ u_2 \end{pmatrix}$:

$$\begin{aligned} 2u_1 + 4u_2 &= 2u_1, \\ 3u_2 &= 2u_2. \end{aligned}$$

The second equation can only be satisfied if $u_2 = 0$, in which case the first equation becomes $u_1 = u_1$, which is always true. Thus, any value of u_1 will satisfy the first equation when $u_2 = 0$. When finding eigenvectors, it will always be the case that the value of one of the elements is arbitrary, because eigenvectors describe a direction, but their length does not matter. Therefore, we arbitrarily choose $u_1 = 1$, giving us the eigenvector $\vec{u} = \begin{pmatrix} 1 \\ 0 \end{pmatrix}$ associated with the eigenvalue $\lambda = 2$. If you choose a larger number for u_1, you will get a longer eigenvector, but it will still point in the same direction. In general,

Rule P2.30:
An eigenvector can be multiplied by any nonzero number, and it will still represent an eigenvector of the same eigenvalue.

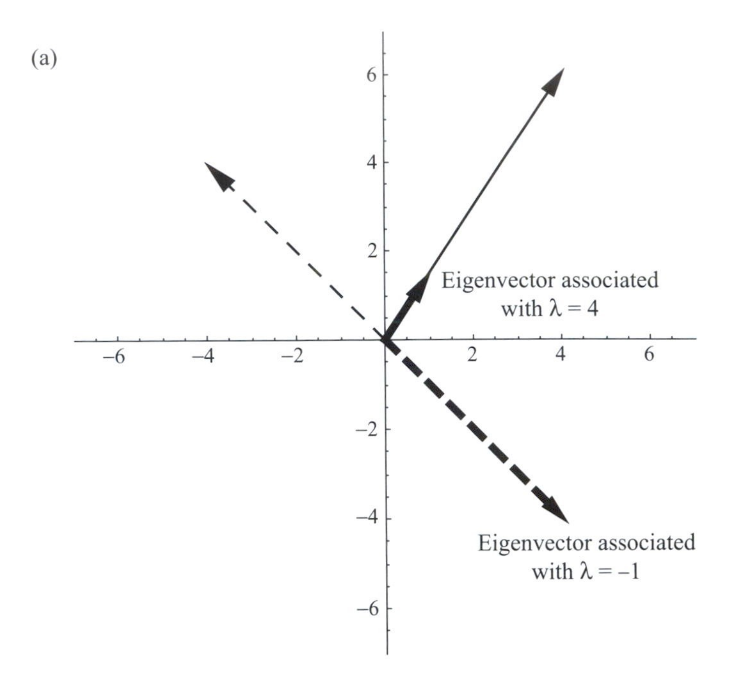

(b)

Figure P2.7: A matrix stretches or shrinks but does not rotate eigenvectors. Here, we use the matrix $\begin{pmatrix} 1 & 2 \\ 3 & 2 \end{pmatrix}$ to multiply original vectors (thick lines) to get new vectors (thin lines). (a) The two original vectors, $\begin{pmatrix} 1 \\ 3/2 \end{pmatrix}$ and $\begin{pmatrix} 4 \\ -4 \end{pmatrix}$, represent two eigenvectors of this matrix. The solid eigenvector is associated with the eigenvalue 4 and becomes stretched by a factor of 4 when multiplied on the left by the matrix. The dashed eigenvector is associated with the eigenvalue −1 and changes sign when multiplied by the matrix. (b) The two original vectors, $\begin{pmatrix} 3 \\ -1 \end{pmatrix}$ and $\begin{pmatrix} -2 \\ -0.5 \end{pmatrix}$, do not point in the direction of the eigenvectors. These vectors are rotated as well as stretched when multiplied by the matrix.

Next, we turn to the second eigenvalue, $\lambda = 3$, and find the eigenvector that satisfies equation (P2.10) with $\lambda = 3$:

$$\begin{pmatrix} 2 & 4 \\ 0 & 3 \end{pmatrix}\begin{pmatrix} u_1 \\ u_2 \end{pmatrix} = 3\begin{pmatrix} u_1 \\ u_2 \end{pmatrix},$$

which can be written out as

$$2u_1 + 4u_2 = 3u_1,$$
$$3u_2 = 3u_2.$$

Again, it is easier to start with the second equation, which is always true, so we arbitrarily choose $u_2 = 1$. Plugging this into the first equation indicates that $u_1 = 4$, so the eigenvector associated with $\lambda = 3$ is $\vec{u} = \binom{4}{1}$.

In the general case of a 2×2 matrix $\mathbf{M} = \begin{pmatrix} a & b \\ c & d \end{pmatrix}$, the eigenvectors must satisfy

$$\begin{pmatrix} a & b \\ c & d \end{pmatrix}\begin{pmatrix} u_1 \\ u_2 \end{pmatrix} = \lambda_i \begin{pmatrix} u_1 \\ u_2 \end{pmatrix},$$

which can be written out as

$$a\,u_1 + b\,u_2 = \lambda_i\,u_1,$$
$$c\,u_1 + d\,u_2 = \lambda_i\,u_2.$$

Using the first equation and arbitrarily setting $u_1 = 1$, u_2 must equal $(\lambda_i - a)/b$. Multiplying this eigenvector by b to avoid fractions, we get a general equation for the two eigenvectors of a 2×2 matrix:

$$\vec{u} = \begin{pmatrix} b \\ \lambda_i - a \end{pmatrix}, \qquad \text{(P2.11)}$$

one for each eigenvalue λ_i given by (P2.7). Alternatively, we could use the second equation to write the eigenvector as $\vec{u}_i = \binom{\lambda_i - d}{c}$, but this points in the same direction as (P2.11) if the correct eigenvalues are used (check this claim for the matrix **A** examined in the previous paragraph).

Let us next calculate the eigenvectors of a 3×3 matrix

$$\mathbf{C} = \begin{pmatrix} 1 & 0 & 0 \\ 3 & 4 & -3 \\ 2 & 2 & -1 \end{pmatrix}.$$

In the last section, we identified the three eigenvalues of this matrix as $\lambda = 1$, $\lambda = 2$, and $\lambda = 1$. We start with the eigenvalue that is not repeated, $\lambda = 2$, and find the eigenvector that satisfies equation (P2.10):

$$\begin{pmatrix} 1 & 0 & 0 \\ 3 & 4 & -3 \\ 2 & 2 & -1 \end{pmatrix}\begin{pmatrix} u_1 \\ u_2 \\ u_3 \end{pmatrix} = 2\begin{pmatrix} u_1 \\ u_2 \\ u_3 \end{pmatrix}.$$

This provides us with three equations:

$$u_1 = 2u_1,$$
$$3u_1 + 4u_2 - 3u_3 = 2u_2,$$
$$2u_1 + 2u_2 - u_3 = 2u_3.$$

The first equation can only be satisfied if $u_1 = 0$, in which case the second and third equations can both be rewritten as $2u_2 - 3u_3 = 0$. Therefore, we arbitrarily choose $u_2 = 1$ so that u_3 must equal 2/3, giving us the eigenvector

$$\vec{u} = \begin{pmatrix} 0 \\ 1 \\ 2/3 \end{pmatrix}$$

associated with the eigenvalue $\lambda = 2$. Next, we turn to the repeated eigenvalue $\lambda = 1$. Our goal is to find two eigenvectors that satisfy equation (P2.10) with $\lambda = 1$:

$$\begin{pmatrix} 1 & 0 & 0 \\ 3 & 4 & -3 \\ 2 & 2 & -1 \end{pmatrix} \begin{pmatrix} u_1 \\ u_2 \\ u_3 \end{pmatrix} = \begin{pmatrix} u_1 \\ u_2 \\ u_3 \end{pmatrix},$$

which can be written out as

$$\begin{aligned} u_1 &= u_1, \\ 3u_1 + 4u_2 - 3u_3 &= u_2, \\ 2u_1 + 2u_2 - u_3 &= u_3. \end{aligned}$$

The first equation is always true, so let us arbitrarily choose $u_1 = 1$. Plugging this into the second and third equations and bringing all the terms to the left-hand side gives us

$$\begin{aligned} 3 + 3u_2 - 3u_3 &= 0, \\ 2 + 2u_2 - 2u_3 &= 0. \end{aligned}$$

These equations are both multiples of $1 + u_2 - u_3 = 0$, so any eigenvector that satisfies this equation will work. For example, we can let $u_2 = 1$ and then $u_3 = 2$, giving us one eigenvector,

$$\vec{u} = \begin{pmatrix} 1 \\ 1 \\ 2 \end{pmatrix}$$

associated with $\lambda = 1$. But can we find a second eigenvector for the second eigenvalue of 1? Let us try another value for u_2, say $u_2 = 0$, in which case $u_3 = 1$ satisfies the equations, giving us a second eigenvector associated with $\lambda = 1$:

$$\vec{u} = \begin{pmatrix} 1 \\ 0 \\ 1 \end{pmatrix}.$$

At this point, however, we must make sure that the eigenvectors that we have chosen are not just multiples of one another. They don't appear to be, but one way to check is to place all the eigenvectors as columns in a new matrix

$$\begin{pmatrix} 0 & 1 & 1 \\ 1 & 1 & 0 \\ 2/3 & 2 & 1 \end{pmatrix}.$$

As long as the determinant of this matrix is not zero, the rows will be linearly independent, and we have made an appropriate choice of eigenvectors that point in different directions. The determinant of this matrix is 1/3, so we are fine.

We could have made many other choices along the way. We will have succeeded in finding an appropriate set of eigenvectors as long as they satisfy equation (P2.10) and are linearly independent. If you are ever unsure of your choices and your algebra, however, remember that you can always plug each eigenvalue and its eigenvector back into equation (P2.10) to make sure that they work. This is a fast check that is worth doing. It is also important to emphasize that each eigenvalue has its own associated eigenvector and that one must keep track of which belong together. If you mix up the eigenvalues and eigenvectors, they will not, in general, satisfy (P2.10).

For most $d \times d$ matrices with repeated eigenvalues, it will be possible to find d eigenvectors that point in different directions. That is, the eigenvectors will be linearly independent (i.e., if placed together in a matrix, they yield a matrix whose determinant is not zero). Sometimes, however, it will prove impossible to choose eigenvectors that point in different directions every time an eigenvalue is repeated (e.g., Exercise P2.15). A $d \times d$ matrix that does not have d linearly independent eigenvectors is called "defective" (only matrices with repeated eigenvalues can be defective). The general solutions described in Chapter 9 are not valid for defective matrices, but fortunately defective matrices do not arise too often in biological models. When they do, other methods can be applied to obtain general solutions (see Further Readings).

The above results illustrate how we can go about calculating the eigenvectors associated with all the eigenvalues of a matrix, but why are these special vectors useful? Once again, graphs can provide us with a better intuitive feel. Consider the vectors in Figure P2.6a. Looking at these vectors, it is hard to "see" how they change when multiplied by the matrix $\begin{pmatrix} 0.5 & 1 \\ -1 & 2 \end{pmatrix}$ to give the vectors in Figure P2.6b. In fact, the changes in the vectors are particularly hard to see because we use the standard horizontal and vertical axes to measure the vectors. But there is no reason why we have to use this standard coordinate system as our yardstick. In fact, if we choose the eigenvectors as our axes, then matrix multiplication results in very simple changes to a vector. In the direction of each eigenvector, a vector is stretched or shrunk by a factor equal to the eigenvalue associated with that eigenvector (Figure P2.8). Eigenvectors are therefore a more natural coordinate system in which to measure changes caused by matrix multiplication, a fact that we will use in Chapter 9 to analyze models. Also notice that the eigenvectors need not be perpendicular to one another (Figure P2.8), although they will be if the matrix is symmetric ($m_{ij} = m_{ji}$).

Before ending our introduction of eigenvectors, we note that the eigenvector defined by equation (P2.10) is more specifically called the *right eigenvector* of the matrix **M**, because the matrix is multiplied on the *right* by the eigenvector. Although the right eigenvector is more commonly encountered (e.g., in Chapters 7–9), the left eigenvector also appears in various applications (e.g., in Chapters 10 and 14). As you might have guessed, the *left eigenvector* of **M** is defined by

$$\vec{v}_i^T \mathbf{M} = \lambda_i \vec{v}_i^T, \tag{P2.12}$$

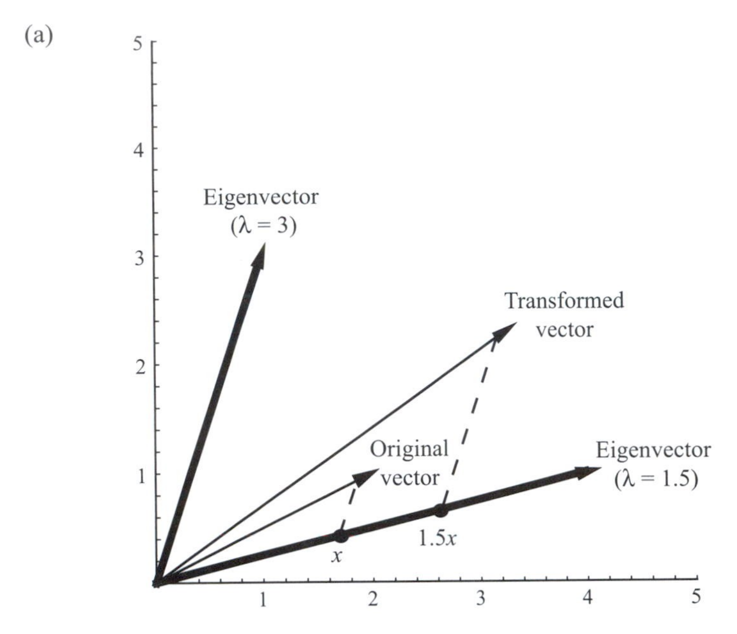

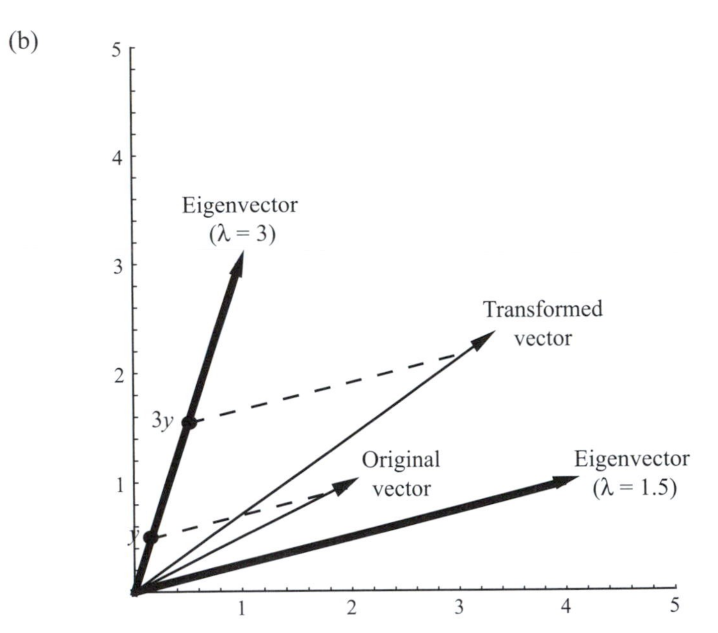

Figure P2.8: Matrix multiplication visualized using eigenvectors as axes. The matrix $\mathbf{M} = \begin{pmatrix} 15/11 & 6/11 \\ -9/22 & 69/22 \end{pmatrix}$ has one eigenvector of $\begin{pmatrix} 4 \\ 1 \end{pmatrix}$ associated with an eigenvalue of 1.5 and a second eigenvector of $\begin{pmatrix} 1 \\ 3 \end{pmatrix}$ associated with an eigenvalue of 3. These eigenvectors are drawn using thick lines. An original vector $\begin{pmatrix} 2 \\ 1 \end{pmatrix}$ is multiplied on the left by $\mathbf{M}$ to get the transformed vector $\begin{pmatrix} 36/11 \\ 51/22 \end{pmatrix}$. This transformation does not appear to follow a pattern. (a) If we project the original and transformed vectors onto the first eigenvector by following the dashed lines, we see, however, that the transformed vector has a length that is $\lambda = 1.5$ times the length of the original vector (x). (b) Similarly, if we project the vectors onto the second eigenvector, we see that the transformed vector has a length that is $\lambda = 3$ times the length of the original vector (y). Thus, multiplication by $\mathbf{M}$ stretches a vector along each eigenvector by an amount proportional to the associated eigenvalue. The dashed lines that allow us to project a vector onto an eigenvector are parallel to the other eigenvector.

where now the matrix is multiplied on the *left* by the eigenvector $\vec{v}_i$, transposed to become a row vector. When it is not specified whether an eigenvector is "right" (P2.10) or "left" (P2.12), then we will take it for granted that we are referring to the right eigenvector.

The right and left eigenvectors of a matrix are not generally the same. To demonstrate this point, we now find the left eigenvectors of $\mathbf{A} = \begin{pmatrix} 2 & 4 \\ 0 & 3 \end{pmatrix}$.

Beginning again with the eigenvalue $\lambda = 2$, the left eigenvector must satisfy equation (P2.12) with $\lambda = 2$:

$$(v_1 \quad v_2)\begin{pmatrix} 2 & 4 \\ 0 & 3 \end{pmatrix} = 2(v_1 \quad v_2).$$

This provides us with two equations that both must be satisfied by the left eigenvector (v_1, v_2):

$$\begin{aligned} 2v_1 &= 2v_1, \\ 4v_1 + 3v_2 &= 2v_2. \end{aligned}$$

These equations are not satisfied by the right eigenvector $\binom{1}{0}$. To find the left eigenvector, we set v_1 to 1. This solves the first equation and causes the second equation to become $4 + 3v_2 = 2v_2$, which is true if $v_2 = -4$. Thus, we have a left eigenvector of $\vec{v}^{\mathrm{T}} = (1, -4)$ associated with the eigenvalue $\lambda = 2$. Next, we find the left eigenvector that satisfies equation (P2.10) with $\lambda = 3$:

$$(v_1 \quad v_2)\begin{pmatrix} 2 & 4 \\ 0 & 3 \end{pmatrix} = 3(v_1 \quad v_2),$$

which can be written out as

$$\begin{aligned} 2v_1 &= 3v_1, \\ 4v_1 + 3v_2 &= 3v_2. \end{aligned}$$

The first equation will never be true unless we choose $v_1 = 0$. Plugging this into the second equation leaves $3v_2 = 3v_2$, which is true for all values of v_2. We can arbitrarily choose $v_2 = 1$ giving the left eigenvector $\vec{v}^{\mathrm{T}} = (0, 1)$ associated with $\lambda = 3$ (the only bad choice for v_2 would be 0 because then all elements of the eigenvector would be zero).

Exercise P2.14: For the following matrices, find right eigenvectors associated with the eigenvalues found in Exercise P2.12:

(a)
$$\begin{pmatrix} \alpha & 0 \\ \beta & 3\delta \end{pmatrix}.$$

(b)
$$\begin{pmatrix} 1 & 2 \\ 3 & 2 \end{pmatrix}.$$

(c)
$$\begin{pmatrix} 2 & 7 & 5 \\ 0 & 19 & 0 \\ 4 & 3 & 1 \end{pmatrix}.$$

Exercise P2.15: Find the eigenvalues, right eigenvectors, and left eigenvectors of the following matrices. Specify which matrices are defective.

(a) $$\begin{pmatrix} 7 & -10 \\ 5 & -8 \end{pmatrix},$$

(b) $$\begin{pmatrix} 3 & 5 \\ 3 & 1 \end{pmatrix},$$

(c) $$\begin{pmatrix} 1 & 1 \\ -1 & 3 \end{pmatrix}.$$

(d) $$\begin{pmatrix} 3 & -2 \\ 2 & -1 \end{pmatrix}.$$

Further Readings

The following books offer more information on linear algebra at an introductory level with many exercises:

- Lay, D. C. 2003. *Linear Algebra and Its Applications*, 3rd ed. Addison-Wesley-Longman, New York.
- Lay, D. C. 2003. *Student's Study Guide*, 3rd ed. Addison-Wesley-Longman, New York.
- Lipschutz, S., and M. Lipson. 2000. *Schaum's Outline of Linear Algebra*, 3rd ed. Schaum's Outline Series.

The following books offer information on linear algebra at an advanced level, describing powerful methods and proofs in linear algebra:

- Gantmacher, F. R. 1990. *Matrix Theory*, Vol. 1, 2nd ed., Chelsea Publications, London.
- Gantmacher, F. R. 2000. *Theory of Matrices*, Vol. 2, 2nd ed., Chelsea Publications, London.

Answers to Exercises

Exercise P2.1

(a) Two-dimensional row vector (or a 1×2 matrix) whose transpose is $\begin{pmatrix} 2 \\ 4 \end{pmatrix}$. (b) Three-dimensional column vector (or a 3×1 matrix) whose transpose is $(1 \quad 2 \quad 3)$. (c) A 2×2 matrix whose transpose is $\begin{pmatrix} 6 & 2 \\ -4 & 0 \end{pmatrix}$. (d) A 2×3 matrix whose transpose is $\begin{pmatrix} 1+x & 5 \\ 2+x & 7 \\ 3+x & 9 \end{pmatrix}$. (e) A 3×2 matrix whose transpose is $\begin{pmatrix} a & c & e \\ b & d & f \end{pmatrix}$.

Exercise P2.2

Only (b), (c), and (e) are valid. (b)$\begin{pmatrix} 7 \\ 11 \end{pmatrix}$. (c)$\begin{pmatrix} 5 \\ 7 \\ 9 \end{pmatrix}$. (e) $(0.5 + a \quad 1.3 + b \quad 2 + c)$.

Exercise P2.3

(a), (c), (d), and (e) are valid. (a) $\begin{pmatrix} 7 & 6 \\ 5 & 4 \end{pmatrix}$. (c) $\begin{pmatrix} a-1 & b-2 \\ c-3 & d-4 \end{pmatrix}$. (d) $\begin{pmatrix} 2a & b+a & c+a \\ 2d & e+d & f+d \end{pmatrix}$.
(e) $\begin{pmatrix} 0 & 0 & 0 \\ 0 & 0 & 0 \\ 0 & 0 & 0 \end{pmatrix}$.

Exercise P2.4

Scalar multiplication is always valid. (a) $(6 \quad 12)$. (b) $\begin{pmatrix} -4 \\ -5 \\ -6 \end{pmatrix}$. (c) $\begin{pmatrix} 2 & 4 \\ 6 & 8 \end{pmatrix}$.
(d) $\begin{pmatrix} a & 2a & 3a \\ 5a & 7a & 9a \end{pmatrix}$. (e) $\begin{pmatrix} 2+a & 4+2a \\ 6+3a & 8+4a \end{pmatrix}$.

Exercise P2.5

(a) $\mathbf{AB} = \begin{pmatrix} ae+bg & af+bh \\ ce+dg & cf+dh \end{pmatrix}$, which does not equal $\mathbf{BA} = \begin{pmatrix} ea+fc & eb+fd \\ ga+hc & gb+hd \end{pmatrix}$.

Exercise P2.6

The trace of matrices **A** and **B** is 5. The trace of matrix **C** is −5. All of these matrices have the same determinant, namely, −2.

Exercise P2.7

Because the determinant of matrices **D** and **F** is zero, their rows are not independent. For matrix **D**, the second row is −3 times the first row. For matrix **F**, the second row is x times the first row. Only matrix **E** has linearly independent rows, because its determinant is not zero but −6.

Exercise P2.8

Using Definition P2.6, the determinant of **A** is 6, which equals the product of the diagonal elements (2×3) because this is an upper triangular matrix. Using Rule P2.16, the determinant of **B** is also 6, which equals the product of the diagonal elements ($2 \times 3 \times 1$) because this is an upper triangular matrix.

Exercise P2.9

(a) $\begin{pmatrix} 1 & 0 \\ 0 & 1 \end{pmatrix}$. (b) $\begin{pmatrix} 1 & 0 \\ 0 & 1/3 \end{pmatrix}$. (c) This matrix is noninvertible because its determinant is zero. (d) $\begin{pmatrix} 2 & -1 \\ -1/2 & 1/2 \end{pmatrix}$. (e) $\begin{pmatrix} -2 & 1 \\ 3/2 & -1/2 \end{pmatrix}$.

Exercise P2.11

$\mathbf{M} = \begin{pmatrix} -\beta-\delta & \gamma \\ \beta & -\gamma-\delta \end{pmatrix}$, $\vec{v} = \begin{pmatrix} -\alpha \\ 0 \end{pmatrix}$, and

$$\mathbf{M}^{-1} = \begin{pmatrix} \dfrac{-\gamma-\delta}{\delta(\beta+\gamma+\delta)} & \dfrac{-\gamma}{\delta(\beta+\gamma+\delta)} \\ \dfrac{-\beta}{\delta(\beta+\gamma+\delta)} & \dfrac{-\beta-\delta}{\delta(\beta+\gamma+\delta)} \end{pmatrix}.$$

Thus,

$$\vec{n} = \begin{pmatrix} \dfrac{\gamma\alpha + \delta\alpha}{\delta(\beta + \gamma + \delta)} \\ \dfrac{\beta\alpha}{\delta(\beta + \gamma + \delta)} \end{pmatrix}$$

is the equilibrium number of unbound (top row) and bound (bottom row) tRNAs.

Exercise P2.12

(a) $\lambda_1 = \alpha$ and $\lambda_2 = 3\delta$ using Rule P2.26. (b) $\lambda_1 = 4$ and $\lambda_2 = -1$ using Definition P2.7. (c) $\lambda_1 = 6$, $\lambda_2 = -3$, and $\lambda_3 = 19$ using Rule P2.28 and Definition P2.7. (d) $\lambda_1 = 4$, $\lambda_2 = -1$, and $\lambda_3 = z$ using Rule P2.27 or P2.28 and the answer to (b). (e) $\lambda_1 = \alpha + \beta$, $\lambda_2 = \alpha - \beta$, $\lambda_3 = \gamma$, and $\lambda_4 = 2\delta - \gamma$ using Rule P2.27 and Definition P2.7.

Exercise P2.13

(a) The characteristic polynomial is $-\lambda\,(a\,b + a\,c + b\,c + (a + b + c)\,\lambda + \lambda^2) = 0$. (b) $\lambda_1 = 0$, $\lambda_2 = \left(-(a + b + c) + \sqrt{(a + b + c)^2 - 4(ab + ac + bc)}\right)/2$, and $\lambda_3 = \left(-(a + b + c) - \sqrt{(a + b + c)^2 - 4(ab + ac + bc)}\right)/2$. Note that λ_2 and λ_3 can be complex numbers. For example, if all the transition rates equal a, these eigenvalues become $-3/2\,a \pm \sqrt{3}ai/2$, where $i = \sqrt{-1}$. As described in Chapter 8, complex eigenvalues are typical of models with cyclic dynamics.

Exercise P2.14

The eigenvector that you choose is correct if it is a constant multiple of the following choices (i.e., if it points along the same line). (a) $\vec{u} = \binom{0}{1}$ associated with $\lambda_1 = 3\delta$ and $\vec{u} = \binom{1}{\beta/(\alpha - 3\delta)}$ associated with $\lambda_2 = \alpha$. (b) $\vec{u} = \binom{2}{3}$ associated with $\lambda_1 = 4$ and $\vec{u} = \binom{1}{-1}$ associated with $\lambda_2 = -1$. (c) $\vec{u} = \begin{pmatrix} 1 \\ 0 \\ 4/5 \end{pmatrix}$ associated with $\lambda_1 = 6$, $\vec{u} = \begin{pmatrix} 1 \\ 0 \\ -1 \end{pmatrix}$ associated with $\lambda_2 = -3$, and $\vec{u} = \begin{pmatrix} 141 \\ 286 \\ 79 \end{pmatrix}$ associated with $\lambda_3 = 19$.

Exercise P2.15

The eigenvector that you choose is correct if it is a constant multiple of the following choices (i.e., if it points along the same line). (a) Left $\vec{v}^{\mathrm{T}} = (1, -2)$ and right $\vec{u} = \binom{1}{1}$ associated with $\lambda_1 = -3$; left $\vec{v}^{\mathrm{T}} = (1, -1)$ and right $\vec{u} = \binom{2}{1}$ associated with $\lambda_2 = 2$. (b) Left $\vec{v}^{\mathrm{T}} = (1, 1)$ and right $\vec{u} = \binom{5}{3}$ associated with $\lambda_1 = 6$; left $\vec{v}^{\mathrm{T}} = (3, -5)$ and right $\vec{u} = \binom{1}{-1}$ associated with $\lambda_2 = -2$. (c) Left

$\vec{v}^{\mathrm{T}} = (1, -1)$ and right $\vec{u} = \binom{1}{1}$ associated with $\lambda_1 = 2$, and there are no other independent choices for $\lambda_2 = 2$ (the matrix is defective). (d) Left $\vec{v}^{\mathrm{T}} = (1, -1)$ and right $\vec{u} = \binom{1}{1}$ associated with $\lambda_1 = 1$ and there are no other independent choices for $\lambda_2 = 1$ (the matrix is defective).

CHAPTER 7

Equilibria and Stability Analyses—Linear Models with Multiple Variables

Chapter Goals:

- To construct dynamical models involving linear combinations of multiple variables
- To find equilibria of these models
- To analyze the stability of these equilibria

Chapter Concepts:

- Multivariable equilibria
- Eigenvectors
- Multivariable linear models
- Eigenvalues
- Characteristic polynomial
- Leading eigenvalues in continuous- and discrete-time models

7.1 Introduction

At this stage we have a considerable suite of techniques at our disposal for constructing and analyzing models, but there are still some very important tools that are missing. Suppose, for example, that we want to know the vaccination coverage that is required to eradicate a disease. Or suppose we want to decide how best to conserve an endangered elephant population. Or suppose we want to develop theory to explain why females in many birds of paradise have evolved mating preferences for males with very extravagant and gaudy-looking tails. To construct an adequate model for any of these questions requires that we keep track of more than one variable. To model vaccination programs, we need to track the numbers of infected and susceptible hosts, at the very least. In the case of endangered elephants, we need to track the numbers of individuals of each age, because age strongly influences survival and reproduction. Finally, in the case of birds of paradise, we need to track the evolutionary dynamics of both female preference and male trait. All these examples require us to analyze models with two or more variables whose dynamics depend upon one another.

In some ways, multivariable models pose no new conceptual issues. Just as with one-variable models, equilibria are found by using the dynamical equations, together with the condition that none of the variables are changing (Definition 5.1). Similarly, the stability of an equilibrium point is determined by considering the dynamics of populations near the equilibrium (Definition 5.2). The difference is that, with more than one variable, there are now many different directions in which the system can move, and therefore we need to develop an appropriate measure of whether a population is approaching or moving away from an equilibrium point.

This chapter is devoted to the analysis of these issues for linear models involving multiple variables. Nonlinear models with multiple variables are treated in Chapter 8. In this chapter we work primarily with two-variable models because they are easiest to portray graphically. Techniques for models with more than two variables are analogous and are summarized in the Rules. Many of the techniques introduced in Chapters 7 and 8 are justified by the general solution to linear models in section 9.2. Therefore, those readers who prefer to understand the mathematical underpinnings before learning a technique should read section 9.2 before this chapter.

In section 7.2, we introduce multivariable models and describe, conceptually, why eigenvectors and eigenvalues (Primer 2) play such an important role in their analysis. In section 7.3, we detail how to find equilibria and determine their stability in continuous-time models. Section 7.4 then presents the discrete-time counterpart.

7.2 Models with More than One Dynamic Variable

The simplest one-variable model is that of exponential growth. In continuous time, it is described by the differential equation

$$\frac{dn}{dt} = rn. \tag{7.1}$$

This model was termed linear because the right-hand side of equation (7.1) is a linear function of the dynamic variable n. The model is simple because we can immediately determine its behavior from the growth rate r. If the growth rate is greater than zero then n will grow indefinitely, whereas if r is less than zero then n will decay to zero. Thus the value of r determines the stability of the equilibrium, $\hat{n} = 0$, which is the only equilibrium of this model. In fact, we even know the general solution for (7.1): $n(t) = n(0)\, e^{rt}$ (see Chapter 6).

To introduce methods for determining equilibria and stability in multivariable models, we start with a model that is a simple extension of (7.1). Consider a bacterial species with two different strains, each growing exponentially at rates r_1 and r_2. If n_1 and n_2 are their population sizes, then we have

$$\frac{dn_1}{dt} = r_1 n_1, \tag{7.2a}$$

$$\frac{dn_2}{dt} = r_2 n_2. \tag{7.2b}$$

Equation (7.2) represents two independent population-dynamic processes. The population dynamics of n_1 do not affect the dynamics of n_2, or vice versa. Because the two populations grow independently and exponentially, the general solution to this model is $n_1(t) = n_1(0)\, e^{r_1 t}$ and $n_2(t) = n_2(0)\, e^{r_2 t}$.

The first step in analyzing one-variable models is to identify the equilibria (Chapter 5). Because the present model represents two independent processes of exponential growth or decay, there is only one possible *joint* equilibrium for the system; namely, $\hat{n}_1 = 0$, $\hat{n}_2 = 0$. Formally, this equilibrium can be found by setting $dn_1/dt = 0$ and $dn_2/dt = 0$. Only when both of these equilibrium conditions are satisfied is there no change in the variables over time. Using (7.2), these equilibrium conditions become $r_1 n_1 = 0$ and $r_2 n_2 = 0$, which are satisfied at the equilibrium $\hat{n}_1 = \hat{n}_2 = 0$. Model (7.2) is not at equilibrium when $\hat{n}_1 = 0$ but $n_2 > 0$ (or when $\hat{n}_2 = 0$ but $n_1 > 0$), because change still occurs in one of the strains. Only when both $\hat{n}_1 = 0$ and $\hat{n}_2 = 0$ does the entire system remain unchanged over time. As emphasized in Chapter 5, an

At a *multivariable equilibrium,* all variables must remain unchanged.

equilibrium describes the values of the variables, not the parameters, at which the system remains constant. Thus, $r_1 = r_2 = 0$ does not represent an equilibrium, but rather a special case of the parameters.

Now that we know the only equilibrium, we proceed to determine its stability, as in Chapter 5. Let us first plot the vector field for this model (see section 4.4.3). Although we are only interested in non-negative values of n_1 and n_2 in a model of population growth, we plot negative values as well because these might be of interest in other models using the same differential equations (Figure 7.1). Three different qualitative behaviors are observed depending on the values of r_1 and r_2: (i) both growth rates are positive ($r_1 > 0$, $r_2 > 0$), (ii) one growth rate is positive and the other is negative ($r_1 > 0$, $r_2 < 0$ or the reverse), and (iii) both growth rates are negative ($r_1 < 0$, $r_2 < 0$).

There are a couple of important points to take away from these plots. First, for the joint equilibrium to be stable to a perturbation in any direction, *both* r_1 and r_2 must be negative (e.g., Figure 7.1c). Otherwise, once perturbed from the origin (i.e., the equilibrium), the system moves away from it indefinitely. Second, if we happened to start the model with initial conditions on the horizontal axis (e.g., $n_1(0) = 100$, $n_2(0) = 0$) then the system would remain on this axis forever. This is because strain 2 is missing from the culture and it is not produced by strain 1. The system would move toward the equilibrium if $r_1 < 0$ (i.e., strain 1 has a negative growth rate) and away from it if $r_1 > 0$ (i.e., strain 1 has a positive growth rate). An analogous situation holds if we started instead on the vertical axis.

7.2.1 Eigenvalues and Eigenvectors

Lines in the phase plane from which the system never leaves, such as the vertical and horizontal axes in the above example, play a very important role in analyzing the behavior of models with multiple variables. These lines are called the *eigenvectors* of the system. By definition, the system grows or decays along an eigenvector, but it never rotates away from the eigenvector. For the two eigenvectors in the above example, movement occurs at rates r_1 and r_2. These growth rates are known as the *eigenvalues* associated with the eigenvectors (the prefix "eigen" roughly translates from German to mean "characteristic"). It is easy to get confused at this stage about the distinction between eigenvectors and the null clines introduced in Chapter 4, so we summarize the key differences in Box 7.1.

> A system grows or decays in the directions of its *eigenvectors*.

Eigenvectors and their associated eigenvalues provide key information about the stability of an equilibrium. This was apparent from the vector-field plots of Figure 7.1. Only if there is movement toward the equilibrium along both eigenvectors is the equilibrium stable. With movement away from the equilibrium along one or both of the eigenvectors, the equilibrium is unstable. What makes this example particularly straightforward is that the eigenvectors happen to coincide with the coordinate axes, and the eigenvalues are the growth rates of each strain considered separately, r_1 and r_2. Thus, stability of the equilibrium $\hat{n}_1 = 0$, $\hat{n}_2 = 0$ requires that both eigenvalues be negative because then (and only then) will the movement along both eigenvectors be toward the equilibrium.

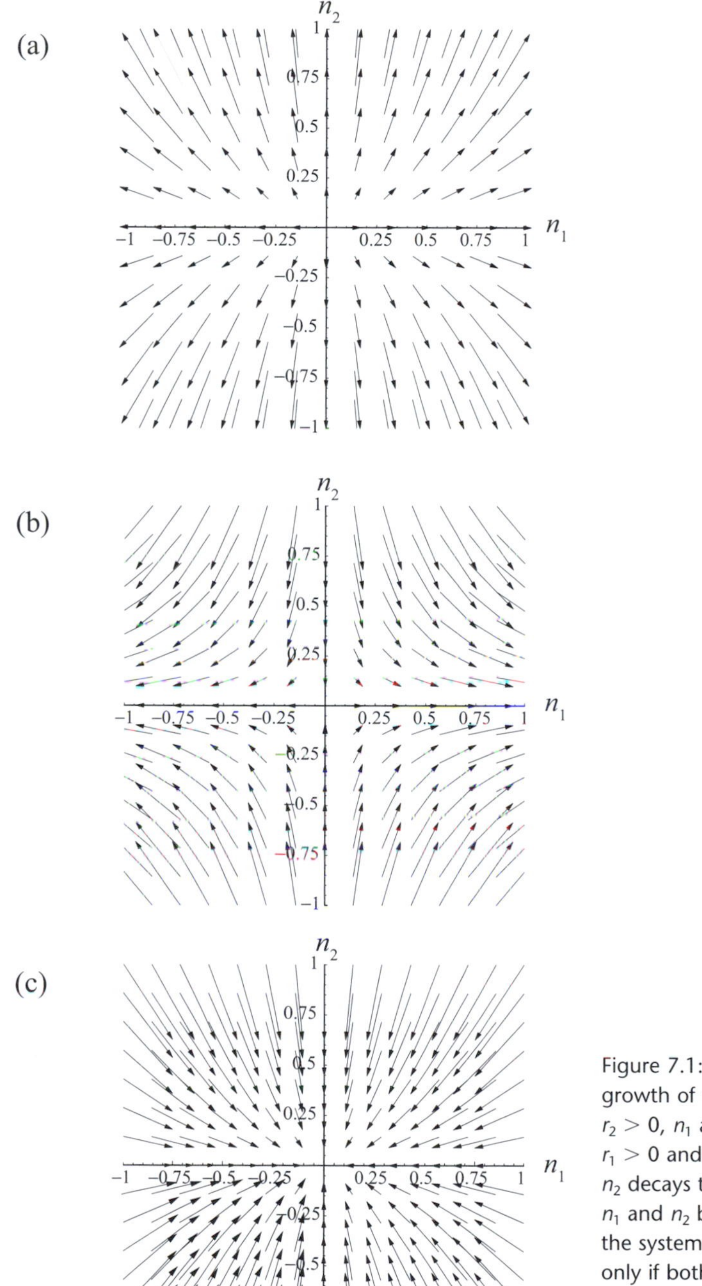

Figure 7.1: Vector-field plots for independent growth of two strains. (a) When $r_1 > 0$, and $r_2 > 0$, n_1 and n_2 grow exponentially. (b) When $r_1 > 0$ and $r_2 < 0$, n_1 grows exponentially whereas n_2 decays to zero. (c) When $r_1 < 0$ and $r_2 < 0$, n_1 and n_2 both decay to zero. Over the long term, the system moves toward the equilibrium (0,0) only if both r_1 and r_2 are negative, as in panel (c). The eigenvectors of model (7.2) are marked as thick black lines, corresponding to the horizontal and vertical axes.

For most linear models involving n variables, we can find n eigenvectors and their associated eigenvalues. Most of the time the eigenvectors will not coincide with the coordinate axes. Typically, linear models have only one possible equilibrium (e.g., $\hat{n}_1 = 0$, $\hat{n}_2 = 0$ in the above example), and the eigenvalues determine its stability. Given this fact, we do not even need to calculate the

Box 7.1: Null Clines versus Eigenvectors

Null clines and eigenvectors are two of the most useful things that can be plotted on a phase-plane diagram, and it is important to understand the distinction between the two. Null clines represent combinations of the dynamic variables that cause one of the dynamic variables to remain constant. An equilibrium occurs wherever the null clines for all variables intersect. In contrast, eigenvectors represent combinations of the dynamic variables for which the relative values of the variables remain constant. The dynamic variables still change through time, but they do so in a way that preserves their relative values.

By definition, when a linear dynamical system is on a null cline, one of the variables remains constant, but the other variables can change, moving the dynamical system off the null cline. In contrast, when a dynamical system is on an eigenvector, it never leaves the eigenvector. When viewed on a phase-plane diagram, the main distinction between null clines and eigenvectors is that, on a null cline, movement is either exactly vertical or exactly horizontal (because one of the variables is unchanging) whereas, on an eigenvector, movement is either inward or outward (Figure 7.1.1).

To come to terms with these two different concepts, try adding the null clines of model (7.2) to Figure 7.1 using the techniques of Chapter 4. For model (7.2) there are two null clines: n_1 remains constant (i.e., $dn_1/dt = 0$) when $n_1 = 0$, and n_2 remains constant (i.e., $dn_2/dt = 0$) when $n_2 = 0$. Thus, in this model, the null clines of model (7.2) happen to coincide with the eigenvectors (the horizontal and vertical axes).

In the simple model of equation (7.2) both the eigenvectors and the null clines happened to coincide with the coordinate axes and with each other. Generally, however, the eigenvectors of a model will be distinct from the null clines, and neither of these will typically coincide with the coordinate axes. Although the null clines of a model provide important information about the dynamics of the variables in different regions of the phase plane (as seen in Chapter 4) as well as important information about the equilibria of the model, it is the eigenvectors (and in particular their associated eigenvalues) that provide important information about the stability of equilibria.

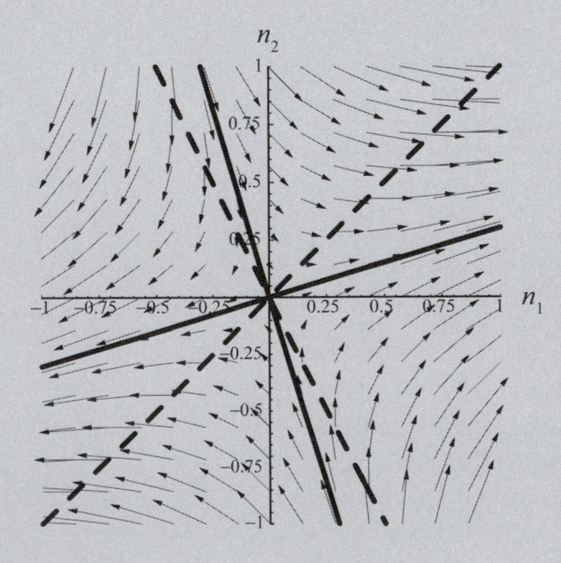

Figure 7.1.1: Adding null clines and eigenvectors to a vector field plot. Each arrow illustrates the direction of movement predicted by the continuous-time model (7.6) with $a = 2$, $b = 1$, $c = 1$, $d = -1$. One null cline represents the line along which n_1 is constant, and one represents the line along which n_2 is constant (dashed lines). Try to determine which is which according to the figure. One eigenvector has an associated eigenvalue that is negative (implying exponential decay toward the origin), and one eigenvector has an associated eigenvalue that is negative (implying exponential growth away from the origin) (solid lines). Again, try to determine which is which.

eigenvectors to determine stability per se; instead, we can calculate the eigenvalues directly to determine whether the system approaches the equilibrium or grows away from it.

Before proceeding to a more general context, it is worth writing the above model in matrix form. Matrix notation is an invaluable aid for simplifying and analyzing multivariable models, and we present some of the most useful concepts of matrices and vectors in Primer 2. If you are not already familiar with matrix operations, you should now look over sections P2.1–P2.6. Using boldface notation to signify matrices (e.g., $\mathbf{M}$) and a half arrow to denote vectors (e.g., $\vec{n}$), we can write equations (7.2) as

$$\begin{pmatrix} \frac{dn_1}{dt} \\ \frac{dn_2}{dt} \end{pmatrix} = \mathbf{M}\begin{pmatrix} n_1 \\ n_2 \end{pmatrix}, \tag{7.3a}$$

where

$$\mathbf{M} = \begin{pmatrix} r_1 & 0 \\ 0 & r_2 \end{pmatrix}. \tag{7.3b}$$

(Check that multiplying out the right-hand side of (7.3) produces (7.2) using Rule P2.9b.)

The matrix $\mathbf{M}$ in (7.3b) does not contain the variables, so it is constant over time. Any model that can be written in the form of (7.3) using a matrix $\mathbf{M}$ with only constant elements is a *linear model*. A model involving multiple variables is linear provided that each term in the equations is proportional to only one variable in the model (e.g., n_1 but not $n_1 n_2$).

A *multivariable linear model* can be written as a matrix of constants times a vector of variables.

Incidentally, writing system (7.3) entirely in matrix notation helps to reveal the connection to linear, one-variable models. Introducing the vectors

$$\frac{d\vec{n}}{dt} = \begin{pmatrix} \frac{dn_1}{dt} \\ \frac{dn_2}{dt} \end{pmatrix} \text{ and } \vec{n} = \begin{pmatrix} n_1 \\ n_2 \end{pmatrix},$$

we can write

$$\frac{d\vec{n}}{dt} = \mathbf{M}\vec{n}. \tag{7.4}$$

Provided that $\mathbf{M}$ is a matrix of constants, equation (7.4) is a multivariable model describing exponential growth, just as equation (7.1) is a one-variable model of exponential growth.

As mentioned in section 6.3, there is also a second type of linear model that includes additional constant terms (so-called affine models). Any model that can be written in the form $d\vec{n}/dt = \mathbf{M}\vec{n} + \vec{c}$, where $\vec{c}$ is a vector of constants and $\mathbf{M}$ is a matrix of constants, is an *affine* model. As in one-variable models, multivariable affine models can be transformed into nonaffine models of the form (7.4) (Box 7.2). We therefore focus this chapter on nonaffine models.

Box 7.2: Affine Models Involving Multiple Variables

In Chapter 6, we discussed the fact that there are two types of linear models in continuous time: those having a nonaffine form $dn/dt = r\,n$ and those having an affine form $dn/dt = r\,n + c$. Here, the parameter c represents a constant input into the system (e.g., migration) or a constant outflow (e.g., through harvesting). Affine models also arise in multivariable models and have the form

$$\frac{d\vec{n}}{dt} = \mathbf{M}\,\vec{n} + \vec{c}, \qquad (7.2.1)$$

where $\mathbf{M}$ is a matrix of constants and $\vec{c}$ is a vector of constants, representing the rate of inputs and outflows to each variable. Because each variable must be unchanging at equilibrium, the equilibrium of (7.2.1) must satisfy the equilibrium condition $\vec{0} = \mathbf{M}\hat{\vec{n}} + \vec{c}$, where $\vec{0}$ is a vector of zeros. Subtracting $\vec{c}$ from both sides and then multiplying both sides on the left by $\mathbf{M}^{-1}$ (the inverse of matrix $\mathbf{M}$), we can solve for the equilibrium of (7.2.1): $\hat{\vec{n}} = -\mathbf{M}^{-1}\vec{c}$ (see section P2.7, Primer 2). If $\vec{c}$ is a vector of zeros, the equilibrium is the origin, $\hat{\vec{n}} = \vec{0}$, as expected for a nonaffine linear model (Rule 7.1). Otherwise, the equilibrium will not be at the origin.

Affine models also arise in discrete-time models and take the form

$$\vec{n}(t+1) = \mathbf{M}\,\vec{n}(t) + \vec{c}, \qquad (7.2.2)$$

where again $\mathbf{M}$ is a matrix of constants and $\vec{c}$ is a vector of constants. The equilibrium of (7.2.2) must satisfy the equilibrium condition $\hat{\vec{n}} = \mathbf{M}\hat{\vec{n}} + \vec{c}$. Subtracting $\mathbf{I}\hat{\vec{n}} + \vec{c}$ from both sides gives $\hat{\vec{n}} - \mathbf{I}\,\hat{\vec{n}} - \vec{c} = \mathbf{M}\,\hat{\vec{n}} - \mathbf{I}\,\hat{\vec{n}}$, which simplifies to $-\vec{c} = (\mathbf{M} - \mathbf{I})\hat{\vec{n}}$. Multiplying both sides on the left by the inverse of $\mathbf{M} - \mathbf{I}$ gives the equilibrium solution $\hat{\vec{n}} = -(\mathbf{M} - \mathbf{I})^{-1}\vec{c}$.

The behavior of affine models can be understood using the techniques for linear models, provided that we measure the variables of an affine model in the right way. For models with one variable, we showed in Chapter 6 that a transformation could be used to write the variable n in terms of the distance to its equilibrium, $\delta = n - \hat{n}$. The dynamic equations for the new variable δ then become linear and nonaffine (see Recipes 6.1 and 6.3). The same transformation can be used to convert affine models involving multiple variables of the form (7.2.1) or (7.2.2) into nonaffine models of the form (7.6a) or (7.23b), by rewriting each variable in terms of the deviation from its equilibrium, $\vec{\delta} = \vec{n} - \hat{\vec{n}}$. Thus, any results for linear nonaffine models can be applied directly to affine models, once the affine model has been transformed.

7.3 Linear Multivariable Models

Equation (7.2) was a particularly simple two-variable linear model, because changes in the variables did not depend on one another. Now consider a more general two-variable linear model, where changes in the variables n_1 and n_2 do depend on one another. To help motivate this section biologically, let us extend our simple model of a bacterial species to allow strain 1 individuals to become strain 2 individuals by mutation, and vice versa. If a is the per capita rate at which strain 1 individuals produce strain 1 daughter cells, and b is the per capita rate at which strain 2 individuals produce strain 1 daughter cells by mutation, then the rate of change of strain 1 is

$$\frac{dn_1}{dt} = a\, n_1 + b\, n_2. \tag{7.5a}$$

Similarly, if c is the per capita rate at which strain 1 individuals produce strain 2 daughter cells, and d is the per capita rate at which strain 2 individuals produce strain 2 daughter cells, then the rate of change of strain 2 is

$$\frac{dn_2}{dt} = c\, n_1 + d\, n_2. \tag{7.5b}$$

We can write this model in matrix notation as

$$\frac{d\vec{n}}{dt} = \mathbf{M}\,\vec{n}, \tag{7.6a}$$

where now

$$\mathbf{M} = \begin{pmatrix} a & b \\ c & d \end{pmatrix}. \tag{7.6b}$$

Importantly, while we have provided a biological motivation for (7.6), this is nevertheless a completely general description that would apply to *any* two-variable linear model with appropriate interpretations of the parameters a, b, c, and d.

7.3.1 Finding Equilibria

The first step in analyzing a continuous-time model is to determine the equilibria by setting $dn_1/dt = 0$ and $dn_2/dt = 0$, giving

$$0 = a\, n_1 + b\, n_2, \tag{7.7a}$$
$$0 = c\, n_1 + d\, n_2. \tag{7.7b}$$

These are the *equilibrium conditions* for the two-variable model; both equilibrium conditions must be satisfied for a point to be an equilibrium. We can find the equilibria of (7.7) in a number of ways. A straightforward way is described in Recipe P2.1 of Primer 2. In short, we solve equation (7.7a) for one of the variables, e.g., $n_2 = (-a/b)\, n_1$, and then plug this solution into the other equation (7.7b). This gives $0 = c\, n_1 + d\,(-a/b)\, n_1$. This equation involves n_1 only and is satisfied at the single point $\hat{n}_1 = 0$. We then plug this solution back into the previous equation $n_2 = (-a/b)\, n_1$. There is only one value of n_2 that satisfies this equation when $\hat{n}_1 = 0$; namely, $\hat{n}_2 = 0$. Thus, for this linear model, there is only one equilibrium point, $\hat{n}_1 = \hat{n}_2 = 0$. As a check, if we plug this equilibrium into the equilibrium conditions (7.7), we confirm that both conditions are satisfied.

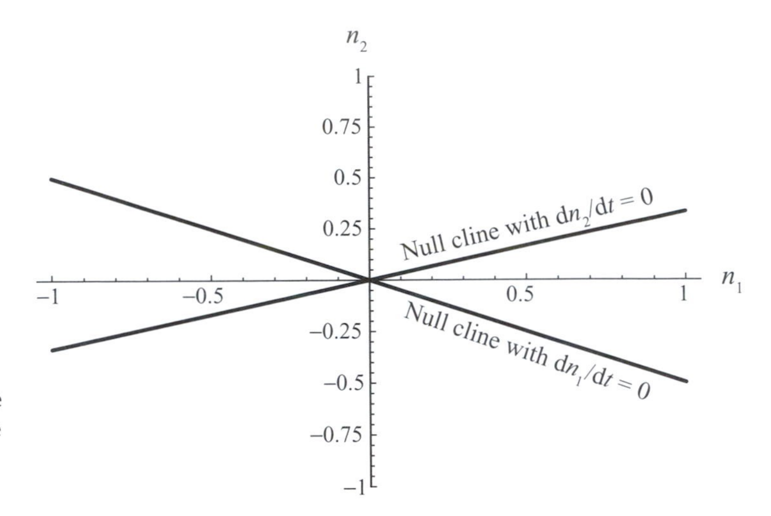

Figure 7.2: Null clines in a two-variable model. The null clines for model (7.7) are given by the lines $n_2 = -a\, n_1/b$ for the $dn_1/dt = 0$ null cline and $n_2 = -c\, n_1/d$ for the $dn_2/dt = 0$ null cline ($a = 1$, $b = 2$, $c = -1$, $d = 3$).

There is a second way to determine the equilibrium of (7.7) that is also worth describing, because it clarifies the relationship between an equilibrium and null clines. We start by solving both equilibrium conditions (7.7a) and (7.7b) for one variable $\hat{n}_2$ in terms of the other variable, obtaining $n_2 = (-a/b)\, n_1$ and $n_2 = (-c/d)\, n_1$. These equations represent the null clines of the model, and they indicate when species 1 and when species 2 remain constant, respectively. These equilibrium conditions are simultaneously satisfied by the same value of n_2 only when $(-a/b)\, n_1 = (-c/d)\, n_1$. For arbitrary parameter values this requires that $\hat{n}_1 = 0$. Plugging this result back into either $n_2 = (-a/b)\, n_1$ or $n_2 = (-c/d)\, n_1$ then tells us that $\hat{n}_2$ must be zero at equilibrium. Graphically, this method is equivalent to determining where the null clines for this model cross (Figure 7.2; see section 4.4). In a linear nonaffine model, these null clines are straight lines that pass through the origin, indicating that the origin is the only possible equilibrium.

An exception to the above result occurs if the two null clines happen to lie exactly on top of one another. In this special case of the parameters we must have $(-a/b) = (-c/d)$, and there are then an infinite number of equilibria, all lying along the common null cline. Interestingly, the condition $(-a/b) = (-c/d)$ is also the condition that the determinant of the matrix $\mathbf{M}$ is zero (Problem 7.11). In fact this results holds for linear models with any number of variables: when the determinant of the matrix $\mathbf{M}$ is zero, there are an infinite number of equilibria. Having a determinant of 0 is a special case of the parameters, analogous to $r = 0$ in the one-variable model.

Rule 7.1: Equilibrium of a Linear Multivariable Model in Continuous Time

A linear model in continuous time has only one equilibrium regardless of the number of variables, provided that the determinant of $\mathbf{M}$ is not zero.

(continued)

Rule 7.1 *(continued)*

- For a linear model of the form $d\vec{n}/dt = \mathbf{M}\vec{n}$, the equilibrium point is the origin, $\hat{\vec{n}} = \vec{0}$, where $\vec{0}$ is a vector of zeros.
- For an affine model of the form $d\vec{n}/dt = \mathbf{M}\vec{n} + \vec{c}$, the equilibrium point is $\hat{\vec{n}} = -\mathbf{M}^{-1}\vec{c}$ (see Box 7.2).

If the determinant of **M** is zero, there then are an infinite number of equilibria.

7.3.2 Determining the Stability of the Equilibrium—Real Eigenvalues

Our next step is to determine the stability of the equilibrium that we have identified. Let us begin as above, by plotting the vector field, along with the eigenvectors for the general two-variable model (7.6). The same three possibilities arise as in the model without mutation, except that now the eigenvectors no longer lie on the coordinate axes (Figure 7.3). Nevertheless, the system either moves away from the equilibrium along both eigenvectors (Figure 7.3a), toward the equilibrium along one eigenvector but away from it along the other (Figure 7.3b), or toward the equilibrium along both eigenvectors (Figure 7.3c). In either of the first two cases, the equilibrium is unstable because the system moves away from the equilibrium in at least one direction. Only when movement occurs toward the equilibrium along both eigenvectors is the equilibrium stable. For the moment, we shall ignore a fourth possibility, that the eigenvectors are not real numbers, but we will return to this in section 7.3.3.

We have just seen that movement along the eigenvectors provides important information about the stability of an equilibrium, but what do these eigenvectors represent? By definition, if we start the model on one of the eigenvectors, then the system will remain on that vector forever, either moving outward or inward. In model (7.3), the eigenvectors coincided with the coordinate axes, representing the absence of strain 1 (vertical axis) or strain 2 (horizontal axis). If a strain is initially absent in model (7.6), however, it will soon appear by mutation. The eigenvectors in this model indicate when the relative numbers of the two strains remain constant (see Figure 7.3). If we choose the relative starting population sizes for the two strains such that the system lies on an eigenvector, the *absolute* population size of each strain will grow or decay exponentially, but the population size of one strain *relative* to the other will remain the same.

Let us now explore the rate at which the system grows or decays along an eigenvector. Suppose that $\vec{v}$ is one of the eigenvectors. If we start the system on this eigenvector, the system must remain on $\vec{v}$, moving inward or outward at a rate determined by the eigenvalue r. In other words, all components of the vector $\vec{v}$ must grow or shrink at the same rate r:

A system grows or decays in the direction of an eigenvector at a rate given by its *eigenvalue*.

$$\frac{d\vec{v}}{dt} = r\vec{v}. \tag{7.8a}$$

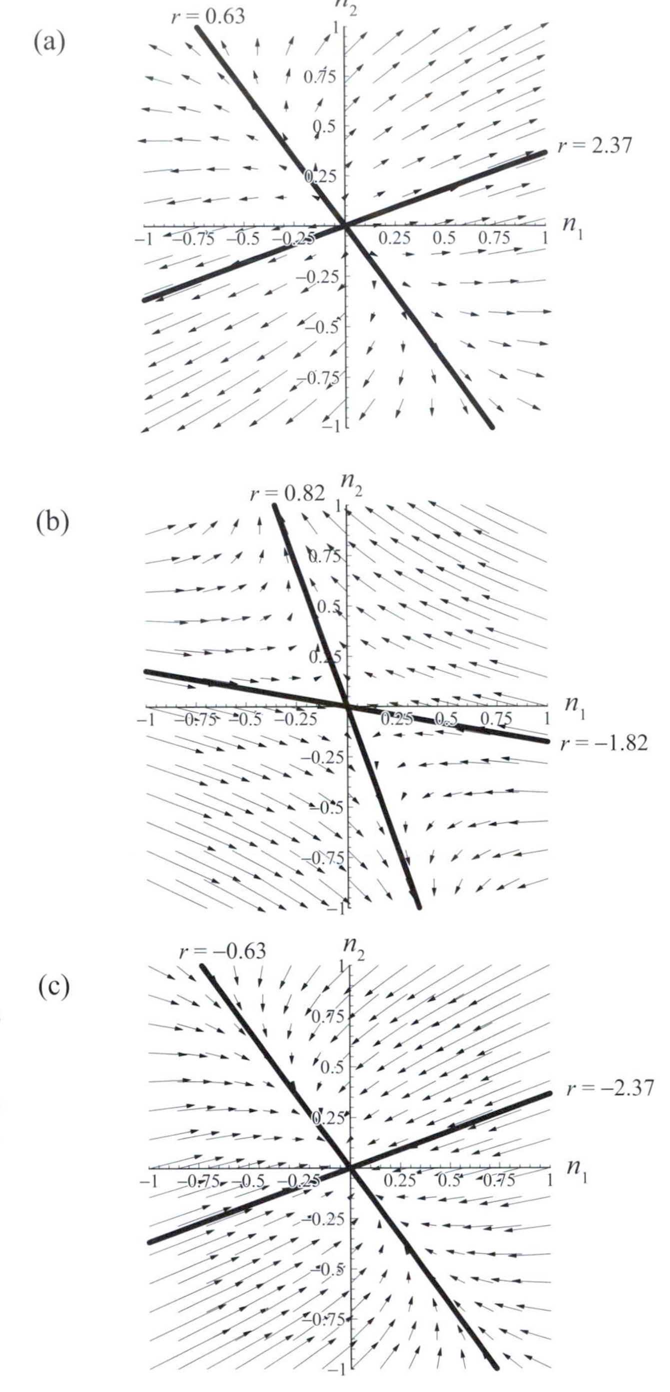

Figure 7.3: Eigenvalues, eigenvectors, and vector-field plots in a two-variable model. The eigenvalues give the growth or decay rates along the corresponding eigenvectors (marked as thick black lines) and can be calculated for model (7.6) using equations (7.12) (also see section P2.8, Primer 2). (a) $a = 2$, $b = 1$, $c = 1/2$, $d = 1$, resulting in two positive eigenvalues. (b) $a = -2$, $b = -1$, $c = 1/2$, $d = 1$, resulting in one positive and one negative eigenvalue. (c) $a = -2$, $b = -1$, $c = -1/2$, $d = -1$, resulting in two negative eigenvalues. Over the long term, the system moves toward the equilibrium (0,0) only if both eigenvalues are negative (Panel c).

Regardless of where we start, however, the dynamics must also satisfy equation (7.6):

$$\frac{d\vec{v}}{dt} = \mathbf{M}\,\vec{v}. \tag{7.8b}$$

Combining equations (7.8a) and (7.8b), the eigenvalue and associated eigenvector must satisfy

$$\mathbf{M}\,\vec{v} = r\,\vec{v}. \qquad (7.9)$$

Mathematically, equation (7.9) *defines* the eigenvalues and eigenvectors of the matrix **M**.

In sections P2.8 and P2.9 of Primer 2, we describe how to use equation (7.9) to find the eigenvalues and eigenvectors of a matrix. If you are not already familiar with calculating eigenvalues and eigenvectors, you should take time to work through these sections. In short, the eigenvalues of the matrix **M** satisfy the characteristic polynomial.

Definition 7.1: Characteristic Polynomial
The characteristic polynomial of a matrix **M** is given by the equation

$$\text{Det}(\mathbf{M} - \mathbf{I}\,r) = 0. \qquad (7.10)$$

Equation (7.10) represents a polynomial equation in r, referred to as the *characteristic polynomial* of the matrix **M**. The number of variables in the model determines the degree of the characteristic polynomial. For example, in the two-variable model (7.6), equation (7.10) represents a quadratic equation in r; the two solutions to this equation are the two eigenvalues of (7.6). If we wanted to determine the eigenvectors, we would then substitute each eigenvalue into (7.9) and solve for $\vec{v}$ (see section P2.9 of Primer 2). To assess the stability of an equilibrium, however, we only need to know the eigenvalues. For the origin to be a stable equilibrium, the variables must decay toward zero along every eigenvector, which requires that all eigenvalues be negative (see equation (7.8a)). This assumes that the eigenvalues are real; in section 7.3.3, we generalize this result to include the possibility of complex eigenvalues.

Eigenvalues are found by determining the roots of a *characteristic polynomial.*

Solving the characteristic polynomial to determine the eigenvalues can be a difficult task, particularly when the number of variables (i.e., the dimension of **M**) is large. The reason is twofold. First, it is tedious to calculate the determinant of a large matrix (Rule P2.17; Primer 2). In such cases we usually use a computer program like *Mathematica* or Maple to do the task. Second, even when we do calculate the determinant in a model with n variables, the resulting nth-degree polynomial equation can be difficult or even impossible to solve for r.

Nevertheless, for a model involving two variables the calculations are quite easy. Following section P2.8 of Primer 2, equation (7.10) becomes

$$r^2 - (a + d)\,r + (a\,d - b\,c) = 0. \qquad (7.11)$$

The quadratic formula (A1.10, Appendix 1) can be used to give the two solutions:

$$r_1 = \frac{a + d + \sqrt{(a + d)^2 - 4(a\,d - b\,c)}}{2} \quad \text{and}$$
$$r_2 = \frac{a + d - \sqrt{(a + d)^2 - 4(a\,d - b\,c)}}{2}. \tag{7.12}$$

Equations (7.12) are the two eigenvalues for the general linear model (7.6) involving two variables. The equilibrium will be stable if (and only if) both r_1 and r_2 are negative. Only then will the system move toward the equilibrium along both eigenvectors.

Example: Metastasis of Malignant Tumors (adapted from Glass and Kaplan 1995)

As an example, let us apply this method to understand the spread of cancer cells. Metastasis is a process by which cancer cells spread throughout the body. Sometimes cancer cells move via the bloodstream and become lodged in the capillaries of different organs. Some of these cells then move across the capillary wall, where they initiate new tumors. We will construct a model for the dynamics of the number of cancer cells lodged in the capillaries of an organ, C, and the number of cancer cells that have actually invaded that organ, I. Suppose that cells are lost from the capillaries by dislodgement or death at a per capita rate δ_1 and that they invade the organ from the capillaries at a per capita rate β. Once cells are in the organ they die at a per capita rate δ_2, and the cancer cells replicate at a per capita rate ρ (Figure 7.4). All of these parameters are assumed positive. These processes result in two differential equations governing the dynamics of C and I:

$$\frac{dC}{dt} = -\delta_1 C - \beta C,$$
$$\frac{dI}{dt} = \beta C - \delta_2 I + \rho\, I, \tag{7.13a}$$

or

$$\begin{pmatrix} \frac{dC}{dt} \\ \frac{dI}{dt} \end{pmatrix} = \mathbf{M} \begin{pmatrix} C \\ I \end{pmatrix}, \tag{7.13b}$$

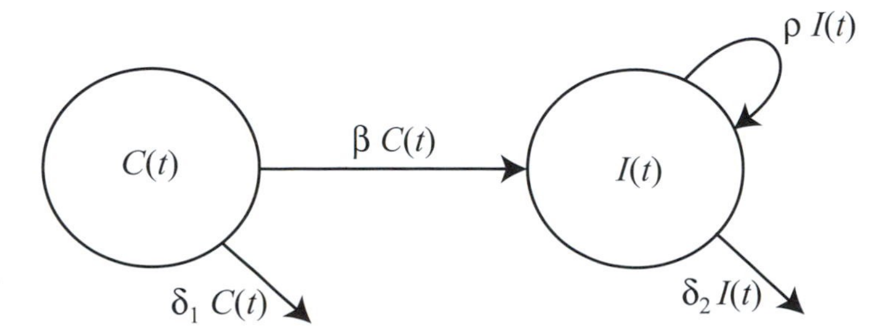

Figure 7.4: A flow diagram for the metastasis of malignant tumors.

where

$$\mathbf{M} = \begin{pmatrix} -(\delta_1 + \beta) & 0 \\ \beta & \rho - \delta_2 \end{pmatrix}.$$

To obtain (7.13b) from (7.13a), gather together terms in equations (7.13a) that involve the same variable: $dC/dt = -(\delta_1 + \beta)\, C$ and $dI/dt = \beta C + (\rho - \delta_2)\, I$. The elements of the matrix **M** are then easier to identify.

Now the question of interest is this: Will a new tumor be able to take hold and grow or will it disappear? The only equilibrium of this system is $C = 0, I = 0$ (i.e., no cancer cells in the capillaries or the organ), so the tumor being able to take hold and grow is represented mathematically by this equilibrium being *unstable*. Using equation (7.12) or, more easily, Rule P2.26, we get the two eigenvalues

$$r_1 = -(\delta_1 + \beta), \qquad r_2 = \rho - \delta_2. \tag{7.14}$$

Because the parameters are positive by the way that they are defined, the first eigenvalue r_1 is always negative. Therefore, the stability of the equilibrium is entirely determined by the sign of the second eigenvalue r_2. The equilibrium is unstable, and the tumor will take hold and grow provided that $r_2 > 0$, meaning that $\rho > \delta_2$. This condition tells us that the growth rate in the organ, ρ, must be larger than the death rate in the organ, δ_2, for the tumor to grow. This makes a lot of intuitive sense. Surprisingly, however, the death rate of cancer cells in the capillaries, δ_1, and the rate at which cancer cells pass from the capillaries to the organ, β, do not affect whether or not the tumor takes hold (although these rates can influence the speed of this process).

Incidentally, it is always a good idea to look at the matrix **M** to determine if it has a special form (e.g., diagonal, upper triangular, lower triangular, block triangular). For such matrices, Rules P2.26–P2.28 make it very easy to calculate eigenvalues. For instance, in this example it is much easier to calculate the eigenvalues of

$$\mathbf{M} = \begin{pmatrix} -(\delta_1 + \beta) & 0 \\ \beta & \rho - \delta_2 \end{pmatrix}$$

by recognizing that it is a lower triangular matrix (whose eigenvalues are the diagonal elements; Rule P2.26) than it is to evaluate (7.12).

7.3.3 Determining the Stability of the Equilibrium—Complex Eigenvalues

There is one important issue that remains to be addressed. What if the eigenvalues in a model are not real? In the general two-variable model, if $(a + d)^2 - 4(a\,d - b\,c)$ is negative in equation (7.12), we are faced with taking the square root of a negative number. This means that the eigenvalues will be complex; i.e., they will involve the imaginary number $i = \sqrt{-1}$ (Box 7.3). We need to

Box 7.3: The Importance of *i*

Many people become uncomfortable when complex numbers are mentioned. Certainly, there is no obvious real-world interpretation of what it means to take the square root of -1. Nevertheless, complex numbers, or imaginary numbers as they are sometimes called, often arise in models describing real-world processes. To understand and appreciate their importance, it is helpful to consider why complex numbers were first introduced into mathematics.

Algebra is an incredibly important and useful area of mathematics, which we have used extensively throughout this book. For our purposes, algebra is essentially the branch of mathematics that involves solving equations, more specifically, equations having the form $a_0 + a_1 x + a_2 x^2 + \cdots + a_n x^n$ (termed polynomial equations). Now if you were asked to solve the equation $x + 3 = 10$, you would have little trouble arriving at the solution $x = 7$. If asked to solve the equation $x + 10 = 3$, you would also have little trouble arriving at the solution $x = -7$, despite the fact that this latter example required you to use negative numbers. But what are negative numbers? Positive numbers have familiar interpretations in the sense that we know what it means to have 5 apples or 14 dollars. But what possible interpretation could a negative quantity have? While this probably seems like a naive question, historically there was considerable apprehension about such negative numbers. It would have been fine if we could simply ignore negative numbers and only "allow" those that are positive, but the above example would then pose a problem. The equation $x + 10 = 3$ involves only positive numbers, yet we are forced to "look outside" these numbers if we are to find a solution. Consequently, people eventually accepted the inevitability (and utility) of negative numbers.

The difficulties do not end there, however. Consider the equation $x^2 + 5 = -5$, which involves only positive and negative numbers. It is easy to see that x must satisfy $x^2 = -10$ meaning that, whatever x is, if we multiply it by itself the result must be -10. But of course squaring a real number should always produce a positive result, so we are left with the seemingly nonsensical solutions $x = \sqrt{-10}$ and $x = -\sqrt{-10}$. Historically, the square root of a negative number caused considerable apprehension. Yet, the quantity $i = \sqrt{-1}$ was eventually integrated into the mathematician's repertoire, because it was the only way to solve some equations, even if the equation involved only positive and negative numbers. The introduction of i broadened the number system to what is termed *complex numbers*. A complex number has the form $A + Bi$ where A and B are real numbers and $i = \sqrt{-1}$. The numbers you are most familiar with—called real numbers—are special cases of complex numbers in which $B = 0$. Thus A is usually called the "real part" of the complex number, and B is called the "imaginary part."

At this stage you might begin to worry that we are chasing a carrot on a stick. If we write an equation in terms of complex numbers, might there be yet another type of number that we need to introduce in order to obtain a solution? Will we ever have a self-contained numbering system? Fortunately, complex numbers do the trick. Around 1800, Carl Friedrich Gauss proved (in his doctoral dissertation no less!) that all solutions of polynomial equations that are written in terms of complex numbers can be written in terms of complex numbers. Moreover, if the highest power of x in a polynomial equation is n, then there are exactly n solutions, some of which might be complex, and some of which might be repeated. This incredible result is now known as the Fundamental Theorem of Algebra. Interested readers should peruse the fascinating book *The Mathematical Universe* (Dunham 1994) to gain a better appreciation for this and other milestones in mathematics.

confront this possibility and determine what it means to the dynamics of a model.

In general, complex numbers come in two parts: a real part and an imaginary part. The above eigenvalues provide a good example. Whenever $(a + d)^2 - 4(a\,d - b\,c) < 0$, the two eigenvalues have the form $r_1 = A + Bi$ and $r_2 = A - Bi$ where $A = (a + d)/2$, $B = \sqrt{|(a + d)^2 - 4(a\,d - b\,c)|}/2$, and $i = \sqrt{-1}$. Here A is called the *real part* of the eigenvalue, and B is called the *imaginary part*. It is also apparent from these expressions that either both eigenvalues are complex or neither of them are. In fact this holds more generally: complex eigenvalues always come in pairs (sometimes called complex conjugates).

In the three cases considered in Figure 7.3 (where the eigenvalues are real), movement along each eigenvector corresponds to either exponential growth or decay. Specifically, movement is described by e^{rt} where r is the corresponding eigenvalue. What happens when the eigenvalues are complex? Let us proceed by supposing that this continues to hold (justified in chapter 9). Then, the movement along one of the eigenvectors would be described by $e^{(A+Bi)t} = e^{At}\,e^{Bit}$, and the movement along the other would be described by $e^{(A-Bi)t} = e^{At}e^{-Bit}$. Now there is a very famous (and quite sophisticated) mathematical equation relating $e^{i\theta}$ to the trigonometric functions sine and cosine:

$$e^{i\theta} = \cos(\theta) + i\sin(\theta). \tag{7.15}$$

We are not going to try to prove this equation (Box 7.4) but we will use it! Applying (7.15), movement along the eigenvectors can be rewritten as

$$e^{(A+Bi)\,t} = e^{At}(\cos(Bt) + i\sin(Bt)) \tag{7.16a}$$

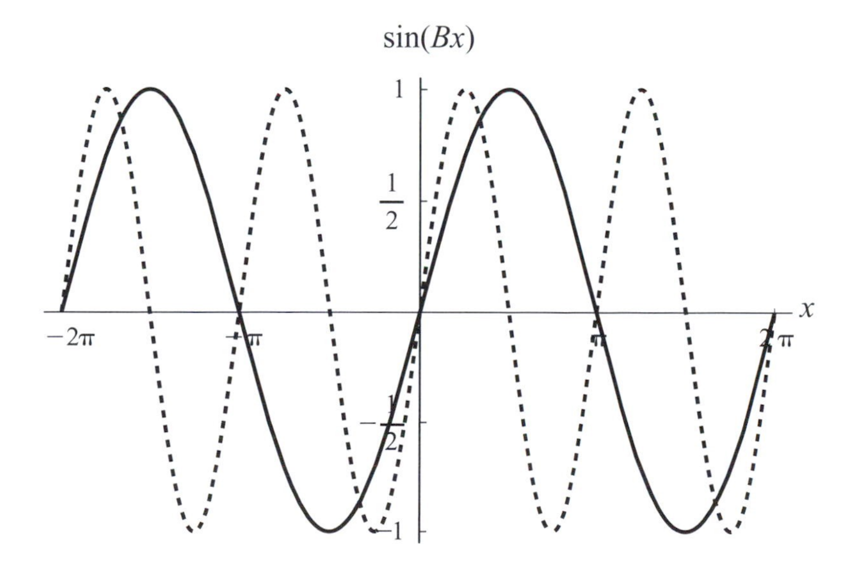

Figure 7.5: Sine functions. A plot of sin(Bx), which fluctuates periodically from −1 to 1. The period of the oscillation (i.e., the time to complete one full cycle) is $2\pi/B$. The solid curve has $B = 1$, giving a period of 2π. The dashed curve has $B = 2$, giving a period of π.

Box 7.4: *e* and Euler's Extraordinary Equation

Much of modern mathematical notation was systematized by the famous Swiss mathematician Leonhard Euler (1707–1783). Today, Euler is viewed as one of the greatest mathematicians of modern times. Readers who go on to study more mathematics will undoubtedly be awed by the breadth and depth of Euler's contributions to the subject. It is estimated that he averaged around 800 published pages of predominately mathematical analysis per year! Even more astounding is the fact that a great deal of this volume was instrumental in laying the foundations of modern mathematical analysis.

For example, Euler introduced the notation "*e*" to represent the only number whose natural logarithm is one, $\ln(e) = 1$. Because natural logarithms are defined by the integral $\ln(x) = \int_1^x (1/t)dt$, *e* is implicitly defined by the relationship $1 = \int_1^e (1/t)dt$. While this definition does not allow us to calculate *e* directly, *e* can also be written as an infinite series $1 + 1/1! + 1/2! + 1/3! + \cdots$, which can be calculated to any degree of accuracy desired. Euler himself used this series to calculate *e* as 2.71828182845904523536028 . . . and proved that *e* is an irrational number. This series can also be expressed in the more general relationship $e^x = 1 + x/1! + x^2/2! + x^3/3! + \cdots$, which we have already seen as the Taylor series of e^x (equation (P1.14); Primer 1). The series $1 + x/1! + x^2/2! + x^3/3! + \cdots$ can be shown to equal the limit of the expression $(1 + x/n)^n$ as n goes to infinity, establishing yet another important relationship involving *e*, that is, $e^x = \lim_{n\to\infty} (1 + x/n)^n$.

The quantity *e* arises in a variety of situations, but none is perhaps as surprising as the equation $e^{i\theta} = \cos(\theta) + i\sin(\theta)$. This incredibly elegant relationship between imaginary numbers and the trigonometric functions sine and cosine was derived by Euler and is known as Euler's equation. The importance of Euler's equation can be further appreciated by considering the special case in which we set $\theta = \pi$. Upon doing so, Euler's equation simplifies to $e^{i\pi} = -1$. This result links four of the most important quantities in mathematics, *e*, *i*, π, and -1. It is astounding enough that such a simple relationship exists among these important quantities, but the fact that these quantities were all introduced into mathematics at different points in history adds a further mystique to the discovery of this relationship. The book *Calculus Gems* (Simmons 1992) provides interesting historical accounts of Euler and other important mathematicians.

and

$$e^{(A-Bi)t} = e^{At}(\cos(B\,t) - i\sin(B\,t)). \tag{7.16b}$$

Equations (7.16) still involve i, but the fact that sines and cosines are involved suggests that the dynamics will cycle. This is exactly the case, as we shall describe in more detail in Chapter 9 (see Box 9.3). The period of the oscillations (the time it takes for a cycle to repeat itself) is given by $2\pi/B$ (Figure 7.5). Although the sine and cosine functions oscillate over time, they are multiplied by a term e^{At} that either decays to zero if $A < 0$ or grows over time if $A > 0$. Consequently, if $A > 0$, the equilibrium is unstable, whereas if $A < 0$, the equilibrium is stable.

(a)

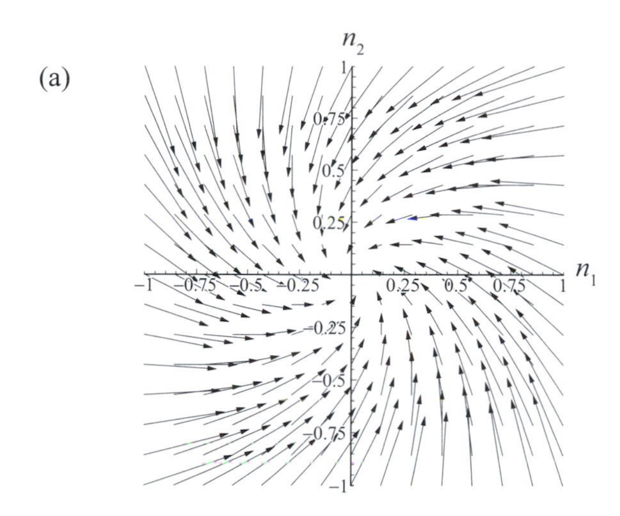

(b)

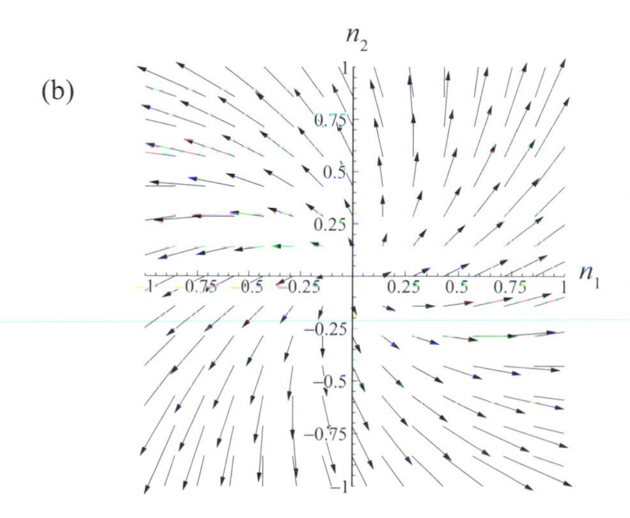

Figure 7.6: Spiraling dynamics with complex eigenvalues. Vector field plots of model (7.6) are shown for cases where the eigenvalues are complex. (a) $a = -2$, $b = -1$, $c = 1$, $d = -2$, resulting in the complex eigenvalues $-2 \pm \sqrt{i}$, with negative real parts. The system spirals in toward the equilibrium (0,0), which is stable. (b) $a = 2$, $b = -1$, $c = 1$, $d = 2$, resulting in the complex eigenvalues $2 \pm \sqrt{i}$ with positive real parts. The system spirals out from equilibrium (0,0), which is unstable.

When the eigenvalues are complex, the eigenvectors are themselves complex (i.e., their elements are complex numbers). Consequently, even though we can still talk of movement along these vectors, and even though decay still results in stability whereas growth still results in instability, the eigenvectors can no longer be represented in the phase plane (which is a plot of real numbers). That is, there is no real line in the phase plane along which the system remains once started on the line. Instead, we see spiraling behavior. The system spirals out from the origin if the real part of the eigenvalues, $A = (a + d)/2$, is positive, and it spirals in to the origin if the real part is negative (Figure 7.6).

We conclude that the real part of all eigenvalues must be negative for an equilibrium to be stable in continuous time. If we order the eigenvalues of a matrix according to their *real* parts, from the most negative to the most positive, the stability of an equilibrium depends only on the eigenvalue with the most positive real value. This is referred to as the *leading eigenvalue* (r_L) and it

The *leading eigenvalue* of a continuous-time model is the one with the largest real part. The *leading eigenvalue* dominates the long-term dynamics of a system.

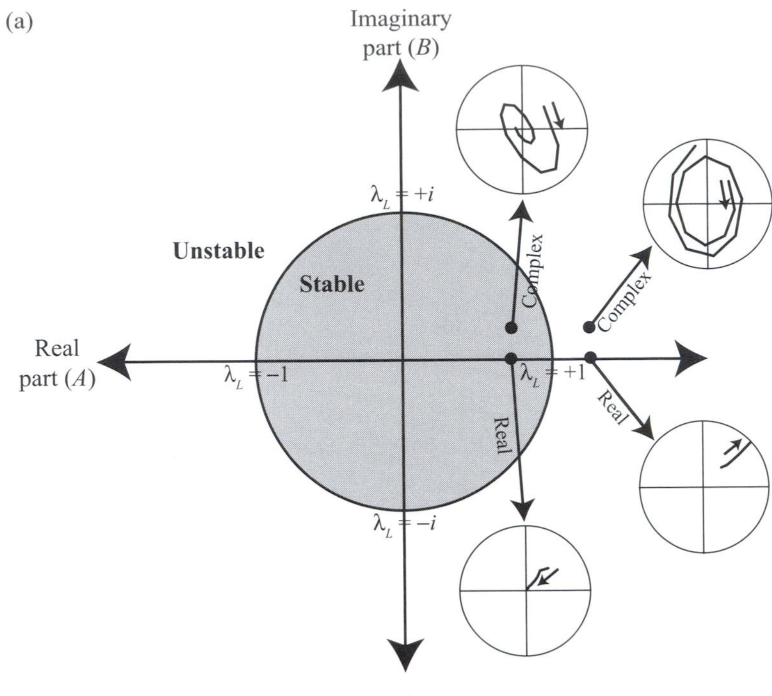

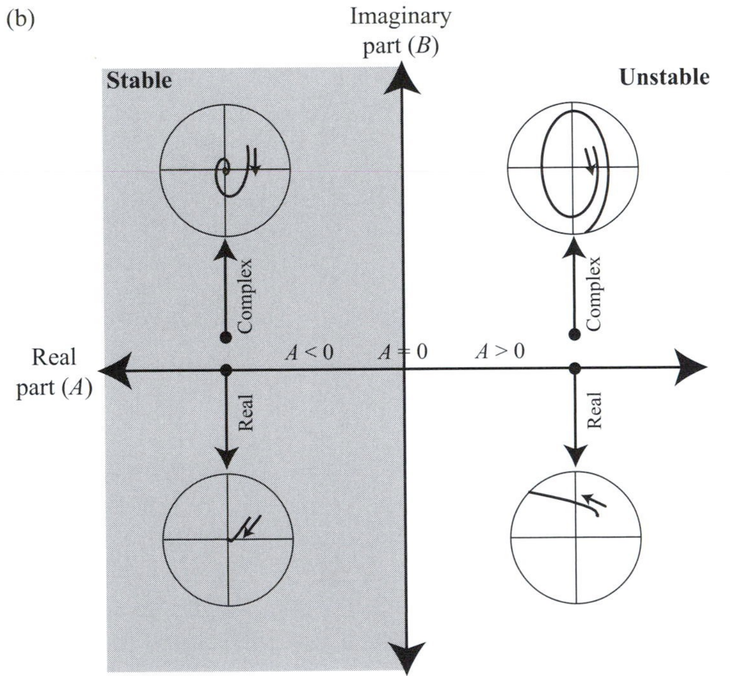

Figure 7.7: Assessing the stability of an equilibrium. Any eigenvalue can be written as $A + Bi$, where A is the real part (horizontal axis) and B is the imaginary part (vertical axis). (a) In discrete-time models, an equilibrium is stable only if the absolute values of all eigenvalues are less than one (within shaded circle). (b) In continuous-time models, an equilibrium is stable only if the real parts of all eigenvalues are negative (shaded half). Cycling is expected if the eigenvalues have imaginary parts (points off the horizontal axes). The inset phase-plane diagrams illustrate dynamics in a two-variable model given different leading eigenvalues.

determines the long-term behavior of the system (Box 9.1). When the real part of the leading eigenvalue is positive, a linear system of equations will move away from its equilibrium. When the real part of the leading eigenvalue is negative, a linear system of equations will move toward its equilibrium. This result, along with the corresponding result for discrete-time models, is illustrated in Figure 7.7.

Rule 7.2: Stability of an Equilibrium in a Linear Continuous-Time Model

- For an equilibrium to be stable in continuous time, the real parts of all eigenvalues must be negative.
- Equivalently, the real part of the *leading eigenvalue* must be negative (see Figure 7.7). The leading eigenvalue of a continuous-time model is the eigenvalue with the largest real part.

Example—Evolution by Sexual Selection

In many species, males exhibit greatly exaggerated, extravagant traits. Perhaps the most familiar examples are the incredibly showy tails of male peacocks. Although this might seem like an anomaly, the males of many species display this sort of gaudy extravagance, both in morphology and often in behavior. Presumably such extreme traits come at a cost to males by reducing foraging efficiency or increasing the likelihood of being killed by a predator. How can we explain the evolution of such characteristics?

Darwin (1871) was the first to recognize that such traits, despite having costs to males, might still evolve through what he termed sexual selection. There are different forms of sexual selection (Andersson 1994), but an important form involves female choice. If females prefer mating with males that have a large tail, then such males will have a mating advantage that might offset the survival cost associated with bearing such an extravagant trait.

While this provides a satisfactory answer on one level, on another level it pushes the problem one step further away. In particular, assuming females do have such mating preferences, we are then led to ask, why do preferences for such bizarre traits evolve? There have been a number of hypotheses put forward to explain the simultaneous evolution of male display traits and female preferences for them, and here we will focus on one known as "Fisher's Runaway Hypothesis." This hypothesis basically states that female preference and male trait exaggeration jointly coevolve in a runaway fashion owing to an evolutionary feedback between them. As females evolve stronger preferences, the fitness of males with more extreme traits rises, which selects for females with even stronger preferences, who produce more sexy sons.

Here we consider a continuous-time model for the evolution of a single male display trait (say tail length) and the level of female preference for this trait. Suppose z represents tail length, and p represents female preference for tails. A female with $p > 0$ prefers males with large tails while one with $p < 0$ prefers males with small tails (having $p = 0$ implies no preference). For simplicity, we measure tail length as a deviation from the tail length that gives the highest survival. In other words, $z = 0$ is the best tail length from the perspective of survival, and larger ($z > 0$) or smaller ($z < 0$) tails are associated with lower survival. We also assume that very choosy females have a lower fitness because they waste time looking for extravagant males. Thus, natural selection acting on females favors a female preference of zero. Finally, we use $\bar{z}$ and $\bar{p}$ to represent the average values of the male trait and female preference in the population. Based on similar models in the literature (see Lande 1982; Iwasa et al. 1991), a model for the evolutionary dynamics of the average tail length and preference is

$$\frac{d\bar{z}}{dt} = G_z a \bar{p} - G_z c \bar{z}, \tag{7.17a}$$

$$\frac{d\bar{p}}{dt} = -G_p b \bar{p}. \tag{7.17b}$$

This type of model is often referred to as a quantitative genetic model. The specific form of this model of evolution is derived in more detail in Supplementary Material 7.1. Here, we justify equations (7.17) on intuitive grounds.

Tail length evolves in response to two different evolutionary forces: sexual and natural selection. Sexual selection on tail length is measured by the term $G_z a \bar{p}$ in (7.17a). The parameter G_z is the additive genetic variance in male tail length—without genetic variation in tail length, tail length does not evolve (see Primer 3 for the definition of variance). The parameter a determines the strength of sexual selection resulting from the mean female preference within the population, $\bar{p}$. Both G_z and a are assumed to be positive. Thus, sexual selection through female choice favors longer tails when $\bar{p} > 0$ ($G_z a \bar{p}$ is positive) but shorter tails when $\bar{p} < 0$ ($G_z a \bar{p}$ is negative). Natural selection on tail length is measured by the term $-G_z c \bar{z}$ in (7.17a). The parameter c determines the strength of natural selection resulting from survival differences among males with different tail lengths and is also assumed to be positive. Therefore, natural selection favors shorter tails when tails are longer than the optimum ($\bar{z} > 0$ so that $-G_z c \bar{z}$ is negative) but longer tails when tails are shorter than the optimum ($\bar{z} < 0$ so that $-G_z c \bar{z}$ is positive).

Next, consider the evolution of female preference. In (7.17b), female preference evolves in response to only one evolutionary force, natural selection, as measured by the term $-G_p b \bar{p}$. G_p denotes the additive genetic variance in female preference—female preference does not evolve if there is no genetic variation for preferences in the population. The parameter b determines the strength of selection acting on female preference. Both G_p and b are assumed to be positive. Thus, natural selection favors preferences for shorter tails when females prefer long tails ($\bar{p} > 0$ so that $-G_p b \bar{p}$ is negative) but preferences for longer tails when females prefer short tails ($\bar{p} < 0$ so that $-G_p b \bar{p}$ is positive).

System (7.17) is a two-dimensional, linear system in continuous time. We can now ask whether this model predicts the evolution of exaggerated male tails and female preferences for them. Mathematically, we can answer this question by finding the equilibria and determining their stability. The only equilibrium of this model is $\hat{\bar{z}} = 0$, $\hat{\bar{p}} = 0$ (no tail exaggeration and no female preference). Therefore, for this model to predict the evolution of exaggerated tails and preferences, this equilibrium must be unstable. We can write (7.17) in matrix form as

$$\begin{pmatrix} \frac{d\bar{z}}{dt} \\ \frac{d\bar{p}}{dt} \end{pmatrix} = \mathbf{M}\begin{pmatrix} \bar{z} \\ \bar{p} \end{pmatrix}, \tag{7.18a}$$

where

$$\mathbf{M} = \begin{pmatrix} -G_z c & G_z a \\ 0 & -G_p b \end{pmatrix}. \tag{7.18b}$$

Because $\mathbf{M}$ is a triangular matrix, its eigenvalues are $r_1 = -G_z c$ and $r_2 = -G_p b$ (Rule P2.26 of Primer 2). As all parameters are assumed to be positive, both eigenvalues are negative. Therefore the equilibrium is stable, and we expect neither male trait exaggeration nor female preference to evolve.

The above model lacks an important feature of evolution, namely, that evolutionary change in one trait can occur as a correlated response to selection acting on other traits (Lande 1979; Lande and Arnold 1983). For example, let us imagine that an allele coding for a positive preference (in females) and an allele coding for a longer tail (in males) are very often found together on the same chromosome. Parents carrying such chromosomes are then very likely to have daughters who prefer longer tails and sons with longer tails. If longer tails are sexually selected, then as tail length increases, so too will female preferences for longer tails, just because the two types of alleles tend to be found together on the same chromosome (Figure 7.8). In the quantitative-genetic lexicon, we expect a correlated evolutionary response in female preference to selection on male tail length.

But why might chromosomes with alleles for a positive preference also carry the alleles for long tails? There are many reasons why such a genetic correlation might occur. This type of correlation is expected to arise with sexual selection because females with alleles generating a preference for large tails will tend to mate with males having long tails. As a result, their offspring (both male and female) will carry the alleles for both. A completely satisfying mathematical model would therefore have an additional dynamic variable that describes the magnitude of this correlation (e.g., Kirkpatrick 1982). For simplicity however, we follow several published treatments of this hypothesis that suppose this correlation is constant (e.g., Pomiankowski et al. 1991).

To incorporate the effect of correlated responses to selection into the above model, we include an extra term $-b\bar{p}B$ in equation (7.17a), where $-b\bar{p}$ is the strength of selection acting on female preference and B denotes the genetic covariance between mail tail length and female preference (see Primer 3 for the definition of covariance). (This is more fully justified in Supplementary

(a)

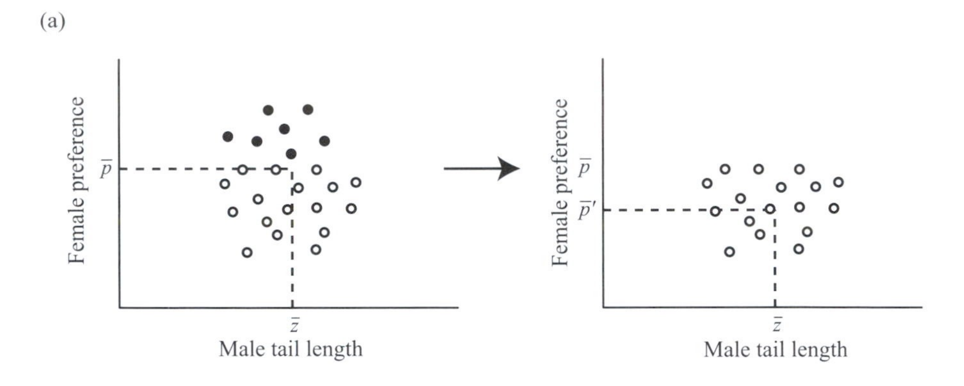

(b)

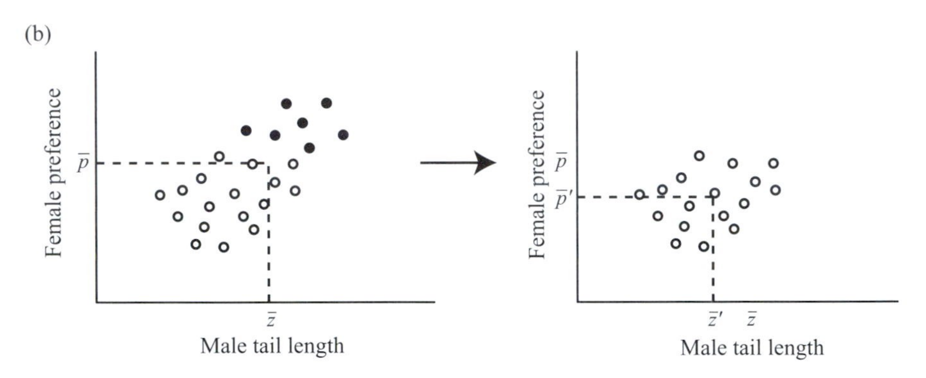

Figure 7.8: The effect of genetic covariance on the outcome of selection. Circles represent genotypes for female preference and mail tail length. Here, we assume that selection acts on female preference only and that those females with large preferences have lower survival. Black circles denote those genotypes that do not survive the bout of selection on female preference. Panels on the left represent the population prior to selection on female preference and those on the right represent the population after selection. (a) No genetic covariance between female preference and male tail length. After selection on female preference, the average female preference is reduced but the average male tail length remains unchanged. (b) A positive relationship between preference and tail lengths is illustrated, as expected under sexual selection. After selection on female preference, the average female preference is reduced and the average male tail length is reduced as well, because of the positive association between preference and tail length. This is referred to as "indirect" or "correlated" selection on male tail length.

Material 7.1; see also Lande 1982.) We expect B to be positive because alleles for positive preferences tend to be found with alleles for long tails. Thus, the product $-b\bar{p}B$ gives the evolutionary change in tail length resulting from indirect selection on female preference. In the same way, males that carry alleles for long tails tend to carry (unexpressed) alleles for positive female preferences that they pass on to their daughters. This association is again measured by the genetic covariance B. As a result, selection on males (i.e., the term $a\bar{p} - c\bar{z}$ multiplying G_z in (7.17a)) creates an indirect force of selection on female preference, adding an extra term $B(a\,\bar{p} - c\,\bar{z})$ to (7.17b). Including these indirect selective forces, we expect male trait and female preference to evolve according to

$$\frac{d\bar{z}}{dt} = G_z(a\,\bar{p} - c\,\bar{z}) - B\,b\,\bar{p},$$
$$\frac{d\bar{p}}{dt} = B\,(a\bar{p} - c\bar{z}) - G_p\,b\,\bar{p}. \tag{7.19}$$

Our revised model is still linear and two dimensional. Again, the only equilibrium is $\hat{\bar{z}} = 0$, $\hat{\bar{p}} = 0$. Thus we need to determine the stability of this point. The matrix in (7.18a) is now

$$\mathbf{M} = \begin{pmatrix} -G_z c & G_z a - bB \\ -Bc & -G_p b + aB \end{pmatrix}. \tag{7.20}$$

The eigenvalues of this matrix can be calculated using equations (7.12), giving

$$\frac{1}{2}\left\{-b\,G_p - c\,G_z + a\,B \pm \sqrt{\left(b\,G_p + c\,G_z - a\,B\right)^2 - 4\,b\,c\left(G_pG_z - B^2\right)}\right\}. \tag{7.21}$$

To make it easier to interpret the eigenvalues, let's simplify the notation by defining $\gamma = -b\,G_p - c\,G_z + a\,B$ and rewriting the eigenvalues as:

$$\frac{1}{2}\left\{\gamma \pm \sqrt{\gamma^2 - 4\,b\,c\left(G_pG_z - B^2\right)}\right\}. \tag{7.22}$$

Eigenvalues (7.22) can be complex numbers, depending on the values of the parameters. Thus, to determine whether the equilibrium with no trait exaggeration and no female preference is unstable, we must consider cases where the term inside the square root is positive and where it is negative. These cases are considered in Table 7.1, where we assume that the genetic covariance B is not as strong as the genetic variances themselves, so that $G_z\,G_p - B^2 > 0$. When $\gamma < 0$ (i.e., $a\,B < b\;G_p + c\;G_z$), the real part of all eigenvalues is negative, as required for the stability of a continuous-time model. Consequently, we predict no exaggeration of male display and no female preferences in Cases 3 and 4. Conversely, the equilibrium is unstable and we expect male tail exaggeration and female preference to evolve when $\gamma > 0$ (i.e., $a\,B > b\,G_p + c\,G_z$; Cases 1 and 2). This requires that the total response to direct selection on both male tail length sexual and female preference (i.e., $b\,G_p + c\,G_z$) is smaller than the indirect effect of selection on female preference arising from its genetic correlation with male tail length (i.e., $a\,B$).

Incidentally, there is a mathematically equivalent way to evaluate the eigenvalues of a 2×2 matrix, which is often easier to apply. The eigenvalues of a two-dimensional matrix $\mathbf{M}$ have negative real parts only if the determinant of

TABLE 7.1:
Analyzing the eigenvalues in the model of sexual selection

Case 1: $\gamma > 0$ and $\gamma^2 - 4bc(G_pG_z - B^2) > 0$
In this case the eigenvalues (7.22) are real, and the largest one is $\frac{1}{2}\{\gamma + \sqrt{\gamma^2 - 4bc(G_pG_z - B^2)}\}$. Because the quantity under the squareroot is positive in this case, it must be true that $\frac{1}{2}\{\gamma + \sqrt{\gamma^2 - 4bc(G_pG_z - B^2)}\}$ is greater than $\gamma/2$, which is positive. This implies that the equilibrium is unstable.

Case 2: $\gamma > 0$ and $\gamma^2 - 4bc(G_pG_z - B^2) < 0$
In this case the eigenvalues (7.22) are complex. Because the real part of the eigenvalues, $\gamma/2$, is positive, the equilibrium is unstable.

Case 3: $\gamma < 0$ and $\gamma^2 - 4bc(G_pG_z - B^2) > 0$
In this case the eigenvalues (7.22) are real, and the largest one is $\frac{1}{2}\{\gamma + \sqrt{\gamma^2 - 4bc(G_pG_z - B^2)}\}$. Now the quantity under the squareroot is positive, but it is less than γ^2 under our assumptions. Therefore, we know that $\sqrt{\gamma^2 - 4bc(G_pG_z - B^2)}$ is a positive number whose magnitude is less than the negative quantity γ. As a result, $\frac{1}{2}\{\gamma + \sqrt{\gamma^2 - 4bc(G_pG_z - B^2)}\}$ is negative, which implies that the equilibrium is stable.

Case 4: $\gamma < 0$ and $\gamma^2 - 4bc(G_pG_z - B^2) < 0$
In this case the eigenvalues are complex. Because the real part of the eigenvalues, $\gamma/2$, is negative, the equilibrium is stable.

M is positive and the trace of **M** is negative (Rule P2.25 in Primer 2). For the sexual selection model (7.19), the determinant is $b\,c(G_p\,G_z - B^2)$. This determinant is positive if we assume that there is more genetic variance for the traits than genetic covariance between them ($G_p\,G_z > B^2$). Making this realistic assumption, the stability of the equilibrium then depends on the sign of the trace. Summing the diagonal elements, the trace of (7.20) is $\gamma = -bG_p - cG_z + aB$. Thus, this alternative method leads to the same conclusion: the equilibrium with no trait exaggeration and no female preference is unstable when $\gamma > 0$ (i.e., $aB > bG_p + cG_z$).

Incorporating genetic correlations in this more realistic model thus confirms that exaggerated male traits can coevolve with female preferences. Indeed, when $\gamma > 0$, a "runaway process" results, with more extreme preferences evolving over time and driving the evolution of more extreme traits. In fact, because this model is linear in the variables $\bar{z}$ and $\bar{p}$, the growth away from the origin is exponential when $\gamma > 0$ (possibly cycling along the way if the eigenvalues are complex; Figure 7.9). Yet, clearly, there must be an upper bound to female preferences and male tail length. Eventually something must halt the runaway process, either the depletion of genetic variance or stronger counteracting natural selection. To incorporate such effects and to determine the ultimate level of male traits and female preferences requires that we consider a nonlinear version of model (7.19). We examine just such a model in Chapter 11.

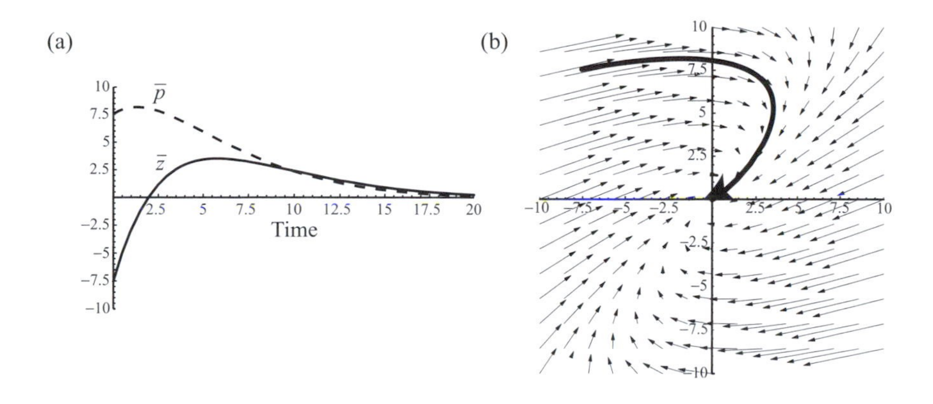

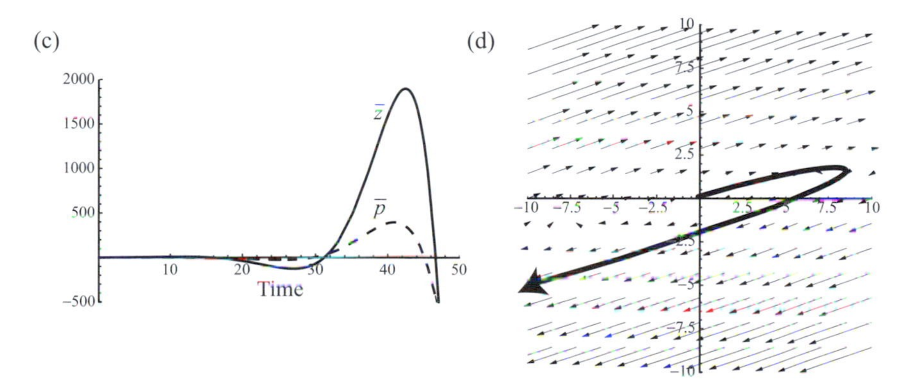

Figure 7.9: Dynamics of the sexual selection model with indirect selection. (a) A plot of the variables against time and (b) a phase-plane diagram for model (7.19) with $a = 0.8$, $b = 0.5$, $c = 0.9$, $B = 0.2$, $G_z = 0.5$, $G_p = 0.4$, and initial conditions $\bar{p}(0) = 7.55$ and $\bar{z}(0) = -7.55$. For these parameter values, the equilibrium is stable ($\gamma < 0$). (c) A plot of the variables against time and (d) a phase-plane diagram with $a = 5$, $b = 0.5$, $c = 0.9$, $B = 0.2$, $G_z = 0.5$, $G_p = 0.4$, and initial conditions $\bar{p}(0) = 0.1$ and $\bar{z}(0) = -0.1$. For these parameter values, the equilibrium is unstable ($\gamma > 0$). The trajectories from panels (a) and (c) are drawn as thick curves in panels (b) and (d).

7.4 Equilibria and Stability for Linear Discrete-Time Models

We have now seen how to find equilibria and determine stability in continuous-time models. While these techniques are extremely useful, some biological phenomena are better described by discrete-time models. This is particularly true for models of species that inhabit seasonal environments or that have nonoverlapping generations. In this section we present the analogous techniques for multivariable discrete-time models. Conceptually, the techniques for finding equilibria and determining their stability are very similar to those of continuous-time models, but the details differ in critical ways.

As in section 7.2, we start with the simplest possible discrete-time model of two variables that are independently growing over time:

$$x_1(t+1) = \lambda_1\, x_1(t),$$
$$x_2(t+1) = \lambda_2\, x_2(t). \tag{7.23a}$$

Writing these recursions in matrix form,

$$\vec{x}(t+1) = \mathbf{M}\,\vec{x}(t) \tag{7.23b}$$

where

$$\vec{x}(t) = \begin{pmatrix} x_1(t) \\ x_2(t) \end{pmatrix} \text{ and } \mathbf{M} = \begin{pmatrix} \lambda_1 & 0 \\ 0 & \lambda_2 \end{pmatrix} \tag{7.23c}$$

M is sometimes referred to as a *transition matrix*, because it quantifies the transitions in the variables from one time step to the next.

Because the recursion for one variable does not depend on the state of the other variable, we can iterate the two equations in (7.23a) separately to obtain the general solution $x_1(t) = x_1(0)\,\lambda_1^t$ and $x_2(t) = x_2(0)\,\lambda_2^t$ (see equation 4.1 of Chapter 4). Consequently, $x_1(t)$ will shrink to zero provided that the absolute value of λ_1 is less than one (i.e., $|\lambda_1| < 1$), and $x_2(t)$ will shrink to zero provided that the absolute value of λ_2 is less than one (i.e., $|\lambda_2| < 1$). This means that the equilibrium of the system, $\hat{x}_1 = 0,\ \hat{x}_2 = 0$, is stable provided that both $|\lambda_1| < 1$ and $|\lambda_2| < 1$. As in the stability analysis of one-variable discrete-time models (Chapter 5), it is the absolute values of the growth factors λ_i relative to one that matters, not the sign of these growth factors.

In this example, the eigenvectors are again the coordinate axes and the eigenvalues are the growth factors λ_1 and λ_2. If the system is started on one of these axes, it remains on that axis forever, and growth (or decay) occurs along this axis by a factor equal to the corresponding eigenvalue. To determine the stability of the equilibrium in a discrete-time model, we need only determine whether the flow along the eigenvectors is toward or away from the equilibrium, just as in the continuous-time model. Thus, the equilibrium will be stable if (and only if) the absolute values of both eigenvalues are less than one.

Now we consider a general discrete-time, two-variable linear model:

$$x_1(t+1) = a\,x_1(t) + b\,x_2(t),$$
$$x_2(t+1) = c\,x_1(t) + d\,x_2(t), \tag{7.24a}$$

or, in matrix notation,

$$\vec{x}(t+1) = \mathbf{M}\,\vec{x}(t), \tag{7.24b}$$

where the transition matrix is

$$\mathbf{M} = \begin{pmatrix} a & b \\ c & d \end{pmatrix}. \tag{7.24c}$$

The equilibrium for this model must satisfy the *equilibrium condition*

$$\begin{pmatrix}\hat{x}_1\\ \hat{x}_2\end{pmatrix} = \mathbf{M}\begin{pmatrix}\hat{x}_1\\ \hat{x}_2\end{pmatrix} \tag{7.25a}$$

or

$$\vec{0} = (\mathbf{M} - \mathbf{I})\,\hat{\vec{x}}. \tag{7.25b}$$

To go from (7.25a) to (7.25b), first write (7.25a) as $\hat{\vec{x}} = \mathbf{M}\,\hat{\vec{x}}$, where $\hat{\vec{x}} = \begin{pmatrix}\hat{x}_1\\ \hat{x}_2\end{pmatrix}$. Next, bring all of the $\hat{\vec{x}}$ terms to one side, $\vec{0} = \mathbf{M}\hat{\vec{x}} - \hat{\vec{x}}$, where $\vec{0}$ is a vector of zeros. The final step is to factor this expression, which requires that we first multiply the second term by $\mathbf{I}$ (so that the dimensions are correct; see section P2.7 of Primer 2).

To solve equation (7.25b) we can again use the results of section P2.7 in Primer 2. There are two possibilities: (i) the determinant of the matrix $\mathbf{M} - \mathbf{I}$ is nonzero, in which case the only equilibrium is $\hat{x}_1 = 0, \hat{x}_2 = 0$, and (ii) the determinant of the matrix $\mathbf{M} - \mathbf{I}$ is zero, in which case there are an infinite number of equilibria:

Rule 7.3: Equilibrium of Linear Multivariable Models in Discrete Time

As long as the determinant of $\mathbf{M} - \mathbf{I}$ is not zero, a linear model in discrete time has only one equilibrium, regardless of the number of variables.

- For a linear model of the form $\vec{n}(t+1) = \mathbf{M}\,\vec{n}(t)$, the equilibrium point is the origin, $\hat{\vec{n}} = \vec{0}$, where $\vec{0}$ is a vector of zeros.
- For an affine model of the form $\vec{n}(t+1) = \mathbf{M}\,\vec{n}(t) + \vec{c}$, the equilibrium point is $\hat{\vec{n}} = -(\mathbf{M} - \mathbf{I})^{-1}\vec{c}$ (see Box 7.2).

If the determinant of $\mathbf{M} - \mathbf{I}$ is zero, then there are an infinite number of equilibria.

Rules 7.1 and 7.3 differ slightly in that it is the determinant of $\mathbf{M}$ alone that matters in the continuous-time case, but the determinant of $\mathbf{M} - \mathbf{I}$ that matters in the discrete-time case.

Focusing on the case where the determinant of $\mathbf{M} - \mathbf{I}$ is not zero, we now describe how to determine the stability of the equilibrium $\hat{x}_1 = 0, \hat{x}_2 = 0$. Once again, the key is to determine the eigenvalues of $\mathbf{M}$. For a 2×2 matrix, the eigenvalues are still given by expressions (7.12). The only difference for a discrete-time model lies in how we use the eigenvalues to determine stability. As we shall demonstrate in Chapter 9, we must now compare the absolute values of the eigenvalues to one, regardless of whether the eigenvalues are real or complex.

In general, the eigenvalues of a discrete-time linear model will have the form $\lambda_j = A_j + B_j i$, where A_j and B_j are real numbers. The absolute value (sometimes termed the *modulus*) of such an eigenvalue is $|\lambda_j| = \sqrt{A_j^2 + B_j^2}$. Stability of the equilibrium requires that all eigenvalues have absolute value less than one ($|\lambda_j| < 1$; Figure 7.7). As with continuous-time models, we can again reorder the eigenvalues, but now in terms of their absolute values. The stability of the equilibrium depends only on the eigenvalue with largest absolute value. This eigenvalue is termed the *leading eigenvalue* of a discrete-time model, λ_L, and again the leading eigenvalue determines the long-term behavior of the system (Box 9.1). When the leading eigenvalue is greater than one, a linear system of equations in discrete time moves away from its equilibrium. When the leading eigenvalue is less than one, a linear system of equations will move toward its equilibrium (see Figure 7.7):

The *leading eigenvalue* of a discrete-time model is the one with the largest absolute value. The leading eigenvalue dominates the long-term dynamics of a system.

Rule 7.4: Stability of an Equilibrium in a Linear Discrete-Time Model

For stability of an equilibrium in discrete time, all eigenvalues must be less than one in abolute value. Equivalently, the absolute value of the leading eigenvalue must be less than one (see Figure 7.7).

- For complex eigenvalues $\lambda = A \pm Bi$, stability requires that the absolute value $\sqrt{A^2 + B^2}$ be less than one.
- For real eigenvalues, stability requires that both $\lambda < 1$ and $-1 < \lambda$.

Example—Host Use of Sea Anemones by Crabs

Some crab species colonize and live within sea anemones. These anemones provide the crab with shelter and protection from potential predators. One such species, *Allopetrolisthes spinifrons*, colonizes a variety of anemones of different colors, including reddish-green anemones, green anemones, and blue anemones. The crab itself is reddish-green, and therefore it is believed that reddish-green anemones are the preferred host because they offer the greatest camouflage (Baeza and Stotz 2001).

A study by Baeza and Stotz (2003) demonstrated that crabs in the wild tend to be found more often than expected by chance on reddish-green anemones and that they tend to be found less often than expected by chance on blue anemones. Their occurrence on green anemones was near that expected by chance (Figure 7.10). This suggests that crabs might have a preference for reddish-green anemones over all others and that they actively avoid blue anemones. Baeza and Stotz (2003) conducted a number of experiments to test this idea, giving crabs a choice between different-colored anemones. Overall they found that crabs avoided the blue anemones but did not strongly discriminate among the other colors (Table 1, Baeza and Stotz 2003). Consequently, it was suggested that other factors besides host color preference must be important to explain the fact that reddish-green hosts are more often inhabited than green hosts.

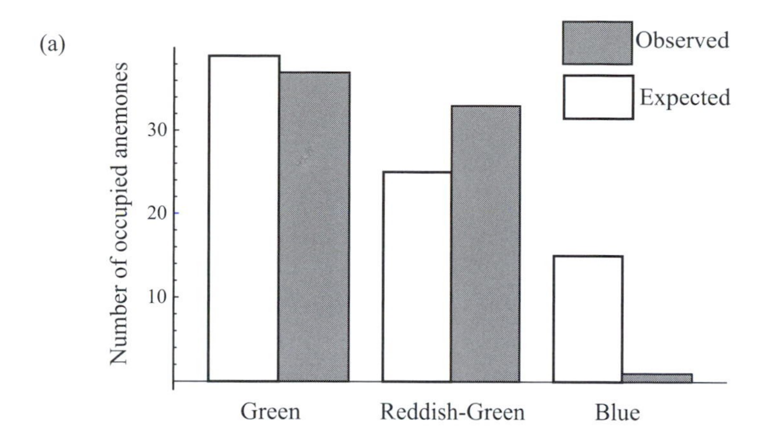

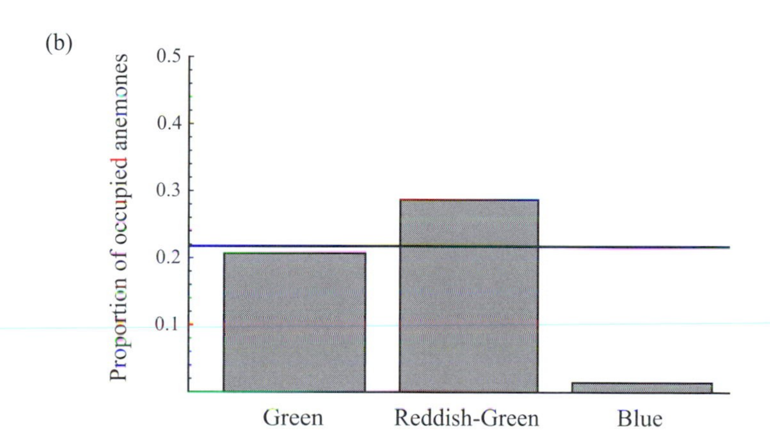

Figure 7.10: Host use of anemones by crabs. (a) The number of anemones of each color that were occupied by crabs in the field, along with how many were expected to be occupied if the hosts were utilized at random. (b) Same data as in (a) but expressed as the proportion of hosts of each color that were occupied. Horizontal line indicates the overall probability that a host of any color is occupied. Data are adapted from Baeza and Stotz (2003).

Let us use a model to examine another possible explanation for the discrepancy between the patterns of host use seen in nature and the host color preferences that were documented in the experiments. It is possible that differential mortality of crabs on different-colored hosts plays an important role in explaining the observed patterns of host use. For example, even if crabs have no color preferences, mortality rates are expected to be highest on blue anemones, followed by green, and then reddish-green anemones. This differential mortality might thereby contribute to the observed patterns of host use. We would like to be able to predict the pattern of host use that is expected to occur, given different levels of host preference and host-specific mortality.

Let us construct a discrete-time model of crab survival and colonization on anemones of different colors. Nearly all colonized anemones contain only a single crab, so we keep track of the number of blue, reddish-green, and green anemones that are occupied by a crab at time t, as denoted by the three variables $B(t)$, $R(t)$, and $G(t)$. At each time step, we suppose that a fraction of crabs die, where this fraction depends on the degree of camouflage. Then there is recruitment of new crabs to unoccupied hosts at a rate that depends on the color preferences of the crabs. This yields the recursions

$$\begin{aligned} B(t+1) &= B(t) - \mu_B B(t) + \text{recruits}_B, \\ R(t+1) &= R(t) - \mu_R R(t) + \text{recruits}_R, \\ G(t+1) &= G(t) - \mu_G G(t) + \text{recruits}_G, \end{aligned} \tag{7.26}$$

where μ_B, μ_R, and μ_G are the fractions of crabs dying each time step, for crabs living on blue, reddish-green, and green anemones, respectively. We would expect that $\mu_B > \mu_G > \mu_R$ because the crabs are least camouflaged on blue anemones and most camouflaged on reddish-green anemones.

The number of recruits to each host color will depend on the preference of crabs for each color, as well as on the number of new crabs that are produced each time step. To complete equations (7.26), we must decide how to model the number of recruits to each host color in each time step. Not much is known about the exact details of recruitment, and therefore, we will consider three different possibilities.

(i) Constant per capita recruitment

We start by assuming that each individual produces a constant number of offspring, b, per time step. These offspring then colonize new anemones, which we assume are always in plentiful supply. The total number of recruiting crabs per time step will then be $(B(t) + R(t) + G(t))\, b$. If we further suppose that a crab chooses a blue, reddish-green, or green host with probability p_B, p_R, or p_G, then model (7.26) becomes

$$\begin{aligned} B(t+1) &= B(t) - \mu_B B(t) + (B(t) + R(t) + G(t))\, b\, p_B, \\ R(t+1) &= R(t) - \mu_R R(t) + (B(t) + R(t) + G(t))\, b\, p_R, \\ G(t+1) &= G(t) - \mu_G G(t) + (B(t) + R(t) + G(t))\, b\, p_G. \end{aligned} \tag{7.27a}$$

At this stage it is a good idea to group all terms involving the same variables. Doing so for (7.27a) gives

$$\begin{aligned} B(t+1) &= B(t)(1 - \mu_B + b\, p_B) + R(t)\, b\, p_B + G(t)\, b\, p_B, \\ R(t+1) &= B(t)\, b\, p_R + R(t)(1 - \mu_R + b\, p_R) + G(t)\, b\, p_R, \\ G(t+1) &= B(t)\, b\, p_G + R(t)\, b\, p_G + G(t)(1 - \mu_G + b\, p_G), \end{aligned} \tag{7.27b}$$

which is a linear system of recursion equations that can be written in matrix form:

$$\begin{pmatrix} B(t+1) \\ R(t+1) \\ G(t+1) \end{pmatrix} = \mathbf{M} \begin{pmatrix} B(t) \\ R(t) \\ G(t) \end{pmatrix}, \tag{7.28a}$$

where

$$\mathbf{M} = \begin{pmatrix} 1 - \mu_B + b p_B & b p_B & b p_B \\ b p_R & 1 - \mu_R + b p_R & b p_R \\ b p_G & b p_G & 1 - \mu_G + b p_G \end{pmatrix}. \tag{7.28b}$$

According to Rule 7.3, the only equilibrium of this model is $\hat{B} = 0$, $\hat{R} = 0$, $\hat{G} = 0$, because the determinant of $\mathbf{M} - \mathbf{I}$ is not zero for reasonable parameter choices. This equilibrium will be stable provided that all eigenvalues of matrix (7.28b) are less than one in absolute value; otherwise it will be unstable.

The eigenvalues can be calculated by determining the characteristic polynomial of matrix (7.28b) and then solving for its three roots (see Primer 2), but these calculations turn out to be quite tedious. Before getting too mired in these details, however, let us first consider what conclusions we could draw if we had the expressions for these eigenvalues. Because model (7.28) is linear and nonaffine, we know that it will predict that either no anemones are colonized or an ever-increasing number of crabs occur on each host type. Obviously neither of these predictions is borne out in real populations, and therefore this model must not be an accurate description of the crab-anemone system. The problem lies in the assumption of a constant per capita recruitment at every time step. Ultimately, there must be some limit to recruitment as the population size of crabs gets large and fewer anemones are available for colonization. Thus we must revise the way that we have modeled recruitment.

(ii) Constant total recruitment

Instead of supposing that there is a constant number of offspring per parent, let us assume that there is a constant total number of offspring produced per time step. This assumption might be realistic if there is some other process regulating the number of juvenile crabs (e.g., density-dependent competition for food), such that a roughly constant number of crabs, b, seek out anemones at each time step. Using the same definitions for the p's as above, equations (7.26) become

$$\begin{aligned} B(t+1) &= B(t) - \mu_B B(t) + b\,p_B, \\ R(t+1) &= R(t) - \mu_R R(t) + b\,p_R, \\ G(t+1) &= G(t) - \mu_G G(t) + b\,p_G. \end{aligned} \tag{7.29}$$

Model (7.29) is an affine model in discrete time (Box 7.2), which can be written in matrix notation as

$$\begin{pmatrix} B(t+1) \\ R(t+1) \\ G(t+1) \end{pmatrix} = \mathbf{M} \begin{pmatrix} B(t) \\ R(t) \\ G(t) \end{pmatrix} + \vec{c}, \tag{7.30a}$$

where

$$\mathbf{M} = \begin{pmatrix} 1-\mu_B & 0 & 0 \\ 0 & 1-\mu_R & 0 \\ 0 & 0 & 1-\mu_G \end{pmatrix} \quad \text{and} \quad \vec{c} = \begin{pmatrix} b\,p_B \\ b\,p_R \\ b\,p_G \end{pmatrix}. \tag{7.30b}$$

All of the techniques that we have learned so far apply to linear models of the form (7.24). Therefore, we first need to transform the affine model (7.30) into a model of this type. Following the approach of Box 7.2, equation (7.30) can be transformed into a linear model by first calculating its equilibrium and then defining a new set of three variables that represent the deviations of each of the original variables from their respective equilibrium values. The equilibrium of an affine model is $-(\mathbf{M} - \mathbf{I})^{-1}\vec{c}$ (Box 7.2), which in this case equals

$$\begin{pmatrix} \hat{B} \\ \hat{R} \\ \hat{G} \end{pmatrix} = \begin{pmatrix} \mu_B & 0 & 0 \\ 0 & \mu_R & 0 \\ 0 & 0 & \mu_G \end{pmatrix}^{-1} \begin{pmatrix} b\,p_B \\ b\,p_R \\ b\,p_G \end{pmatrix}. \tag{7.31a}$$

Because the matrix in (7.31a) is diagonal, its inverse is easy to calculate using Rule P2.24:

$$\begin{pmatrix} \mu_B & 0 & 0 \\ 0 & \mu_R & 0 \\ 0 & 0 & \mu_G \end{pmatrix}^{-1} = \begin{pmatrix} \mu_B^{-1} & 0 & 0 \\ 0 & \mu_R^{-1} & 0 \\ 0 & 0 & \mu_G^{-1} \end{pmatrix}.$$

Using this fact, we can multiply out (7.31a) to get the equilibrium

$$\hat{B} = \frac{b\,p_B}{\mu_B}, \quad \hat{R} = \frac{b\,p_R}{\mu_R}, \quad \hat{G} = \frac{b\,p_G}{\mu_G}. \tag{7.31b}$$

Equations (7.31b) give the equilibrium number of anemones of each color that are occupied by crabs. We can then define the new variables $\delta_B(t) = B(t) - \hat{B}$, $\delta_R(t) = R(t) - \hat{R}$, and $\delta_G(t) = G(t) - \hat{G}$, and rewrite the original model (7.30a) as

$$\begin{pmatrix} \delta_B(t+1) \\ \delta_R(t+1) \\ \delta_G(t+1) \end{pmatrix} = \mathbf{M} \begin{pmatrix} \delta_B(t) \\ \delta_R(t) \\ \delta_G(t) \end{pmatrix} \tag{7.32}$$

(Problem 7.6). The equilibrium of the transformed system (7.32) is $\hat{\vec{\delta}} = \vec{0}$. Because (7.32) is a nonaffine linear model, we can now evaluate the stability of this equilibrium by determining the eigenvalues of $\mathbf{M}$. Because $\mathbf{M}$ is a diagonal matrix, its eigenvalues are the diagonal elements: $\lambda_1 = 1 - \mu_B$, $\lambda_2 = 1 - \mu_R$, and $\lambda_3 = 1 - \mu_G$ (see Rule P2.26). For a discrete-time model, stability requires that the absolute value of all eigenvalues be less than one, which they are in this case because the μ's represent the fraction of the crab population that dies. Thus, the equilibrium $\hat{\vec{\delta}} = \vec{0}$ of the transformed system (7.32) is stable, as is the equilibrium (7.31b) of the original model.

Our revised model predicts the equilibrium numbers of crabs inhabiting blue, reddish-green, and green anemones, as given by equations (7.31b). This equilibrium makes some intuitive sense. At equilibrium, more crabs will inhabit anemones of a certain color if the crabs' preference for that color, p, is high and if the mortality rate of crabs on that color, μ, is low. The experimental data suggest that the preferences for green and for reddish-green anemones are the same (i.e., $p_G = p_R$), and therefore we predict the abundances of each type to be $\hat{B} = bp_B/\mu_B$, $\hat{R} = bp_R/\mu_R$, and $\hat{G} = bp_R/\mu_G$. Therefore, reddish-green anemones will be used more than blue anemones provided that $\hat{R} > \hat{B}$, or $p_R/\mu_R > p_B/\mu_B$. The available data suggest that this condition will be met since $p_R > p_B$ (from the experiments mentioned earlier) and $\mu_B > \mu_R$ (mortality of crabs is greater on blue than on reddish-green anemones). Similarly, reddish-green anemones will be used more than green anemones provided that $\hat{R} > \hat{G}$, or $\mu_G > \mu_R$. Again this is expected to hold because mortality on green anemones is likely to be greater than that on reddish-green anemones.

The above analysis goes some way toward illustrating that the observed pattern of host use by crabs might be explained solely by host color preference and differential mortality, but there is still a troublesome problem with the model. The equilibrium in (7.31b) suggests that the relative abundance of anemones of different colors does not affect the patterns of host use. Surely this cannot be true in the real world. For example, if blue anemones are extremely common relative to reddish-green anemones, then presumably a greater total number of blue anemones will be colonized than reddish-green anemones, even if crabs prefer reddish-green anemones. Indeed, the field data of Baeza and Stotz (2003) control for this issue by comparing the observed patterns of host use with what would be expected by chance (Figure 7.10). Ideally we would like our model to account for the abundance of the different types of anemones, and so we must return to the drawing board and revise our model.

(iii) Constant total recruitment and variable host availability

To account for the fact that hosts of different colors are present at different abundances, we suppose that recruitment to a host of a given color is composed of two steps. First, the crab must choose an anemone, say a green anemone, over the other colors, which happens with probability p_G. Then, the green host that this crab attempts to colonize must be unoccupied, otherwise the juvenile is injured or killed. If the total number of green anemones is N_G and if $G(t)$ are currently occupied, then the fraction of unoccupied green anemones is $1 - G(t)/N_G$. We thus let the recruitment rate to green anemones be $b\,p_G(1 - G(t)/N_G)$ at time step t, where again we assume that the total number of juveniles seeking out anemones is b at every time step. Similar calculations hold for the other two host colors. Equations (7.26) then become

$$\begin{aligned} B(t+1) &= B(t) - \mu_B B(t) + bp_B(1 - B(t)/N_B), \\ R(t+1) &= R(t) - \mu_R R(t) + bp_R(1 - R(t)/N_R), \\ G(t+1) &= G(t) - \mu_G G(t) + bp_G(1 - G(t)/N_G)\,. \end{aligned} \tag{7.33a}$$

Grouping terms involving the same variables gives

$$\begin{aligned} B(t+1) &= B(t)(1 - \mu_B - bp_B/N_B) + bp_B, \\ R(t+1) &= R(t)(1 - \mu_R - bp_R/N_R) + bp_R, \\ G(t+1) &= G(t)(1 - \mu_G - bp_G/N_G) + bp_G\,. \end{aligned} \tag{7.33b}$$

Model (7.33b) is again an affine model in discrete time that can be written in matrix form as

$$\begin{pmatrix} B(t+1) \\ R(t+1) \\ G(t+1) \end{pmatrix} = \mathbf{M} \begin{pmatrix} B(t) \\ R(t) \\ G(t) \end{pmatrix} + \vec{c}, \tag{7.34a}$$

where

$$\mathbf{M} = \begin{pmatrix} 1 - \mu_B - \dfrac{b p_B}{N_B} & 0 & 0 \\ 0 & 1 - \mu_R - \dfrac{b p_R}{N_R} & 0 \\ 0 & 0 & 1 - \mu_G - \dfrac{b p_G}{N_G} \end{pmatrix} \text{ and } \vec{c} = \begin{pmatrix} b\,p_B \\ b\,p_R \\ b\,P_G \end{pmatrix}. \tag{7.34b}$$

We can now follow the same approach that we did in the previous section to find the equilibrium of model (7.34):

$$\hat{B} = \frac{b p_B}{\mu_B + b p_B/N_B}, \quad \hat{R} = \frac{b p_R}{\mu_R + b p_R/N_R}, \quad \hat{G} = \frac{b p_G}{\mu_G + b p_G/N_G}. \tag{7.35a}$$

Alternatively, we can express the equilibrium abundances as fractions of the total number of available hosts of each color by dividing each of the solutions in (7.35a) by the total number of hosts of the corresponding color (e.g., $\hat{f}_B = \hat{B}/N_B$):

$$\hat{f}_B = \frac{b p_B}{\mu_B N_B + b p_B}, \quad \hat{f}_R = \frac{b p_R}{\mu_R N_R + b p_R}, \quad \hat{f}_G = \frac{b p_G}{\mu_G N_G + b p_G}. \tag{7.35b}$$

In Problem 7.7, you are asked to transform model (7.34) into a nonaffine model and then to show that the equilibrium (7.35) is stable as long as we assume that the number of juvenile crabs recruiting to a particular type of anemone does not exceed the number of such anemones available (i.e., $b\,p_i < N_i$). If we do not make this assumption, then, according to (7.33a), it is possible to have more occupied anemones of a particular color than there are anemones of that color (e.g., if $B(t) = 0$ and $b\,p_G > N_G$ then (7.33a) predicts that $G(t+1) = b\,p_G > N_G$). As this is unreasonable, we proceed under the assumption that $b\,p_i < N_i$ and that the expected pattern of host use at equilibrium is given by (7.35).

We can evaluate the plausibility of this model using the data of Baeza and Stotz. Their survey data suggest that $N_G = 179$, $N_R = 115$, and $N_B = 69$, and their experimental data suggest that $p_G \approx p_R \approx 0.43$ and $p_B \approx 0.14$. Therefore, the equilibrium fractions of occupied anemones of each color are predicted by (7.35b) to be

$$\hat{f}_B = \frac{0.14 b}{69\mu_B + 0.14 b}, \quad \hat{f}_R = \frac{0.43 b}{115\mu_R + 0.43 b}, \quad \hat{f}_G = \frac{0.43 b}{179\mu_G + 0.43 b},$$

or

$$\hat{f}_B = \frac{0.14}{69\frac{\mu_B}{b} + 0.14}, \quad \hat{f}_R = \frac{0.43}{115\frac{\mu_R}{b} + 0.43}, \quad \hat{f}_G = \frac{0.43}{179\frac{\mu_R}{b} + 0.43}.$$

If we assume that the observed fractions of occupied anemones have reached equilibrium ($\hat{f}_B = 0.014$, $\hat{f}_R = 0.29$, $\hat{f}_G = 0.21$; Figure 7.10b), we can solve for the mortality rate on each type of anemone relative to the birth rate: $\mu_B \approx 0.14\, b$, $\mu_R \approx 0.0092\, b$, and $\mu_G \approx 0.0090\, b$. These mortality rates are consistent with the expectation that the reddish-green crabs should be most susceptible to predation on blue anemones, against which they are least camouflaged. On the other hand, the model predicts that the mortality rate should be higher on reddish-green anemones than on green anemones, which is opposite to what is believed. The difference in the predicted mortality rates is small, however, and could easily reflect sampling error.

Before concluding this example it is worth considering one more revision to the model. Closer examination of model (7.34) reveals that the dynamics of the number of occupied anemones of a particular color (e.g., blue) is unaffected by the number of occupied anemones of the other colors (e.g., reddish-green and green). This independence of the dynamics explains why the transition matrix **M** is diagonal for these models. The validity of this assumption is an empirical question, but we might expect the number of births to depend on the number of occupied anemones rather than being a fixed quantity b. To describe this possibility requires a nonlinear model (see Problem 7.8) of the sort that we shall analyze in the next chapter. Interestingly, this nonlinear model and model (7.34) generate very similar predictions for the mortality rates on the three types of anemones (see Problem 8.14).

7.5 Concluding Message

In this chapter, we have shown how to find an equilibrium and to analyze its stability for linear models involving multiple variables. In both continuous-time and discrete-time models, an equilibrium is found by solving for the values of the variables that cause the system to remain constant. For linear models, this equilibrium point corresponds to the origin (unless the model is affine; see Box 7.2). For any linear model, the dynamics can be written in matrix form, and the stability of the equilibrium is entirely determined by the eigenvalues of this matrix. For continuous-time models, stability requires that the real part of all eigenvalues be negative. For discrete-time models, stability requires that the absolute value of all eigenvalues be less than one. In the next chapter, we turn to nonlinear models involving multiple variables. Just as in one-variable models (Chapter 5), our approach will be to find the equilibria of nonlinear models and then to determine the stability of each equilibrium by approximating the nonlinear model with a linear model in the vicinity of the equilibrium. The techniques introduced in this chapter will again play a central role.

Problems

Problem 7.1: Sea anemones sometimes provide refuge to small fish called anemone fish. Suppose that $O(t)$ and $U(t)$ denote the numbers of occupied and unoccupied anemones at time t. Suppose that, at each time step, an empty anemone is occupied by a fish with probability κ, and an occupied anemone loses its fish through mortality and becomes unoccupied with probability μ. (a) Derive the pair of recursion equations in discrete time for this model. (b) What are the equilibria of this model? (Hint: the determinant of $\mathbf{M} - \mathbf{I}$ is zero in this case.) (c) In (b) you will have found an infinite number of equilibria, characterized by a relationship between $\hat{O}$ and $\hat{U}$ that must be satisfied for neither variable to change over time. What does this infinite number of equilibria mean from a biological standpoint? (d) Calculate the eigenvalues of the transition matrix from part (a). (e) Suppose that there are a total of N anemones. Rewrite the pair of recursion equations that you obtained in part (a) as a single recursion equation, by writing $U(t) = N - O(t)$. (f) Find the equilibrium of the model in (e) and determine its stability. How does the eigenvalue compare with the two eigenvalues of the transition matrix that you calculated in part (d)?

Problem 7.2: Consider a population in which individuals can be classified as either "juveniles" or "adults", where juveniles are not yet reproductively mature. Use $J(t)$ and $A(t)$ to denote the number of juveniles and adults at time t. Suppose that adults each produce b juvenile offspring that survive to the next time step and that each adult survives to the next time step with probability p_a. Furthermore, suppose that juveniles survive one time step and become adults with probability p_j. (a) Derive the recursion equations for $J(t)$ and $A(t)$ in discrete time. (b) What is the equilibrium of the model? (c) Derive a condition involving the parameters of the model that must be satisfied in order for the population to avoid extinction.

Problem 7.3: Figure 7.11 is a flow diagram for a model of migration between two patches, where R_1 is the number of individuals in patch 1, R_2 is the number of individuals in patch 2, and θ is the rate of arrival of individuals to the first patch. (a) Write down the pair of differential equations that governs the dynamics of the number of individuals in each patch. (b) Find all possible equilibria of the model obtained in part (a). (c) Transform the model you obtained in part (a) into a linear

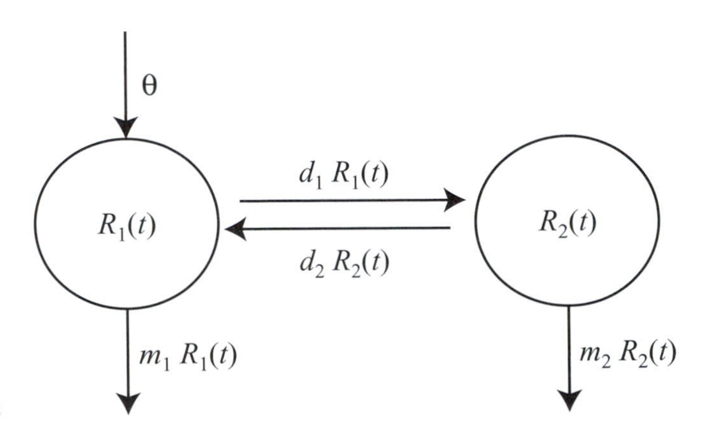

Figure 7.11: Migration between two patches

model. (d) Find all possible equilibria of the model obtained in part (c). (e) Conduct a stability analysis of any equilibrium that you identified in part (d) to determine the conditions under which it is stable. (f) What do the results of the stability analysis in part (e) for the transformed model tell you about the stability properties of the equilibrium of the original model in part (a)?

Problem 7.4: In the text, we described a model of sexual selection under the assumption that females having zero preference have the highest survival probability (see equation 7.19). Instead suppose that the probability of survival of a female is highest if she prefers the males that are currently most common. We can describe this scenario by letting natural selection against female preferences be proportional to $-b(\bar{p} - \bar{z})$ instead of just $-b\bar{p}$:

$$\frac{d\bar{z}}{dt} = G_z(a\bar{p} - c\bar{z}) - Bb(\bar{p} - \bar{z}),$$
$$\frac{d\bar{p}}{dt} = B(a\bar{p} - c\bar{z}) - G_p b(\bar{p} - \bar{z}).$$

(a) What are the equilibria/ium of the revised model? (b) Write the model in matrix form and determine the conditions under which the equilibrium at (0,0) is unstable, and thus trait and preference evolution will occur. You can assume that $G_p G_z > B^2$, $G_z > B$, and $G_p > B$. (c) In the model of the text we found the condition for instability to be $aB > G_z c + G_p b$. Compare this condition with your answer in part (b). Which model is the most conducive to the evolution of exaggerated male traits and female preferences (i.e., predicts nonzero trait and preferences over the broadest range of parameters)? (d) Provide a biological interpretation for your answer to part (c).

Problem 7.5: In the model of sexual selection, the optimum value of the male trait might be at a value $z = \theta$ instead of at $z = 0$. In this case, equation (7.19) becomes

$$\frac{d\bar{z}}{dt} = G_z(a\bar{p} - c(\bar{z} - \theta)) - Bb\bar{p},$$
$$\frac{d\bar{p}}{dt} = B(a\bar{p} - c(\bar{z} - \theta)) - \hat{G}_p b\bar{p}.$$

(a) What are the equilibria/ium of the revised model? (b) Write the model in matrix form and determine the conditions under which the equilibrium is unstable, and thus trait and preference evolution will occur. You can assume that $G_pG_z > B^2$, $G_z > B$, and $G_p > B$.

Problem 7.6: For the anemone model of the text, demonstrate that the dynamics of the transformed variables $\delta_B(t) = B(t) - \hat{B}$, $\delta_R(t) = R(t) - \hat{R}$, and $\delta_G(t) = G(t) - \hat{G}$ in model (ii) are given by equation (7.32).

Problem 7.7: (a) Transform the affine model (7.34) into a linear model using the equilibrium solutions (7.35a). (b) Determine the eigenvalues of the model. (c) Explain why equilibrium (7.35a) is stable if we assume that $b\,p_i < N_i$ for i equal to B, R, or G.

Problem 7.8: Suppose that the total number of juvenile crabs at time t is proportional to the number of colonized anemones, i.e., $b\,(B(t) + R(t) + G(t))$, as in model (i). Suppose also that a juvenile crab that attempts to colonize an occupied anemone is killed, as in model (iii). Specifically, assume that the probability of successfully colonizing a blue, reddish-green, or green anemone is $1 - B(t)/N_B$, $1 - R(t)/N_R$, or $1 - G(t)/N_G$, respectively. Derive the recursion equations for this new model, and explain why it is no longer linear.

Problem 7.9: Many more plant species have an even number of chromosomes at their gamete stage than an odd number. The simplest explanation for this phenomenon is that genomes occasionally double in size over evolutionary time ("polyploidization"), which always leads to an even number of chromosomes. A model that describes the number of species with an even, $E(t)$, versus odd, $O(t)$, number of chromosomes is shown in Figure 7.12,

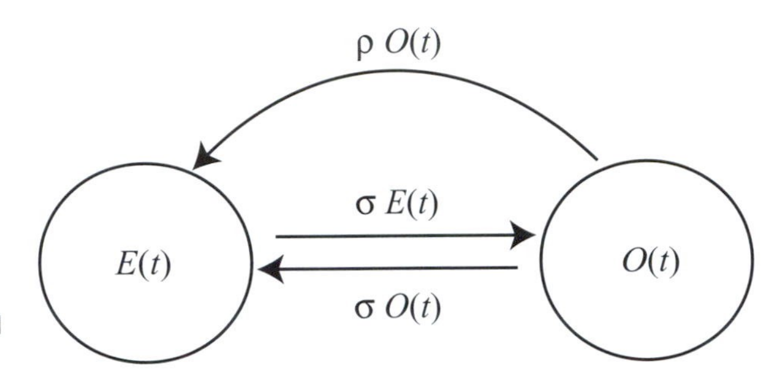

Figure 7.12: Transitions between an even and odd number of chromosomes

where σ denotes the probability of gain or loss of a single chromosome and ρ denotes the probability of doubling of the entire set of chromosomes in a time step. (a) Write down a pair of discrete-time recursion equations for the number of species with an even chromosome number and for the number of species with an odd chromosome number. (b) At equilibrium, what is the relationship between the number of species with even chromosome numbers and the number of species with odd chromosome numbers? (Hint: the determinant of $\mathbf{M} - \mathbf{I}$ is zero in this case.) (c) Find the eigenvalues and eigenvectors of this model. (d) Show that the equilibrium found in (b) is the same as the relationship between the two variables along the eigenvector associated with the largest eigenvalue.

Problem 7.10: In the general two-variable linear model (7.6), show that the system remains at whatever value of $cn_1 - an_2$ is initially present when the determinant of $\mathbf{M}$ is zero. That is, show that $d(cn_1 - an_2)/dt = 0$ when $ad - bc = 0$.

Problem 7.11: In the general two-variable linear model (7.6), prove that the condition for the two null clines to lie on top of one another, $(-a/b) = (-c/d)$, is equivalent to the condition that the determinant of $\mathbf{M}$ equals zero (see Definition P2.6)

Further Reading

For a fascinating introduction to the history of mathematics, consult

- Dunham, W. 1994. *The Mathematical Universe: An Alphabetical Journey Through the Great Proofs, Problems, and Personalities*. John Wiley & Sons, New York.
- Simmons, G. F. 1992. *Calculus Gems: Brief Lives and Memorable Mathematics*. McGraw-Hill, New York.

For a comprehensive treatment of discrete-time linear models, consult

- Caswell, H. 2000. *Matrix Population Models; Construction, Analysis, and Interpretation.* Sinauer Associates, Sunderland, Mass.

References

Andersson, M. 1994. *Sexual Selection.* Princeton University Press, Princeton, N.J.

Baeza, J. A., and W. Stotz. 2001. Host-use pattern and host selection during ontogeny of the commensal crab *Allopetrolisthes spinifrons* (H. M. Edwards, 1837) (Decapoda: Anomura: Porcellanidae). *J. Nat. Hist.* 35:341–355.

Baeza, J.A., and W. Stotz. 2003. Host-use and selection of differently colored sea anemones by the symbiotic crab *Allopetrolisthes spinifrons. J. Exp. Mar. Biol. Ecol.* 284:25–39.

Darwin, C. 1871. *The Descent of Man, and Selection in Relation to Sex.* John Murray, London.

Dunham, W. 1994. *The Mathematical Universe: An Alphabetical Journey Through the Great Proofs, Problems, and Personalities.* John Wiley & Sons, New York.

Glass, L., and D. Kaplan. 1995. *Understanding Nonlinear Dynamics.* Springer-Verlag, New York.

Iwasa, Y., A. Pomiankowski, and S. Nee. 1991. The evolution of costly female preferences. II. The "handicap" principle. *Evolution* 45:1431–1442.

Kirkpatrick, M. 1982. Sexual selection and the evolution of female choice. *Evolution* 36:1–12.

Lande, R. 1979. Quantitative genetic analysis of multivariate evolution applied to brain: body size allometry. *Evolution* 33:402–416.

Lande, R. 1982. Models of speciation by sexual selection on polygenic traits. *Proc. Natl. Acad. Sci. U.S.A.* 78:3721–3725.

Lande, R., and S. J. Arnold. 1983. The measurement of selection on correlated characters. *Evolution* 37:1210–1226.

Pomiankowski, A., Y. Iwasa, and S. Nee. 1991. The Evolution of Costly Mate Preferences .1. Fisher and Biased Mutation. *Evolution* 45:1422–1430.

Simmons, G. F. 1992. *Calculus Gems: Brief Lives and Memorable Mathematics.* McGraw-Hill, New York.

CHAPTER 8

Equilibria and Stability Analyses—Nonlinear Models with Multiple Variables

Chapter Goals:

- To construct dynamical models involving nonlinear combinations of multiple variables
- To find equilibria of these models
- To analyze the stability of these equilibria

Chapter Concepts:

- Multivariable equilibria
- Linearization near equilibria
- Stability analyses with multiple equilibria
- Linkage disequilibrium
- Approximate stability analyses

8.1 Introduction

Chapter 7 discussed methods for determining equilibria and their stability properties for models involving multiple variables, but it only considered cases where the dynamical equations were linear functions of the variables. Most interesting biological models are not linear, because any interaction among individuals requires a nonlinear model. For example, Phillips' (1996) model of within-human HIV dynamics involved interactions: virus particles contact and infect CD4+ cells (note the nonlinear $R\,V$ term in (2.4.1)). Similarly, the model of Blower et al. (2000) describing the spread of HIV involved interactions: unprotected sexual contact between infected and uninfected males (note the nonlinear $\lambda\,X$ terms in (2.5.1)). In this chapter, we describe how the methods of Chapter 7 can be extended to nonlinear models.

In section 8.2 we present nonlinear models with multiple variables in continuous time. Section 8.3 then presents the discrete-time counterpart. Again, we develop the techniques in the context of two variable models for simplicity, and then present general recipes for handling models with an arbitrary number of variables. Finally, section 8.4 illustrates how perturbation techniques can be used in nonlinear models with multiple variables to obtain useful approximations when conducting stability analyses.

8.2 Nonlinear Multiple-Variable Models

We begin with an example to illustrate the process of finding equilibria in nonlinear models with multiple variables. Consider a two-variable model for the spread of a disease that tracks the dynamics of the number of susceptible and infected individuals in a population (denoted by S and I):

$$\begin{aligned}\frac{\mathrm{d}S}{\mathrm{d}t} &= \theta - dS - \beta S I + \gamma I,\\ \frac{\mathrm{d}I}{\mathrm{d}t} &= \beta S I - (d + \nu + \gamma) I.\end{aligned} \tag{8.1}$$

For simplicity, we have assumed that the host population is replenished by immigration at total rate, θ, and that recovered individuals immediately become susceptible again (i.e., there is no immunity). In equation (8.1), β denotes the transmission rate of the disease (i.e., $\beta = c\,a$ where c is the rate of contact between susceptible and infected hosts and a is the probability of transmission given that a contact occurs; Chapter 3), d denotes the per capita background mortality rate of the host, ν denotes the additional mortality that is caused by infection, and γ denotes the rate of clearance of disease through host defense mechanisms. Given these definitions, all parameters in the model are positive.

8.2.1 Finding Equilibria

To identify the equilibria $\hat{S}$, $\hat{I}$ of the model, we set $dS/dt = 0$ and $dI/dt = 0$. This gives two equilibrium conditions:

$$0 = \theta - d\hat{S} - \beta \hat{S}\,\hat{I} + \gamma \hat{I}, \tag{8.2a}$$

$$0 = \beta \hat{S}\,\hat{I} - (d + \nu + \gamma)\hat{I}. \tag{8.2b}$$

There can be several equilibria in a nonlinear model, unlike a linear model, and we would like to obtain explicit expressions for all of them.

In general, finding all equilibria can be a difficult task because the equations that these equilibria must satisfy might be complicated functions of the dynamic variables. One good place to start is to factor these equations. If we can do this, then we can look for values of the dynamic variables that make any of the factors zero (because the entire expression will then be zero). For the present model we can see that equation (8.2a) cannot be factored, but equation (8.2b) can, giving

$$0 = \hat{I}\{\beta \hat{S} - (d + \nu + \gamma)\}. \tag{8.2c}$$

This simplifies our task because it is much clearer from (8.2c) that dI/dt equals zero only if $\hat{I} = 0$ or $0 = \beta\hat{S} - (d + \nu + \gamma)$. To find an equilibrium of the model as a whole, however, we also require that the dynamic variable S is unchanging (i.e., that equation (8.2a) holds). Therefore, for each of the different ways in which the variable I can be unchanging, we must determine the conditions required for S to be unchanging as well. Specifically, when $\hat{I} = 0$, we must determine if there are conditions under which S is also constant. Similarly, when $0 = \beta\hat{S} - (d + \nu + \gamma)$, we must determine if there are conditions under which S is again constant.

We begin with the case where $\hat{I} = 0$. Substituting this into equation (8.2a), we see that the equation, $0 = \theta - d\hat{S}$ must hold for S to remain constant. This implies that $\hat{S} = \theta/d$. As a result, one equilibrium of this model is

$$\hat{S} = \frac{\theta}{d}, \quad \hat{I} = 0 \tag{8.3a}$$

Similarly, we can substitute $\hat{S} = (d + \nu + \gamma)/\beta$ into equation (8.2a) to obtain another equilibrium (try this)

$$\hat{S} = \frac{d + \nu + \gamma}{\beta}, \quad \hat{I} = \frac{\theta - \frac{d}{\beta}(d + \nu + \gamma)}{d + \nu} \tag{8.3b}$$

Equilibrium (8.3a) corresponds to the case where the disease is absent, and equilibrium (8.3b) corresponds to the case where the disease is present, which is often referred to as the *endemic* equilibrium.

Before proceeding, we should step back for a moment and consider how the above procedure works generally. Model (8.1) is based on a specific set of assumptions about how the population of susceptible hosts gets replenished (i.e., by immigration) as well as how susceptible hosts become infected. A general model involving two variables can be written as

$$\frac{dS}{dt} = f(S, I),$$
$$\frac{dI}{dt} = g(S, I), \tag{8.4}$$

where $f(S,I)$ and $g(S,I)$ can be any functions of the variables S and I. To find the equilibria of this model, we again need to identify values of the variables $\hat{S}$ and $\hat{I}$ that result in no change in either variable. That is, we set $dS/dt = 0$ and $dI/dt = 0$ to get two equilibrium conditions, which we must solve for the two unknowns $\hat{S}$ and $\hat{I}$:

$$0 = f(\hat{S}, \hat{I}), \tag{8.5a}$$

$$0 = g(\hat{S}, \hat{I}). \tag{8.5b}$$

The equilibrium conditions (8.5) describe the null clines of the model (Figure 8.1). If (8.5a) holds, then S remains constant. If (8.5b) holds, then I remains constant. But unless both equilibrium conditions hold, changes to one variable will typically lead to changes in the other. Therefore, for both variables to remain constant, an equilibrium must simultaneously satisfy both equilibrium conditions. Graphically, this means that any equilibrium must lie on null clines for every variable in a model (see filled circles in Figure 8.1).

> At a *multivariable equilibrium,* all variables must remain unchanged. There can be multiple equilibria in nonlinear models.

Depending on the functions f and g, identifying the possible equilibria of a model can be straightforward, difficult, or even impossible. A good strategy is to factor each of the equilibrium conditions, identify which one is easiest to solve for its null clines, and then plug in these null clines, one by one, into the other equilibrium condition to see if it can be solved as well.

These procedures can be generalized for models involving any number of variables:

Definition 8.1: General Nonlinear Models in Continuous Time

A general, nonlinear, continuous-time model with n dynamic variables $x_1, \ldots, x_n$ can be written as

$$\frac{dx_1}{dt} = f_1(x_1, x_2, \ldots, x_n),$$
$$\frac{dx_2}{dt} = f_2(x_1, x_2, \ldots, x_n),$$
$$\vdots$$
$$\frac{dx_n}{dt} = f_n(x_1, x_2, \ldots, x_n),$$

where $f_1, f_2, \ldots, f_n$ denote different functions specifying the rate of change of each variable.

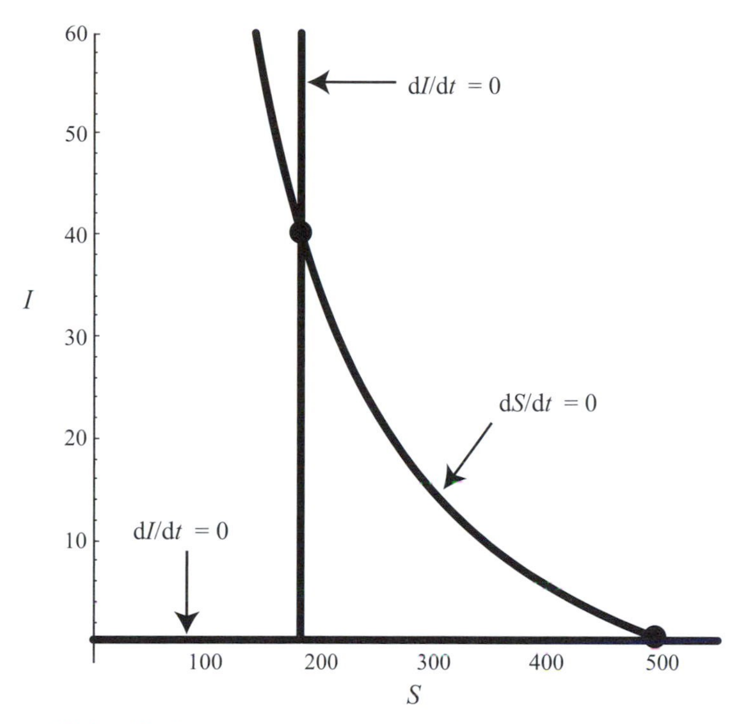

Figure 8.1: Null clines for the disease transmission model. For model (8.1), the equilibrium condition (8.2a) defines the null cline along which the number of susceptible individuals does not change ($dS/dt = 0$). The equilibrium condition (8.2b) defines two null clines along which the number of infected individuals does not change ($dI/dt = 0$). An equilibrium must lie on the null clines for both variables, which occurs at $\hat{S} = 180$, $\hat{I} = 40$ and at $\hat{S} = 500$, $\hat{I} = 0$ (filled circles). Parameter values are $v = 0.7$, $d = 0.1$, $\gamma = 0.1$, $\beta = 0.005$, $\theta = 50$.

Recipe 8.1 then describes how to find equilibria of models of the form in Definition 8.1:

Recipe 8.1: Equilibria of Nonlinear Multivariable Models in Continuous Time

Equilibria are found by determining the values of the variables that cause all of the variables to remain constant: $dx_1/dt = 0$, $dx_2/dt = 0, \ldots, dx_n/dt = 0$. This results in n equations in n unknowns: $f_1(\hat{x}_1, \hat{x}_2, \ldots, \hat{x}_n) = 0$, $f_2(\hat{x}_1, \hat{x}_2, \ldots, \hat{x}_n) = 0, \ldots, f_n(\hat{x}_1, \hat{x}_2, \ldots, \hat{x}_n) = 0$. Any point that satisfies *all* of these conditions simultaneously is an equilibrium. To identify the equilibria:

Step 1: Factor each equation, $f_i(\hat{x}_1, \hat{x}_2, \ldots, \hat{x}_n) = 0$.

Step 2: Identify all possible solutions to one of these equations (start with the simplest one).

Step 3: Plug each possible solution into the remaining equations, and repeat the above steps until equilibrium values for all variables are identified.

For nonlinear models, there can be more than one equilibrium. Depending on the complexity of the model, it may or may not be possible to identify all of the equilibria explicitly.

8.2.2 Determining the Stability of the Equilibria

Once the equilibria have been identified, our next task is to determine the behavior of the model if we move the system slightly away from one of these points. In other words, we want to know when each equilibrium is *locally stable* (Definition 5.2). Stability conditions can be used to address very fundamental biological questions. For example, in our example model, can the disease spread and become established within a population? We can answer this question by asking if the equilibrium without the disease, $\hat{S} = \theta/d$, $\hat{I} = 0$, is unstable following the introduction of a small number of infected individuals.

To address this question, we imagine starting the model very close to an equilibrium. We define ε_S and ε_I to be the *displacements* of the starting values of the variables from the equilibrium $\hat{S}$, $\hat{I}$. That is, $\varepsilon_s = S - \hat{S}$ and $\varepsilon_I = I - \hat{I}$ where S and I are the numbers of susceptible and infected individuals, respectively. The terms ε_S and ε_I are dynamic variables that change over time as S and I change. These displacements will get larger or smaller depending on whether the system moves away from the equilibrium (an unstable equilibrium) or toward the equilibrium (a stable equilibrium). Thus, to determine whether the displacements grow or decay we need equations describing how the displacements change through time.

Differential equations for the displacements can be obtained by differentiating ε_S and ε_I with respect to time:

$$\begin{aligned}\frac{\mathrm{d}\varepsilon_S}{\mathrm{d}t} &= \frac{\mathrm{d}}{\mathrm{d}t}(S - \hat{S}) \\ &= \frac{\mathrm{d}S}{\mathrm{d}t} \\ &= \theta - dS - \beta\, S\, I + \gamma I\end{aligned} \tag{8.6a}$$

and

$$\begin{aligned}\frac{\mathrm{d}\varepsilon_I}{\mathrm{d}t} &= \frac{\mathrm{d}}{\mathrm{d}t}(I - \hat{I}) \\ &= \frac{\mathrm{d}I}{\mathrm{d}t} \\ &= \beta SI - (d + \nu + \gamma)\, I.\end{aligned} \tag{8.6b}$$

These results follow from the fact that $\hat{S}$ and $\hat{I}$ are constants and hence their derivatives with respect to time are zero. Consequently, the dynamics of the displacements are governed by the same equations that govern the dynamics of the variables S and I. We have not quite finished our derivation, however, because the right-hand side of each equation is written in terms of the original variables S and I. To have a self-contained model for the displacements ε_S and ε_I, we need to rewrite equations (8.6) in terms of ε_S and ε_I alone. This can be accomplished by using the fact that $S = \hat{S} + \varepsilon_S$ and $I = \hat{I} + \varepsilon_I$, giving

$$\begin{aligned}\frac{\mathrm{d}\varepsilon_S}{\mathrm{d}t} &= \theta - d(\hat{S} + \varepsilon_S) - \beta(\hat{S} + \varepsilon_S)(\hat{I} + \varepsilon_I) + \gamma(\hat{I} + \varepsilon_I) \\ &= \theta - d\hat{S} - d\,\varepsilon_S - \beta\hat{S}\,\hat{I} - \beta\,\hat{I}\,\varepsilon_S - \beta\,\hat{S}\varepsilon_I - \beta\,\varepsilon_S\,\varepsilon_I + \gamma\,\hat{I} + \gamma\varepsilon_I,\end{aligned} \tag{8.7a}$$

$$\frac{d\varepsilon_I}{dt} = \beta(\hat{S} + \varepsilon_S)(\hat{I} + \varepsilon_I) - (d + \nu + \gamma)(\hat{I} + \varepsilon_I)$$

$$= \beta\hat{S}\,\hat{I} + \beta\hat{I}\varepsilon_S + \beta\hat{S}\varepsilon_I + \beta\varepsilon_S\varepsilon_I - d\hat{I} - \nu\hat{I} \tag{8.7b}$$

$$- \gamma\hat{I} - d\,\varepsilon_I - \nu\varepsilon_I - \gamma\varepsilon_I.$$

We are almost done, but we can now simplify (8.7) by using the fact that the displacements ε_S and ε_I are small. As a consequence, higher powers in these terms (e.g., ε_S^2, ε_I^2, and $\varepsilon_S\varepsilon_I$) will be extremely small and thus we can ignore them. Doing so, and grouping terms together that involve the displacements ε_S and ε_I, we get

$$\frac{d\varepsilon_S}{dt} = (\theta - d\hat{S} - \beta\hat{S}\,\hat{I} + \gamma\hat{I}) + (-d\varepsilon_S - \beta\hat{S}\varepsilon_I - \beta\hat{I}\varepsilon_S + \gamma\varepsilon_I), \tag{8.8a}$$

$$\frac{d\varepsilon_I}{dt} = (\beta\hat{S}\hat{I} - d\hat{I} - \nu\hat{I} - \gamma\hat{I}) + (-d\varepsilon_I - \nu\varepsilon_I - \gamma\varepsilon_I + \beta\hat{S}\varepsilon_I + \beta\hat{I}\varepsilon_S). \tag{8.8b}$$

The beauty of writing the equations in this way is that the first parenthetical term in each equation is identical to one of the equilibrium conditions, and therefore it must be zero. The equilibrium condition (8.2a) tells us that $(\theta - d\hat{S} - \beta\hat{S}\hat{I} + \gamma\hat{I})$ is zero, while the equilibrium condition (8.2b) tells us that $(\beta\hat{S}\hat{I} - d\hat{I} - \nu\hat{I} - \gamma\hat{I})$ is zero. As a result, the dynamics of the small displacements ε_S and ε_I near any equilibrium point of the original model are governed by the equations

$$\frac{d\varepsilon_S}{dt} = -d\varepsilon_S - \beta\hat{S}\varepsilon_I - \beta\hat{I}\varepsilon_S + \gamma\varepsilon_1, \tag{8.9a}$$

$$\frac{d\varepsilon_I}{dt} = -d\,\varepsilon_I - \nu\varepsilon_I - \gamma\varepsilon_I + \beta\hat{S}\varepsilon_I + \beta\hat{I}\varepsilon_S. \tag{8.9b}$$

The most important point to take away from the above calculations is that we started with a *nonlinear* model and we have arrived at a *linear* system of equations that governs the dynamics of the displacements from an equilibrium. We have been able to do this because we allow only small displacements, and therefore we can approximate the nonlinear model with a linear one near the equilibrium by ignoring higher-powered terms in the ε's. This process is called a *local stability analysis* or *linearization* near an equilibrium. Given that we have a linear system of equations, we can then use the techniques from Chapter 7 to determine whether the equilibrium is stable or unstable. If the displacements get smaller with time then the equilibrium is locally stable.

The stability of an equilibrium is determined by a local stability analysis, which is a *linearization* of the nonlinear model near the equilibrium of interest.

As in Chapter 7, we can write equations (8.9) in matrix form:

$$\begin{pmatrix} \frac{d\varepsilon_S}{dt} \\ \frac{d\varepsilon_I}{dt} \end{pmatrix} = \begin{pmatrix} -\beta\hat{I} - d & -\beta\hat{S} + \gamma \\ \beta\hat{I} & \beta\hat{S} - (d + \nu + \gamma) \end{pmatrix} \begin{pmatrix} \varepsilon_S \\ \varepsilon_I \end{pmatrix}. \tag{8.10}$$

Matrix equation (8.10) describes the dynamics of small displacements from any equilibrium of the original model. According to Rule 7.2 for continuous-time linear models, the equilibrium is stable provided that all eigenvalues have negative real parts. This ensures that the system moves toward the equilibrium along all eigenvectors.

To use the above results for determining local stability properties, we must first specify which equilibrium we are near. Let us begin with the equilibrium where the disease is absent. The stability of this equilibrium is determined by the eigenvalues of the matrix in (8.10). For the disease-absent equilibrium, $\hat{S} = \theta/d, \hat{I} = 0$, and this matrix therefore simplifies to

$$\begin{pmatrix} -d & -\beta\dfrac{\theta}{d} + \gamma \\ 0 & \beta\dfrac{\theta}{d} - (d + \nu + \gamma) \end{pmatrix}. \tag{8.11}$$

Matrix (8.11) is upper triangular, and therefore its eigenvalues can be read directly from the diagonal (Rule P2.26):

$$r_1 = \beta\frac{\theta}{d} - (d + \nu + \gamma) \quad \text{and} \quad r_2 = -d. \tag{8.12}$$

Because d is a positive constant, the stability of the disease-absent equilibrium is completely determined by the sign of the first eigenvalue $\beta\,(\theta/d) - (d + \nu + \gamma)$. If this is negative, then the equilibrium will be stable and the disease will not spread. If it is positive, then the disease will spread into the population.

The requirement for a disease to spread, $\beta\,(\theta/d) - (d + \nu + \gamma) > 0$, can be rewritten in terms of the number of susceptible individuals $\hat{S} = \theta/d$. Making this substitution and rearranging, we get the condition $\beta\hat{S}/(d + \nu + \gamma) > 1$ for a disease to spread. The quantity on the left-hand side is sometimes called R_0, the reproductive number, and represents the expected number of new infections produced per infected host when a disease is introduced into a susceptible population. For R_0 to be greater than one, the population size in the absence of the disease, $\hat{S}$, must be large enough that an infected individual encounters enough susceptible individuals to ensure infection of at least one other member of the population before the infected individual dies or recovers.

When there are *multiple equilibria*, the *stability* of each must be evaluated separately.

We have just completed a local stability analysis of the $\hat{S} = \theta/d, \hat{I} = 0$ equilibrium, but this model has a second equilibrium: the endemic equilibrium $\hat{S} = (d + \nu + \gamma)/\beta, \hat{I} = (\theta - (d/\beta)\,(d + \nu + \gamma))/(d + \nu)$. Each equilibrium of a model has its own stability properties, and we must identify these properties by performing a local stability analysis for each equilibrium separately.

Before we begin, we should determine the conditions under which the second equilibrium is biologically feasible (i.e., biologically valid) by asking when $\hat{S}$ and $\hat{I}$ are both positive. Given that the parameters are assumed to be positive, $\hat{S}$ will always be positive, but $\hat{I}$ need not be. For $\hat{I}$ to be positive, we require that $\theta > (d/\beta)\,(d + \nu + \gamma)$, which can be reorganized as $\beta\,(\theta/d) - (d + \nu + \gamma) > 0$. This is exactly the same condition required for the disease-absent equilibrium to be unstable. Therefore, the endemic equilibrium is biologically feasible only when the disease can spread when rare.

Once again, we use the linear version of the model (8.10) that describes the dynamics near an equilibrium for our stability analysis, but we now plug in the second equilibrium. After factoring, equation (8.10) becomes

$$\begin{pmatrix} \dfrac{d\varepsilon_S}{dt} \\ \dfrac{d\varepsilon_I}{dt} \end{pmatrix} = \begin{pmatrix} \dfrac{-\beta\theta + \gamma d}{d + \nu} & -\nu - d \\ \dfrac{\beta\theta - d(d + \nu + \gamma)}{d + \nu} & 0 \end{pmatrix} \begin{pmatrix} \varepsilon_S \\ \varepsilon_I \end{pmatrix}. \tag{8.13}$$

Again, according to Rule 7.2, the equilibrium of (8.13) will be stable only if all of the eigenvalues of the matrix have negative real parts. We could calculate the eigenvalues of (8.13) directly and then examine their real parts, but it is easier to use the trace and determinant conditions in Rule P2.25 of Primer 2. According to Rule P2.25, for the real part of both eigenvalues to be negative, the determinant of a 2×2 matrix must be positive and its trace must be negative.

The determinant of the matrix in (8.13) is $\beta\theta - d(d + \nu + \gamma)$. The sign of this expression might not be obvious at first, but recall that we require that $\beta(\theta/d) - (d + \nu + \gamma) > 0$ for $\hat{I}$ to be positive. This implies that the determinant is positive whenever the endemic equilibrium is biologically feasible. The trace of (8.13) is $(-\beta\theta + \gamma d)/(d + \nu)$, and because all the parameters are positive, the sign of the trace depends on the sign of $(-\beta\theta + \gamma d)$. The fact that $\hat{I}$ must be positive implies that $\beta\theta$ must be greater than $d^2 + d\nu + d\gamma$ (from our condition for biological feasibility), which in turn implies that $\beta\theta$ is greater than $d\gamma$. Therefore, $(-\beta\theta + \gamma d)$ must be negative, meaning that the trace must be negative. Thus, according to Rule P2.25, the real parts of both eigenvalues are negative, and we conclude that the endemic equilibrium is locally stable whenever it is biologically feasible.

The above example illustrates how we can use the techniques of Chapter 7 for linear models to conduct a local stability analysis of our nonlinear epidemiological model. The approach rests on the assumption that our linear approximation adequately captures the dynamics of the nonlinear model near equilibria. Figure 8.2 shows that, indeed, this linearization provides a remarkably good approximation to the nonlinear model (8.1) near both equilibria. In fact, mathematicians have proven that such approximations will typically work for local stability analyses of general nonlinear models as well. These general results are presented next.

8.2.3 The General Approach for Determining the Local Stability of Equilibria

In the above derivation, we approximated the dynamics near an equilibrium with linear equations by first writing out nonlinear equations (8.7) and then dropping higher-order terms in the displacement, like ε_S^2, ε_I^2, and $\varepsilon_S\varepsilon_I$. In other models, however, it might not be so obvious which terms should be kept and which can be dropped. For example, what should we do with a term like $(1 + \varepsilon_I)/(1 - \varepsilon_S)$? Fortunately, there is a much simpler route to reaching matrix (8.10b), both conceptually and computationally, which automatically drops the correct terms.

We describe the procedure using the more general two-variable model (8.4). Having found the equilibria of a model, we can again define the displacements ε_S and ε_I from one of the equilibria and derive equations for their dynamics.

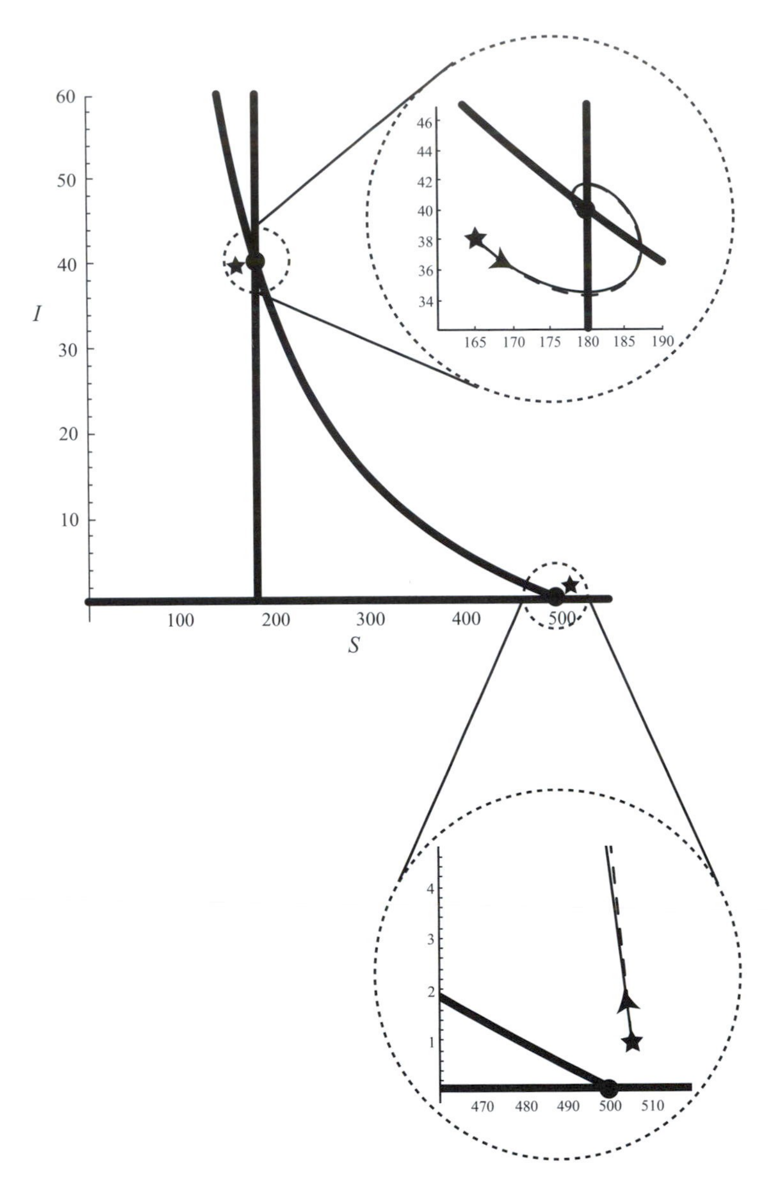

Figure 8.2: Dynamics near the equilibria of the disease transmission model. The inset figure below the graph shows a numerical solution to the differential equations using *Mathematica* (see Chapter 4) starting from an initial number of susceptible and infected individuals (stars) near the disease-absent equilibrium with $S(0) = 505$, $I(0) = 1$; the solid thin curve uses the exact differential equations (8.1), while the dashed curve uses the linear differential equations (8.9) that are approximately correct near the equilibrium. The inset figure to the right of the graph is similarly drawn, but starting near the endemic equilibrium with $S(0) = 165$, $I(0) = 38$. In the region of each equilibrium, the non-linear dynamics are well approximated by linear differential equations. Parameters as in Figure 8.1.

The same procedure used in the specific model can be followed to obtain the more general version of (8.7a) and (8.7b):

$$\frac{d\varepsilon_S}{dt} = f(\hat{S} + \varepsilon_S, \hat{I} + \varepsilon_I), \tag{8.14a}$$

$$\frac{d\varepsilon_I}{dt} = g(\hat{S} + \varepsilon_S, \hat{I} + \varepsilon_I). \tag{8.14b}$$

The next step is to come up with a linear approximation to these equations.

In the model of section 8.2.2, we first multiplied out the equations and then discarded terms involving higher powers of the ε's. We can accomplish the same task more generally by linearizing the functions f and g using the Taylor series for each function (Primer 1) near the equilibrium point $(\hat{S}, \hat{I})$. This is exactly what we did for one-variable nonlinear models in Chapter 5. Box 8.1 generalizes the Taylor series for functions of more than one variable.

Box 8.1: The Taylor Series Approximation with Multiple Variables

The Taylor series was introduced in Primer 1 as a method to rewrite functions as a series of power terms, $f(x) = b_0 + b_1 x + b_2 x^2 + \cdots$. Although the functions we considered in Primer 1 involved only one variable, x, the Taylor series can also be used for functions involving multiple variables. The Taylor series of a function of d variables, $f(x_1, x_2, \ldots, x_d)$, around the point $x_i = a_i$ is given by

$$f(x_1, x_2, \ldots, x_d) = f(a_1, a_2, \ldots, a_d) + \sum_{j=1}^{d} \left(\left. \frac{\partial f}{\partial x_j} \right|_{x_1 = a_1,\, x_2 = a_2, \ldots,\, x_d = a_d} \right) (x_j - a_j) + \text{residual}. \tag{8.1.1}$$

Before going any further, we must take a short mathematical excursion to discuss exactly what we mean by $\partial f / \partial x_j$.

Mathematical aside: Partial derivatives

The term $\partial f / \partial x_j$ is known as a "partial derivative" and is written with curly ∂'s rather than the d's normally used in more familiar derivatives (see Appendix 2). A partial derivative is a straightforward extension of the normal derivative to functions involving multiple variables. Specifically, $\partial f / \partial x_j$ involves taking the normal derivative of the function f with respect to the variable x_j treating all of the other variables as constants.

For example, consider the function, $f(x,y) = x + e^{xy}$. If we treat y as if it were a constant, we get the partial derivative with respect to x: $\partial f / \partial x = 1 + y e^{xy}$. Similarly, if we treat x as a constant, we get the partial derivative with respect to y: $\partial f / \partial y = x e^{xy}$. As another example, consider the function, $f(x,y) = x^2 + 3\,x\,y + y^2$. If we treat y as a constant, we get the partial derivative with respect to x: $\partial f / \partial x = 2x + 3y$. If we treat x as a constant, we get the partial derivative with respect to y: $\partial f / \partial y = 3x + 2y$.

In the Taylor series (8.1.1), each $\partial f / \partial x_j$ term involves taking the partial derivative of $f(x_1, x_2, \ldots, x_d)$ with respect to only one variable, x_j. This partial derivative is then evaluated at the point $x_1 = a_1, x_2 = a_2, \ldots, x_d = a_d$. Thus, once again, the derivative terms are no longer functions of the variables x_j.

The "residual" term in (8.1.1) contains second-order terms (double derivatives $\partial f^2 / \partial x_j \partial x_k$ evaluated at $x_i = a_i$ multiplied by $(x_j - a_j)(x_k - a_k)/2$ for all variable pairs j and k), plus third-order terms, etc. As long as the higher-order derivatives are not too large, the residual term may be dropped, giving a linear approximation to the multivariable function for points near enough to $(a_1, a_2, \ldots, a_d)$.

(continued)

Box 8.1 *(continued)*

For example, let us approximate the function $f(x,y) = e^x \sin(y)$ as a linear function of the variables x and y around the point $x = y = 0$. The constant term is $f(0,0) = e^0 \sin(0) = 0$. The linear term consists of two parts,

$$\left(\frac{\partial f}{\partial x}\bigg|_{x=0,\,y=0}\right)(x - 0) \text{ and } \left(\frac{\partial f}{\partial y}\bigg|_{x=0,\,y=0}\right)(y - 0).$$

The two partial derivatives that we need are $\partial f/\partial x = e^x \sin(y)$ and $\partial f/\partial y = e^x \cos(y)$, which become 0 and 1, respectively, when evaluated at $x = y = 0$. Plugging these terms into (8.1.1)

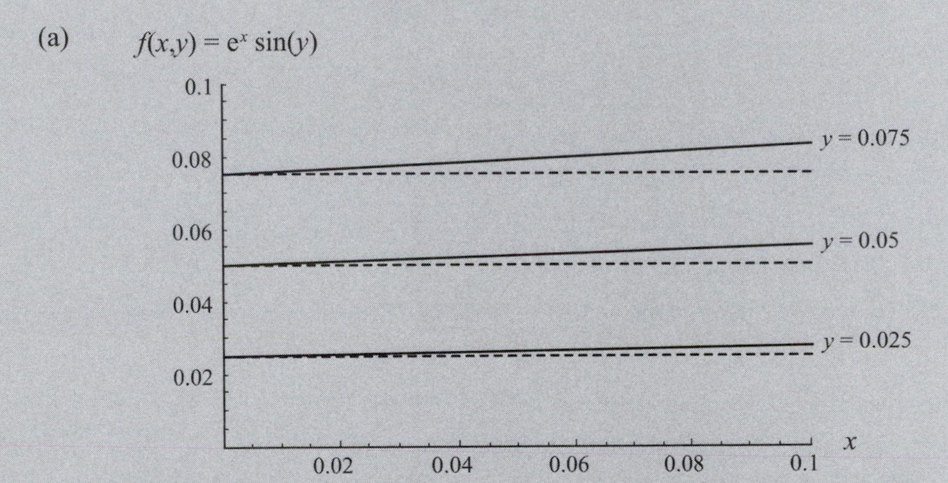

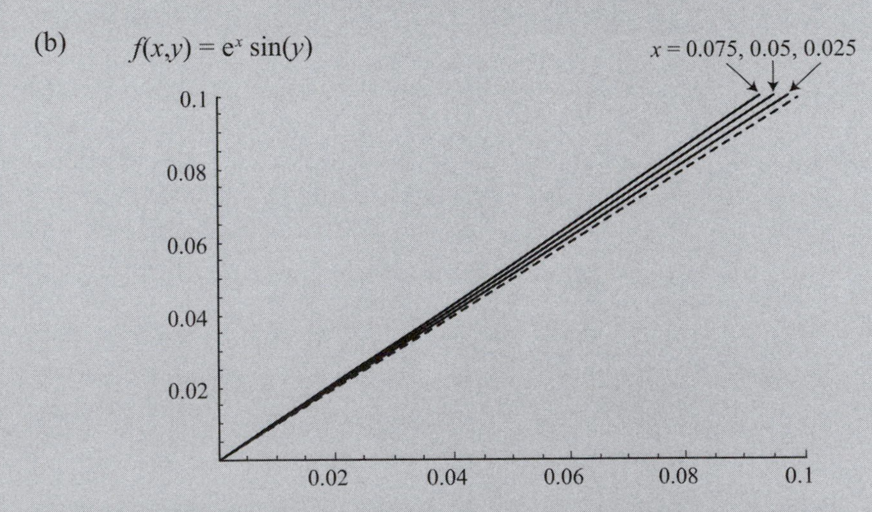

Figure 8.1.1: The Taylor series of a function of two variables. The function, $f(x,y) = e^x \sin(y)$, plotted as a function of x and y for various values of the other variable. For small values of x and y, $f(x,y)$ is (a) fairly insensitive to x but (b) almost proportional to y, as predicted by the linear Taylor series approximation, $f(x, y) \approx y$ (illustrated for comparison by dashed lines).

(continued)

Box 8.1 (*continued*)

gives the linear approximation $e^x \sin(y) \approx y$. This approximation is certainly not obvious, but we can see how well it works when we plot $f(x,y)$ as a function of both x and y (Figure 8.1.1).

For more practice, use the Taylor series to derive the following linear approximations for the given functions when x and y are both near zero:

$$f(x,y) = e^x e^y \approx 1 + x + y, \tag{8.1.2}$$

$$\frac{y}{1+x} \approx y, \tag{8.1.3}$$

$$\frac{e^y}{1+e^x} \approx \frac{1}{2} + \frac{y}{2} - \frac{x}{4}. \tag{8.1.4}$$

The Taylor series of the function f near the equilibrium $(\hat{S}, \hat{I})$ is

$$\frac{d\varepsilon_S}{dt} = f(\hat{S}, \hat{I}) + \left.\frac{\partial f}{\partial S}\right|_{S=\hat{S}, I=\hat{I}} (S - \hat{S}) + \left.\frac{\partial f}{\partial I}\right|_{S=\hat{S}, I=\hat{I}} (I - \hat{I}) \tag{8.15a}$$
$$+ \text{ higher-power terms}$$

$$= \left.\frac{\partial f}{\partial S}\right|_{S=\hat{S}, I=\hat{I}} \varepsilon_S + \left.\frac{\partial f}{\partial I}\right|_{S=\hat{S}, I=\hat{I}} \varepsilon + \text{higher-power terms}, \tag{8.15b}$$

where $(\partial f/\partial S)\Big|_{S=\hat{S}, I=\hat{I}}$ denotes the partial derivative of f with respect to the variable S, evaluated at the equilibrium point $S = \hat{S}$, $I = \hat{I}$ (see Box8.1). Expression (8.15b) follows from (8.15a) using the definitions of the deviations ε_S and ε_I, and from the fact that $f(\hat{S}, \hat{I}) = 0$ at equilibrium (from equation (8.5a)). Importantly, the derivatives $\partial f/\partial S$ and $\partial f/\partial I$ are evaluated at the equilibrium point $(\hat{S}, \hat{I})$ meaning that they are constants and not functions of S and I. Finally, the "higher-power terms" in expressions (8.15) are terms that involve higher-powers of ε_S and ε_I. Ignoring these, we arrive at the linear approximation

$$\frac{d\varepsilon_S}{dt} = \left.\frac{\partial f}{\partial S}\right|_{S=\hat{S}, I=\hat{I}} \varepsilon_S + \left.\frac{\partial g}{\partial I}\right|_{S=\hat{S}, I=\hat{I}} \varepsilon_I. \tag{8.16a}$$

The same calculations can be done for the function g, giving

$$\frac{d\varepsilon_I}{dt} = \left.\frac{\partial g}{\partial S}\right|_{S=\hat{S}, I=\hat{I}} \varepsilon_S + \left.\frac{\partial g}{\partial I}\right|_{S=\hat{S}, I=\hat{I}} \varepsilon_I. \tag{8.16b}$$

As a result, we can write the system of differential equations in matrix form as

$$\begin{pmatrix} \frac{d\varepsilon_S}{dt} \\ \frac{d\varepsilon_I}{dt} \end{pmatrix} = \begin{pmatrix} \frac{\partial f}{\partial S} & \frac{\partial f}{\partial I} \\ \frac{\partial g}{\partial S} & \frac{\partial g}{\partial I} \end{pmatrix}_{S=\hat{S}, I=\hat{I}} \begin{pmatrix} \varepsilon_S \\ \varepsilon_I \end{pmatrix}. \tag{8.17}$$

Fortunately, to perform a local stability analysis or linearization, we need not bother with the above derivation every time. Instead, we can just determine the matrix of derivatives in (8.17), which is known as the *Jacobian matrix*. For example, the matrix in (8.17) can be calculated for our epidemiological model using the functions $f(S,I) = \theta - d\,S - \beta\,S\,I + \gamma\,I$ and $g(S,I) = \beta\,S\,I - (d + \nu + \gamma)\,I$. Doing so reduces the Jacobian matrix in equation (8.17) to that of (8.10).

Definition 8.2: The Jacobian Matrix

Given n functions $f_1(x_1, x_2, \ldots, x_n), f_2(x_1, x_2, \ldots, x_n), \ldots, f_n(x_1, x_2, \ldots, x_n)$ describing the dynamics of n variables $x_1, \ldots, x_n$, the *Jacobian matrix* $\mathbf{J}$ is defined as

$$\mathbf{J} = \begin{pmatrix} \frac{\partial f_1}{\partial x_1} & \frac{\partial f_1}{\partial x_2} & \cdots & \frac{\partial f_1}{\partial x_n} \\ \frac{\partial f_2}{\partial x_1} & \frac{\partial f_2}{\partial x_2} & \cdots & \frac{\partial f_2}{\partial x_n} \\ \vdots & \vdots & \ddots & \vdots \\ \frac{\partial f_n}{\partial x_1} & \frac{\partial f_n}{\partial x_2} & \cdots & \frac{\partial f_n}{\partial x_n} \end{pmatrix}.$$

CAUTION: The order of the functions determines the order in which the derivatives must be taken. If the ith row uses the function describing the change over time in x_i, then the ith column must contain derivatives with respect to the variable x_i.

The Jacobian matrix, evaluated at an equilibrium, provides a linear approximation of the non-linear model near that equilibrium. In the vicinity of an equilibrium, the trajectories predicted using the full general model (8.4) will be very nearly the same as those of the linear approximation (Figure 8.2). The accuracy of the linearization allows us to use the techniques of Chapter 7 to determine stability, by using the linear model as a proxy for the more complicated nonlinear model. This approximation breaks down as the system moves farther away from the equilibrium, but it provides an excellent way to determine how the model behaves near any equilibrium point.

A local stability analysis (Recipe 8.2) must be performed separately on each equilibrium of interest.

Recipe 8.2
Local Stability of Equilibria for Nonlinear Multivariable Models in Continuous Time

To determine whether an equilibrium of interest is locally stable:

Step 1: Evaluate the Jacobian matrix (Definition 8.2) at the equilibrium of interest, $\mathbf{J}|_{\hat{x}_1, \hat{x}_2, \ldots, \hat{x}_n}$. This matrix is often called the *local stability matrix*.

Step 2: Solve the characteristic polynomial $\text{Det}(\mathbf{J}|_{\hat{x}_1, \hat{x}_2, \ldots, \hat{x}_n} - r\,\mathbf{I}) = 0$, which is an nth-degree polynomial.

(continued)

Recipe 8.2 ***(continued)***

Step 3: The n solutions ("roots") to this characteristic polynomial are the n eigenvalues $r_1, r_2, \ldots, r_n$.

Step 4: The real parts of all n eigenvalues must be negative for the equilibrium to be locally stable. Equivalently, the equilibrium is locally stable if the real part of the eigenvalue with the largest real part (the *leading eigenvalue*) is negative. (If the real part of the leading eigenvalue is zero, the local stability analysis is inconclusive, and higher order terms must be considered.)

Although the real part completely determines stability, if the eigenvalues have complex parts, then the system will spiral around the equilibrium along some axes (for details, see Box 9.3).

Example: Predator-Prey Dynamics

Let us now get some practice by analyzing the predator-prey model introduced in equations (3.17) of Chapter 3. In these equations, the prey species is replenished by immigration rather than by reproduction, and the predator species has its birth rate solely determined by its rate of prey intake:

$$\frac{dn_1}{dt} = \theta - a c n_1 n_2, \qquad \frac{dn_2}{dt} = \varepsilon a c n_1 n_2 - \delta n_2. \tag{8.18}$$

In this model, θ is the prey immigration rate, c is the rate of contact between predator and prey, a is the probability that a predator consumes a prey given a contact, ε is the conversion efficiency of predators into prey, and δ is the per capita death rate of the predator. We used this model to explore how the population dynamics of a predator affects the population dynamics of a prey species and vice versa. Now we shall conduct a local stability analysis of its equilibria.

Following Recipe 8.1, the equilibria of this model are obtained by solving the equations that result from setting $dn_1/dt = 0$ and $dn_2/dt = 0$. This gives $0 = \theta - ac\hat{n}_1\hat{n}_2$ and $0 = \varepsilon ac\hat{n}_1\hat{n}_2 - \delta\hat{n}_2$. The second equation can be factored into $0 = \hat{n}_2(\varepsilon ac\hat{n}_1 - \delta)$, and if either factor is zero, then the population size of the predator remains constant. This occurs if either $\hat{n}_2 = 0$ or $\hat{n}_1 = \delta/(\varepsilon ac)$. We first consider the case where $\hat{n}_2 = 0$. Substituting this into the other equilibrium condition gives $0 = \theta$. This equilibrium condition can never be met because we have assumed that the immigration rate θ is positive. Therefore, there is no possible equilibrium when $\hat{n}_2 = 0$. Biologically, if predators are absent (i.e., $\hat{n}_2 = 0$) then the prey population will grow indefinitely through immigration. Now consider the second possibility, in which $\hat{n}_1 = \delta/(\varepsilon\, a\, c)$. Substituting this into the equilibrium equation for $dn_1/dt = 0$ gives $0 = \theta - ac\delta/(\varepsilon\; a\; c)\,\hat{n}_2$, or $\hat{n}_2 = \theta\,\varepsilon/\delta$. Therefore, the only equilibrium is

$$\hat{n}_1 = \frac{\delta}{\varepsilon ac}, \quad \hat{n}_2 = \frac{\theta\varepsilon}{\delta}. \tag{8.19}$$

To determine the local stability of this equilibrium, we first calculate the Jacobian matrix (Definition 8.2)

$$\mathbf{J} = \begin{pmatrix} -a\,c\,\hat{n}_2 & -a\,c\,\hat{n}_1 \\ \varepsilon\,a\,c\,\hat{n}_2 & \varepsilon\,a\,c\,\hat{n}_1 - \delta \end{pmatrix}. \tag{8.20}$$

Evaluating (8.20) at the equilibrium (8.19) then gives

$$\mathbf{J} = \begin{pmatrix} -a\,c\,\varepsilon\,\theta/\delta & -\delta/\varepsilon \\ \varepsilon^2\,a\,c\,\theta/\delta & 0 \end{pmatrix} \tag{8.21}$$

To determine if the equilibrium is stable (i.e., if the real parts of both eigenvalues are negative), we can use the determinant and trace conditions for a 2×2 matrix (Rule P2.25). The determinant of (8.21) is $\varepsilon\, a\, c\, \theta$, which is always positive because all parameters are positive. Furthermore, the trace of (8.21) is $-a\, c\, \varepsilon\, \theta/\delta$, which is always negative. Therefore, the real parts of both eigenvalues are negative according to Rule P2.25, and the equilibrium (8.19) is always locally stable.

Although the equilibrium is locally stable, we do not yet know whether the system approaches it smoothly or whether it displays cycles. To answer this question, we can calculate the eigenvalues of (8.21) explicitly. Using equations (P2.7) of Primer 2, we have

$$r = \frac{-\varepsilon\,a\,c\,\theta \pm \sqrt{\varepsilon\,a\,c\,\theta(\varepsilon\,a\,c\,\theta - 4\delta^2)}}{2\delta} \tag{8.22}$$

The eigenvalues will be real if $\varepsilon\, a\, c\, \theta - 4\,\delta^2 > 0$, which occurs if the rate of conversion of prey to predator offspring, multiplied by the prey immigration rate, is high relative to the predator's mortality rate. In this case, the equilibrium is reached smoothly without oscillations (Figure 8.3a). On the other hand, the eigenvalues will be complex if $\varepsilon\, a\, c\, \theta - 4\delta^2 < 0$, in which case, the equilibrium population size is reached in an oscillatory fashion (Figure 8.3b).

Finally, we can go one step further, by first writing the complex eigenvalues as $r = \alpha \pm \beta i$, where α and β are real and $i = \sqrt{-1}$. The results of Box 9.3 in the next chapter then show that dynamics of the variables are described by sinusoidal cycles with a period of $\tau = 2\pi/\beta$. Thus, when our predator-prey model has complex eigenvalues, the population sizes of each species will approach their equilibrium values while cycling, with a period $\tau = 4\,\pi\,\delta/\sqrt{\varepsilon\,a\,c\,\theta(4\delta^2 - \varepsilon\,a\,c\,\theta)}$. For the parameters used in Figure 8.3b ($\delta = 2$, $\theta = 10$, $\varepsilon = 1$, and $a\,c = 0.1$), this predicts a period of 6.5 generations, which is consistent with the observed cycles.

Example: Phillips' Model of HIV Dynamics

The above example illustrates a linear stability analysis for a two-variable model. It would be a bit misleading, however, for us to imply that stability analyses are always so straightforward. The calculations can sometimes be very tedious or even impossible for some models, particularly those having more than two dynamic variables. Thankfully some additional techniques are available, and here we introduce one known as the Routh-Hurwitz conditions (Box 8.2). Phillips' model of HIV infection from Chapter 1 is used as an example to illustrate the approach.

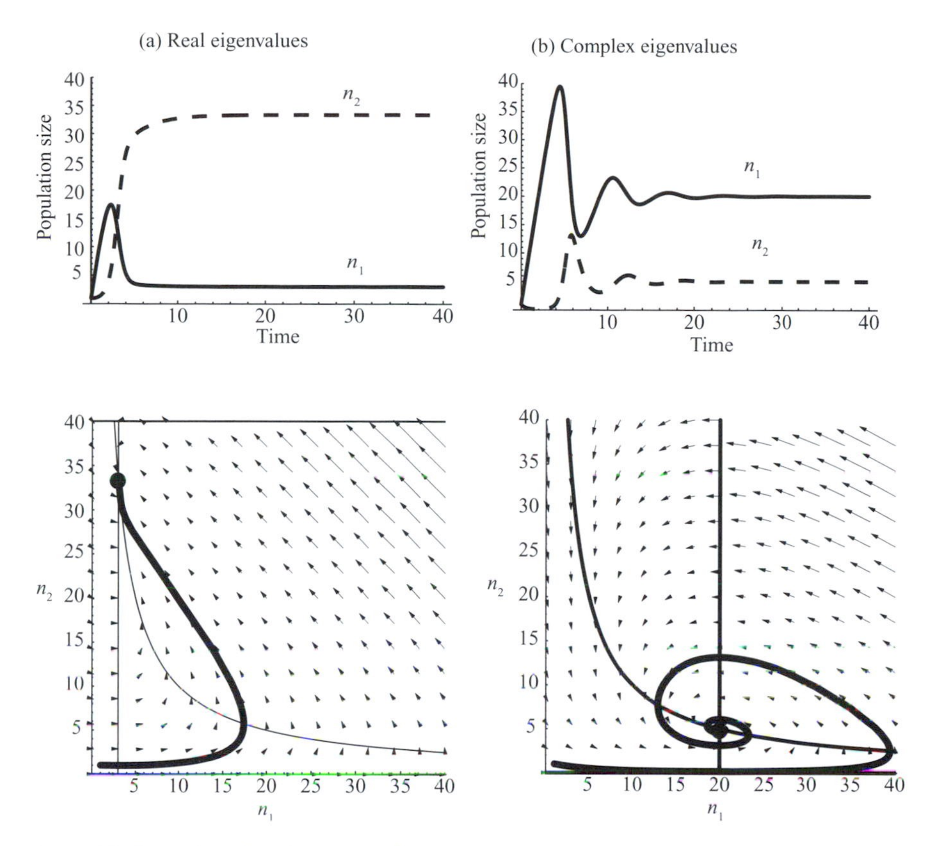

Figure 8.3: Examples of predator-prey dynamics. Dynamics over time (top) and phase plane diagrams (bottom) with (a) real eigenvalues ($\delta = 0.3$) or (b) complex eigenvalues ($\delta = 2$). In both cases, the dynamics are illustrated for 40 generations with $\theta = 10$, $\varepsilon = 1$, and $a\,c = 0.1$.

Box 8.2: Routh-Hurwitz Conditions for Local Stability

The Routh-Hurwitz conditions can be used to determine whether an equilibrium is stable without having to calculate the eigenvalues explicitly. Here we describe the Routh-Hurwitz conditions for models with different numbers of variables. We begin by assuming that the model is a continuous-time model, which is the most common context in which the Routh-Hurwitz conditions are used. Nevertheless, under certain circumstances, the Routh-Hurwitz conditions can be used to determine stability in discrete-time models, as we describe at the end of this box.

Two-Dimensional Matrices: To begin, we consider a model with two variables. In this case, the Jacobian matrix is a 2×2:

$$\mathbf{J} = \begin{pmatrix} j_{11} & j_{12} \\ j_{21} & j_{22} \end{pmatrix}, \tag{8.2.1}$$

whose characteristic polynomial is a quadratic equation in r (see equations (7.11) in Chapter 7):

$$r^2 - (j_{11} + j_{22})\, r + (j_{11} j_{22} - j_{12} j_{21}) = 0. \tag{8.2.2a}$$

(continued)

Box 8.2 *(continued)*

We can simplify the notation by defining the coefficients $a_1 = -(j_{11} + j_{22})$ and $a_2 = j_{11}j_{22} - j_{12}j_{21}$. Then we can rewrite (8.2.2a) as

$$r^2 + a_1 r + a_2 = 0 \tag{8.2.2b}$$

The eigenvalues of (8.2.1) are the values of r that satisfy equation (8.2.2b).

For a continuous-time model, the equilibrium is locally stable only if both eigenvalues have negative real parts. According to Rule P2.25, this will be true if the trace of (8.2.1) is negative ($j_{11} + j_{22} < 0$) and the determinant is positive ($j_{11}j_{22} - j_{12}j_{21} > 0$). We can rewrite these trace and determinant conditions in terms of the coefficients of the characteristic polynomial (8.2.2b):

$$\begin{aligned} a_1 &> 0, \\ a_2 &> 0. \end{aligned} \tag{8.2.3}$$

These are the Routh-Hurwitz conditions for a second-degree characteristic polynomial.

Three-Dimensional Matrices: Next we consider a 3×3 Jacobian matrix:

$$\mathbf{J} = \begin{pmatrix} j_{11} & j_{12} & j_{13} \\ j_{21} & j_{22} & j_{23} \\ j_{31} & j_{32} & j_{33} \end{pmatrix}. \tag{8.2.4}$$

The characteristic polynomial for (8.2.4) can be written in the form

$$r^3 + a_1 r^2 + a_2 r + a_3 = 0 \tag{8.2.5}$$

where the coefficients a_1, a_2, and a_3 are calculated from the elements of the matrix (8.2.4), just as the coefficients in (8.2.2b) were calculated from the elements of the matrix (8.2.1). The conditions on the coefficients for local stability are no longer easily identified as involving the trace or the determinant of (8.2.4). Nonetheless, it has been proven that the eigenvalues of (8.2.4) all have negative real parts if (and only if) the following three conditions are met (Edelstein-Keshet 1988):

$$\begin{aligned} a_1 &> 0, \\ a_3 &> 0, \\ a_1 a_2 &> a_3. \end{aligned} \tag{8.2.6}$$

If these conditions are met for the characteristic polynomial of a continuous-time model, then we are guaranteed that an equilibrium is locally stable. If one or more condition fails, however, the equilibrium will be unstable.

Four-Dimensional Matrices: For a 4×4 Jacobian matrix, the characteristic polynomial is a fourth-order polynomial, having the general form

$$r^4 + a_1 r^3 + a_2 r^2 + a_3 r + a_4 = 0. \tag{8.2.7}$$

The Routh-Hurwitz conditions on the coefficients of (8.2.7) that must be met for local stability are then (Edelstein-Keshet 1988)

$$\begin{aligned} a_1 &> 0, \\ a_3 &> 0, \\ a_4 &> 0, \\ a_1 a_2 a_3 &> a_3^2 + a_1^2 a_4. \end{aligned} \tag{8.2.8}$$

Again, all of these conditions must be met for an equilibrium to be locally stable in a continuous-time model.

***n*-Dimensional Matrices:** There is a general recipe that can be followed for matrices of any size to obtain the Routh-Hurwitz conditions that must be met for local stability. As the dimension of the matrix gets large, however, the recipe becomes more and more difficult to apply (see Edelstein-Keshet 1988).

Consider the characteristic polynomial for an $n \times n$ matrix. It will be an nth-order polynomial, having the form

$$r^n + a_1 r^{n-1} + a_2 r^{n-2} + \cdots + a_n = 0, \tag{8.2.9}$$

where, again, the a_i's will be functions of the elements of the $n \times n$ matrix. To obtain the desired conditions on the coefficients of (8.2.9) we first need to construct the following n matrices:

$$\mathbf{H}_1 = (a_1), \tag{8.2.10a}$$

$$\mathbf{H}_2 = \begin{pmatrix} a_1 & 1 \\ a_3 & a_2 \end{pmatrix}, \tag{8.2.10b}$$

$$\mathbf{H}_3 = \begin{pmatrix} a_1 & 1 & 0 \\ a_3 & a_2 & a_1 \\ a_5 & a_4 & a_3 \end{pmatrix}, \tag{8.2.10c}$$

$$\mathbf{H}_4 = \begin{pmatrix} a_1 & 1 & 0 & 0 \\ a_3 & a_2 & a_1 & 1 \\ a_5 & a_4 & a_3 & a_2 \\ a_7 & a_6 & a_5 & a_4 \end{pmatrix}, \tag{8.2.10d}$$

etc., where the i,j element in the $k \times k$ matrix $\mathbf{H}_k$ is given by

$$\begin{array}{ll} a_{2i-j} & \text{for } 0 < 2i - j < k, \\ 1 & \text{for } 2i = j, \\ 0 & \text{for } 2i < j \text{ or } 2i > k + j. \end{array} \tag{8.2.11}$$

An equilibrium will be stable in a continuous-time model (i.e., all eigenvalues will have negative real parts) if and only if the determinants of all n matrices are positive:

$$\text{Det}(\mathbf{H}_1) > 0, \ \text{Det}(\mathbf{H}_2) > 0, \ \ldots, \ \text{Det}(\mathbf{H}_n) > 0 \tag{8.2.12}$$

(continued)

Box 2.1 ***(continued)***

(Problem 8.13). In principle, the Routh-Hurwitz conditions can be used for Jacobian matrices of any size. In practice, even these conditions become cumbersome and difficult to use when the size of the Jacobian matrix gets large.

Discrete-Time Models: While the Routh-Hurwitz conditions can be directly applied to determine the stability of continuous-time models, they can also be used in special circumstances to assess the stability of equilibria in discrete-time models. We illustrate this point using a two-variable model, but the principle holds generally.

If the stability matrix for a discrete-time model with two variables is given by (8.2.1), its eigenvalues λ will solve the characteristic polynomial

$$\lambda^2 - (j_{11} + j_{22})\,\lambda + (j_{11}\,j_{22} - j_{12}\,j_{21}) = 0. \tag{8.2.13}$$

Stability of an equilibrium requires that each eigenvalue lies between -1 and $+1$. If all of the elements of the stability matrix are non-negative, however, then the Perron-Frobenius theorem (Appendix 3) guarantees that there is a real and positive eigenvalue larger in magnitude than (or equal to) all other eigenvalues. This eigenvalue will be the leading eigenvalue. For such non-negative matrices, we need only focus on determining whether the leading eigenvalue falls above or below one. Equivalently, if we write λ as $1 + r$, we need only focus on determining whether r is positive or negative. While the Routh-Hurwitz conditions only tell us whether the real part of r is positive or negative, we are guaranteed by the Perron-Frobenius theorem that the leading eigenvalue of a non-negative matrix is real, and hence its real part must be larger than or equal to the real parts of all remaining eigenvalues.

Making the substitution $\lambda = 1 + r$, we can rewrite (8.2.13) as

$$\begin{aligned} &(1 + r)^2 - (j_{11} + j_{22})(1 + r) + (j_{11}\,j_{22} - j_{12}\,j_{21}) = 0, \\ &r^2 + (2 - j_{11} - j_{22})r + (1 - j_{11} - j_{22} + j_{11}\,j_{22} - j_{12}\,j_{21}) = 0. \end{aligned} \tag{8.2.14}$$

Defining the coefficients $a_1 = (2-j_{11}-j_{22})$ and $a_2 = 1-j_{11}-j_{22} + j_{11}\,j_{22}-j_{12}\,j_{21}$, we are guaranteed that the solutions for r are negative and that the leading eigenvalue will be less than one in magnitude if the matrix satisfies the Perron-Frobenius theorem and the Routh-Hurwitz conditions (8.2.3) are met.

The same procedure can be followed for matrices of any size. First define λ as $1 + r$, then rewrite the characteristic polynomial in terms of r, and apply the Routh-Hurwitz conditions. Only if all of these conditions hold will the equilibrium be locally stable. This assumes that you know that the leading eigenvalue is real and positive, as is true for non-negative matrices (Appendix 3). These conditions come in handy for discrete-time models because non-negative matrices often arise in biological applications (e.g., the matrix (8.48b)).

We introduced Phillips' model for HIV dynamics within the body in Chapter 1 and derived equations for its four dynamic variables in Chapter 2 (Box 2.4). In Chapter 4 we explored the dynamical behavior of the model using graphical techniques. One of our findings was that there appeared to be a critical value of the transmission parameter β above which HIV infection takes hold and below which it is cleared from the body. This makes sense because there must be some level of transmission β ensuring that viruses infect new cells and replicate faster than they die off. But what is this critical value of β? We can identify this critical value in terms of the parameters of the model by performing a local stability analysis. These results then allow us to determine which factors most increase the critical value of β. Targeting these factors soon after exposure to HIV could boost the chances that the body is able to clear the infection.

The first step is to calculate the equilibria of the model (see equations 2.4.1). You can carry out these calculations yourself, but our main interest here is the possibility that there is an equilibrium at which $\hat{E} = 0$, $\hat{L} = 0$, and $\hat{V} = 0$ (i.e., the infection is absent from the body). Substituting these values into the system of equations (2.4.1) reveals that all variables remain constant at this point as long as the number of uninfected cells is $\hat{R} = \Gamma\tau/\mu$.

We now want to determine the conditions on the transmission parameter β under which this equilibrium is stable, and HIV cannot become established within the body. To do so, we first calculate the Jacobian matrix using Definition 8.2. For Phillips' model, we obtain

$$\mathbf{J} = \begin{pmatrix} -V\beta-\mu & 0 & 0 & -R\beta \\ pV\beta & -\alpha-\mu & 0 & pR\beta \\ (1-p)V\beta & \alpha & -\delta & (1-p)R\beta \\ 0 & 0 & \pi & -\sigma \end{pmatrix}. \tag{8.23}$$

The above Jacobian matrix must now be evaluated at the equilibrium of interest, $\hat{R} = \Gamma\tau/\mu$, $\hat{E} = 0$, $\hat{L} = 0$, and $\hat{V} = 0$, giving

$$\mathbf{J} = \begin{pmatrix} -\mu & 0 & 0 & -\Gamma\tau\beta/\mu \\ 0 & -\alpha-\mu & 0 & p\Gamma\tau\beta/\mu \\ 0 & \alpha & -\delta & (1-p)\Gamma\tau\beta/\mu \\ 0 & 0 & \pi & -\sigma \end{pmatrix}. \tag{8.24}$$

Next, we need to determine the eigenvalues of matrix (8.24). The eigenvalues are the roots r of the characteristic polynomial equation given by $\text{Det}(\mathbf{J}-r\,\mathbf{I}) = 0$, which can be calculated using Rule P2.17 (we used *Mathematica*). For a 4×4 matrix, the characteristic polynomial is a fourth-degree polynomial in r and yields four roots (four eigenvalues). Unfortunately, identifying the eigenvalues from such fourth-order polynomial equations is difficult unless the polynomial factors. Fortunately, in this example, the characteristic polynomial *can* be factored into two terms:

$$(r + \mu)(r^3 + a_1 r^2 + a_2 r + a_3) = 0, \tag{8.25}$$

where

$$a_1 = \alpha + \delta + \mu + \sigma, \tag{8.26a}$$

$$a_2 = \mu\sigma + \alpha(\delta + \sigma) + \delta(\mu + \sigma) - \frac{(1-p)\beta\Gamma\tau\pi}{\mu}, \tag{8.26b}$$

$$a_3 = \delta\sigma(\alpha + \mu) - \frac{(\alpha + \mu(1-p))\beta\Gamma\tau\pi}{\mu}. \tag{8.26c}$$

One eigenvalue is easy to identify: $r = -\mu$. Because Phillips defined his parameters so that they are all positive (e.g., μ is the rate at which cells die or are eliminated from the body), this eigenvalue will always have a negative real part. But the other three eigenvalues are the roots of a cubic, which are harder to interpret. At times like these, the Routh-Hurwitz conditions (Box 8.2) come in very handy. These conditions are generalizations of the trace and determinant conditions (Rule P2.25) for 2×2 matrices, and tell us when the real parts of all eigenvalues are negative. Consequently, the Routh-Hurwitz conditions are often used to determine whether an equilibrium in a continuous-time model is locally stable.

To determine when the real parts of the remaining three eigenvalues are negative, we can use the Routh-Hurwitz conditions given by (8.2.6). The process of examining the three conditions required for stability (8.2.6) is somewhat tedious for this model, but we will go through it in detail to provide an illustration of how these conditions can be applied. Remember that we are trying to find the critical value of the transmission parameter β above which the equilibrium loses stability (and the virus takes hold in the body).

For the equilibrium without HIV to be stable requires four conditions to be met: $-\mu < 0$ (from the first eigenvalue), $a_1 > 0$, $a_3 > 0$, and $a_1 a_2 - a_3 > 0$ (from the Routh-Hurwitz conditions (8.2.6) for the remaining three eigenvalues). Biologically, it makes sense that the infection should not spread if β is low enough, in which case we expect each of these conditions to hold. Now imagine increasing β but holding all other parameters fixed. At some point, one of the conditions should fail, and the HIV-absent equilibrium should become unstable. We need to determine which of these conditions fails first.

We begin by examining the first Routh-Hurwitz condition: $a_1 > 0$. From (8.26a) we have $a_1 = \alpha + \delta + \mu + \sigma$, which does not involve β at all. In fact, under the assumption that all parameters are positive, a_1 is positive. Therefore, the first Routh-Hurwitz condition is always satisfied, and we can proceed to the next condition.

From (8.26c), the second Routh-Hurwitz condition, $a_3 > 0$, requires that

$$\delta\,\sigma\,(\alpha + \mu) > \frac{(\alpha + \mu\,(1 - p))\,\beta\Gamma\tau\pi}{\mu} \tag{8.27a}$$

Rearranging (8.27a) as a condition on β, we find that this condition on stability is satisfied only for β values below

$$\beta_{a_3} = \frac{\delta\mu\sigma\,(\alpha + \mu)}{\Gamma\tau\pi\,(\alpha + \mu(1 - p))}. \tag{8.27b}$$

If β is greater than β_{a_3}, then the condition $a_3 > 0$ fails and the equilibrium with HIV absent will be unstable.

We now proceed to the third and final Routh-Hurwitz condition, $a_1 a_2 - a_3 > 0$. Factoring terms involving, and not involving, β we get

$$\begin{aligned}&(\alpha + \delta + \mu)(\delta + \sigma)(\alpha + \mu + \sigma)\\ &-((\delta + \sigma)(1 - p) - p\,\alpha)\frac{\beta\Gamma\tau\pi}{\mu} > 0.\end{aligned} \tag{8.28a}$$

Whether or not (8.28a) holds is not obvious. To proceed, consider the two halves on the left-hand side of (8.28a). The first half is positive under the assumption that each parameter is positive. The second half is also positive if $(\delta + \sigma)(1-p) - p\,\alpha$ is negative, in which case we are guaranteed that $a_1 a_2 - a_3 > 0$. If, however, $(\delta + \sigma)(1-p) - p\,\alpha$ is positive, we have an additional requirement for stability: β must be less than

$$\beta_{a_1a_2-a_3} = \frac{\mu(\alpha + \delta + \mu)(\delta + \sigma)(\alpha + \mu + \sigma)}{\Gamma\tau\pi((\delta + \sigma)(1 - p) - p\alpha)} \tag{8.28b}$$

If β is larger than $\beta_{a_1a_2-a_3}$, then the condition $a_1a_2 - a_3 > 0$ fails, and the equilibrium will be unstable.

Now we need to determine which cutoff, $\beta_{a_1a_2-a_3}$ or β_{a_3}, is smaller, because it is the smaller cutoff that determines when the HIV-absent equilibrium first becomes unstable. To do this, we can try to factor $\beta_{a_1a_2-a_3} - \beta_{a_3}$ to see if the result is clearly positive or negative. In this case, the result does factor:

$$\beta_{a_1a_2-a_3} - \beta_{a_3} =$$

$$\frac{\mu(\alpha + \delta + \mu + \sigma)(\alpha\delta\sigma p + (\alpha\delta + \delta\mu + \alpha\sigma + \mu\sigma)(\alpha + \mu(1 - p)))}{\Gamma\tau\pi(\alpha + \mu(1 - p))((\delta + \sigma)(1 - p) - p\alpha)} \tag{8.29}$$

The numerator in (8.29) is positive because all of the parameters are positive and because p is a proportion. The denominator is also positive as long as $(\delta + \sigma)(1-p) - p\alpha$ is positive. In this case, $\beta_{a_1a_2-a_3}$ will be greater than β_{a_3}. If $(\delta + \sigma)(1-p) - p\alpha$ is negative then $a_1a_2 - a_3 > 0$ is necessarily satisfied, and we are left with the single condition β_{a_3}. Thus, either $\beta_{a_1a_2-a_3}$ is larger than β_{a_3} or there is only one critical value β_{a_3}. In either case, β_{a_3} is the value of β at which stability is lost.

We must admit that it took some effort to write (8.29) in a way that was clearly positive. When you suspect that an equation is positive, the first thing to do is to check numerically that it is positive over a range of values for each parameter. There is no point in trying to prove an equation is positive if it isn't! A helpful technique is to identify terms that are positive but that include negative terms within them. For example, $(1 - p)$ is positive if p is a proportion. You can then try to rewrite negative parts of your equation (e.g., involving $-p$) in terms of these positive quantities. Determining the sign of terms is often the most time-consuming part of a stability analysis; success requires persistence and a lot of trial and error.

In summary, the Routh-Hurwitz conditions have shown us that HIV can take hold in the body only if the transmission parameter is greater than β_{a_3}; i.e.,

$$\beta > \frac{\delta\mu\sigma\,(\alpha + \mu)}{\Gamma\pi\tau\,(\alpha + (1 - p)\mu)}. \tag{8.31}$$

In our numerical analysis, we found that the HIV-absent equilibrium first became unstable around $\beta = 0.000034$ per susceptible cell per day. Using (8.31), we now have a more precise estimate of the critical value of the transmission parameter: $\beta = 0.0000331$.

More importantly, we now have a general formula that specifies how large the transmission parameter must be in order for HIV to establish itself within the body. With such a general condition in hand, we can determine how each parameter affects the critical transmission rate (Figure 8.4). For example,

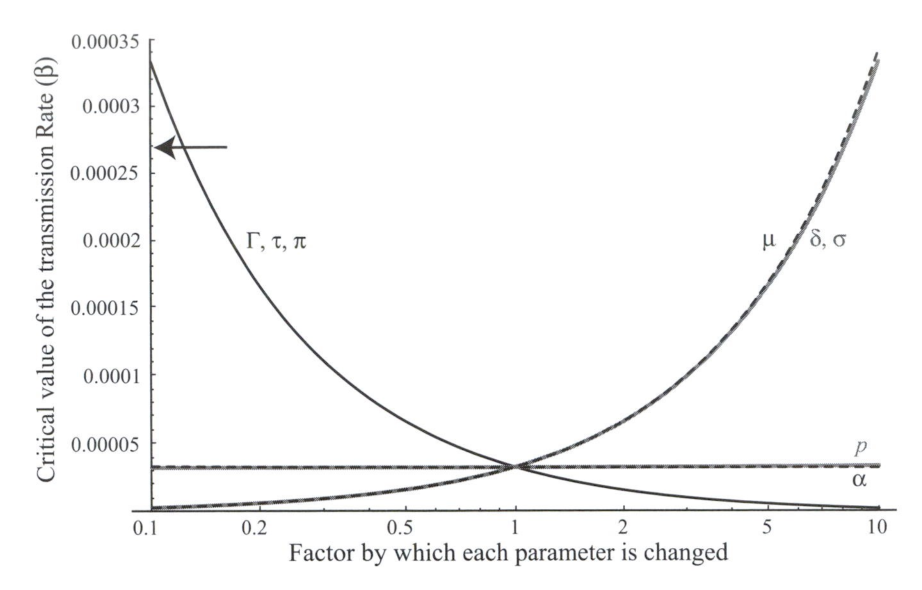

Figure 8.4: The critical value of the transmission rate for HIV to become established. The critical value of β (8.31) above which viral counts are expected to increase is shown as each parameter is varied by a factor from 0.1 to 10 (horizontal axis). Except for the parameter being varied, the parameters are those used by Phillips (1996): $\Gamma = 1.36$, $\mu = 1.36 \times 10^3$, $\tau = 0.2$, $p = 0.1$, $\alpha = 3.6 \times 10^{-2}$, $\sigma = 2$, $\delta = 0.33$, $\pi = 100$. His choice of $\beta = 0.00027$ is marked by the arrow. The critical value is fairly insensitive to p, the probability that HIV becomes latent within a newly infected cell, and to α, the rate at which latently infected cells become activated (these parameters affect the speed of invasion, but not whether invasion occurs). In contrast, the critical value rises rapidly, making it more difficult for HIV to become established within the body, if the death rate of infected (δ) or uninfected (α) CD4+ cells increases or if the death rate of viruses in the bloodstream (σ) increases. Conversely, the critical value declines and becomes easier to satisfy if the birth rate (Γ) or maturation rate (τ) of CD4+ cells increases or if the rate of production of virus particles per actively infected cell increases (π).

increasing the daily death rate of viruses in the bloodstream (σ) makes it harder for the virus to invade (a higher critical value of the transmission rate is required). More surprising, HIV is less likely to establish within the body if CD4+ cells die at a higher rate (higher δ and μ) or are produced at a lower rate (lower Γ and τ). Although a rigorous immune system and a high CD4+ count are necessary for long-term survival, in the short term, a healthy CD4+ population provides a large reservoir of potential cells for HIV to attack, and it therefore makes it easier for HIV to invade. This suggests a counterintuitive drug strategy; drugs that target and remove CD4+ cells might help the body ward off HIV in the first few days after exposure.

8.3 Equilibria and Stability for Nonlinear Discrete-Time Models

We now turn our attention to nonlinear models in discrete time. We focus on models with two variables and generalize these results to models with more variables in Recipes 8.3 and 8.4. The method is very similar to that used in continuous time, with the main difference being the way in which the eigenvalues are interpreted to determine stability.

Consider a general, two-variable, nonlinear model having the form

$$\begin{aligned} x_1(t+1) &= f(x_1(t), x_2(t)), \\ x_2(t+1) &= g(x_1(t), x_2(t)), \end{aligned} \tag{8.32}$$

where f and g are arbitrary functions of the variables x_1 and x_2. As with continuous-time models, the first thing we must do is find the equilibria of the model, and again there can be several. At each equilibrium the variables must be unchanging, and therefore $x_1(t+1) = x_1(t) = \hat{x}_1$ and $x_2(t+1) = x_2(t) = \hat{x}_2$. This gives two equilibrium conditions, $f(\hat{x}_1, \hat{x}_2) = \hat{x}_1$ and $g(\hat{x}_1, \hat{x}_2) = \hat{x}_2$, which must both be satisfied simultaneously at any equilibrium.

Now we would like to know what happens to the system if we start it very close to an equilibrium. Will it move in toward the equilibrium (and hence that equilibrium is locally stable), or will it move away (and hence that equilibrium is unstable)? Again, we will derive the method used to assess stability, but this method results in a recipe (Recipe 8.3) that can be used without rederiving it every time.

Let us start the system a small amount ε_1 from $\hat{x}_1$, and a small amount ε_2 from $\hat{x}_2$, so that $\varepsilon_1 = x_1(t) - \hat{x}_1$ and $\varepsilon_2 = x_2(t) - \hat{x}_2$ are the deviations of x_1 and x_2 from their equilibrium values. If the equilibrium point $(\hat{x}_1, \hat{x}_2)$ is locally stable, then these deviations ε_1 and ε_2 must decay to zero as time passes (meaning that $x_1(t) \to \hat{x}_1$ and $x_2(t) \to \hat{x}_2$ as time passes). To determine whether or not an equilibrium is stable, we need a pair of equations that tell us the dynamics of ε_1 and ε_2. These can be obtained as

$$\begin{aligned} \varepsilon_1(t+1) &= x_1(t+1) - \hat{x}_1 \\ &= f(x_1(t), x_2(t)) - \hat{x}_1 \\ &= f(\hat{x}_1 + \varepsilon_1(t), \hat{x}_2 + \varepsilon_2(t)) - \hat{x}_1 \end{aligned} \tag{8.33a}$$

and

$$\begin{aligned} \varepsilon_2(t+1) &= x_2(t+1) - \hat{x}_2 \\ &= g(x_1(t), x_2(t)) - \hat{x}_2 \\ &= g(\hat{x}_1 + \varepsilon_1(t), \hat{x}_2 + \varepsilon_2(t)) - \hat{x}_2. \end{aligned} \tag{8.33b}$$

We can again write the multiple-variable Taylor series (Box 8.1) for each of the functions f and g near the equilibrium point $(\hat{x}_1, \hat{x}_2)$, just as we did for continuous-time models. The Taylor series of (8.33a) with respect to ε_1 and ε_2 near (0,0) is

$$\begin{aligned} \varepsilon_1(t+1) = \Bigg(& f(\hat{x}_1, \hat{x}_2) = \frac{\partial f}{\partial x_1}\bigg|_{x_1=\hat{x}_1,\ x_2=\hat{x}_2} \varepsilon_1(t) + \frac{\partial f}{X_2}\bigg|_{x_1=\hat{x}_1,\ x_2=\hat{x}_2} \varepsilon_2(t) \\ & + \text{higher-power terms} \Bigg) - \hat{x}_1 \end{aligned} \tag{8.34a}$$

$$= \frac{\partial f}{\partial x_1}\bigg|_{x_1=\hat{x}_1,\ x_2=\hat{x}_2} \varepsilon_1(t) + \frac{\partial f}{\partial x_2}\bigg|_{x_1=\hat{x}_1,\ x_2=\hat{x}_2} \varepsilon_2(t) + \text{higher-power terms}. \tag{8.34b}$$

Expression (8.34b) follows from (8.34a) and from the fact that $f(\hat{x}_1, \hat{x}_2) = \hat{x}_1$ at equilibrium. "Higher-power terms" in the above expression are terms that involve higher powers of ε_1 and ε_2, and the derivatives in this expression are evaluated at the equilibrium (hence they are constants). Consequently, if we start the system very close to the equilibrium, so that ε_1 and ε_2 are very small, then all the terms having higher powers in these deviations will be extremely small and can be ignored. This gives

$$\varepsilon_1(t+1) = \left.\frac{\partial f}{\partial x_1}\right|_{x_1=\hat{x}_1,\ x_2=\hat{x}_2} \varepsilon_1(t) + \left.\frac{\partial f}{\partial x_2}\right|_{x_1=\hat{x}_1,\ x_2=\hat{x}_2} \varepsilon_2(t). \tag{8.34c}$$

These calculations can be repeated to obtain a linear approximation for the function g in (8.33b) near the equilibrium point $(\hat{x}_1, \hat{x}_2)$:

$$\varepsilon_2(t+1) = \left.\frac{\partial g}{\partial x_1}\right|_{x_1=\hat{x}_1,\ x_2=\hat{x}_2} \varepsilon_1(t) + \left.\frac{\partial g}{\partial x_2}\right|_{x_1=\hat{x}_1,\ x_2=\hat{x}_2} \varepsilon_2(t). \tag{8.34d}$$

Finally, we can write this pair of recursions equations in matrix form as

$$\begin{pmatrix} \varepsilon_1(t+1) \\ \varepsilon_2(t+1) \end{pmatrix} = \left.\begin{pmatrix} \frac{\partial f}{\partial x_1} & \frac{\partial f}{\partial x_2} \\ \frac{\partial g}{\partial x_1} & \frac{\partial g}{\partial x_2} \end{pmatrix}\right|_{x_1=\hat{x}_1,\ x_2=\hat{x}_2} \begin{pmatrix} \varepsilon_1(t) \\ \varepsilon_2(t) \end{pmatrix}, \tag{8.35}$$

The matrix in (8.35) is again a Jacobian matrix and is sometimes referred to as a *stability matrix*. The system of equations is linear in the two deviations ε_1 and ε_2. Thus, (8.35) allows us to describe the nonlinear dynamics in terms of linear equations near an equilibrium of interest. Having done so, we can apply the techniques described in Chapter 7 for linear models. If the deviations ε_1 and ε_2 decay to zero over time, then the equilibrium is locally stable. Conversely, if the deviations ε_1 and ε_2 grow over time, then the equilibrium is unstable. Specifically, the equilibrium is stable if the absolute value of every eigenvalue is less than one ($|\lambda_1| < 1$ and $|\lambda_2| < 1$; see section 7.4). And, again, we must remember to repeat this stability analysis for every one of the equilibria of interest.

We summarize the above methods and generalize them to models with multiple variables in the following definitions and rules:

Definition 8.3: A General Nonlinear Model in Discrete Time

A general, nonlinear, discrete-time model with n dynamic variables $x_1, \ldots, x_n$ can be written as

$$\begin{aligned} x_1(t+1) &= f_1(x_1(t), x_2(t), \cdots, x_n(t)), \\ x_2(t+1) &= f_2(x_1(t), x_2(t), \cdots, x_n(t)), \\ &\vdots \\ x_n(t+1) &= f_n(x_1(t), x_2(t), \cdots, x_n(t)), \end{aligned}$$

where $f_1, f_2, \ldots, f_n$ denote different functions that map the variables from one time step to the next.

Recipe 8.3 then describes how to find equilibria of such models:

Recipe 8.3

Equilibria of a Nonlinear Multivariable Model in Discrete Time

Equilibria are found by determining the values of the variables that cause all of the variables to be the same in the next time step: $f_1(\hat{x}_1, \hat{x}_2, \ldots, \hat{x}_n) = \hat{x}_1$, $f_2(\hat{x}_1, \hat{x}_2, \ldots, \hat{x}_n) = \hat{x}_2, \ldots, f_n(\hat{x}_1, \hat{x}_2, \ldots, \hat{x}_n) = \hat{x}_n$. This results in n equations in n unknowns. Any point that satisfies *all* of these conditions simultaneously is an equilibrium. To identify the equilibria:

Step 1: Factor each equation $f_i(\hat{x}_1, \hat{x}_2, \ldots, \hat{x}_n) - \hat{x}_i = 0$.

Step 2: Identify all possible solutions to one of these equations (start with the simplest one).

Step 3: Plug each possible solution into the remaining equations and repeat the above steps until equilibrium values for all variables are identified.

For nonlinear models, there may be more than one equilibrium. Depending on the complexity of the model, it may or may not be possible to identify all of the equilibria explicitly.

Recipe 8.4 describes how to assess the stability of equilibria:

Recipe 8.4

Stability of a Nonlinear Multivariable Model in Discrete Time

To determine whether an equilibrium of interest is stable:

Step 1: Evaluate the Jacobian matrix (Definition 8.2) at the equilibrium of interest $\mathbf{J}|_{\hat{x}_1, \hat{x}_2, \ldots, \hat{x}_n}$. This matrix is often called the *local stability matrix*.

Step 2: Solve the characteristic polynomial Det $(\mathbf{J}|_{\hat{x}_1, \hat{x}_2, \ldots, \hat{x}_n} - \lambda \mathbf{I}) = 0$, which is an nth-degree polynomial.

Step 3: The n solutions ("roots") to this characteristic polynomial are the n eigenvalues $\lambda_1, \lambda_2, \ldots, \lambda_n$.

Step 4: The equilibrium will be locally stable if the absolute values of all n eigenvalues are less than one.

- For complex eigenvalues $\lambda = A \pm Bi$, stability requires that the absolute value $\sqrt{A^2 + B^2}$ be less than one.
- For real eigenvalues, stability requires that both $\lambda < 1$ and $-1 < \lambda$.

Equivalently, the equilibrium is locally stable if the eigenvalue with the largest absolute value (*the leading eigenvalue*) has an absolute value less than one. (If the leading eigenvalue equals one exactly, the local stability analysis is inconclusive, and higher-order terms must be considered.)

(continued)

Recipe 8.4 *(continued)*

Once again, whether the eigenvalues are real or complex provides information about the behavior near the equilibrium. If they are complex, then the system will spiral around the equilibrium along some axes (for details, see Box 9.2).

Example: Density-Dependent Natural Selection

We now apply the above methods to investigate how ecological interactions within a species might evolve. Specifically, we will ask when a new allele that experiences competition in a different manner from a resident allele can invade a population. To answer this question, we use the recursions developed in Problem 3.17 of Chapter 3 that track the dynamics of population size and the frequency of an allele that affects reproductive success. This two-variable model has the form

$$
\begin{aligned}
N(t+1) &= \overline{W}\,N(t), \\
p(t+1) &= \frac{p(t)\,W_{AA}(N(t)) + (1-p(t))\,W_{Aa}(N(t))}{\overline{W}}\,p(t).
\end{aligned}
\tag{8.36}
$$

Here $N(t)$ and $p(t)$ are the population size and the frequency of allele A at time t. $\overline{W}$ is the mean fitness of the population and is a function of the current population size and allele frequency: $\overline{W} = p^2\,W_{AA}(N) + 2p(1-p)\,W_{Aa}(N) + (1-p)^2\,W_{aa}(N)$.

The $W_{ij}(N)$ are the fitnesses of individuals with genotype ij, which are assumed to be decreasing functions of the population size due to competition for resources. We assume that some genotypes are more sensitive to competition than others; for example, some individuals might be better able to switch to alternative resources in the face of competition, or they might be less likely to engage in lethal battles over resources. There are many possible ways in which we could model how fitness declines with the population size (some are listed in Problem 5.1). Here, we use the functions $W_{AA}(N) = (1+r)e^{-\alpha_{AA}N}$, $W_{Aa}(N) = (1+r)e^{-\alpha_{Aa}N}$, and $W_{aa}(N) = (1+r)e^{-\alpha_{aa}N}$, which imply that fitness declines exponentially with population size. This is similar to the Ricker model but now the rate of decline depends on genotype (a large α_{ij} means that the fitness of genotype ij decreases very quickly with increases in population size; Figure 8.5). The term r represents the intrinsic growth rate, which we assume is positive and the same for all genotypes.

We begin by finding equilibria by setting $N(t+1) = N(t) = \hat{N}$ and $p(t+1) = p(t) = \hat{p}$, giving the equations

$$
\begin{aligned}
\hat{N} &= \left(\hat{p}^2\,W_{AA}(\hat{N}) + 2\hat{p}(1-\hat{p})\,W_{Aa}(\hat{N}) + (1-\hat{p})^2\,W_{aa}(\hat{N})\right)\hat{N}, \\
\hat{p} &= \frac{\hat{p}\,W_{AA}(\hat{N}) + (1-\hat{p})\,W_{Aa}(\hat{N})}{\hat{p}^2\,W_{AA}(\hat{N}) + 2\hat{p}(1-\hat{p})\,W_{Aa}(\hat{N}) + (1-\hat{p})^2\,W_{aa}(\hat{N})}\,\hat{p}.
\end{aligned}
\tag{8.37}
$$

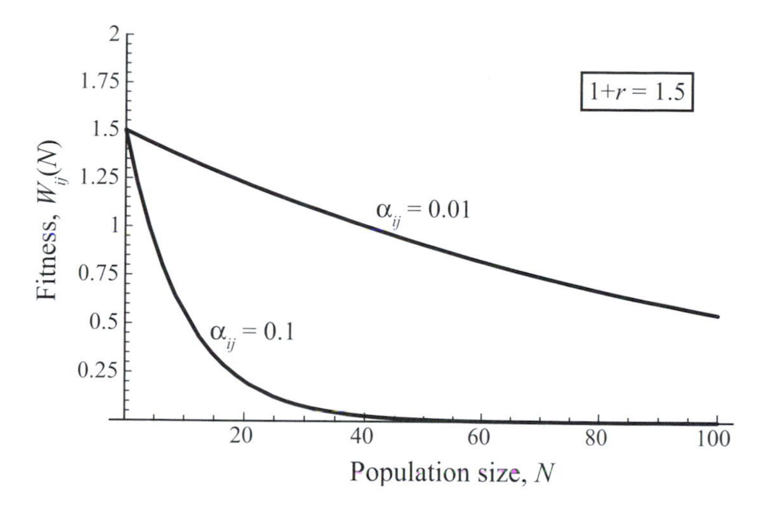

Figure 8.5: Fitness as a function of population size. The fitness function, $W_{ij}(N) = (1 + r)\, e^{-\alpha_{ij} N}$, is plotted against population size, N. Fitness declines more quickly with population size when α_{ij} is larger.

We have not yet substituted the specific forms of W_{ij} into the equilibrium conditions, but doing so produces a pair of complicated equations. We could analyze these expressions to obtain all of the possible equilibria (Recipe 8.3), but instead we will search for a specific equilibrium that is relevant to our biological question (Problem 8.2 gives more practice obtaining equilibria using the more formal route).

We are interested in knowing whether a rare allele can invade a resident population with a different competition coefficient, so let us focus on finding an equilibrium where the resident allele is fixed; i.e., $\hat{p} = 1$. Because there is no mutation in the model, the population must remain at $p = 1$ (allele A fixed) if it ever reaches this point. Substituting $\hat{p} = 1$ into the equilibrium condition for the population size gives $\hat{N} = W_{AA}(\hat{N})\,\hat{N}$, where $W_{AA}(\hat{N}) = (1 + r)e^{-\alpha_{AA}\hat{N}}$. One equilibrium of this model is $\hat{N} = 0$, representing extinction, and a second equilibrium occurs at $\hat{p} = 1$ and $\hat{N} = \ln(1 + r)/\alpha_{AA}$. We shall focus on the equilibrium with a positive population size. It is the stability of this equilibrium that interests us, and we need not search for any remaining equilibria.

We now perform a stability analysis of the equilibrium with $\hat{p} = 1$ and $\hat{N} = \ln(1 + r)/\alpha_{AA}$. Biologically, if this equilibrium is unstable, then either the population can be invaded by the a allele, or the population size moves away from $\hat{N} = \ln(1 + r)/\alpha_{AA}$ (or both). The general Jacobian matrix is rather large, but when evaluated at $\hat{p} = 1$, $\hat{N} = \ln(1 + r)/\alpha_{AA}$ it simplifies to

$$\mathbf{J} = \begin{pmatrix} 1 - \ln(1 + r) & \dfrac{2(1 - (1 + r)^{1-(\alpha_{Aa}/\alpha_{AA})})\ln(1 + r)}{\alpha_{AA}} \\ 0 & (1 + r)^{1-(\alpha_{Aa}/\alpha_{AA})} \end{pmatrix}. \tag{8.38}$$

Because (8.38) is upper triangular, its eigenvalues are $\lambda_1 = 1 - \ln(1 + r)$ and $\lambda_2 = (1 + r)^{1-(\alpha_{Aa}/\alpha_{AA})}$ (Rule P2.12).

According to Recipe 8.4, the equilibrium will be stable if both eigenvalues are less than one in absolute value. Because we assumed that the intrinsic growth rate is positive, $\ln(1 + r)$ will always be positive, so that $\lambda_1 = 1 - \ln(1 + r) < 1$. Even so, we must also check to make sure that $-1 < \lambda_1$ for the equilibrium to be stable. This condition requires that $\ln(1 + r) < 2$, implying that r must be less than $e^2 - 1$. This indicates that the rate of population growth must be small enough if the equilibrium is to be stable; the same condition was found in Problem 5.7, where we analyzed the local stability of the standard Ricker model with only one genotype. If the intrinsic growth rate is too large, the population overshoots the carrying capacity by such a large degree that the equilibrium becomes unstable.

Stability of the equilibrium with $\hat{p} = 1$ and $\hat{N} = \ln(1 + r)/\alpha_{AA}$ also requires that the second eigenvalue λ_2 be less than one in absolute value. Because $(1 + r)$ is positive, $\lambda_2 = (1 + r)^{1-(\alpha_{Aa}/\alpha_{AA})}$ must be positive. Therefore, we need only check that $(1 + r)^{1-(\alpha_{Aa}/\alpha_{AA})} < 1$ for stability. Taking the natural logarithm of both sides, stability requires that $(1 - \alpha_{Aa}/\alpha_{AA}) \ln(1 + r) < 0$. Given the assumption of a positive intrinsic growth rate, $\ln(1 + r) > 0$, and this condition is met only if $\alpha_{AA} < \alpha_{Aa}$. When $\alpha_{AA} < \alpha_{Aa}$, the fitness of *AA* individuals declines less rapidly with population size than the fitness of *Aa* individuals. Furthermore, according to our fitness functions, this stability condition implies that $W_{AA}(N)$ is greater than $W_{Aa}(N)$ at any population size. This condition is thus analogous to the requirement that $W_{AA} > W_{Aa}$ for stability in the one-gene diploid model of selection in a population of constant size (see discussion after equation (5.22)).

Overall, the resident equilibrium is locally stable only if both $r < e^2 - 1$ and $\alpha_{AA} < \alpha_{Aa}$ (Figure 8.6). The first condition depends only on the intrinsic growth rate and not on attributes of each genotype, while the second condition depends only on how sensitive each genotype is to population size. Intuitively, the first condition determines whether the system is ecologically stable at the genetic equilibrium ($\hat{p} = 1$), while the second condition determines whether the system is genetically stable at the ecological equilibrium ($\hat{N} = \ln(1 + r)/\alpha_{AA}$). Accurate predictions about stability can be made only by considering both the displacement from the ecological equilibrium *and* the displacement from the genetic equilibrium whenever growth rates depend on genotype.

Assuming that the ecological equilibrium is stable in the resident population before the introduction of the new allele, we can use these stability conditions to answer our original question. A new allele will invade a population whenever it is less sensitive to competition than the resident.

Example: The Evolutionary Dynamics of Two Genes

The genomes of most organisms contain thousands of genes, yet the evolutionary models we have considered so far have tracked changes at one gene in isolation of the rest of the genome. Is it reasonable to ignore neighboring genes? Do the evolutionary dynamics of one gene depend on the evolutionary dynamics of other genes in important ways? Ideally, to answer these questions, we would model a whole genome (see, for example, Barton 1995; Barton and Turelli 1991; Kirkpatrick et al. 2002; Turelli and Barton 1990). Yet we can get a lot of insight by just moving up from one gene to two (see also Bürger 2000;

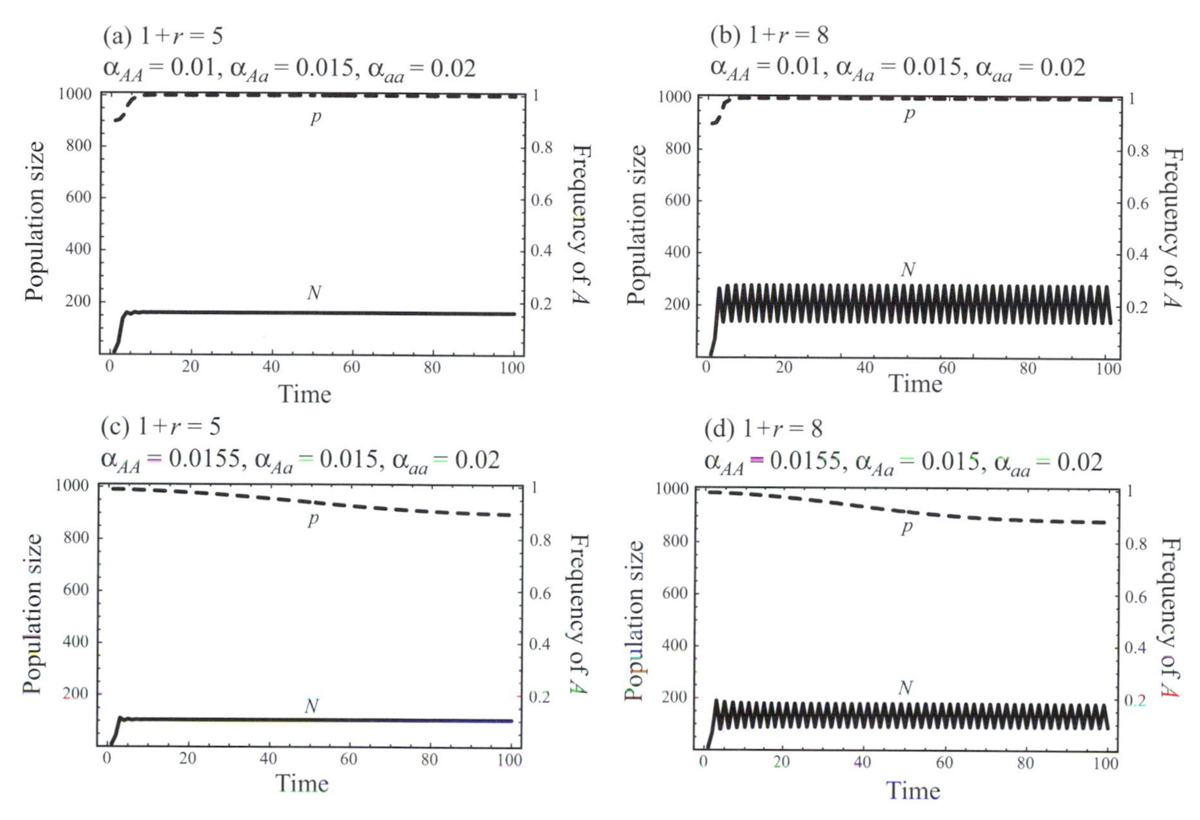

Figure 8.6: The evolution of competition coefficients. Plots of the population size (solid) and allele frequency (dashed) dynamics for model (8.27) with initial conditions near the equilibrium point, $\hat{p} = 1$, $\hat{N} = \ln(1 + r)/\alpha_{AA}$. (a) The equilibrium is stable because both $1 + r < e^2$ and $\alpha_{AA} < \alpha_{Aa}$. (b) The equilibrium is unstable because $1 + r > e^2$. (c) The equilibrium is unstable because $\alpha_{AA} > \alpha_{Aa}$, (d) The equilibrium is unstable because both $1 + r > e^2$ and $\alpha_{AA} > \alpha_{Aa}$.

Christiansen 2000; Crow and Kimura 1970; Karlin 1975). If adding a second gene causes major changes to the results, then we know that one-gene models are potentially misleading.

We denote the two genes by **A** and **B** and assume that there are two alleles at each gene, which we denote A_1, A_2 and B_1, B_2. There are thus four possible combinations of these alleles that can be found on any chromosome: A_1B_1 (with frequency x_1), A_1B_2 (with frequency x_2), A_2B_1 (with frequency x_3), and A_2B_2 (with frequency x_4). The frequency of these four types must sum to one (i.e., $x_1 + x_2 + x_3 + x_4 = 1$).

Let us now consider the life cycle given in Figure 8.7. If we assume that gametes (sperm and eggs) come together at random, then we can census the population at the gamete stage where there are only four types of chromosomes (as opposed to all the possible combinations of diploid individuals: A_1B_1/A_1B_1, A_1B_1/A_1B_2, etc.). Developing the recursions for this model requires a lot of bookkeeping as we go around the life cycle, which is best organized in the form of a table (Table 8.1). In the gamete pool, there are male gametes (sperm) and female gametes (eggs), each of which contains one of the four possible chromosome types. As a result, there are $4 \times 4 = 16$ different kinds of unions that produce diploid individuals (given in the first column of Table 8.1).

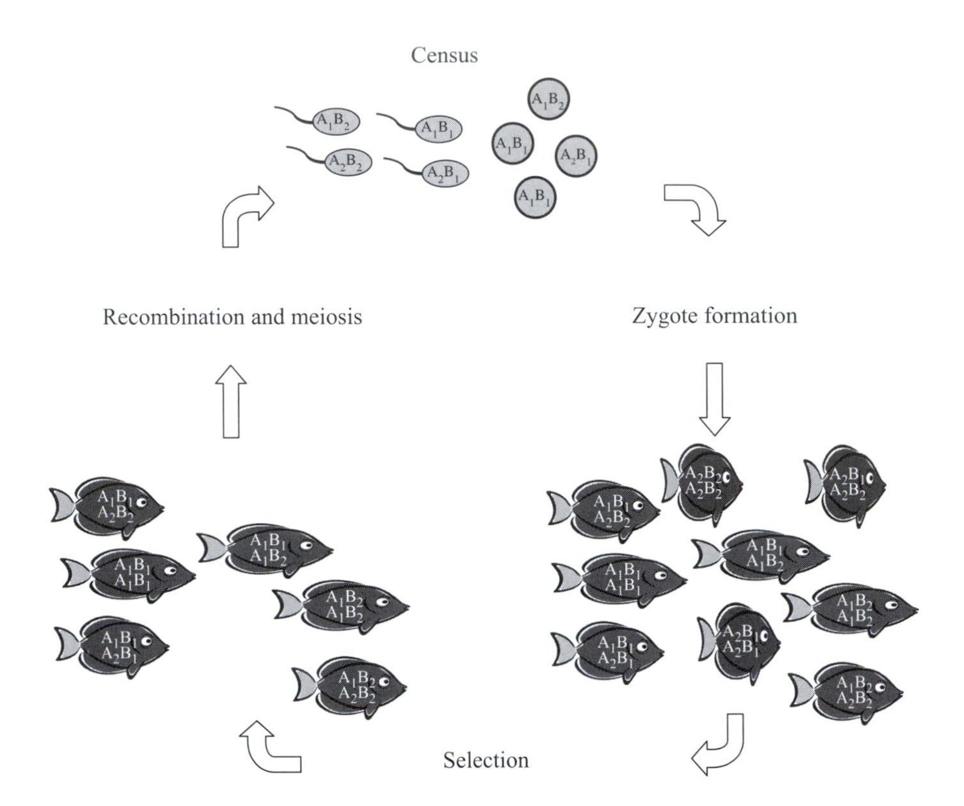

Figure 8.7: Life cycle diagram for the two-gene model

If eggs and sperm unite at random to form diploid individuals, then the frequency of these different unions is given in the second column of Table 8.1.

The next phase in the life cycle involves natural selection, and we suppose that certain diploid genotypes survive better than others. There are many different ways in which we might assign fitnesses to the different genotypes. Table 8.2 describes a very general scheme in which the fitness of an individual depends on whether it inherited a particular chromosome from its mother or father. For most genes, this parent-of-origin effect is absent (in which case $w_{ij} = w_{ji}$) but we allow for this possibility here to keep the model more general. The third column of Table 8.1 specifies the frequency of the different possible diploid individuals after selection has acted.

At this point, gametes are produced. In our previous models involving a single gene, we did not worry about where, in the genome, the gene of interest was found. Now we must be more explicit because the types of gametes that can be produced by a parent depend on the distance between the two genes within the genome. If the two genes are on different chromosomes (or very far apart on the same chromosome), then the alleles will assort independently of one another according to Mendel's second law (Figure 8.8). If they are on the same chromosome, a chromosome could be passed intact to a gamete, carrying the same combination of alleles as found on a parental chromosome. Alternatively, recombination can occur in the parent, producing a chromosome with a different combination of alleles (Figure 8.9). We can develop a

TABLE 8.1:
Life cycle table with mating and selection

Female × male	Frequency of union	Frequency after selection[a]	Gamete frequencies			
			A_1B_1	A_1B_2	A_2B_1	A_2B_2
$A_1B_1 \times A_1B_1$	x_1x_1	$x_1x_1 \frac{w_{11}}{\bar{w}}$	1	0	0	0
$A_1B_1 \times A_1B_2$	x_1x_2	$x_1x_2 \frac{w_{12}}{\bar{w}}$	1/2	1/2	0	0
$A_1B_1 \times A_2B_1$	x_1x_3	$x_1x_3 \frac{w_{13}}{\bar{w}}$	1/2	0	1/2	0
$A_1B_1 \times A_2B_2$	x_1x_4	$x_1x_4 \frac{w_{14}}{\bar{w}}$	$(1-r)/2$	$r/2$	$r/2$	$(1-r)/2$
$A_1B_2 \times A_1B_1$	x_2x_1	$x_2x_1 \frac{w_{21}}{\bar{w}}$	1/2	1/2	0	0
$A_1B_2 \times A_1B_2$	x_2x_2	$x_2x_2 \frac{w_{22}}{\bar{w}}$	0	1	0	0
$A_1B_2 \times A_2B_1$	x_2x_3	$x_2x_3 \frac{w_{23}}{\bar{w}}$	$r/2$	$(1-r)/2$	$(1-r)/2$	$r/2$
$A_1B_2 \times A_2B_2$	x_2x_4	$x_2x_4 \frac{w_{24}}{\bar{w}}$	0	1/2	0	1/2
$A_2B_1 \times A_1B_1$	x_3x_1	$x_3x_1 \frac{w_{31}}{\bar{w}}$	1/2	0	1/2	0
$A_2B_1 \times A_1B_2$	x_3x_2	$x_3x_2 \frac{w_{32}}{\bar{w}}$	$r/2$	$(1-r)/2$	$(1-r)/2$	$r/2$
$A_2B_1 \times A_2B_1$	x_3x_3	$x_3x_3 \frac{w_{33}}{\bar{w}}$	0	0	1	0
$A_2B_1 \times A_2B_2$	x_3x_4	$x_3x_4 \frac{w_{34}}{\bar{w}}$	0	0	1/2	1/2
$A_2B_2 \times A_1B_1$	x_4x_1	$x_4x_1 \frac{w_{41}}{\bar{w}}$	$(1-r)/2$	$r/2$	$r/2$	$(1-r)/2$
$A_2B_2 \times A_1B_2$	x_4x_2	$x_4x_2 \frac{w_{42}}{\bar{w}}$	0	1/2	0	1/2
$A_2B_2 \times A_2B_1$	x_4x_3	$x_4x_3 \frac{w_{43}}{\bar{w}}$	0	0	1/2	1/2
$A_2B_2 \times A_2B_2$	x_4x_4	$x_4x_4 \frac{w_{44}}{\bar{w}}$	0	0	0	1

[a] $\bar{w}$ is chosen such that all elements in the column "Frequency after selection" sum to 1.

general model that allows for any genomic arrangement by letting r equal the probability that recombination occurs between the two genes, where $r = 1/2$ corresponds to the case of independent assortment (see Figure 8.9).

The final four columns of Table 8.1 specify the fraction of each type of gamete produced by a diploid individual. For example, consider an A_1B_1/A_2B_2

TABLE 8.2
Genotypic fitnesses in a diploid model of selection

Chromosome from mother	Chromosome from father:			
	A_1B_1 (freq. x_1)	A_1B_2 (freq. x_2)	A_2B_1 (freq. x_3)	A_2B_2 (freq. x_4)
A_1B_1 (freq. x_1)	w_{11}	w_{12}	w_{13}	w_{14}
A_1B_2 (freq. x_2)	w_{21}	w_{22}	w_{23}	w_{24}
A_2B_1 (freq. x_3)	w_{31}	w_{32}	w_{33}	w_{34}
A_2B_2 (freq. x_4)	w_{41}	w_{42}	w_{43}	w_{44}

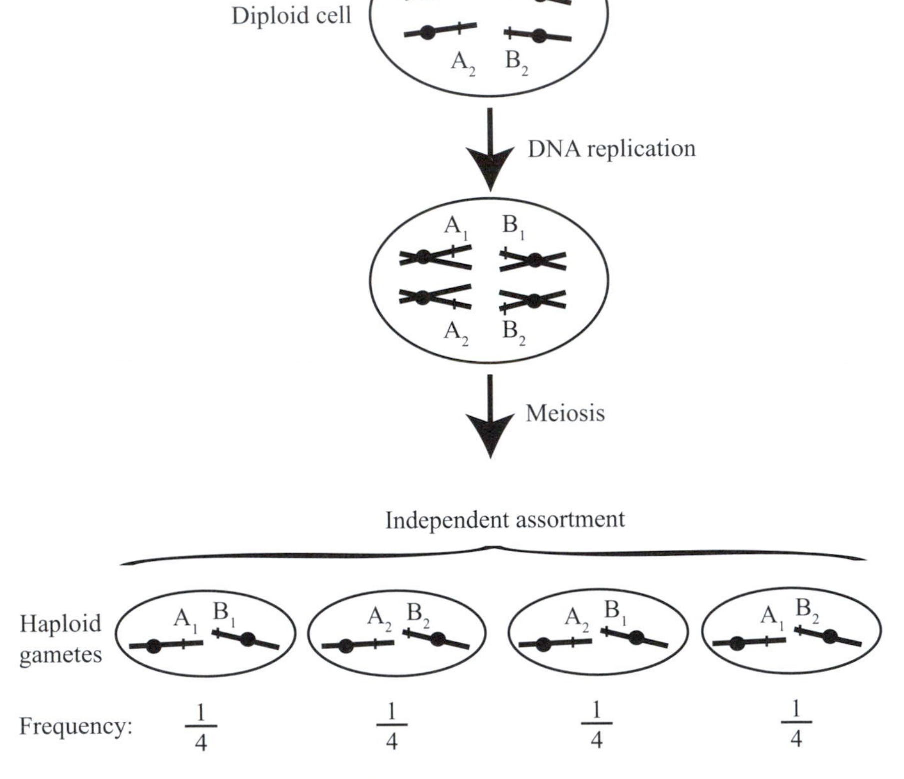

Figure 8.8: Independent assortment of chromosomes. Two chromosomes are shown in the nucleus of a diploid cell (each line represents a double helix of DNA, with a centromere located at the circle). During the production of gametes, the DNA replicates (producing X-shaped pairs of chromosomes) and then undergoes two meiotic divisions to produce four haploid cells. Because assortment of chromosomes is independent across different chromosomes, each of the four gamete types is equally likely.

parent (row 4 of Table 8.1). With probability $1-r$, recombination does not occur between the two genes (see Figure 8.9), in which case half of the gametes carry the A_1B_1 chromosome and half carry the A_2B_2 chromosome. Thus, we place $(1-r)/2$ in both the fourth and seventh columns of row 4. With probability r,

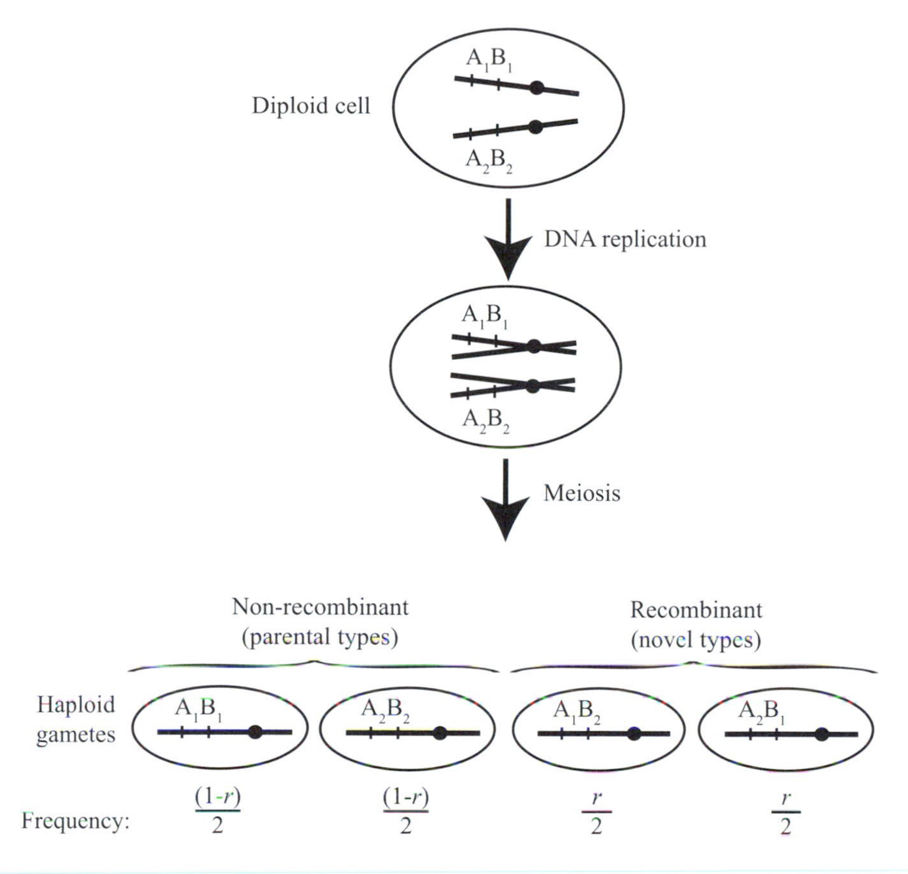

Figure 8.9: Recombination and meiosis. Two genes (**A** and **B**) on the same chromosome are illustrated. After DNA replication but before cell division, recombination can occur between the chromosomes, creating chromosomes with different combinations of the alleles than the parental chromosomes. The rate of recombination between two genes, r, measures the probability that a chromosome carries an **A** allele from one parental chromosome and a **B** allele from the other parental chromosome. When $r = 1/2$, all gamete types are equally like to be produced, whether recombinant or non-recombinant, as was the case for genes on different chromosomes (Figure 8.8).

recombination does occur between the two genes, in which case half of the gametes carry the A_1B_2 chromosome and half carry the A_2B_1 chromosome. Thus, we place $r/2$ in both the fifth and sixth columns of row 4. The calculations for the other rows are analogous, but the recombination rate only matters for some of the genotypes. For example, an A_1B_1/A_1B_2 parent can produce only A_1B_1 and A_1B_2 gametes regardless of the rate of recombination. Only when both genes are heterozygous does recombination affect the array of gametes produced. This property serves as a useful check when deriving models with recombination. Another important check is to make sure that the sum of the gamete frequencies produced by any one genotype is one. Indeed, across any row, the last four columns do sum to one.

Once gamete production is complete, the life cycle then begins again. To tabulate the frequency of the A_1B_1 gamete in the next generation (i.e., $x_1(t + 1)$) as a function of the frequency of the four types in the current generation, we multiply column 3 (which gives the postselection frequency of each parental genotype) by column 4 (which gives the probability that this parent produces an A_1B_1 gamete) and then sum these products over all rows. We get

$$\begin{aligned} x_1(t+1) = {} & x_1x_1\frac{w_{11}}{\overline{w}} + \frac{1}{2}x_1x_2\frac{w_{12}}{\overline{w}} + \frac{1}{2}x_1x_3\frac{w_{13}}{\overline{w}} + \frac{1-r}{2}x_1x_4\frac{w_{14}}{\overline{w}} \\ & + \frac{1}{2}x_2x_1\frac{w_{21}}{\overline{w}} + \frac{r}{2}x_2x_3\frac{w_{23}}{\overline{w}} + \frac{1}{2}x_3x_1\frac{w_{31}}{\overline{w}} + \frac{r}{2}x_3x_2\frac{w_{32}}{\overline{w}} \\ & + \frac{1-r}{2}x_4x_1\frac{w_{41}}{\overline{w}}, \end{aligned} \tag{8.39}$$

where the gamete frequencies on the right are measured at time t and where $\overline{w}$ is the population mean fitness and is equal to $\overline{w} = \sum_{i=1}^{4}\sum_{j=1}^{4} x_i(t)\, x_j(t)\, w_{ij}$.

To simplify this recursion, we bring together the terms that do not involve the recombination rate:

$$\begin{aligned} x_1\Bigg(& x_1\frac{w_{11}}{\overline{w}} + \frac{1}{2}x_2\frac{w_{12}}{\overline{w}} + \frac{1}{2}x_3\frac{w_{13}}{\overline{w}} + \frac{1}{2}x_4\frac{w_{14}}{\overline{w}} + \frac{1}{2}x_2\frac{w_{21}}{\overline{w}} \\ & + \frac{1}{2}x_3\frac{w_{31}}{\overline{w}} + \frac{1}{2}x_4\frac{w_{41}}{\overline{w}}\Bigg), \end{aligned} \tag{8.40}$$

which can be written more compactly as

$$x_1(t)\sum_{j=1}^{4} x_j(t)\left(\frac{w_{1j}+w_{j1}}{2\overline{w}}\right).$$

The remaining terms involve the recombination rate and are

$$\frac{-r}{2}x_1x_4\frac{w_{14}}{\overline{w}} + \frac{r}{2}x_2x_3\frac{w_{23}}{\overline{w}} + \frac{r}{2}x_3x_2\frac{w_{32}}{\overline{w}} + \frac{-r}{2}x_4x_1\frac{w_{41}}{\overline{w}}. \tag{8.41}$$

We can rewrite (8.41) as $-r\,D^*$, where

$$D^* = x_1x_4\left(\frac{w_{14}+w_{41}}{2\overline{w}}\right) - x_2x_3\left(\frac{w_{23}+w_{32}}{2\overline{w}}\right). \tag{8.42}$$

A nonrandom association between alleles at different genes is referred to as *linkage disequilibrium*. A random association is referred to as linkage equilibrium.

D^* is known as the *linkage disequilibrium*, measured after selection in (8.42). We use an asterisk to distinguish this measure from the linkage disequilibrium before selection, $D = x_1x_4 - x_2x_3$. If you have not encountered the concept of linkage disequilibrium before, the main point to remember is that it describes the association between alleles carried at gene **A** and at gene **B**. When linkage disequilibrium is positive, a chromosome is more likely to carry a B_1 allele if it carries an A_1 allele (and more likely to carry a B_2 allele if it carries A_2) than you would predict based on random associations among the alleles within the population. In other words, there is a correlation between the alleles carried at the two genes when disequilibrium is present. Recall that we encountered this idea of a genetic correlation in the example of sexual selection in Chapter 7 (see also Figure 7.8).

Following the same procedure for the remaining three gamete types, we can write the four recursion equations in a nice compact form as

$$\begin{aligned}
x_1(t+1) &= x_1(t)\left(\sum_{j=1}^{4} x_j(t)\left(\frac{w_{1j}+w_{j1}}{2\bar{w}}\right)\right) - r\,D^*,\\
x_2(t+1) &= x_2(t)\left(\sum_{j=1}^{4} x_j(t)\left(\frac{w_{2j}+w_{j2}}{2\bar{w}}\right)\right) + r\,D^*,\\
x_3(t+1) &= x_3(t)\left(\sum_{j=1}^{4} x_j(t)\left(\frac{w_{3j}+w_{j3}}{2\bar{w}}\right)\right) + r\,D^*,\\
x_4(t+1) &= x_4(t)\left(\sum_{j=1}^{4} x_j(t)\left(\frac{w_{4j}+w_{j4}}{2\bar{w}}\right)\right) - r\,D^*,
\end{aligned} \tag{8.43}$$

Model (8.43) is a very complicated system of nonlinear equations. Even finding all of the equilibria of this model is an impossible task. Here we ask only one question with this model that illustrates a stability analysis without being overly complicated. Suppose that the two alleles A_1 and B_1 are fixed in the population (i.e., $x_1 = 1$). Under what conditions can the alternative alleles, A_2 and/or B_2 invade (Crow and Kimura 1965)? To answer this question we first need to calculate the stability matrix for (8.43) and evaluate it at the equilibrium $\hat{x}_1 = 1$, $\hat{x}_2 = 0$, $\hat{x}_3 = 0$, $\hat{x}_4 = 0$. This task is greatly aided by mathematical software such as Maple or *Mathematica* and results in the matrix

$$\mathbf{J} = \begin{pmatrix}
0 & -\frac{w_{12}+w_{21}}{2w_{11}} & -\frac{w_{13}+w_{31}}{2w_{11}} & -(1-r)\frac{w_{14}+w_{41}}{2w_{11}}\\
0 & \frac{w_{12}+w_{21}}{2w_{11}} & 0 & r\frac{w_{14}+w_{41}}{2w_{11}}\\
0 & 0 & \frac{w_{13}+w_{31}}{2w_{11}} & r\frac{w_{14}+w_{41}}{2w_{11}}\\
0 & 0 & 0 & (1-r)\frac{w_{14}+w_{41}}{2w_{11}}
\end{pmatrix} \tag{8.44}$$

The stability matrix (8.44) is upper triangular, and therefore its four eigenvalues are given by the diagonal elements (Rule P2.26) $\lambda_1 = 0$, $\lambda_2 = (w_{12} + w_{21})/(2w_{11})$, $\lambda_3 = (w_{13} + w_{31})/(2w_{11})$, and $\lambda_4 = (1-r)(w_{14} + w_{41})/(2w_{11})$. For an equilibrium to be stable in a discrete-time model, we must show that the absolute values of all eigenvalues are less than one. None of the eigenvalues can be negative (because fitness cannot be negative), and therefore we need only check whether the eigenvalues are less than one.

The first eigenvalue $\lambda_1 = 0$ reveals that there is some direction in which the system always goes to zero. In this case, the genotype frequencies are constrained to satisfy a particular relationship: they must sum to one. Because of this constraint, there are effectively only three dimensions in which evolution occurs, even though there are four equations in (8.43).

For λ_2 to be less than one requires that $w_{11} > (w_{12} + w_{21})/2$. That is, the fitness of an A_1B_1/A_1B_1 individual must be larger than the average fitness of individuals that are heterozygous at the **B** gene and homozygous for the A_1 allele at the **A** gene. When this condition holds, the A_1B_2 chromosome is, on average, selected against. Similarly, for λ_3 to be less than one requires that $w_{11} > (w_{13} + w_{31})/2$. That is, the fitness of an A_1B_1/A_1B_1 individual must be larger than the average fitness of individuals that are heterozygous at the **A** gene and homozygous for the B_1 allele at the **B** gene. When this condition holds, the A_2B_1 chromosome is, on average, selected against. Finally, for λ_4 to be less than one requires that $w_{11} > (1-r)(w_{14} + w_{41})/2$. This inequality is more interesting and involves the recombination rate. For stability of the equilibrium with A_1 and B_1 fixed, the fitness of an A_1B_1/A_1B_1 individual must be larger than the average fitness of individuals that are heterozygous at both genes times the probability that no recombination occurs.

How does the presence of two genes affect the stability of the equilibrium? The equilibrium is always unstable if the A_1B_2 or A_2B_1 chromosome has higher fitness than the resident ($\lambda_2 > 1$ or $\lambda_3 > 1$), just as we would predict from the one-gene model (section 5.3.2). Unlike with one-gene models, even if the A_1B_2 and A_2B_1 chromosomes are less fit than the residents ($\lambda_2 < 1$ and $\lambda_3 < 1$), the equilibrium can still be unstable if the fitness of A_1B_1/A_2B_2 individuals is higher than the fitness of A_1B_1/A_1B_1 individuals (and the recombination rate is low enough, $(1-r)(w_{14} + w_{41})/2 > w_{11}$).

When the recombination rate is very low, the new A_2B_2 chromosome spreads if it has higher fitness than the resident. Recombination with the resident A_1B_1 breaks apart the good A_2B_2 combination of alleles, hindering the spread of the A_2B_2 chromosome, even if A_2B_2 causes its carriers to have higher fitness. Interestingly, this demonstrates that recombination can prevent the spread of gene combinations that are selectively favorable by breaking them apart.

If the A_2 and B_2 alleles decrease fitness on their own but increase fitness when combined, the two-gene model predicts that the new alleles can invade a population of A_1B_1/A_1B_1 individuals if $\lambda_4 > 1$ (Figure 8.10). This requires that the two genes interact to affect fitness, a phenomenon known as *epistasis*. As we shall discuss in greater depth in the next chapter, a one-gene model is generally not sufficient to predict the outcome of evolution whenever there are such fitness interactions among alleles at different genes.

8.4 Perturbation Techniques for Approximating Eigenvalues

Although we can often calculate and interpret the eigenvalues of a model, this is certainly not always true. When we can't, different techniques can be tried to determine the stability of an equilibrium. We have already mentioned one technique — using the Routh-Hurwitz conditions (Box 8.2). Another technique involves a graphical analysis of the characteristic polynomial (Sup. Mat. 8.1), which we use to analyze the discrete-time Lotka-Volterra model of competition given by equation (3.14). Alternatively, it is often possible to approximate the leading eigenvalue using the perturbation techniques introduced in Chapter 5.

In Box 5.1, we introduced *perturbation analysis* for approximating the solution to an equation. In that context, we were interested in obtaining an expression

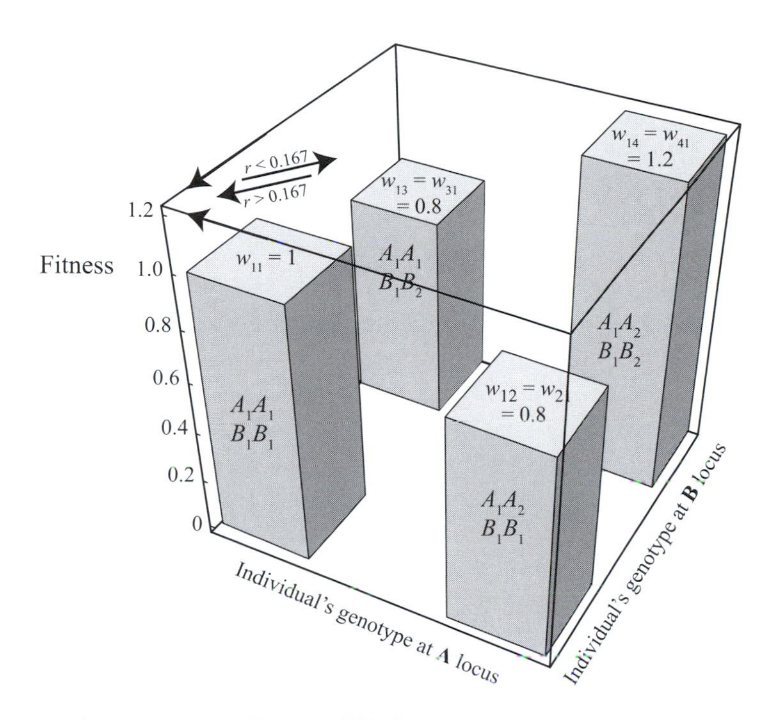

Figure 8.10: The bars give the fitnesses of the four relevant genotypes when alleles A_2 and B_2 are first introduced into a population of A_1B_1 chromosomes. The resident $A_1A_1B_1B_1$ individuals are more fit than $A_1A_1B_1B_2$ and $A_1A_2B_1B_1$ individuals, so that neither A_2 nor B_2 could spread on its own (arrows along the axes). Carrying both genetic changes, however, confers a 20% fitness advantage to $A_1A_2B_1B_2$ individuals. Even so, the A_2 and B_2 alleles spread only if they are sufficiently linked ($r < 0.167$) such that $\lambda_4 = (1-r)(w_{14} + w_{41})/(2w_{11}) > 1$ (diagonal arrow).

that provided an approximation to an equilibrium. But perturbation analysis can be useful for finding approximate solutions to any equation, including the characteristic polynomial of a matrix. In this latter context, it thereby provides an approximation for the eigenvalues of a matrix.

As in any perturbation analysis, we must first identify a parameter (or a set of parameters) that is small. For example, we might be willing to assume that the mutation rate or the effect of a new allele is small in an evolutionary model, or we might assume that the difference between two species is small in an ecological model. For the perturbation method to be useful in a stability analysis, it must be possible to determine the eigenvalues when the small parameter is set to zero. One situation in which perturbation analysis is particularly useful is when the leading eigenvalue falls on the boundary between stability and instability (i.e., equals one in a discrete-time model or zero in a continuous-time model). In this case, introducing the small parameter with some non-zero value might cause a critical shift in the leading eigenvalue. If the leading eigenvalue increases, then including the small parameter causes the equilibrium to become unstable. If the leading eigenvalue decreases, then including the small parameter causes the equilibrium to become stable.

An *approximate stability analysis* can be performed when a parameter is thought to be near a special value, using perturbation methods.

To begin, we assume that we have a characteristic polynomial that we want to solve for the leading eigenvalue λ. We then rewrite the characteristic polynomial as a function of the small parameter ζ; i.e., $f(\zeta) = 0$ (see Box 5.1). Next, we write the leading eigenvalue λ as a sum of terms involving powers of the small parameter ζ:

$$\lambda = \lambda_0 + \lambda_1 \zeta + \lambda_2 \zeta^2 + \lambda_3 \zeta^3 + \cdots. \tag{8.45}$$

You can think of each successive ζ^i term as providing a more refined estimate for the eigenvalue. Although there will be more than one eigenvalue in a multiple-variable model, we are particularly interested in the behavior of the leading eigenvalue. Thus, we set λ_0 equal to the value of the leading eigenvalue when the small parameter is set to zero. Next, we plug (8.45) into the characteristic polynomial $f(\zeta)$ and take the Taylor series of the characteristic polynomial with respect to ζ, around the point $\zeta = 0$ (Primer 1). Finally, we determine the values of the λ_i in (8.45) that cause each term in the Taylor series to equal zero. Such values ensure that the characteristic polynomial does indeed equal zero (see Box 5.1 for more details).

Knowing λ_1 is typically sufficient to determine the direction in which the eigenvalue changes with the addition of the small parameter. When λ_0 in (8.45) lies on the border between stability and instability (i.e., $\lambda_0 = 1$ in discrete time and $\lambda_0 = 0$ in continuous time), the sign of λ_1 in (8.45) determines whether the balance tips toward stability or instability for nonzero values of the small parameter ζ.

Example: The Evolution of Haploid and Diploid Organisms

We saw in Chapter 3 that an important genetic distinction between some organisms is whether they are haploid (i.e., they carry only one copy of each gene) or diploid (i.e., they carry two copies of each gene). For example, humans are primarily diploid because the majority of their life cycle is carried out in the diploid state, even though they have a haploid stage (sperm and eggs). In contrast, many fungi, algae, and unicellular organisms are primarily haploid. Why have some organisms evolved to become diploid while others have evolved to become haploid? Let us construct a model to gain some insight into this question.

Consider an organism that reproduces sexually and is capable of growth and development in either the haploid or diploid stage. Because the organism is sexual, it will necessarily pass through a haploid stage after meiosis and a diploid stage after the union of gametes (Figure 8.11). To allow the proportion of time spent in each state to evolve, we suppose that there is a gene that alters the life cycle. At this gene, allele C_1 causes meiosis to occur early in life (before natural selection has acted), resulting in a predominantly haploid life cycle (left pathway, Figure 8.11). In contrast, allele C_2 causes a delay in meiosis until after selection, resulting in a predominantly diploid life cycle (right pathway, Figure 8.11). To simplify matters, we assume that C_2 is dominant, so that C_1C_2 individuals are also diploid.

We also suppose that there are no intrinsic costs or benefits to being haploid or diploid, and therefore the frequency of allele C_2 would remain constant over

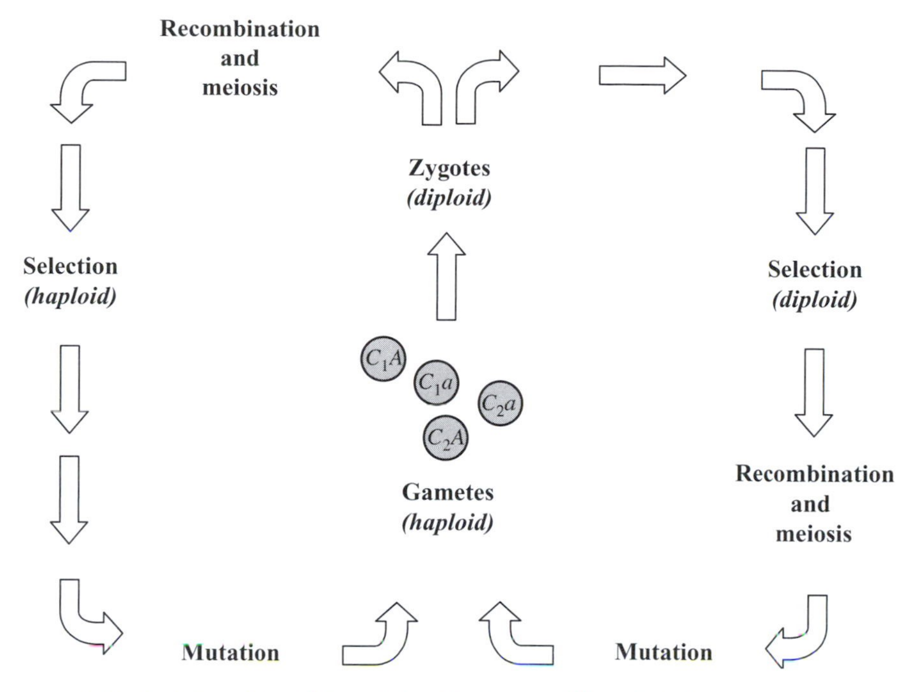

Figure 8.11: Evolution of haploidy and diploidy. A general life cycle is shown in which an organism either undergoes meiosis early and experiences selection as a haploid (left pathway) or undergoes meiosis late and experiences selection as a diploid (right pathway).

time if it were the only gene considered in the model. There are certainly other genes in the genome under selection, however, and we thus include a second gene under selection with two alleles, where allele A is the most fit but mutates regularly to a less fit allele, a, at rate μ. Specifically, we assume that the fitnesses of individuals are

Haploid fitness		Diploid fitness		
A	a	AA	Aa	aa
1	$1 - s$	1	$1 - h\,s$	$1 - s$

where h is the coefficient of dominance and s is the selective disadvantage. Both h and s are assumed positive so that the a allele is deleterious. We also assume that, at the end of the life cycle, all individuals produce haploid gametes that unite at random to begin the next generation (Figure 8.11).

We census the population at the gamete stage immediately after mutation, letting the frequency of C_1A, C_1a, C_2A, and C_2a gametes equal x_1, x_2, x_3, and x_4, respectively (Figure 8.11). Using a mating table like Table 8.1, it is possible to track the frequency of these genotypes as they unite, undergo early meiosis (in C_1C_1 individuals), undergo selection, and then undergo late meiosis (in C_1C_2 and C_2C_2 individuals). The only difference from Table 8.1 is that C_1C_1 individuals experience selection only after meiosis. Working through such a table, the frequencies of the four chromosome types after meiosis but before mutation are

$$x_1' = (x_1^2 + x_1x_2 + x_1x_3 + (1 - r)\,x_1x_4\,(1 - h\,s) + r\,x_2x_3\,(1 - h\,s))/\overline{W},$$

$$
\begin{aligned}
x_2' &= (x_1x_2\,(1-s) + r\,x_1x_4\,(1-h\,s) + x_2^2\,(1-s) \\
&\quad + (1-r)\,x_2x_3\,(1-h\,s) + x_2x_4\,(1-s))/\overline{W}, \\
x_3' &= (x_1x_3 + x_1x_4\,r\,(1-h\,s) + x_2x_3\,(1-r)\,(1-h\,s) + x_3^2 \\
&\quad + x_3x_4\,(1-h\,s))/\overline{W} \\
x_4' &= (x_1x_4(1-r)\,(1-h\,s) + x_2x_3\,(1-h\,s)\,r + x_2x_4\,(1-s) \\
&\quad + x_3x_4\,(1-h\,s) + x_4^2\,(1-s))/\overline{W},
\end{aligned}
\tag{8.46a}
$$

where r is the rate of recombination between the two genes, and $\overline{W}$ is the mean fitness of the population (the sum of the numerators on the right-hand side). Finally, we allow mutations to occur from A to a (mutations occurring in the reverse direction are assumed to be very rare and are ignored), which gives us the frequency of the four types of gametes in the next generation:

$$
\begin{aligned}
x_1'' &= (1-\mu)x_1', \\
x_2'' &= \mu\,x_1' + x_2', \\
x_3'' &= (1-\mu)x_3', \\
x_4'' &= \mu\,x_3' + x_4'.
\end{aligned}
\tag{8.46b}
$$

If we want to understand how life cycles evolve, one approach is to assume that the population has a certain life cycle and then determine when a mutation altering the life cycle can invade. To this end, let's consider a haploid population, with allele C_1 fixed, into which we will introduce the C_2 allele, thereby generating diploid individuals.

First, we must determine the equilibrium reached by a haploid population when only allele C_1 is present ($x_3 = x_4 = 0$). In Problem 8.3, you are asked to find the equilibrium where selection is balanced by mutation. This equilibrium occurs at $\hat{x}_1 = 1 - \mu/s$ and $\hat{x}_2 = \mu/s$, as we found in the one-gene model (Problem 5.4).

Next, we explore what happens after C_2 arises at some small frequency, causing its carriers to remain diploid throughout selection. The allele will ultimately decrease in frequency if the equilibrium $\hat{x}_1 = 1 - \mu/s$, $\hat{x}_2 = \mu/s$, $\hat{x}_3 = 0$, $\hat{x}_4 = 0$ is locally stable, and the population will remain haploid. Therefore, we can determine when we expect the allele to die out by performing a local stability analysis of this equilibrium. The stability matrix for this model is

$$
\mathbf{J} = \left.\begin{pmatrix}
\frac{\partial x_1''}{\partial x_1} & \frac{\partial x_1''}{\partial x_2} & \frac{\partial x_1''}{\partial x_3} & \frac{\partial x_1''}{\partial x_4} \\
\frac{\partial x_2''}{\partial x_1} & \frac{\partial x_2''}{\partial x_2} & \frac{\partial x_2''}{\partial x_3} & \frac{\partial x_2''}{\partial x_4} \\
\frac{\partial x_3''}{\partial x_1} & \frac{\partial x_3''}{\partial x_2} & \frac{\partial x_3''}{\partial x_3} & \frac{\partial x_3''}{\partial x_4} \\
\frac{\partial x_4''}{\partial x_1} & \frac{\partial x_4''}{\partial x_2} & \frac{\partial x_4''}{\partial x_3} & \frac{\partial x_4''}{\partial x_4}
\end{pmatrix}\right|_{\substack{x_1 = 1-\mu/s,\, x_2 = \mu/s, \\ x_3 = 0, x_4 = 0}}
\tag{8.47}
$$

where the x_i'' are given by equations (8.46). Analyzing this matrix is aided by the fact that the 2×2 submatrix on the bottom left is full of zeros:

$$\left.\begin{pmatrix} \frac{\partial x_3''}{\partial x_1} & \frac{\partial x_3''}{\partial x_2} \\ \frac{\partial x_4''}{\partial x_1} & \frac{\partial x_4''}{\partial x_2} \end{pmatrix}\right|_{\substack{x_1=1-\mu/s,\, x_2=\mu/s, \\ x_3=0, x_4=0}} = \begin{pmatrix} 0 & 0 \\ 0 & 0 \end{pmatrix}.$$

Thus, matrix (8.47) has a block-triangular form (see Primer 2), and its four eigenvalues are given by the eigenvalues of the two submatrices along the diagonal (Rule P2.27), which we label **A** and **B**:

$$\mathbf{A} = \left.\begin{pmatrix} \frac{\partial x_1''}{\partial x_1} & \frac{\partial x_1''}{\partial x_2} \\ \frac{\partial x_2''}{\partial x_1} & \frac{\partial x_2''}{\partial x_2} \end{pmatrix}\right|_{\substack{x_1=1-\mu/s,\, x_2=\mu/s, \\ x_3=0, x_4=0}} = \begin{pmatrix} \frac{\mu(1-s)}{(1-\mu)s} & -\frac{(s-\mu)(1-s)}{(1-\mu)s} \\ -\frac{\mu(1-s)}{(1-\mu)s} & \frac{(s-\mu)(1-s)}{(1-\mu)s} \end{pmatrix} \tag{8.48a}$$

$$\mathbf{B} = \left.\begin{pmatrix} \frac{\partial x_3''}{\partial x_3} & \frac{\partial x_3''}{\partial x_4} \\ \frac{\partial x_4''}{\partial x_3} & \frac{\partial x_4''}{\partial x_4} \end{pmatrix}\right|_{\substack{x_1=1-\mu/s,\, x_2=\mu/s, \\ x_3=0, x_4=0}}$$

$$= \begin{pmatrix} 1-\frac{\mu\,(r+hs-rhs)}{s} & \frac{(s-\mu)(1-hs)\,r}{s} \\ \frac{\mu\,(r+s-rhs)-\mu^2\,(r+hs-rhs)}{(1-\mu)\,s} & 1-\frac{(s-\mu)(hs+(1-\mu)(1-hs)\,r)}{(1-\mu)\,s} \end{pmatrix}. \tag{8.48b}$$

From the partial derivatives contained in the submatrix **A**, we can tell that this submatrix describes the sensitivity of the recursions for x_1 and x_2 to displacements in x_1 and x_2. The eigenvalues of this submatrix thus describe the stability of the equilibrium in the absence of the new mutant C_2 allele. Using Rule P2.20 and factoring, these eigenvalues are given by $\lambda = 0$ and $\lambda = (1-s)/(1-\mu)$. The zero eigenvalue indicates that the model has only three effective dimensions, and again this occurs because all four gamete frequencies must sum to one. The eigenvalue of $(1-s)/(1-\mu)$ is positive and less than one as long as the mutation rate is small relative to selection ($\mu < s$), which is both a reasonable assumption and necessary for $\hat{x}_1 = 1-\mu/s$ to be a valid equilibrium frequency. Thus, when the C_2 allele is absent, the mutation-selection balance equilibrium is stable when it exists.

The real question of interest is whether this equilibrium is stable if we perturb it by introducing the C_2 allele (i.e., if we have x_3 and/or x_4 not equal to zero). As seen by the partial derivatives contained in submatrix **B**, the eigenvalues of **B** address this question. These eigenvalues can be calculated, but they are ugly and difficult to interpret. To obtain interpretable results, we use the perturbation method under the assumption that the mutation rate is very small, $\mu = \zeta$.

First, we must find λ_0, the leading term in the eigenvalue (8.45), by setting the mutation rate to zero in submatrix **B**. The submatrix is then triangular and has eigenvalues equal to the diagonal elements; $\lambda_0 = 1$ and $\lambda_0 = (1-hs)(1-r)$. Because we have assumed that all of the parameters are positive, $\lambda_0 = 1$ is the leading eigenvalue (i.e., the one with the largest absolute value).

Because the leading term $\lambda_0 = 1$ falls on the border between stability and instability, we must seek out the next-order term λ_1 in (8.45). Plugging (8.45) in for λ in the characteristic polynomial of **B**, taking the Taylor series with respect to ζ, and keeping only terms to first order in ζ, we get an equation that λ_1 must satisfy:

$$-\mu\,(r\,(1-2h)(1-hs) - h^2 s - \lambda_1\,(r+s-rhs)) = 0, \tag{8.49}$$

which we have written in terms of the original parameter μ (see Problem 8.9). Solving (8.49) for λ_1 and plugging λ_0 and λ_1 into (8.45), we get an approximation for the leading eigenvalue:

$$\lambda = 1 + \frac{r\,(1-2h)(1-hs)-h^2 s}{r+s-rhs}\,\mu + O(\mu^2), \tag{8.50}$$

which is extremely close to the exact numerical value (Figure 8.12). The denominator of (8.50) is always positive, but the numerator is more difficult to interpret. If $h > 1/2$ then all terms in the numerator are negative and λ is less than one, indicating that diploids cannot invade a haploid population. If $h < 1/2$, then the two terms in the numerator are of opposite sign, and diploidy can invade ($\lambda > 1$) as long as the first term is larger. This requires that $r > h^2 s/((1-2h)(1-hs))$.

What do these results mean biologically? Intuitively, one might expect diploidy to be favored because mutant alleles can be "masked" by the good copy of the allele in heterozygotes. This does provide diploidy with an advantage, but this advantage is counterbalanced by the fact that a diploid individual has two chances of carrying a deleterious mutation. Thus, only when masking is strong enough ($h < 1/2$) would the average fitness of diploids be better than haploids, assuming that they have the same frequency of deleterious alleles.

The final twist to our result, however, is that diploidy spreads only if recombination rates are high enough, even when mutations are better masked in diploid individuals ($h < 1/2$). Because of masking among diploid ancestors, mutant alleles are more likely to survive and persist among the descendants carrying the diploid allele, C_2. Consequently, chromosomes with the diploid allele C_2 are more loaded with deleterious mutations, while chromosomes with the haploid allele C_1 are more effectively purged of deleterious mutations. The

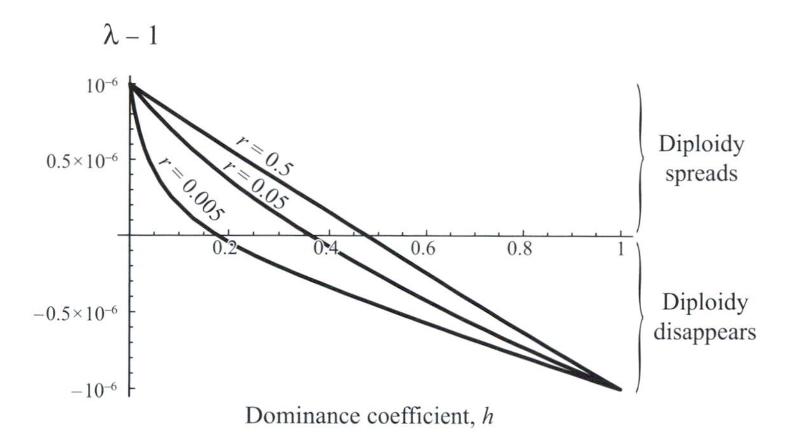

Figure 8.12: Evolution of diploidy. The leading eigenvalue minus one ($\lambda - 1$) is shown as a function of the dominance coefficient. A mutation causing diploid life cycles spreads within a haploid population only when the curve lies above the horizontal axis. This requires that h is small enough ($\lambda - 1 > 0$); the larger the recombination rate between the genes, the broader the range of dominance conditions under which diploidy evolves. For the parameters chosen ($s = 0.1$, $\mu = 10^{-6}$), the curves for the leading eigenvalue based on the approximation (8.50) or a numerical evaluation of the eigenvalues of the matrix (8.48b) cannot be distinguished.

tighter the recombination rate between the ploidy and selected genes, the greater the difference in mutant allele frequency expected between C_1-bearing and C_2-bearing chromosomes, and the less likely diploids are to invade a population. While diploidy protects the individual from its burden of deleterious mutations, it does so at the expense of future generations, which are more likely to inherit deleterious mutations. Thus, diploidy is favored only when there is enough genetic mixing among the chromosomes in a population (see also Otto 1994; Otto and Goldstein 1992; Otto and Marks 1996; Perrot et al. 1991).

8.5 Concluding Message

In this chapter we have presented the general techniques for finding equilibria and determining their stability properties in non-linear models with multiple variables. This concludes our tour of the main techniques for analyzing equilibria of dynamical models. To summarize, we write a nonlinear model involving n dynamic variables $x_1, \ldots, x_n$, using n differential equations of the form $dx_i/dt = f_i(x_1, x_2, \ldots, x_n)$ (continuous time) or n recursion equations of the form $x_i(t+1) = f_i(x_1(t), x_2(t), \ldots, x_n(t))$ (discrete time).

At any equilibrium of such a model, none of the variables changes over time. In continuous time, the equilibria are found by determining the values of the variables that simultaneously cause $dx_1/dt = 0$, $dx_2/dt = 0, \ldots, dx_n/dt = 0$. In discrete time, the equilibria are found by determining the values of the variables that simultaneously cause $f_1(\hat{x}_1, \hat{x}_2, \ldots, \hat{x}_n) = \hat{x}_1$, $f_2(\hat{x}_1, \hat{x}_2, \ldots, \hat{x}_n) = \hat{x}_2, \ldots,$ $f_n(\hat{x}_1, \hat{x}_2, \ldots, \hat{x}_n) = \hat{x}_n$. Nonlinear models can have multiple equilibria, each of which has its own stability properties.

For each equilibrium, we can determine whether it is stable or unstable by performing a local stability analysis. A local stability analysis or linearization approximates the nonlinear equations with linear equations that are accurate near the equilibrium and that can be analyzed using the techniques of Chapter 7. First, we find the Jacobian matrix (Definition 8.2), which consists of the partial derivatives of each function with respect to each variable. Next, we determine the eigenvalues of this matrix evaluated at an equilibrium. For a continuous-time model, the equilibrium is stable as long as the real parts of all eigenvalues are negative (Recipe 8.2). For a discrete-time model, the equilibrium is stable as long as the absolute values of all eigenvalues are less than one (Recipe 8.4).

For some models with multiple variables, it is possible to go beyond analyzing the stability of equilibrium points. In the next chapter, we will see how to obtain general solutions for some such models. As with the one-variable models considered in Chapter 6, the general solution to multiple-variable models provides us with complete information about the model's behavior.

Problems

Problem 8.1: For each pair of hypothetical eigenvalues from a stability analysis, specify whether the equilibrium is stable or unstable; also specify when cycles are expected around the equilibrium because the eigenvalues are complex (you can assume that p is a proportion, and therefore lies between 0 and 1). (a) $r_1 = 0.5$ and $r_2 = -2/3$ in a continuous-time model. (b) $\lambda_1 = 0.5$ and $\lambda_2 = -2/3$ in a discrete-time model. (c) $r_1 = (1 + \sqrt{1+p})/2$ and $r_2 = (1 - \sqrt{1+p})/2$ in a continuous-time model. (d) $\lambda_1 = (1 + \sqrt{1+p})/2$ and $\lambda_2 = (1 - \sqrt{1+p})/2$ in a discrete-time model. (e) $r_1 = (1 + \sqrt{-p})/2$ and $r_2 = (1 - \sqrt{-p})/2$ in a continuous-time model. (f) $\lambda_1 = (1 + \sqrt{-p})/2$ and $\lambda_2 = (1 - \sqrt{-p})/2$ in a discrete-time model.

Problem 8.2: In the text, we explored a diploid version of density-dependent natural selection (8.36). A haploid version of this model is

$$N(t+1) = \overline{W}(N(t), p(t))\, N(t),$$

$$p(t+1) = \frac{W_A(N(t))}{\overline{W}(N(t), p(t))}\, p(t),$$

where $\overline{W}(N(t), p(t)) = p(t)\, W_A(N(t)) + (1 - p(t))\, W_a(N(t))$, $W_A(N) = (1 + r)e^{-\alpha_A N}$, and $W_a(N) = (1 + r)e^{-\alpha_a N}$. Identify all of the equilibria of this model.

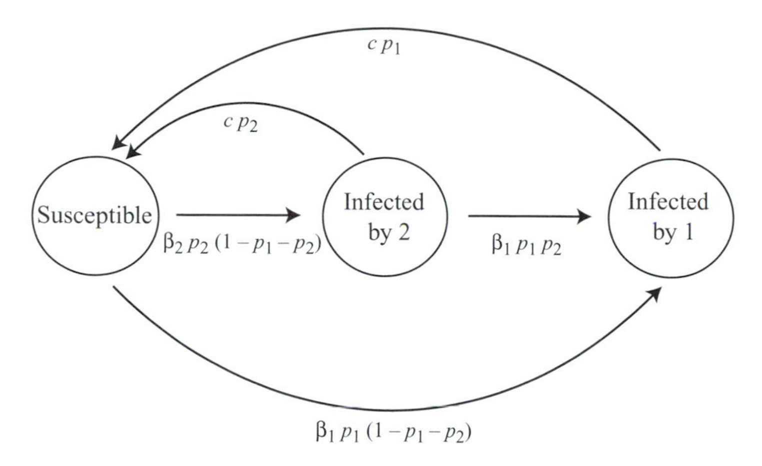

Figure 8.13: A flow diagram of infection and clearance.

Problem 8.3: Find the two equilibria of (8.46) under the assumption that the allele C_2 is absent. Say why your answer does not depend on the recombination rate r or the dominance coefficient h.

Problem 8.4: Here we construct an epidemiological model in which there are two parasite types (labeled 1 and 2). To simplify matters, we assume that the total host population is constant in size, and we track the frequency of the host population that is infected by each type of parasite in continuous time. We further assume that a host can be infected by at most one type of parasite and that there is an asymmetry such that parasite 1 can infect and immediately take over a host infected with parsite 2 but not vice versa (this is sometimes referred to as superinfection: Levin and Pimentel 1981; Nowak and May 1994). At any given time, some fraction of the hosts will be susceptible $(1-p_1-p_2)$, some fraction will be infected by parasite 1 (p_1), and some fraction will be infected by parasite 2 (p_2) (Figure 8.13).

Assuming a mass-action infection process, susceptible hosts are infected with parasite 2 at rate $\beta_2 p_2 (1-p_1-p_2)$ (i.e., proportional to the number of hosts currently infected with parasite 2 and to the number of susceptible hosts). Hosts infected with parasite 2 clear the infection at rate $c\, p_2$ or can be superinfected by parasite 1 at rate $\beta_1 p_1 p_2$. Parasite 1 infects both susceptible hosts and hosts infected by parasite 2 at a total rate of $\beta_1 p_1 (1-p_1-p_2) + \beta_1 p_1 p_2 = \beta_1 p_1 (1-p_1)$. Hosts infected with parasite 1 also clear the infection at rate $c\, p_1$. These assumptions generate the model

$$\frac{dp_1}{dt} = \beta_1 p_1 (1 - p_1) - c\, p_1,$$

$$\frac{dp_2}{dt} = \beta_2 p_2 (1 - p_1 - p_2) - \beta_1 p_1 p_2 - c\, p_2.$$

In the following, the quantities β_1/c and β_2/c will arise frequently. Whenever possible, simplify the notation by defining $R_1 = \beta_1/c$ and $R_2 = \beta_2/c$, which represent the infectivities of the parasites relative to their clearance rates. Intuitively we might expect to find four equilibria in this model: (i) one with both parasites absent, (ii) one with parasite 1 present and parasite 2 absent, (iii) one with parasite 2 present and parasite 1 absent, and (iv) one with both parasites present.

(a) Find the four equilibria and express them in terms of the quantities, R_1 and R_2. (b) For equilibria (i)–(iv), determine the conditions under which they are biologically feasible (i.e., represent fractions between 0 and 1), expressed in terms of the quantities R_1 and R_2. (c) Calculate the general Jacobian matrix for the differential equations, without plugging in a particular equilibrium. (d) Suppose $R_1 < 1$ and $R_2 < 1$. Which equilibria are biologically feasible? For each feasible equilibrium, determine if it is locally stable or unstable. Provide a biological interpretation of the assumption that $R_1 < 1$ and $R_2 < 1$ to explain your findings. (e) Suppose $R_1 > 1$ and $R_2 < 1$. Which equilibria are biologically feasible? For each feasible equilibrium, determine if it is locally stable or unstable. Provide a biological interpretation of the assumption that $R_1 > 1$ and $R_2 < 1$ to explain your findings. (f) Suppose $R_1 > 1$ and $R_2 > 1$ and $R_2 < R_1^2$. Which equilbria are biologically feasible? For each feasible equilibrium, determine if it is locally stable or unstable. Provide a biological interpretation of the assumption that $R_1 > 1$ and $R_2 > 1$ and $R_2 < R_1^2$ to explain your findings. (g) Suppose $R_1 > 1$ and $R_2 > 1$ and $R_2 > R_1^2$. Which equilibria are biologically feasible? For each feasible equilibrium, determine if it is locally stable or unstable. Provide a biological interpretation of the assumption that $R_1 > 1$ and $R_2 > 1$ and $R_2 > R_1^2$ to explain your findings.

Problem 8.5: Consider the following predator-prey model:

$$\frac{dR}{dt} = rR\left(1 - \frac{R}{K}\right) - P\frac{cR}{a + R}$$

$$\frac{dP}{dt} = \frac{\varepsilon c R}{a + R}P - dP.$$

R and P represent the number of prey and predators respectively, r is the per capita growth rate of the prey when their numbers are small and the predator is absent, K is the carrying capacity of the prey in the absence of the predator, a and c are parameters governing the shape of the functional response, ε is the conversion efficiency of consumed prey into new predators, and d is the per capita mortality rate of the predator. All parameters are positive.

(a) Provide a biological interpretation for the above model. In particular, discuss the nature of the prey dynamics in the absence of the predator and the predator dynamics in the absence of the prey. Also, discuss whether the consumption of prey by any given predator individual is affected by prey and predator population sizes. If it is affected by these population sizes, discuss how and provide a reason why this might be the case in reality. (b) Find all equilibria of the model. You should find three of them, only one of which has both species present. Label the equilibrium with neither species present as (i), with one species as (ii), and with both species as (iii). Give the conditions on the parameters under which each equilibrium is biologically feasible. (c) Calculate the Jacobian matrix for the above model but do not evaluate it at any particular equilibrium. (d) Determine the conditions under which equilibrium (i) is *unstable* and provide a biological interpretation of the result. (e) Determine the conditions under which equilibrium (ii) is *unstable* and provide a biological interpretation of the result. (f) Show that equilibria (i) and (ii) are both unstable when equilibrium (iii) is feasible. (g) Given that equilibrium (iii) is biologically feasible, determine the conditions under which it is *stable*. Use the fact that the equilibrium is stable if the determinant of the 2×2 Jacobian matrix is positive and if its trace is negative. (h) As the carrying capacity of the prey, K, increases, there comes a point at which the equilibrium (iii) loses its stability. Determine this critical value of K in terms of the other parameters. (For K larger than this critical value, predator-prey cycles occur.)

Problem 8.6: Consider the following discrete-time model for the population dynamics of a host species and a parasitoid (a parasitoid is an insect whose larvae develop inside other insects):

$$N(t+1) = R\,N(t)\exp(-a\,P(t)),$$

$$P(t+1) = a\,N(t)\,(1 - \exp(-a\,P(t))).$$

Here, N is the host population size and P is the parasitoid population size, and a and R are positive parameters. (a) Find all equilibria of model (2). (b) Calculate the Jacobian matrix and evaluate it at each equilibrium obtained in (a). (c) Determine the conditions under which the equilibrium where both species are absent is unstable. (d) Provide a biological interpretation for your answer to (c).

Problem 8.7: Here, we analyze the dynamics of a source-sink model of population dynamics. Consider two populations of a single species, one in a good environment (the "source" population 1) and one in a poor environment (the "sink" population 2). The number of individuals in the two populations is n_1 and n_2, respectively. In the good environment, the population is able to grow logistically with a carrying capacity K_1 and an intrinsic rate of growth r_1. Each generation, a proportion m of the adults migrate from the good to the poor environment. In the poor environment, each individual has R offspring. R is less than one and the population is unable to replace itself. Thus, the sink population is maintained by the constant input of migrants every generation from the source population. Equations describing this model are

$$n_1(t+1) = (1-m)\,n_1(t) + n_1(t)\,r_1\left(1 - \frac{n_1(t)}{K}\right),$$

$$n_2(t+1) = R\,n_2(t) + m\,n_1(t).$$

(a) What are the equilibrium population sizes for the two populations in this model? (b) Under what conditions is the source-sink metapopulation stable?

Problem 8.8: Here we analyze the Volterra model of predator-prey dynamics in discrete time. We let $P(t)$ equal the number of predators at time t and $H(t)$ equal the number of prey. The parameters of the model are: the per capita growth of the prey in the absence of the predator (r), the per capita probability that a predator contacts and kills a prey (β), the per capita growth of the predator following the consumption of prey (c), and the death rate of predators (δ). With these definitions, the discrete time predator-prey model is

$$H(t+1) = H(t) + r\,H(t) - \beta\,H(t)\,P(t),$$

$$P(t+1) = P(t) + c\,H(t)\,P(t) - \delta\,P(t).$$

(a) Determine the two equilibria of these equations. Double check that the numbers of *both* predators and prey do not change over time when started at an equilibrium. (b) Determine the local stability matrix that approximates these equations near the equilibrium with both species absent. Repeat, finding the local stability matrix near the equilibrium with both species present. (c) Find the eigenvalues for the two matrices in (b). (d) From these eigenvalues, determine whether each equilibrium is stable or unstable, assuming that every parameter is positive. [Recall that an equilibrium in a discrete-time model is stable if all of its eigenvalues are less than one in magnitude, where the magnitude of a complex eigenvalue $A + B\,i$ is $\sqrt{A^2 + B^2}$.]

Problem 8.9: Here we derive equation (8.50) of the text, which gives the eigenvalue near one in the model of haploid-diploid evolution using the stability matrix (8.48b),

$$\mathbf{B} = \begin{pmatrix} 1 - \dfrac{\mu\,(r + hs - rhs)}{s} & \dfrac{(s-\mu)(1-hs)\,r}{s} \\ \dfrac{\mu\,(r + s - rhs) - \mu^2\,(r + hs - rhs)}{(1-\mu)\,s} & 1 - \dfrac{(s-\mu)(hs + (1-\mu)(1-hs)\,r)}{(1-\mu)\,s} \end{pmatrix}.$$

(a) Find the characteristic polynomial $\text{Det}(\mathbf{B} - \lambda\mathbf{I}) = 0$. (b) Replace μ with ζ, replace λ with $\lambda_0 + \lambda_1\,\zeta + \lambda_2\,\zeta^2 + \lambda_3\,\zeta^3$, and replace λ_0 with 1 (based on the case where $\mu = 0$). (c) Perform a Taylor series approximation of this characteristic polynomial with respect to ζ around the point $\zeta = 0$ to linear order in ζ (Primer 1). (d) Undo the transformation by replacing ζ with μ, and show that your result is equivalent to (8.50). Note that your result does not depend on λ_2 or λ_3, nor would it depend on any higher-order term in (8.45). [Use a mathematical software package if available.]

Problem 8.10: The Lotka-Volterra model of competition between two species in discrete-time can be described by the recursion equations

$$n_1(t+1) = n_1(t) + r_1\,n_1(t)\left(1 - \frac{n_1(t) + \alpha_{12}\,n_2(t)}{K_1}\right),$$

$$n_2(t+1) = n_2(t) + r_2\,n_2(t)\left(1 - \frac{n_2(t) + \alpha_{21}\,n_1(t)}{K_2}\right).$$

(a) Determine the four equilibria and confirm that $\hat{n}_1 = (K_1 - \alpha_{12}\,K_2)/(1 - \alpha_{12}\,\alpha_{21})$ and $\hat{n}_2 = (K_2 - \alpha_{21}\,K_1)/(1 - \alpha_{12}\,\alpha_{21})$ is the only equilibrium with both species present. (b) Determine the local stability matrices for the three equilibria in which both species are *not* maintained (stability of the equilibrium with both species present is analyzed in Supplementary Material 8.1). (c) Use Rule P2.26 to find the eigenvalues for these three equilibria. (d) [Challenging] Interpret the stability conditions for these three equilibria using the fact that, near an equilibrium,

$$1 + r_i\left(1 - \frac{\hat{n}_i + \alpha_{ij}\,\hat{n}_j}{K_i}\right)$$

is the number of offspring per parent of species i, which can never be negative.

Problem 8.11: The Lotka-Volterra model of competition in continuous time can be described by the differential equations

$$\frac{dn_1}{dt} = r_1\,n_1(t)\left(1 - \frac{n_1(t) + \alpha_{12}\,n_2(t)}{K_1}\right),$$

$$\frac{dn_2}{dt} = r_2\,n_2(t)\left(1 - \frac{n_2(t) + \alpha_{21}\,n_1(t)}{K_2}\right).$$

(a) Show that $\hat{n}_1 = (K_1 - \alpha_{12}K_2)/(1 - \alpha_{12}\alpha_{21})$ and $\hat{n}_2 = (K_2 - \alpha_{21}K_1)/(1 - \alpha_{12}\alpha_{21})$ is an equilibrium of the continuous-time model as well as the discrete-time model (Problem 8.10) and that it is the only equilibrium with both species present. (b) Determine the local stability matrix for this equilibrium. How does this matrix differ from the discrete-time equivalent (S8.1.5) of Supplemental Material 8.1? (c) Use the Routh-Hurwitz conditions to determine when the equilibrium with both species is stable. Contrast these conditions to the discrete-time model analyzed in Supplemental Material 8.1.

Problem 8.12: In Box 2.5, we described the model of HIV infection developed by Blower et al. (2000) to predict changes in HIV incidence in the San Francisco community of gay males. Before treatment begins, the model describing changes to the number of uninfected (X) and infected (Y) individuals simplifies to

$$\frac{dX}{dt} = \pi - c\,\beta\,\frac{Y}{Y + X}X - \mu X,$$

$$\frac{dY}{dt} = c\,\beta\,\frac{Y}{Y + X}X - (\mu + \nu)Y,$$

where π describes the immigration rate of males, c is the number of partners per male, β is the rate of infection per sexual contact with an infected male, μ is the normal mortality or emigration rate, and ν is the additional mortality rate suffered by infected individuals. (a) Determine the equilibria of this model and describe the conditions under which each equilibrium is biologically feasible. (b) Use a local stability analysis to determine the conditions under which HIV will spread when rare within a population. (c) Use the trace and determinant conditions (Rule

P2.25) to determine whether the equilibrium with both infected and uninfected individuals present is stable. Would you expect to observe cycling around this equilibrium?

Problem 8.13: Show that the general Routh-Hurwitz conditions (8.2.12) of Box 8.2 are equivalent to the conditions for a 3×3 matrix (8.2.6) of Box 8.2.

Problem 8.14: In this problem, we analyze a nonlinear crab-anemone model that allows the number of juvenile crabs to depend on the current population size and incorporates competition among these juveniles for available sea anemones. The recursions for this model are

$$B(t+1) = B(t) - \mu_B B(t) + (B(t) + R(t) + G(t))\, b\, p_B\, (1 - B(t)/N_B),$$

$$R(t+1) = R(t) - \mu_R R(t) + (B(t) + R(t) + G(t))\, b\, p_R\, (1 - R(t)/N_R),$$

$$G(t+1) = G(t) - \mu_G G(t) + (B(t) + R(t) + G(t))\, b\, p_G\, (1 - G(t)/N_G),$$

(see Problem 7.8). (a) Write down the equilibrium conditions for these recursions and solve for the mortality rates (do *not* solve them for the equilibrium). That is, determine what μ_B, μ_R, and μ_G are as functions of the other parameters and $\hat{B}$, $\hat{R}$, and $\hat{G}$. (b) Using the data of Baeza and Stotz (2001), solve for the mortality rates relative to the per capita birth rate b. Set $N_B = 69$, $N_R = 115$, $N_G = 179$, $p_B \approx 0.14$, $p_G \approx p_R \approx 0.43$, and assume that the observed numbers of occupied anemones in Figure 7.10a are near the equilibrium so that $\hat{B} \approx 1$, $\hat{R} \approx 33$, $\hat{G} \approx 38$. (c) Compare your predictions from (b) to the predictions $\mu_B \approx 0.14b$, $\mu_R \approx 0.0092b$, and $\mu_G \approx 0.0090b$ obtained from model (iii) in Section 7.4. Remember that b represents the per capita birth rate in this model but the total birth rate for the whole population in (7.33a).

Problem 8.15: [Challenging]. Consider model (8.36) but where the fitness expressions $W_{ij}(N)$ are left unspecified. Here you will again focus on the equilibrium with only the resident allele A. (a) Set $\hat{p} = 1$ in equation (8.37) and show that the equilibrium population size must then satisfy the equation $W_{AA}(\hat{N}) = 1$. Provide a biological interpretation of this result. (b) Even though we do not have an explicit expression for $\hat{N}$, we can still use the conditions $\hat{p} = 1$ and $W_{AA}(\hat{N}) = 1$ to simplify the Jacobian matrix at the equilibrium with allele A fixed. Show that this matrix can be written as

$$\mathbf{J} = \begin{pmatrix} W_{AA}(\hat{N}) + \hat{N}\left.\dfrac{dW_{AA}}{dN}\right|_{\hat{N}} & -2\hat{N}\,(W_{Aa}(\hat{N}) - W_{AA}(\hat{N})) \\ 0 & W_{Aa}(\hat{N})/W_{AA}(\hat{N}) \end{pmatrix}.$$

(c) What are the eigenvalues of the Jacobian matrix in (b) (use the fact that $W_{AA}(\hat{N}) = 1$ to simplify these expressions as much as possible). (d) Provide an argument that, as with model (8.36) of the text, these two eigenvalues correspond, respectively, to a condition for ecological stability of the equilibrium with allele A fixed and to a condition for the stability of the genetic equilibrium with the population size fixed.

Further Reading

For a more complete mathematical treatment of population genetics models consult

- Bürger, R. 2000. *The Mathematical Theory of Selection, Recombination, and Mutation*. Wiley, Chichester.
- Christiansen, F. B. 1999. *Population Genetics of Multiple Loci*. Wiley, Chichester.
- Crow, J. F., and M. Kimura. 1970. *An Introduction to Population Genetics Theory*. Harper & Row, New York.
- Hartl. D. L., and A. G. Clark. 1989. *Principles of Population Genetics*. Sinauer Associates, Sunderland. Mass.

References

Baeza, J. A., and W. Stotz. 2003. Host-use and selection of differently colored sea anemones by the symbiotic crab *Allopetrolisthes spinifrons*. *J. Exp. Mar. Biol. Ed.* 284:25–39.

Barton, N. H. 1995. A general model for the evolution of recombination. *Genet. Res.* 65:123–144.

Barton, N. H., and M. Turelli. 1991. Natural and sexual selection on many loci. *Genetics* 127:229–255.

Blower, S. M., H. B. Gershengorn, and R. M. Grant. 2000. A tale of two futures: HIV and antiretroviral therapy in San Francisco. *Science* 287:650–654

Bürger, R. 2000. *The Mathematical Theory of Selection, Recombination, and Mutation*. Wiley, Chichester.

Christiansen, F. B. 2000. *Population Genetics of Multiple Loci*. Wiley, Chichester.

Crow, J. F., and M. Kimura. 1965. Evolution in sexual and asexual populations. *Amer. Nat.* 99:439–450.

Crow, J. F., and M. Kimura. 1970. *An Introduction to Population Genetics Theory*. Harper & Row, New York.

Edelstein-Keshet, L. 1988. *Mathematical Models in Biology*. McGraw-Hill, New York.

Hastings, A. 1997. *Population Biology: Concepts and Models*. Springer, New York.

Karlin, S. 1975. General two-locus selection models: some objectives, results and interpretations. *Theoretical Population Biology* 7:364–398.

Kirkpatrick, M., T. Johnson, and N. Barton. 2002. General models of multilocus evolution. *Genetics* 161:1727–1750.

Levin, S., and D. Pimentel. 1981. Selection of intermediate rates of increase in parasite-host systems. *Am. Nat.* 117:308–315.

Nowak, M. A., and R. M. May. 1994. Superinfection and the evolution of parasite virulence. *Proc. R. Soc. London, Ser. B* 255:81–89.

Otto, S. P. 1994. The role of deleterious and beneficial mutations in the evolution of ploidy levels. *Lectures on Mathematics in the Life Sciences* 25:69–96.

Otto, S. P., and D. B. Goldstein. 1992. Recombination and the evolution of diploidy. *Genetics* 131:745–751.

Otto, S. P., and J. Marks. 1996. Mating systems and the evolutionary transition between haploidy and diploidy. *Biol. J. Linnean Soc.* 57:197–218.

Perrot, V., S. Richerd, and M. Valero. 1991. Transition from haploidy to diploidy. *Nature* 351:315–317.

Phillips, A. N. 1996. Reduction of HIV concentration during acute infection: Independence from a specific immune response. *Science* 271:497–499.

Turelli, M., and N. H. Barton. 1990. Dynamics of polygenic characters under selection. *Theor. Popul. Biol.* 38:1–57.

CHAPTER 9

General Solutions and Transformations—Models with Multiple Variables

Chapter Goals:

- To derive the general solution to linear models involving multiple variables
- To examine methods for solving nonlinear models with multiple variables
- To introduce transformations for nonlinear models with multiple variables
- To obtain approximate general solutions using a separation of time scales

Chapter Concepts:

- Long-term dynamics
- Proportional transformation
- Separation of time scales
- Quasi-linkage equilibrium

9.1 Introduction

In this chapter, we describe recipes and techniques for finding general solutions of models with multiple variables. As we shall see, there is a recipe for finding the general solution to any linear model, regardless of the number of variables. General solutions can also sometimes be obtained for nonlinear models using transformations, but there is no single recipe. When a general solution cannot be found using transformations, we can sometimes use another technique known as a *separation of time scales* to approximate the dynamics. Although this technique is only approximate, it often provides a more thorough understanding of a model's behavior.

In section 9.2 we derive the general solution to linear models with multiple variables. Section 9.3 then considers nonlinear models with multiple variables. There we revisit the use of transformations (Chapter 6) in the context of models with multiple variables. Then we describe how a separation of times scales can be used to approximate the general solution.

9.2 Linear Models Involving Multiple Variables

Linear equations involving multiple variables commonly arise in models of age- or class-structured populations (Chapters 7 and 10), and they are also used to approximate the behavior of nonlinear models in the vicinity of equilibria (Chapter 8). Here, we derive the exact general solution to linear models with multiple variables. We focus on discrete-time models and provide the general solution for continuous-time models in Box 9.3 below. General solutions allow us to generate exact predictions when accuracy matters, and they also provide greater insight into why a model behaves the way that it does.

Consider a set of linear recursion equations describing the state of a population in the next generation, $n_i(t+1)$, as a function of its state in the current generation, $n_i(t)$. As with single-variable models, linear multivariable models have one of two forms. The model might involve d equations in d variables without constant terms; i.e.,

$$\begin{aligned} n_1(t+1) &= m_{11}\,n_1(t) + m_{12}\,n_2(t) + \cdots + m_{1d}\,n_d(t), \\ n_2(t+1) &= m_{21}\,n_1(t) + m_{22}\,n_2(t) + \cdots + m_{2d}\,n_d(t), \\ &\vdots \\ n_d(t+1) &= m_{d1}\,n_1(t) + m_{d2}\,n_2(t) + \cdots + m_{dd}\,n_d(t). \end{aligned} \tag{9.1}$$

In Chapter 7, we explored several models of this form, such as model (7.27) describing the number of occupied sea anemones of different colors.

Alternatively, the model might be *affine* and involve d equations in d variables with the addition of constant terms. As in the one-variable case, we can apply Steps 1–3 of Recipe 6.1 to every variable in a multiple-variable affine model, transforming it into the form (9.1) (see Problem 9.2). For example, we can write a discrete-time affine model involving two variables as

$$\begin{aligned} n_1(t+1) &= a\,n_1(t) + b\,n_2(t) + \alpha, \\ n_2(t+1) &= c\,n_1(t) + d\,n_2(t) + \beta. \end{aligned} \tag{9.2}$$

The equilibrium of (9.2) is

$$\hat{n}_1 = \frac{b\beta + (1-d)\,\alpha}{(1-a)(1-d) - b\,c} \text{ and } \hat{n}_2 = \frac{(1-a)\beta + c\alpha}{(1-a)(1-d) - b\,c}.$$

If we transform the system into the new variables $\delta_1 = n_1 - \hat{n}_1$ and $\delta_2 = n_2 - \hat{n}_2$, describing the distance of the system from this equilibrium, the recursions then become

$$\begin{aligned} \delta_1(t+1) &= a\delta_1(t) + b\delta_2(t), \\ \delta_2(t+1) &= c\delta_1(t) + d\delta_2(t) \end{aligned} \tag{9.3}$$

(Recipe 6.1). These recursions are linear and no longer contain constant terms. Therefore, they can be analyzed in the same way that we analyze nonaffine models (9.1).

The recursion equations (9.1) can be written in matrix notation as $\vec{n}(t+1) = \mathbf{M}\,\vec{n}(t)$, where

$$\vec{n} = \begin{pmatrix} n_1 \\ n_2 \\ \vdots \\ n_d \end{pmatrix} \quad \text{and} \quad \mathbf{M} = \begin{pmatrix} m_{11} & m_{12} & \cdots & m_{1d} \\ m_{21} & m_{22} & \cdots & m_{2d} \\ \vdots & \vdots & \vdots & \vdots \\ m_{d1} & m_{d2} & \cdots & m_{dd} \end{pmatrix}. \tag{9.4}$$

Our goal is to find the *general solution* to such matrix recursions, describing the state of the system, $\vec{n}(t)$, at any time t as a function of its initial state $\vec{n}(0)$. To find the general solution, we can apply the matrix recursion repeatedly as we did in the one-variable case:

$$\vec{n}(t) = \mathbf{M}\,\vec{n}(t-1) = \mathbf{M}^2\,\vec{n}(t-2) = \cdots = \mathbf{M}^t\,\vec{n}(0). \tag{9.5}$$

This gives the general solution, and it has the same form as the one-variable model of exponential growth: $n(t) = R^t\,n(0)$. Now $\mathbf{M}$ takes the place of the reproductive factor R, and the vector $\vec{n}$ takes the place of the single variable n. When the elements of $\mathbf{M}$ are just numbers, then $\mathbf{M}^t$ can be calculated relatively easily (e.g., try calculating $\mathbf{M}^3$ where $\mathbf{M} = \left(\begin{smallmatrix} 1 & 2 \\ 2 & 5 \end{smallmatrix}\right)$). When the elements of $\mathbf{M}$ are parameters, however, the calculations become much more cumbersome (e.g., try calculating $\mathbf{M}^3$ where $\mathbf{M} = \left(\begin{smallmatrix} a & b \\ c & d \end{smallmatrix}\right)$). Therefore, to make the general solution (9.5) more useful, we need a simpler way to calculate powers of the matrix $\mathbf{M}$.

To begin, we first recall the equation defining the right eigenvectors of the matrix **M** and their associated eigenvalues (equation (P2.4a)):

$$\mathbf{M}\,\vec{u}_i = \lambda_i \vec{u}_i. \tag{9.6}$$

With the exception of "defective" matrices (see Primer 2) any $d \times d$ matrix will have d right eigenvectors $\vec{u}_i$, each of which is associated with an eigenvalue λ_i. Therefore, we can use matrix notation to write all d equations for the eigenvectors and eigenvalues of **M** as

$$\mathbf{M}(\vec{u}_1 \quad \vec{u}_2 \quad \cdots \quad \vec{u}_d) = (\lambda_1 \vec{u}_1 \quad \lambda_2 \vec{u}_2 \quad \cdots \quad \lambda_d \vec{u}_d) \tag{9.7a}$$

or

$$\mathbf{M}\,\mathbf{A} = \mathbf{A}\,\mathbf{D}. \tag{9.7b}$$

Here **A** is a $d \times d$ matrix whose *columns* are the right eigenvectors of **M**: i.e.,

$$\mathbf{A} = (\vec{u}_1 \quad \vec{u}_2 \quad \cdots \quad \vec{u}_d) = \begin{pmatrix} u_{11} & u_{12} & \cdots & u_{1d} \\ u_{21} & u_{22} & \cdots & u_{2d} \\ \vdots & \vdots & \vdots & \vdots \\ u_{d1} & u_{d2} & \cdots & u_{dd} \end{pmatrix} \tag{9.8a}$$

and **D** is a diagonal matrix with eigenvalues λ_i arrayed along the diagonal, in the same order as their associated eigenvectors were arrayed in the matrix **A**; i.e.,

$$\mathbf{D} = \begin{pmatrix} \lambda_1 & 0 & \cdots & 0 \\ 0 & \lambda_2 & \cdots & 0 \\ \vdots & \vdots & \vdots & \vdots \\ 0 & 0 & \cdots & \lambda_d \end{pmatrix} \tag{9.8b}$$

(convince yourself that multiplying out the right-hand side of (9.7b) does, in fact, give the right-hand side of (9.7a)). We can then multiply equation (9.7b) on the right by $\mathbf{A}^{-1}$ (the inverse of **A**; Primer 2, section P2.6) to obtain

$$\mathbf{M} = \mathbf{A}\,\mathbf{D}\,\mathbf{A}^{-1}. \tag{9.9}$$

(Recall from Primer 2 that a matrix times its inverse, e.g., $\mathbf{A}\,\mathbf{A}^{-1}$, equals the identity matrix **I**, and that multiplying a matrix by **I** has no effect, so that $\mathbf{M}\,\mathbf{A}\,\mathbf{A}^{-1} = \mathbf{M}$.)

Equation (9.9) gives us a way of writing the matrix **M** in terms of its right eigenvectors (the columns of **A**), its eigenvalues (the entries of **D**), and the inverse matrix $\mathbf{A}^{-1}$. Interestingly, the rows of matrix $\mathbf{A}^{-1}$ are nothing other than *left* eigenvectors of the matrix **M**. To see this, first recall the equation (P2.12) that defines the left eigenvectors: $\vec{v}_i^{\mathrm{T}}\mathbf{M} = \lambda_i \vec{v}_i^{\mathrm{T}}$. Now, if we start by defining $\mathbf{A}^{-1}$ as a $d \times d$ matrix whose *rows* are the left eigenvectors of **M**, then all d of these equations can be written in matrix notation as

$$\mathbf{A}^{-1}\mathbf{M} = \mathbf{D}\,\mathbf{A}^{-1}. \tag{9.10}$$

If we multiply both sides of equation (9.10) on the left by **A**, we once again get equation (9.9). Thus, we arrive at the same result if and only if the columns of **A** contain the right eigenvectors and the rows of $\mathbf{A}^{-1}$ contain the left eigenvectors of the original matrix **M**. (Although the lengths of eigenvectors are typically arbitrary, setting the left eigenvectors to the rows of $\mathbf{A}^{-1}$ constrains their lengths in a manner that depends on the lengths chosen for the right eigenvectors in the columns of **A**.)

By rewriting the matrix **M** in terms of its eigenvalues and their right and left eigenvectors, we can greatly simplify the general solution (9.5). Substituting (9.9) into (9.5), we obtain

$$\vec{n}(t) = \mathbf{M}^t\,\vec{n}(0) = (\mathbf{A}\,\mathbf{D}\,\mathbf{A}^{-1})^t\,\vec{n}(0). \tag{9.11}$$

Expanding $(\mathbf{A}\,\mathbf{D}\,\mathbf{A}^{-1})^t$ is straightforward. Multiplying $\mathbf{A}\,\mathbf{D}\,\mathbf{A}^{-1}$ by itself once, we get $\mathbf{A}\,\mathbf{D}\,\mathbf{A}^{-1}\mathbf{A}\,\mathbf{D}\,\mathbf{A}^{-1} = \mathbf{A}\,\mathbf{D}\,\mathbf{I}\,\mathbf{D}\,\mathbf{A}^{-1} = \mathbf{A}\,\mathbf{D}^2\,\mathbf{A}^{-1}$. We can repeat this operation any number of times, and each time the exponent of **D** just increases by one. Thus, we can rewrite (9.11) as

$$\vec{n}(t) = \mathbf{A}\,\mathbf{D}^t\,\mathbf{A}^{-1}\,\vec{n}(0). \tag{9.12}$$

The tremendous advantage of (9.12) is that, while $\mathbf{M}^t$ is difficult to compute, $\mathbf{D}^t$ is easy to compute; it is just the diagonal matrix with each eigenvalue raised to the tth power:

$$\mathbf{D}_t = \begin{pmatrix} \lambda_1^t & 0 & \cdots & 0 \\ 0 & \lambda_2^t & \cdots & 0 \\ \vdots & \vdots & \vdots & \vdots \\ 0 & 0 & \cdots & \lambda_d^t \end{pmatrix} \tag{9.13}$$

Now imagine what happens to (9.12) after a long period of time. If all of the eigenvalues are less than one in magnitude, then the diagonal matrix shrinks toward a matrix of zeros, causing the vector $\vec{n}(t)$ to approach the origin. If the absolute value of any of the eigenvalues of **M** is greater than one, however, then the system will grow away from the origin as time passes because at least one of the elements of $\mathbf{D}^t$ grows. Only if there is an eigenvalue equal to one, and only if this is the largest eigenvalue, will the system equilibrate at a point other than zero or infinity.

With the general solution (9.12) in hand, we can also develop an approximate general solution by focusing only on the leading eigenvalue and its associated eigenvector (Box 9.1). This method is commonly used in demographic models to predict the future state of a population that is structured into age or stage classes (Chapter 10). The method works wells when one eigenvalue is much larger in magnitude than any other eigenvalue, but it fails when the leading eigenvalue is close or equal in magnitude to any of the other eigenvalues. In such cases the exact general solution should be used.

> The *long-term dynamics* of a linear model with multiple variables is dominated by its leading eigenvalue.

Equation (9.12) also gives some insight into why local stability analyses work (Chapter 8). In a local stability analysis, we approximate a system of nonlinear equations with a set of linear equations that describes the dynamics of a model near an equilibrium. By linearizing the recursions, we obtain a matrix equation of the form $\vec{n}(t+1) = \mathbf{M}\,\vec{n}(t)$, where $\vec{n}$ represents a vector of perturbations from the equilibrium (see (8.35)). Because we can rewrite this matrix equation as $\vec{n}(t) = \mathbf{A}\,\mathbf{D}^t\,\mathbf{A}^{-1}\vec{n}(0)$, we can now understand exactly why the equilibrium is stable if and only if every eigenvalue is less than one in magnitude—only then will $\mathbf{D}^t$ shrink, causing the perturbations to disappear over time. Furthermore, the general solution describes exactly what happens when the eigenvalues are complex, and it explains why complex eigenvalues result in oscillatory behavior (Box 9.2).

We now summarize all of the above results in a single recipe:

Recipe 9.1
Solving Linear Recursion Equations Involving Multiple Variables

Step 1: Write the d equations in matrix form, giving a $d \times d$ matrix $\mathbf{M}$. Order the variables in a consistent way such that the jth row of $\mathbf{M}$ describes the recursion equation for the jth variable, and the jth column of $\mathbf{M}$ describes how each recursion equation depends on the jth variable.

Step 2: Determine the d eigenvalues of $\mathbf{M}$.

Step 3: Make a diagonal matrix $\mathbf{D}$ with one eigenvalue in each of the diagonal positions.

Step 4: Determine the right eigenvector associated with each eigenvalue.

Step 5: Make a matrix $\mathbf{A}$ whose columns are the eigenvectors (placed in the same order as the eigenvalues in matrix $\mathbf{D}$).

Step 6: The general solution of the linear recursion equations is then

$$\vec{n}(t) = \mathbf{A}\,\mathbf{D}^t\,\mathbf{A}^{-1}\,\vec{n}(0).$$

Before leaving the topic, it is worth considering a graphical interpretation of the above recipe. As mentioned in Primer 2, matrix multiplication by $\mathbf{M}$ can be seen as an operation that moves a vector $\vec{n}(t)$ to a new vector $\vec{n}(t+1)$. Most of the time, matrix multiplication rotates and stretches vectors (e.g., Figures P2.3 and P2.7). For certain special vectors, however, the matrix does not rotate the vector, but only stretches or shrinks it. These special vectors are the eigenvectors of $\mathbf{M}$. Now, imagine what would happen if we used the right eigenvectors as the axes along which we tracked the dynamics of a model, instead of the original axes (e.g., instead of axes describing the population size of each species). In this new coordinate system, the dynamics would appear simple: any point is just stretched or shrunk along the ith axis by a factor λ_i in each generation (Figure 9.1). But how do we accomplish this change of coordinate system? The answer involves the matrices $\mathbf{A}$ and $\mathbf{A}^{-1}$.

We can think of the matrices $\mathbf{A}$ and $\mathbf{A}^{-1}$ as *transformation* matrices. In general, transformation matrices can be used to change from one set of coordinate axes to another. Specifically, the matrix $\mathbf{A}^{-1}$ transforms a point $\vec{n}(t)$ in the original coordinate system to a point $\vec{y}(t)$ in the new coordinate system defined by the right eigenvectors; that is, $\vec{y}(t) = \mathbf{A}^{-1}\,\vec{n}(t)$. Similarly, the reverse transformation from the new to the original coordinate system can be accomplished by multiplying by $\mathbf{A}$; that is, $\vec{n}(t) = \mathbf{A}\,\vec{y}(t)$. We now apply these transformations to a system of recursion equations.

Box 9.1: Long-Term Dynamics and the Role of the Leading Eigenvalue

In this box, we describe an approximation to the dynamics of a linear discrete-time model in multiple variables. Consider a transition matrix **M** whose eigenvalues are known and placed along the diagonal of a matrix **D**, and whose eigenvectors are known and placed in the columns of a matrix **A**. We are free to place the eigenvalues in **D** in any order (as long as we place their associated eigenvectors in the same order in **A**), so let us place the leading eigenvalue (the eigenvalue with the largest magnitude, λ_1) in the first row and first column of **D**. To simplify the presentation further, we adjust the length of the eigenvector associated with the leading eigenvalue $\vec{u}_1$, so that its elements sum to one. As discussed in Primer 2, eigenvectors point in a particular direction but can be of any length, and therefore we are free to choose whatever length is convenient.

We start by factoring out λ_1^t from $\mathbf{D}^t$:

$$\mathbf{D}^t = \lambda_1^t \begin{pmatrix} 1 & 0 & \cdots & 0 \\ 0 & \left(\frac{\lambda_2}{\lambda_1}\right)^t & \cdots & 0 \\ \vdots & \vdots & \vdots & \vdots \\ 0 & 0 & \cdots & \left(\frac{\lambda_d}{\lambda_1}\right)^t \end{pmatrix}. \tag{9.1.1}$$

As long as λ_1 is larger in magnitude than all other eigenvalues, (λ_i/λ_1) will be less than one in magnitude and $(\lambda_i/\lambda_1)^t$ will approach zero over time. Consequently, over time, $\mathbf{D}^t$ becomes more and more similar to

$$\widetilde{\mathbf{D}}^t = \lambda_1^t \begin{pmatrix} 1 & 0 & \cdots & 0 \\ 0 & 0 & \cdots & 0 \\ \vdots & \vdots & \vdots & \vdots \\ 0 & 0 & \cdots & 0 \end{pmatrix}. \tag{9.1.2}$$

Whether or not $\widetilde{\mathbf{D}}^t$ provides a sufficiently accurate approximation depends on how much time has passed and on the magnitude of the other eigenvalues relative to the leading eigenvalue. If, for example, the eigenvalues of a 4×4 matrix are 3/2, –3/2, 1/3, and $-1/2$, two eigenvalues are equally large in magnitude ($\lambda = 3/2$ and $-3/2$), and the approximation (9.1.2) should be avoided in favor of the exact general solution (9.12) (see Appendix 3 for a description of which matrices can and cannot have more than one eigenvalue equal to the leading eigenvalue). On the other hand, if the eigenvalues are 5/2, $-3/2$, 1/3, and $-1/2$, then after only five time steps (9.1.1) becomes

$$\mathbf{D}^3 = \left(\frac{5}{2}\right)^5 \begin{pmatrix} 1 & 0 & 0 & 0 \\ 0 & \left(-\frac{3}{5}\right)^5 & 0 & 0 \\ 0 & 0 & \left(\frac{2}{15}\right)^5 & 0 \\ 0 & 0 & 0 & \left(-\frac{1}{5}\right)^5 \end{pmatrix} = \left(\frac{5}{2}\right)^5 \begin{pmatrix} 1 & 0 & 0 & 0 \\ 0 & -0.078 & 0 & 0 \\ 0 & 0 & 0.000042 & 0 \\ 0 & 0 & 0 & 0.00032 \end{pmatrix},$$

(continued)

Box 9.1 *(continued)*

which is very close to (9.1.2). In this case, it would be reasonable to approximate $\mathbf{D}^t$ with $\widetilde{\mathbf{D}}^t$ unless you needed short-term or extremely accurate predictions.

Using (9.1.2), we can approximate the general solution (9.12) by

$$\vec{n}(t) \approx \mathbf{A}\,\widetilde{\mathbf{D}}^t\,\mathbf{A}^{-1}\,\vec{n}(0). \tag{9.1.3}$$

The first two terms in this equation can be multiplied together to give

$$\mathbf{A}\,\widetilde{\mathbf{D}}^t = \lambda_1^t \begin{pmatrix} u_{11} & 0 & \cdots & 0 \\ u_{21} & 0 & \cdots & 0 \\ \vdots & \vdots & \vdots & \vdots \\ u_{d1} & 0 & \cdots & 0 \end{pmatrix}. \tag{9.1.4}$$

Let $\widetilde{v}_{ij}$ stand for the element in the i_{th} row and j_{th} column of $\mathbf{A}^{-1}$. Multiplying $\mathbf{A}\,\widetilde{\mathbf{D}}^t$ by $\mathbf{A}^{-1}$ on the right then gives

$$\mathbf{A}\,\widetilde{\mathbf{D}}^t\,\mathbf{A}^{-1} = \lambda_1^t \begin{pmatrix} u_{11}\widetilde{v}_{11} & u_{11}\widetilde{v}_{12} & \cdots & u_{11}\widetilde{v}_{1d} \\ u_{21}\widetilde{v}_{11} & u_{21}\widetilde{v}_{12} & \cdots & u_{21}\widetilde{v}_{1d} \\ \vdots & \vdots & \vdots & \vdots \\ u_{d1}\widetilde{v}_{11} & u_{d1}\widetilde{v}_{12} & \cdots & u_{d1}\widetilde{v}_{1d} \end{pmatrix}. \tag{9.1.5}$$

Finally, we can multiply this matrix by the vector describing the initial state of the population, $\vec{n}(0)$, to get an approximation for $\vec{n}(t)$:

$$\vec{n}(t) \approx \mathbf{A}\,\widetilde{\mathbf{D}}^t\,\mathbf{A}^{-1}\,\vec{n}(0) = \lambda_1^t \begin{pmatrix} u_{11}(\widetilde{v}_{11}\,n_1(0) + \widetilde{v}_{12}\,n_2(0) + \cdots \widetilde{v}_{1d}\,n_d\,(0)) \\ u_{21}(\widetilde{v}_{11}\,n_1(0) + \widetilde{v}_{12}\,n_2(0) + \cdots \widetilde{v}_{1d}\,n_d\,(0)) \\ \vdots \\ u_{d1}(\widetilde{v}_{11}\,n_1(0) + \widetilde{v}_{12}\,n_2(0) + \cdots \widetilde{v}_{1d}\,n_d\,(0)) \end{pmatrix}. \tag{9.1.6}$$

This approximation can be rewritten as

$$\vec{n}(t) \approx \lambda_1^t\,c\,\vec{u}_1, \tag{9.1.7}$$

where $c = \widetilde{v}_{11}\,n_1(0) + \widetilde{v}_{12}\,n_2(0) + \ldots + \widetilde{v}_{1d}\,n_d(0) = \vec{\widetilde{v}}_1\,\vec{n}(0)$. The constant c can be thought of as the initial size of the system, adjusted by the left eigenvector $\vec{\widetilde{v}}$ (which is found in the first row of $\mathbf{A}^{-1}$; see Chapter 10).

The approximation (9.1.7) indicates that the system will eventually grow at a rate equal to λ_1^t. Furthermore, because we have adjusted $\vec{u}_1$ so that its elements sum to one, the proportion of the system that is of type i is given by the ith element of $\vec{u}_1$ (i.e., by the ith element of the right eigenvector of $\mathbf{M}$ associated with the eigenvalue λ_1). This result helps us to understand why the leading eigenvalue determines the stability of an equilibrium: eventually it will come to dominate the recursions. It also forms the basis for important results in demography, as described in Chapter 10.

Box 9.2: General Solution of a Discrete-Time Linear Model with Complex Eigenvalues

In equation (9.12), we showed that the general solution for a discrete-time linear model can be written as $\vec{n}(t) = \mathbf{A}\,\mathbf{D}^t\mathbf{A}^{-1}\,\vec{n}(0)$. How do we interpret this solution if the eigenvalues are complex? Here, we show how the solution for a two-variable model can be written in terms of sine and cosine functions containing only real terms. To keep things general, we will work with the two-dimensional matrix $\mathbf{M} = \left(\begin{smallmatrix} a & b \\ c & d \end{smallmatrix}\right)$. The eigenvalues and associated eigenvectors of this matrix are

$$\lambda_1 = \frac{a + d + \sqrt{(a-d)^2 + 4bc}}{2} \quad \text{with } \vec{u} = \left(1, \frac{-a + d + \sqrt{(a-d)^2 + 4bc}}{2b}\right), \tag{9.2.1a}$$

$$\lambda_2 = \frac{a + d - \sqrt{(a-d)^2 + 4bc}}{2} \quad \text{with } \vec{u} = \left(1, \frac{-a + d - \sqrt{(a-d)^2 + 4bc}}{2b}\right). \tag{9.2.1b}$$

If $(a-d)^2 + 4bc$ is positive, then the eigenvalues and associated eigenvectors are real, and (9.12) can be multiplied out to describe the system at any future point in time without any difficulties. If $(a-d)^2 + 4bc$ is negative, however, then the eigenvalues and eigenvectors are complex numbers involving $i = \sqrt{-1}$. How can we interpret these complex numbers? And how do we raise a complex eigenvalue to the tth power to evaluate $\mathbf{D}^t$?

Insights from geometry help. When an eigenvalue is complex, we can write it as

$$\lambda_1 = \alpha + \beta i$$

where

$$\alpha = \frac{a + d}{2} \text{ and } \beta = \frac{\sqrt{-(a-d)^2 - 4bc}}{2}.$$

By definition, α and β are real numbers, and βi is the imaginary part of the eigenvalue. Using Figure 9.2.1 as a guide, we can also write $\alpha + \beta i$ in terms of sine and cosine functions. This figure illustrates that

$$\alpha + \beta i = R(\cos(\theta) + i\sin(\theta)),$$

where the magnitude (i.e., absolute value) of the complex number is $R = \sqrt{\alpha^2 + \beta^2}$, which simplifies to $\sqrt{a\,d - b\,c}$, and θ is the angle between the number plotted on the complex plane and the horizontal axis;

$$\theta = \arctan\left(\frac{\beta}{\alpha}\right) = \arctan\left(\frac{\sqrt{-(a-d)^2 - 4bc}}{a + d}\right).$$

We have already encountered $(\cos\theta + i\sin\theta)$ in Euler's formula (Box 7.4). Euler showed that $(\cos\theta + i\sin\theta) = e^{i\theta}$. Consequently, we can rewrite $\alpha + \beta i$ as $R\,e^{i\theta}$. This is helpful because we

(*continued*)

Box 9.2 *(continued)*

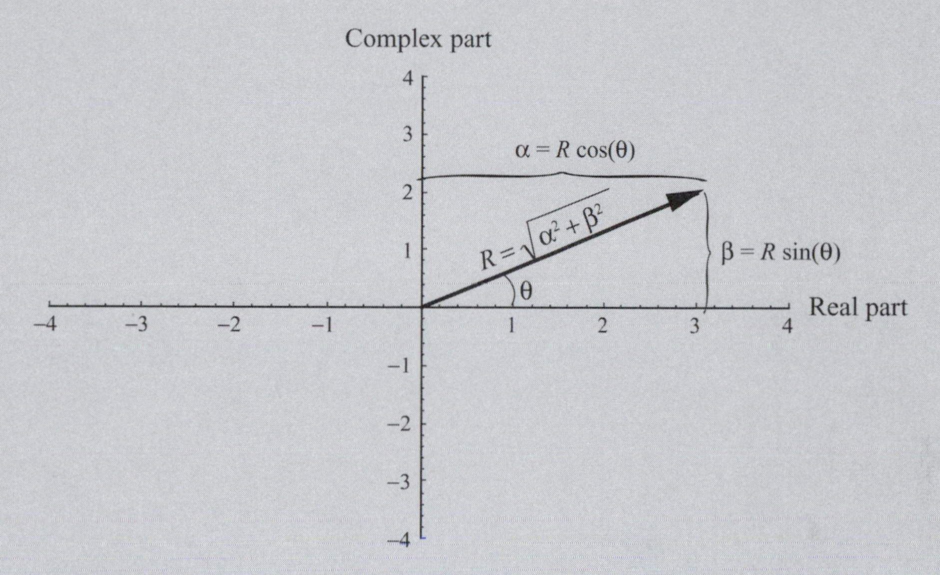

Figure 9.2.1: A complex number represented on the real-complex plane. Any complex number can be written as $\alpha + \beta\, i$ where both α and β are real numbers and $i = \sqrt{-1}$. Any number can thus be represented as a vector on a plot where the real part of the number (α) gives the position along the horizontal axis and the part multiplying i (β) gives the position along the vertical axis. For example, we illustrate the number $3 + 2\,i$. From trigonometry, α must equal $R \cos(\theta)$, where θ is the angle between the vector and the horizontal axis and R is the length of the vector. Similarly, β must equal $R \sin(\theta)$. Thus, we can write $\alpha + \beta\, i$ in terms of sines and cosines as $R \cos(\theta) + R \sin(\theta)\, i$. The angle, θ, can be found using the trigonometric relationship, $\theta = \arctan(\beta/\alpha)$. The total length of the vector (its "magnitude") can be found from the theorem of Pythagoras, $R = \sqrt{\alpha^2 + \beta^2}$. For $3 + 2\,i$, $\theta = \arctan(2/3) = 33.7°$ and $R = \sqrt{2^2 + 3^2} = 3.6$.

can easily raise $R\, e^{i\,\theta}$ to the tth power, to get $\lambda^t = R^t e^{i\theta t}$. We can then apply Euler's formula again to write λ_1^t as $R^t(\cos(\theta t) + i \sin(\theta t))$. We can repeat this procedure for the second eigenvalue as well, the only difference being that the sign of β (and thus the sign of θ) changes.

This procedure allows us to raise a complex number to the tth power, but the eigenvalue still involves a complex number. The beauty of using Euler's transformation is that when we multiply out $\mathbf{A}\,\mathbf{D}^t\,\mathbf{A}^{-1}$ and simplify the answer, we get a real matrix

$$\begin{pmatrix} n_1(t) \\ n_2(t) \end{pmatrix} = R^t \begin{pmatrix} \cos(\theta t) + \dfrac{(a-d)}{2\beta}\sin(\theta t) & \dfrac{b}{\beta}\sin(\theta t) \\ \dfrac{c}{\beta}\sin(\theta t) & \cos(\theta t) - \dfrac{(a-d)}{2\beta}\sin(\theta t) \end{pmatrix} \begin{pmatrix} n_1(0) \\ n_2(0) \end{pmatrix}, \tag{9.2.2}$$

where we have factored out $R^t = (a\,d - b\,c)^{t/2}$, which was present in every term in the matrix. This general solution tells us that the system cycles, with the matrix in (9.2.2) returning to the same value after a period of $\tau = 2\pi/\theta$. At the same time, the system expands or shrinks by a factor R every time step.

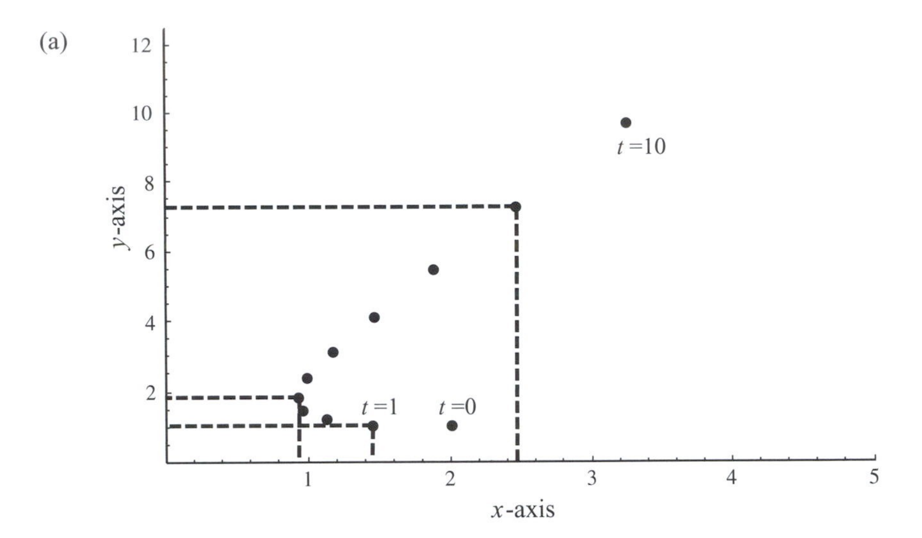

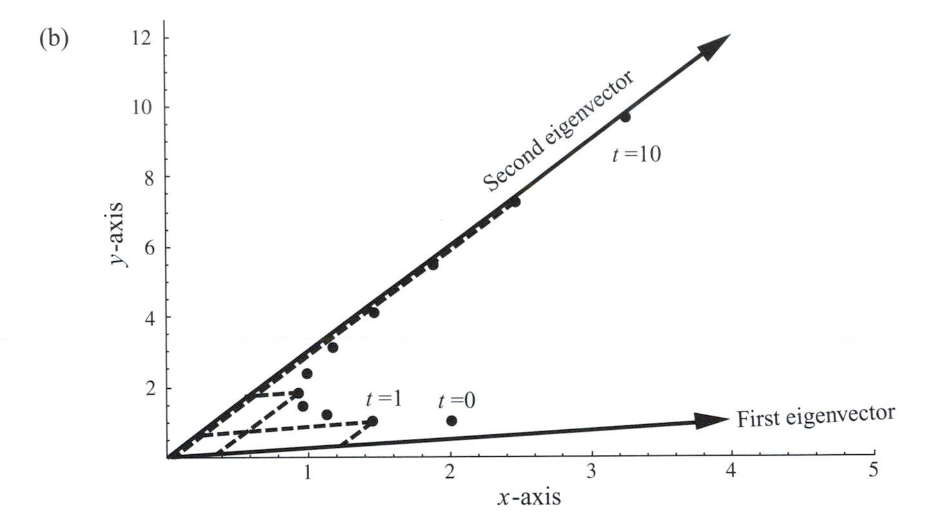

Figure 9.1: Observing dynamics in a new coordinate system based on the eigenvectors. We illustrate how a two-dimensional system, initially at $\vec{n}(t) = \binom{2}{1}$, changes over ten time steps when multiplied by the matrix $\mathbf{M} = \begin{pmatrix} 20/33 & 8/33 \\ -2/11 & 46/33 \end{pmatrix}$. $\mathbf{M}$ has an eigenvector of $\binom{4}{1}$ associated with an eigenvalue of $\lambda = 2/3$ and a second eigenvector of $\binom{4}{12}$ associated with an eigenvalue of $\lambda = 4/3$ (the eigenvectors are drawn as arrows). (a) The position of the system does not change in an obvious fashion when viewed on the standard horizontal and vertical axes. In fact, the position along the horizontal axis initially decreases and then starts to increase after $t = 4$. (b) If we change the axes along which we view the dynamics to the eigenvectors, however, the dynamics are easier to interpret. Along these new axes, the state of the system at time t can be found by following the dashed lines (parallel to the eigenvectors). Each generation, the distance along the first eigenvector shrinks by a factor $\lambda = 2/3$, while the distance along the second eigenvector is stretched by a factor $\lambda = 4/3$. Because the leading eigenvalue is $\lambda = 4/3$, the system eventually moves in the direction of its associated eigenvector, expanding by a factor 4/3 at each time step.

At time $t + 1$, if the state of the system is $\vec{n}(t+1)$ in the original coordinate system (e.g., counts of each species), then the same state can be described in the new coordinate system based on the right eigenvectors as $\vec{y}(t+1) = \mathbf{A}^{-1}\vec{n}(t+1)$. Then, using the recursion $\vec{n}(t+1) = \mathbf{M}\vec{n}(t)$ and the transformations $\vec{y}(t+1) = \mathbf{A}^{-1}\vec{n}(t+1)$ and $\vec{n}(t) = \mathbf{A}\vec{y}(t)$, we get $\vec{y}(t+1) = \mathbf{A}^{-1}\mathbf{M}\vec{n}(t) = \mathbf{A}^{-1}\mathbf{M}\mathbf{A}\vec{y}(t)$. Finally, we can use the fact that $\mathbf{M} = \mathbf{A}\ \mathbf{D}\ \mathbf{A}^{-1}$ to get $\vec{y}(t+1) = \mathbf{A}^{-1}\mathbf{A}\ \mathbf{D}\ \mathbf{A}^{-1}\mathbf{A}\vec{y}(t)$. Remarkably, canceling out the terms that equal the identity matrix ($\mathbf{A}^{-1}\mathbf{A} = \mathbf{A}\mathbf{A}^{-1} = \mathbf{I}$) gives $\vec{y}(t+1) = \mathbf{D}\vec{y}(t)$. Thus, in the new coordinate system, a simple diagonal transition matrix $\mathbf{D}$ describes changes to the system from one generation to the next. This means that the state of the system changes by a constant factor λ_i along each axis in each generation (Figure 9.2). Pretty nice!

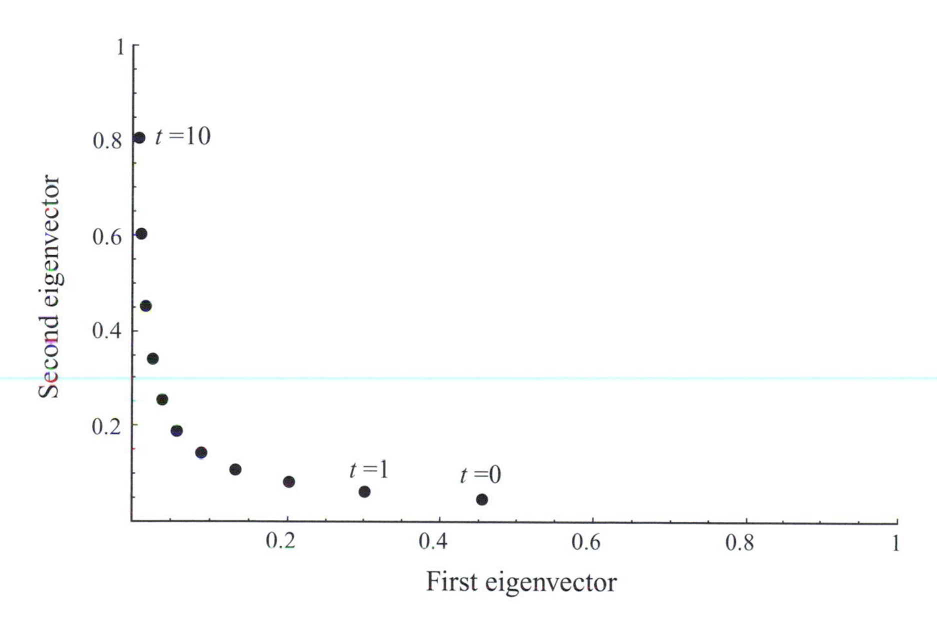

Figure 9.2: Observing dynamics using the eigenvectors as horizontal and vertical axes. Here, we redraw the dynamics illustrated in Figure 9.1 using the eigenvectors as our horizontal and vertical axes. To do this, we must change coordinates using the transformation $\vec{y}(t) = \mathbf{A}^{-1}\vec{n}(t)$, where the eigenvectors are arrayed in columns to get $\mathbf{A} = \begin{pmatrix} 4 & 4 \\ 1 & 12 \end{pmatrix}$, which is inverted to get $\mathbf{A}^{-1} = \begin{pmatrix} 3/11 & -1/11 \\ -1/44 & 1/11 \end{pmatrix}$. At $t = 0$, the initial position is given by the vector $\vec{y}(0) = \mathbf{A}^{-1}\vec{n}(0) = \binom{5/11}{1/22}$, indicating that the system starts at 5/11 along the first eigenvector (horizontal axis) and at 1/22 along the second eigenvector (vertical axis). In this coordinate system, the dynamics are described by $\vec{y}(t) = \mathbf{D}^t\vec{y}(0)$, where $\mathbf{D} = \begin{pmatrix} 2/3 & 0 \\ 0 & 4/3 \end{pmatrix}$. The state of the system now changes in more obvious fashion each time step, decreasing by a factor $\lambda = 2/3$ along the horizontal axis and increasing by a factor $\lambda = 4/3$ along the vertical axis.

In this new coordinate system, the general solution is $\vec{y}(t) = \mathbf{D}^t \vec{y}(0)$, which involves taking the power of each eigenvalue along the diagonal of **D**. The general solution in the original coordinate system can then be obtained by back transformation, to get $\vec{n}(t) = \mathbf{A}\,\vec{y}(t) = \mathbf{A}\,\mathbf{D}^t\,\vec{y}(0) = \mathbf{A}\,\mathbf{D}^t\mathbf{A}^{-1}\vec{n}(0)$, which is exactly the general solution we obtained earlier (9.12). Thus, the general solution in Recipe 9.1 can be interpreted as first transforming the coordinates to a new "vantage point" from which to view the variables, then solving this transformed model, and finally transforming back to the original variables. A similar method can be used for linear continuous-time models as well (Box 9.3).

Example: Describing DNA Sequence Evolution

Here we apply Recipe 9.1 to predict how DNA sequences should evolve over time in the presence of mutations. We will use Kimura's (1980) two-parameter model of mutation (Figure 9.3), which takes into account the fact that mutations altering a nucleotide's ring structure (*transversion* mutations from purines to pyrimidines or from pyrimidines to purines) are less common than mutations that retain the same ring structure (*transition* mutations between purines or between pyrimidines). We use α to describe the fraction of sites that experience a transition mutation in a time step, and β to describe the fraction that

---

Box 9.3: General Solution to a Continuous-Time Linear Model Involving Multiple Variables

Here we describe the solution to a system of linear differential equations in continuous time:

$$\begin{aligned}
\frac{dn_1}{dt} &= m_{11}\,n_1 + m_{12}\,n_2 + \cdots + m_{1d}\,n_d,\\
\frac{dn_2}{dt} &= m_{21}\,n_1 + m_{22}\,n_2 + \cdots + m_{2d}\,n_d,\\
&\vdots\\
\frac{dn_d}{dt} &= m_{d1}\,n_1 + m_{d2}\,n_2 + \cdots + m_{dd}\,n_d.
\end{aligned} \tag{9.3.1}$$

If the equations have constant terms representing inflow or outflow, then we must first transform the equations into the form of (9.3.1) by applying steps in Box 7.2.

We can represent (9.3.1) in matrix form as $d\vec{n}/dt = \mathbf{M}\,\vec{n}$. The first step is to find the eigenvalues and eigenvectors of **M**. We write the eigenvalues as r instead of λ to emphasize the fact that the eigenvalues are for a continuous-time model. The eigenvectors of **M** are then placed in the columns of a new matrix **A**, and the eigenvalues are placed along the diagonal elements of a diagonal matrix **D**. Again, we must make sure to correctly order the columns of **A** and **D**: the eigenvector in the jth column of **A** must be associated with the eigenvalue in the jth column of **D** according to the relationship $\mathbf{M}\,\vec{u} = r\,\vec{u}$.

The general solution to (9.3.1) is $\vec{n}(t) = e^{\mathbf{M}t}\,\vec{n}(0)$, which is analogous to the one-variable solution in continuous time, $n(t) = e^{rt}n(0)$. The problem with this solution, however, is that it

(continued)

Box 9.3 *(continued)*

requires us to find the exponential of $\mathbf{M}t$, which is defined as $e^{\mathbf{M}t} = \sum_{i=0}^{\infty} (\mathbf{M}^i t^i / i!)$. Matrix exponentiation is clearly not a straightforward operation! Fortunately, however, we can rewrite this solution as

$$\vec{n}(t) = \mathbf{A}\, e^{\mathbf{D}t} \mathbf{A}^{-1}\, \vec{n}(0), \tag{9.3.2}$$

which is an enormous help because diagonal matrices are easy to exponentiate:

$$e^{\mathbf{D}t} = \begin{pmatrix} e^{r_1 t} & 0 & \cdots & 0 \\ 0 & e^{r_2 t} & \cdots & 0 \\ \vdots & \vdots & \vdots & \vdots \\ 0 & 0 & \cdots & e^{r_d t} \end{pmatrix}. \tag{9.3.3}$$

There are two issues that might arise when calculating (9.3.2). The first is that the $d \times d$ matrix $\mathbf{M}$ might not have d independent eigenvectors (i.e., $\mathbf{M}$ might be defective). In such cases, $\mathbf{A}$ cannot be inverted and (9.3.2) cannot be applied. Fortunately, there are extensions of this technique for solving differential equations involving defective matrices. These techniques involve writing matrices in "Jordan form" and are described in the Further Reading.

The second issue is that the eigenvalues might be complex numbers (Box 7.3) even if all of the elements of $\mathbf{M}$ are real. Complex eigenvalues imply that the system cycles. As a concrete example, we describe the solution to the general two-variable linear model in continuous time:

$$\begin{pmatrix} \dfrac{dn_1}{dt} \\ \dfrac{dn_2}{dt} \end{pmatrix} = \begin{pmatrix} a & b \\ c & d \end{pmatrix} \begin{pmatrix} n_1 \\ n_2 \end{pmatrix}. \tag{9.3.4}$$

Let $\binom{v_{11}}{v_{21}}$ and $\binom{v_{12}}{v_{22}}$ represent the eigenvectors of $\mathbf{M}$ associated with the eigenvalues r_1 and r_2, respectively. The general solution is then

$$\begin{pmatrix} n_1(t) \\ n_2(t) \end{pmatrix} = \begin{pmatrix} v_{11} & v_{12} \\ v_{21} & v_{22} \end{pmatrix} \begin{pmatrix} e^{r_1 t} & 0 \\ 0 & e^{r_2 t} \end{pmatrix} \begin{pmatrix} \dfrac{v_{22}}{\mathrm{Det}(\mathbf{A})} & -\dfrac{v_{12}}{\mathrm{Det}(\mathbf{A})} \\ -\dfrac{v_{21}}{\mathrm{Det}(\mathbf{A})} & \dfrac{v_{11}}{\mathrm{Det}(\mathbf{A})} \end{pmatrix} \vec{n}(0), \tag{9.3.5}$$

where $\mathrm{Det}(A) = v_{11}v_{22} - v_{12}v_{21}$. The eigenvalues and associated eigenvectors of this matrix are again

$$r_1 = \frac{a + d + \sqrt{(a-d)^2 + 4bc}}{2} \quad \text{with } \vec{u} = \begin{pmatrix} 1 \\ \dfrac{-a + d + \sqrt{(a-d)^2 + 4bc}}{2b} \end{pmatrix}, \tag{9.3.6a}$$

$$r_2 = \frac{a + d - \sqrt{(a-d)^2 + 4bc}}{2} \quad \text{with } \vec{u} = \begin{pmatrix} 1 \\ \dfrac{-a + d - \sqrt{(a-d)^2 + 4bc}}{2b} \end{pmatrix}. \tag{9.3.6b}$$

(continued)

Box 9.3 *(continued)*

As in Box 9.2, if $(a-d)^2 + 4bc$ is positive, then the eigenvalues and eigenvectors are real, and (9.3.5) can be multiplied out to describe the system at any future point in time. If $(a-d)^2 + 4bc$ is negative, however, (9.3.5) will involve the exponential function of complex terms. Again, we will use Euler's formula to help us interpret these terms.

We first rewrite the eigenvalues as $r_1 = \alpha + \beta i$ and $r_2 = \alpha - \beta i$, where α is the real part and βi is the imaginary part. Again, $\alpha = (a+d)/2$ and $\beta = \sqrt{-(a-d)^2 - 4bc}/2$. We can then use the fact that $e^{(\alpha+\beta i)t} = e^{\alpha t}e^{\beta t i}$ along with Euler's formula $e^{\beta t i} = \cos(\beta t) + i\sin(\beta t)$ to write $e^{\mathbf{D}t}$ as

$$e^{\mathbf{D}t} = e^{\alpha t}\begin{pmatrix} \cos(\beta t) + i\sin(\beta t) & 0 \\ 0 & \cos(\beta t) - i\sin(\beta t) \end{pmatrix}.$$

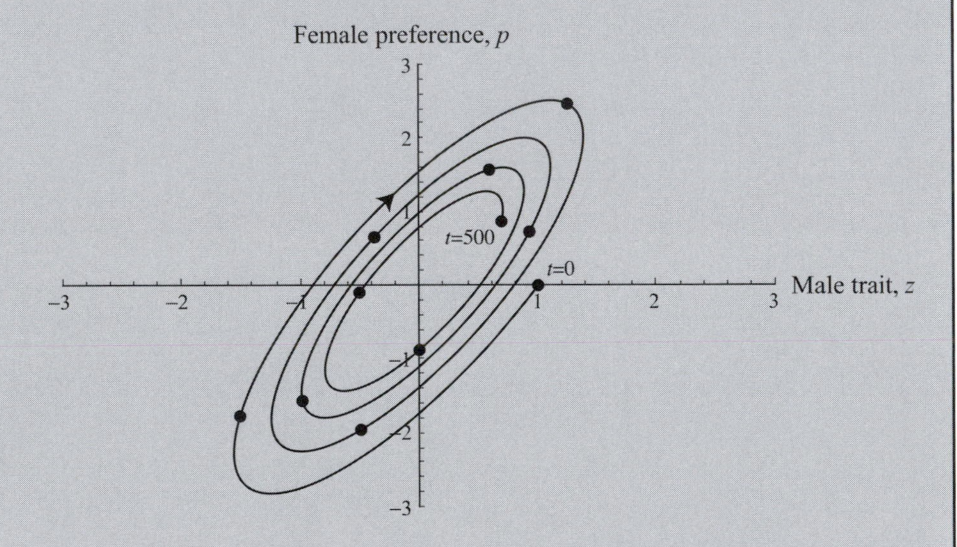

Figure 9.3.1: Sexual selection on male traits and female preferences. The co-evolution of the mean male trait ($\bar{z}$) and mean female preference for the male trait ($\bar{p}$). The differential equations (9.6.8) were used with the parameters $G_z = 0.15$, $G_p = 0.8$, $c = 0.45$, $b = 0.3$, $a = 0.95$, and $B = 0.32$. This model was rewritten in the general matrix form, (9.3.4), whose general solution is (9.3.7). The general solution was then plotted for 500 generations starting from a population where the mean male trait was $\bar{z}(0) = 1$ and the mean female preference was $\bar{p}(0) = 0$. Each dot represents the passage of 50 generations. In this example, cycling is observed with a period of $\tau = 129$ generations, and the system spirals in towards the origin at a per generation rate $r = -0.0175$. Notice that the cycles occur in a clock-wise direction. Initially, this occurs because the male's trait is very exaggerated (+1) despite the fact that sexual and natural selection favor no trait (0). Consequently, the male trait declines over time, which causes a correlated response in females (because $B > 0$), resulting in a preference for males with negative trait values. Female preferences evolve rapidly, however, and eventually the fitness costs of these preferences become so great that natural selection prohibits the further evolution of preferences. This causes the dynamics to curve upwards, propagating the cycle.

(continued)

Box 9.3 ***(continued)***

Multiplying out the general solution and simplifying, we then get

$$\begin{pmatrix} n_1(t) \\ n_2(t) \end{pmatrix} = e^{(a+d)t/2} \begin{pmatrix} \cos(\beta t) + \frac{(a-d)}{2\beta}\sin(\beta t) & \frac{b}{\beta}\sin(\beta t) \\ \frac{c}{\beta}\sin(\beta t) & \cos(\beta t) - \frac{(a-d)}{2\beta}\sin(\beta t) \end{pmatrix} \begin{pmatrix} n_1(0) \\ n_2(0) \end{pmatrix}, \quad (9.3.7)$$

where now β describes how fast the system cycles. The matrix in (9.3.7) returns to the same value after a period equal to $\tau = 2\pi/\beta$, and the system expands or contracts exponentially at rate equal to the real part of the eigenvalue $(a + d)/2$.

As a concrete example, we consider the general solution to the sexual selection model considered in Chapter 7. In that example, the differential equations describing the coevolution of male traits ($\bar{z}$) and female preferences ($\bar{p}$) were given by equation (7.19) as

$$\begin{pmatrix} \frac{d\bar{z}}{dt} \\ \frac{d\bar{p}}{dt} \end{pmatrix} = \begin{pmatrix} -G_z c & G_z a - b B \\ -B c & -G_p b + a B \end{pmatrix} \begin{pmatrix} \bar{z} \\ \bar{p} \end{pmatrix}. \quad (9.3.8)$$

The eigenvalues are complex when $\gamma^2 - 4bc\,(G_p G_z - B^2)$ is negative, where $\gamma = -b\,G_p - c\,G_z + a\,B$. In this case, we can rewrite (9.3.7) using the equivalent terms in this model ($n_1 = \bar{z}$, $n_2 = \bar{p}$, $a = -G_z c$, $b = G_z a - b B$, $c = -B c$, and $d = -G_p b + a B$), which immediately gives us the general solution. In particular, we can plot (9.3.7) and watch how the female preferences and male traits cycle over time (Figure 9.3.1) with a period equal to

$$\tau = \frac{4\pi}{\sqrt{-\gamma^2 + 4bc\,(G_p\,G_z - B^2)}}$$

experience each of the two types of possible transversions (Figure 9.3). We assume that both transitions and transversions occur (α, $\beta > 0$) and that the fraction of sites that experience a mutation is less than one ($\alpha + 2\beta < 1$).

We represent the number of adenines, guanines, cytosines, and thymines in the sequence as $n_A(t)$, $n_G(t)$, $n_C(t)$, and $n_T(t)$ at time t. From Figure 9.3, the numbers of each nucleotide in the next generation are (Step 1 of Recipe 9.1)

$$\begin{pmatrix} n_A(t+1) \\ n_G(t+1) \\ n_C(t+1) \\ n_T(t+1) \end{pmatrix} = \begin{pmatrix} 1-\alpha-2\beta & \alpha & \beta & \beta \\ \alpha & 1-\alpha-2\beta & \beta & \beta \\ \beta & \beta & 1-\alpha-2\beta & \alpha \\ \beta & \beta & \alpha & 1-\alpha-2\beta \end{pmatrix} \begin{pmatrix} n_A(t) \\ n_G(t) \\ n_C(t) \\ n_T(t) \end{pmatrix} \quad (9.14)$$

As a check, notice that the columns sum to one, which reflects the fact that any particular base pair at time t must fall into one of the four categories at time

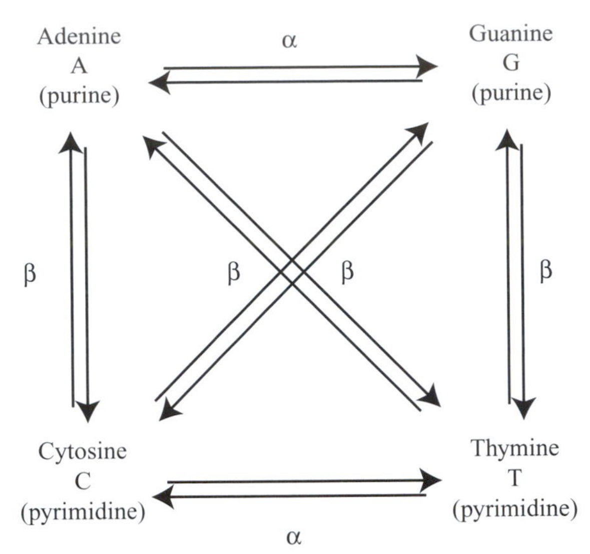

Figure 9.3: The Kimura two-parameter model. Nucleotides within a DNA sequence are classified as purines (adenine (A) and guanine (G)) or pyrimidines (cytosine (C) and thymine (T)), according to whether they contain two rings or one ring of carbon and nitrogen atoms. In the Kimura two-parameter model of mutation, the probability of mutation is lower for mutations that alter the ring structure (transversion mutations with probability β, A,G $\leftrightarrow$ C,T) than for mutations that do not change the number of rings (transition mutations with probability α, A $\leftrightarrow$ G, and C $\leftrightarrow$ T).

$t + 1$. Using Rule P2.17 and Definition P2.7 (or *Mathematica* or Maple), the four eigenvalues of (9.14) are $\lambda = 1, 1 - 4\beta, 1 - 2\alpha - 2\beta$, and $1 - 2\alpha - 2\beta$ (Step 2). The fact that one eigenvalue equals one suggests that there is some aspect to the system that remains constant over time. Here, it is the total number of nucleotides that remains constant and equal to the sequence length. We can now place the eigenvalues along the diagonal of a matrix (Step 3):

$$\mathbf{D} = \begin{pmatrix} 1 & 0 & 0 & 0 \\ 0 & 1 - 4\beta & 0 & 0 \\ 0 & 0 & 1 - 2\alpha - 2\beta & 0 \\ 0 & 0 & 0 & 1 - 2\alpha - 2\beta \end{pmatrix}, \tag{9.15}$$

and raise this matrix to the power t:

$$\mathbf{D}^t = \begin{pmatrix} 1 & 0 & 0 & 0 \\ 0 & (1 - 4\beta)^t & 0 & 0 \\ 0 & 0 & (1 - 2\alpha - 2\beta)^t & 0 \\ 0 & 0 & 0 & (1 - 2\alpha - 2\beta)^t \end{pmatrix}. \tag{9.16}$$

Eigenvectors associated with these eigenvalues are (Step 4)

$$\begin{pmatrix} 1 \\ 1 \\ 1 \\ 1 \end{pmatrix} \text{for } \lambda = 1; \begin{pmatrix} -1 \\ -1 \\ 1 \\ 1 \end{pmatrix} \text{for } \lambda = 1 - 4\beta;$$

$$\begin{pmatrix} 0 \\ 0 \\ -1 \\ 1 \end{pmatrix} \text{for } \lambda = 1 - 2\alpha - 2\beta; \text{ and} \begin{pmatrix} -1 \\ 1 \\ 0 \\ 0 \end{pmatrix} \text{for } \lambda = 1 - 2\alpha - 2\beta.$$

Notice that while one of the eigenvalues is repeated ($\lambda = 1 - 2\alpha - 2\beta$), it is possible to find two eigenvectors associated with this eigenvalue (Primer 2). Hence, we can write a matrix **A** whose columns are the four independent eigenvectors (Step 5):

$$\mathbf{A} = \begin{pmatrix} 1 & -1 & 0 & -1 \\ 1 & -1 & 0 & 1 \\ 1 & 1 & -1 & 0 \\ 1 & 1 & 1 & 0 \end{pmatrix}. \tag{9.17}$$

The order of these columns must match the order of the eigenvalues placed in **D**.

The general solution is then given by $\vec{n}(t) = \mathbf{A}\,\mathbf{D}^t\,\mathbf{A}^{-1}\vec{n}(0)$, where $\mathbf{A}\,\mathbf{D}^t\,\mathbf{A}^{-1}$ can be calculated by hand or by using a mathematical software package:

$$\begin{pmatrix} \frac{1}{4}+\frac{(1-4\beta)^t}{4}+\frac{(1-2\alpha-2\beta)^t}{2} & \frac{1}{4}+\frac{(1-4\beta)^t}{4}-\frac{(1-2\alpha-2\beta)^t}{2} & \frac{1}{4}-\frac{(1-4\beta)^t}{4} & \frac{1}{4}-\frac{(1-4\beta)^t}{4} \\ \frac{1}{4}+\frac{(1-4\beta)^t}{4}-\frac{(1-2\alpha-2\beta)^t}{2} & \frac{1}{4}+\frac{(1-4\beta)^t}{4}+\frac{(1-2\alpha-2\beta)^t}{2} & \frac{1}{4}-\frac{(1-4\beta)^t}{4} & \frac{1}{4}-\frac{(1-4\beta)^t}{4} \\ \frac{1}{4}-\frac{(1-4\beta)^t}{4} & \frac{1}{4}-\frac{(1-4\beta)^t}{4} & \frac{1}{4}+\frac{(1-4\beta)^t}{4}+\frac{(1-2\alpha-2\beta)^t}{2} & \frac{1}{4}+\frac{(1-4\beta)^t}{4}-\frac{(1-2\alpha-2\beta)^t}{2} \\ \frac{1}{4}-\frac{(1-4\beta)^t}{4} & \frac{1}{4}-\frac{(1-4\beta)^t}{4} & \frac{1}{4}+\frac{(1-4\beta)^t}{4}-\frac{(1-2\alpha-2\beta)^t}{2} & \frac{1}{4}+\frac{(1-4\beta)^t}{4}+\frac{(1-2\alpha-2\beta)^t}{2} \end{pmatrix} \tag{9.18}$$

We should now perform a couple of checks to make sure that we have made no errors. First, the columns each add to one, which they must because one of the four base pairs must be present at each sequence position at any point in time. Second, when $t = 0$, $(1 - 4\beta)^t = 1$, $(1 - 2\alpha - 2\beta)^t = 1$, and the above matrix reduces to the identity matrix **I**. This predicts that $\vec{n}(0) = \mathbf{I}\,\vec{n}(0)$, which is correct. Therefore, our result does give correct predictions under these scenarios.

What does the model predict in the long term? As t goes to infinity, both $(1 - 4\beta)^t$ and $(1 - 2\alpha - 2\beta)^t$ go to zero, because both terms are numbers less than one that are raised to a large power. Consequently, every element in the matrix goes to 1/4 over time. Multiplying by the initial vector $\vec{n}(0)$ indicates that the number of each type of nucleotide becomes $(n_A(0) + n_G(0) + n_C(0) + n_T(0))/4$. Thus, eventually, we expect each type of nucleotide to comprise 25% of the sequence, regardless of the initial sequence composition. This prediction might seem counterintuitive at first. Because transitions are more common than transversions, you might expect adenines and guanines to predominate if you started with an AG-rich sequence. Nevertheless, the symmetry of the model (with mutations to and from each nucleotide occurring at the same total rate, Figure 9.3) ensures that each type of nucleotide ultimately reaches the same equilibrium level.

Of course, this model ignores several important constraints on actual DNA sequences. In particular, it ignores the fact that selection will eliminate many mutations, especially those that alter amino acids or regulatory regions of the gene. Nevertheless, the Kimura two-parameter model has played an important role in biology. This model has been used to describe how gene sequences change over time, allowing us to predict how rapidly the genomes of two species should diverge. Conversely, given the number of differences between two DNA sequences, the model can be used to estimate the total evolutionary time t separating the sequences. If two sequences differ at a site, the model also predicts how likely it is that the difference is a transition versus a transversion, with transitions being more likely for closely related sequences and transversions predominating over longer evolutionary time (assuming $\alpha > \beta$; Figure 9.4). Most importantly, the Kimura two-parameter model and various extensions form the basis of likelihood methods for reconstructing phylogenetic trees. In this context, the general solution (9.18) is used to determine how likely it is that the sequence data would be observed, given a set of parameters describing the mutation rates, the phylogeny, and the amount of evolutionary time along each branch in the phylogeny (see Li 1997 and Felsenstein 2004 for more information on models of molecular evolution and phylogenetic reconstruction).

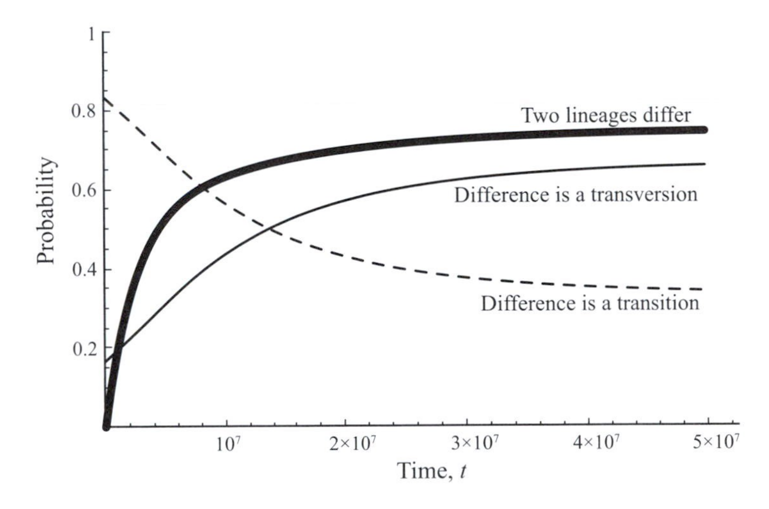

Figure 9.4: The probability that two sequences differ at a site. Consider a sequence that gives rise to two lineages at time $t = 0$. Initially, the two lineages carry the same nucleotide. The probability that a site differs between the two lineages ($D = n_A n_G + n_A n_C + n_A n_T + n_G n_C + n_G n_T + n_C n_T$, all at time t) rises over time as mutations accumulate (thick curve). Because transition mutations are more common, the chance that the two lineages differ by a transition mutation, $(n_A n_G + n_C n_T)/D$, is initially high (dashed curve). Eventually, however, it becomes more likely to observe a transversion difference between the two lineages, $(n_A n_C + n_A n_T + n_G n_C + n_G n_T)/D$, because more pairs of nucleotides differ by a transversion mutation (Figure 9.3). Parameters used are $\alpha = 10^{-7}$, $\beta = 10^{-8}$.

9.3 Nonlinear Models Involving Multiple Variables

Only rarely is there a general solution for nonlinear equations involving multiple variables. Occasionally, it might be possible to transform the model so that each equation becomes linear. For example, consider the multiallele extension to the haploid model of selection, where there are alleles $A_1, A_2, \ldots, A_d$ at frequencies $p_1, p_2, \ldots, p_d$. The recursion equation for allele j is given by an equation similar to (6.4):

$$p_j(t+1) = \frac{W_{A_j}\, p_j(t)}{W_{A_1} p_1(t) + W_{A_2} p_2(t) + \cdots + W_{A_d} p_d(t)}. \tag{9.19}$$

Thus, each of the alleles changes in frequency according to a nonlinear equation that involves all of the other variables. But let us try the same sort of transformation that we used in (6.7), by defining new variables that measure the allele frequencies relative to the frequency of A_1: $f_j(t) = p_j(t)/p_1(t)$. Equation (9.19) then becomes

$$f_j(t+1) = \frac{p_j(t+1)}{p_1(t+1)} = \frac{W_{A_j}\, p_j(t)}{W_{A_1}\, p_1(t)} = \frac{W_{A_j}}{W_{A_1}} f_j(t). \tag{9.20}$$

Not only is (9.20) a linear equation, but it also does not depend on any of the variables besides $f_j(t)$. We can thus iterate (9.20) to determine the general solution: $f_j(t) = (W_{A_j}/W_{A_1})^t\, f_j(0)$.

In other cases, it is possible to transform a nonlinear model into a familiar model whose solution is already known. This is true, for example, when modeling a pool of resources (n_1) that is being used by a consumer species (n_2):

$$\begin{aligned} \frac{dn_1}{dt} &= -a\, c\, n_1 n_2, \\ \frac{dn_2}{dt} &= \varepsilon\, a\, c\, n_1 n_2. \end{aligned} \tag{9.21}$$

Equation (9.21) makes the mass-action assumption that consumers contact (c) and use up (a) resources at a rate that is proportional to the amount of resources available (n_1) and the number of consumers around (n_2) (also called a type-I functional response; Section 3.4). It also assumes that the resources are nonrenewable (i.e., there is no input in the differential equation for n_1) and that the consumers do not die over the short-time frame during which the model is applicable (see equation (3.17) with $\theta = \delta = 0$). Finally, the model assumes that consumers convert resources into progeny according to a conversion rate of ε consumer offspring per unit of resources eaten.

The above model might be a reasonable description for the growth of yeast feeding on liquid media in a vial. As illustrated in Figure 4.5, the number of yeast cells rises over time in a manner reminiscent of logistic growth curves (Figure 4.3). In fact, as it turns out, equations (9.21) are identical to the differential equation for logistic growth (3.5). To see this, let us step back and think about the fundamental process encapsulated in the model. The model describes how energy moves between two pools, a resource pool and a consumer pool, where the total amount of energy in the system remains constant (no input or outflow to the system). This suggests an important insight into this model: if we were to total up all of the energy E in the system, it should remain constant.

We can put this insight into the form of an equation relating E to the amount of resources, n_1, and consumers, n_2. First, we have to choose a unit of energy. We focus our attention on the predicted number of consumers and count energy in units of consumer-equivalents. Specifically, E equals the number of consumers, n_2, plus the number of consumers that could be produced given the amount of resources currently available, εn_1. Next, we can check to make sure that the total energy in the system does actually remain constant, by deriving a differential equation for $E = \varepsilon n_1 + n_2$. We get

$$\frac{\mathrm{d}E}{\mathrm{d}t} = \frac{\mathrm{d}(\varepsilon\, n_1 + n_2)}{\mathrm{d}t} = \varepsilon\frac{\mathrm{d}n_1}{\mathrm{d}t} + \frac{\mathrm{d}n_2}{\mathrm{d}t}. \tag{9.22}$$

Plugging in (9.21) confirms that E does not change over time:

$$\frac{\mathrm{d}E}{\mathrm{d}t} = -\varepsilon\, a\, c\, n_1 n_2 + \varepsilon\, a\, c\, n_1 n_2 = 0. \tag{9.23}$$

How can we use this insight to simplify the model? The relationship $E = \varepsilon\, n_1 + n_2$ can be used to replace one of the variables in the model in terms of the other variable. Focusing on the number of consumers, we can rewrite the amount of resources as $n_1 = (E - n_2)/\varepsilon$. We then plug this formula into the differential equation for the number of consumers:

$$\begin{aligned}\frac{\mathrm{d}n_2}{\mathrm{d}t} &= \varepsilon\, a\, c\, n_1 n_2 \\ &= \varepsilon\, a\, c\left(\frac{E - n_2}{\varepsilon}\right)n_2 \\ &= a\, c\, E\, n_2 - a\, c\, n_2^2.\end{aligned} \tag{9.24}$$

Equation (9.24) involves only the variable n_2. Furthermore, equation (9.24) contains only two terms, one proportional to n_2 and the other proportional to n_2^2, which is exactly the same form as the logistic model. By defining the coefficients multiplying these two terms as $a\, c\, E \equiv r$ and $a\, c \equiv r/K$, equation (9.21) is converted into the logistic equation. We can also use these relationships to show that the new parameter E equals K, the carrying capacity in the logistic model. With these relationships in hand, we can immediately use the general solution for the logistic model (6.14b) to solve for the number of consumers over time, n_2. We can then plug this solution into $n_1 = (E - n_2)/\varepsilon$ to solve for the level of resources over time.

Now we can understand why the two-variable model of resource competition behaves so much like the logistic model—they are mathematically equivalent.

Importantly, this does not mean that the models are biologically equivalent. The logistic model assumes that resources are constantly renewed, and these resources sustain a reproducing population of constant size K. In contrast, using equations (9.21), the resources are eventually depleted, producing a total of K individuals that no longer reproduce once the resources are used up.

The above model illustrates an important lesson: insight matters in mathematics. Here we used the insight that energy was conserved to simplify the dynamical equations from two variables down to one. This insight allowed us to apply everything that we already know about the logistic model in continuous time to equations (9.21). So pay attention to any insights that you have, and think about whether you can use these to simplify your model. We should add, however, that such insights do not always come easily or rapidly when working with a model—they often pop into your mind when you are stewing over some surprising results or when you least expect it. While nobody can really teach the creative process by which such insights are made, it helps to step back from the equations, to let your mind wander over the problem, and to think about the model in general terms.

9.3.1. Proportional Transformations (Nondimensionalization)

The examples considered in the previous section demonstrate that nonlinear models involving multiple variables aren't always as complicated as they might seem. Working with the model and trying various transformations can help to determine whether there is any hope for a general solution. Nevertheless, even without obtaining a general solution, transformations can be used to simplify a model and allow you to clarify the fundamental nature of the interactions among the variables. A particularly helpful transformation in this regard is known as a *proportional transformation* or sometimes *nondimensionalization*. The idea behind a proportional transformation is to measure each variable on some sort of natural scale, which helps to simplify the model.

As an example, consider a predator-prey model where N and P denote the number of prey and predators, respectively. Suppose that, in the absence of predation, the prey population grows logistically to carrying capacity, K, and that the predator consumes prey in a way described by a type-I functional response (Section 3.4 of Chapter 3). In continuous time, the model is

$$\begin{aligned} \frac{dN}{dt} &= r\,N\left(1 - \frac{N}{K}\right) - a\,c\,N\,P, \\ \frac{dP}{dt} &= \varepsilon\,a\,c\,N\,P - \mu\,P. \end{aligned} \tag{9.25}$$

Model (9.25) contains six parameters: r, the intrinsic growth rate of the prey when at low abundance; K, the carrying capacity of the prey; c, the contact rate of predator with prey; a, the probability that a predator consumes a prey, given it makes a contact; ε, the conversion coefficient of captured prey into new predators; and μ, the per capita mortality rate of the predator. To use the proportional transformation technique, begin by defining the new variables $N^* = N/\tilde{N}, P^* = P/\tilde{P}$ as well as the new measurement of time $t^* = t/\tilde{t}$, where $\tilde{N}$, $\tilde{P}$, and $\tilde{t}$ are constants (to be chosen shortly).

We now derive an equation for the dynamics of N^* with respect to the new time variable t^*, i.e., dN^*/dt^*. Applying the chain rule of calculus twice (Appendix 2, A2.14), we get

$$\frac{dN^*}{dt^*} = \frac{dN^*}{dN}\frac{dN}{dt^*} = \frac{dN^*}{dN}\frac{dN}{dt}\frac{dt}{dt^*}. \tag{9.26}$$

Using the definitions for N^* and t^* to evaluate dN^*/dN and dt/dt^* gives

$$\frac{dN^*}{dt^*} = \frac{\tilde{t}}{\tilde{N}}\frac{dN}{dt}. \tag{9.27}$$

Plugging in equation (9.25) for dN/dt then produces

$$\frac{dN^*}{dt^*} = \frac{\tilde{t}}{\tilde{N}}\left(r\,N\left(1 - \frac{N}{K}\right) - a\,c\,N\,P\right), \tag{9.28}$$

which we can rewrite in terms of the new variables as

$$\frac{dN^*}{dt^*} = \frac{\tilde{t}}{\tilde{N}}\left(r\,N^*\tilde{N}\left(1 - \frac{N^*\tilde{N}}{K}\right) - a\,c\,N^*\tilde{N}P^*\tilde{P}\right)$$

$$= r\tilde{t}N^*\left(1 - \frac{N^*\tilde{N}}{K}\right) - a\,c\,\tilde{t}N^*P^*\tilde{P} \tag{9.29}$$

We can perform the same type of calculation for P^*, arriving at

$$\frac{dP^*}{dt^*} = \frac{\tilde{t}}{\tilde{P}}\frac{dP}{dt}$$

$$= \frac{\tilde{t}}{\tilde{P}}(\varepsilon\,a\,c\,N^*\,\tilde{N}\,P^*\,\tilde{P} - \mu\,P^*\tilde{P}) \tag{9.30}$$

$$= \varepsilon\,a\,c\,\tilde{t}\,N^*\tilde{N}P^* - \mu\tilde{t}\,P^*.$$

The real power of this technique comes from a judicious choice of the constants $\tilde{N}$, $\tilde{P}$, and $\tilde{t}$. There is no "correct" choice for these—you are free to choose them in any way you please; but a good pair of guiding principles is to (i) make choices that are biologically meaningful (often this involves making choices so that the new variables are dimensionless; hence the name "nondimensionalization"), and (ii) make choices that reduce the number of parameters in the new model as much as possible. For example, from a biological standpoint, a natural choice is to measure the population size of the prey species relative to its carrying capacity by defining $\tilde{N} = K$. With this choice, the transformed prey variable is $N^* = N/K$, meaning that N^* is a dimensionless measure of prey abundance (i.e., it is the proportion of the carrying capacity). The advantage of this choice is that $(1 - N^*\tilde{N}/K)$ in (9.29) simplifies to $(1 - N^*)$. Substituting $\tilde{N} = K$ into the above equations yields

$$\frac{dN^*}{dt^*} = r\,\tilde{t}\,N^*(1 - N^*) - a\,c\,\tilde{t}\,N^*P^*\,\tilde{P},$$

$$\frac{dP^*}{dt^*} = \varepsilon\,a\,c\,K\,\tilde{t}\,N^*\,P^* - \mu\tilde{t}\,P^*. \tag{9.31}$$

But now how should we choose $\tilde{P}$ and $\tilde{t}$? This is less obvious, but let's use our second guiding principle and try to simplify the model as much as possible. If we choose $\tilde{t} = 1/r$ and $\tilde{P} = r/(a\,c)$, we can cancel out the parameters in dN^*/dt^* and get a very simple equation:

$$\frac{dN^*}{dt^*} = N^*(1 - N^*) - N^*P^*. \tag{9.32a}$$

We can think of $\tilde{t} = 1/r$ as scaling time in terms of the growth rate of the prey population when prey and predators are rare. Similarly, we can think of $\tilde{P} = r/(a\,c)$ as the predator population size that would maintain a constant prey population when prey are rare (so that $\frac{dN}{dt} = r\,N(1 - N/K) - a\,c\,N\,\tilde{P} \approx r\,N - a\,c\,N\,\tilde{P} = 0$). With these choices, dP^*/dt^* equals

$$\begin{aligned}\frac{dP^*}{dt^*} &= \frac{\varepsilon\,a\,c\,K}{r}N^*P^* - \frac{\mu}{r}P^* \\ &= \alpha\,N^*\,P^* - \beta\,P^*,\end{aligned} \tag{9.32b}$$

which we can write in terms of only two parameters, defined as $\alpha \equiv \varepsilon acK/r$ and $\beta \equiv \mu/r$.

What has this transformation accomplished? It has simplified our original model, for which there were the six parameters r, K, a, c, ε, and μ, to a transformed model of the system, for which there are only two effective parameters α and β. This transformation does not alter the basic dynamical structure of the model. For example, equations (9.32) involve N^* and P^* in the same way that equations (9.25) involve N and P. But it has revealed that, from a mathematical standpoint, the dynamical behavior of the number of predators and the number of prey (P and N) is really determined by only two composite parameters, α and β, rather than the six separate parameters that appeared in the original model.

Proportional transformations can reduce the number of parameters in a model.

Recipe 9.2
Proportional Transformations (Nondimensionalization)

Step 1: For each variable in the model, define a new variable that is the original variable divided by a constant. This is where the term "proportional" comes from; each new variable is proportional to the corresponding original variable (i.e., it is the original variable divided by a constant)

Step 2: If the model is a continuous-time model, define a new time variable that is the original time variable divided by a constant as well.

Step 3: Derive a new system of equations for these transformed variables.

Step 4: Choose the constants for the transformations of each variable so that the new model is as simple as possible. Make choices that (i) are biologically meaningful (e.g., cause the new variables to be dimensionless), and (ii) reduce the number of parameters in the transformed model as much as possible.

In general, the benefit of performing a proportional transformation is that many models contain more parameters than are required to specify the model's behavior. Looking at the model in the right way, through the use of this transformation, allows you to find ways in which the parameters can be lumped together. This is extremely useful for three reasons. First, this procedure often allows you to "see" the dynamical structure of the model more easily because it is less cluttered with extraneous parameters. Second, it usually makes all subsequent calculations less tedious and more transparent because you no longer need to manipulate all of these parameters (which has the added advantage of reducing the risk of a calculation error!). And third, if you want to simulate the model on a computer, reducing the number of parameters is helpful because it reduces the number of different cases that you have to examine. For example, suppose you want to simulate the original model (9.25) for ten different values of each parameter in order to develop a fairly comprehensive description of the model's behavior. This results in a total of 10^6 different simulations that you would have to run! If you used the transformed model (9.32), however, you would only have to run 100 simulations because there are only two effective parameters. A related advantage is that it is possible to describe the behavior of the system on a graph with only two axes (α and β) whereas doing so on a graph with six axes is impossible. Clearly a small amount of time spent transforming your model can result in a huge savings in time when simulating the model! Moreover, a much deeper understanding of the model can result from such transformations because they reveal particular relationships among the parameters that are fundamental to the dynamics of the system.

9.3.2. Separation of Time Scales

For most nonlinear models involving multiple variables, there is no general solution that can be used to predict the state of the system at any future point in time. To understand the dynamics of such models, one typically combines numerical simulations (Chapter 4) with stability analyses of specific equilibria (Chapter 8). For models with many parameters and variables, however, it can be nearly impossible to get an overall picture of the dynamics of the system from this approach. In this case, there is another technique that can often be used to obtain an approximate description of the general solution. This technique is known as a separation of time scales.

The underlying rationale behind a separation of time scales is that biological processes can sometimes be divided into *fast* and *slow* processes. When such a division is possible, you can then try to determine where the fast processes lead the system, assuming that the slow processes are unchanging. For example, if you were to analyze a model that involves ecological and evolutionary processes, then you might be willing to assume that the ecological time scale is faster. In this case, you could determine where the system heads while the current gene frequencies are held constant. This state is known as a *quasi-equilibrium*. It is not a true equilibrium because the gene frequencies will change, albeit slowly, and therefore alter the ecological dynamics. But it represents the state toward which the system tends over the short term. Once the quasi-equilibrium has been determined, it can be used to understand how the slow process affects the system, by assuming that the system always remains near the quasi-equilibrium.

A *separation of time scales* analyzes rapidly and slowly changing variables separately.

As an example, let us consider the model of density-dependent natural selection introduced in Chapter 8:

$$N(t+1) = \overline{W}\big(N(t), p(t)\big)\, N(t),$$

$$p(t+1) = \frac{p(t)\, W_{AA}\big(N(t)\big) + \big(1 - p(t)\big)\, W_{Aa}\big(N(t)\big)}{\overline{W}\big(N(t), p(t)\big)}\, p(t), \tag{9.33}$$

where $\overline{W} = p^2\, W_{AA}(N) + 2p(1-p)\, W_{Aa}(N) + (1-p)^2\, W_{aa}(N)$. We assume that competition causes fitness to decrease linearly with increasing population size as in the logistic model (3.5a) and that genotypes differ in their sensitivity to this competition. The fitness function $W_j(N) = 1 + r\,(1 - \alpha_j\, N)$ captures these assumptions, where $1 + r$ is the fitness when the population size is very small, and α_j is the competition coefficient of genotype j.

In this example, we might be interested in exploring the dynamics under the assumption that the ecological time scale over which the population size $N(t)$ changes is much shorter than the evolutionary time scale over which the allele frequency $p(t)$ changes. Informally, we might expect such a separation of time scales to be valid when the intrinsic growth rate r is large relative to the fitness differences among genotypes.

Assuming that the population size equilibrates rapidly, we can treat the allele frequency as fixed and determine the quasi-equilibrium population size (i.e., the equilibrium population size if we hold allele frequency fixed). We call this value $\hat{N}_Q$ to emphasize that it is like an equilibrium (hence the caret) but is only a quasi-equilibrium (hence the Q). To find this quasi-equilibrium, we set $N(t+1) = N(t) = \hat{N}_Q$ and solve the first recursion in (9.33) for $\hat{N}_Q$, treating $p(t)$ as a constant. Substituting $W_j(N) = 1 + r\,(1 - \alpha_j N)$ into (9.33), we must solve the following for $\hat{N}_Q$:

$$\hat{N}_Q = \left(1 + r\left(p^2\left(1 - \alpha_{AA}\hat{N}_Q\right) + 2p(1-p)\left(1 - \alpha_{Aa}\hat{N}_Q\right) + (1-p)^2\left(1 - \alpha_{aa}\hat{N}_Q\right)\right)\right)\hat{N}_Q. \tag{9.34}$$

Equation (9.34) has two solutions, $\hat{N}_Q = 0$ and

$$\hat{N}_Q = \frac{1}{\overline{\alpha}}, \tag{9.35}$$

where $\overline{\alpha}$ is the average strength of density dependence at time t:

$$\overline{\alpha} = p(t)^2\alpha_{AA} + 2p(t)(1-p(t))\,\alpha_{Aa} + (1-p(t))^2\alpha_{aa}. \tag{9.36}$$

We assume that the population is able to grow when rare ($r > 0$) and focus on the quasi-equilibrium (9.35) where the population is present.

At this point, we can now ask what happens over the long term in response to the slow process of evolution. Assuming that the population is always near quasi-equilibrium, we can replace N with $\hat{N}_Q$ from (9.35) in the fitnesses $W_j(N) = 1 + r\,(1 - \alpha_j N)$. Incorporating these fitnesses into (9.33) and simplifying, we can describe the allele frequency dynamics most compactly in terms of the difference equation

$$\begin{aligned}\Delta p &= p(t+1) - p(t)\\ &= -r\,\frac{p(t)\left(1-p(t)\right)\left(p(t)\left(\alpha_{AA}-\alpha_{Aa}\right) + (1-p(t))\left(\alpha_{Aa}-\alpha_{aa}\right)\right)}{\overline{\alpha}}.\end{aligned} \tag{9.37}$$

This result is interesting. The fraction on the right-hand side is equivalent to the standard difference equation (3.3.2c) describing the per generation change in allele frequency in a diploid model of selection, but with α_j taking the place of fitness. But there is an additional factor, $-r$, in equation (9.37) as well. The negative sign indicates that selection drives the allele frequency in the opposite direction from that expected if fitness were given by α_j. This makes sense because the genotype with the smallest value of α has the greatest fitness. The factor r indicates that the speed of allele frequency change depends on the growth rate of the population. Although you might think that increasing r just acts to increase the strength of selection, this is not precisely true. If (9.37) were a standard evolutionary model with r affecting the strength of selection, then the denominator (i.e., the mean fitness) would contain r as well. Thus, this density-dependent model of natural selection gives rise to allele frequency dynamics that are not exactly equivalent to the dynamics under natural selection in a population of constant size.

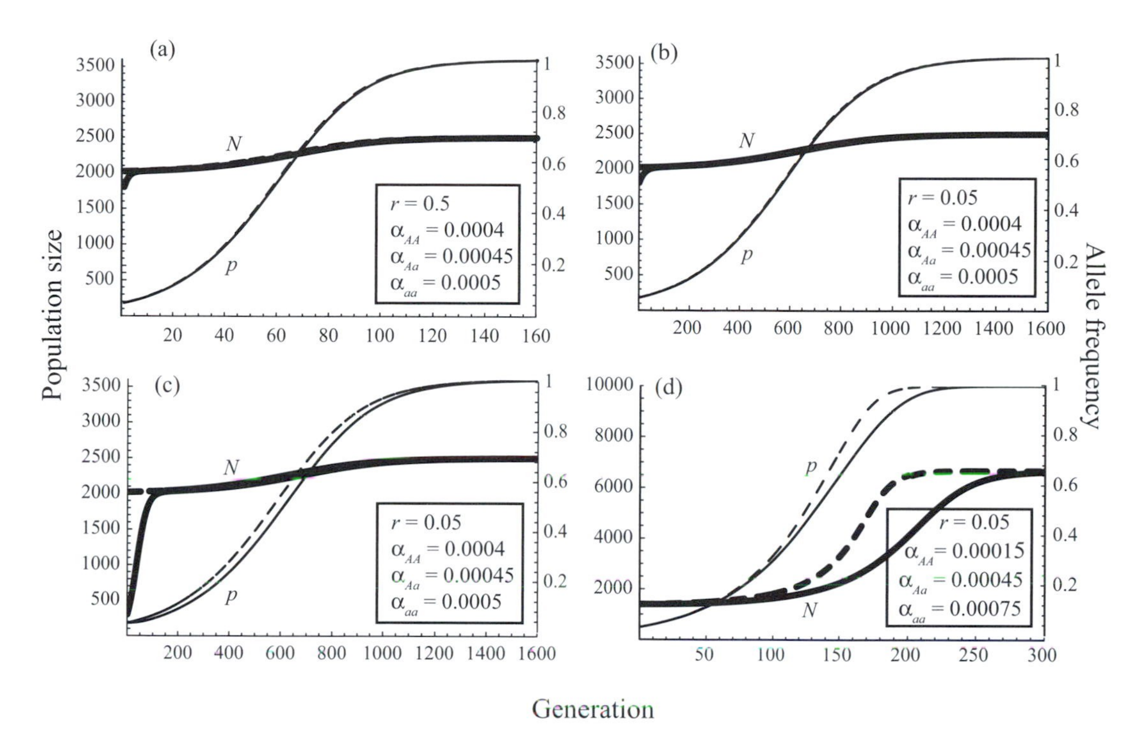

Figure 9.5: Dynamics of the density-dependent model of natural selection. Population size (thick curves) and allele frequency (thin curves) changes are graphed over time. The solid curves illustrate the exact dynamics (9.33) and the dashed curves (often indistinguishable) illustrate the quasi-equilibrium approximations (9.35) and (9.37). In calculating the quasi-equilibrium population size, we use the approximate allele frequencies given by (9.37); the fit improves if we use the actual allele frequencies at any point in time. In cases (a)–(c), the α_j values are similar across genotypes and the quasi-equilibrium approximation provides a good fit; case (c) fits less well because we start the population size far from its quasi-equilibrium (it is otherwise identical to case b). In case (d), the α_j values differ more among genotypes, and the quasi-equilibrium approximation breaks down.

The sign of Equation (9.37) predicts that allele A should rise in frequency to fixation whenever it is less sensitive to population size: $\alpha_{AA} < \alpha_{Aa} < \alpha_{aa}$. Figure 9.5 confirms this prediction and shows that the population size rises as a consequence, as we would expect from (9.35). More importantly, Figure 9.5 demonstrates that the quasi-equilibrium predictions for $\hat{N}_Q$ (9.35) and $p(t)$ (9.37) are excellent. Problems arise only during the initial time period as the population approaches the quasi-equilibrium (Figure 9.5c) or when the differences among genotypes are large (Figure 9.5d).

In summary:

Recipe 9.3
Separation of Time Scales

Step 1: Identify variables that you expect to change rapidly (fast-changing variables) and variables that you expect to change over longer time scales (slow-changing variables).

(continued)

Recipe 9.3 (*continued*)

Step 2: Treating the slow-changing variables as constants, determine the "equilibrium" for the fast-changing variables using Recipe 8.1 for continuous-time models or Recipe 8.3 for discrete-time models. This "equilibrium" is known as the *quasi-equilibrium*.

Step 3: Assume that the fast-changing variables are always at quasi-equilibrium and plug their quasi-equilibrium values into the equations for the slow-changing variables. Analyze the dynamics of the slow-changing variables using these new equations.

Step 4: Check to make sure that the fast-changing variables reach their quasi-equilibrium at a faster rate than the slow-changing variables, at least under appropriate values of the parameters.

Before proceeding to another example, we first examine Step 4 in Recipe 9.3 in more detail. This is related to the question of how and when we can justify the use of a quasi-equilibrium approximation. The key is to identify a parameter related to the slow process that you can assume is smaller in magnitude than the parameters associated with the fast process. In this example, we assume that proportional differences among the α_j are small, because this corresponds to assuming that selection is weak. The best way to describe the α_j is to write each one in terms of how different they are from the average value of α_j:

$$\begin{aligned} \alpha_{AA} &= \bar{\alpha}\,(1 + \delta_{AA}\,\varepsilon), \\ \alpha_{Aa} &= \bar{\alpha}\,(1 + \delta_{Aa}\,\varepsilon), \\ \alpha_{aa} &= \bar{\alpha}\,(1 + \delta_{aa}\,\varepsilon). \end{aligned} \tag{9.38}$$

We can now formalize our assumption that selection is weak compared to the growth rate of the population by requiring that ε be small relative to r.

Assuming that ε is small, we can expand the recursions using a Taylor series in ε (Primer 1). The leading-order terms are

$$\begin{aligned} N(t+1) &= (1 + r\,(1 - \bar{\alpha}\,N(t)))\,N(t) + O(\varepsilon), \\ p(t+1) &= p(t) + O(\varepsilon). \end{aligned} \tag{9.39}$$

If we drop terms of order ε, then the allele frequency remains roughly constant while the population size changes, and the equilibrium population size predicted by (9.39) is precisely the quasi-equilibrium identified in (9.35). This indicates that our calculations based on a separation of times scales should provide a reasonable approximation provided that ε is smaller than r. If r were also small (order ε), however, the equilibrium population size would be zero according to (9.39); to obtain a nonzero population size, we would then have to keep the next-order terms in the recursions, in which case the allele frequency would also change. Therefore, our separation of times scales approximation is valid provided that the population growth rate r is of a larger magnitude than the proportional differences among the competition coefficients.

Example: The Evolutionary Dynamics of Two Genes

Let us use a separation of time scales to gain insight into the dynamics of genetic models with more than one gene. We focus on the two-gene model of natural selection described by equation (8.43) in the previous chapter, but the methods described below can be applied more generally to an arbitrarily large numbers of genes and the genetic associations among them (Barton 1995; Barton and Turelli 1991; Kirkpatrick et al. 2002; Turelli and Barton 1990). In the context of genetic models, the separation of time scales described below is known as a *quasi-linkage equilibrium* or QLE approximation.

Even though (8.43) represents the simplest model with more than one gene, we cannot derive its general solution. In fact, we do not even know all of its equilibria. A separation of time scales can, however, provide some insight into its behavior. To begin, we must identify those processes that are slow and those that are fast. Again, we assume that selection is weak (the slow process), with small fitness differences among genotypes. The other process in the model is recombination, and we assume that the recombination rate (denoted by r) is large relative to selection.

This example is not yet as straightforward as the previous example because we cannot clearly separate the variables (i.e., the chromosome frequencies x_1, x_2, x_3, and x_4) into ones that are affected by the slow process and ones that are affected by the fast process. Rather, all four variables are affected by both selection and recombination. In contrast, with density-dependent natural selection, there was one variable (the allele frequency) that was mainly affected by the slow process, and a second variable (the population size) that was mainly affected by the fast process.

The best place to start is to first transform the model into a new set of variables from which we can readily distinguish fast-changing and slow-changing variables. In the two-gene model of selection, there is a natural choice of new variables: the frequencies of alleles A_1 and B_1 ($p_A = x_1 + x_2$ and $p_B = x_1 + x_3$) and the *linkage disequilibrium* between the two genes ($D = x_1 x_4 - x_2 x_3$). Linkage disequilibrium measures whether certain alleles are found together at different genes more often than expected by chance. When $D > 0$, alleles A_1 and B_1 are more often found together, as are alleles A_2 and B_2; when $D < 0$, alleles A_1 and B_2 are more often found together, as are alleles A_2 and B_1. The allele frequencies should be driven primarily by the slow process of selection, whereas the genetic associations between the two genes (D) should be primarily driven by the fast process of recombination.

In population-genetic models, *quasi-linkage equilibrium* refers to a separation of time scales in which allele frequencies at multiple loci change more slowly than the asociations among them.

Thus, we want to transform model (8.43) into one with the three new variables:

$$p_A = x_1 + x_2 \qquad \text{(the frequency of allele } A_1\text{)},$$
$$p_B = x_1 + x_3 \qquad \text{(the frequency of allele } B_1\text{)},$$
$$D = x_1 x_4 - x_2 x_3 \qquad \text{(the linkage disequilibrium)}.$$

These equations, plus the fact that $x_1 + x_2 + x_3 + x_4 = 1$, provide four equations that can be solved to write the old variables (the chromosome frequencies) in terms of the new variables:

$$\begin{aligned}
x_1 &= p_A\, p_B + D && \text{(the frequency of chromosome } A_1B_1\text{)},\\
x_2 &= p_A\,(1 - p_B) - D && \text{(the frequency of chromosome } A_1B_2\text{)},\\
x_3 &= (1 - p_A)\, p_B - D && \text{(the frequency of chromosome } A_2B_1\text{)},\\
x_4 &= (1 - p_A)(1 - p_B) + D && \text{(the frequency of chromosome } A_2B_2\text{)}.
\end{aligned}$$

Using these equations, we can determine the recursions for p_A, p_B, and D using equation (8.43). To simplify the presentation, we focus on the case of selection acting in the haploid phase (immediately following the census in Figure 8.8), allowing us to replace the diploid fitness w_{ij} with $w_i\, w_j$, that is, with the product of the fitness of haploids carrying chromosome i and the fitness of haploids carrying chromosome j. After a bit of algebra, these new recursions can be written as

$$p_A(t+1) = \frac{(p_A\, p_B + D)\, w_1 + (p_A\,(1 - p_B) - D)\, w_2}{\overline{w}_h}, \tag{9.40a}$$

$$p_B(t+1) = \frac{(p_A\, p_B + D)\, w_1 + ((1 - p_A)\, p_B - D)\, w_3}{\overline{w}_h}, \tag{9.40b}$$

$$D(t+1) = (1 - r)\frac{w_1 w_4 (p_A p_B + D)((1 - p_A)(1 - p_B) + D) - w_2 w_3 (p_A(1 - p_B) - D)((1 - p_A)p_B - D)}{\overline{w}_h^2}, \tag{9.40c}$$

where the right-hand sides involve the allele frequencies and disequilibria at time t, and

$$\begin{aligned}
\overline{w}_h = {} & (p_A p_B + D)\, w_1 + (p_A\,(1 - p_B) - D)\, w_2 + ((1 - p_A)\, p_B - D)\, w_3\\
& + ((1 - p_A)(1 - p_B) + D)\, w_4
\end{aligned}$$

is the mean fitness among haploids.

We now have a three-variable model, and we might expect the dynamics of the linkage disequilibrium, D, to occur on a time scale that is shorter than that of p_A and p_B as long as recombination rates are large relative to the strength of selection. As in the previous example, we start by calculating the quasi-equilibrium value for the variable with fast dynamics. Here, we set $D(t+1) = D(t) = \hat{D}_Q$ and solve the following for $\hat{D}_Q$:

$$\hat{D}_Q = (1 - r)\frac{w_1 w_4 (p_A p_B + \hat{D}_Q)((1 - p_A)(1 - p_B) + \hat{D}_Q) - w_2 w_3 (p_A(1 - p_B) - \hat{D}_Q)((1 - p_A)p_B - \hat{D}_Q)}{((p_A p_B + \hat{D}_Q)w_1 + (p_A(1 - p_B) - \hat{D}_Q)w_2 + ((1 - p_A)p_B - \hat{D}_Q)w_3 + ((1 - p_A)(1 - p_B) + \hat{D}_Q)w_4)^2} \tag{9.41}$$

We would then substitute the quasi-equilibrium solution into equations (9.40a,b) to obtain the dynamics of the allele frequencies, under the assumption that the disequilibrium always remains near its quasi-equilibrium value. If you try to do this, however, you immediately run into a problem: equation (9.41) cannot be solved for $\hat{D}_Q$ because it yields a complicated cubic polynomial in $\hat{D}_Q$.

What to do next? This issue should seem vaguely familiar; we faced the same problem in Chapter 5 when we encountered models whose equilibrium conditions could not be solved explicitly (e.g., the Ricker model with migration (5.10)). In section 5.4, we introduced perturbation techniques to obtain an approximate expression for the equilibrium in such cases. Similarly, we can use perturbation techniques to obtain an approximate expression for the quasi-equilibrium value $\hat{D}_Q$ from condition (9.41).

As described in Box 5.1 and Recipe 5.4, Step 1 of a perturbation analysis requires that we identify a parameter, or set of parameters, that is small and can be written in terms of a small quantity ζ ("zeta"). As we have already assumed that selection is weak, we can write the haploid fitnesses as

$$\begin{aligned} w_1 &= 1 + \tilde{s}_1 \zeta, \\ w_2 &= 1 + \tilde{s}_2 \zeta, \\ w_3 &= 1 + \tilde{s}_3 \zeta, \\ w_4 &= 1 + \tilde{s}_4 \zeta, \end{aligned} \tag{9.42}$$

where the selection coefficients s_i, are written as $\tilde{s}_i$ times the small term ζ. This makes explicit our assumption that selection is weak.

Step 2 of the perturbation analysis requires that we write our desired solution as a sum of increasing powers in ζ. In this example, we desire a quasi-equilibrium solution for the linkage disequilibrium, which we write as $\hat{D}_Q = \hat{D}_0 + \hat{D}_1 \zeta + \hat{D}_2 \zeta^2 + \cdots$. This solution must satisfy a condition that, according to the notation of Recipe 5.4, we write as $f(\zeta) = 0$. Here, $f(\zeta)$ is obtained by subtracting $\hat{D}_Q$ from both sides of (9.41):

$$f(\zeta) = (1 - r)\frac{w_1 w_4 (p_A p_B + \hat{D}_Q)((1 - p_A)(1 - p_B) + \hat{D}_Q) - w_2 w_3 (p_A(1 - p_B) - \hat{D}_Q)((1 - p_A)p_B - \hat{D}_q)}{((p_A p_B + \hat{D}_Q)w_1 + (p_A(1 - p_B) - \hat{D}_Q)w_2 + ((1 - p_A)p_B - \hat{D}_Q)w_3 + ((1 - p_A)(1 - p_B) + \hat{D}_Q)w_4)^2} - \hat{D}_Q. \tag{9.43}$$

We then plug $\hat{D}_Q = \hat{D}_0 + \hat{D}_1 \zeta + \hat{D}_2 \zeta^2 + \cdots$ and the fitnesses (9.42) into expression (9.43) and use the Taylor series to obtain as many terms in our approximation for $\hat{D}_Q$ as desired (Primer 1).

The terms in the approximation for $\hat{D}_Q$ are found by setting each term in the Taylor series of $f(\zeta)$ with respect to ζ to zero and solving for the $\hat{D}_i$. Specifically, we set $\left.\frac{d^i f(\zeta)}{d\zeta^i}\right|_{\zeta = 0} = 0$ and solve for $\hat{D}_i$, starting with $i = 0$ and working up until we have a sufficiently accurate approximation (Steps 3–4 of Recipe 5.4). Starting with $i = 0$, our solution must satisfy $f(0) = 0$. Plugging $\zeta = 0$ into (9.43) and simplifying, $f(0)$ equals zero only if $\hat{D}_0 = 0$. Next, we set $i = 1$ and solve $\left.\frac{df(\zeta)}{d\zeta}\right|_{\zeta = 0} = 0$ for $\hat{D}_1$. After some simplification, we obtain

$$\hat{D}_1 = \frac{(1 - r)\, p_A (1 - p_A)\, p_B (1 - p_B)}{r} (\tilde{s}_1 - \tilde{s}_2 - \tilde{s}_3 + \tilde{s}_4). \tag{9.44}$$

Plugging $\hat{D}_0 = 0$, (9.44), and (9.42) into $\hat{D}_Q = \hat{D}_0 + \hat{D}_1 \zeta + \hat{D}_2 \zeta^2 + \cdots$, we get a first-order approximation for the linkage disequilibrium:

$$\hat{D}_Q = \frac{(1 - r)\, p_A (1 - p_A)\, p_B (1 - p_B)}{r} (w_1 - w_2 - w_3 + w_4) + O(\zeta^2). \tag{9.45}$$

Equation (9.45) suggests that the sign of linkage disequilibrium is determined by the quantity $w_1 - w_2 - w_3 + w_4$. This quantity measures the degree to which fitnesses depart from an *additive* expectation. If alleles A_1 and B_1 have selection coefficients equal to s_A and s_B, respectively, an additive fitness scheme can be written as

$$\begin{aligned} w_1 &= 1 + s_A + s_B, \\ w_2 &= 1 + s_A, \\ w_3 &= 1 + s_B, \\ w_4 &= 1. \end{aligned} \tag{9.46}$$

(Check for yourself that $w_1 - w_2 - w_3 + w_4$ equals zero under an additive fitness scheme.) Before concluding that disequilibrium always depends on departures from an additive fitness scheme, however, we must remember that (9.45) was derived under the assumption that selection is weak (it ignores terms of order ζ^2). Furthermore, this approximation also implicitly assumes that the quantity $w_1 - w_2 - w_3 + w_4$ is not very small (i.e., fitnesses are not close to additive). Otherwise, if the fitnesses were close to additive and $w_1 - w_2 - w_3 + w_4$ was very small (e.g., of order ζ^2), then we would have to continue with the perturbation analysis to find the next-order term $\hat{D}_2$ in the approximation. Therefore, we can use approximation (9.45) only when the fitnesses are *not* close to additive (i.e., when $w_1 - w_2 - w_3 + w_4$ is not small). A more general expression can be obtained that allows the fitnesses to be nearly additive (Supplementary Material 9.1), but for our purposes approximation (9.45) is perfectly adequate. We mention this here only because the population genetics literature typically measures fitness as a departure from a *multiplicative* expectation, as suggested by the more general analysis in Supplementary Material 9.1. Because these scales are very similar, nothing is changed in the following discussion if we were to replace (9.45) with the more general version in Supplementary Material 9.1.

Given that the fast process of recombination drives the disequilibrium in the system toward (9.45), we must now examine how the allele frequencies change under the slow process of selection. At this point, we plug the quasi-equilibrium expression (9.45) and the fitness expressions (9.42) into (9.40a) and (9.40b), to obtain recursion equations for the allele frequencies. Because we have ignored terms of order ζ^2 in the disequilibrium, we must also ignore such terms in the recursions for the allele frequencies. Performing a Taylor series with respect to ζ, the change in frequency of allele A_1 $(p(t+1) - p(t))$ at quasi-linkage equilibrium factors into

$$\begin{aligned} \Delta p_A = p_A(t)\,(1 - p_A(t))\,(w_1\,p_B(t) + w_2\,(1 - p_B(t)) - w_3\,p_B(t) \\ - w_4\,(1 - p_B(t))) + O(\zeta^2). \end{aligned} \tag{9.47a}$$

Equation (9.47a) is easier to interpret when written in terms of the *marginal fitnesses* of the A_1 and A_2 alleles. The marginal fitness of an allele is just the average fitness of all individuals carrying that allele. Here, we average over those individuals carrying the B_1 allele and those carrying the B_2 allele, to get the marginal fitnesses of the A_1 and A_2 alleles: $W_{A_1} = p_B w_1 + (1 - p_B) w_2$ and $W_{A_2} = p_B w_3 + (1 - p_B) w_4$. The change in frequency of allele A_1 is then

$$\Delta p_A = p_A(t)\,(1 - p_A(t))\,(W_{A_1} - W_{A_2}) + O(\zeta^2). \tag{9.47b}$$

We encountered a similar equation in Chapter 3. Indeed, the one-gene haploid recursion equation, $p_A(t + 1) = p_A W_{A_1}/(p_A W_{A_1} + (1 - p_A) W_{A_2})$, gives us the same approximation (9.47b) if we assume weak selection (equation P1.7). Analogous results apply to the dynamics at gene **B**.

According to (9.47), allele frequency change is not affected by the linkage disequilibrium (to leading order) as long as the system is near the quasi-equilibrium value of the disequilibrium. Thus, assuming that selection is weak relative to recombination, the main impact that selected genes have on one another is through their marginal fitnesses, because the fitness effect of any allele depends on the alleles currently present at other genes (Problem 9.6). Figure 9.6 compares the exact dynamics of the frequency of allele A_1 to the quasi-equilibrium approximation (9.47). The result is quite accurate over a broad range of conditions, as long as the recombination rate is large relative to the differences in fitness among the chromosome types.

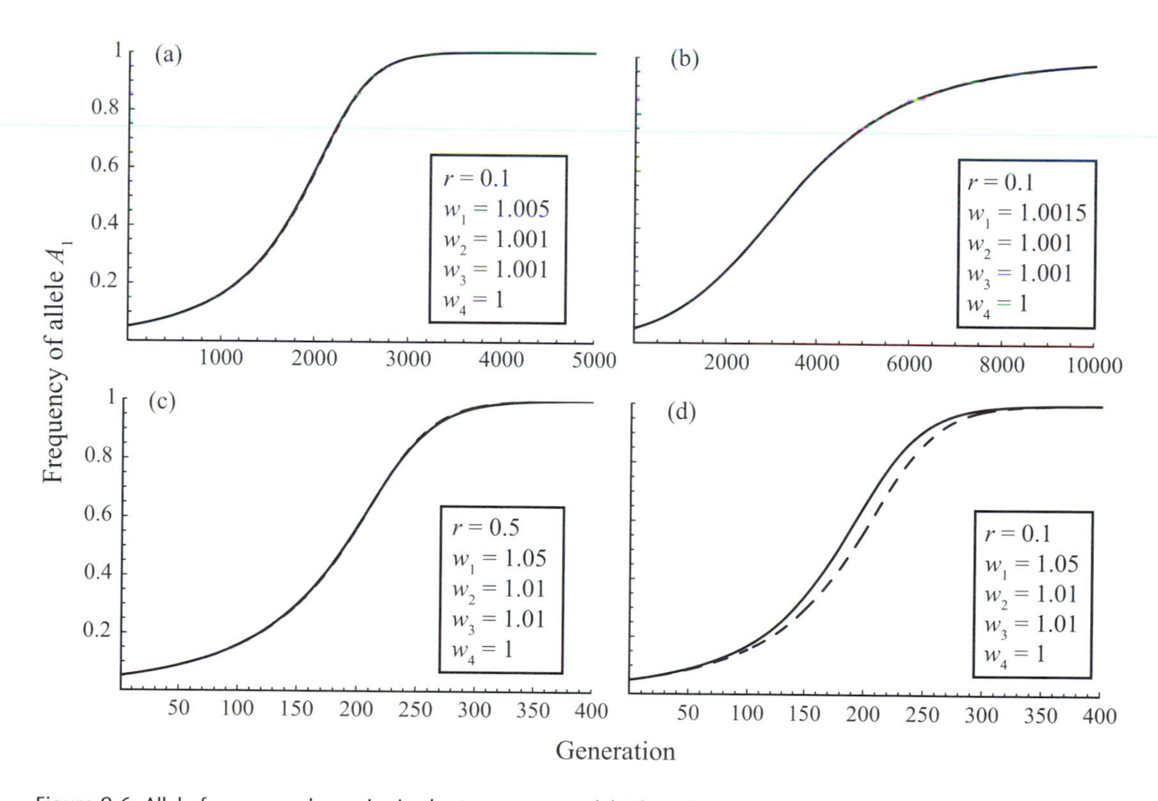

Figure 9.6: Allele frequency dynamics in the two-gene model. The solid curves illustrate the exact dynamics (9.40) and the dashed curves (often indistinguishable) illustrate the quasi-equilibrium approximation (9.47) for the frequency of allele A_1. In every case except (d), recombination is frequent relative to the fitness differences among genotypes, and the quasi-equilibrium approximation is excellent.

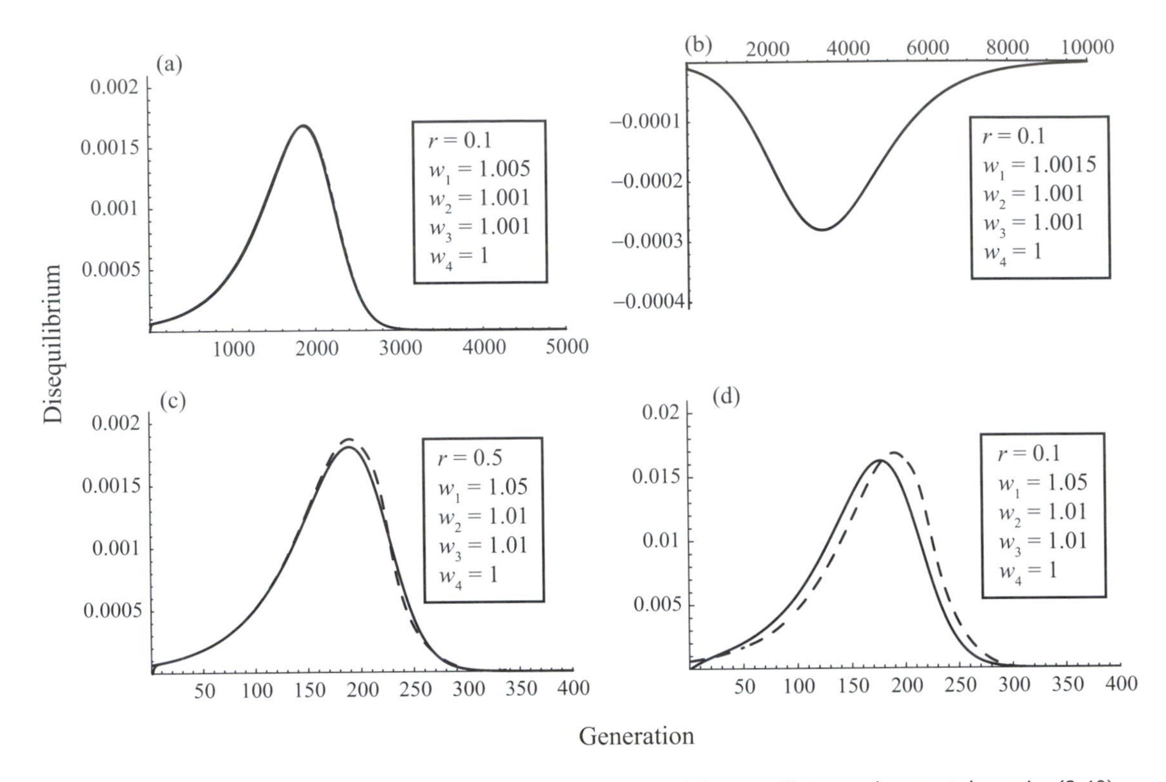

Figure 9.7: Disequilibrium dynamics in the two-gene model. The solid curves illustrate the exact dynamics (9.40), and the dashed curves (often indistinguishable) illustrate the quasi-equilibrium approximation (9.45). Even though we use the approximate allele frequencies from Figure 9.6 rather than the exact allele frequencies in (9.45), the match is excellent except when recombination and selection are similar in magnitude (case d).

Given the predicted changes in allele frequency, we can now turn our attention back to the linkage disequilibrium and describe how the quasi-equilibrium value of D changes in response to selection over the longer time scale. In each generation, we can plug the predicted allele frequencies using (9.47) into equation (9.45) to determine the disequilibrium (Figure 9.7). We can see that our approximation for the disequilibrium is also very accurate over a broad range of conditions.

Before concluding this example, let us perform one final calculation and check that our assumptions are consistent with the claim that the dynamics of the disequilibrium, D, are faster than the dynamics of the allele frequencies. To leading order in ζ, the recursions (9.40) with the fitnesses (9.42) are

$$p_A(t+1) = p_A(t) + O(\zeta) \tag{9.48a}$$

$$p_B(t+1) = p_B(t) + O(\zeta) \tag{9.48b}$$

$$D(t+1) = (1-r)\,D(t) + O(\zeta). \tag{9.48c}$$

Furthermore, we can subtract $\hat{D}_Q$ from both sides of (9.48c) to determine how far the system is from the quasi-equilibrium. Again, to leading order,

$$\left\{D(t+1) - \hat{D}_Q\right\} = (1-r)\left\{D(t) - \hat{D}_Q\right\} + O(\zeta). \tag{9.48d}$$

Equations (9.48) indicate that, over the short term, the allele frequencies change very slowly, by an amount proportional to the small quantity ζ, whereas the disequilibrium decays toward its quasi-equilibrium value by a factor $(1 - r)$ each generation. Therefore, provided that selection is considerably weaker than recombination (i.e., provided that ζ is considerably smaller than r), it is reasonable to approximate the dynamics of the allele frequencies by supposing that the disequilibrium D is at its quasi-equilibrium value.

In closing, many recent theoretical advances in ecology and evolution have relied on a separation of time scales. A separation of time scales underlies results from adaptive dynamics (Geritz et al. 1998; Metz et al. 1996), QLE analyses (Barton 1995; Barton and Turelli 1991; Kirkpatrick et al. 2002; Turelli and Barton 1990), and coalescent theory involving structured populations (Nordborg 1997; Nordborg and Krone 2002; Wakeley 1999). For many problems, a separation of time scales provides the only analytically tractable method to explore how a biological system behaves. Clearly, however, the method is not a panacea—it can only be used when there is reason to believe that some processes occur more quickly than others. The advantage, however, is that we get an analytical prediction as well as insight into the fundamental forces at work. We can say how ecological patterns evolve, how disequilibria develop, and how multigene systems behave. With complex models, the main alternative—simulations—is also useful, but it is often difficult to generalize beyond a limited set of parameters, and difficult to develop a deeper understanding of why a model behaves as it does.

9.4 Concluding Message

In this chapter, we have explored various routes by which a general solution can be obtained for models involving more than one variable. For a linear model, there is a recipe (Recipe 9.1 and Box 9.3) that can be followed to obtain a general solution. This recipe requires that we determine all of the eigenvalues and eigenvectors of a model explicitly, however, and therefore it isn't guaranteed to help. Sometimes, even if you are able to write down a general solution, the resulting expression is too complex to readily interpret. As a rule of thumb, if the eigenvalues of a linear model cannot be found (or are complicated), then the general solution cannot be found (or will be complicated) as well.

Even when obtaining a general solution is elusive, the concepts presented in this chapter allow us to understand some of the mathematical rules used in previous chapters. We can now understand why the leading eigenvalue is so important in stability analysis. The general solution of a linear model (or a linear approximation to a nonlinear model) is dominated by the leading eigenvalue as time passes (Box 9.1), and this therefore determines the fate of small perturbations. The general solution also helps us to interpret more complicated situations, such as when the eigenvalues are complex (Boxes 9.2 and 9.3) or when there is more than one eigenvalue with the magnitude of the leading eigenvalue.

As with Chapter 6, however, the picture is not so rosy for nonlinear models and there is no single recipe to follow that is guaranteed to yield a general solution. Various techniques, from brute force iteration to transformations (e.g., section 9.3.1), must be tried with the hope of hitting upon a general solution. When a general solution cannot be obtained, it can still be fruitful to search for an approximate solution. The separation of time scales technique introduced in section 9.3.2 is one of the most commonly used and successful ways to gain an approximate analytical solution to nonlinear models involving multiple variables.

Problems

Problem 9.1: In addition to being colonized by crabs, sea anemones are also sometimes colonized by specialized fish species. Let the variables $O(t)$ and $U(t)$ denote the number of occupied and unoccupied anemones in generation t. Suppose that, in each generation, an empty anemone becomes occupied with probability κ, and an occupied anemone loses its fish through mortality and becomes unoccupied with probability μ. This results in the discrete-time model:

$$\begin{pmatrix} O(t+1) \\ U(t+1) \end{pmatrix} = \begin{pmatrix} 1-\mu & \kappa \\ \mu & 1-\kappa \end{pmatrix} \begin{pmatrix} O(t) \\ U(t) \end{pmatrix}.$$

(a) Find the general solution to this model describing the long-term dynamics of occupied and unoccupied anemones (simplify your answer as much as possible). (b) Describe in words what this general solution tells you about the sea anemone dynamics.

Problem 9.2: Many eukaryotic genes have both exons and introns, where only exons code for protein sequence. mRNAs transcribed from such genes initially include the introns, which must be spliced out before the mRNAs can be translated. Let $n_1(t)$ equal the number of unspliced "pre-mRNAs," which are produced by the cell at a rate, c, per minute, and let $n_2(t)$ equal the number of spliced "processed mRNAs" within a cell at time t. If a is the fraction of the pre-mRNAs that are spliced per minute and if a proportion d of processed mRNAs degrade per minute, the numbers of pre-mRNA and processed mRNA change over time according to the equations:

$$n_1(t+1) = (1-a)\, n_1(t) + c,$$
$$n_2(t+1) = a\, n_1(t) + (1-d)\, n_2(t).$$

(a) Find the one equilibrium of this model, i.e., the point at which *both* types of mRNA remain constant in number over time. (b) Perform a transformation of this model. Define the new variables $\varepsilon_1(t) = n_1(t) - \hat{n}_1$ and $\varepsilon_2(t) = n_2(t) - \hat{n}_2$. This transformation represents how far each type of mRNA is from its equilibrium. Determine recursion equations for $\varepsilon_1(t+1)$ and $\varepsilon_2(t+1)$ in terms of $\varepsilon_1(t)$ and $\varepsilon_2(t)$. (c) Show that the recursions developed in part (b) can be written in matrix form:

$$\begin{pmatrix} \varepsilon_1(t+1) \\ \varepsilon_2(t+1) \end{pmatrix} = \mathbf{M} \begin{pmatrix} \varepsilon_1(t) \\ \varepsilon_2(t) \end{pmatrix}$$

and specify the transition matrix **M**. [If you cannot write the recursions in this form or if the matrix **M** looks ugly, then go back and check your answers to (a) and (b).] (d) Calculate the eigenvalues of **M**. Given that both a and d represent proportions, can there be an eigenvalue whose magnitude is greater than one? Whenever there is an eigenvalue greater than one, the equilibrium is unstable. Conversely, as long as all eigenvalues are less than one in magnitude, the equilibrium will be stable. (d) Calculate the eigenvectors of **M**. (e) Write a diagonal matrix **D** containing the two eigenvalues along the diagonal. Find $\mathbf{D}^t$ by raising these eigenvalues to the tth power. (f) Write a transformation matrix **A** containing the eigenvector associated with the first eigenvalue of **D** in the first column and the eigenvector associated with the second eigenvalue of **D** in the second column. (g) Find the inverse matrix $\mathbf{A}^{-1}$. (h) With these calculations in hand, predict the numbers of pre-mRNA and processed mRNA at any point in time (the general solution) by multiplying out and simplifying:

$$\begin{pmatrix} \varepsilon_1(t) \\ \varepsilon_2(t) \end{pmatrix} = \mathbf{A}\,\mathbf{D}^t\,\mathbf{A}^{-1} \begin{pmatrix} \varepsilon_1(0) \\ \varepsilon_2(0) \end{pmatrix}.$$

You can check your answer by setting $t = 0$ to regain the initial condition and $t = \infty$ to regain the equilibrium. (i) Transform the general solution obtained in (h) back into the original variables $n_1(t)$ and $n_2(t)$ to predict how the numbers of pre-mRNAs and processed mRNAs change over time.

Problem 9.3: You are presented with a series of linear equations that you solve by matrix manipulation. In your general solution, you let time go to infinity and find that the system reaches a specific point. Is this point an equilibrium? Is it unstable or stable? If stable, is it locally or globally stable? Should you do a local stability analysis?

Problem 9.4: Perform a proportional transformation of the logistic model $dn/dt = r\,n\,(1 - n/K)$. What choices of $\tilde{n}$ and $\tilde{t}$ lead to the simplified differential equation $dn^*/dt^* = n^*\,(1 - n^*)$? What do these choices of $\tilde{n}$ and $\tilde{t}$ represent biologically?

Problem 9.5: In the text, we performed a separation of time scales using a diploid version of density-dependent natural selection. Using the haploid version of the model,

$$N(t+1) = \overline{W}(N(t), p(t))\,N(t),$$

$$p(t+1) = \frac{W_A(N(t))}{\overline{W}(N(t), p(t))}\,p(t),$$

where $\overline{W}(N(t), p(t)) = p(t)\,W_A(N(t)) + (1 - p(t))\,W_a(N(t))$, $W_A(N) = 1 + r(1 - \alpha_A N)$, and $W_a(N) = 1 + r(1 - \alpha_a N)$, perform a separation of time scales. As in the text, assume that ecological changes occur over a faster time scale than allele frequency changes. Compare your answers to (9.35) and (9.37).

Problem 9.6: In the two-locus model of selection, we performed a separation of time scales and found that the change in frequency of allele A_1 is the same as the one-locus recursion,

$$p_A(t+1) = \frac{p_A\,W_{A_1}}{p_A\,W_{A_1} + (1 - p_A)W_{A_2}},$$

to leading order in the selection coefficients. Here, $W_{A_1} = p_B\, w_1 + (1 - p_B)\, w_2$ and $W_{A_2} = p_B\, w_3 + (1 - p_B)\, w_4$ represent the marginal fitnesses of alleles A_1 and A_2, averaged over the genetic states at the **B** locus. Using this QLE approximation, show that the dynamics at locus **A** do not depend on the allele frequencies at locus **B** under a *multiplicative* fitness regime:

$$\begin{aligned} w_1 &= (1 + s_A)(1 + s_B), \\ w_2 &= 1 + s_A, \\ w_3 &= 1 + s_B, \\ w_4 &= 1, \end{aligned}$$

but not in the case of the additive fitness regime (9.46). This result supports the idea that it is natural to measure fitnesses as a departure from a multiplicative expectation (Supplementary Material 9.1).

Problem 9.7: Consider the following consumer-resource model with two consumers and two resources:

$$\begin{aligned} \frac{dR_1}{dt} &= r_1 R_1 \left(1 - \frac{R_1}{K_1}\right) - N_1\, a_{11}\, R_1 - N_2\, a_{21}\, R_1, \\ \frac{dR_2}{dt} &= r_2 R_2 \left(1 - \frac{R_2}{K_2}\right) - N_1\, a_{12}\, R_2 - N_2\, a_{22}\, R_2, \\ \frac{dN_1}{dt} &= N_1\,(a_{11} R_1 + a_{12} R_2 - \mu_1), \\ \frac{dN_2}{dt} &= N_2\,(a_{21} R_1 + a_{22} R_2 - \mu_2). \end{aligned}$$

The resources have abundances R_1 and R_2 and the consumers have abundances N_1 and N_2. The resources are assumed to grow logistically in the absence of consumption, and the consumers eat these resources according to a type-1 functional response. This model provides a mechanistic description of the dynamics of resource competition between the two consumer species. Use a separation of time scales in which the dynamics of the resources are assumed to occur much faster than those of the consumer to demonstrate that the model can be reduced to a two-variable model having the form

$$\begin{aligned} \frac{dN_1}{dt} &= \rho_1 N_1 \left(1 - \frac{N_1 + \alpha_{12} N_2}{\chi_1}\right), \\ \frac{dN_2}{dt} &= \rho_2 N_2 \left(1 - \frac{N_2 + \alpha_{21} N_1}{\chi_2}\right), \end{aligned}$$

where you should specify how ρ_1, ρ_2, α_{12}, α_{21}, χ_1 and χ_2 relate to the parameters of the original model. This two-variable model is identical in form to the Lotka-Volterra competition equations (3.15).

Further Reading

For further information on using a separation of time scales in population-genetic models, consult

- Kirkpatrick, M., T. Johnson, and N. Barton. 2002. General models of multilocus evolution. *Genetics* 161:1727–1750.

For further information on general solutions in discrete and continuous time, consult

- Edelstein-Keshet, L. 1988. *Mathematical Models in Biology*. Birkhäuser Mathematics Series, McGraw-Hill, New York.

References

Barton, N. H. 1995. A general model for the evolution of recombination. *Genet. Res.* 65:123–144.

Barton, N. H., and M. Turelli. 1991. Natural and sexual selection on many loci. *Genetics* 127:229–255.

Felsenstein, J. 2004. *Inferring Phylogenies*. Sinauer Associates, Sunderland, Mass.

Geritz, S.A.H., É. Kisdi, G. Meszéna, and J.A.J. Metz. 1998. Evolutionarily singular strategies and the adaptive growth and branching of the evolutionary tree. *Evol. Ecol.* 12:35–57.

Kimura, M. 1980. A simple method for estimating evolutionary rates of base substitutions through comparative studies of nucleotide sequences. *J. Mol. Evol.* 16:111–120.

Kirkpatrick, M., T. Johnson, and N. Barton. 2002. General models of multilocus evolution. *Genetics* 161:1727–1750.

Li, W. H. 1997. *Molecular Evolution*. Sinauer Associates, Sunderland, Mass.

Metz, J.A.J., S.A.H. Geritz, G. Meszéna, F.J.A. Jacobs, and J. S. Van Heerwaarden. 1996. Adaptive dynamics: A geometrical study of the consequences of nearly faithful reproduction. Pp. 183–231 in S. J. van Strien and S. M. Verduyn Lunel, eds. *Stochastic and Spatial Structures of Dynamical Systems*. North-Holland Elsevier, Amsterdam.

Nordborg, M. 1997. Structured coalescent processes on different time scales. *Genetics* 146:1501–1514.

Nordborg, M., and S. Krone. 2002. Separation of time scales and convergence to the coalescent in structured populations. Pp. 194–232 in M. Slatkin and M. Veuille, eds., *Modern Developments in Theoretical Population Genetics*. Oxford University Press, Oxford.

Turelli, M., and N. H. Barton. 1990. Dynamics of polygenic characters under selection. *Theor. Popul. Biol.* 38:1–57.

Wakeley, J. 1999. Nonequilibrium migration in human history. *Genetics* 153:1863–1871.

CHAPTER 10

Dynamics of Class-Structured Populations

Chapter Goals:

- To describe the dynamics of age-structured and class-structured populations
- To introduce concepts and techniques important in demography

Chapter Concepts:

- Demography
- Vital statistics
- Long-term growth rate
- Stable class distribution
- Reproductive value
- Sensitivity of the leading eigenvalue
- Leslie matrix

10.1 Introduction

One simplification in the models considered so far is that all individuals in the population (or at least all individuals of a particular genotype) are identical in terms of their probability of survival and reproductive output. In this chapter we present the most commonly used techniques for constructing models of "class-structured populations." A class-structured model is one in which individuals fall into different classes, and where each class has its own survival and reproductive characteristics. For example, we might classify individuals by age and specify age-specific survival probabilities or reproductive outputs. Alternatively, we might classify individuals by size, sex, developmental stage, or physiological condition.

Some class variables are continuous (e.g., age, size) whereas others are discrete (e.g., sex, reproductively mature/immature). To be completely faithful to the biology, we would ideally use different techniques for models with continuous and discrete variables (see Box 10.1). Nevertheless, many class-structured models in the literature treat the class variable as discrete regardless of its true form, because such models are typically easier to construct, analyze, and interpret. Furthermore, discrete classes usually provide an adequate approximation for continuous variables as long as they are broken into a sufficiently large number of classes (e.g., measuring human age structure by rounding everyone's age to the nearest year). Therefore, we consider only discrete-class (and discrete-time) models. In addition, we restrict attention to linear models because most of the unique and useful properties of class-structured models apply only to these.

With the above restrictions, you might now be asking yourself why we need to devote a separate chapter to these models. After all, Recipe 9.1 provides the complete general solution to discrete-time linear models with multiple classes (i.e., multiple variables). While in principle this is true, in practice Recipe 9.1 becomes unworkable for models with several classes (i.e., several variables). For such models, the eigenvalues are often difficult (or impossible) to calculate, and even when they are calculable, the resulting general solution is long and cumbersome. This complexity obscures some simple patterns and useful properties of many models, especially if we are primarily interested in the long-term dynamics. Therefore we need additional techniques to draw out these long-term predictions. Such long-term predictions have played an important role in

Box 10.1: Modeling Class-Structured Populations

Constructing a class-structured population model involves identifying a variable of interest that categorizes individuals in the population. Let us consider "sex" and "size" as two specific examples. The variable "sex" is a discrete variable, and therefore we simply need to model the dynamics of the number of individuals in each of the two discrete classes, males (n_m) and females (n_f). This can be done in either discrete time or continuous time, resulting in either a pair of coupled recursion equations,

$$n_m(t+1) = f(n_m(t), n_f(t)),$$

$$n_f(t+1) = g(n_m(t), n_f(t)), \tag{10.1.1a}$$

or a pair of coupled ordinary differential equations,

$$\frac{\mathrm{d}n_m}{\mathrm{d}t} = f(n_m(t), n_f(t)),$$

$$\frac{\mathrm{d}n_f}{\mathrm{d}t} = g(n_m(t), n_f(t)). \tag{10.1.1b}$$

We have considered these general sorts of models in Chapters 7 and 8, and their construction and analysis therefore poses no new challenges.

"Size," on the other hand, is a continuous variable, and therefore we might want to model the dynamics of the number of individuals over the continuum of possible sizes. Again this can be handled in discrete or continuous time, resulting in either a recursion equation that applies to individuals of any size, s,

$$n(s,t+1) = F[n] \tag{10.1.2a}$$

or a *partial differential equation*,

$$\frac{\partial n(s,t)}{\partial t} = F[n]. \tag{10.1.2b}$$

The term "partial differential equation" stems from the fact that the variable n in equation (10.1.2b) is now a function of the two independent variables s and t. Although differentiation can be carried out with respect to either variable, here we are interested in changes to the population over time.

Also, notice the unusual notation that we have used in (10.1.2) involving $F[n]$. We used this notation to signal that something quite different is going on with the right-hand side. In particular, the function on the right can involve any operation on the function n, including derivatives or integrals. For example, the recursion equation (10.1.2a) might be $n(s,t+1) = \int_0^\infty n(\sigma,t)p(s,\sigma,t)\mathrm{d}\sigma$,

(continued)

Box 10.1 ***(continued)***

where $p(s,\sigma,t)$ represents the probability that an individual of size σ becomes an individual of size s during the time period t. This integration "sums up" all of the individuals that move into size class s from all other size classes during the time period in question. Similarly, using the chain rule, the partial differential equation (10.1.2b) might take the form $\partial n(s,t)/\partial t = -(\partial n(s,t)/\partial s)(ds/dt)$. The key point to recognize is that models such as (10.1.2) call for some new mathematical techniques. The required techniques are typically more sophisticated than those dealt with so far, and they are beyond the scope of this book (see Further Readings). Importantly, however, most continuous class variables can be reasonably well approximated by discrete variables with a sufficiently large number of classes. Therefore models such as (10.1.1) will usually suffice.

The study of population age structure or size structure is known as *demography*.

demography, which is the study of natural populations (often human populations), focusing on statistics such as size and age structure.

In section 10.2 below we illustrate the construction of class-structured models. Then we explore four main types of questions that can be addressed without a full general solution. In section 10.3 we ask: (i) What is the long-term growth rate of a population? and (ii) what is the long-term class structure of a population? These questions can be important for ensuring the persistence of endangered species or for assessing the risk of invasion of a non-native species. Section 10.4 then asks: (iii) Which classes contribute most to the long-term growth rate of a population? This question is important in conservation programs that can target and protect particular classes of individuals. It will also turn out to be important in later chapters when we develop models of evolution in class structure populations. Section 10.5 then asks: (iv) Which parameters have the greatest impact on the long-term growth rate of a population? Again, this question is important in deciding among intervention strategies that alter fertility and survival parameters in a conservation program. Finally, in section 10.6 we apply these concepts to age-structured populations, which represents a particularly important type of class structure.

10.2 Constructing Class-Structured Models

We begin by considering two examples of class-structured models. First, recall the model that you were asked to develop in Problem 7.2 of Chapter 7. This model tracked the dynamics of a population containing the two classes; "juveniles" and "adults." The discrete-time recursion is

$$\begin{pmatrix} J(t+1) \\ A(t+1) \end{pmatrix} = \begin{pmatrix} 0 & b \\ p_j & p_a \end{pmatrix} \begin{pmatrix} J(t) \\ A(t) \end{pmatrix} \tag{10.1}$$

where $J(t)$ and $A(t)$ denote the number of juveniles and adults at time t, b is the fecundity of an adult per time step, and p_j and p_a are the juvenile and adult survival probabilities. Equation (10.1) is a linear system of recursion equations. In demography, the matrix of coefficients, $\begin{pmatrix} 0 & b \\ p_j & p_a \end{pmatrix}$, is referred to as the transition matrix or the projection matrix, because it projects the number of individuals in each class at time step t into the number in each class at time step $t + 1$. The parameters b, p_j, and p_a are often referred to as vital statistics.

In demography, age-specific mortality rates and birth rates are known as *vital statistics*.

For our second example, we focus on a more specific biological situation, and construct a model for the recovery of the right whale population in the North Atlantic. Right whales of the North Atlantic Ocean were once abundant, but whaling pressure in the 1800s and early 1900s drove this species near to extinction. With the end of commercial whaling, however, the species is now slowly beginning to recover.

Females control the rate of offspring production and are critical to the growth and recovery of the species. Therefore, we confine our model to the female portion of the population. Right whales are an excellent example of a species where accounting for class structure is important, because individuals are long lived and their survival probability and reproductive output changes significantly over a female's lifespan. We will incorporate this in a relatively simple fashion, dividing the female population into four different stages: (i) newborn calves, (ii) immature females, (iii) mature females, and (iv) females that are actively reproducing (Figure 10.1).

Using $n_C(t)$, $n_I(t)$, $n_M(t)$, and $n_R(t)$ to denote the population size of these four different classes in year t, we can proceed to add up all of the ways these numbers change in one year from Figure 10.1. Calves are produced by reproductively active females only, and their fecundity per year is b. Thus, the number of newborn calves in the next year is

$$n_C(t+1) = n_R(t)\, b. \tag{10.2a}$$

The number of immature females in the next year will be the number of calves in this year that mature by next year (with probability s_{IC}), plus the number of

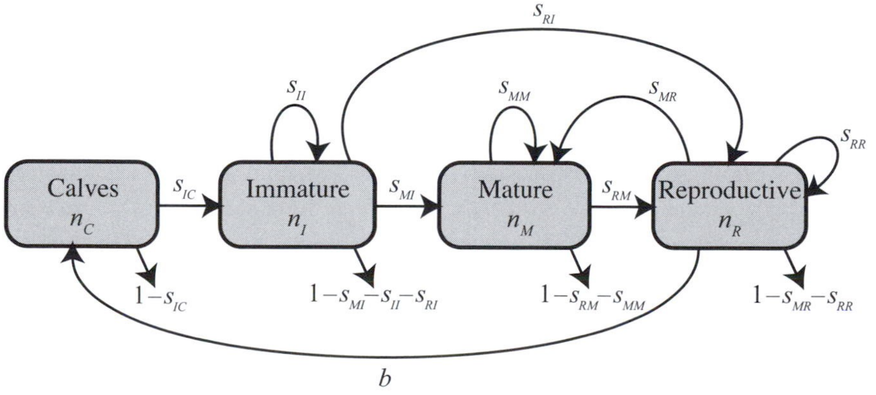

Figure 10.1: A flow diagram for the discrete-time model of right whales. The parameters, s_{ij} give the probabilities of an individual moving from j to i, and b is the fecundity of a reproducing female.

immature females this year that survive one year and remain immature (with probability s_{II}):

$$n_I(t+1) = n_C(t)\ s_{IC} + n_I(t)\ s_{II}. \tag{10.2b}$$

The number of mature (but nonreproductive) females in the next year will be the number of immature females that survive one year and become mature (with probability s_{MI}), plus the number of mature females that survive one year and remain mature (with probability s_{MM}), plus the number of reproductive females that finish their current bout of reproduction and move back into the mature nonreproductive class until they gain enough energy to become reproductive again (with probability s_{MR}). Altogether, we have

$$n_M(t+1) = n_I(t)\ s_{MI} + n_M(t)\ s_{MM} + n_R(t)\ s_{MR}. \tag{10.2c}$$

Finally, the number of reproductive females in the next year will be the number of immature females that survive and move directly to the reproductive class (with probability s_{RI}), plus the number of mature female that survive and become reproductive (with probability s_{RM}), plus the number of reproductive females that survive and remain reproductive (with probability s_{RR}):

$$n_R(t+1) = n_I(t)\ s_{RI} + n_M(t)\ s_{RM} + n_R(t)\ s_{RR}. \tag{10.2d}$$

Equations (10.2a–10.2d) can be written more compactly in matrix notation as

$$\begin{pmatrix} n_C(t+1) \\ n_I(t+1) \\ n_M(t+1) \\ n_R(t+1) \end{pmatrix} = \begin{pmatrix} 0 & 0 & 0 & b \\ s_{IC} & s_{II} & 0 & 0 \\ 0 & s_{MI} & s_{MM} & s_{MR} \\ 0 & s_{RI} & s_{RM} & s_{RR} \end{pmatrix} \begin{pmatrix} n_C(t) \\ n_I(t) \\ n_M(t) \\ n_R(t) \end{pmatrix}. \tag{10.3}$$

Because the elements of the transition matrix are constants (i.e., not functions of any of the dynamic variables n_i), system (10.3) is a linear model.

Before proceeding to the new ideas and techniques of this chapter, let us explore simulations to get a better sense of the behavior of these two class-structured models, starting with model (10.1). Figure 10.2 displays the total population size as a function of time, as well as its growth rate per time step, and the proportion of the population that is made up of juveniles and adults. For the parameter values used, the population eventually grows at a constant rate, and the proportion of the population that is in each of the two classes settles down to a constant value. A consideration of the phase-plane diagram for this model (Figure 10.2) provides a nice way to understand this long-term behavior. At first, the system moves in different directions, but eventually it settles down and moves in a direction parallel to the eigenvector associated with the leading eigenvalue. From the definition of an eigenvector (a line representing values of the dynamic variables along which their relative values remain constant), we thus expect the relative number of individuals in each class to settle down to a constant number, given by the elements of this eigenvector.

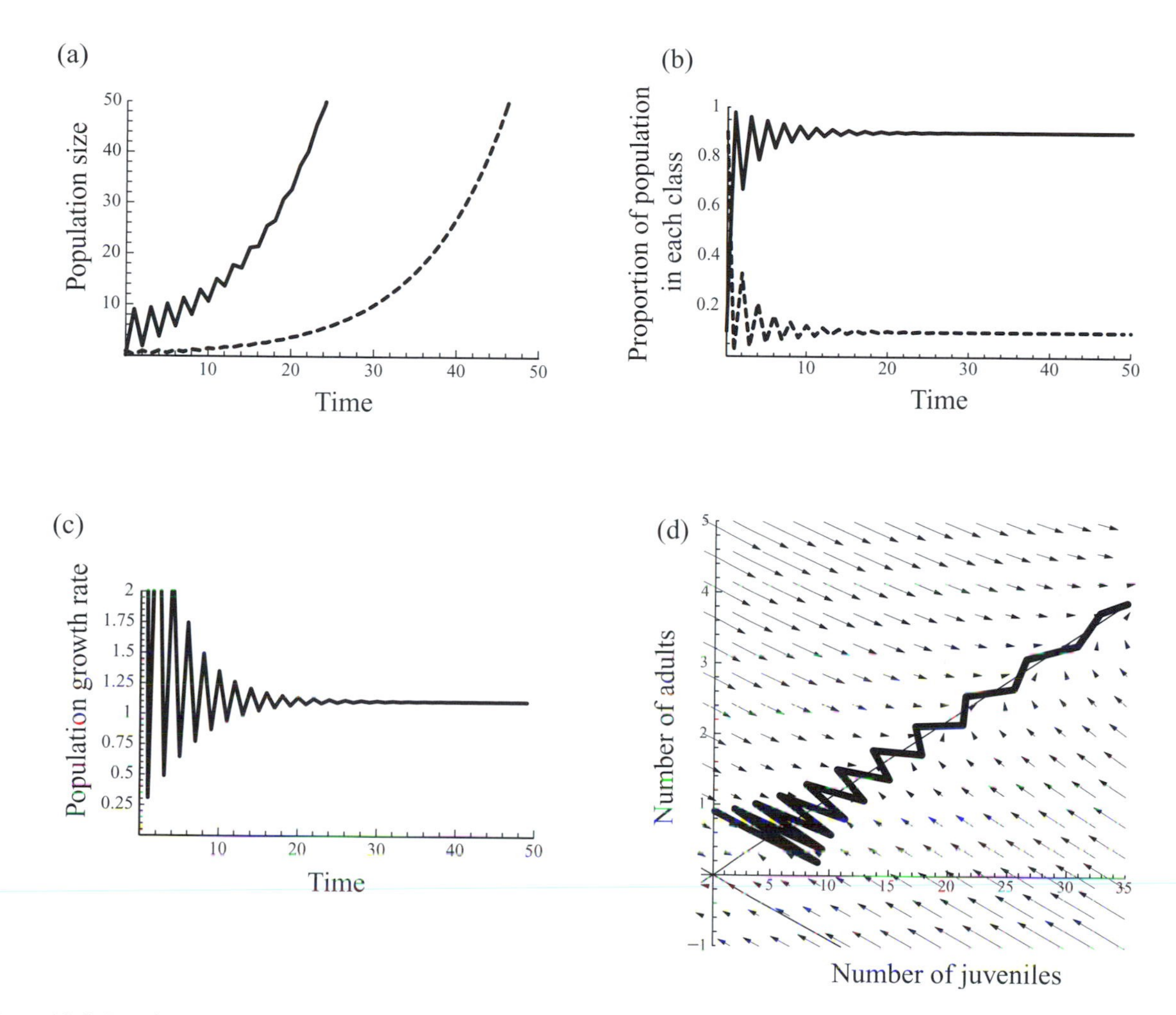

Figure 10.2: Population dynamics in an age-structured model with juveniles and adults. Model (10.1) is iterated with $b = 10$, $p_j = 0.1$, $p_a = 0.2$, and starting conditions $J(0) = 0.1$ and $A(0) = 0.9$. (a) Number of juveniles (solid line) and adults (dashed line) at each time step. (b) The proportion of the population that are juveniles (solid line) and adults (dashed line) at each time step. (c) The growth rate of the population in each time step. It eventually reaches a constant value given by the leading eigenvalue of the transition matrix. (d) The phase plane with the eigenvectors plotted as thin lines. Thick line connects the points in the phase plane that represent the trajectory of the system. The system approaches an eigenvector, at which point the relative proportion of the population in each class reaches a stable value given by the elements of the eigenvector.

Furthermore, once this has occurred, we expect the population as a whole to grow at a rate that is given by the leading eigenvalue.

In fact, in Problem 7.2 you were asked to determine the condition under which the population grows indefinitely, by determining when the equilibrium $\hat{J} = 0, \hat{A} = 0$ is unstable. This requires that at least one of the eigenvalues of the transition matrix be greater than one in absolute value. The two eigenvalues are

$$\lambda_1 = \frac{1}{2}\left(p_a + \sqrt{p_a^2 + 4bp_j}\right), \qquad \lambda_2 = \frac{1}{2}\left(p_a - \sqrt{p_a^2 + 4bp_j}\right). \quad (10.4)$$

These eigenvalues are both real valued. Furthermore, λ_1 is always larger in magnitude than λ_2, assuming that there is some adult survival, $p_a > 0$. Therefore, the entire population ultimately grows in size if $\lambda_1 > 1$, and then it does so by a factor λ_1 each generation. With these considerations we can now be confident that this model exhibits the long-term behavior depicted in Figure 10.2 regardless of the parameter values (provided $\lambda_1 > 1$). Furthermore, if we are interested in knowing how this long-term growth rate changes as a result of changes in different parameters, we can simply examine how λ_1 changes, without having to resort to more simulations.

Analogous simulation results appear to hold for the model of the right whale population (Figure 10.3). The proportion of whales in each class and the population growth rate all settle down to constant values over time. Now, however, we cannot draw a phase plot because the model is four-dimensional. We cannot obtain simple expressions for the eigenvalues of this model either, because the

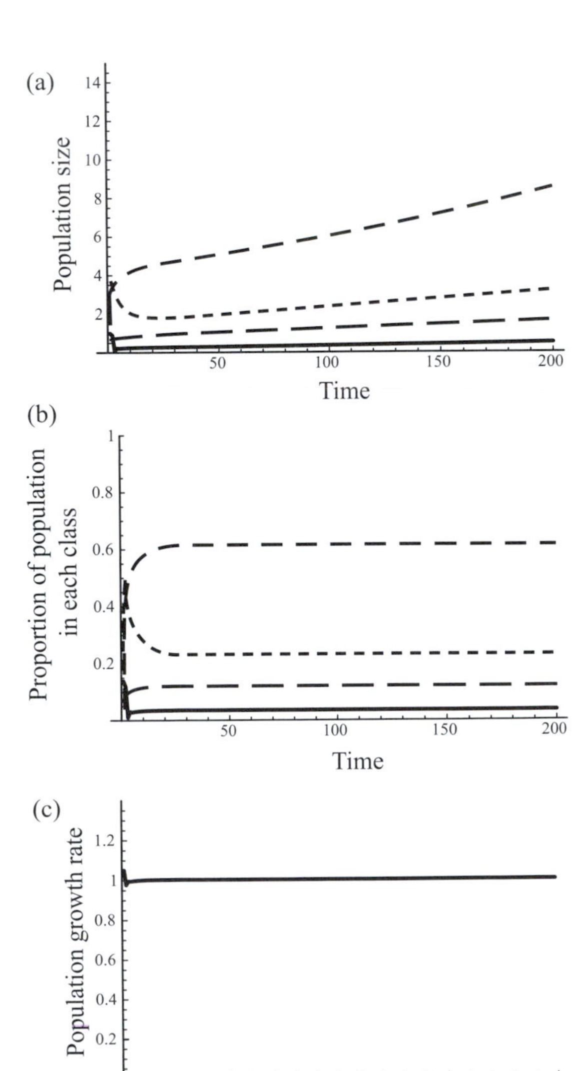

Figure 10.3: Population growth of a right whale population. Model (10.4) is iterated using parameter values for a right whale population taken from estimates in Fujiwara and Caswell (2001): $s_{IC} = 0.92$, $s_{II} = 0.86$, $s_{MI} = 0.08$, $s_{MM} = 0.8$, $s_{MR} = 1 - \mu$, $s_{RI} = 0.02$, $s_{RM} = 0.19$, $s_{RR} = 0$, where μ is an unknown mortality rate of reproducing females and b is the unknown fecundity; all parameters are per year. Here, we assume that $b = 0.3$ and $\mu = 0.12$. Solid lines represent the calf class, and increasingly large dashed lines represent the immature, mature, and reproductive classes respectively. (a) Number of individuals in each class. (b) Proportion of the population in each class. (c) Population growth rate.

characteristic equation is now a fourth-order polynomial. Therefore, we are left wondering if the simulation results of this Figure 10.3 are characteristic of all parameter values, or if other types of qualitative dynamics sometimes occur. In the next section we provide a general way to answer this question for any class-structured model.

10.3 Analyzing Class-Structured Models

In Chapter 9 we described the general solution for any linear multivariable model. For example, the two-class model with juveniles and adults has a general solution of the form

$$\begin{pmatrix} J(t) \\ A(t) \end{pmatrix} = \mathbf{A}\,\mathbf{D}^t\mathbf{A}^{-1}\begin{pmatrix} J(0) \\ A(0) \end{pmatrix}, \tag{10.5}$$

where $\mathbf{D}$ is the 2×2 diagonal matrix containing the two eigenvalues λ_1 and λ_2 from (10.4), $\mathbf{A}$ is a matrix whose first column is the eigenvector associated with λ_1 and whose second column is the eigenvector associated with λ_2, and $\mathbf{A}^{-1}$ is the inverse of $\mathbf{A}$. Once the eigenvalues and eigenvectors have been calculated, equation (10.5) can be used to predict the number of juveniles or adults at any time of interest. Similarly, the general solution for the model of the right whale population dynamics is

$$\begin{pmatrix} n_C(t) \\ n_I(t) \\ n_M(t) \\ n_R(t) \end{pmatrix} = \mathbf{A}\,\mathbf{D}^t\mathbf{A}^{-1}\begin{pmatrix} n_C(0) \\ n_I(0) \\ n_M(0) \\ n_R(0) \end{pmatrix}, \tag{10.6}$$

where $\mathbf{D}$ is now the 4×4 diagonal matrix containing the four eigenvalues of the transition matrix in (10.4), $\mathbf{A}$ is a matrix whose columns are the eigenvectors of this transition matrix, and $\mathbf{A}^{-1}$ is the inverse of $\mathbf{A}$. Although we can calculate the required eigenvalues and eigenvectors to use solution (10.5), we cannot do so for solution (10.6). Fortunately, the results of Box 9.1 can be used to obtain important information about the long-term behavior of either model without actually calculating these quantities.

The first main question of interest concerns the long-term growth rate of a population. In Box 9.1 we showed that the long-term dynamics of a linear model can be approximated by ignoring all eigenvalues in $\mathbf{D}$ other than the leading eigenvalue (assuming that there is such an eigenvalue). Each variable then grows in the long term by a constant factor given by the leading eigenvalue λ_1. As a result, the total population size eventually grows at this rate. Figures 10.2 and 10.3 demonstrate exactly this behavior for the juvenile/adult model and for the right whale model, respectively. Importantly, this long-term growth rate does not depend on the initial population size or the initial number of individuals in each class.

The *long-term growth rate* of a population is given by the leading eigenvalue.

The second main question of interest concerns the long-term proportion of the population found in each class. Again, provided that there is a leading eigenvalue larger than all other eigenvalues, Box 9.1 indicates that the proportion of individuals in each class eventually settles down to a constant over time. The relative frequencies in each class are given by the elements of the eigenvector corresponding to the leading eigenvalue of the transition matrix. This is the so-called "stable class distribution," and the eigenvector describing it is referred to as the *dominant* or *leading* eigenvector.

The *stable class distribution* describes the long-term proportion of individuals in each class; these proportions are given by the leading eigenvector.

In general, provided that the leading eigenvalue is (i) real, (ii) positive, and (iii) larger than all other eigenvalues, the qualitative predictions from our simulations will hold for any linear, class-structured model. The population will eventually grow or decay at a rate given by this leading eigenvalue, and the proportion of the population in each class will stabilize at values given by the leading eigenvector. Therefore, we can make important qualitative predictions for any linear class-structured model provided that these three conditions are met.

Fortunately, class-structured models are special in a way that ensures that the leading eigenvalue is neither negative nor complex. Class-structured models have transition matrices that are *non-negative*, meaning that every entry in the matrix is either positive or zero. Indeed, this has to be the case for any matrix that contains only probabilities of moving among classes or numbers of new offspring, neither of which can be negative. There is a group of very useful mathematical results for non-negative matrices that is collectively referred to as the Perron-Frobenius Theorem (Appendix 3). In short, for matrices that are non-negative, the Perron-Frobenius Theorem states that the eigenvalue with the largest magnitude will never be negative and will always be real. Furthermore, the elements of the eigenvector associated with this leading eigenvalue will always be non-negative and real as well. Therefore, the first two of the three conditions are guaranteed to hold for class-structure models.

The third condition requires that the leading eigenvalue be strictly larger than all other eigenvalues. As discussed in Box 9.1, the approximation (9.1.3) assumes that $(\lambda_i/\lambda_1)^t$ eventually goes to zero for every eigenvalue λ_i, except for the leading eigenvalue λ_1. But if there is more than one eigenvalue with the same magnitude as λ_1, then these terms will no longer go to zero. Here again, the Perron-Frobenius Theorem can tell us whether to worry about this problem. According to this theorem, if we can show that all classes have descendants in every class at some future point in time (i.e., the matrix is *primitive*; see Appendix 3), then we are guaranteed that all other eigenvalues will be less than λ_1. If the matrix is not primitive, the long-term dynamics will not necessarily settle down, and our qualitative predictions about the model no longer hold.

One final issue that we must deal with is exactly what is meant by "long term." How long must we wait until the leading eigenvalue and its eigenvector dominate the dynamics? According to Box 9.1, the effects of the other eigenvalues can be ignored in the long term because the quantity $(\lambda_i/\lambda_1)^t$ decays to zero as t increases. Consequently, if the next largest eigenvalue λ_2 is very close in magnitude to the leading eigenvalue, λ_1, then we will have to wait a very long time because $(\lambda_i/\lambda_1)^t$ decays to zero very slowly. Conversely, if the leading eigenvalue is much larger in magnitude than all other eigenvalues, then the "long-term" behavior will be reached after only a few time steps. Therefore, the amount of time that must pass before the variables grow according to the leading eigenvalue depends on how similar the remaining eigenvalues are to the leading eigenvalue.

Rule 10.1: The Long-Term Dynamics of a Class-Structured Population

Class-structured models have transition matrices with non-negative elements describing births and survival probabilities. As a result, their leading eigenvalue and its associated eigenvector are guaranteed to be non-negative and real.

If we can further show that there is a single leading eigenvalue that is greater than all other eigenvalues, then, in the long term,

- The population will grow by a factor given by the leading eigenvalue of its transition matrix.
- The population will have stable proportions in each class, given by the leading right eigenvector, whose elements have been standardized to sum to one (see Box 9.1).

The leading eigenvalue of the transition matrix will be greater than all other eigenvalues provide that all classes have descendants in every other class at some future point in time. Mathematically, this requires that the elements of $\mathbf{M}^t$ are all positive after some number of time steps, t. $\mathbf{M}$ is then a primitive matrix (see Appendix 3 for more details).

We already know that the requirements of Rule 10.1 hold for the model of juveniles and adults, as we discussed after finding the eigenvalues (10.4). (It can also be checked that the transition matrix for this model has all positive elements when raised to the power 2, assuming that $p_a > 0$, thereby satisfying the conditions of Rule 10.1.) Therefore, let us now apply this recipe to the long-term dynamics of the right whale model. The transition matrix in system (10.4) is clearly non-negative as we would expect in a class-structured model, but can each class of whale give rise to descendants in every class at some point in the future? Let us power up the transition matrix, with the hope that some low power will give a result with positive elements. This can be done by hand, but software such as *Mathematica* makes the task trivial. You can check that raising the transition matrix to the power of 3 (i.e., multiplying it by itself 3 times) does the job; the matrix is primitive. As a result, the leading eigenvalue is real, positive, and strictly greater than all other eigenvalues. This eigenvalue represents the long-term growth rate of the whale population. The associated eigenvector gives the stable class distribution of individuals within the population.

It is now also easy to explore how the long-term growth rate changes as we alter parameters in the right whale model, without requiring extensive simulations

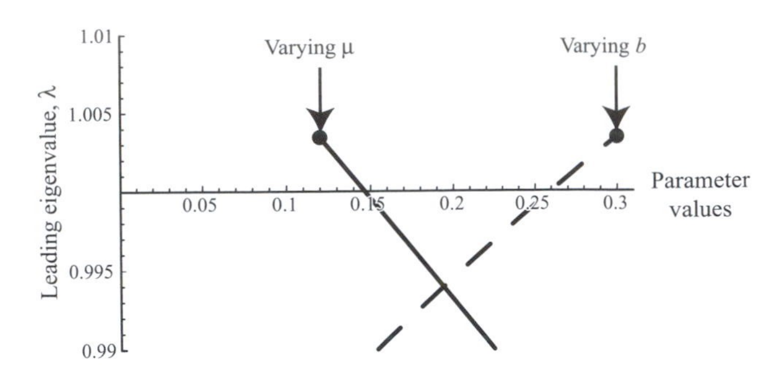

Figure 10.4: The influence of the unknown parameters on the growth rate of the right whale population. The right whale population grows over the long term as long as the leading eigenvalue is greater than one (above the horizontal axis). The solid and dashed lines show how the leading eigenvalue depends on the unknown parameters, μ and b, respectively, holding all other parameters at their values in Figure 10.3. The dots show the leading eigenvalue when $\mu = 0.12$ and $b = 0.3$ as in Figure 10.3.

(Figure 10.4). We can simply calculate the leading eigenvalue numerically using estimates for the parameters and then look at the effect of changing unknown parameters of interest. If, for example, our estimate of the birth rate used in Figure 10.3 is incorrect, then the population will still grow over the long term as long as $b > 0.26$ (holding μ at 0.12). But if the birth rate were lower, the leading eigenvalue would be less than one, and the population will ultimately be unable to sustain itself. Similarly, as long as the probability of reproductive females dying per year is small enough, $\mu < 0.15$ (holding b at 0.3), the population will grow over the long term because the leading eigenvalue remains greater than one. This analysis provides insight into how large these unknown parameters must be for the right whale population to persist over the long term. In section 10.5, we will return to the question of how parameters affect the long-term growth rate of a population.

We turn next to a very different sort of biological question, exploring patterns of methylation in genomic sequences, to see the broad applicability of Rule 10.1.

Example: Methylation of DNA

In vertebrates, a very large proportion of sequences with cytosine followed by guanine (CG sites) are methylated, with a methyl group ($-CH_3$) attached to the cytosine. Methylation appears to play an important role in gene expression, and therefore it is of interest to understand levels of methylation throughout the genome. Newly replicated DNA is always unmethylated, and methylase enzymes recognize sites where the parental strand of DNA is methylated but the daughter strand is not. These enzymes then methylate the daughter strand. This provides a mechanism for the maintenance of methylation from one cell division to the next. In addition, a small fraction of previously unmethylated sites become methylated each cell division through a process known as *de novo* methylation.

We construct a class-structured model that tracks the proportion of CG sites in the genome that are methylated through time (ignoring any selection that might act on methylation levels). We want to track proportions in this model rather than numbers, and we require a class-structured model because each site can be in one of two possible classes: methylated or unmethylated.

We denote the proportion of methylated sites in cell generation t by $p(t)$ and the proportion of unmethylated sites by $q(t)$ (with $p(t)+q(t)=1$). If we use α to denote the probability that any given site in the DNA remains methylated after a cell division (through the action of methylase enzymes) and β to denote the probability that an unmethylated site becomes newly methylated, we obtain the following recursions:

$$\begin{aligned} p(t+1) &= p(t)\,\alpha + q(t)\,\beta, \\ q(t+1) &= p(t)(1-\alpha) + q(t)(1-\beta), \end{aligned} \tag{10.7}$$

or in matrix notation,

$$\begin{pmatrix} p(t+1) \\ q(t+1) \end{pmatrix} = \begin{pmatrix} \alpha & \beta \\ 1-\alpha & 1-\beta \end{pmatrix} \begin{pmatrix} p(t) \\ q(t) \end{pmatrix}. \tag{10.8}$$

Given that the methylases are active (α and $\beta > 0$), the transition matrix in (10.8) is strictly positive. For such matrices, each class can move to any other class in a single time step. (Positive matrices are always primitive; Appendix 3.) Therefore, the conditions laid out in Rule 10.1 are guaranteed. The leading eigenvalue is real, positive, and strictly greater than all other eigenvalues, and the leading eigenvector has strictly positive elements. Indeed, you can calculate the eigenvalues of the matrix (Problem 10.3) to show that the leading eigenvalue is $\lambda_1 = 1$.

It might seem surprising that the leading eigenvalue turns out to be exactly one in this model, regardless of the parameter values α and β. This result is found in any class-structured model that tracks the *proportion* of individuals in different classes rather than the *numbers* of individuals. The reason is that the sum of the proportions in each class must remain constant at one, even if the total number of individuals increases or decreases through time. Therefore, a good check that you have formulated the transition matrix correctly is that the leading eigenvalue should be one in a model tracking proportions in each class.

The stable class distribution (i.e., the dominant right eigenvector, scaled so that the sum of its elements is one) can also be calculated using equation (P2.11) from Primer 2, giving

$$\vec{u}_1 = \begin{pmatrix} \beta/(1-\alpha+\beta) \\ (1-\alpha)/(1-\alpha+\beta) \end{pmatrix}. \tag{10.9}$$

The proportion of sites in the genome that are methylated at the stable class distribution is determined by the spontaneous methylation probability, β, relative to the sum of this and the probability of loss of methylation, $1-\alpha$; i.e., $\beta/(1-\alpha+\beta)$. This ratio represents the rate of flow into the methylated class, relative to the total rate of flow into and out of the methylated class. Therefore, if the spontaneous methylation probability is high and/or the maintenance probability of methylation is high, the long-term proportion of sites in the genome that is methylated will also be high. This makes good intuitive sense, but it also reveals a slightly subtle prediction that might not have been obvious beforehand: we expect the proportion of sites in the genome that are methylated to be very high, even if the flow into the methylated class, β, is very small, provided that the maintenance of methylation, α, is also very high.

10.4 Reproductive Value and Left Eigenvectors

A third type of question about the long-term dynamics of populations that remains to be answered has to do with the contribution of individuals of different classes to the future population size. For example, suppose you are a conservation biologist in charge of the recovery of right whale populations. Your goal is to invest the resources at your disposal in such a way as to make the population size in the future as large as possible. Budgets are typically limited, and therefore you are faced with difficult decisions about how best to invest. You could focus your efforts on newborn calves, immature females, mature females, or reproductive females (or some combination of these). But which would yield the best outcome?

We again start with simulations and ask how the long-term population size is affected by a program that supplements the number of individuals in each of the different classes. If we are trying to conserve the species and want to focus our efforts on one class of the whale population, Figure 10.5 indicates that focusing on reproductive females would yield the greatest benefit in terms of future population size. Again, however, while these simulation results give us some insight into the question, it would be better to have general results that can be guaranteed to hold, regardless of the parameter values chosen. This requires that we develop another

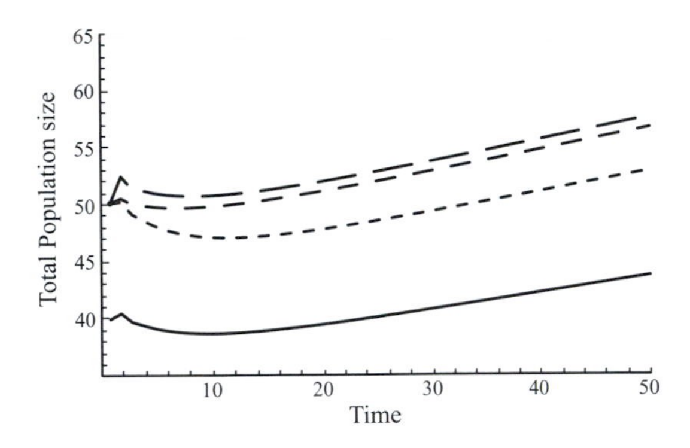

Figure 10.5: The total size of the right whale population over time. Solid curve is for a population starting with ten individuals in each class (for a total initial population size of $N(0) = 40$), using the parameter values from Figure 10.3. Other curves are for a total initial population size of 50, but with ten additional individuals placed in one of the four different classes at time zero: as calves (dotted curve), as immature females (short dashed curve), as mature females (intermediate dashed curve), or as reproductive females (long dashed curve). The effects of each of these options can be quantified numerically by calculating the difference in population size at time 50 between each of the four cases when ten extra individuals were added and when they were not (i.e., $N_C(50)-N(50)$, $N_I(50)-N(50)$, $N_M(50)-N(50)$, and $N_R(50)-N(50)$, where N is the total population size, and the subscript denotes the class in which the ten extra individuals were initially placed). Dividing these numbers by $N_C(50)-N(50)$, we obtain estimates of the reproductive values of the four age classes relative to the reproductive value of calves: (1, 1.1, 1.5, 1.7). We conclude that reproductive females are most valuable in terms of increasing the population size.

important concept for class-structured population models: *reproductive value*. As we shall see, the reproductive value of each class is related to the left eigenvector $\vec{v}$ associated with the leading eigenvalue of the matrix **M**. Recall (see equation (P2.12)) that the left eigenvector is defined as a vector that satisfies

$$\vec{v}^{\mathrm{T}}\mathbf{M} = \lambda\,\vec{v}^{\mathrm{T}}. \tag{10.10}$$

To understand why left eigenvectors are of interest, it is first useful to provide a conceptual description of what is meant by reproductive value. Before considering the more complex model of the right whale population, let us consider the juvenile/adult model (10.1). Imagine that we plan to introduce an extra individual from a captive breeding program to the population, and we need to decide whether it should be a juvenile or an adult. Which of these would be most effective for increasing future population size? To answer this question we need some way to quantify the value of juveniles relative to adults in terms of their contribution to future generations. Such quantities are termed *reproductive values*.

We denote the reproductive values of a juvenile and an adult by ρ_J and ρ_A, respectively. Suppose that the population is in the stable class distribution and is therefore growing at its long-term rate given by the leading eigenvalue, λ_1. The first thing to notice is that adding a juvenile at time t will result in λ_1 times more future individuals than adding a juvenile at $t+1$. Thus, by definition, we say that the reproductive value of a juvenile at time t is λ_1 times the reproductive value of a juvenile at $t+1$. This is the same principle as the compounding of money in a bank account (Box 6.1).

Now let us consider the value of a juvenile relative to an adult. With probability p_j the juvenile will become an adult in the next time step, at which point it will have the value of an adult, discounted by the fact that the population has grown by a factor, λ_1. Conversely, with probability $1 - p_j$, it will die and have no reproductive value. Therefore, the expected reproductive value of a juvenile is $\rho_J = (p_j\ \rho_A/\lambda_1)$. Similarly, if we consider an adult, it will produce b juveniles in the next time step, each of which will have a reproductive value of ρ_J, discounted again by the fact that the population has grown by a factor λ_1. These juveniles thus contribute $b\ (\rho_J/\lambda_1)$ to the reproductive value of an adult. The adult might also survive with probability p_a, contributing further to its reproductive value at time t by an amount $p_a\ (\rho_A/\lambda_1)$. Therefore, the total reproductive value of an adult is $\rho_A = b\ (\rho_J/\lambda_1) + p_a\ (\rho_A/\lambda_1)$. Both of these expressions can be written more compactly using matrix notation as

$$\begin{pmatrix}\rho_J & \rho_A\end{pmatrix} = \begin{pmatrix}\rho_J & \rho_A\end{pmatrix}\begin{pmatrix}0 & b\\ p_j & p_a\end{pmatrix}\frac{1}{\lambda_1}. \tag{10.11}$$

Multiplying both sides of (10.11) by λ_1 and defining the vector $\vec{v}^{\mathrm{T}} = \begin{pmatrix}\rho_J & \rho_A\end{pmatrix}$, we can write this as

$$\begin{aligned}\lambda_1\vec{v}^{\mathrm{T}} &= \vec{v}^{\mathrm{T}}\begin{pmatrix}0 & b\\ p_j & p_a\end{pmatrix}\\ &= \vec{v}^{\mathrm{T}}\mathbf{M}.\end{aligned} \tag{10.12}$$

The *reproductive value* of each class is proportional to the left eigenvector associated with the leading eigenvalue.

Comparison of (10.12) with (10.10) reveals the key result that we are after. The relative reproductive values of each class are given by the elements of the left eigenvector corresponding to the leading eigenvalue of the transition matrix. This is referred to as the dominant or leading left eigenvector. The same considerations can be applied to any model of a class-structured population, with analogous results. Therefore, given a leading eigenvalue, its right eigenvector corresponds to the proportions of the stable class distribution, whereas its left eigenvector corresponds to the relative reproductive values of individuals in each class (see Supplementary Material 10.1 for a more formal derivation of reproductive value). Because the length of an eigenvector is arbitrary, we typically choose a class against which the reproductive values of all other classes are compared. Alternatively, we could choose the left eigenvector so that the average reproductive value within a population at the stable age distribution is one, that is, $\vec{v}^T \vec{u} = 1$ (this is equivalent to setting the left eigenvector to the first row in $\mathbf{A}^{-1}$ as we did on Box 9.1). As the first choice is more common in demographic models, we will follow this convention.

CAUTION: The stable class distribution is given by the elements of the right eigenvector, while the reproductive value of each class is given by the elements of the left eigenvector.

Let us now apply these results to the model of right whales. For the parameters used in Figure 10.3, the left eigenvector corresponding to the leading eigenvalue $\lambda_1 = 1.004$ is (using *Mathematica*) $\vec{v}^{\mathrm{T}} = (1, 1.1, 1.5, 1.7)$, where we have standardized the reproductive value of a calf to be one. These values agree perfectly with our simulation results (Figure 10.5). Thus, the reproductive values tell us that investing in reproductive females will have the greatest effect on the future population size of whales, because it is these females that contribute most to future generations. Importantly, these results also give us quantitative information about the relative value of investing in different classes. For example, although investing in reproductive females is the best option, investing in mature but non-reproductive females provides nearly as much of a benefit (i.e., $v_M = 1.5$ versus $v_R = 1.7$).

10.5 The Effect of Parameters on the Long-Term Growth Rate

With the concepts of the stable class distribution and reproductive value in hand, we can now address the fourth main question of interest. Suppose that you can affect the future population size by altering one (or more) of the life history parameters of the individuals that are already present in the population. For example, you might be able to conserve the breeding ground of an endangered species to increase its birth rate, or you might be able to limit hunting of mature individuals to increase survival. Which strategy would be most effective in terms of the long-term growth of the population? We can answer this question mathematically by altering one (or more) of the elements in the transition matrix and determining the effect on the future population size.

This question is subtly different than that considered in the previous section. There, we imagined changing the initial number of individuals in one of the classes, while leaving the life history parameters unchanged. As a result, the long-term growth rate, the stable class distribution, and the reproductive values for the population were the same in both the manipulated and unmanipulated populations. If we change some of the elements of the transition matrix, however, then the long-term growth rate, the stable class distribution, and the reproductive values of individuals in each class will change as well. In principle we could determine which interventions are most effective by calculating the long-term growth rate (i.e., the leading eigenvalue) for each set of parameters. This is the most direct approach, but it can be difficult to determine the best possible intervention, especially if there are multiple parameters and we have to calculate the leading eigenvalue numerically. Fortunately, if we consider only small changes in the parameter values, then there is a very simple relationship between the change in the long-term growth rate and the reproductive values in the unmanipulated population.

Suppose that z is a parameter of interest that affects one or more of the elements of the transition matrix $\mathbf{M}$ and that z^* is the value of this parameter in the unmanipulated population. We would like to know how the leading eigenvalue, λ_1, changes with an increase in z; specifically, we would like to know $d\lambda_1/dz$, which we call the *sensitivity* of the leading eigenvalue to z. If the eigenvalue can be determined analytically, then this derivative is straightforward to calculate. Often, however, λ_1 is impossible to calculate or is too complex to interpret. As shown in Box 10.2, the sensitivity of the eigenvalue to changes in z near z^* can be related to the stable class distribution (given by the right eigenvector $\vec{u}$) and the reproductive values (given by the left eigenvector $\vec{v}$):

The *sensitivity of the leading eigenvalue* to a parameter describes the impact of changing the parameter on the long-term growth of a population.

$$\left.\frac{d\lambda}{dz}\right|_{z=z^*} = \frac{\vec{v}^T\left(\left.\frac{d\mathbf{M}}{dz}\right|_{z=z^*}\right)\vec{u}}{\vec{v}^T\vec{u}}. \tag{10.13}$$

Here, $\vec{u}$ and $\vec{v}$ are calculated from the transition matrix $\mathbf{M}$ for the unmanipulated population (at z^*) and $(d\mathbf{M}/dz)|_{z=z^*}$ is the matrix obtained by differentiating all of the elements of the transition matrix with respect to z at z^*.

At this point, you might wonder what we gain by using (10.13): if the eigenvalues are impossible to calculate or difficult to interpret, the same will be true of the eigenvectors. There are three main advantages to this method. First, although the general expressions for the eigenvalues and eigenvectors of the matrix $\mathbf{M}$ can be quite complicated, the eigenvalues and eigenvectors are sometimes simpler for the special case where $z = z^*$, which is all that we need in (10.13). Second, equation (10.13) can be used even without specifying the elements of the eigenvectors, if all that we want is to know is the effect of changing a parameter in terms of the reproductive values of each class. Third, this method can be used when the eigenvalues and eigenvectors have been calculated numerically for a given set of parameter estimates, because the only derivative in (10.14) involves the transition matrix, which can be easily differentiated.

Box 10.2: Determining the Influence of Parameters on the Long-Term Growth Rate

Suppose we have a model of a structured population and that one or more of the elements of the transition matrix $\mathbf{M}$ are functions of some parameter z. We write $\mathbf{M}(z)$ to denote this fact. When we vary z, the elements of the matrix $\mathbf{M}$ will vary, and therefore its leading eigenvalue λ as well as the associated left and right eigenvectors, $\vec{v}$ and $\vec{u}$, will vary. From Primer 2 we know that the eigenvalue and its right eigenvector must satisfy the equation $\mathbf{M}\vec{u} = \lambda\vec{u}$ (where each of these is a function of z). A similar equation must hold for the left eigenvector; $\vec{v}^{\mathrm{T}}\mathbf{M} = \vec{v}^{\mathrm{T}}\lambda$. Multiplying this latter equation on the right by $\vec{u}$, we also know that $\vec{v}^{\mathrm{T}}\mathbf{M}\vec{u} = \vec{v}^{\mathrm{T}}\lambda\vec{u}$. Now differentiate both sides with respect to z using the chain rule (A2.15):

$$\frac{d\vec{v}^{\mathrm{T}}}{dz}\mathbf{M}\vec{u} + \vec{v}^{\mathrm{T}}\frac{d\mathbf{M}}{dz}\vec{u} + \vec{v}^{\mathrm{T}}\mathbf{M}\frac{d\vec{u}}{dz} = \frac{d\vec{v}^{\mathrm{T}}}{dz}\lambda\vec{u} + \vec{v}^{\mathrm{T}}\frac{d\lambda}{dz}\vec{u} + \vec{v}^{\mathrm{T}}\lambda\frac{d\vec{u}}{dz} \tag{10.2.1}$$

where $d\mathbf{M}/dz$ represents the matrix whose elements are the derivatives (with respect to z) of the elements of $\mathbf{M}$. Because $\mathbf{M}\vec{u} = \lambda\vec{u}$ and $\vec{v}^{\mathrm{T}}\mathbf{M} = \vec{v}^{\mathrm{T}}\lambda$, (10.2.1) can be simplified to

$$\vec{v}^{\mathrm{T}}\frac{d\mathbf{M}}{dz}\vec{u} = \frac{d\lambda}{dz}\vec{v}^{\mathrm{T}}\vec{u}. \tag{10.2.2a}$$

Furthermore, because $\vec{v}^{\mathrm{T}}\vec{u}$ is a scalar, we can rearrange this equation to obtain the sensitivity of the eigenvalue to z:

$$\frac{d\lambda}{dz} = \frac{\vec{v}^{\mathrm{T}}\frac{d\mathbf{M}}{dz}\vec{u}}{\vec{v}^{\mathrm{T}}\vec{u}} \tag{10.2.2b}$$

As long as $\vec{v}^{\mathrm{T}}\vec{u}$ is positive (which it will be as long as transitions can eventually occur among all classes, i.e., $\mathbf{M}$ is irreducible; see Appendix 3), $d\lambda/dz$ will have the same sign as $\vec{v}^{\mathrm{T}}(d\mathbf{M}/dz)\vec{u}$. Equation (10.2.2b) holds for any value of z, but we often evaluate the sensitivity of λ to z at a particular value z^* so that we can calculate the eigenvectors, which leads to equation (10.13).

To illustrate these ideas, let us suppose that you are trying to increase the future population size of right whales and that you must choose between improving protection of immature whales or of calves. In terms of the long-term population growth rate, which is better: increasing the rate at which immature females become mature, s_{MI}, or increasing the rate at which calves reach the immature stage, s_{IC}? Assuming that you can cause only small changes in either of these parameters, we can use equation (10.13) to evaluate the consequences of each. For s_{MI}, all entries of the matrix $d\mathbf{M}/ds_{MI}$ are zero except for the entry in the third row and second column, which equals one because $ds_{MI}/ds_{MI} = 1$. Thus we get $d\lambda_1/ds_{MI} = u_I v_M/(\vec{v}^{\mathrm{T}}\vec{u})$ where the subscripts refer to the class. This result has a very nice interpretation. The benefit of increasing the maturation rate of immature females (s_{MI}) depends on the fraction of the population that is affected (u_I) as well as the reproductive value of the mature individuals that are produced by increasing the maturation rate of immature females (v_M). Performing the same calculation for s_{IC} yields $\mathrm{d}\lambda_1/\mathrm{d}s_{IC} = u_C v_I/(\vec{v}^{\mathrm{T}}\vec{u})$. Thus, the relative value of the two options is $(\mathrm{d}\lambda_1/\mathrm{d}s_{MI})/(\mathrm{d}\lambda_1/\mathrm{d}s_{IC}) = (u_I v_M)/(u_C v_I)$. Here, we have obtained general insights without actually specifying the elements of the left and right eigenvectors.

We can proceed further numerically, using the reproductive values obtained earlier along with the stable class distribution, $\vec{u} = (0.04, 0.23, 0.61, 0.12)$. In this case,

$$\left(\frac{\mathrm{d}\lambda_1/\mathrm{d}s_{MI}}{\mathrm{d}\lambda_1/\mathrm{d}s_{IC}}\bigg|_{\substack{s_{MI}=0.08\\ s_{IC}=0.92}}\right) = \frac{(0.23)(1.5)}{(0.04)(1.1)} \approx 7.8.$$

Thus, increasing the maturation rate of immature females (s_{MI}) is nearly eight times more effective at increasing the long-term growth rate of the population than increasing the maturation rate of calves. This makes sense because the maturation rate is so low ($s_{MI} = 0.08$) that it represents a more severe bottleneck to the growth of the population.

10.6 Age-Structured Models—The Leslie Matrix

Age structure is a special type of class structure that is important enough to warrant a separate section of its own. Nearly all populations have some important component of age structure, and the transition matrices of all age-structured models have the same special form. This type of matrix is referred to as a Leslie matrix, after P. H. Leslie (see Caswell 2001 for a discussion), who was one of the first to analyze this form of class-structured model.

10.6.1 Construction of Age-Structured Models

Let us construct an age-structured model for the females of the three-spine stickleback, a freshwater fish. Sticklebacks are temperate fish with a distinctly seasonal reproductive pattern, and therefore it makes sense to measure age in this species in years. Female sticklebacks tend not to live more than four years, and therefore we will build a model that keeps track of the number of individuals that are 1, 2, 3, and 4 years old (i.e., $n_1(t)$, $n_2(t)$, $n_3(t)$, and $n_4(t)$). We will census the population at the beginning of the season, before reproduction occurs. (This is an arbitrary choice, and many demographic models census immediately after reproduction.)

The number of 1-year-old sticklebacks at the beginning of the next season is equal to the number of offspring produced by all individuals this year, multiplied by the probability that an offspring survives to reach age 1:

$$n_1(t+1) = n_1(t)\, m_1 + n_2(t)\, m_2 + n_3(t)\, m_3 + n_4(t)\, m_4, \qquad (10.14\text{a})$$

where m_i is the expected number of female offspring produced by an i-year-old female that survive to become 1-year-olds in the next season. Similarly, the number of 2-year-old sticklebacks at the beginning of next season is equal to the number of 1-year-olds this year, multiplied by the probability that an individual survives year 1 to become a 2-year-old, p_1:

$$n_2(t+1) = n_1(t)\, p_1. \tag{10.14b}$$

This same principle applies for 3- and 4-year old sticklebacks; i.e.,

$$n_3(t+1) = n_2(t)\, p_2, \tag{10.14c}$$
$$n_4(t+1) = n_3(t)\, p_3, \tag{10.14d}$$

where p_i is the probability that an individual survives age class i.

Equations (10.14a)–(10.14d) can be written in matrix form as

$$\begin{pmatrix} n_1(t+1) \\ n_2(t+1) \\ n_3(t+1) \\ n_4(t+1) \end{pmatrix} = \mathbf{L} \begin{pmatrix} n_1(t) \\ n_2(t) \\ n_3(t) \\ n_4(t) \end{pmatrix}, \tag{10.15}$$

where

$$\mathbf{L} = \begin{pmatrix} m_1 & m_2 & m_3 & m_4 \\ p_1 & 0 & 0 & 0 \\ 0 & p_2 & 0 & 0 \\ 0 & 0 & p_3 & 0 \end{pmatrix} \tag{10.16}$$

is the Leslie matrix. Figure 10.6 illustrates an example of the dynamics of the stickleback population.

A *Leslie matrix* describes an age-structured population; surviving individuals always move to the next age class and can give birth to individuals in the first age class.

Leslie matrices always have the form of (10.16), where the top row of the matrix is the effective fecundity of individuals of each age class, and all other rows contain one element p_i. With age-structure, all individuals must pass through each class in succession, and assuming that time is measured in terms of the age classes, an individual must move on to the next class every time step. Thus, the survival probabilities p_i enter the matrix in the cells immediately below the diagonal. The elements of Leslie matrices are often referred to as the life-history parameters of the model.

10.6.2 Analysis of Age-Structured Models

The special form of Leslie matrices allows us to write its characteristic polynomial in a pleasingly simple form (Box 10.3):

$$1 = \sum_{i=1}^{n} \frac{l_i m_i}{\lambda^i}, \tag{10.17}$$

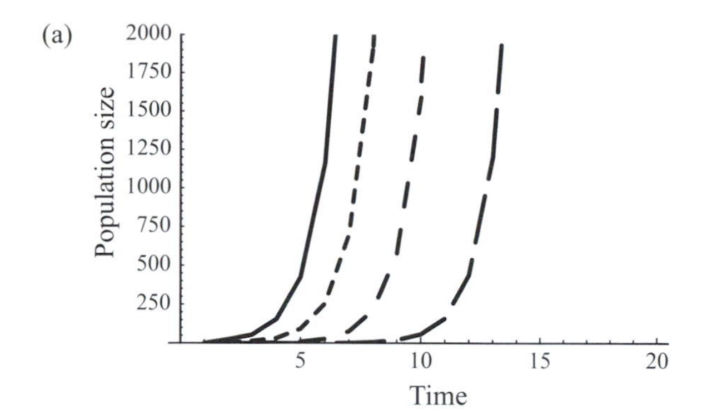

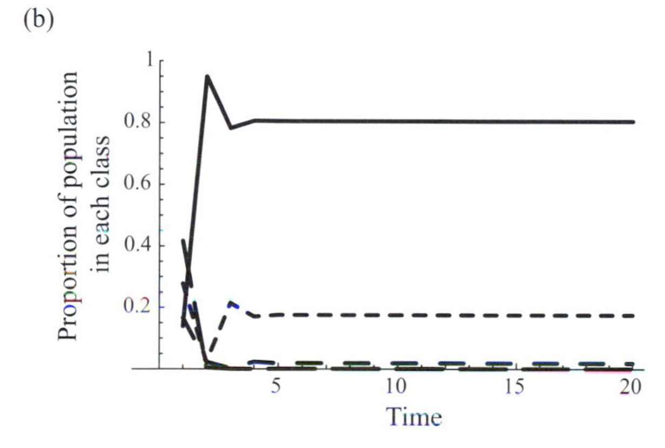

Figure 10.6: Growth of each age class of sticklebacks. Model (10.16) is iterated for a stickleback population. The 1-year old class is depicted by the solid line, and 2-, 3-, and 4-year old classes are depicted by lines with increasingly long dashes. (a) Population size of each age class. (b) Proportion of the population in each age class (the 4-year-old class is too rare to appear in the figure). Parameter values: $m_1 = 2$, $m_2 = 3$, $m_3 = 4$, $m_4 = 4$, $p_1 = 0.6$, $p_2 = 0.3$, and $p_3 = 0.1$.

where n is the number of age classes, $l_i = p_1 p_2 \ldots p_{i-1}$ is the probability that an individual survives until age class i, and $l_1 = 1$ because we defined the fecundities m_i as the number of offspring that survive to age 1. Regardless of the number of age classes, the n eigenvalues λ of a Leslie matrix are the n roots of equation (10.17).

Equation (10.17) is sometimes referred to as the Euler-Lotka equation. It features prominently in life-history theory, which is devoted to understanding the evolution of age-specific patterns of fecundity and survival (Roff 1992; Stearns 1992; Charlesworth 1994). The Euler-Lotka equation also provides an alternative way of calculating the long-term growth of a population, by solving (10.17) for the largest root λ_1. If the population were not growing in size ($\lambda_1 = 1$), then the expected lifetime reproductive success of a newborn individual, $\Sigma_{i=1}^{n} l_i m_i$, would be one—that is, each individual would exactly replace itself. The Euler-Lotka equation shows us how to generalize this statement for populations that are growing or shrinking in size. Now, the expected lifetime reproductive success of a newborn individual, discounted by the amount the population has grown from when it was born until it reaches age i, $\Sigma_{i=1}^{n} l_i m_i / \lambda_1^i$, will equal one.

For our model of sticklebacks, equation (10.17) can be written out as

$$1 = \lambda^{-1} l_1 m_1 + \lambda^{-2} l_2 m_2 + \lambda^{-3} l_3 m_3 + \lambda^{-4} l_4 m_4, \tag{10.18a}$$

or

$$1 = \lambda^{-1} m_1 + \lambda^{-2} p_1 m_2 + \lambda^{-3} p_1 p_2 m_3 + \lambda^{-4} p_1 p_2 p_3 m_4. \tag{10.18b}$$

Box 10.3: The Characteristic Polynomial of Leslie Matrices

Suppose we have a Leslie matrix with n age classes:

$$\mathbf{L} = \begin{pmatrix} m_1 & m_2 & m_3 & \cdots & m_n \\ p_1 & 0 & 0 & \cdots & 0 \\ 0 & p_2 & 0 & \cdots & 0 \\ \vdots & \vdots & \ddots & \ddots & 0 \\ 0 & 0 & 0 & p_{n-1} & 0 \end{pmatrix}. \tag{10.3.1}$$

From Definition (P2.7) of Primer 2, any eigenvalue λ of the matrix $\mathbf{L}$ must satisfy

$$\text{Det}(\mathbf{L} - \mathbf{I}\,\lambda) = 0. \tag{10.3.2}$$

Because of the relatively simple form of Leslie matrices we can explicitly calculate (10.3.2) using the general definition for determinants (P2.18) in Primer 1. We get

$$\begin{aligned}(-1)^0(m_1-\lambda)(-\lambda)^{n-1} + (-1)^1 m_2 p_1(-\lambda)^{n-2} + (-1)^2 m_3 p_1 p_2(-\lambda)^{n-3} + \cdots \\ + (-1)^{n-1} m_n p_1\, p_2 \ldots p_{n-1}(-\lambda)^0 = 0.\end{aligned} \tag{10.3.3a}$$

We can define the probability of surviving from birth to age class i as $l_i = p_1\, p_2 \ldots p_{i-1}$, with $l_1 = 1$, allowing us to write equation (10.3.3) as

$$(-\lambda)^n + m_1(-\lambda)^{n-1} + (-1)^1 m_2 l_2(-\lambda)^{n-2} + (-1)^2 m_3 l_3(-\lambda)^{n-3} + \cdots + (-1)^{n-1} m_n l_n = 0. \tag{10.3.3b}$$

Factoring out $(-\lambda)^n$ then gives

$$1 - m_1\,\lambda^{-1} - m_2\, l_2\,\lambda^{-2} - m_3\, l_3\,\lambda^{-3} - \cdots - m_n\, l_n\,\lambda^{-n} = 0, \tag{10.3.3c}$$

or

$$1 = \sum_{i=1}^{n} \frac{l_i m_i}{\lambda^i}. \tag{10.3.3d}$$

In some demographic models, individuals survive in the last age class for multiple censuses, which adds an entry p_n to the last row and column of the Leslie matrix:

$$\mathbf{L} = \begin{pmatrix} m_1 & m_2 & m_3 & \cdots & m_n \\ p_1 & 0 & 0 & \cdots & 0 \\ 0 & p_2 & 0 & \cdots & 0 \\ \vdots & \vdots & \ddots & \ddots & 0 \\ 0 & 0 & 0 & p_{n-1} & p_n \end{pmatrix} \tag{10.3.4}$$

(continued)

Box 10.3 *(continued)*

Revising the calculations in (10.3.3), the eigenvalues now solve

$$1 = \left(\sum_{i=1}^{n-1} \frac{l_i m_i}{\lambda^i}\right) + \left(\frac{l_n m_n}{\lambda^n}\right)\left(\frac{\lambda}{\lambda - p_n}\right). \tag{10.3.5}$$

If the last age class has zero fecundity (m_n=0), then the long-term growth rate is unaffected by individuals in the oldest age class, and equation (10.3.5) reduces to (10.3.3d).

It is not possible to obtain any meaningful, explicit expression for the population growth rate by solving equation (10.18) for λ, but we can easily evaluate it for certain parameter values. For example, suppose that 1-year-olds produce an average of two female offspring that survive to the next year, 2-year-olds produce three such offspring, and 3- and 4-year-olds both produce four surviving female offspring (i.e., $m_1 = 2$, $m_2 = 3$, $m_3 = 4$, and $m_4 = 4$). Further, suppose that the age-specific survival probabilities are $p_1 = 0.6$, $p_2 = 0.3$, and $p_3 = 0.1$. Equation (10.18b) is then

$$1 = 2\lambda^{-1} + 1.8\lambda^{-2} + 0.72\lambda^{-3} + 0.072\lambda^{-4}. \tag{10.18c}$$

All eigenvalues of the Leslie matrix must satisfy (10.18c). *Mathematica* is readily able to solve (10.18c), giving the four eigenvalues $\lambda_1 = 2.75$, $\lambda_2 = -0.3 + 0.3i$, $\lambda_3 = -3 - 0.3i$, and $\lambda_4 = -0.14$. The long-term growth rate of this stickleback population is therefore $\lambda_1 = 2.75$.

The special form of Leslie matrices results in the dominant right and left eigenvectors (which represent the stable age distribution and the age-specific reproductive values respectively) having a very specific form as well, regardless of the number of age classes. Box 10.4 shows that the proportion of the population in each of the age classes at the stable age distribution (i.e., the elements of the dominant right eigenvector) is

$$u_x = \frac{l_x \lambda_1^{-(x-1)}}{\sum_{i=1}^{n} l_i \lambda_1^{-(i-1)}}. \tag{10.19}$$

For our stickleback population, equation (10.19) predicts 80.4% 1-year olds, 17.5% 2-year olds, 2% 3-year-olds, and 0.06% 4-year-olds over the long term, using the leading eigenvalue $\lambda = 2.75$ and the above parameter values. These values agree perfectly with simulation results (Figure 10.6b). The approach of Box 10.4 can also be used (see Problem 10.4) to show that the reproductive value of each age class (i.e., the elements of the dominant left eigenvector) measured relative to the reproductive value of newborns (v_1) satisfies

$$\frac{v_x}{v_1} = \frac{\lambda_1^{x-1}}{l_x} \sum_{i=x}^{n} \frac{l_i m_i}{\lambda_1^i}. \tag{10.20}$$

Box 10.4: The Right Eigenvector Associated with the Leading Eigenvalue of a Leslie Matrix

Multiplying the n-dimensional Leslie matrix (10.3.1) in Box 10.3 on the right by the right eigenvector $\tilde{\vec{u}}$ gives $L\,\tilde{\vec{u}} = \lambda\,\tilde{\vec{u}}$ (the reason for labeling the eigenvector as $\tilde{\vec{u}}$ instead of $\vec{u}$ will become apparent shortly). This describes a system of n equations, and except for the first equation, these equations have the form:

$$\tilde{u}_{x-1}\, p_{x-1} = \lambda \tilde{u}_x \tag{10.4.1}$$

Solving (10.4.1) for $\tilde{u}_x$ gives

$$\tilde{u}_x = \frac{\tilde{u}_{x-1} p_{x-1}}{\lambda}. \tag{10.4.2}$$

We can also use (10.4.2) to define $\tilde{u}_{x-1}$ in terms of $\tilde{u}_{x-2}$, which we can plug into (10.4.2) to write $\tilde{u}_x$ in terms of $\tilde{u}_{x-2}$. Continuing to work recursively backward, we can write a general formula for $\tilde{u}_x$ in terms of $\tilde{u}_1$:

$$\tilde{u}_x = \tilde{u}_1 l_x \lambda^{-(x-1)}. \tag{10.4.3}$$

Expression (10.4.3) gives all the elements of the stable age distribution, relative to age 1. We usually prefer to have the elements of the right eigenvector represent the proportion of the population in age class x, and we can obtain this scaled version of (10.4.3) by dividing (10.4.3) by the sum of all the elements. Using $\vec{u}$ to denote this scaled version of (10.4.3), we obtain

$$u_x = \frac{l_x \lambda^{-(x-1)}}{\sum_{i=1}^{n} l_i \lambda^{-(i-1)}}. \tag{10.4.4}$$

If individuals survive in the last age class for multiple censuses according to the Leslie matrix (10.3.4), then the last in the system of equations must be revised to:

$$\tilde{u}_{n-1} p_{n-1} + \tilde{u}_n p_n = \lambda \tilde{u}_n, \tag{10.4.5}$$

whose solution is

$$\tilde{u}_n = \frac{\tilde{u}_{n-1} p_{n-1}}{\lambda - p_n}. \tag{10.4.6}$$

Plugging in (10.4.3) and dividing by the sum of the $\tilde{u}_x$ gives the right eigenvector in terms of the proportion in each age class:

$$u_x = \frac{l_x \lambda^{-(x-1)}}{\left(\sum_{i=1}^{n-1} l_i \lambda^{-(i-1)}\right) + l_n \lambda^{-(x-2)} (\lambda - p_n)^{-1}} \quad \text{for } x < n \tag{10.4.7}$$

(continued)

Box 10.4 *(continued)*

and

$$u_n = \frac{l_n \lambda^{-(x-2)}(\lambda - p_n)^{-1}}{\left(\sum_{i=1}^{n-1} l_i \lambda^{-(i-1)}\right) + l_n \lambda^{-(x-2)}(\lambda - p_n)^{-1}}.$$

For our stickleback population, equation (10.20) predicts reproductive values of $v_1=1$, $v_2=1.25$, $v_3=1.5$, and $v_4=1.45$. From these numbers we can see that a 3-year-old stickleback contributes more to the growth of the population than a stickleback of any other age. It makes sense that 3-year-olds have a higher reproductive value than 1- or 2-year-olds, because 3-year-olds have survived to the age of highest fecundity. It also makes sense that 3-year-olds have a higher reproductive value than 4-year-olds, because 3-year-olds have some chance of surviving to age 4 and reproducing again, whereas all 4-year-olds are assumed to die over the following year.

Rule 10.2: Long-Term Growth of an Age-Structured Population

An age-structured population with n age classes is described by a Leslie matrix:

$$\mathbf{L} = \begin{pmatrix} m_1 & m_2 & m_3 & \cdots & m_n \\ p_1 & 0 & 0 & \cdots & 0 \\ 0 & p_2 & 0 & \cdots & 0 \\ \vdots & \vdots & \ddots & \ddots & 0 \\ 0 & 0 & 0 & p_{n-1} & 0 \end{pmatrix}.$$

- The leading eigenvalue of a Leslie matrix is the largest root of $1 = \Sigma_{i=1}^{n} l_i m_i / \lambda^i$, which describes the long-term growth of the population.
- In the long term, the proportion of individuals in age class x is

$$u_x = \frac{l_x \lambda_1^{-(x-1)}}{\sum_{i=1}^{n} l_i \lambda_1^{-(i-1)}}$$

(the elements of the dominant right eigenvector), which describes the stable age distribution.

(continued)

Rule 10.2 ***(continued)***

- In the long term, the reproductive value of individuals in age class x relative to the youngest age class is

$$\frac{v_x}{v_1} = \frac{\lambda_1^{x-1}}{l_x} \sum_{i=x}^{n} \frac{l_i m_i}{\lambda_1^i}$$

(the elements of the dominant left eigenvector).

10.6.3 The Effect of Life-History Parameters on Population Growth in Age-Structured Models

Now suppose that we want to know how a change in some life-history parameter in the Leslie matrix affects the population growth rate. We examined this sort of question in section 10.5, where we found that equation (10.13) describes the effect of changing a parameter, z, on the long-term growth rate. Because of the very special form of Leslie matrices, equation (10.13) can be greatly simplified. In fact, there are only two kinds of life-history parameters in age-structured models: age-specific survival probabilities and age-specific fecundities. Here we simplify equation (10.13) for each.

Suppose first that the parameter of interest is the age-specific survival probability p_i. The matrix of derivatives in (10.13) contains zeros everywhere except for a 1 in column i of row $i+1$. Carrying out the matrix multiplication in (10.13) gives

$$\frac{d\lambda}{dp_i} = \frac{u_i v_{i+1}}{E[v_k]}, \tag{10.21}$$

where $E[v_k] = \vec{v}^{\mathrm{T}}\vec{u} = u_1 v_1 + u_2 v_2 + \cdots + u_n v_n$ is the frequency of each age class at the stable age distribution (u_i) times its reproductive value (v_i), which is just the long-term average reproductive value of the population. Equation (10.21) has a simple interpretation. A proportion u_i of the population is in age class, i, and increasing their survival probability creates more age $i+1$ individuals, each of which has reproductive value v_{i+1}. The total effect on the growth rate is therefore the product of the two, standardized by the average reproductive value of the population.

Suppose next that the parameter of interest is the age-specific fecundity m_i. The matrix of derivatives in (10.13) now contains zeros everywhere except for a one in column i of row 1. Carrying out the matrix multiplication in (10.13) then gives

$$\frac{d\lambda}{dm_i} = \frac{u_i v_1}{E[v_k]}. \tag{10.22}$$

TABLE 10.1
The effect of life history parameters on stickleback population growth. $\mathrm{d}\lambda/\mathrm{d}x$ describes the sensitivity of the eigenvalue to the parameter x. If the parameter x is increased by a small amount, Δx, the long-term growth rate is altered by $\Delta x\,(\mathrm{d}\lambda/\mathrm{d}x)$.

$$\frac{\mathrm{d}\lambda}{\mathrm{d}p_1} = 0.18, \quad \frac{\mathrm{d}\lambda}{\mathrm{d}p_2} = 0.27, \quad \frac{\mathrm{d}\lambda}{\mathrm{d}p_3} = 0.31$$

$$\frac{\mathrm{d}\lambda}{\mathrm{d}m_1} = 0.14, \quad \frac{\mathrm{d}\lambda}{\mathrm{d}m_2} = 0.18, \quad \frac{\mathrm{d}\lambda}{\mathrm{d}m_3} = 0.22, \quad \frac{\mathrm{d}\lambda}{\mathrm{d}m_4} = 0.21$$

Equation (10.22) has an analogous interpretation to (10.21). A proportion u_i of the population is in age class i, and increasing their fecundity will create more age 1 individuals, each of which has reproductive value v_1. The total effect on the growth rate is again the product of the two, standardized by the average reproductive value of the population.

With equations (10.21) and (10.22), we can now return to our sticklebacks and use the right and left eigenvectors that we obtained earlier to examine the effect of all life history parameters on the growth rate of the population. Table 10.1 presents these results, illustrating that an increase in the survival probability from age 3 to age 4 (i.e., p_3) has the greatest impact on the population growth rate. Even so, increasing p_3 by 1% is expected to lead to only a 0.31% increase in the growth rate of the population. This result suggests that it would be difficult to substantially improve the growth rate of the population by targeting conservation efforts on the survival of only one age class.

Before considering another example, it is worth mentioning that equations (10.21) and (10.22) feature prominently in life-history theory. Life-history theory has been developed to explain how natural selection has shaped the way individuals invest their resources in fecundity and survival at different ages. The underlying assumption is that an individual has limited resources at its disposal, and therefore it must trade off fecundity and/or survival at each age with fecundity and/or survival at other ages. We might expect that natural selection would favor the best balance in this tradeoff. Much of life-history theory has assumed that the "best" balance is the one that maximizes the long-term population growth rate λ_1 (we'll have more to say on this in Chapter 12). To predict the best balance, equations (10.21) and (10.22) are used to evaluate the effect of investing in age-specific survival versus reproduction. These equations are often written in a more explicit form, using equations (10.19) and (10.20) for the elements of the right and left eigenvectors:

$$\frac{\mathrm{d}\lambda_1}{\mathrm{d}p_i} = \frac{\sum_{j=i+1}^{n} \lambda_1^{-(j-1)} l_j m_j}{p_i T} \tag{10.23}$$

and

$$\frac{d\lambda_1}{dm_i} = \frac{\lambda_1^{-(i-1)} l_i}{T}, \tag{10.24}$$

where $T = \sum_{x=1}^{n} \sum_{j=x}^{n} l_j m_j \lambda_1^{-j} = \sum_{x=1}^{n} x\, l_x m_x \lambda_1^{-x}$.

Example: Demography of the Canadian Population

Let us now see how Leslie matrices can be used to understand more complex age-structured populations by modeling the demographic trends in the human population of Canada (data are from Statistics Canada, catalogues 84–210, 84–211, 91–213-XPB, and from the U.S. Census Bureau). Figure 10.7 presents census data for the total population size of Canada at five-year intervals, from 1950 to 2005. The population has more than doubled in size over this period, from 14.0 million in 1950 to 32.8 million in 2005. If we use the discrete-time equation $n(2005) = n(1950)\, \lambda^{55}$ as a model of population growth, where λ is the growth rate per year, then we can estimate the annual growth rate as $\lambda = 1.0156$ (i.e., the population has grown by approximately 1.56% per year).

The above estimate for the population growth rate can be used to predict the total population size at different times in the future. This approach, however, ignores the pronounced differences in age structure of the population over the last half of the twentieth century (Figure 10.8). 1951 was near the beginning of the "baby boom" (1946–1964), and more individuals were in the (0–5)-year-old age class than in any other class. In contrast, the most common age class in 1991 was the (30–35)-year-old class. Therefore, it would be much more accurate to have a model that predicts the future age distribution, in addition to the total population size. Leslie matrix models are perfect for this task.

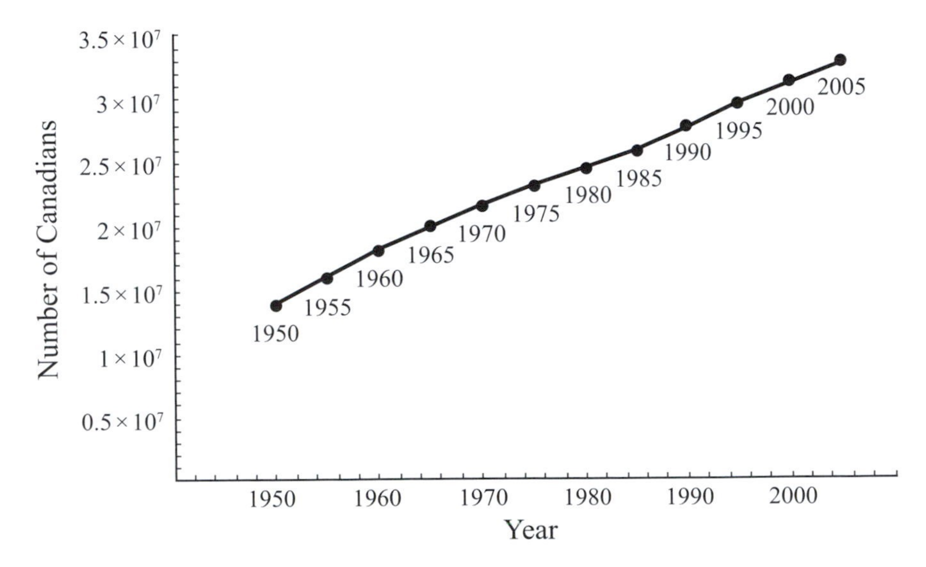

Figure 10.7: The total population size of Canada over time. Based on Table 001 in U.S. Census Bureau, Total Midyear Population, International Data Base (IDB), (http://www. census.gov/ipc/www/idbprint.html release; date April 26, 2005).

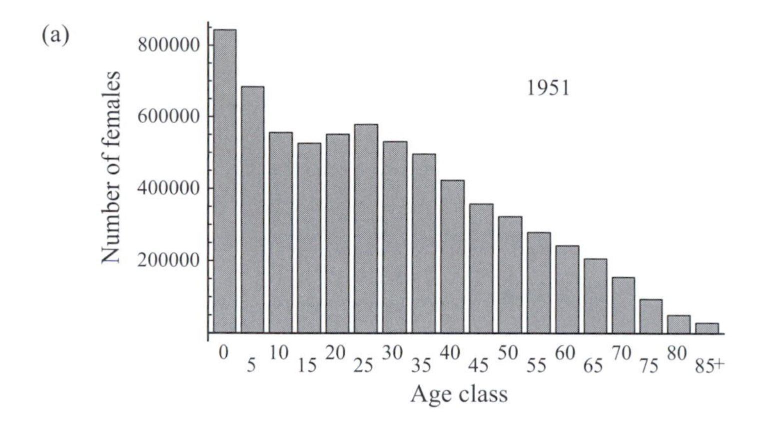

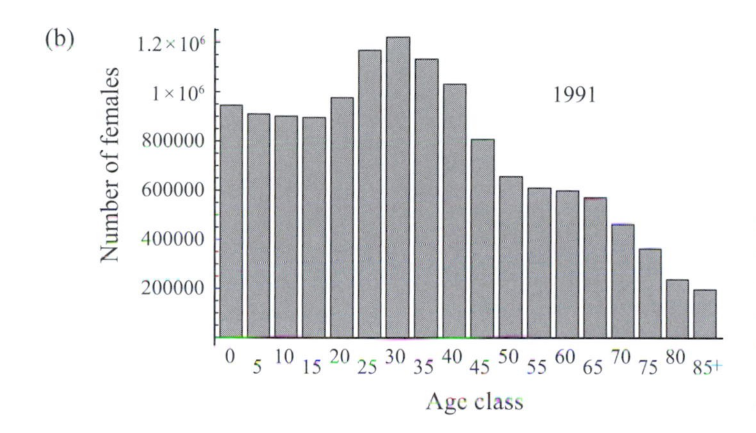

Figure 10.8: The number of females in Canada by age class. (a) 1951 (from Table 004 in U.S. Census Bureau, Enumerated and Adjusted Census Data, International Data Base (IDB), http://www.census.gov/ipc/www/idbprint.html, release date: April 26, 2005). (b) 1991 (see Table 10.2b). Age classes are grouped into five-year intervals.

We will break down the age classes of the Canadian population into five-year intervals. We chose five-year intervals because the census takes place every five years. Choosing age categories whose lengths match the time between censuses ensures that all individuals either die or move to the next age class between censuses. This causes the transition matrix to have the form of a Leslie matrix (10.3.1).

In our model, we keep track of females only. The number of females in each age class is expected to change according to the matrix recursion

$$\begin{pmatrix} n_{0-4}(t+5) \\ n_{5-9}(t+5) \\ n_{10-14}(t+5) \\ \vdots \\ n_{85+}(t+5) \end{pmatrix} = \mathbf{L} \begin{pmatrix} n_{0-4}(t) \\ n_{5-9}(t) \\ n_{10-14}(t) \\ \vdots \\ n_{85+}(t) \end{pmatrix}, \tag{10.25a}$$

where

$$\mathbf{L} = \begin{pmatrix} m_{0-4} & m_{5-9} & m_{10-14} & \cdots & m_{85+} \\ p_{0-4} & 0 & 0 & \cdots & 0 \\ 0 & p_{5-9} & 0 & \cdots & 0 \\ \vdots & \vdots & \ddots & \ddots & \vdots \\ 0 & 0 & 0 & p_{80-84} & p_{85+} \end{pmatrix} \tag{10.25b}$$

The elements of the Leslie matrix (10.25b) are the survival probabilities and fecundities over five-year intervals.

The structure of (10.25b) is slightly different from that of (10.3.1) in that it has an additional element in the bottom right corner. This stems from the fact that the Canadian census groups all individuals over 85 into one class, and such individuals can continue to live through more than one census. This element causes only a slight adjustment to the methods, as discussed in Boxes 10.3 and 10.4.

We now need to obtain estimates for the values of all these elements. We do so using data from the 1991 census and treat the parameters in matrix (10.25b) as constants. In reality, the rates of reproduction and probabilities of survival will change over time as social mores, disease prevalence, health care, and economic conditions change.

First consider the survival probabilities. Table 10.2a gives the number of females that died in each age class in the year 1991. From this data we need to

TABLE 10.2:
Canadian mortality records. (a) The total number of female deaths by age class in 1991 (from Table 3 in Statistics Canada, *Deaths, 1991*, Catalogue 84-211). (b) The total number of females in 1991 (p. 41 in Statistics Canada, *Deaths, 1991*, Catalogue 84-211). (c) The mortality rate of females per census; column (c) was estimated from columns (a) and (b) as $1 - p_i = 1 - (1 - (a)/(b))^5$. (d) Probability of surviving from birth to age class i, $l = p_1 p_2 \dots p_{i-1}$.

Age class	(a) Number deaths (per year)	(b) Total number of females	(c) Mortality rate (per census)	(d) Survival to age i
0–4	1344	945400	0.00709	1
5–9	155	909700	0.000852	0.993
10–14	152	901500	0.000843	0.992
15–19	345	895900	0.00192	0.991
20–24	361	976800	0.00185	0.989
25–29	556	1168800	0.00238	0.987
30–34	681	1221800	0.00278	0.985
35–39	939	1133000	0.00414	0.982
40–44	1337	1031500	0.00646	0.978
45–49	1659	807500	0.0102	0.972
50–54	2142	656600	0.0162	0.962
55–59	3275	610100	0.0266	0.946
60–64	5032	599400	0.0413	0.921
65–69	7763	571800	0.0661	0.883
70–74	9757	461700	0.101	0.825
75–79	12949	362600	0.166	0.741
80–84	14261	237900	0.266	0.618
85+	27415	197100	0.527	0.454

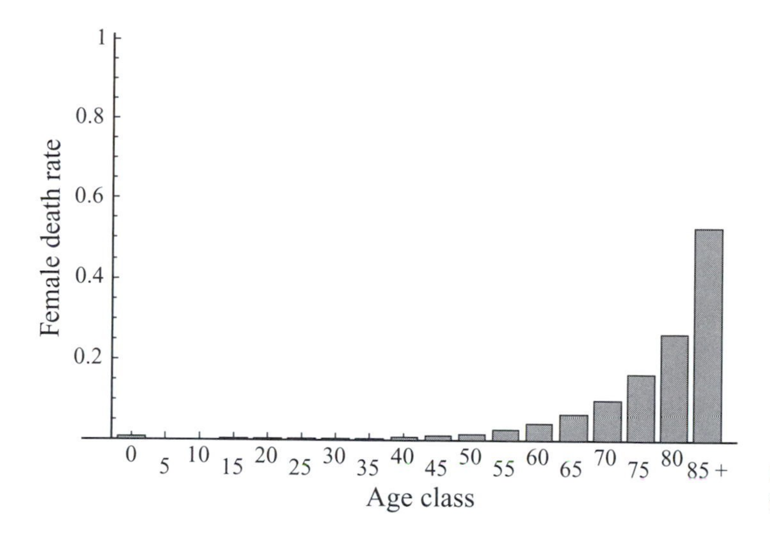

Figure 10.9: The death rate per female per five-year census period. See Table 10.2c.

calculate the probability of an individual surviving through each five-year age class. This is done by calculating the annual probability of surviving, which equals one minus the number that died divided by the number of individuals in each age class (using the columns in Table 10.2, this is 1 – (a)/(b)). The probability of surviving over a census, p_i, is given by this survival probability raised to the power five (the length of the census). The age-specific probability of dying, $1 - p_i$, over the five-year interval between censuses is illustrated in Figure 10.9 (see also Table 10.2c).

Now consider fecundity. Table 10.3 gives the total number of babies produced by Canadian females in the various age classes in 1991. From these data we need to calculate the age-specific expected number of daughters per female, over the five-year census period. This is done by first multiplying the annual number of births in Table 10.3a by the fraction of daughters and by five to obtain the total number of daughters over the census period. We then divide by the number of females in each age class to convert this into a fecundity per female. This gives the results of Table 10.3b and Figure 10.10.

Our model is now completely parameterized. The leading eigenvalue of the Leslie matrix (10.25b) can be calculated using the standard methods of Primer 2 or, more directly, by solving (10.3.5). Either way, the leading eigenvalue is $\lambda = 0.977$. Similarly, the right eigenvector can be solved using standard methods or by plugging the parameters into (10.4.7), giving $\vec{u}^{\mathrm{T}} = (0.0495, 0.0503, 0.0515, 0.0526, 0.0537, 0.0549, 0.0560, 0.0572, 0.0583, 0.0592, 0.0600, 0.0604, 0.0602, 0.0590, 0.0564, 0.0519, 0.0443, 0.0645)$.

These results are very informative. Over the long term, the Canadian population is predicted to decrease in size by a factor of 0.977 every five-year period. This amounts to a 2.3% decline every census (in the absence of immigration), rather than the increase that we predicted by extrapolating from past growth (Figure 10.7). Furthermore, at the stable age distribution, the fraction of the population over age 60 is expected to equal $0.0602 + 0.0590 + 0.0564 + 0.0519 + 0.0443 + 0.0645$, or 33.6%. Similarly, 21.7% are expected to be over the age of

TABLE 10.3
Canadian birth records. (a) The total number of births by age class in 1991 (from Table 11 in Statistics Canada, *Births, 1991*, Catalogue 84-210). (b) The birth rate of daughters per census, obtained by multiplying (a) by the sex ratio (48.67% daughters; from Table 1 in Statistics Canada, *Births, 1991*, Catalogue 84-210), times the census period (×5), divided by the number of females (column b in Table 10.2).

Age class	(a) Total births (per year)	(b) Female birth rate (per census)
0–4	0	0
5–9	0	0
10–14	265	0.000715
15–19	24180	0.0657
20–24	80723	0.201
25–29	150024	0.312
30–34	107560	0.214
35–39	33107	0.0711
40–44	4124	0.00973
45–49	138	0.000416
50–54	0	0
55–59	0	0
60–64	0	0
65–69	0	0
70–74	0	0
75–79	0	0
80–84	0	0
85+	0	0

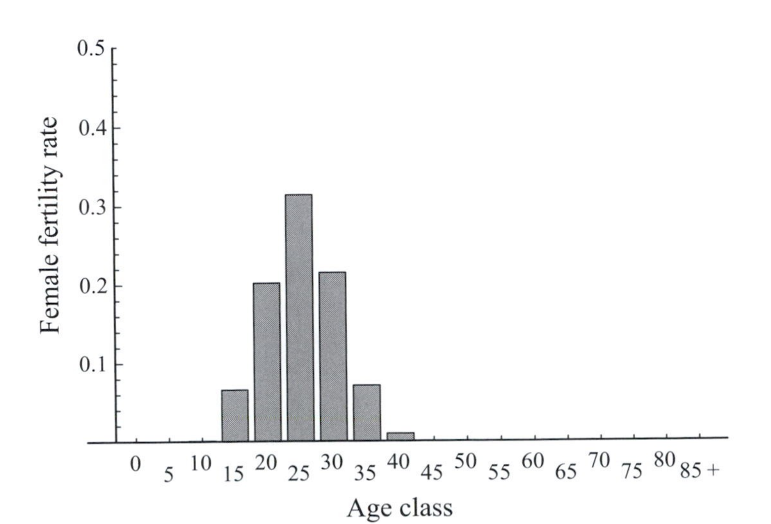

Figure 10.10: The number of daughters born per female per five-year census period. See Table 10.3b.

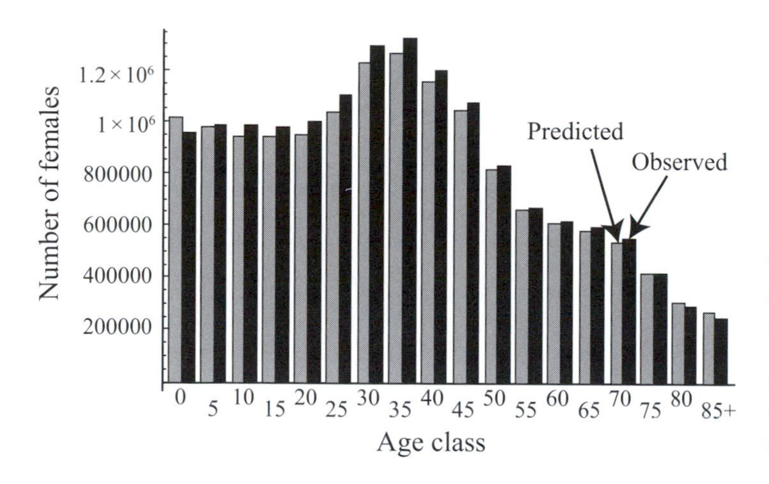

Figure 10.11: Observed and predicted age distribution. The observed age distribution of females in Canada in 1996 (black bars, from Table 1.4 in Statistics Canada, *Annual Demographic Statistics, 1996*, Catalogue 91-213-XPB) is compared with that predicted by multiplying the Leslie matrix (10.25b) with the 1991 data from Table 10.2b (gray bars).

70. If we compare these statistics with the 1991 proportions of 17.76% and 9.20%, respectively, we expect roughly twice the fraction of senior citizens at the stable age distribution. Such predictions have led to a call for the reallocation of government resources, in preparation for the medical and social infrastructure required by these age classes.

We can also iterate the Leslie matrix in *Mathematica*, starting with the number of females in each age class in the 1991 census data (Figure 10.8b). From this we can see that the Leslie matrix does reasonably well at predicting the 1996 census data, but it slightly underestimates the number in most age classes (Figure 10.11). Moreover, the predicted age distribution does differ slightly from that observed. These discrepancies are due to three main factors: (1) immigration and emigration, (2) changing values in the Leslie matrix, and (3) overly coarse age categories. In fact, the net migration rate into Canada over this time period was nearly 500,000 females (Chart 1 in Statistics Canada, *Annual Demographic Statistics, 1996*, Catalogue 91–213-XPB). Furthermore, the birth rate has continued to drop over this period, especially for women under 30.

Sometimes governmental programs aim to increase or decrease the future population size. If such a program were to focus on a particular age group in the Canadian population, which age group would be most effective to target? Interestingly, equation (10.22) reveals that it is best to focus on the most abundant age class, because all u_i are weighted by the same factor $v_1/E[v_k]$. (We can continue to apply equation (10.22) even for this Leslie matrix where p_{85+} is not zero, because the oldest age class does not reproduce; see Box 10.3.) The dominant right eigenvector shows that, in the long term, the most abundant age class will be the (55–59)-year-old age group. This result is disturbing, because females over the age of 55 are typically not physiologically capable of producing offspring. Indeed, Table 10.3b reveals that no such females reproduced in 1991. What has gone wrong? The sensitivity of the leading eigenvalue (10.22) tells us the effect on the long-term growth rate of an *absolute* change in fecundity. But fecundity cannot be changed by the same absolute amount for all age classes. More likely, family planning programs might affect fecundity in a *proportional* manner, by a factor $1 + f$ (that is, $m_i \to m_i\,(1 + f)$). In this case, targeting individuals of age class i would increase the leading eigenvalue by $\Delta m_i\, d\lambda/dm_i = f m_i\, d\lambda/dm_i = f m_i\, u_i v_1/E[v_k]$. As a result, the most effective program would target the age class with the highest product of fecundity and abundance, $m_i\, u_i$, which is the (25–29)-year-old class.

We can also use our general results to explore the likely effects of improvements in medical care on the growth rate of the population. As the medical facilities in a country are enhanced, the population growth rate will increase. Some of the medical improvements result in a greater survival rate of young children, while others result in a greater survival rate of adults. We can use equation (10.21) to determine the different effects that these sorts of medical improvements will have on the population growth rate. To evaluate this equation, we first need to calculate the dominant left eigenvector (i.e., the reproductive value of an individual in each age class), giving $\vec{v}^{T}$ = (1, 0.984, 0.963, 0.941, 0.855, 0.634, 0.306, 0.084, 0.010, 0.00043, 0, 0, 0, 0, 0, 0, 0, 0). The zeroes in the last several elements reveal that females above the age 50 contribute nothing to population growth. You can then evaluate equation (10.21) for all possible i using this eigenvector. Doing so reveals that medical improvements that enhance the survival rate of newborns (i.e., (0–4)-year-olds) cause the greatest increase in the population growth rate.

Example: Calculating the Risk of Infection with HIV Using an Age-Structured Model

Age-structured models are essential in cases where we know that the parameters are likely to change dramatically with age. Although the models of HIV that we considered in previous chapters assumed that all age classes are equivalent, sexual behavior and risk factors depend strongly on age. Thus, accurate predictions of the dynamics of sexually transmitted diseases also require an age-structured model. In Supplementary Materials 10.2, we explore the age-structured model mentioned in Chapter 1 that was used by Williams et al. (2001) to estimate the age-specific risk of contracting HIV in South Africa.

10.7 Concluding Message

In this chapter we have developed techniques for constructing and analyzing a very wide variety of linear models for which the population of interest is class structured. Often, the focus is on the long-term dynamics, and in such cases, the long-term growth rate is given by the leading eigenvalue of the transition matrix. The long-term proportion in each class (the stable class distribution) is given by the right eigenvector. The long-term reproductive value of each class is given by the left eigenvector. We have also described techniques that can be used to assess how the long-term growth rate of a population is affected by a change in the number of individuals or a change in a parameter of the model (the sensitivity of the eigenvalues). Such techniques have been used, for example, to assess the efficacy of programs aimed at conserving a species.

A special form of class structure is age structure. With age structure, individuals move from one age class up to the next age class in subsequent censuses.

Age-structured models are described by a special transition matrix, called the Leslie matrix. The form of the Leslie matrix allows us to develop specific formulas for the characteristic polynomial, the eigenvectors, and the sensitivity of the eigenvalues to changes in parameters.

The range of applications of the techniques for class-structured population models is enormous. Caswell (2001) provides an excellent survey of biological examples, along with more advanced material related to such matrix models. The techniques of this chapter also prove to be invaluable when constructing evolutionary models for populations that are class structured (Charlesworth 1994; Chapter 12).

Problems

Problem 10.1: Figure 10.12 illustrates a model with three patches of plants, arranged from west to east.

Each generation, b seeds are produced, of which a fraction d disperse and $1-d$ remain on the parental patch. Of the dispersing seeds, a fraction f disperse west and $1-f$ disperse east. The parents then die and all seeds that have landed on a patch grow into plants. This model can be described by the following transition matrix:

$$\mathbf{M} = \begin{pmatrix} b\,(1-d) & b\,d\,(1-f) & 0 \\ b\,d\,f & b\,(1-d) & b\,d\,(1-f) \\ 0 & b\,d\,f & b\,(1-d) \end{pmatrix}$$

The eigenvalues and associated eigenvectors of this transition matrix are

$$\begin{array}{lll} (1) & \lambda = b\,(1-d), & \vec{u} = (1-f,\ \ 0,\ \ -f), \\ (2) & \lambda = b\,(1-d) - b\,d\sqrt{2f(1-f)}, & \vec{u} = (1-f,\ \ -\sqrt{2f(1-f)},\ \ f), \\ (3) & \lambda = b\,(1-d) + b\,d\sqrt{2f(1-f)}, & \vec{u} = (1-f,\ \ \sqrt{2f(1-f)},\ \ f), \end{array}$$

(a) At what rate will the population eventually grow? (b) In the long term, what fraction of plants inhabits each patch? (c) Explain why f affects the growth rate of the population.

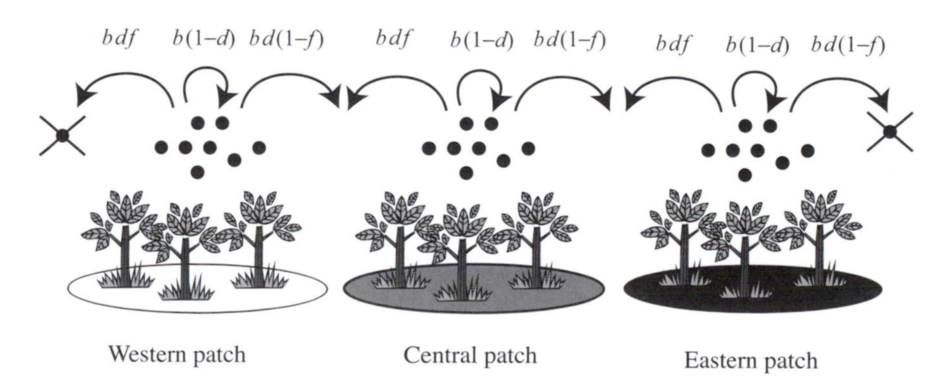

Figure 10.12: A schematic of a three-patch plant population.

Problem 10.2: Consider the model of seed dispersal in Problem 10.1. (a) What are the reproductive values of a plant in each of the three patches? Standardize the reproductive values so that the reproductive value of a plant in the westernmost patch is one. (b) Suppose that a mutant allele appears in the population that causes the fraction of seeds that disperse east, f, to increase slightly. Under what conditions will this mutant allele have a larger long-term growth rate than the wild-type allele? (c) By using the fact that $d\lambda/dz = 0$ at a maximum, find the value of f that yields the largest long-term growth rate.

Problem 10.3: Find the two eigenvalues of the transition matrix in the methylation model,

$$\begin{pmatrix} \alpha & \beta \\ 1-\alpha & 1-\beta \end{pmatrix},$$

using the trace and determinant method (equation (P2.8) in Primer 2). Confirm that the trace equals the sum of the two eigenvalues. [Hint: the algebra is made easier by defining $\rho = \alpha - \beta$.]

Problem 10.4: Here you will prove that

$$\frac{v_x}{v_1} = \frac{\lambda^{x-1}}{l_x} \sum_{i=x}^{n} \frac{l_i m_i}{\lambda^i}.$$

(a) Multiply out $\vec{v}^{\mathrm{T}} L = \lambda \vec{v}^{\mathrm{T}}$ using the matrix $\mathbf{L}$ in (10.3.1) and the left eigenvector $\vec{v}^{\mathrm{T}} = (v_1 \quad v_2 \quad v_3 \cdots v_n)$ (b) Show that the results can be written as $v_x/v_1 = m_x/\lambda + (p_x/\lambda)(v_{x+1}/v_1)$ for x from 1 to $n-1$ and $v_n/v_1 = m_n/\lambda$. (c) Show that the first of these equations is equivalent to

$$\frac{v_x}{v_1} = \frac{\lambda^{x-1}}{l_x}\left(\frac{l_x m_x}{\lambda^x} + \frac{l_{x+1}}{\lambda^x}\frac{v_{x+1}}{v_1}\right),$$

where $l_i = p_1 p_2 \ldots p_{i-1}$ is the probability of surviving from birth to age class i. (d) Use your result from (c) and $v_n/v_1 = m_n/\lambda$ to show that

$$\frac{v_{n-1}}{v_1} = \frac{\lambda^{n-2}}{l_{n-1}} \sum_{i=n-1}^{n} \frac{l_i m_i}{\lambda^i}.$$

You can then replace n–1 with x to infer the desired result

$$\frac{v_x}{v_1} = \frac{\lambda^{x-1}}{l_x} \sum_{i=x}^{n} \frac{l_i m_i}{\lambda^i}.$$

(e) Prove that this result obeys the equation in (c).

Problem 10.5: A scientist is studying a colony of mice. She finds that they produce, on average, one surviving daughter per female in the first year of life and eight in the second year of life. She also finds that they have only a 25% chance of surviving to the second year and do not survive beyond this. She constructs the following Leslie matrix for this population:

$$\mathbf{L} = \begin{pmatrix} 1 & 8 \\ 1/4 & 0 \end{pmatrix}.$$

(a) Find the eigenvalues and right eigenvectors, $\vec{u}$, of $\mathbf{L}$. (b) Using your answers to (a), what is the long-term growth rate and the proportion of the population in each age class at the stable age distribution? (c) Write a matrix $\mathbf{A}$ containing the eigenvectors of $\mathbf{L}$ as columns. Use your answer to (b) to write the eigenvector associated with the leading eigenvalue as proportions and place this in the first column of $\mathbf{A}$. Calculate the inverse matrix $\mathbf{A}^{-1}$, whose first row gives the left eigenvector associated with the leading eigenvalue. (Check: Make sure that your answer does give a left eigenvector by showing that $\vec{v}^{\mathrm{T}}\mathbf{L} = \lambda\vec{v}^{\mathrm{T}}$.) (d) We can approximate the dynamics of the mice colony using $\vec{n}(t) \approx \lambda_L^t \vec{u}_L(\vec{v}_L^{\mathrm{T}}\vec{n}(0))$, where $\vec{u}_L$ is the right eigenvector (written in column format) and $\vec{v}_L^{\mathrm{T}}$ is the left eigenvector (written in row format) associated with the leading eigenvalue. For a population that begins with ten mice, all in the first age class, what is the approximate total population size after two years? After ten years? (e) Compare your answers to (d) to the exact answer obtained by iterating the Leslie matrix, which predicts that the total population size is 32.5 at t=2 and 7682.5 at t=10. Explain why the error is proportionately worse at t=2 and why the approximation works so well given that neither eigenvalue is below one. (g) In the long run, would the colony be larger if it had been started from ten mice of age one or from ten mice of age two? Compose an answer using information from the eigenvectors.

Problem 10.6: Consider a special case of the model of juveniles and adults in equation (10.1) where the transition matrix is given by

$$\mathbf{L} = \begin{pmatrix} 0 & 1.6 \\ 0.6 & 0.4 \end{pmatrix}.$$

(a) What is the long-term growth rate of the population? (b) What is the stable class distribution of the population? (c) Once the population has reached its stable class distribution, prove that adults contribute twice as much to the future population size relative to juveniles. (d) Suppose that you want to increase the future population size as much as possible, and you can do this by either introducing three juveniles or two adults. Which is your best option?

Problem 10.7: Consider the model of Problem 10.6, but now suppose that you can change the population size in the future by permanently altering some of the elements of the transition matrix. In particular, suppose that you can increase the juvenile survival from 0.6 to 0.61 but this necessarily comes at the cost of reducing adult survival from 0.4 to 0.39. (a) Determine the effect of this alteration on the long-term population growth rate by calculating the eigenvalues numerically. (b) Determine the effect of this alteration on the long-term population growth rate by using the formula (10.13) of the text, where the effect on the leading eigenvalue, $\Delta\lambda$, is approximately $\Delta\lambda \approx d\lambda/dz\ \Delta z$.

Further Reading

For further information and examples of class-structured models see

- Charlesworth, B. 1994. *Evolution in Age-Structured Populations*. Cambridge University Press, Cambridge.

- Edelstein-Keshet, L. 1988. *Mathematical Models in Biology*. Random House, New York.
- Gurney, W. S. C. and R. M. Nisbet. 1998. *Ecological Dynamics*. Oxford University Press, Oxford.
- Kot, M. 2001. *Elements of Mathematical Ecology*. Cambridge University Press, Cambridge.
- Murray, J. D. 1993. *Mathematical Biology*. Springer, New York.
- Okubo, A., and S. A. Levin. 2001. *Diffusion and Ecological Problems*, 2nd ed. Springer, New York.

References

Caswell, H. 2001. *Matrix Population Models*, 2nd ed. Sinauer Associates. Sunderland, Mass.

Charlesworth, B. 1994. *Evolution in Age-Structured Populations*. Cambridge University Press, Cambridge.

Fujiwara, M., and H. Caswell. 2001. Demography of the endangered North Atlantic right whale. *Nature* 414:537–541.

Roff, D. A. 1992. *The Evolution of Life Histories*. Chapman and Hall, New York.

Stearns, S. C. 1992. *The Evolution of Life Histories*. Oxford University Press, Oxford.

CHAPTER 11

Techniques for Analyzing Models with Periodic Behavior

Chapter Goals:

- To describe periodic dynamics around an equilibrium
- To predict parameter values that allow for periodic dynamics
- To determine the stability of periodic dynamics

Chapter Concepts:

- Composite mapping
- Stable limit cycle
- Hopf bifurcation theorem
- Constants of motion

11.1 Introduction

A major focus of the book so far has been identifying equilibria and characterizing their stability. While these techniques are of great importance, many biological processes are characterized by nonequilibrium dynamics—a situation in which the values of the variables of interest never approach a steady state. In fact, there are two different reasons why nonequilibrium dynamics might occur in biological systems. First, there is undoubtedly an important component of randomness or stochasticity to many biological systems, thereby preventing them from ever remaining constant for very long. Incorporating such stochasticity into mathematical models in ecology and evolution will be discussed later in Chapters 13–15. Second, many biological processes generate persistent fluctuations or oscillations. In other words, the biological processes themselves generate self-sustaining nonequilibrium dynamics even in the absence of any stochasticity. It is this second class of processes that we study in this chapter.

There are many sorts of deterministic nonequilibrium dynamics that can occur, including recurring (usually termed periodic) dynamics and chaos (Edelstein-Keshet 1988; Hofbauer and Sigmund 1988; Kaplan and Glass 1995; Gurney and Nisbet 1998; Chapter 4). While there are a few mathematical techniques available to analyze models that exhibit these behaviors, many techniques are quite sophisticated and difficult to apply in general. Indeed computer simulations become an invaluable aid to analyzing models with such complicated behaviors. There are, however, a few relatively straightforward techniques for models exhibiting periodic behavior. We explore three of these in this chapter.

In section 11.2 we begin with a precise definition of periodic dynamics. Section 11.3 then presents a technique for discrete-time models, which we call composite mapping. Section 11.4 presents a technique known as the Hopf bifurcation theorem. Finally, section 11.5 presents a technique based on identifying constants of motion. Despite these somewhat intimidating sounding names, these techniques are relatively straightforward to apply.

11.2 What Are Periodic Dynamics?

Perhaps one of the best-known biological examples of periodic dynamics is the predator-prey cycle in ecology (Hofbauer and Sigmund 1988; Begon et al. 1996; Case 2000). Roughly speaking, when the predator population is large, it has a very strong negative effect on the prey population causing it to crash to low numbers. As a result, the predator's food supply becomes insufficient to maintain so many predators, and the predator population subsequently crashes. Once the negative effect of the predator on the prey has diminished, the prey population is free to grow in numbers again, and the cycle starts anew. Thus the ecological interaction between predator and prey can generate sustained fluctuating dynamics (Figure 4.17). This sort of behavior is quite common in models of other biological processes, especially those involving antagonistic interactions.

From a mathematical standpoint, a system is said to display *periodic* behavior if the values of its variables change through time but repeatedly return to the same state at fixed time intervals.

Definition 11.1: Periodic Dynamics

A variable x is said to display periodic dynamics with period τ if it varies over time, but at any point in time t, its value is the same as its value at time $t + \tau$; i.e., $x(t) = x(t + \tau)$ (Figure 11.1). By convention the period τ is taken to be the shortest interval of time over which all variables return to the same values.

As an example, if a predator-prey model exhibits periodic cycles with a period of 10 years, then although the predator numbers would be continually changing over time, if there were 500 predators in year 2000, then there would also be 500 predators in years 2010, 2020, 2030, etc. Moreover, if there were 750 predators in year 2005, there would also be 750 predators in year 2015, 2025, 2035, etc.

Periodic dynamics are a form of cyclic behavior, but it is important to distinguish periodic dynamics from other types of oscillatory and cyclic behavior.

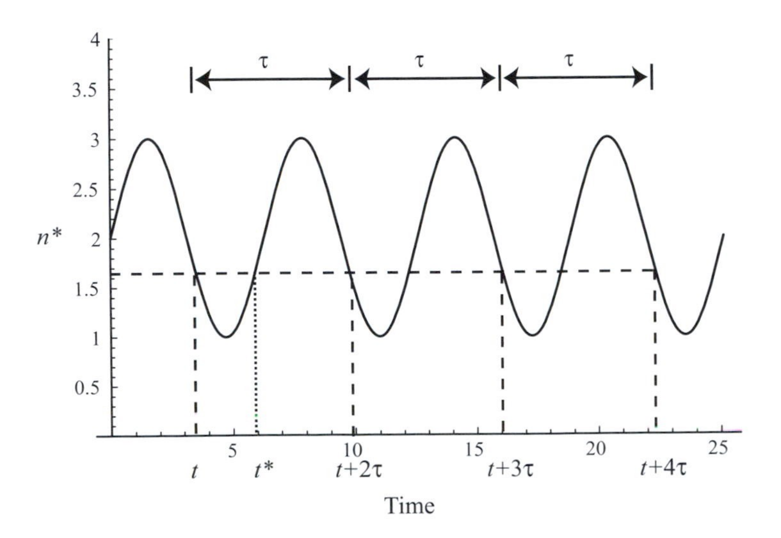

Figure 11.1: An example of periodic dynamics in continuous time. The period length (i.e., the time to complete a single cycle) is τ. Even though the value of n at time t and time t^* is the same, this does not constitute a cycle of period $t^* - t$ because other values of n do not recur over that time period. You can see from the plot that any value of n recurs with period τ and that this is the smallest time interval for which this is true.

Oscillations or cycles that increase or decrease in magnitude can occur around any equilibrium, as we saw in the logistic model with $r = 1.8$ and $r = 2.7$ (Figure 4.2a,c). But these do not represent periodic dynamics, because different values of the variable are reached during every fluctuation. Similarly, spirals away from or toward an equilibrium, as seen in Figure 7.6, also represent cyclic dynamics, but they do not represent periodic dynamics. Only when a cycle passes through the very same points at the interval τ do we say that the dynamics are periodic. For example, we saw periodic dynamics in Figure 4.2a,b for the logistic model with $r = 2.1$ (giving a two-point cycle) and $r = 2.5$ (giving a four-point cycle).

In the three techniques that follow, we will address two main questions about periodic dynamics: How can we predict whether a model will exhibit periodic dynamics, and how do the periodic dynamics depend on the parameters of the model?

11.3 Composite Mappings

We start with a technique that is useful for analyzing periodic dynamics in discrete-time models. We refer to this technique as "composite mappings" although there is no standard name for the approach. Rather than analyzing equations that describe the change in a system over one time step, we analyze composite mappings, which are equations that describe the change in a system over a particular number of time steps. There is a very nice correspondence between equilibria in the composite mapping of a discrete-time model and periodic dynamics. Consequently, we can use composite mappings to determine whether a model will display periodic behavior, and whether or not this cycling will be stable (i.e., whether or not, the behavior approaches the same cyclic pattern if started nearby). In principle, this technique can be applied to any discrete-time model, although in practice it is difficult to use for most multiple-variable models. We therefore focus on single-variable models, using a model of population growth for illustration.

Suppose that $N(t)$ is the population size of an insect species in year t and that the population size changes from one year to the next according to the recursion

$$N(t + 1) = f(N(t)). \tag{11.1}$$

As we saw in Chapters 4 and 5, this type of model (which includes the exponential growth and logistic models) can exhibit a variety of behaviors. These include stable equilibria, periodic or cyclic dynamics, as well as chaos (see Figures 4.1 and 4.2). As with continuous-time models, periodic behavior means that the same temporal pattern of population sizes repeats itself over and over again. Therefore, for some value of τ, we have $N(t + \tau) = N(t)$.

Here we consider the logistic model (3.5a), which exhibits periodic behavior for certain values of the intrinsic growth rate (Chapter 4). In terms of equation (11.1), $f(N(t)) = N(t)\,(1 + r\,(1 - N(t)/K))$. As described in Chapter 5, this model has two equilibria, $\hat{N} = 0$ and $\hat{N} = K$. The first of these is unstable whenever $r > 0$, whereas the second is stable when $0 < r < 2$.

To begin, let us suppose that we want to know whether it is possible to have cycles with a period of two generations; i.e., $N(t + 2) = N(t)$. We can obtain an explicit expression for $N(t + 2)$ in terms of $N(t)$ by iterating equation (11.1) twice; i.e.,

$$\begin{aligned} N(t+2) &= f(N(t+1)) \\ &= f(f(N(t))). \end{aligned} \tag{11.2}$$

A function f applied to itself is referred to as a *composite mapping* or a composition of the function f.

Expressions such as $f(f(N(t)))$ are termed composite mappings or compositions of the function f as they denote the function f applied to itself (possibly more than once). In fact, we can define a new function g as

$$g(N(t)) = f(f(N(t))). \tag{11.3}$$

which equals

$$\begin{aligned} g(N(t)) &= f(N(t))\left(1 + r\left(1 - \frac{f(N(t))}{K}\right)\right) \\ &= \{N(t)\,(1 + r\,(1 - N(t)/K))\} \\ &\quad \left(1 + r\left(1 - \frac{\{N(t)\,(1 + r\,(1 - N(t)/K))\}}{K}\right)\right). \end{aligned} \tag{11.4}$$

The function g can then be used to define the "composite" model

$$N(i + 1) = g(N(i)) \tag{11.5}$$

where a single time step i in this composite model corresponds to two time steps t in our original model. The dynamics of this composite model are exactly those that would be observed if we looked at the dynamics of the original model (11.1) at every other time step. If we start on an odd time step,

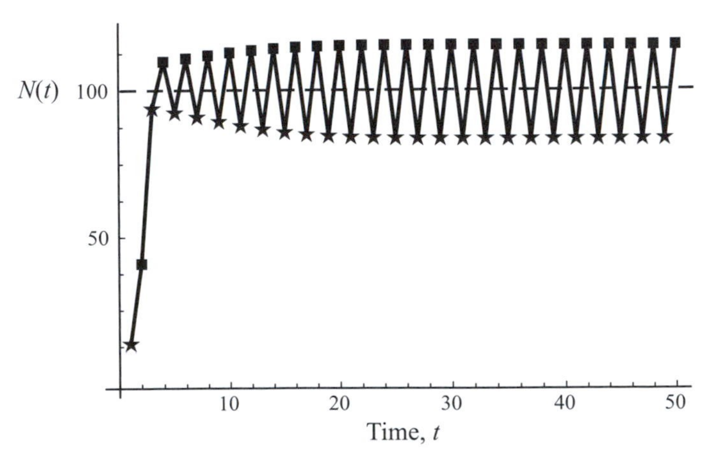

Figure 11.2: A period 2 cycle in the discrete-time logistic model. The dynamics of the logistic model, $N(t + 1) = f(N(t)) = N(t)\,(1 + r\,(1 - N(t)/K))$ (solid line) is compared to the dynamics of the composite model, $N(i + 1) = g(N(i)) = f(f(N(i)))$. There are two different patterns depending upon whether we start observing these dynamics in time step 1 of the original model (stars) or in time step 2 (squares). Here $r = 2.1$ and $K = 100$ (dashed line). Equation (11.6) thus predicts a period 2 cycle alternating between $\hat{N}_3 = 112.9$ and $\hat{N}_4 = 82.4$, as observed.

equation (11.5) describes the set of transitions among odd time steps, whereas if we start on an even time step, it describes the set of transitions among even time steps (Figure 11.2).

Because a cycle with a period of two generations must return to the same point every other time step, a two-point cycle will appear as an equilibrium in the "composite" model (Figure 11.2). This is a very powerful observation, because it allows us to use all of our techniques for finding equilibria and determining their stability to find cycles and determine their stability properties. In particular, solving the equation $\hat{N} = g(\hat{N})$ using equation (11.4) yields four potential solutions:

$$\begin{aligned} \hat{N}_1 &= 0, \\ \hat{N}_2 &= K, \\ \hat{N}_3 &= K\frac{2 + r + \sqrt{r^2 - 4}}{2r}, \\ \hat{N}_4 &= K\frac{2 + r - \sqrt{r^2 - 4}}{2r}. \end{aligned} \tag{11.6}$$

The first two solutions are the equilibria of the original model. We should expect this result because if $\hat{N}$ is an equilibrium of the original model, then it will also be an equilibrium of the composite model. For this reason, the first two solutions do not represent cycles with a period of two times steps (a "period-2 cycle").

When the last two solutions in (11.6) are real, the original model will display period-2 cycles. Assuming r is positive, the solutions are real if $r > 2$. Combining this with our earlier results, we see that the logistic model will have a period-2 cycle whenever the equilibrium $\hat{N} = K$ is unstable. In the original model, this period-2 cycle alternates back and forth between the two population densities $\hat{N}_3$ and $\hat{N}_4$ in (11.6) (Figure 11.2). At this stage, however, we do not yet know if the period-2 cycle is stable. If we started the model near one of the population sizes that makes up this cycle, do the dynamics of the system eventually attain this periodic pattern, or do they move away?

In a discrete-time model, this question can be answered by asking whether the equilibria $\hat{N}_3$ and $\hat{N}_4$ are locally stable in the composite model defined by (11.4). If they are, then we know that the period-2 cycle of the original model is locally stable. Using (11.4) we can check that the stability condition (Recipe 5.3) for both of these equilibria of the composite model is the same: the period-2 cycle is stable if $-1 < 5 - r^2 < 1$. This requires that $2 < r < \sqrt{6} \approx 2.4495$. Therefore, as r increases past 2, the equilibrium at the carrying capacity K loses stability, and the model then displays a locally stable cycle in population size with period 2. The cycle remains locally stable as r increases (although the amplitude of the cycle changes) until r reaches approximately 2.4495. This period-2 cycle then loses stability as well. This is entirely consistent with the bifurcation plot for the logistic model (Figure 4.1.2), but now we know exactly when the period-2 cycle loses stability.

Recipe 11.1
Composite Mappings

Composite mappings can be used to determine whether periodic dynamics are possible in discrete-time models, and when such dynamics are stable. The technique is most useful for one-variable models.

Consider the discrete-time recursion $N(t + 1) = f(N(t))$, and suppose that we want to know whether periodic dynamics with period τ are possible.

Step 1: Define the composite function, $g_\tau(N(i)) \equiv f(f(f(\ldots f(N(i)))))$, which represents the original recursion function $f(N)$ applied to itself τ times. Define analogous composite functions $g_2, g_3, \ldots, g_{\tau-1}$ for all periods less than τ as well.

Step 2: Beginning with $g_2(N(i)) \equiv f(f(N(i)))$, define the composite recursion $N(i + 1) = g_2(N(i))$; one time step i in this composite recursion corresponds to two time steps in the original recursion.

Step 3: Equilibria of this composite recursion that are not also equilibria of the original model correspond to cycles of period 2 in the original recursion.

Step 4: Determine the local stability properties of any period-2 cycles found, using the stability techniques from Recipes 5.2 and 8.4 on the composite recursion.

Step 5: Steps 2–4 can be applied to composite recursions of larger and larger period τ. Equilibria of the period-τ composite recursion that are not also equilibria of the original model or any of the composite recursions for shorter periods correspond to cycles of period τ in the original recursion.

Recipe 11.1 describes the general approach for identifying periodic dynamics using composite mappings in discrete-time models. In principle this approach can be used to explore cycles of any period τ by constructing the composite mapping $N(t + \tau) = f(f(f(\ldots f(N(t)))))$. Then one looks for equilibria of this composite model (ignoring equilibria that are already known to correspond to cycles of a period less than τ) and determines their stability properties. In practice, however, as the period length τ increases, analytical solutions become more difficult to obtain.

11.4 Hopf Bifurcations

In this section we introduce a technique for analyzing periodic dynamics known as the Hopf bifurcation theorem (Hopf 1941; Marsden and McCracken 1976; Edelstein-Keshet 1988). A Hopf bifurcation (named after the mathematician Eberhard Hopf) occurs when an equilibrium's stability properties change and periodic dynamics appear around the equilibrium as a parameter of the model is changed. The Hopf bifurcation theorem specifies conditions under which such periodic dynamics occur. We introduce it through an example from evolutionary biology.

Example: Evolution by Sexual Selection

In Chapter 7 we constructed and analyzed a model of Fisher's Runaway Process of sexual selection. Our model was a continuous-time linear model with two variables—one for the average male tail length in the population, $\bar{z}$, and one for the average female mate preference, $\bar{p}$. Recall that tail length was standardized so that $z = 0$ was the optimal tail length from the perspective of natural selection, and $p = 0$ represented no female mate preference. We examined the question of whether male tail exaggeration and female preference would evolve by determining whether the equilibrium $\hat{\bar{z}} = 0, \hat{\bar{p}} = 0$ was unstable, and we found that instability was predicted under certain parameter values. Because that model was linear, it predicted that the magnitude of male tail length and female preference should continue to increase forever. Here we explore a nonlinear version of this model that allows selection to act more strongly against tail length and female preference as they get farther from zero.

To do so, we first review the form of the model in Chapter 7. In that model, the average male tail length had two sources of selection acting on it: (i) sexual selection and (ii) natural selection. Sexual selection favored longer tails than are optimal for survival if $\bar{p} > 0$ or smaller tails than are optimal if $\bar{p} < 0$. This was modeled by having the differential equation for $d\bar{z}/dt$ contain a term involving $a\,\bar{p}$ (which is positive if $\bar{p} > 0$ and negative if $\bar{p} < 0$; see equation (7.19)). Natural selection favored an optimal tail length ($z = 0$), and this was modeled by having the differential equation for $d\bar{z}/dt$ contain a term involving $-c\,\bar{z}$ (which is negative if $\bar{z}$ is above zero and positive if it is below zero; see equation (7.19)). Similarly, natural selection on the female trait favored a preference value of zero, and the differential equation for $d\bar{p}/dt$ therefore contained a term involving $-b\,\bar{p}$ (see equation (7.19)).

Let us now alter the form of natural selection so that the strength of selection increases as $\bar{z}$ gets further from the optimum. To ensure that the direction of natural selection remains the same (favoring the optimal tail length, $z = 0$), we must multiply $-c\,\bar{z}$ in model (7.19) by a strictly positive factor that rises as $\bar{z}$ becomes more positive or more negative. We chose $1 + \bar{z}^2$ as one of the simplest factors with these properties. Similarly, we alter the form of natural selection on female preference by changing the term $-b\,\bar{p}$ in model (7.19) to $-b\,\bar{p}\,(1 + \bar{p}^2)$. Model (7.19) then becomes

$$\frac{d\bar{z}}{dt} = G_z(a\,\bar{p} - c\,\bar{z}\,(1 + \bar{z}^2)) - B\,b\,\bar{p}\,(1 + \bar{p}^2),$$

$$\frac{d\bar{p}}{dt} = B\,(a\,\bar{p} - c\,\bar{z}\,(1 + \bar{z}^2)) - G_p\,b\,\bar{p}\,(1 + \bar{p}^2) \qquad (11.7)$$

(see Problem S7.4 in Supplementary Material 7.1). The form of sexual selection (i.e., the terms involving the parameter a) is unchanged from the model in Chapter 7 (see equations (7.19)).

Model (11.7) is nonlinear, and we can begin to analyze it using the techniques from Chapter 8. The first step is to determine the equilibria. Setting $d\bar{z}/dt = 0$ and $d\bar{p}/dt = 0$, you can see that every term is multiplied by either $\bar{z}$ or $\bar{p}$. Therefore, $\hat{\bar{z}} = 0$, $\hat{\bar{p}} = 0$ must be an equilibrium. (Indeed, this is the only equilibrium, as you can show by solving $d\bar{z}/dt = 0$ for $\bar{z}(1 + \bar{z}^2)$ and using this to solve $d\bar{p}/dt = 0$). The stability of the origin is particularly interesting, as it will determine whether or not female preferences and exaggerated male displays can evolve. To determine stability, we follow Recipe 8.2 and first find the Jacobian matrix (Definition 8.2) at this equilibrium:

$$\mathbf{J} = \begin{pmatrix} -G_z c & G_z a - b B \\ -B c & -G_p b + a B \end{pmatrix}. \tag{11.8}$$

This matrix is identical to the matrix of coefficients (7.20) of the linear model, because the nonlinear terms in the model are neglected in a linearization around the equilibrium $\hat{\bar{z}} = 0$, $\hat{\bar{p}} = 0$. In Chapter 7, we found that the equilibrium with no exaggeration or preference was unstable if $\gamma > 0$, where $\gamma = -b\,G_p - c\,G_z + a\,B$ (which is the trace of the Jacobian matrix; see Definition P2.5 in Primer 2). Therefore, the same conclusion applies here as well. For the linear model, the general solution (Box 9.4) predicts that the system will grow away from the equilibrium whenever the equilibrium is unstable. Thus, the linear model will not exhibit a periodic pattern, returning to the same values over regular intervals of time (although it might exhibit oscillations as the variables move away from zero). What is not clear, however, is what will happen to the dynamics in the nonlinear model when the equilibrium is unstable.

One of the best places to start exploring the dynamics of such nonlinear models is with simulations. Figure 11.3a illustrates the dynamics of female preference and male tail length for one set of parameter values where the equilibrium is unstable. In this example, both variables grow over time and appear to reach a stable cycling pattern. The phase plane for this example reveals that the dynamics spiral out from the origin $\hat{\bar{z}} = 0$, $\hat{\bar{p}} = 0$, and approach a cycle around this unstable equilibrium (Figure 11.3b). Additionally, if we start the model with large values of $\bar{z}$ and $\bar{p}$, then the dynamics spiral in toward the same cycle around the unstable equilibrium (Figure 11.3c). This suggests that the dynamics are attracted toward a particular type of periodic pattern. Such periodic patterns are termed *stable limit cycles* (Box 11.1).

A *stable limit cycle* is a trajectory that is periodic and attracting when the system starts away from the cycle.

One drawback with simulations is that it is difficult to know when periodic dynamics will occur and how they depend on the parameter values chosen. The Hopf bifurcation theorem can address these questions. Generally speaking, the term *bifurcation* (as applied to dynamic models) refers to a situation in which the behavior of a model changes from one qualitative type to another as a parameter of the model is changed. For example, as we increased the per capita growth rate in the discrete-time logistic model of Chapter 4, we saw that its behavior changed from having a stable equilibrium to cycling between two values, then to four values, etc. (see the bifurcation diagram 4.1.2 in Box 4.1 of Chapter 4). As this parameter was increased even further the qualitative behavior changed again, resulting in chaotic dynamics. Each one of these qualitative

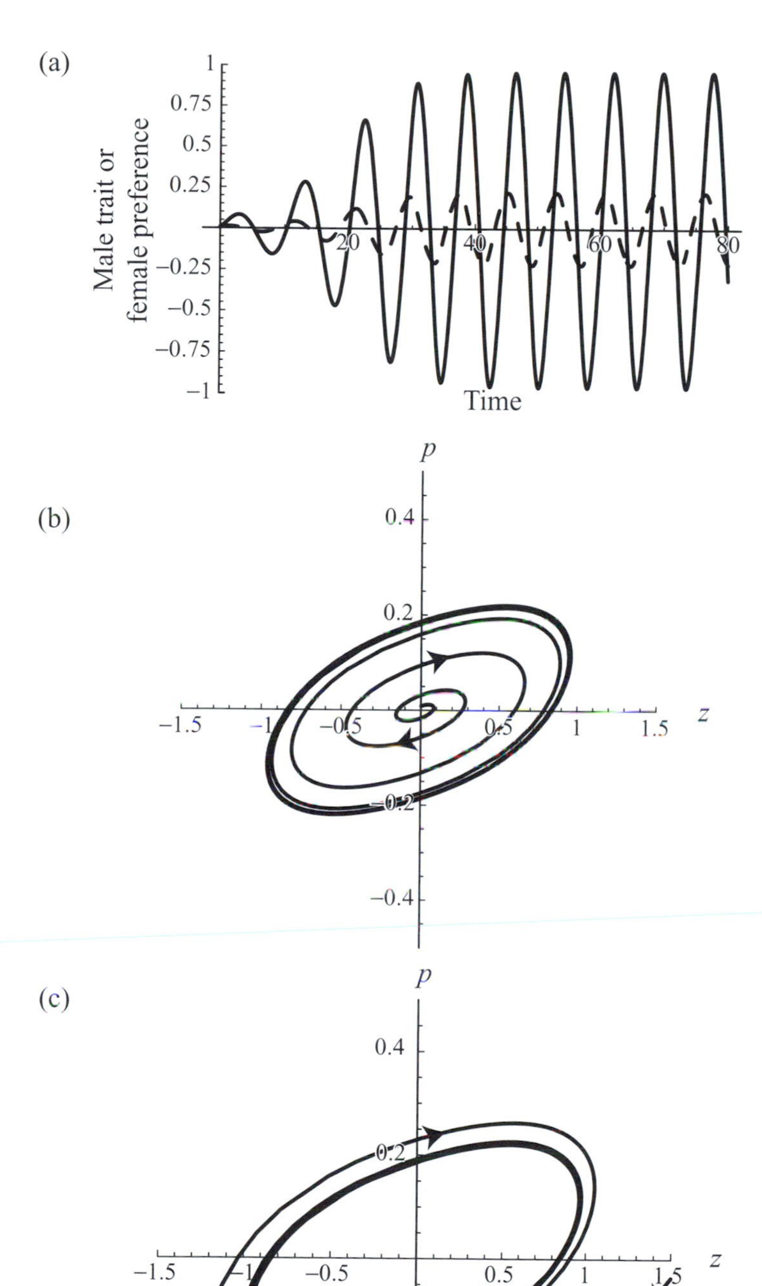

Figure 11.3: The evolutionary dynamics of male tail length and female preference in the nonlinear model. Parameter values are $G_z = 1, G_p = 1, B = 0.5, c = 0.25, b = 2, a = 5$. With these parameter values, $\gamma = 0.125$. (a) Male tail length (solid curve) and female preference (dashed curve) over time. Both increase in magnitude while oscillating and approach a stable, periodic pattern. (b) The corresponding phase plane. (c) The phase plane with the same parameter values but with initial conditions that lie 'outside' of the periodic cycle.

Box 11.1: Definitions of Limit Cycles and Attractors

Limit cycles are best described by first defining the notion of an attractor in dynamic models. An *attractor* is a set of values that the variables approach over time. An attractor can be a local or a global attractor.

The stable equilibria of the previous chapters are attractors, and they are sometimes referred to as "point attractors" because they are points that are eventually approached as time passes. It might seem odd, but unstable equilibrium points are also called attractors. This surprising convention follows from the fact that such equilibria are attractors (in the colloquial sense of the word) if time runs backward! Nevertheless, it is often more intuitive to refer to such points as *repellors*.

Although an equilibrium is the simplest type of attractor, there are other interesting attractors. Two frequently mentioned attractors are chaotic attractors and periodic attractors. *Chaotic attractors* (also called *strange attractors*) are sets of values of the variables (sometimes very complicated) that the system eventually approaches. We have already seen a chaotic attractor in Chapter 4 where we explored the dynamics of the logistic model for different values of the population growth rate. As mentioned in Box 4.1, chaotic attractors are characterized by sensitivity to initial conditions.

Periodic attractors are sets of points that the system exactly retraces over and over again and are often referred to as *limit cycles*. As with equilibria, a limit cycle is termed locally stable if the system approaches the cycle as time passes for all values of the state variables close to the cycle. Limit cycles can be globally stable as well. Similarly, *unstable limit cycles* (or *repelling limit cycles*) can also occur. The system will remain on an unstable limit cycle if it is started exactly on the cycle, but once perturbed off the cycle, it will move away. Finally, neutrally stable cycles, called *neutral limit cycles*, are also possible. The neutral cycle that a system exhibits is determined by the starting conditions; if perturbed, the system does not return to the original neutral cycle but exhibits a new neutral cycle. When there is a neutral limit cycle, infinitely many such cycles exist.

changes in behavior is referred to as a bifurcation, and the parameter that caused the change (in that case the per capita growth rate) is referred to as the *bifurcation parameter* (see Box 4.1). There are different kinds of bifurcations that are possible and their mathematical study is quite sophisticated. The Hopf bifurcation, where a transition occurs between a stable equilibrium and periodic dynamics, arises frequently in biological models and can be identified using standard mathematical techniques.

The *Hopf bifurcation theorem* determines whether periodic dynamics arise near an equilibrium; the theorem applies for parameters close to a critical value at which an equilibrium becomes unstable.

We expect periodic cycles to occur around an equilibrium if the dynamical equations satisfy the conditions of the Hopf bifurcation theorem. The newly formed cycle can then be a stable limit cycle, an unstable limit cycle, or a neutrally stable cycle (see Box 11.1). The Hopf bifurcation theorem specifies conditions under which such periodic dynamics occur, as well as their stability properties. We focus on two-dimensional models in continuous time, although the method can be extended to higher dimensions and to discrete time (see Marsden and McCracken 1976; Hofbauer and Sigmund 1988; Edelstein-Keshet 1988).

Recipe 11.2
Hopf Bifurcation Theorem

Suppose we have a continuous-time model of the form

$$\frac{dx}{dt} = f(x, y, \mu)$$
$$\frac{dy}{dt} = g(x, y, \mu)$$

where x and y are the dynamic variables, and μ is the bifurcation parameter. Let us suppose that $(\hat{x}_\mu, \hat{y}_\mu)$ is the equilibrium point of interest, which may or may not be a function of μ. We also assume that this equilibrium is stable when $\mu < \mu^*$ and unstable when $\mu > \mu^*$ (we can always redefine the parameter of interest, μ, so that this is true). Therefore, μ^* is the critical value of the parameter μ that delineates the boundary between stability and instability.

Step 1: Write the eigenvalues of the linearized model at this equilibrium as

$$r_1 = \alpha(\mu) + i\,\beta(\mu),$$
$$r_2 = \alpha(\mu) - i\,\beta(\mu),$$

where $\alpha(\mu)$ and $\beta(\mu)$ are both real quantities. These represent the real and the imaginary part of the eigenvalues, respectively. Because μ^* marks the transition from stability to instability, we know that $\alpha(\mu^*) = 0$ (see Chapter 7).

Step 2: The Hopf bifurcation theorem states that periodic cycles around the equilibrium $(\hat{x}_\mu, \hat{y}_\mu)$ will occur for values of μ close to μ^*, provided that the following two conditions are met:

(i) $\beta(\mu)$ is real for values of μ near μ^*; in particular, $\beta(\mu^*) \neq 0$.

(ii) $\left.\frac{d\alpha}{d\mu}\right|_{\mu=\mu^*} > 0.$

Condition (i) states that the eigenvalues must be complex for values of μ around μ^*. Condition (ii) states that the real part of the eigenvalue must rise with a "nonzero speed" as μ increases through the value μ^*. If condition (ii) does not hold, there might still be periodic behavior around the equilibrium, but more sophisticated techniques are required to resolve this case (Marsden and McCracken 1976).

(continued)

Recipe 11.2 ***(continued)***

If conditions (i) and (ii) are both satisfied, a model is guaranteed to exhibit periodic dynamics. The type of cycle that occurs will have one of three forms:

- At $\mu = \mu^*$ an infinite number of neutrally stable periodic solutions surround the equilibrium point $(\hat{x}_\mu, \hat{y}_\mu)$.
- For all values of μ slightly larger than μ^* a single stable limit cycle surrounds the equilibrium point, $(\hat{x}_\mu, \hat{y}_\mu)$. This is referred to as a *supercritical bifurcation* because it takes place for values of μ above μ^*.
- For all values of μ slightly smaller than μ^* a single unstable limit cycle surrounds the equilibrium point, $(\hat{x}_\mu, \hat{y}_\mu)$. This is referred to as a *subcritical bifurcation* because it takes place for values of μ below μ^*.

With this classification, we can now add a third condition:

(iii) If $(\hat{x}_\mu, \hat{y}_\mu)$ is stable at $\mu = \mu^*$, then the cyclic behavior will consist of a unique stable limit cycle for all values of μ in an interval between μ^* and $\mu^* + a$ where a is a small positive constant (a supercritical Hopf bifurcation). Conversely, if $(\hat{x}_\mu, \hat{y}_\mu)$ is unstable at $\mu = \mu^*$, the cyclic behavior will consist of a unique unstable limit cycle for all values of μ in an interval between $\mu^* - a$ and μ^* where a is a small positive constant (a subcritical Hopf bifurcation).

Condition (iii) is the most difficult to check (because a local stability analysis is inconclusive when $\alpha(\mu^*) = 0$). We provide an on-line supplementary *Mathematica* package to automate the calculations.

We now apply the Hopf bifurcation theorem to the sexual selection model (11.7). First we must decide which parameter we will treat as the bifurcation parameter. From a mathematical standpoint it does not matter which parameter we use, but we certainly want to choose one that affects the stability of the equilibrium. An obvious choice would be the composite parameter $\gamma = -b\,G_p - c\,G_z + a\,B$, because the equilibrium is stable only when γ is negative. Therefore, we take γ as the bifurcation parameter (represented by μ in Recipe 11.2). The critical value of the parameter, which is where the equilibrium switches from stable to unstable, is then $\gamma^* = 0$ (i.e., the real parts of the eigenvalues are zero at $\gamma = 0$).

Next we need to identify the real and imaginary parts of the eigenvalues at this equilibrium. The eigenvalues for this equilibrium are

$$\frac{1}{2}\left\{\gamma \pm \sqrt{\gamma^2 - 4\,b\,c\,(G_p\,G_z - B^2)}\right\} \tag{11.9}$$

These eigenvalues can be written as $\alpha(\gamma) \pm \sqrt{-1}\,\beta(\gamma)$, where $\alpha(\gamma) = \gamma/2$ and $\beta(\gamma) = \sqrt{4\,b\,c\,(G_p\,G_z - B^2) - \gamma^2/2}$. We now proceed to check the conditions of Recipe 11.2.

If the eigenvalues are real, neither spiraling behavior nor periodic dynamics will occur in the vicinity of the equilibrium. Thus, condition (i) of Recipe 11.2 tests whether the eigenvalues are complex for values of γ near γ^*, including γ^*. In the formulation of this model (Chapter 7), we assumed that the genetic covariance B is not as strong as the genetic variances themselves (i.e., $G_z G_p - B^2 > 0$). Therefore, unless bc = 0, $\beta(\gamma)$ will be real for small enough values of γ near $\gamma^* = 0$. Therefore condition (i) of Recipe 11.2 is met.

To determine whether periodic behavior will occur, we proceed to condition (ii) of Recipe 11.2, which holds because $(\mathrm{d}\alpha/\mathrm{d}\gamma|_{\gamma=\gamma^*} = 1/2 > 0$. Therefore, the Hopf bifurcation theorem tells us that, for some values of γ near the critical value of $\gamma^* = 0$, periodic dynamics will occur around the equilibrium with no female preference and no male display exaggeration.

What we do not yet know is whether these periodic dynamics are observed when $\gamma < 0$ (in which case they would be unstable limit cycles), when $\gamma = 0$ (neutrally stable cycles) or when $\gamma > 0$ (stable limit cycles). Our simulations (Figure 11.3) suggest that the cycling dynamics approach a stable limit cycle whenever the equilibrium is unstable ($\gamma > 0$) but to be sure that this is generally true we need to check condition (iii) of Recipe 11.2. If you carry out the calculations in the accompanying on-line *Mathematica* notebook, you can verify that this condition is indeed satisfied.

This technique also takes us some way toward understanding when evolutionary cycles can be expected in this model. In particular, the above eigenvalues reveal that the conditions of the Hopf bifurcation theorem are not satisfied unless $G_p G_z - B^2 > 0$ holds. If this inequality does not hold, then the equilibrium can still be unstable but the eigenvalues no longer have imaginary parts when stability is lost. Therefore, Recipe 11.2 reveals that no periodic behavior is exhibited. In fact simulation results suggest that indefinite evolutionary escalation of male trait and female preference then occurs just as in the linear sexual selection model of Chapter 7 (Figure 7.9). The generality of this finding can be confirmed using the facts about two-variable continuous-time models presented in Box 11.2.

In Supplementary Material 11.1, we work through a second example of a Hopf bifurcation using the Lotka-Volterra predator-prey model.

Box 11.2: Special Facts about Two-Dimensional Continuous-Time Models

As mentioned in Box 4.1, chaotic dynamics are not possible in a continuous-time model when the number of dynamic variables is less than three. In fact, the list of possible behaviors that can be exhibited by a continuous-time model with two variables is restricted. This fact is often

(continued)

Box 11.2 *(continued)*

extremely useful for determining the global behavior of such models. One main result is referred to as the Poincaré-Bendixon theorem (see Edelstein-Keshet 1988). This theorem states that, as time passes, any trajectory of such models must do one of the following four things:

(i) The variables approach a steady state (i.e., an equilibrium).
(ii) One or both of the variables grows indefinitely to plus or minus infinity.
(iii) The variables approach a periodic cycle.
(iv) The variables approach a "cycle graph" (a set of equilibria that are connected to one another by trajectories that form a closed loop; if the system is perturbed off one of these equilibria in the direction of this trajectory, then it moves to the next equilibrium in the loop).

Behaviors (i)–(iv) are not mutually exclusive in the sense that a single model can exhibit more than one of these behaviors depending upon the starting conditions. One implication is the following fact:

> If the state space of a two-dimensional continuous-time model is bounded, so that a given trajectory never leaves a specified region, and if the trajectory within this region never approaches an equilibrium (including an equilibrium on a cycle graph), then it must exhibit sustained periodic behavior.

In light of this fact, let's consider the sexual selection model (11.7). In section 11.4, we used the Hopf bifurcation theorem to determine that a newly unstable equilibrium gives rise to periodic dynamics for some parameter values. We also mentioned that, for other parameter values, the unstable equilibrium gives rise to trajectories in which the variables (male tail length and female preference) grow indefinitely to plus or minus infinity. The above results reveal that, when the equilibrium is unstable, these are the only two possibilities, because cases (i) and (iv) are then excluded.

11.5 Constants of Motion

A *constant of motion* is some function of the variables that remains constant over time.

The last technique that we examine involves constants of motion. As the name suggests, a *constant of motion* for a dynamic model is a function of the dynamic variables that remains constant over time. For example, in physics, the total energy of a closed system remains constant even though the different energy components might change through time. Thus, the total energy is a constant of motion. This constraint is sometimes useful for understanding and making predictions about a system. For example, we used the conservation of energy in equation (9.23) to simplify the model of yeast consuming resources in a vial. As another trivial example, the sum of allele frequencies is a constant of motion; regardless of how the frequencies change over time, their sum is constrained to equal one.

In this section, we explore two less obvious examples, where a function of the heterozygosity in two species remains constant over time while allele frequencies cycle. Through these examples, we shall discuss how to identify such constants of motion and see how constants of motion can be used to gain a more complete understanding of a model's dynamics.

Example: Host-Parasite Coevolution (Matching-Alleles Model)

Interactions between hosts and parasites have received a great deal of attention by theoreticians, particularly the evolutionary consequences of such interactions (see Frank 2002 for an excellent treatment). One common type of model is the "matching-alleles" model. This name refers to the way in which the genotypes of the host and parasite affect their interaction. A parasite can successfully attack the host only if it carries an allele that matches a corresponding allele in the host. For example, many microparasites have surface molecules that must attach to a receptor molecule on a host's cell, and different parasite and host genotypes have different surface and receptor molecules, respectively.

In the simplest situation, we can model this by supposing that the host and parasite are both haploid and that each host interacts with one parasite at any point in time. The outcome of this interaction depends on a "surface molecule" gene in the parasite (with two alleles A and a) and a "receptor" gene in the host (with two alleles B and b). A parasite with genotype A can successfully attack only hosts of genotype B, and a parasite with genotype a can successfully attack only hosts of genotype b. We will construct a population-genetic model for this phenomenon that tracks evolutionary changes in the frequency of allele A in the parasite (denoted by p) and allele B in the host (denoted by q).

Let us suppose that time can be scaled such that a parasite's growth rate is one when it successfully attacks a host and that a host's growth rate is one when it successfully escapes parasitism. A parasite that fails to successfully attack a host will have a growth rate $1 - s_p$, where s_p is the cost to the parasite of failing to attack a host. Similarly, we model the growth rate of infected hosts as $1 - s_h$, where s_h is the cost to the host of being parasitized.

Table 11.1 specifies the outcome for each of the four different host-parasite combinations, and these can be used to calculate the average growth rate of each allele. For example, a B host will interact with an A parasite a proportion p of the time (in which case its growth rate is $1 - s_h$), and it will interact with an a parasite a proportion $(1 - p)$ of the time (in which case its growth rate is 1). Therefore, the average growth rate of the B allele in hosts is

$$r_B(p) = p\,(1 - s_h) + (1 - p) = 1 - p\, s_h. \tag{11.10a}$$

Likewise, the average growth rate of the b allele in hosts is

$$r_b(p) = p + (1 - p)(1 - s_h) = 1 - s_h\,(1 - p). \tag{11.10b}$$

TABLE 11.1
Matching-alleles model

Host genotype	Parasite genotype: A	Parasite genotype: a
B	Infection Parasite: 1, Host: $1-s_h$	No Infection Parasite: $1-s_p$, Host: 1
b	No Infection Parasite: $1-s_p$, Host: 1	Infection Parasite: 1, Host: $1-s_h$

$r_A(q) = q + (1-q)(1-s_p) = 1 - s_p(1-q)$
$r_a(q) = q(1-s_p) + (1-q) = 1 - q\,s_p$
$r_B(p) = p(1-s_h) + (1-p) = 1 - p\,s_h$
$r_b(p) = p + (1-p)(1-s_h) = 1 - s_h(1-p)$

Finally, analogous calculations show that the average growth rates of the A and a alleles in the parasite are

$$r_A(q) = q + (1-q)(1-s_p) = 1 - s_p(1-q), \tag{11.10c}$$

$$r_a(q) = q(1-s_p) + (1-q) = 1 - q\,s_p. \tag{11.10d}$$

Using equation 3.1.3 from Chapter 3 to describe the change in allele frequency, our two-dimensional matching-allele model is then

$$\begin{aligned}\frac{dq}{dt} &= q(1-q)(r_B(p) - r_b(p)),\\ \frac{dp}{dt} &= p(1-p)(r_A(q) - r_a(q)).\end{aligned} \tag{11.11a}$$

Plugging the growth rates (11.10) into (11.11a), we get

$$\begin{aligned}\frac{dq}{dt} &= -s_h q(1-q)(2p-1),\\ \frac{dp}{dt} &= s_p p(1-p)(2q-1).\end{aligned} \tag{11.11b}$$

It is clear from equations (11.11) that the allele frequency dynamics in each species depend on the allele frequencies in the other species.

A quick simulation of this model reveals some interesting behavior (Figure 11.4). It appears to undergo sustained evolutionary fluctuations in allele frequency in both the host and the parasite, and the height of the cycles (i.e., the amplitude) appears to depend on initial conditions. From a biological standpoint, cycling behavior of some kind makes sense because, roughly speaking, there is selection for the host genotype that evades the most common

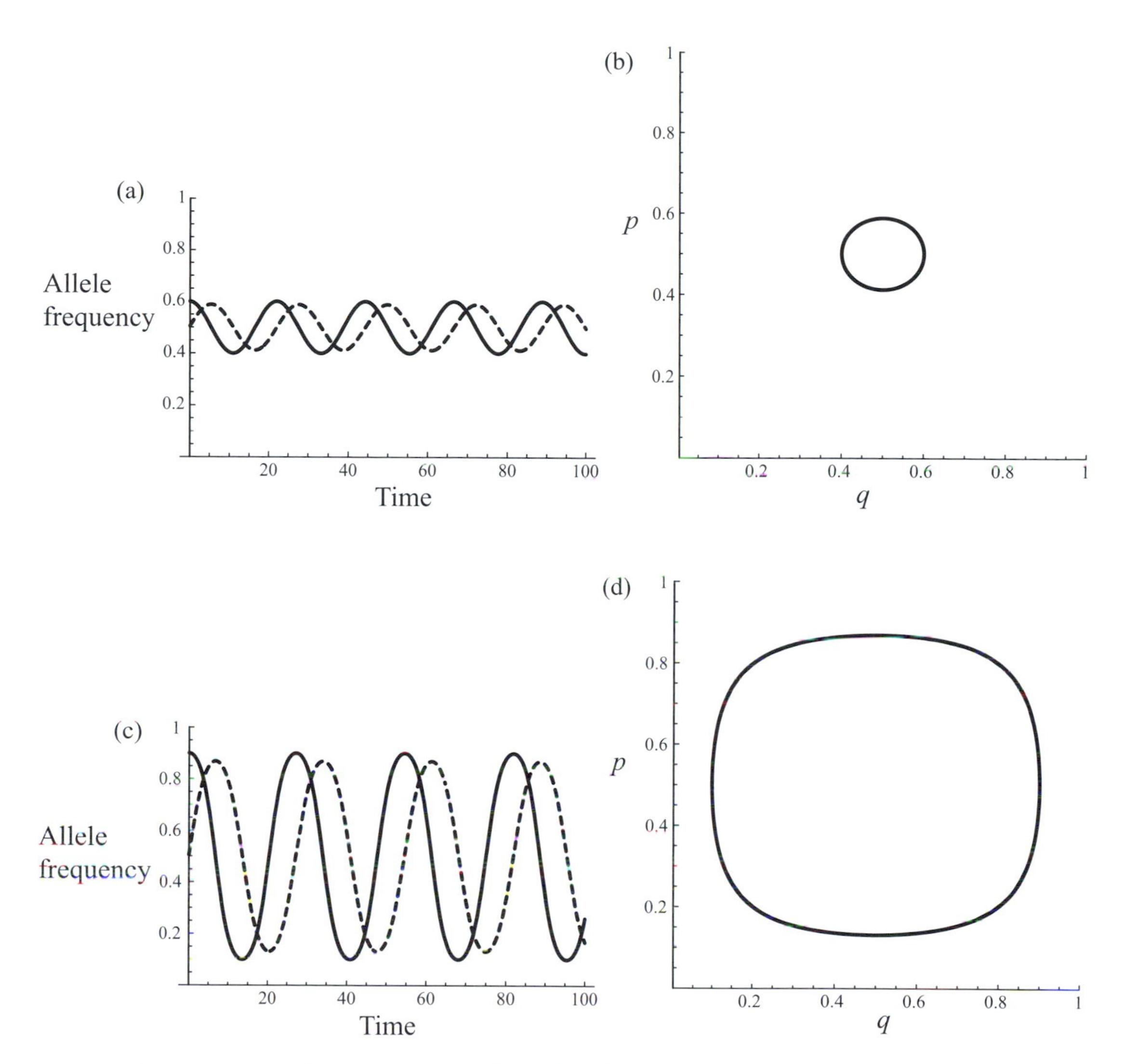

Figure 11.4: Simulations of the matching-alleles model. Parameter values are $s_h = 0.65$, $s_p = 0.5$. (a) A plot of the frequency of allele A in the parasite (dashed) and allele B in the host (solid) over time when $q(0) = 0.6$ and $p(0) = 0.505$. (b) The corresponding phase plane. (c) A plot of the frequency of allele A (dashed) and allele B (solid) when $q(0) = 0.9$ and $p(0) = 0.505$. (d) The corresponding phase plane.

parasite genotype but there is also selection for the parasite genotype that matches the most common host genotype.

Model (11.11) has five equilibria, four of which are the four different combinations of allelic fixation that might occur (e.g., alleles A and b fixed, alleles A and B fixed, etc.). All four of these equilibria are locally unstable. The fifth equilibrium is $\hat{q} = 1/2$, $\hat{p} = 1/2$, and a calculation of the Jacobian matrix at this equilibrium shows that the two eigenvalues are purely imaginary. Because there is no real part, our linear stability analysis for this equilibrium is inconclusive, and we cannot apply the Hopf bifurcation theorem (condition (ii) of Recipe 11.2 fails).

Our simulations suggest that this model exhibits a type of neutral cycling, much like what we saw in the Lotka-Volterra predator-prey model in Figure 4.17.

Neutrally stable cycles hint that the system is constrained in some way—constrained to remain on whatever cycle it starts on. If a system appears to be constrained (or if certain variables always change in parallel or always sum to the same quantity), it is worth looking for a *constant of motion*, defined as a function of the variables that remains constant over time. Constants of motion can be used to simplify dynamical equations, and they often allow us to answer general questions about a model. For example, are the cycles observed in Figure 11.4 truly neutral or do they slowly spiral in or out? Does the model always exhibit neutrally stable cycles?

But how do we come up with a constant of motion? For two-variable models in continuous time, there is a relatively straightforward recipe that often works:

Recipe 11.3
Deriving Constants of Motion

Suppose you have the following two-variable model:

$$\frac{dx}{dt} = f(x, y),$$
$$\frac{dy}{dt} = g(x, y).$$

Step 1: For $h(x,y)$ to be a constant of motion, $dh(x,y)/dt$ must be zero. Using the chain rule of calculus, this requirement implies that

$$0 = \frac{\partial h}{\partial x}\frac{dx}{dt} + \frac{\partial h}{\partial y}\frac{dy}{dt}$$
$$= \frac{\partial h}{\partial x} f(x, y) + \frac{\partial h}{\partial y} g(x, y).$$

Step 2: If possible, multiply or divide the above equation by a factor so that it has the form

$$0 = \frac{\partial h}{\partial x} F(y) + \frac{\partial h}{\partial y} G(x),$$

where $F(y)$ is a function of y only and $G(x)$ is a function of x only. This is the crucial step. If you cannot find a way to do this, then the only way to obtain a constant of motion (if one exists) is by educated guesswork and biological intuition.

Step 3: The above equation will be satisfied by any function $h(x,y)$ for which $\partial h/\partial x = -G(x)$ and $\partial h/\partial y = F(y)$. Using a separation of variables (Recipe 6.2) and summing the results, a constant of motion is

$$h(x, y) = -\int G(x)dx + \int F(y)dy.$$

(continued)

Recipe 11.3 *(continued)*

Step 4: If $h(x,y)$ is a constant of motion, so too is any function of the constant of motion, $g(h)$. Use this fact to look for a constant of motion that is easier to interpret, by considering functions $g(h)$ that simplify $h(x,y)$.

Not all models will have a constant of motion. But when a constant of motion can be found, it will not be unique. As indicated in Step 4, any function that depends on the variables only through the constant of motion will itself be a constant of motion. For example, if $h(x,y)$ is a constant of motion, then so too are $h(x,y)^2$ and $h(x,y) + 1$. This fact can be useful because some constants of motion are biologically more meaningful than others.

In Box 11.3, we use Recipe 11.3 to derive the constant of motion for model (11.11b):

$$h(q, p) = [q\,(1 - q)]^{s_p}\,[p\,(1 - p)]^{s_h} \tag{11.12}$$

The quantity $q\,(1 - q)$ is a measure of genetic diversity and is proportional to the expected heterozygosity of a population (in this case, the host population). Therefore, the function h involves the genetic diversity of the host multiplied by the genetic diversity of the parasite, where each is raised to some positive power (s_p and s_h, respectively).

Box 11.3: Deriving the Constant of Motion for the Matching-Alleles Model

A function $h(q,p)$ is a constant of motion of model (11.11b) if

$$\frac{\mathrm{d}}{\mathrm{d}t}h(q, p) = \frac{\partial h}{\partial q}\frac{\mathrm{d}q}{\mathrm{d}t} + \frac{\partial h}{\partial p}\frac{\mathrm{d}p}{\mathrm{d}t} = 0 \tag{11.3.1}$$

(Step 1 of Recipe 11.3). Using the equations for $\mathrm{d}q/\mathrm{d}t$ and $\mathrm{d}p/\mathrm{d}t$ from equations (11.11b), we can write (11.3.1) more explicitly as

$$-\frac{\partial h}{\partial q}s_h\,q(1 - q)(2p - 1) + \frac{\partial h}{\partial p}s_p\,p(1 - p)(2q - 1) = 0. \tag{11.3.2}$$

Equation (11.3.2) represents a condition that the derivatives $\partial h/\partial q$ and $\partial h/\partial p$ must satisfy for h to be a constant of motion. Equation (11.3.2) has the form $-(\partial h/\partial q)\,X + (\partial h/\partial p)\,Y = 0$. One way for this equation to be satisfied is if we find a function h whose derivatives are $\partial h/\partial q = Y$ and $\partial h/\partial p = X$ (because $-Y\,X + X\,Y = 0$ is always true). In principle, then, all we have to do is find a

(continued)

Box 11.3 ***(continued)***

function h whose derivatives with respect to q and p equal $\partial h/\partial q = s_p\, p\,(1-p)(2q-1)$ and $\partial h/\partial p = s_h\, q\,(1-q)(2p-1)$. This is not so easy, however, as these equations depend on both variables q and p.

Matters are considerably simpler, however, if we can make Y a function of q only and X a function of p only. Examining equation (11.3.2), we can achieve this goal by dividing through by $p(1-p)\,q\,(1-q)$:

$$-\frac{\partial h}{\partial q} s_h \frac{2p-1}{p(1-p)} + \frac{\partial h}{\partial p} s_p \frac{2q-1}{q(1-q)} = 0 \qquad (11.3.3)$$

(Step 2 of Recipe 11.3).

Equation (11.3.3) is satisfied by any function h whose derivatives are $\partial h/\partial q = s_p(2q-1)/(q(1-q))$ and $\partial h/\partial p = s_h(2p-1)/(p(1-p))$. These equations can be solved using a separation of variables (Recipe 6.2):

$$\begin{aligned} h(q,p) &= -s_p \ln(q(1-q)) + c_1, \\ h(q,p) &= -s_h \ln(p(1-p)) + c_2, \end{aligned} \qquad (11.3.4)$$

where c_1 is a constant of integration that does not depend on q and c_2 is a constant of integration that does not depend on p. One solution that is consistent with both equations (11.3.4) is the sum

$$h(q,p) = -s_p \ln(q(1-q)) - s_h \ln(p(1-p)) \qquad (11.3.5)$$

(Step 3 of Recipe 11.3).

We could stop here, but recall that any function of $h(q,p)$ is also a constant of motion (Step 4 of Recipe 11.3). We can get rid of the natural logarithms and obtain a more intuitive and biologically meaningful constant of motion by taking $e^{-h(q,p)}$ (along with Rule A1.11). Doing so gives expression (11.12):

$$h(q,p) = [q(1-q)]^{s_p}\,[p(1-p)]^{s_h}. \qquad (11.3.6)$$

Both (11.3.5) and (11.3.6) are perfectly legitimate constants of motion.

To confirm that h is a constant of motion, we can differentiate it with respect to time, using the chain rule (Appendix A2.15), and treating q and p as functions of time whose derivatives are given by equations (11.11). The step-by-step calculations are

$$\begin{aligned} \frac{d}{dt}h(q,p) &= s_p[q(1-q)]^{s_p-1}\,[p(1-p)]^{s_h}\,\frac{d\{q(1-q)\}}{dt} \\ &\quad + s_h[q(1-q)]^{s_p}\,[p(1-p)]^{s_h-1}\,\frac{d\{p(1-p)\}}{dt} \end{aligned}$$

$$
\begin{aligned}
&= s_p[q(1-q)]^{-1}h(q,p)\frac{\mathrm{d}\{q(1-q)\}}{\mathrm{d}t} \\
&\quad + s_h[p(1-p)]^{-1}h(q,p)\frac{\mathrm{d}\{p(1-p)\}}{\mathrm{d}t} \qquad (11.13) \\
&= h(q,p)\left\{s_p[q(1-q)]^{-1}(1-2q)\frac{\mathrm{d}q}{\mathrm{d}t} + s_h[p(1-p)]^{-1}(1-2p)\frac{\mathrm{d}p}{\mathrm{d}t}\right\} \\
&= h(q,p)\{-s_p s_h(1-2q)(2p-1) + s_h s_p(1-2p)(2q-1)\} \\
&= 0.
\end{aligned}
$$

Because $h(q, p)$ is a constant of motion (i.e., it does not change over time) it is always equal to its initial value, $h(q_0,p_0)$. This does not mean that the variables do not change. Figure 11.4 demonstrates that the allele frequencies do vary over time in both the host and parasite. Nevertheless, $h(q,p)$ remains constant, at $h(q,p) = 0.199$ in Figure 11.4b and at $h(q,p) = 0.122$ in Figure 11.4d.

Let us now see why a constant of motion is so useful. To begin, notice that h always lies between zero and some finite maximum because both s_p and s_h are positive. Additionally, h is "well behaved" in the sense that it never jumps abruptly as p or q changes (i.e., h is a continuous function of p and q). As a consequence, the equilibrium point $\hat{q} = 1/2$, $\hat{p} = 1/2$ cannot be locally stable (recall that our linear stability analysis was inconclusive). The reason is that, if it were locally stable, then the value of h along any trajectory in the vicinity of the equilibrium would have to approach $h(1/2,1/2) = [1/4]^{s_p}[1/4]^{s_h}$. But this is not possible because h takes on some other constant value depending on the starting values of q and p. By a similar argument, the corner equilibria cannot be locally stable ($\hat{q} = 0$ or; $\hat{p} = 0$ or 1).

In fact, there is a general and useful statement that can be made for models with a constant of motion:

Rule 11.1: Constants of Motion and the Stability of an Equilibrium

The existence of a constant of motion rules out the possibility that an equilibrium is locally stable as long as

- the constant of motion is a continuous function of the variables near the equilibrium,
- the constant of motion takes on different values in a small region around the equilibrium (i.e., it is not just a constant like $h = 1$).

The logic behind Rule 11.1 is as follows. Locally stable equilibria have a region around them such that any initial conditions lying in that region are on trajectories leading to the equilibrium. Therefore, if the constant of motion is continuous at the equilibrium, then the value of the constant of motion at this point must be equal to the value of the constant of motion along all trajectories leading to the equilibrium. But if the constant of motion differs between any two initial conditions in this region, then it must also differ along the two

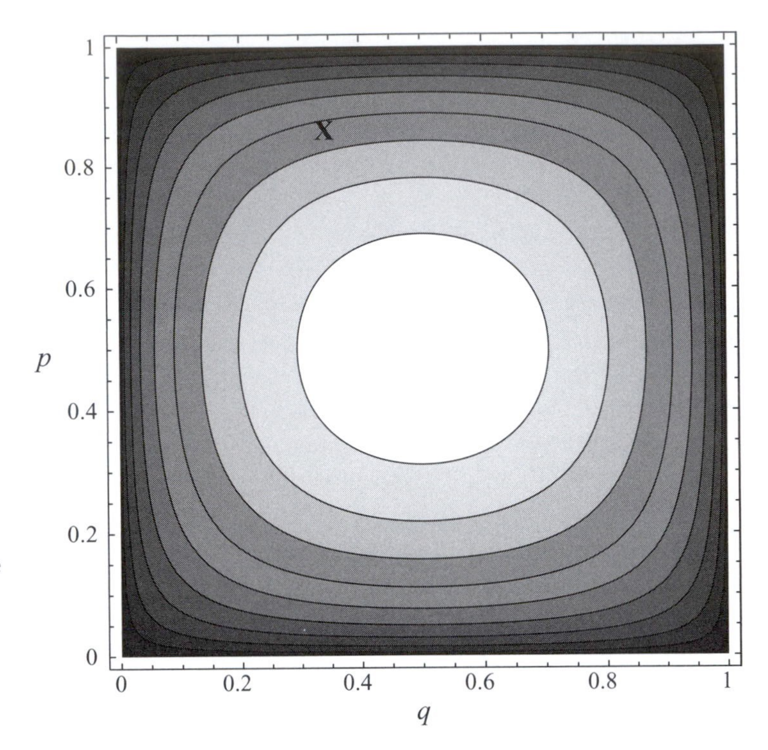

Figure 11.5: Contour plot of the constant of motion for the matching-alleles model. A contour plot shows curves along which some function is constant. Here each curve shows a particular value for the constant of motion, $h(q,p)$. Thus, the system remains on whichever contour it starts on (e.g., the X marks the trajectory from Figure 11.4d). All contours form concentric circles around the neutrally stable equilibrium at $\hat{q} = 1/2$, $\hat{p} = 1/2$. Here, $s_h = 0.65$, $s_p = 0.5$.

trajectories that begin at these two sets of initial conditions. As a result, both trajectories cannot reach the same point, and the equilibrium cannot be locally stable.

We can also get some very general information about the global behavior of the model by graphing the constant of motion, h. In Figure 11.5, we plot curves ("contours") along which h has the same value. Model (11.11) remains on the contour on which it is started because $\mathrm{d}h/\mathrm{d}t = 0$. By looking at the form of the function h, we can also answer our question about the generality of such neutral cycles. The contours of h always form concentric rings around the equilibrium point $\hat{q} = 1/2$, $\hat{p} = 1/2$ whose shape is determined by the parameters s_p and s_h. Therefore this model exhibits neutral cycles for all parameter values. This illustrates the power of finding constants of motion; they allow us to deduce global properties of a model.

We can also infer some general predictions from the shapes of these contours. For example, if the cost to a host of being parasitized is very small relative to the cost to a parasite of failing to successfully attack (which means that $s_h << s_p$), then the function h does not change much over most of the range of values of p (Figure 11.6). The contour lines are virtually vertical in this case, and therefore different initial values of p place the system on virtually the same contour. From a biological standpoint, this boxlike phase portrait means that the allele frequency in the parasite fluctuates from one extreme to the other whereas the allele frequency in the host has a much smaller amplitude. This makes good sense. If $s_h << s_p$, then selection is much stronger on the parasite than on the host. As a result, we would expect the allele frequencies in the parasite to change much more quickly. Of course the converse argument would hold if s_p

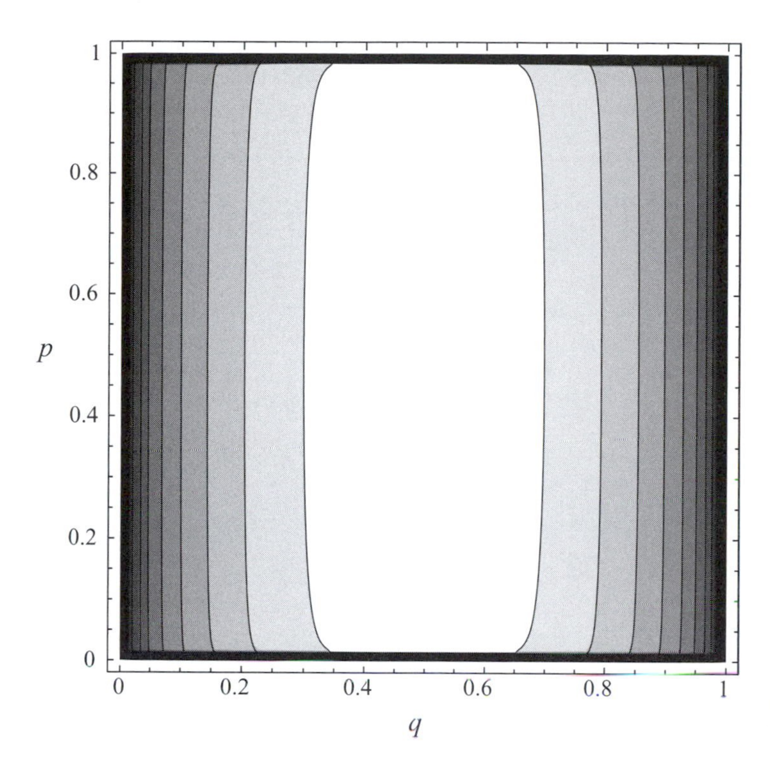

Figure 11.6: Contour plot of the constant of motion for the matching-alleles model. As in Figure 11.5, but now $s_h = 0.009$ is very small relative to $s_p = 0.4$.

were very small relative to s_h (i.e., the cost to the parasite is smaller than the cost to the host).

Example: Host-Parasite Coevolution (Gene-for-Gene Model)

The matching-alleles model is not the only model for host-parasite coevolution. Another model is the so-called gene-for-gene model, which assumes that there are parasite genotypes that are, in general, better at parasitizing, and host genotypes that are more resistant to parasitism. Evidence for such host-parasite interactions is found in plant-pathogen interactions (e.g., Whalen et al., 1991; Kunkel 1996). Thus we might expect an evolutionary escalation in parasitic ability in the parasite and resistance in the host. In many cases, however, there are costs to resistance in the host (Bergelson and Purrington 1996) and potentially costs to increased parasitic ability in the parasite as well, which might halt such evolutionary arms races.

Again we suppose that the host and parasite are both haploid and that there are two genotypes for each (A and a in the parasite, and B and b in the host). Host genotype B is "resistant," whereas genotype b is "susceptible." Parasite genotype A is termed "virulent" and genotype a is termed "avirulent." The virulent parasite genotype A can successfully attack both host genotypes, whereas the avirulent genotype a can successfully attack only the susceptible host genotype b. We will also assume that the resistant host genotype incurs a fixed cost relative to the susceptible genotype and the virulent parasite genotype incurs a fixed cost relative to the avirulent genotype. Again we denote the frequency of the A allele in the parasite population by p, and the frequency of the B allele in the host population by q.

TABLE 11.2
Gene-for-gene model. We have made the specific (and arbitrary) assumption that the host's growth rate when it carries the resistant allele but yet is parasitized (by a "virulent" pathogen) is given by the sum of the two costs; i.e., $-s_h - c_h$.

Host genotype	Parasite genotype: *A* (virulent)	Parasite genotype: *a* (avirulent)
B (resistant)	Infection Parasite: $1 - c_p$ Host: $1 - s_h - c_h$	No Infection Parasite: $1 - s_p$ Host: $1 - c_h$
b (susceptible)	Infection Parasite: $1 - c_p$ Host: $1 - s_h$	Infection Parasite: 1 Host: $1 - s_h$

$r_A(q) = q\,(1 - c_p) + (1 - q)\,(1 - c_p) = 1 - c_p$

$r_a(q) = q\,(1 - s_p) + (1 - q) = 1 - q\,s_p$

$r_B(p) = p\,(1 - s_h - c_h) + (1 - p) = 1 - ps_h - c_h$

$r_b(p) = p\,(1 - s_h) + (1 - p)\,(1 - s_h) = 1 - s_h$

We model growth rates as a function of whether or not infection occurs, just as in the matching-alleles model. Now, however, we also include a cost c_p to the virulent genotype and a cost c_h to the resistant genotypes. Table 11.2 gives the outcome of host-parasite interactions for each of the four possible combinations as well as the average growth rates of the different alleles. Plugging these growth rates into equation (11.11a) and simplifying, we get

$$\begin{aligned} \frac{dq}{dt} &= q(1 - q)\{s_h(1 - p) - c_h\}, \\ \frac{dp}{dt} &= p(1 - p)\{s_p q - c_p\}. \end{aligned} \tag{11.14}$$

Without costs of virulence and resistance, equation (11.14) confirms that the resistant and virulent alleles would rise in frequency if they are not already fixed ($dq/dt \geqslant 0$; $dp/dt \geqslant 0$).

With costs, it appears from simulations that this gene-for-gene model also exhibits a type of neutral cycling (Figure 11.7a, b). Unlike the matching-alleles model, however, there are some parameter values for which stable equilibria also occur (Figure 11.7c). As with the matching-allele model, this model has five equilibria: four that occur at the various fixation points, and one interior equilibrium with the host at $\hat{q} = c_p/s_p$ and the parasite at $\hat{p} = 1 - c_h/s_h$.

A linear stability analysis reveals that both equilibria with the parasite population fixed for the virulent allele (i.e., $\hat{p} = 1$) are always unstable. The equilibrium with susceptible hosts and avirulent parasites, $\hat{q} = \hat{p} = 0$, is locally stable if the costs of resistance are larger than the costs of being parasitized ($c_h > s_h$). This makes sense because resistant hosts are then never more fit than susceptible hosts. Similarly, the equilibrium with resistant hosts and avirulent parasites, $\hat{q} = 1$, $\hat{p} = 0$, is locally stable if the costs of resistance are weaker than the costs of being parasitized ($c_h < s_h$) and if the costs of virulence are

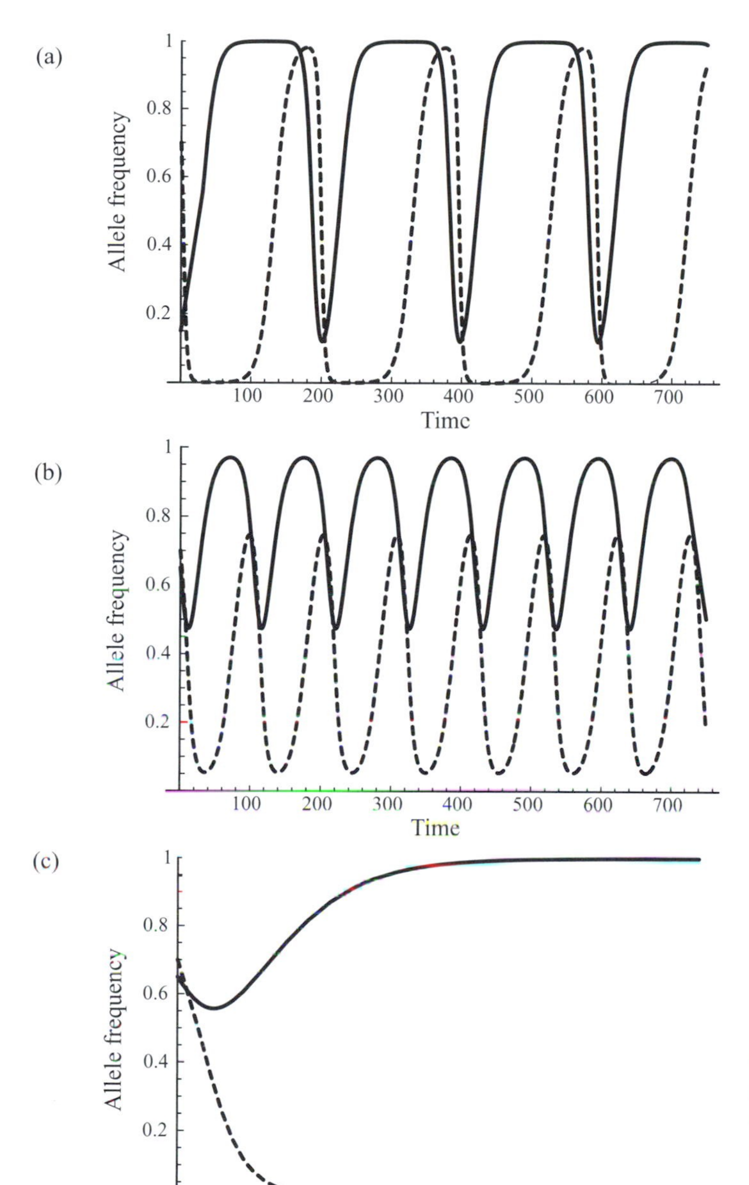

Figure 11.7: Simulations of the gene-for-gene model. The frequency of the virulent parasite allele A (dashed) and the resistant host allele B (solid) over time with $s_h = 0.3, s_p = 0.5$, $c_h = 0.2$, $c_p = 0.4$. (a) $q(0) = 0.15$ and $p(0) = 0.7$. (b) $q(0) = 0.65$ and $p(0) = 0.7$. (c) As in (b) but with $s_p = 0.3$.

greater than the costs to a parasite of failing to attack a host ($s_p < c_p$). Again this makes sense because there is then an advantage to resistance in the presence of avirulent parasites, but there is never an advantage to virulence in the parasite. Finally, a linear stability analysis of the interior equilibrium reveals that, when it is feasible, it is unstable whenever one of the boundary equilibria is stable. Otherwise, the stability analysis is inconclusive, giving two purely imaginary eigenvalues.

The fact that Figures 11.7a,b exhibit neutral cycling again suggests that the system is constrained in some way, making it worth looking for a constant of

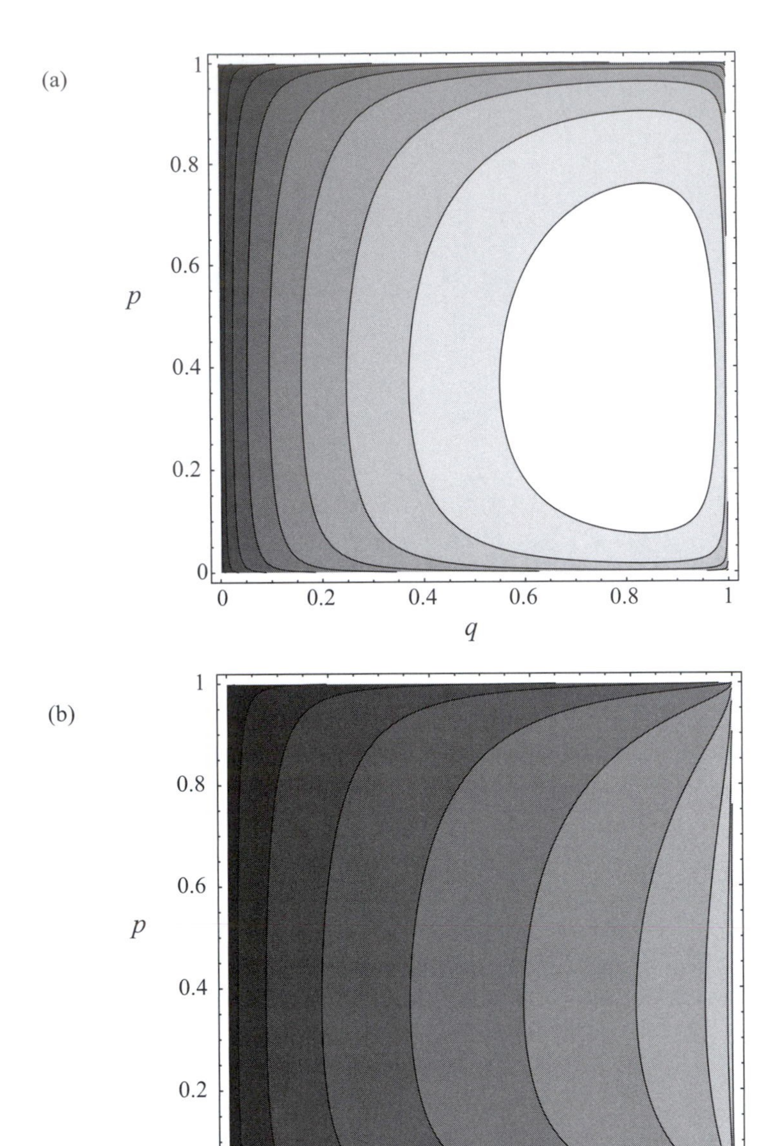

Figure 11.8: Contour plots of the constant of motion for the gene-for-gene model. (a) Parameter values $s_h = 0.3$, $s_p = 0.5$, $c_h = 0.2$, $c_p = 0.4$, corresponding to Figures 11.7a, b (b) Parameter values $s_h = 0.3$, $s_p = 0.3$, $c_h = 0.2$, $c_p = 0.4$, corresponding to Figure 11.7c.

motion. Indeed, the gene-for-gene model (11.14) does have a constant of motion, which we found by following Recipe 11.3:

$$h(q, p) = q^{c_p} (1 - q)^{s_p - c_p} p^{s_h - c_h} (1 - p)^{c_h} \tag{11.15}$$

(Problem 11.5). In light of Rule 11.1, this seems a bit puzzling—there is a constant of motion even when some equilibria are locally stable (as in Figure 11.7c). But remember, Rule 11.1 excludes the possibility of stable equilibria only when the constant of motion is continuous in the variables around each equilibrium. The constant of motion for the gene-for-gene model is not always continuous. The expression in (11.15) is continuous for all values of p and q only if $s_h > c_h$ and $s_p > c_p$. In this case, no equilibrium can be locally stable, and the system must exhibit neutrally stable cycles (Figure 11.8a). If, however, $s_h < c_h$, then expression (11.15) is discontinuous at $p = 0$ (it increases to infinity at this point), and we cannot rule out the possibility of a locally stable equilibrium at $\hat{p} = 0$ in this case. Similarly, if $s_p < c_p$ then expression (11.15) is discontinuous at $q = 1$ (it increases to infinity at this point), and we cannot rule out the possibility of a locally stable equilibrium at $\hat{q} = 1$. Figure 11.8b shows an example where $s_p < c_p$, in which case $\hat{q} = 1$, $\hat{p} = 0$ is a stable equilibrium. The system still remains on a contour of the constant of motion, but now the contours converge at the locally stable equilibrium.

Before concluding this section, it is important to emphasize again that a model need not have a constant of motion. Even if it does, it might be difficult to find. As the number of variables in the model increases, finding constants of motion becomes progressively more difficult. Often they are found through educated guesses, where biological intuition can play a large role.

11.6 Concluding Message

In this chapter we have introduced three techniques that are often useful and relatively easy to apply when modeling biological processes that exhibit periodic behavior. Composite mappings can be used in discrete-time models to examine the state of a system over specific intervals of time, allowing cycles and their stability to be determined. The Hopf bifurcation theorem can be used to identify periodic dynamics that arise near equilibria in continuous-time models. Finally, for some models, constants of motion can be identified, providing a description of the trajectory followed as each variable changes over time.

These approaches work well for many models, but general analytical techniques tend to be much more sophisticated and difficult to apply when models exhibit such nonequilibrium dynamics. Nevertheless, this is an important area of study because many biological processes involve antagonistic interactions that naturally lead to fluctuations in the variables of interest. We encourage the interested reader to pursue these issues further with more advanced texts such as Edelstein-Keshet (1988), Hofbauer and Sigmund (1988), Murrary (1989), and Gurney and Nisbet (1998).

Problems

Problem 11.1: Consider the following predator-prey model:

$$\frac{dR}{dt} = r\,R\,e^{-\alpha R} - R - P\frac{R}{1 + R},$$

$$\frac{dP}{dt} = P\frac{R}{1 + R} - dP.$$

In Supplementary Matarial 11.1, we analyze a similar model, except that here the resource dynamics take the form of a Ricker model in the absence of the predator. (a) Show that one equilibrium of this model is given by

$$\hat{R} = \frac{d}{1-d}, \quad \hat{P} = \frac{e^{-d\alpha/(1-d)}r - 1}{1-d}.$$

(b) What are the conditions under which this equilibrium is biologically feasible? (c) Perform a local stability analysis for this equilibrium and demonstrate that, provided it is feasible, it is locally stable when $r < r^*$, where $r^* = (1-d)e^{d\alpha/(1-d)}/(1-d-\alpha)$. (d) When the parameter r increases above the cutoff r^*, the equilibrium of (a) loses stability. Using conditions (i) and (ii) of Recipe 11.2, determine when we would expect periodic dynamics in the variables R and P to arise as r rises above r^*.

Problem 11.2: In equations (3.18) of Chapter 3, we presented a consumer-resource model having the following form:

$$\frac{dn_1}{dt} = r\,n_1 - a\,c\,n_1 n_2,$$

$$\frac{dn_2}{dt} = \varepsilon\,a\,c\,n_1\,n_2 - \delta\,n_2,$$

where n_1 denotes the abundance of the resource and n_2 denotes the abundance of the consumer. Figure 4.17 depicts the dynamics of this model in the phase plane. (a) Derive a constant of motion for this model. (b) When $r = \delta = 0$, we get model (9.21) for the number of yeast growing in a vial. In this case, use your answer to (a) to show that the total amount of energy $E = \varepsilon\,n_1 + n_2$ measured in units of consumer-equivalents is a constant of motion. [Hint: You might need to use the fact that any function of a constant of motion, $g(h)$, is also a constant of motion.]

Problem 11.3: Model (11.11b) of the text is a special case of a more general population-genetic model having the form

$$\frac{dq}{dt} = q(1-q)(a\,p + b\,(1-p)),$$

$$\frac{dp}{dt} = p(1-p)(c\,q + d\,(1-q)).$$

Derive a general expression for a constant of motion for the above model in terms of the parameters, a, b, c, and d.

Problem 11.4: Conduct a local stability analysis of all five equilibria for model (11.14).

Problem 11.5: Use Recipe 11.3 to derive the constant of motion given by equation (11.15) for model (11.14).

Problem 11.6: Using the results of Box 11.2, provide a logical argument for why the predator-prey model (S11.1.1) of Supplementary Material 11.1 must exhibit periodic behavior when the equilibrium

$$\hat{R} = \frac{a\,d}{\varepsilon\,c - d}, \quad \hat{P} = \frac{\varepsilon\,a\,r(K(\varepsilon\,c - d) - a\,d)}{K(\varepsilon\,c - d)^2}$$

is unstable. [Hint: Consider whether it is possible to draw a box in the phase plane within which the system must remain for initial conditions within the box.]

Problem 11.7: Consider two types, whose densities are x and y, that interact destructively (e.g., each type consumes the other). In this problem, we consider the dynamics of these two types over the short term, ignoring births and other sources of mortality. Depending on how the types interact, two appropriate sets of differential equations might be (a) $\mathrm{d}x/\mathrm{d}t = -a\,y$, $\mathrm{d}y/\mathrm{d}t = -b\,x$, which assumes that each type seeks out and consumes the other at a rate that does not depend on the density of the victim, and (b) $\mathrm{d}x/\mathrm{d}t = -a\,x\,y$, $\mathrm{d}y/\mathrm{d}t = -b\,x\,y$, which assumes that the destructive interactions increase in intensity as either type becomes more common. For each set of differential equations, derive a constant of motion using the method outlined in Recipe 11.3.

Problem 11.8: Whenever females of a species mate with multiple males, the reproductive interests of males have the potential to differ from those of females. For example, if a female has multiple mating partners over her life, then her reproductive output will likely be highest by spreading out her maternal investment over the course of her lifespan. The reproductive output of any given male with whom she mates, however, will be highest if he is able to have the female devote all of her maternal investment into the mating in which he took part. This sexual conflict has the potential to result in evolutionary arms races between male "persistence" traits that induce females to have higher investment in the current mating, and female "resistance" traits that allow the female to counter these male adaptations. A model for this process tracks the average value of a single female resistance trait $\bar{r}$ and the average value of a single male persistence trait $\bar{p}$ (Gavrilets et al. 2001; Rowe et al. 2005):

$$\frac{\mathrm{d}\bar{r}}{\mathrm{d}t} = G_r(-a\,\bar{p}\,(\bar{p} - \bar{r} - \theta) - s_r\,\bar{r}),$$

$$\frac{\mathrm{d}\bar{p}}{\mathrm{d}t} = G_p(b\,\bar{r} - s_p\,\bar{p}).$$

Here, G_r and G_p are the genetic variances of the two traits, and the terms $-s_r\,\bar{r}$ and $-s_p\,\bar{p}$ reflect an assumption that both traits are selected to have a value of zero in the absence of mating interactions. The term $(\bar{p} - \bar{r} - \theta)$ in the equation for the dynamics of resistance reflects an assumption that the optimal balance between the two traits, $\bar{p} - \bar{r}$, for females is θ. The term $b\,\bar{r}$ in the equation for the dynamics of persistence reflects the assumption that higher persistence is always beneficial for males. (a) Find the equilibria of this model. (b) Assume that the relationships $a\,b\,\theta > s_r s_p$ and $b > s_p$ hold. Derive the condition under which the equilibrium with nonzero resistance and persistence is stable. (c) The condition you found in (b) can be written as a condition on the ratio of the genetic variances G_r/G_p. Define a new parameter $\mu = G_r/G_p$. What is the critical value of μ at which the equilibrium with nonzero resistance and persistence loses stability? (d) Using the new parameter μ as a bifurcation parameter, demonstrate that conditions (i) and (ii) of Recipe 11.2 hold. This proves that, when the equilibrium loses stability, periodic behavior in the two variables results.

Problem 11.9: The Ricker model of density-dependent growth (equation 5.7) can be written as:

$$n(t+1) = n(t)\exp\left[r\left(1 - \frac{n(t)}{K}\right)\right],$$

which has two equilibria $\hat{n} = 0$ and $\hat{n} = K$. (a) Determine the conditions under which the nonzero equilibrium $\hat{n} = K$ is stable. (b) When $\hat{n} = K$ becomes unstable, the possibility exists that a cycle of period two occurs. Derive a composite recursion equation whose equilibria correspond to a cycle of period two in the above model. You should be able to write this recursion in the form $n(i+1) = n(i)\, f(n(i))$, for some function $f(n(i))$, where i measures time in two time steps. (c) The nonzero equilibria of the composite recursion must satisfy $f(\hat{n}) = 1$ for the population size to return to the same value every two generations. As $f(\hat{n})$ involves exponential and polynomial functions of n, it cannot be solved explicitly. Nevertheless, you can use the following logic to determine when a two-point cycle exists. First, check that $\hat{N} = K$ satisfies $f(\hat{n}) = 1$ (think about why this must be true). This fact can be seen from where $f(\hat{n}) = 1$ crosses 1 in Figure 11.9. This figure further suggests that for low values of r (solid curve) no other points satisfy $f(\hat{n}) = 1$, but two additional solutions exist for high values of r (dashed curve). Only in this latter case can there be a two-point cycle, which involves a population size below the carrying capacity and a population size above the carrying capacity. By examining the shape of the curves near $\hat{n} = K$, determine the mathematical condition that must be satisfied for a two-point cycle to exist. [Hint: Consider the behavior of the slope.]

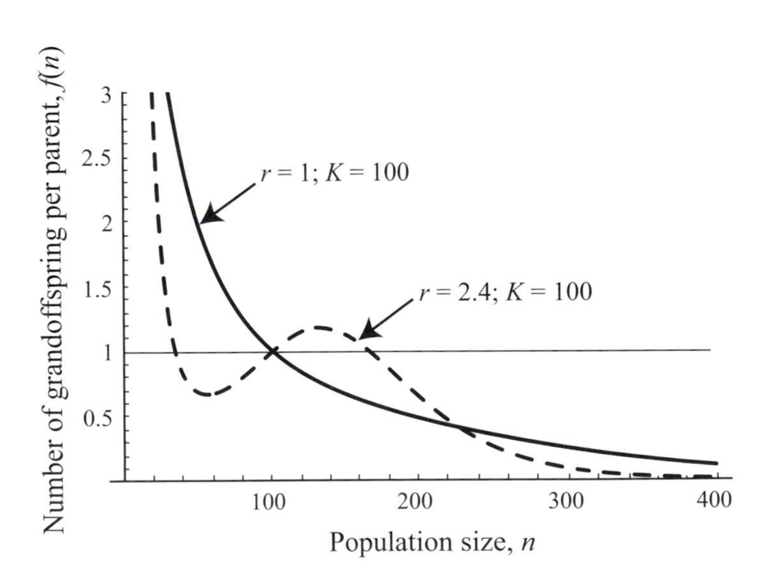

Figure 11.9: Number of grand-offspring in the Ricker model. Equilibria of the composite recursion statisfy $f(n) = 1$ (thin horizontal line).

Further Reading

For further information on cyclic dynamics see

- Glass, L., and M. C. Mackey. 1988. *From Clocks to Chaos: The Rhythms of Life*. Princeton University Press, Princeton, N.J.
- Keener, J., and J. Sneyd. 1998. *Mathematical Physiology*. Springer-Verlag, Berlin.

References

Begon, M., J. L. Harper, and C. R. Townsend. 1996. *Ecology: Individuals, Populations and Communities*. Blackwell Science, Oxford.

Bergelson, J. and C. B. Purrington. 1996. Surveying patterns in the cost of resistance in plants. *Am. Nat.* 148:536–558.

Case, T. J. 2000. *An Illustrated Guide to Theoretical Ecology*. Oxford University Press, Oxford.

Edelstein-Keshet, L. 1988. *Mathematical Models in Biology*. Random House, New York.

Frank, S. A. 2002. *Immunology and Evolution of Infectious Disease*. Princeton University Press, Princeton, N.J.

Gavrilets, S., G. Arnqvist, and U. Friberg. 2001. The evolution of female mate choice by sexual conflict. *Proc. R. Soc. London, Ser. B* 268:531–539.

Gurney, W.S.C., and R. M. Nisbet. 1998. *Ecological Dynamics*. Oxford University Press, Oxford.

Hofbauer, J., and K. Sigmund. 1988. *The Theory of Evolution and Dynamical Systems*. Cambridge University Press, Cambridge.

Kaplan, D., and L. Glass. 1995. *Understanding Nonlinear Dynamics*. Springer-Verlag, New York.

Kunkel, B. N. 1996. A useful weed put to work: genetic analysis of disease resistance in *Arabidopsis thaliana*. *Trends Gene.* 12:63–69.

Marsden, J. E., and M. McCracken. 1976. *The Hopf Bifurcation and Its Applications*. Springer-Verlag, New York.

Murrary, J. D. 1989. *Mathematical Biology*. Springer-Verlag, New York.

Rowe, L., E. Cameron, and T. Day. 2005. Escalation, retreat, and female indifference as alternative outcomes of sexually antagonistic coevolution. *Am. Nat.* 165:S5–S18.

Whalen, M. C., R. W. Innes, A. F. Bent, and B. J. Staskawicz. 1991. Identification of *Pseudomonas syringae* pathogens of *Arabidopsis* and a bacterial locus determining avirulence on both *Arabidopsis* and soybean. *Plant Cell,* 3:49–59.

CHAPTER 12

Evolutionary Invasion Analysis

Chapter Goals:

- To explain the general technique of evolutionary invasion analysis
- To determine how predictions from evolutionary invasion analyses depend on parameters
- To extend evolutionary invasion analyses to models of structured populations

Chapter Concepts:

- Resident allele/mutant allele
- Fitness gradient
- Convergence stability
- Evolutionary stability
- Evolutionary branching point
- Modifier model

12.1 Introduction

In previous chapters we have presented a variety of techniques for constructing dynamic models in ecology and evolutionary biology. In this chapter, we build upon these techniques to address longer-term evolutionary questions. For example, consider the question of why some species have high juvenile dispersal rates whereas others have evolved relatively sedentary lifestyles. This cannot be adequately addressed with an evolutionary model limited to two specific alleles, because "dispersal rate" is a continuous trait that can take on many different values. Rather, we would like to address this question allowing for a wide variety of possible alleles, each of which codes for a different dispersal rate.

The approach that we develop in this chapter considers whether a population that is initially fixed for a particular allele can be invaded by a mutant allele that codes for a different trait value. By determining which alleles can invade which populations, we can infer the direction of evolution. Furthermore, we can determine if there is an allele coding for a particular trait value that can resist invasion by all other possible mutant alleles. These analyses tell us important information about which trait values we expect to see after long periods of evolutionary time. This conceptual approach has been used in a great many models in evolutionary biology (see Box 12.5 at the end of this chapter), and we refer to the approach generically as an *evolutionary invasion analysis*.

The allele for which a population is currently fixed is termed the *resident allele*. The newly arisen mutation is termed the *mutant allele*.

In an evolutionary invasion analysis, the allele currently fixed in a population is referred to as the *resident allele*. The alternative allele that has arisen by mutation and that alters the trait of interest is referred to as the *mutant allele*. In most analyses, it is assumed that mutations are infrequent enough that the population dynamics of the species in question reaches some form of long-term behavior (e.g., an equilibrium) prior to the mutation. Evolutionary invasion analyses then address two main questions:

(i) What types of mutant alleles can invade a resident population?
(ii) Is there a resident allele that can resist invasion by all other mutant alleles?

In section 12.2 below, we begin by giving two examples of evolutionary invasion analyses. Section 12.3 then steps back from these specific examples and outlines the general steps of an evolutionary invasion analysis. In section 12.4

we demonstrate how to determine the effect of model parameters on predictions from evolutionary invasion analyses. Finally, in section 12.5, we extend evolutionary invasion analyses to class-structured populations.

12.2 Two Introductory Examples

Example: Evolution of Reproductive Effort in Plants

Consider a plant species that takes one year to become reproductively mature. We use $N(t)$ to denote the number of mature individuals in year t. Every year, each plant reproduces and has probability p of surviving to the following year. We assume that the rate of establishment of new plants is density dependent, having the same form as the Ricker model (5.7): $b\, e^{-\alpha N(t)}$, where b is the birth rate of a plant when the population size is small and the parameter α determines the strength of density dependence. Consequently, the number of plants changes through time as

$$N(t+1) = N(t)\,(b\, e^{-\alpha N(t)} + p). \tag{12.1}$$

Now imagine that every year each plant allocates a proportion of its resources to reproduction and the remainder to survival. If a plant invests a lot of resources in reproduction, the value of b will be large but p will be small. We refer to the proportion of resources invested in reproduction as the *reproductive effort* E. Because fertility and survival both depend on reproductive effort, we write b and p as $b(E)$ and $p(E)$. The exact nature of these functions will depend on the plant species in question.

What level of reproductive effort is expected to evolve? The answer to this question might help explain why some plants have annual life histories (dying after reproduction) while others are perennial (surviving over multiple years of reproduction). Following the general conceptual approach outlined in the introduction, we begin with a population that is fixed for a resident allele coding for the reproductive effort E, and that has reached a stable equilibrium. Using the methods of Chapter 5, recursion equation (12.1) gives two equilibria for the population size:

$$\hat{N} = 0 \qquad \text{and} \qquad \hat{N} = \frac{1}{\alpha} \ln \frac{b(E)}{1 - p(E)}. \tag{12.2}$$

To address the long-term evolution of reproductive effort, we focus on the nonzero equilibrium. The techniques of Chapter 5 can be used to show that this equilibrium is locally stable provided that $0 < (1 - p) \ln(b/(1 - p)) < 2$.

Now imagine that a mutant allele appears that codes for a different reproductive effort E_m. What is the fate of this mutant allele? Can the allele invade the population or will it be driven to extinction? To answer this question we need an explicit model for how the number of individuals carrying the mutant allele changes over time while it is still rare.

We can obtain a recursion for the number of mutant plants by first extending model (12.1) to allow for two alleles and then using a linear stability analysis to

Box 12.1: The Evolution of Reproductive Effort

Here, we derive equation (12.3) describing the number of plants carrying a mutant allele. As a starting place, we first suppose that plants reproduce asexually, so that offspring are identical copies of their parents. Under this assumption, we can extend model (12.1) to track both resident plants, $N(t)$, and mutant plants, $N_m(t)$:

$$N(t+1) = N(t)\left(b\, e^{-\alpha\,(N(t)+N_m(t))} + p\right), \tag{12.1.1a}$$

$$N_m(t+1) = N_m(t)\left(b_m\, e^{-\alpha\,(N(t)+N_m(t))} + p_m\right), \tag{12.1.1b}$$

where we have simplified the notation by writing $b(E_m) = b_m$ and $p(E_m) = p_m$, respectively. Equations (12.1.1) make no assumptions regarding the rarity of the mutant plants. The dynamics of the two types are coupled to one another solely through the density-dependent term $e^{-\alpha\,(N(t)+N_m(t))}$.

One equilibrium of equations (12.1.1) is

$$\hat{N}_m = 0, \quad \hat{N} = \frac{1}{\alpha}\ln\left(\frac{b(E)}{1-p(E)}\right). \tag{12.1.2}$$

This equilibrium corresponds to the nonzero equilibrium (12.2) of the resident population before the mutant is introduced. Under what conditions is this equilibrium unstable after the mutant is introduced?

When the mutant plant is rare, the total population size $N(t) + N_m(t)$ will very nearly equal the number of residents, $N(t)$. Consequently, the recursion equation for the mutant type is approximately

$$N_m(t+1) \approx N_m(t)\left(b_m\, e^{-\alpha\hat{N}} + p_m\right), \tag{12.1.3}$$

which is equation (12.3). Formally, equation (12.1.3) is the linear approximation (Primer 1) to equation (12.1.1b) when $N_m(t)$ is small.

More generally, to determine when the mutant type can increase when rare, we conduct a stability analysis as in Chapter 8 (Recipe 8.4). Calculating the Jacobian matrix for system (12.1.1) and evaluating at $\hat{N}_m = 0$ and $\hat{N} = (1/\alpha)\ln(b(E)/(1-p(E)))$, gives the local stability matrix

$$\mathbf{J} = \begin{pmatrix} 1-(1-p)\ln\dfrac{b}{1-p} & -(1-p)\ln\dfrac{b}{1-p} \\ 0 & \dfrac{b_m(1-p)+b\,p_m}{b} \end{pmatrix}. \tag{12.1.4}$$

The equilibrium will be unstable if at least one of the eigenvalues of (12.1.4) is larger than one in absolute value.

(continued)

Box 12.1 *(continued)*

Because matrix (12.1.4) is upper triangular, its eigenvalues are simply the diagonal elements (Rule P2.26). The first of these eigenvalues is less than one in absolute value when $-1 < 1 + (1-p)\ln(b/(1-p)) < 1$, which can be written as $0 < (1-p)\ln(b/(1-p)) < 2$. This condition involves only characteristics of the resident type and is the same stability condition as in model (12.1) with only the resident allele. This is no coincidence—some of the eigenvalues obtained from an invasion analysis always correspond to the stability conditions of the resident population before the mutant allele appears.

Only the second eigenvalue $(b_m(1-p) + b\,p_m)/b$ involves characteristics of the mutant allele, and it is this eigenvalue that determines whether the mutant invades (see equation (12.4c)). Because this eigenvalue is always positive, invasion will occur if $(b_m(1-p) + b\,p_m)/b > 1$, which can be rearranged to give condition (12.5). While the above derivation assumes asexual reproduction, condition (12.5) also holds for a sexually reproducing diploid species when mating is random (Problem 12.1).

determine the rate of spread of the mutant allele as in Chapter 8 (Box 12.1). When rare, the number of mutant plants changes through time by approximately

$$N_m(t+1) \approx N_m(t)\,(b(E_m)\,e^{-\alpha\hat{N}} + p(E_m)). \tag{12.3}$$

Recursion (12.3) makes good intuitive sense; mutant plants survive each year with probability $p(E_m)$, and they have a yearly reproductive output of $b(E_m)e^{-\alpha N_{\text{Total}}}$. N_{Total} is the total population size, and when the mutant allele is rare, $N_{\text{Total}} \approx \hat{N}$.

Plugging the nonzero equilibrium from (12.2) into (12.3) and simplifying, we get a more explicit recursion:

$$N_m(t+1) = N_m(t)\left(b(E_m)\frac{1-p(E)}{b(E)} + p(E_m)\right). \tag{12.4a}$$

As (12.4a) is a linear recursion equation for $N_m(t)$, we can write it as

$$N_m(t+1) = N_m(t)\,\lambda(E_m,E) \tag{12.4b}$$

where

$$\lambda(E_m, E) = b(E_m)\frac{1-p(E)}{b(E)} + p(E_m). \tag{12.4c}$$

$\lambda(E_m,E)$ represents the *reproductive factor* of the mutant allele (the number of surviving mutant individuals in the next year, per mutant individual in the current year). For invasion to occur, the mutant allele must increase in numbers over time: $N_m(t+1) > N_m(t)$, which requires that $\lambda(E_m, E) > 1$. This condition is easier to interpret if we rearrange the terms so that the left-hand side involves only attributes of the mutant allele and the right-hand side involves only attributes of the resident allele. Invasion is then predicted to occur if

$$\frac{b(E_m)}{1 - p(E_m)} > \frac{b(E)}{1 - p(E)}. \tag{12.5}$$

According to equation (12.5), the mutant allele spreads only if its value of $b/(1-p)$ is larger than that of the resident allele. What does this quantity represent? Consider a plant reproducing for the first time. In the absence of density dependence, its reproductive output in the first year will be b. With probability p it will also survive to the next year, in which case it will have another b offspring. Similarly, with probability p^2 it will survive two years and again have b offspring. In this way, in the absence of density dependence, the plant's *expected lifetime reproductive output* is $b + pb + p^2b + p^3b + \cdots$. This can be written as $b\sum_{i=0}^{\infty} p^i$, and using equation (A1.20) of Appendix 1, this sum simplifies to $b/(1-p)$. Therefore, the mutant allele invades only if it leads to a higher lifetime reproductive output in the absence of density dependence.

Given the above interpretation, our model predicts that evolution will eventually maximize lifetime reproductive output. Only those alleles that increase lifetime reproductive output can spread, and this process of allelic replacement will continue until the population reaches a maximum where no further increase is possible. Keep in mind, however, that this prediction is specific to the assumptions of the model. In Box 12.2, we show how altering the assumptions of the model slightly causes a different quantity, not lifetime reproductive output, to be maximized over evolutionary time. Indeed, sometimes nothing is maximized by evolution (see next example).

To determine the level of reproductive effort that maximizes the lifetime reproductive output we must specify the functions $p(E)$ and $b(E)$. For illustrative purposes, we assume that fecundity increases exponentially with reproductive effort using the function $b(E) = e^E$ (Definition P1.4) and that the probability of survival is a quadratic function, $p(E) = p_0 - E^2$ (Definition P1.3), that decreases with reproductive effort, where p_0 is the baseline survival when reproductive effort is zero. We can then write the lifetime reproductive output explicitly as

$$f(E) = \frac{b(E)}{1 - p(E)} = \frac{e^E}{1 - p_0 + E^2}$$

(Figure 12.1).

To find the reproductive effort that maximizes lifetime reproductive output, we seek the value of E that causes the slope of $f(E)$ to equal zero (i.e., $df/dE = 0$; see tangent line in Figure 12.1). This value is not necessarily a maximum, however, because minimum values also have this property. To ensure that the value of E represents a maximum, we can examine the second derivative, df^2/d^2E, which will be negative at a maximum (see Appendixes 2 and 4).

Box 12.2: The Fitness Measure Maximized by Evolution Depends on Model Assumptions

Here we show that slight changes to the assumptions of a model can fundamentally alter the fitness measure that is maximized over evolutionary time. Let us alter the model of reproductive effort (12.1) to allow density dependence to reduce *both* the birth rate and the probability of survival:

$$N(t+1) = N(t)\,(b+p)\,e^{-\alpha N(t)}. \tag{12.2.1}$$

Model (12.2.1) has two equilibria $\hat{N} = 0$ and $\hat{N} = (1/\alpha)\ln(b+p)$. The nonzero equilibrium is locally stable provided that $0 < \ln(b+p) < 2$. Assuming that this holds and that the total population size is very nearly $\hat{N}$, the number of plants carrying a rare mutant allele changes according to

$$N_m(t+1) = N_m(t)\,(b(E_m) + p(E_m))\,e^{-\alpha \hat{N}}. \tag{12.2.2}$$

Substituting $\hat{N} = (1/\alpha)\ln(b+p)$ into this equation yields an equation having the same form as equation (12.4b), but with

$$\lambda(E_m, E) = \frac{b(E_m) + p(E_m)}{b(E) + p(E)}, \tag{12.2.3}$$

where we must have $\lambda(E_m, E) > 1$ for the mutant allele to invade. Again, we can rearrange (12.2.3) into a condition of the form $f(E_m) > f(E)$; specifically, $b(E_m) + p(E_m) > b(E) + p(E)$. Consequently, the mutant allele invades if its reproductive factor $b + p$ is larger than that of the resident allele in the absence of density dependence. Once again, a measure of reproductive success is maximized by evolution, but now it is the intrinsic growth factor per time step, $b + p$, not the lifetime reproductive output, $b/(1-p)$, that is maximized (also see Mylius and Diekmann 1995).

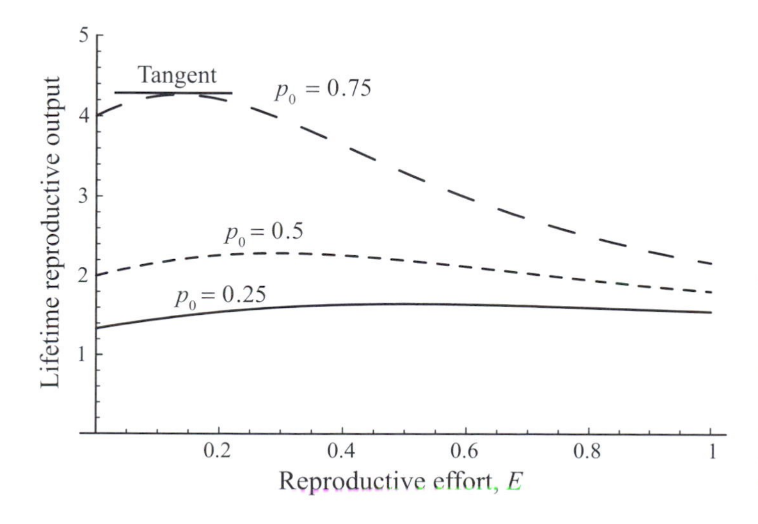

Figure 12.1: Reproductive effort and lifetime reproductive output. The lifetime reproductive output, $f(E) = b(E)/(1 - p(E))$, is plotted as a function of the reproductive effort, E, for different values of the baseline probability of survival, p_0. A short solid line is drawn that is tangent to the curve with $p_0 = 0.75$; the tangent line has a slope of zero at the maximum, $E = 0.13$.

The first derivative of $f(E) = e^E/(1 - p_0 + E^2)$ with respect to E is

$$\frac{\mathrm{d}f}{\mathrm{d}E} = \frac{e^E((1 - E)^2 - p_0)}{(1 - p_0 + E^2)^2}. \tag{12.6a}$$

Equation (12.6a) equals zero when

$$(1 - E)^2 = p_0, \tag{12.6b}$$

which yields two values of E that could potentially maximize lifetime reproductive output:

$$E = 1 - \sqrt{p_0} \quad \text{and} \quad E = 1 + \sqrt{p_0}. \tag{12.6c}$$

As E represents the proportion of resources allocated to reproduction, only $E = 1 - \sqrt{p_0}$ represents a biologically feasible solution.

Next, we examine the second derivative of $f(E)$ at $E = 1 - \sqrt{p_0}$:

$$\left.\frac{\mathrm{d}^2 f}{\mathrm{d}E^2}\right|_{E=1-\sqrt{p_0}} = -\frac{e^{(1-\sqrt{p_0})}\sqrt{p_0}}{2\left(1 - \sqrt{p_0}\right)^2}. \tag{12.6d}$$

Because p_0 is a survival probability that lies between zero and one, expression (12.6d) is always negative. Thus, an allele that codes for a reproductive effort of $E = 1 - \sqrt{p_0}$ maximizes lifetime reproductive output. According to equation (12.5), a mutant allele coding for $E = 1 - \sqrt{p_0}$ will be able to invade any resident allele with a different strategy, because the mutant has a higher lifetime reproductive success. Conversely, a resident allele coding for $E = 1 - \sqrt{p_0}$ can never be invaded, because no other strategy would have a higher lifetime reproductive success. Thus, over the long term, we expect reproductive effort to evolve toward, and then remain at, $E = 1 - \sqrt{p_0}$.

In the above example, evolution maximizes a measure of reproductive success (lifetime reproductive output). As the next example demonstrates, however, there is not always a quantity that is maximized by evolution.

Example: Evolution of Dispersal

Most organisms have offspring that disperse from their place of birth. From an evolutionary standpoint this is puzzling. Given that individuals of the previous generation have successfully produced offspring in that location, it is clearly a suitable site. So why opt for the risky behavior of leaving and moving to a new location? There are many potential explanations, but the one we will model is the possibility that offspring disperse to avoid competing with siblings (Hamilton and May 1977). The trait of interest is the probability that a given offspring disperses. We use d to denote the resident value of this trait and d_m to denote the mutant value.

As before, we suppose that the resident population has reached an equilibrium, and we then derive an equation governing the dynamics of a rare mutant allele. This requires making some assumptions regarding the genetics and population dynamics of the species in question. The simplest scenario assumes that the species is haploid and asexual. We also assume that the population is made up of a very large number of breeding sites, S, each of which supports one and only one breeding individual. Each generation, the individual on a breeding site produces a large number of offspring, the genotype of which determines the probability of dispersal. Offspring that disperse enter a "global dispersal pool" and have a probability of survival of $1 - c$ relative to individuals that do not disperse (c represents the *cost of dispersal*). Individuals in the global dispersal pool then disperse randomly to all sites in the population. Competition occurs among all individuals on a site for access to that breeding site, with all individuals having the same chance of winning that site. The life cycle then starts anew.

If the number of mutants in generation t is $N_m(t)$, then the number in the next generation is $\lambda(d_m,d)\, N_m(t)$, where $\lambda(d_m,d)$ is the reproductive factor of the mutant allele while rare (i.e., the number of successful mutant offspring alleles produced by each mutant allele in the current generation). According to Box 12.3, the mutant reproductive factor is

$$\lambda(d_m,d) = \frac{1 - d_m}{(1 - d_m) + d\,(1 - c)} + \frac{d_m\,(1 - c)}{(1 - d) + d\,(1 - c)}. \tag{12.7}$$

Box 12.3: The Mutant Reproductive Factor in the Dispersal Model

The reproductive factor of a mutant allele can be derived by adding up the number of successful offspring that do not disperse and the number that do disperse per mutant parent:

$$\lambda(d_m,d) = (\text{\# of successful mutant nondispersers}) + (\text{\# of successful mutant dispersers}).$$

By "successful," we mean that the offspring has won a breeding site in the next generation. The number of successful mutant nondispersers equals (# mutant offspring)(probability of not dispersing)(probability that a mutant nondisperser wins the breeding site). On the other hand, the number of successful mutant dispersers equals (# mutant offspring)(probability of dispersing)(probability of surviving dispersal)(probability that a mutant disperser wins a breeding site). Let n be the number of offspring produced by a breeding individual (mutant or resident). We then have

$$\begin{aligned}\lambda(d_m,d) = {} & n\,(1 - d_m) \times (\text{probability that a mutant non-disperser wins the breeding site}) \\ & + n\,d_m\,(1 - c) \times (\text{probability that a mutant disperser wins a breeding site}).\end{aligned}$$

After dispersal, all individuals on a site have an equal chance of becoming the single individual that can breed on the site. Thus, we can rewrite $\lambda(d_m,d)$ as

$$\lambda(d_m, d) = \frac{n\,(1 - d_m)}{N_{\text{nondispersing}}} + \frac{n\,d_m(1 - c)}{N_{\text{dispersing}}}, \tag{12.3.1}$$

(continued)

Box 12.3 ***(continued)***

where $N_{\text{nondispersing}}$ is the total number of individuals competing for the breeding site of a mutant that has not dispersed, and $N_{\text{dispersing}}$ is the total number of individuals competing for the breeding site of a mutant that has dispersed.

Calculating $N_{\text{nondispersing}}$. Because only one individual breeds on each site, a site containing nondispersing mutant offspring must have been occupied by a single mutant individual in the previous generation. Thus, a focal patch containing a mutant nondisperser will contain two types of individuals: (i) mutant types that have not dispersed and (ii) resident types that have immigrated into the patch. We can ignore the possibility that a mutant immigrates in from another patch, because the mutant is rare and mutants migrating out of other patches are exceedingly unlikely to end up in a patch that happens to contain other mutant individuals.

We have already calculated the number of mutant types that do not disperse away from the focal patch: $n\,(1-d_m)$. Next, we must calculate the number of resident types that migrate into the focal patch. Let S be the total number of sites and p be the proportion of these sites occupied by mutants. There are thus $S\,(1-p)$ residents, who produce a total of $n\,S\,(1-p)$ offspring, of which a fraction d disperse. Of these migrants, only a fraction $(1-c)$ survive, and only a fraction $1/S$ land on the focal patch. Thus, the number of resident types that immigrate to the patch is $n\,S\,(1-p)\,d\,(1-c)/S = n\,(1-p)\,d\,(1-c)$. Because we assume that the mutant type is very rare, this number is approximately $n\,d\,(1-c)$, to leading order in p. Overall, the total number of offspring competing for the breeding site of a non-dispersing mutant is thus $N_{\text{nondispersing}} \approx n\,(1-d_m) + n\,d\,(1-c)$.

Calculating $N_{\text{dispersing}}$. A focal patch containing a dispersing mutant offspring will contain three types of individuals: (i) the mutant itself, (ii) resident types that have not dispersed, and (iii) resident types that have dispersed from other sites. We assume that only the one mutant would land on the site, because the mutant is very rare. The number of resident types that have not dispersed will equal $n\,(1-d)$. Finally, the number of resident types that have dispersed from other sites to the focal site is the same as in the previous paragraph: approximately $n\,d\,(1-c)$. Overall, we have $N_{\text{dispersing}} \approx 1 + n\,(1-d) + n\,d\,(1-c)$. If the number of offspring produced per parent, n, is very large (e.g., when a plant produces a large number of seeds), we can make the further approximation that $N_{\text{dispersing}} \approx n\,(1-d) + n\,d\,(1-c)$.

Substituting these expressions into (12.3.1) yields the mutant reproductive factor:

$$\lambda(d_m, d) = \frac{n\,(1-d_m)}{n\,(1-d_m) + n\,d\,(1-c)} + \frac{n\,d_m\,(1-c)}{n\,(1-d) + n\,d\,(1-c)}. \tag{12.3.2}$$

Dividing through by n gives equation (12.7) in the main text.

Unlike (12.4c), equation (12.7) cannot be rearranged in any way that allows us to rewrite the invasion condition $\lambda(d_m, d) > 1$ as $f(d_m) > f(d)$, with mutant parameters on one side and resident parameters on the other. Consequently, there is no function that is strictly maximized over evolutionary time. Rather, the ability of a mutant allele to invade depends on both its own characteristics and those of the resident allele, as described by $\lambda(d_m,d)$. Nevertheless, we can use the mutant reproductive factor (12.7) to predict how dispersal rate evolves by determining which mutations can invade which resident populations (i.e., by determining when $\lambda(d_m,d) > 1$).

Although mathematically equivalent, it is easier to see if $\lambda(d_m,d) - 1$ is positive than it is to see if $\lambda(d_m,d) > 1$, because we can focus on the sign of the result. Subtracting one from (12.7) and factoring yields

$$\lambda(d_m,d) - 1 = \frac{(1-c)(d_m - d)(1 - c\,d - d_m)}{(1 - c\,d)((1-d_m) + d\,(1-c))}. \tag{12.8}$$

We attempted to write each term in equation (12.8) as a quantity whose sign was obviously positive, because we can ignore positive factors when identifying the sign of $\lambda(d_m,d) - 1$. For example, we wrote $((1 - d_m) + d\,(1 - c))$ rather than $(1 + d - d_m - c)$. All terms in $\lambda(d_m,d) - 1$ are necessarily positive except $(d_m - d)(1 - c\,d - d_m)$ because the dispersal probability d and the probability of dying while dispersing, c, are less than one. Therefore, when $(d_m - d)(1 - c\,d - d_m) > 0$, we know that $\lambda(d_m,d) - 1 > 0$, implying that the mutant will spread.

The quantity $(d_m - d)(1 - c\,d - d_m)$ is positive when both terms in parentheses are positive ($d_m > d$ and $1 - c\,d > d_m$) or when both terms are negative ($d_m < d$ and $1 - c\,d < d_m$). In other words, the mutant allele will spread only if its dispersal probability lies between d and $1 - c\,d$. Figure 12.2 illustrates four

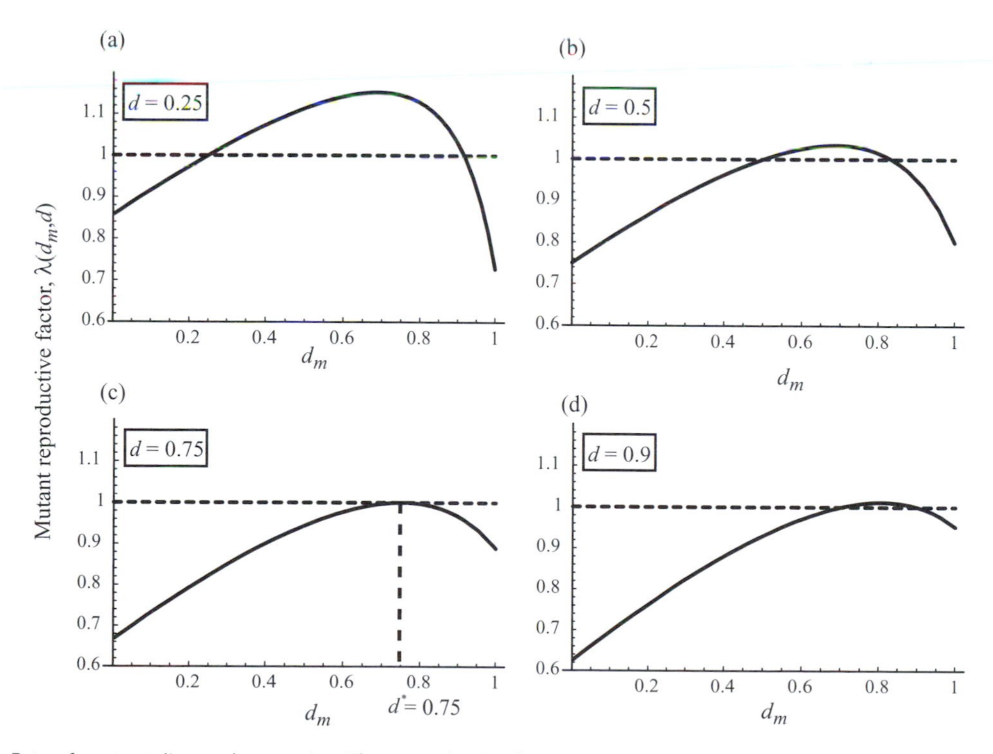

Figure 12.2: Fate of mutant dispersal strategies. The reproductive factor (12.7) of the mutant type, $\lambda(d_m,d)$, is plotted against the mutant dispersal probability, d_m, for different values of the resident dispersal probability, d. In all cases, the cost of dispersal is $c = 1/3$. (a) $d = 0.25$, (b) $d = 0.5$, (c) $d = 0.75$, and (d) $d = 0.9$.

examples of these conditions by plotting the reproductive factor (12.7) as a function of the mutant dispersal rate. Mutants can spread whenever the curve $\lambda(d_m,d)$ lies above the dashed line. In panel (a), mutants with a dispersal rate, d_m, between $d = 0.25$ and $1 - c\,d = 0.917$ can spread. In panel (b), a narrower range of mutant alleles can spread, namely, those with dispersal rates between $d = 0.5$ and $1 - c\,d = 0.833$. In panel (c), the range of mutants that can invade has shrunk to zero, because $d = 1 - c\,d = 0.75$. Finally, in panel (d), mutants can invade if they disperse at a rate between $d = 0.9$ and $1 - c\,d = 0.7$.

Figure 12.2 illustrates that when $d < d_m < 1 - c\,d$, mutants invade that increase the dispersal rate relative to the resident population, whereas when $1 - c\,d < d_m < d$, mutants invade that decrease the dispersal rate relative to the resident population. The resident dispersal rate at which the direction of evolution changes occurs where $d = 1 - c\,d$. Rearranging this equation, we see that this critical resident dispersal rate is given by $d = 1/(1 + c)$. Thus, we now have complete information about the direction in which dispersal rate evolves; it increases when the resident disperses with a probability less than $1/(1 + c)$, and it decreases when the dispersal probability of the resident is higher than this value.

Next we wish to determine if there is a dispersal rate allele that can resist invasion by all possible mutant alleles. You might expect that our above results already answer this question. After all, if dispersal rate increases when it is below $d = 1/(1 + c)$, and if it decreases when above this value, then won't an allele coding for $d = 1/(1 + c)$ resist invasion by all possible mutants? Interestingly, the answer is "not necessarily." The direction of evolution is determined by focusing on a particular resident dispersal rate and asking which mutant alleles can invade when rare. Our analysis shows that, if the resident allele does not code for $d = 1/(1 + c)$, then mutant alleles that are closer to this value (including alleles coding exactly for $d = 1/(1 + c)$) can always invade when rare. What we have not yet checked, however, is whether, a resident allele coding for the critical dispersal rate, $d = 1/(1 + c)$, is able to resist invasion by all possible rare mutant alleles.

To consider the fate of mutants arising in a population where the resident disperses at rate $d = 1/(1 + c)$, we must evaluate the mutant reproductive factor at $d = 1/(1 + c)$. Using equation (12.8) we obtain

$$\lambda(d_m,d)\big|_{d=1/(1+c)} - 1 = -\frac{(1 - c)(d_m + c\,d_m - 1)^2}{(1 - d_m) + (1 - c\,d_m)}. \tag{12.9}$$

The sign of expression (12.9) determines the fate of mutant alleles attempting to invade a resident population at $d = 1/(1 + c)$. Expression (12.9) is zero whenever $d_m + c\,d_m - 1 = 0$, or equivalently, when $d_m = 1/(1 + c)$. This is expected—if the mutant allele codes for the same dispersal rate as the resident allele, then it should be neutral, and neither spread nor disappear. Otherwise, the sign of (12.9) is negative because both d_m and c must lie between zero and one. Therefore, expression (12.9) is negative or zero, and a mutant allele will never be able to spread. Thus, a resident allele that disperses at rate $d = 1/(1 + c)$ can indeed resist invasion by all possible mutants.

This model, analyzed by Hamilton and May (1977), provides an amazingly counterintuitive result about dispersal. Over the long term, dispersal strategies are expected to evolve toward, and remain at, $d = 1/(1 + c)$. Thus, even when dispersing individuals are almost certain to die (c near 1), we expect more than 50% of offspring to disperse! Selection to decrease competition among siblings can thus be a potent evolutionary force.

12.3 The General Technique of Evolutionary Invasion Analysis

With the above two models as examples, we now step back from the calculations and describe the general approach in Recipe 12.1:

Recipe 12.1
Performing an Evolutionary Invasion Analysis

Step 1: Determine the trait of interest and the potential values that it can take (this is sometimes referred to as the "decision variable").

Step 2: Derive an equation (or equations) that specifies the dynamics of a rare mutant allele appearing in a population fixed for some resident allele (this is sometimes referred to as "invasion fitness").

Step 3: Using the equation(s) from Step 2, derive the condition that must be met for the mutant allele to invade.

Step 4: Use the condition from Step 3 to determine how the trait evolves as a function of the resident trait value.

Step 5: Use the condition from Step 3 to identify resident trait values that cannot be invaded by any mutant allele. Such resident strategies are referred to as evolutionarily stable strategies, or ESS's.

We next discuss each of these steps in detail, after which we summarize the mathematical details in Recipe 12.2.

12.3.1 Step 1—The Decision Variable

The decision variable represents any trait that is potentially altered by mutations. It can be a discrete or a continuous trait, but we restrict attention to continuous traits. The evolution of discrete traits is often handled using these same techniques by approximating the trait of interest as continuous. For example, even though the number of eggs that a female produces is a discrete trait (it takes integer values) we can often approximate egg number reasonably well by assuming that a continuum is possible.

12.3.2 Step 2—Invasion Fitness

The most foolproof approach for deriving invasion fitness involves four stages:

(a) Construct a model for the resident population dynamics in the absence of the mutant allele.

(b) Determine the equilibrium of the resident model. More generally, we might allow for other types of long-term dynamics (e.g., periodic cycles), but we focus only on equilibria in this chapter. Also, identify the conditions under which the resident equilibrium is stable. We focus only on parameter values that allow a stable equilibrium in the resident population.

(c) Augment the resident model to allow for two alleles at the gene of interest (i.e., the resident and the mutant allele). How this is done depends on the question and the details of the model (e.g., is the organism haploid or diploid, how does population regulation occur, etc.). As a check, the augmented model should have an equilibrium where the mutant allele is absent and the resident population is at the equilibrium obtained in stage (b) above.

(d) Use the techniques of Chapter 8 to perform a local stability analysis on the augmented model for the equilibrium where the mutant allele is absent. Providing that we group together the equations for the mutant types after the equations for the resident types, the local stability matrix will typically have a block-triangular form:

$$\begin{pmatrix} \mathbf{J}_{\text{res}} & \mathbf{V} \\ \mathbf{0} & \mathbf{J}_{\text{mut}} \end{pmatrix},$$

involving the submatrices $\mathbf{0}$ (a matrix of zeros), $\mathbf{V}$, $\mathbf{J}_{\text{res}}$, and $\mathbf{J}_{\text{mut}}$ (these are potentially 1×1 matrices as in equation (12.1.4)). The eigenvalues of this block triangular matrix are the eigenvalues of $\mathbf{J}_{\text{res}}$ and $\mathbf{J}_{\text{mut}}$ (Rule P2.27). The eigenvalues of $\mathbf{J}_{\text{res}}$ depend only on the parameters of the resident (determining "internal stability" of the resident equilibrium in the absence of the mutant allele), while the eigenvalues of $\mathbf{J}_{\text{mut}}$ typically depend on the parameters of the mutant and the resident (determining "external stability" to the introduction of the mutant allele). The leading eigenvalue of $\mathbf{J}_{\text{mut}}$ is the invasion fitness of the mutant allele.

The block-triangular form of the local stability matrix results from the assumption that the resident type does not continually produce the mutant type (i.e., the mutant is introduced only once). Thus, all elements of the local stability matrix describing transitions from the resident to the mutant type are zero (as in the zero element in equation (12.1.4)).

To ensure that no mistakes are made, it is safest to proceed through stages (a)–(d), as we did in Box 12.1. In practice, however, it is often possible to skip some of these stages. For example, in the dispersal model, we never derived an equation governing the dynamics of the resident population. We could skip stages (a) and (b) because the number of resident individuals was fixed at the number of sites, and the trait value of the resident was fixed at d. We did not need equations to identify this equilibrium. In Box 12.3, we proceeded directly to stage (d) and derived the mutant reproductive factor directly from the life cycle. This is often a useful time-saving approach (see examples in Supplementary Material 12.1). In general, however, it is best to be explicit about the details of a model and proceed through stages (a)–(d) as we did in Box 12.1.

12.3.3 Step 3–Invasion Condition

The leading eigenvalue λ of the mutant subblock $\mathbf{J}_{\text{mut}}$ in a stability analysis is the reproductive factor of the mutant allele when rare. This reproductive factor is typically a function of both the mutant and the resident trait values (see equation (12.4c) for the model of reproductive effort, and equation (12.7) for the model of dispersal). For discrete-time models, the mutant allele will be able to invade if the mutant reproductive factor λ is larger than one in magnitude (Recipe 8.4). Analogously, in a continuous-time model, the mutant allele will be able to invade if the leading eigenvalue from the mutant subblock, r, has a positive real part (Recipe 8.2).

In general, we must consider the possibility that the leading eigenvalue is negative in discrete-time models ($\lambda < -1$) or that the leading eigenvalue is complex. In practice, however, the reproductive factor of the mutant is typically real and positive. This is because the reproductive factor represents the number of surviving mutant offspring per mutant parent when the mutant allele is rare, which should never become negative. That the leading eigenvalue is real and positive can be verified in a discrete-time model using the Perron-Frobenius theorem on the mutant subblock $\mathbf{J}_{\text{mut}}$ (Appendix 3).

Sometimes the invasion condition can be rewritten so that the mutant trait value occurs only in a function on one side of the inequality and the resident trait value occurs in the same function on the other side, as in equation (12.5) in the model of reproductive effort. In this case, mutant alleles spread only if they increase the value of this function; this function is therefore maximized by evolution. In such cases, the direction and ultimate endpoint of evolution is determined by the trait value that maximizes the function (Appendix 4). When there is no function that is maximized, Steps 4 and 5 are applied.

12.3.4 Step 4—Predicting the Direction of Evolution

To predict the direction of evolution, we use the reproductive factor from Step 3 to determine which mutant alleles can spread, for each of the possible values of the resident trait. In the examples of section 12.2, we were able to examine the fate of all possible mutants in the context of all possible resident alleles. For many models, however, the expression for the mutant reproductive factor is too complicated to conduct such a complete analysis.

To make progress when the reproductive factor is complicated, we can focus only on mutant alleles that code for trait values very similar to that of the resident allele. This implies that the mutations that arise are of small effect. The benefit of this restriction is that we can then approximate the mutant reproductive factor using a Taylor series (Recipe P1.2), assuming that the mutant trait value is near the resident trait value:

$$\lambda(d_m,d) \approx \lambda(d,d) + \left.\frac{\partial\lambda}{\partial d_m}\right|_{d_m=d} (d_m - d). \tag{12.10}$$

Mutant alleles that code for the same trait as the resident are neutral, i.e., $\lambda(d,d) = 1$. Therefore, mutant alleles coding for trait values slightly larger than the resident

The slope (derivative) of the reproductive factor of a mutant allele with respect to the mutant trait value is referred to as the *fitness gradient* for that trait.

(i.e., $d_m - d > 0$) invade if $(\partial\lambda/\partial d_m)|_{d_m=d} > 0$. Similarly, mutant alleles coding for trait values slightly less than the resident (i.e., $d_m - d < 0$) invade if $(\partial\lambda/\partial d_m)|_{d_m=d} < 0$. The term $(\partial\lambda/\partial d_m)|_{d_m=d}$ is sometimes referred to as the *fitness gradient*, and its sign determines the direction of evolution. Positive fitness gradients imply that the trait will increase, and negative fitness gradients imply that the trait will decrease as a result of the appearance and spread of mutations of small effect.

A trait value is said to be *convergence stable* if natural selection drives the evolution of the trait toward this value.

According to the above considerations, we expect evolution to lead toward the trait value $\hat{d}$ at which the sign of the fitness gradient changes from positive to negative. In such cases, evolution pushes the trait to higher values when below $\hat{d}$ and to lower values when above $\hat{d}$. At the trait value $d = \hat{d}$, directional selection ceases to act and the quantity $(\partial\lambda/\partial d_m)|_{d_m=d}$ is zero (Figure 12.3). Such trait values are called *convergence stable*. Mathematically, a trait value is convergence stable if $(\partial\lambda/\partial d_m)|_{d_m=d}$ (which is a function of d only) decreases with d, going from positive to negative at $d = \hat{d}$ (see Recipe 12.2).

As an example, if we restrict our attention to mutations of small effect in the dispersal model, the direction of evolution is determined by

$$\left.\frac{\partial\lambda}{\partial d_m}\right|_{d_m=d} = \frac{(1 - c)(1 - d - c\,d)}{(1 - c\,d)^2}. \tag{12.11}$$

Mutant alleles that increase the dispersal rate invade when $(\partial\lambda/\partial d_m)|_{d_m=d}$ is positive, which requires that $1 - d - c\,d > 0$ or, equivalently, $d < 1/(1 + c)$. Similarly, mutant alleles that decrease the dispersal rate invade when $(\partial\lambda/\partial d_m)|_{d_m=d}$ is negative, which requires that $d > 1/(1 + c)$. The direction of evolution changes sign when $(\partial\lambda/\partial d_m)|_{d_m=d} = 0$, which occurs when $\hat{d} = 1/(1 + c)$. Because $(\partial\lambda/\partial d_m)|_{d_m=d}$ is positive below $\hat{d} = 1/(1 + c)$ and negative above it, this point is always convergence stable. These results are consistent with our previous analysis, but here we have restricted our attention to mutants of small effect.

12.3.5 Step 5—Evolutionarily Stable Strategies

The final step in an invasion analysis is to determine if there is a trait value that can resist invasion by all possible rare mutant alleles. This trait value is called an *evolutionarily stable strategy* or ESS. The ESS trait value is often denoted by an asterisk, e.g., d^*. If the resident population is fixed for an allele coding for an ESS, d^*, then the reproductive factor of all rare mutant alleles must be less than a mutant allele coding for the ESS value itself, d^*. This requirement is summarized mathematically by the following condition:

A trait value is said to be *evolutionarily stable* (an ESS) if, once an allele coding for this value has reached fixation, no other mutant allele can increase in number (or frequency).

$$\lambda(d_m,d^*) \le \lambda(d^*,d^*), \tag{12.12}$$

for all mutant strategies d_m, with equality occurring when $d_m = d^*$. For discrete-time models, the right-hand side of equation (12.12) will equal one, because the mutant allele will be neutral. Therefore, equation (12.12) requires that the reproductive factor of any mutant allele in the resident population at d^* be less

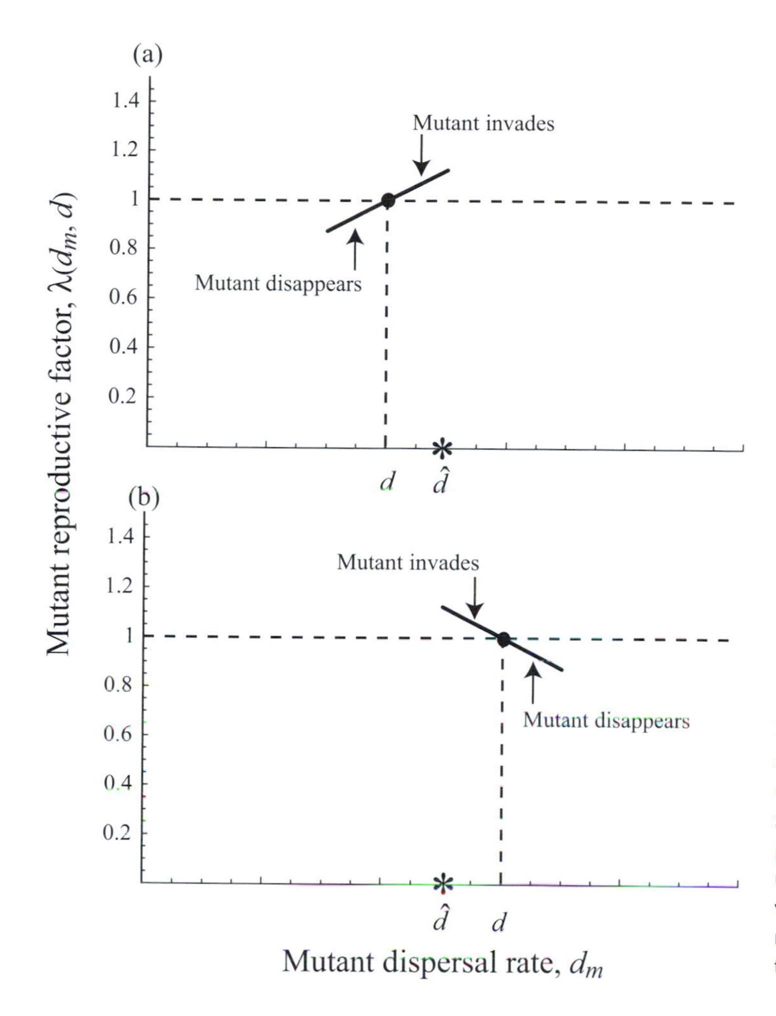

Figure 12.3: A graphical depiction of convergence stability. Suppose that $\hat{d}$ is a convergence stable dispersal rate. (a) When the resident value of d is smaller than $\hat{d}$, the slope $(\partial\lambda/\partial d_m)|_{d_m=d}$ must be positive so that mutant alleles coding for trait values closer to $\hat{d}$ can invade. (b) When the resident value of d is larger than $\hat{d}$, the slope $(\partial\lambda/\partial d_m)|_{d_m=d}$ must be negative so that mutant alleles coding for trait values closer to $\hat{d}$ can invade.

than one. We see a graphical example of this condition in Figure 12.2c, where $\lambda(d_m,d^*)$ falls below one for all mutant strategies when the resident dispersal rate is 0.75. Consequently, $d^* = 0.75$ represents an ESS in this example.

Again, it is often difficult to determine an ESS by examining all possible resident and mutant alleles. As in Step 4, we can simplify the analysis if we restrict attention to mutations of small effect. Considering only mutations of small effect, a strategy d^* will satisfy (12.12) if $\lambda(d_m,d^*)$ is maximized in d_m at $d_m = d^*$. We can thus use the fact that the slope must be zero and the second derivative must be negative at a maximum (Appendix 4):

$$\left.\frac{\partial\lambda(d_m,d)}{\partial d_m}\right|_{\substack{d_m=d^*\\ d=d^*}} = 0, \tag{12.13a}$$

$$\left.\frac{\partial^2\lambda(d_m,d)}{\partial {d_m}^2}\right|_{\substack{d_m=d^*\\ d=d^*}} \le 0. \tag{12.13b}$$

Condition (12.13a) determines trait values at which directional selection ceases. Condition (12.13b) ensures that such points maximize rather than minimize the mutant reproductive factor with respect to d_m. When equations (12.13) are both satisfied, the resident trait value cannot be invaded by any mutant of small effect. It remains possible, however, that mutants of large effect could invade. We thus refer to equations (12.13) as *local* ESS conditions.

As an example, we apply the local ESS conditions to the model of dispersal. We already determined the derivative of $\lambda(d_m,d^*)$ with respect to d_m in equation (12.11), so we can write the first derivative condition (12.13a) by evaluating equation (12.11) at $d = d^*$:

$$\left.\frac{\partial\lambda(d_m,d)}{\partial d_m}\right|_{\substack{d_m=d^* \\ d=d^*}} = \frac{(1-c)(1-d^*-c\,d^*)}{(1-c\,d^*)^2} = 0. \tag{12.14a}$$

Thus, there is a potential ESS when $(1 - d^* - c\,d^*) = 0$, i.e., when

$$d^* = \frac{1}{(1+c)}. \tag{12.14b}$$

Problem 12.4 asks you to confirm that the second derivative condition (12.13b) is satisfied, confirming that d^* is an ESS.

Recipe 12.2 summarizes the local conditions that make it possible to determine the direction of evolution (Step 4) and to characterize the ESS (Step 5) in a wide variety of models. Recipe 12.2 focuses only on cases where the ESS occurs for intermediate values of the decision variable. Results for cases where the ESS occurs on a boundary of allowable values can be derived in a manner similar to that described in Appendix 4 for finding maxima.

Recipe 12.2

The Direction of Evolution and ESS when Mutations Have Small Effects

Specific mathematical conditions can be given for Steps 4 and 5 of Recipe 12.1 if mutant alleles differ from resident alleles by only a small amount. Suppose that the trait of interest is denoted by x and that $\lambda(x_m,x)$ is the discrete-time reproductive factor of a rare mutant allele coding for trait value x_m in a population consisting of a resident allele coding for trait value x. Continuous-time models are handled analogously with $\lambda(x_m,x)$ replaced by the continuous-time growth rate $r(x_m,x)$. As a check, we should have $\lambda(x,x) = 1$ in discrete time or $r(x,x) = 0$ in continuous time, for any trait value x.

Step 4: The trait value x is predicted to evolve in a direction given by the sign of the fitness gradient

(continued)

Recipe 12.2 ***(continued)***

$$\left.\frac{\partial \lambda}{\partial x_m}\right|_{x_m = x}. \tag{12.15a}$$

Directional selection ceases for trait values $x = \hat{x}$ at which equation (12.15a) equals zero. The trait value $\hat{x}$ is convergence stable if equation (12.15a) decreases, changing from positive to negative as x increases through the value $x = \hat{x}$. Mathematically, this requirement for convergence stability is equivalent to

$$\frac{\mathrm{d}}{\mathrm{d}x}\left\{\left.\frac{\partial \lambda(x_m, x)}{\partial x_m}\right|_{x_m = x}\right\}_{x = \hat{x}} < 0. \tag{12.15b}$$

Step 5: An ESS value of x (denoted x^*) must satisfy the following first and second derivative conditions:

$$\left.\frac{\partial \lambda}{\partial x_m}\right|_{\substack{x_m = x^* \\ x = x^*}} = 0, \tag{12.15c}$$

$$\left.\frac{\partial^2 \lambda}{\partial x_m^2}\right|_{\substack{x_m = x^* \\ x = x^*}} \leq 0. \tag{12.15d}$$

Conditions (12.15c) and (12.15d) ensure that a resident population with trait value x^* cannot be invaded by mutant strategies that are similar to x^*, but they do not guarantee that large-effect mutations cannot invade. A trait value that satisfies (12.15) is thus a *local ESS*, while a trait value that also satisfies condition (12.12) is a *global ESS*.

To determine the direction of evolution and to identify ESS in models involving multiple variables, see Supplementary Material 12.2.

As mentioned in the dispersal example, you might expect that a population would always evolve toward an evolutionarily stable state, but this is not necessarily true. There are two other possibilities that arise (Eshel and Motro 1981; Eshel 1983; Taylor 1989; Christiansen 1991; Abrams et al. 1993; Geritz et al. 1998). First, evolution can lead *away* from an ESS, even when the initial trait value of a population is near the ESS. In this case, we say that the ESS is evolutionarily unattainable or *convergence unstable*. Second, evolution can lead to a trait value that is *not* an ESS. A convergence stable trait value that is not an ESS is sometimes referred to as an *evolutionary branching point*. At an evolutionary branching point, any mutant allele of small effect can invade the resident

A trait value is said to be an *evolutionary branching point* if it is convergence stable but not evolutionarily stable.

population, but the mutant allele cannot entirely displace the resident allele because convergence stability ensures that this resident allele is able to spread when rare. Consequently, we expect some form of evolutionary diversification to occur. We now apply Recipes 12.1 and 12.2 to an example illustrating this point.

Example: Competition for Resources

Consider a species of zooplankton such as daphnia that feeds on algal species of different sizes. For simplicity, we assume that there is a continuum of different-sized algal species and that each individual daphnid has a feeding strategy characterized by the algal size it is best able to consume. We assume that a high feeding efficiency on large algal species comes at the cost of a low feeding efficiency on small algal species and vice versa.

The decision variable is the size of algae, s, that the daphnia are best able to consume (Step 1 of Recipe 12.1). Step 2 requires that we formulate a model for the dynamics of a mutant allele that arises in a population fixed for a resident allele (see section 12.3.2). First we formulate a model for the resident population in the absence of the mutant allele. We suppose that the daphnia population is reproducing asexually, and that the population dynamics follow the logistic equation (3.5a) in the absence of the mutant allele:

$$N(t+1) = N(t)\left\{1 + r - r\,\frac{N(t)}{K(s)}\right\} \tag{12.16}$$

where $K(s)$ is the carrying capacity of the daphnia population when the resident allele codes for a feeding preference for algae of size s. For simplicity we assume that the carrying capacity is given by the bell-shaped function $K(s) = \kappa e^{-(s-s_0)^2/a_K}$ (Definition P1.6). The parameter s_0 represents the most abundant algal size class, κ measures the carrying capacity for daphnia that prefer this most abundant size class, and a_K is a positive parameter measuring the breadth of size classes that are present; all of these parameters affect the carrying capacity $K(s)$ of daphnia that prefer size class s (Figure 12.4a). The equilibria of the logistic model (12.16) are $\hat{N} = 0$ and $\hat{N} = K(s)$. We focus on the latter equilibrium, where daphnia are present, which is locally stable provided that $0 < r < 2$ (Chapter 5). We assume that this stability condition holds.

Next we augment model (12.16) to allow for a mutant allele coding for a different feeding strategy. When such a mutant allele arises, we suppose that the joint dynamics of the mutant and resident allele are given by the Lotka-Volterra competition equations (3.14):

$$\begin{aligned} N(t+1) &= N(t)\left\{1 + r - r\,\frac{N(t) + \alpha(s,s_m)\,N_m(t)}{K(s)}\right\}, \\ N_m(t+1) &= N_m(t)\left\{1 + r - r\,\frac{N_m(t) + \alpha(s_m,s)\,N(t)}{K(s_m)}\right\}. \end{aligned} \tag{12.17}$$

The function $\alpha(s_m,s)$ is the competition coefficient, measuring the impact of resident daphnia on mutant daphnia as a function of their feeding preferences.

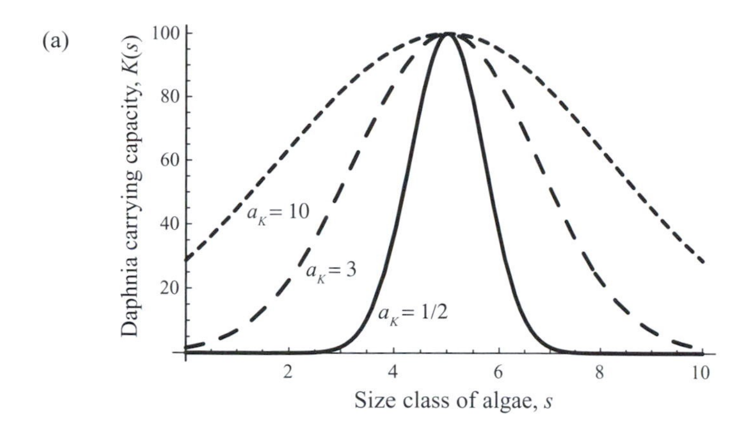

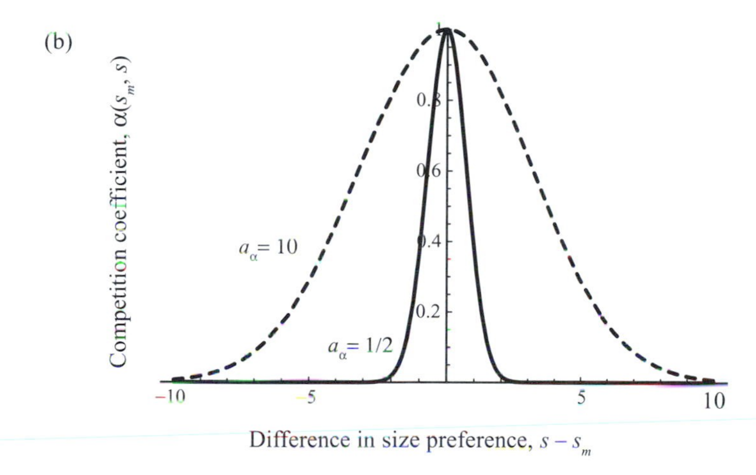

Figure 12.4: Carrying capacity and competition for resources of different sizes. (a) The carrying capacity function $K(s)$ is illustrated as a function of the size class of the resource, s, for different values of the parameter a_K with $s_0 = 5$ and $\kappa = 100$. (b) The strength of competition, $\alpha(s_m, s)$, is illustrated as a function of the difference in size class preferences using different values of the parameter a_α.

We assume that the competition coefficient decreases from a value of one when $s = s_m$ to a value of zero when s and s_m are very different. One such function that obeys these constraints is the bell-shaped function $\alpha(s_m, s) = e^{-(s_m - s)/a_\alpha}$ (Definition P1.6). The parameter a_α is positive and measures the breadth of the competition function (Figure 12.4b), with larger values of a_α corresponding to stronger competition across a broader array of size preferences.

To complete Step 2, we need to use model (12.17) to derive the mutant allele's reproductive factor when rare. This is obtained by conducting a local stability analysis of the equilibrium $\hat{N} = K(s)$ and $\hat{N}_m = 0$ for model (12.17). Calculating the local stability matrix and evaluating at this equilibrium produces an upper triangular matrix whose eigenvalues are the diagonal elements (Chapter 8; Problem 8.10). The first of these eigenvalues is $\lambda = 1 - r$, which determines the stability of the resident equilibrium before the mutant appears; we have already assumed that this eigenvalue is less than one in magnitude (i.e., $0 < r < 2$). Thus, the second eigenvalue is the relevant eigenvalue, giving the reproductive factor of the mutant allele when rare:

$$\lambda(s_m, s) = 1 + r - r\,\frac{\alpha(s_m, s)\,K(s)}{K(s_m)}. \tag{12.18}$$

Moving on to Step 3, we use the mutant reproductive factor to determine the invasion condition. Because this is a discrete-time model, a mutant allele coding for feeding strategy s_m will invade a resident population with feeding strategy s, only if $\lambda(s_m,s) > 1$, which requires that

$$K(s_m) > \alpha(s_m,s)\, K(s) \tag{12.19}$$

(recall that $r > 0$ for stability of the equilibrium before the mutant appears). For the particular functions chosen for $K(s)$ and $\alpha(s_m,s)$, it is not possible to separate all terms involving mutant versus resident trait values on different sides of the inequality (Problem 12.3). Therefore, we cannot obtain a function that is maximized by evolution.

We now proceed to determine the direction of evolution (Step 4). Suppose that the resident allele codes for a feeding preference for algae of size s and that mutations are of small effect. Using equation (12.15a) of Recipe 12.2, s is predicted to evolve in a direction given by the sign of the fitness gradient

$$\left.\frac{\partial\lambda(s_m,s)}{\partial s_m}\right|_{s_m=s} = -2\,r\,\frac{(s - s_0)}{a_K}. \tag{12.20}$$

If s is smaller than s_0 then s is predicted to increase, and vice versa. Once $s = s_0$, directional selection ceases. Therefore, natural selection drives the feeding strategy s to match the most abundant resource (i.e., $\hat{s} = s_0$). In this case, we do not have to evaluate (12.15b) to determine that s_0 is convergence stable, because the sign of equation (12.20) is obvious (changing from positive to negative as s passes through s_0).

Finally, let us determine if there is an evolutionarily stable feeding strategy s^* (Step 5). Again, assuming small mutational effects, we can apply conditions (12.15c) and (12.15d) of Recipe 12.2. The first derivative of (12.18) with respect to s_m was calculated in equation (12.20), which we now evaluate at $s = s^*$ to obtain the first derivative condition:

$$\left.\frac{\partial\lambda(s_m,s)}{\partial s_m}\right|_{\substack{s_m=s^*\\ s=s^*}} = -2\,r\,\frac{(s^* - s_0)}{a_K} = 0, \tag{12.21}$$

Thus, there is a potential ESS when $s^* = s_0$, i.e., when daphnia prefer algae of the size that yields the largest carrying capacity. Next, evaluating the second derivative condition (12.15d) gives

$$\left.\frac{\partial^2\lambda(s_m,s)}{\partial s_m^2}\right|_{\substack{s_m=s^*\\ s=s^*}} = 2\,r\left(\frac{1}{a_\alpha} - \frac{1}{a_K}\right) < 0. \tag{12.22}$$

Interestingly, the feeding strategy $s^* = s_0$ is evolutionarily stable only when $a_K < a_\alpha$. Otherwise, there is no single allele that represents an ESS.

These calculations reveal two possibilities for the evolution of daphnia feeding strategies: (i) $a_K < a_\alpha$, in which case $s^* = s_0$ is a convergence stable ESS, or

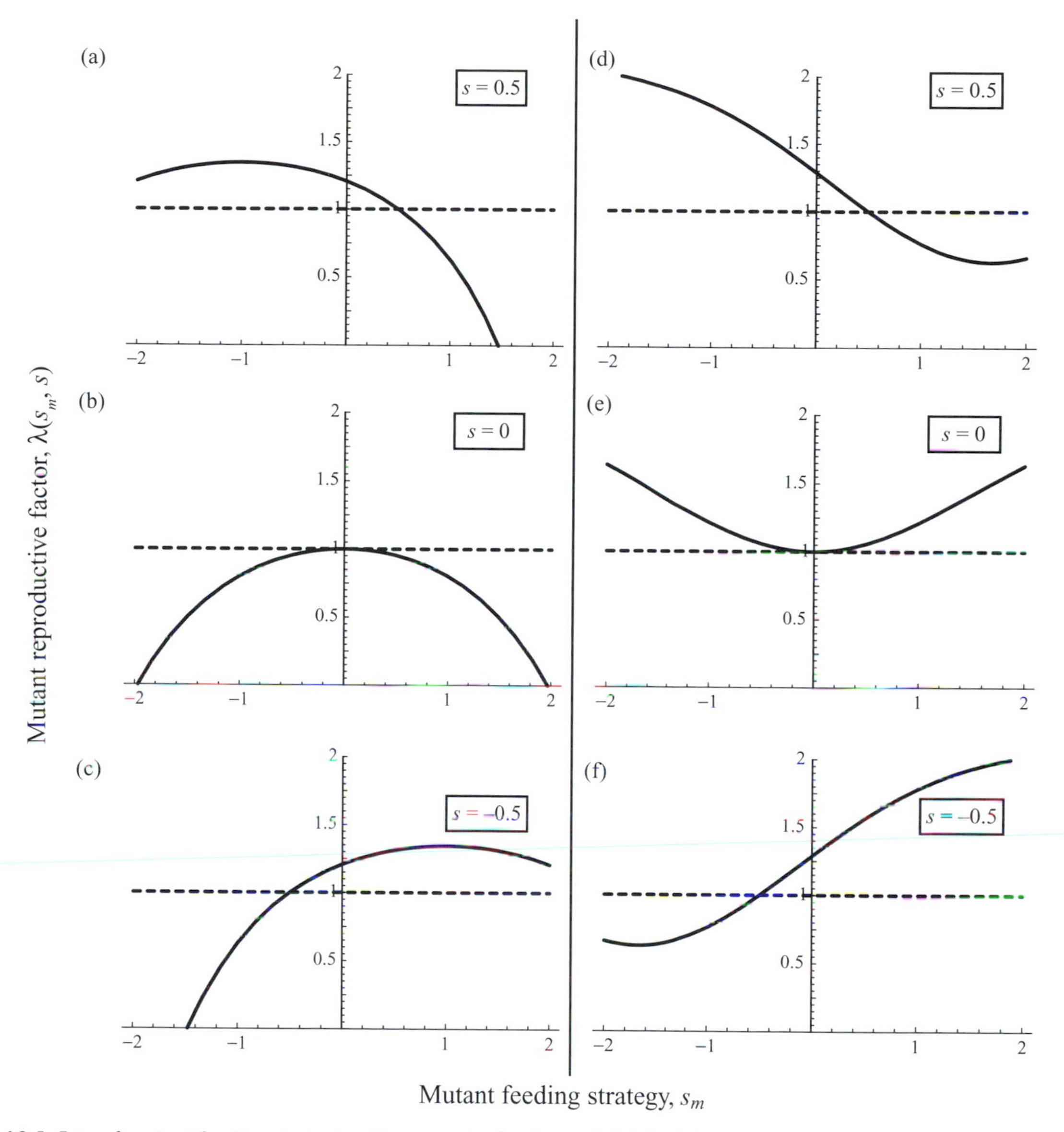

Figure 12.5: Fate of mutant feeding strategies. The reproductive factor (12.18) of the mutant allele, $\lambda(s_m,s)$, is plotted against the mutant feeding strategy s_m using the parameters $s_0 = 0$, $r = 1.1$, and $a_K = 1$. Mutant alleles can invade whenever the curves lie above the dashed line at one. Panels (a), (b), and (c) consider the invasion of mutant alleles in populations with different resident feeding strategies s when $a_\alpha = 1.5$ so that $s = 0$ is a convergence stable ESS. Panels (d), (e), and (f) are similar except that $a_\alpha = 0.7$ so that $s = 0$ is convergence stable but not an ESS. Notice that $\lambda(s_m,s)$ is maximized in panel (b) but minimized in panel (e) at $s = 0$.

(ii) $a_K > a_\alpha$, in which case $s^* = s_0$ is convergence stable but *not* an ESS. In case (i) we expect the population to evolve toward $s^* = s_0$ and to remain there indefinitely (Figure 12.5a–c). In case (ii) we still expect the population to evolve toward $s^* = s_0$ provided that mutational steps are small (Figure 12.5d–f). Once there, however, mutations with any size preference different from $s^* = s_0$ can invade, because $\lambda(s_m,s^*)$ is minimized at $\lambda(s^*,s^*)$ when $a_K > a_\alpha$ (Figure 12.5e).

While such mutations can spread while rare, they will not entirely displace the resident strategy s^*. This is because convergence stability guarantees that an allele coding for size preference s^* would spread if ever it became rare relative to the allele coding for size preference s_m. Thus, once the population reaches s^*, natural selection becomes disruptive and favors the spread and coexistence of more than one type. The strategy $s^* = s_0$ thus corresponds to an evolutionary branching point when $a_K > a_\alpha$. This and other possible evolutionary outcomes can be visualized using *pairwise invasibility plots*, as described in Box 12.4.

Biologically, when $a_K > a_\alpha$, there is a broad array of algal resources that cannot be utilized by daphnia with a single size preference. Thus, it makes sense for evolution to favor diverse size preferences. At the same time, because the carrying capacity is maximized at $s = s_0$, natural selection favors evolution toward this phenotype. When the resident population does not prefer size class s_0, mutants whose size preference is closer to s_0 gain in two ways: they benefit from eating different algae and from having a higher carrying capacity. This is why $s^* = s_0$ is always convergence stable. Once the residents prefer algae of size s_0, however, selection to avoid competition is strong enough to outweigh the costs of preferring less abundant algae. Evolutionary diversification then occurs (Dieckmann and Doebeli

Box 12.4: Pairwise Invasibility Plots

Pairwise invasibility plots (PIP) provide a graphical depiction of the conditions under which a point is an evolutionarily stable strategy (ESS) and/or convergence stable (CS). In a pairwise invasibility plot, possible resident strategies x are plotted along the horizontal axis, and possible mutant strategies x_m are plotted along the vertical axis (Figure 12.4.1). Regions where the number of mutant offspring per mutant parent exceeds one, $\lambda(x_m,x) > 1$, are then shaded ($r(x_m,x) > 0$ in a continuous-time model). Mutants are neutral when the mutant strategy equals the resident strategy (i.e., $\lambda(x,x) = 1$), and such mutants are represented along the diagonal line (dashed) in such a figure. There might be other mutant-resident combinations that yield a value of $\lambda(x_m,x) = 1$, and curves connecting such points are plotted as well (solid curve), because they demarcate the boundary between stability and instability.

A *potential ESS* corresponds to a point where the solid curve and dashed line intersect (circle). Only at such points is the first derivative condition (12.15c) of Recipe 12.2 satisfied (such points are sometimes referred to as evolutionarily "singular" points). By "potential ESS" we mean that, although the first derivative condition (12.15c) of Recipe 12.2 is satisfied, the second derivative condition might not be.

In cases (a) and (b), the circled point is an ESS because no other mutant strategy yields a higher growth rate once the population is at this point (i.e., no mutant strategies along the vertical line passing through the circle fall within a shaded region). Consequently, the second derivative condition (12.15d) of Recipe 12.2 will be satisfied. In cases (c) and (d), on the other hand, the circled point is not an ESS because there are mutant strategies that can invade a population (i.e., there are mutant strategies along the vertical line passing through the circled point that fall within a shaded region). In this case, the potential ESS represents a minimum on the fitness

(continued)

Box 12.4 *(continued)*

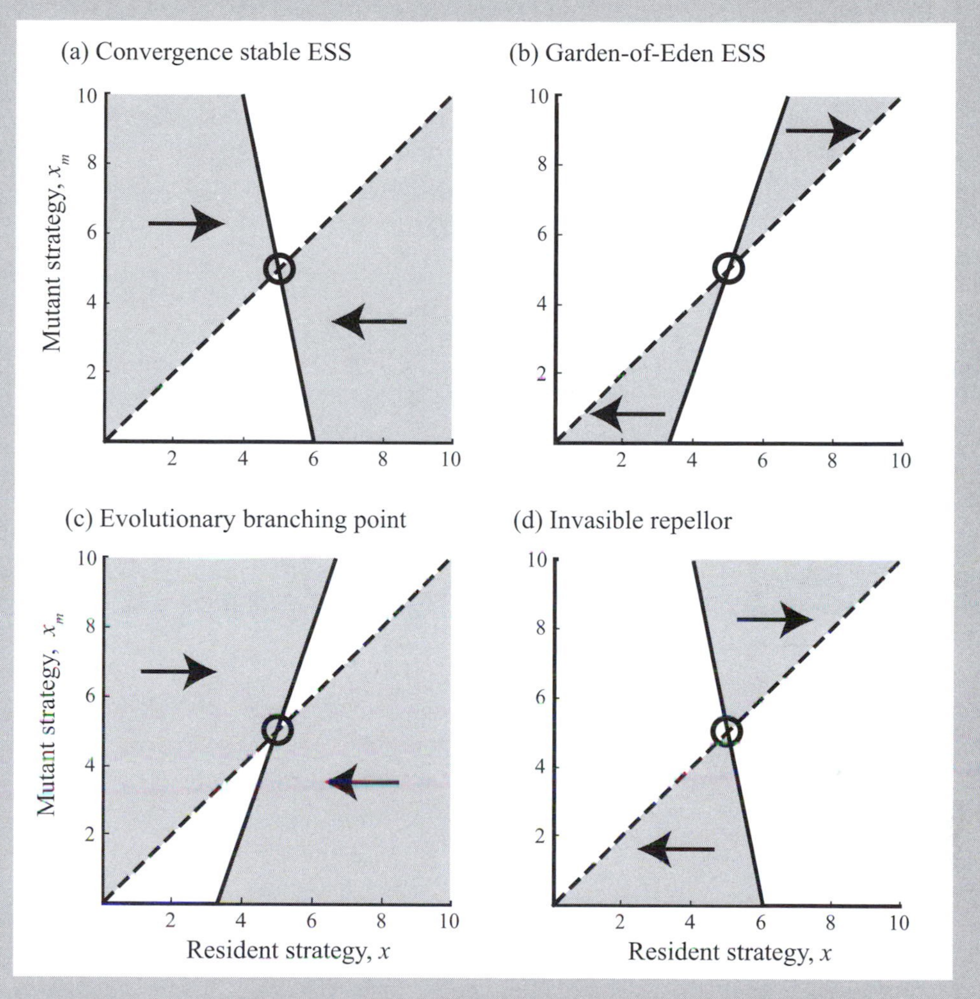

Figure 12.4.1: Pairwise invasibility plots. Mutant strategies invade within the shaded regions, causing the population to evolve in the direction of the arrows. The solid lines indicate where $\lambda(x_m, x) = 1$. Cases (a) with $a_\alpha = 1.5 > a_K = 1$ and (c) with $a_\alpha = 0.5 < a_K = 1$ correspond to the resource competition model (12.18), with s replaced by x and $K(x) = \kappa e^{-(x-x_0)^2/a_K}$, $\alpha(x_m, x) = e^{-(x_m-x)^2/a_\alpha}$, and $x_0 = 5$. In this model, the potential ESS (circle) is always convergence stable; mutants cannot invade a population at the potential ESS in case (a), but they can invade in case (c).

surface given by $\lambda(x_m, x)$, and mutant trait values x_m both above and below the resident trait value x can invade.

As mentioned in the text, a point can be an ESS but not be convergence stable and vice versa. Convergence stability requires that mutant strategies closer to the potential ESS are able to invade populations not at this point. In Recipe 12.2 we presented a local condition for convergence stability assuming that the population was already near the potential ESS (see condition (12.15b) of Recipe 12.2), but pairwise invasibility plots provide an excellent way to examine convergence

(continued)

Box 12.4 *(continued)*

stability globally. Regardless of how far the population is from the potential ESS, convergence stability requires that there be shading immediately above the diagonal to the left of the potential ESS, and immediately below the diagonal to the right of the potential ESS. This ensures that mutants that have a slightly larger x value can invade whenever the resident falls below the potential ESS and vice versa. In Figure 12.4.1, the circled point (i.e., the potential ESS) is convergence stable in cases (a) and (c), but it is not convergence stable in cases (b) and (d).

Altogether there are four classes of potential ESS values:

- Case (a): an ESS that is convergence stable.
- Case (b): an ESS that is not convergence stable.
- Case (c): a point that is not an ESS but that is nevertheless convergence stable.
- Case (d): a point that is neither an ESS nor convergence stable.

Case (a) is called a *convergence-stable ESS*. Case (b) has been termed a *Garden-of-Eden ESS*, because it would be stable if an organism ever had that strategy, but organisms do not evolve toward it (Nowak and Sigmund 1989). Case (c) is known as an *evolutionary branching point*, because a population is expected to evolve toward such a point, but once there, mutant strategies can invade and a polymorphism arises. Finally, case (d) is sometimes called an *invasible repellor*, because any mutant can invade the point and nearby populations evolve away.

Pairwise invasibility plots have two main advantages. For many people, they are much more intuitive than the derivative conditions given in Recipe 12.2. Second, they provide a graphical means of demonstrating that a point is a *global ESS*, which requires that all points along the vertical line through the circle fall within unshaded regions. It should be remembered, however, that pairwise invasibility plots are drawn for particular values of the parameters (although sometimes their general shapes can be inferred) and that they can be represented on a plane only for cases where mutations affect only one trait as in the examples considered in this chapter (see Supplementary Material 12.2 and 12.3 for cases where mutations affect multiple traits). For more information about pairwise invasibility plots and their various possible configurations, see Geritz et al. (1998).

1999). To determine whether more than two feeding preferences can persist simultaneously, the above model must be extended to allow for more than one resident strategy (Taylor 1989; Christiansen 1991; Geritz et al. 1998).

12.4 Determining How the ESS Changes as a Function of Parameters

In many models we are not so interested in the ESS trait value itself but rather in how the ESS varies as a function of parameters. In this section, we present techniques for investigating the relationship between an ESS and a parameter of interest. We do this in the context of a modified SI model of disease, asking how the ESS level of virulence of a pathogen depends on the natural death rate of the host.

Example: Evolution of Pathogen Virulence

Many pathogens infect their hosts without causing substantial damage (e.g., many rhinoviruses that cause the common cold) while others induce higher levels of mortality (e.g., flavivurses that cause dengue fever). One explanation for this variation in pathogen virulence is that the costs and benefits of pathogen-induced mortality vary among pathogens, and this has resulted in the evolution of different levels of virulence. Here we construct a model to predict the level of virulence that we expect to evolve. We work in continuous time to illustrate how Recipes 12.1 and 12.2 can be applied in this setting.

Our decision variable (Step 1, Recipe 12.1) is the pathogen-induced mortality rate (virulence). Step 2 requires that we derive an equation for the dynamics of a mutant pathogen strain with an altered level of virulence. Again, this step is composed of the four stages given in section 12.3.2. We suppose that, in the absence of the mutant strain, the resident pathogen dynamics are given by model (8.1) of Chapter 8, with no recovery. Specifically, the numbers of susceptible and infected individuals change according to

$$\frac{dS}{dt} = \theta - d\,S - \beta(v)\,S\,I,$$
$$\frac{dI}{dt} = \beta(v)\,S\,I - (d + v)\,I. \tag{12.23}$$

The parameter θ is the immigration rate of susceptible hosts, d is the per capita background mortality rate of hosts, v is the increased mortality rate induced by the resident strain (the virulence), and $\beta(v)$ is the transmission rate associated with this level of virulence. Specifically, we suppose that the level of virulence of a pathogen is related to the amount of pathogen in the body, which in turn is related to the transmissibility of the disease. But the exact nature of this relationship is unknown, so we write the transmission rate as an arbitrary function $\beta(v)$ of virulence v. In this example, we are not actually modeling the dynamics of individual pathogens per se, but rather tracking the number of hosts infected by a particular pathogen type.

In Chapter 8 we saw that this model has two equilibria, one in which the pathogen is absent from the population $\hat{S} = \theta/d, \hat{I} = 0$, and the other in which the pathogen is endemic

$$\hat{S} = \frac{d + v}{\beta(v)}, \quad \hat{I} = \frac{\theta}{d + v} - \frac{d}{\beta(v)}.$$

The local stability of these equilibria was also determined in Chapter 8. In the absence of recovery, the local stability matrix in equation (8.17) is

$$\mathbf{J}_{\text{res}} = \begin{pmatrix} -d - \hat{I}\,\beta(v) & -\hat{S}\,\beta(v) \\ \hat{I}\,\beta(v) & -d - v + \hat{S}\,\beta(v) \end{pmatrix}. \tag{12.24}$$

The endemic equilibrium with the disease present is locally stable when $\theta/d > (d + v)/\beta(v)$ (see Chapter 8). We assume that this condition holds.

Next, we must determine the growth rate of a mutant with a different virulence. The joint dynamics of the number of susceptible individuals, individuals infected with the resident strain (I), and individuals infected with the mutant strain (I_m) are

$$\begin{aligned} \frac{dS}{dt} &= \theta - d\,S - \beta(v)\,S\,I - \beta(v_m)\,S\,I_m, \\ \frac{dI}{dt} &= \beta(v)\,S\,I - (d + v)\,I, \\ \frac{dI_m}{dt} &= \beta(v_m)\,S\,I_m - (d + v_m)\,I_m. \end{aligned} \tag{12.25}$$

The parameter v_m is the virulence of the mutant pathogen strain, and $\beta(v_m)$ is its transmission rate. System (12.25) implicitly assumes that the only way in which the two strains interact is through competition for the infection of a common pool of susceptible hosts. As expected, one equilibrium of the augmented system (12.25) is

$$\hat{S} = \frac{d + v}{\beta(v)}, \quad \hat{I} = \frac{\theta}{d + v} - \frac{d}{\beta(v)}, \quad \hat{I}_m = 0,$$

and it is the stability of this equilibrium that determines whether or not the mutant strain can invade.

We now perform a stability analysis on this equilibrium. The local stability matrix for the augmented system (12.25) is

$$\begin{pmatrix} -d - \hat{I}\,\beta(v) & -\hat{S}\,\beta(v) & -\hat{S}\,\beta(v_m) \\ \hat{I}\,\beta(v) & -d - v + \hat{S}\,\beta(v) & 0 \\ 0 & 0 & -d - v_m + \hat{S}\,\beta(v_m) \end{pmatrix}. \tag{12.26a}$$

To appreciate its structure, we can write matrix (12.26a) in a way that emphasizes its block-triangular form:

$$\begin{pmatrix} \mathbf{J}_{\text{res}} & \vec{u} \\ \vec{0} & \mathbf{J}_{\text{mut}} \end{pmatrix} \tag{12.26b}$$

where $\vec{0}$ is $(0 \quad 0)$, $\vec{u}$ is $\begin{pmatrix} -\hat{S}\,\beta(v_m) \\ 0 \end{pmatrix}$, $\mathbf{J}_{\text{res}}$ is given by equation (12.24), and $\mathbf{J}_{\text{mut}} = (-d - v_m + \hat{S}\,\beta(v_m))$. Because (12.26b) is block triangular, its eigenvalues are given by the eigenvalues of the diagonal blocks, $\mathbf{J}_{\text{res}}$ and $\mathbf{J}_{\text{mut}}$ (Rule P2.27).

We have already assumed that the eigenvalues of $\mathbf{J}_{\text{res}}$ have negative real parts to ensure that the endemic equilibrium is locally stable before the mutant appears. Therefore, the relevant eigenvalue is the single element of $\mathbf{J}_{\text{mut}}$: $r(v_m, v) = -d - v_m + \hat{S}\,\beta(v_m)$. Substituting the value of $\hat{S}$ at the endemic equilibrium gives

$$r(v_m, v) = -d - v_m + \frac{d + v}{\beta(v)}\beta(v_m). \tag{12.27}$$

In Step 3 (Recipe 12.1), we use the growth rate (12.27) to obtain the invasion condition. A mutant with virulence v_m will invade a resident population with virulence v provided that (12.27) is positive; i.e.,

$$r(v_m, v) = -d - v_m + \frac{d + v}{\beta(v)}\beta(v_m) > 0 \tag{12.28a}$$

or

$$\frac{\beta(v_m)}{d + v_m} > \frac{\beta(v)}{d + v}. \tag{12.28b}$$

Inequality (12.28b) reveals that mutant alleles will invade provided that they increase the quantity $\beta(v)/(d + v)$. Because infected hosts die at a constant rate $(d + v)$, their expected life span is $1/(d + v)$ time units (as discussed in Primer 3 on probability distributions). Thus, $\beta(v)/(d + v)$ can be thought of as the expected number of times that the pathogen is transmitted before its host dies, per susceptible host. In this model, evolution maximizes this measure of the pathogen's reproductive success (Anderson and May 1982).

With the above example, let us explore how the ESS varies as a function of the parameter values. Specifically, we will determine how an increase in the background host mortality d affects the ESS level of virulence. If we knew the ESS, we could assess the effect of changing d by examining how the ESS varies as a function of d. But to determine the ESS, we need to specify the transmission function $\beta(v)$. For example, suppose that transmissibility increases as a function of virulence but eventually reaches an asymptote at τ. We could model this using the rational function $\beta(v) = \tau\, v/(k + v)$ (Definition P1.5). The level of virulence that maximizes $\beta(v)/(d + v)$ (i.e., the ESS) then equals

$$v^* = \sqrt{d\,k}. \tag{12.29}$$

Therefore, the ESS virulence increases as the background mortality rate of the host, d, increases.

This makes sense; if a host is likely to die soon, the pathogen should evolve a rapid transmission rate and so become more virulent. But how general is the above prediction? If we chose a different function for $\beta(v)$ would it still be true? In models like this, it is worth stepping back and treating $\beta(v)$ as an arbitrary function to see if any general insights are possible. But if we do not specify the functions of a model explicitly, then we cannot obtain an explicit expression for the ESS. Surprisingly, it is sometimes possible to determine how the ESS value would be affected by the parameter of interest even when we do not know the ESS.

To begin, we must first determine the properties of the function $\beta(v)$ that must hold for there to be an ESS level of virulence in the first place. It makes no sense to ask how the ESS value is affected by the parameters if there is no ESS. Furthermore, we focus on ESS values that do not lie on the boundary of allowable trait values.

Leaving $\beta(v)$ unspecified, let us calculate the fitness gradient (12.15a) of Recipe 12.2. Differentiating the mutant growth rate (12.27) with respect to v_m and evaluating at v gives

$$\left.\frac{\partial r(v_m, v)}{\partial v_m}\right|_{v_m = v} = -1 + \frac{d + v}{\beta(v)} \left.\frac{d\beta}{dv_m}\right|_{v_m = v} \tag{12.30a}$$

According to equation (12.15c) of Recipe 12.2, the fitness gradient must be zero at the ESS value v^* and therefore we have

$$\left.\frac{d\beta}{dv_m}\right|_{v_m = v^*} = \frac{\beta(v^*)}{d + v^*}. \tag{12.30b}$$

The second derivative condition (12.15d) of Recipe 12.2 must also hold at an ESS:

$$\left.\frac{\partial^2 r}{\partial v_m^2}\right|_{\substack{v_m = v^* \\ v = v^*}} = \frac{d + v^*}{\beta(v^*)} \left.\frac{d^2\beta}{dv_m^2}\right|_{v_m = v^*} \leq 0. \tag{12.30c}$$

Finally, let us restrict our attention to ESS values that are also convergence stable. Using (12.30a) to evaluate the convergence stability condition (12.15b) of Recipe 12.2 gives

$$\begin{aligned} \frac{d}{dv}\left\{\left.\frac{\partial r(v_m, v)}{\partial v_m}\right|_{v_m = v}\right\}_{v = v^*} &= \frac{d}{dv}\left\{-1 + \frac{d + v}{\beta(v)}\left(\frac{d\beta(v)}{dv}\right)\right\}_{v = v^*} \\ &= \frac{1}{\beta} - \frac{d + v^*}{\beta^2}\left.\frac{d\beta}{dv}\right|_{v = v^*} + \frac{d + v^*}{\beta}\left.\frac{d^2\beta}{dv^2}\right|_{v = v^*} < 0. \end{aligned} \tag{12.30d}$$

From condition (12.30b) we know that $(d\beta/dv_m)|_{v_m = v^*}$ must equal $\beta/(d + v^*)$ at v^*. Furthermore, $(d\beta/dv)|_{v = v^*}$ is the same as $(d\beta/dv_m)|_{v_m = v^*}$, as we are free to use any variable name that we wish in our function. Therefore we can replace $(d\beta/dv)|_{v = v^*}$ with $\beta/(d + v^*)$ in condition (12.30d). The first two terms then cancel, leaving

$$\frac{d + v^*}{\beta(v^*)}\left.\frac{d^2\beta}{dv^2}\right|_{v = v^*} < 0. \tag{12.30e}$$

Condition (12.30e) is easier to interpret than (12.30d), even though they are mathematically equivalent. In particular, it is now easy to see that the ESS condition (12.30c) and the convergence stability condition (12.30e) are identical for this model provided that strict inequality holds in (12.30c).

Conditions (12.30b) and (12.30c) must hold for any intermediate value of virulence that is a convergence stable ESS. Because the right-hand side of equation (12.30b) is positive, the function $\beta(v)$ must therefore increase with virulence if there is to be such an ESS. Furthermore, in order for equation (12.30c) to hold, the second derivative of $\beta(v)$ must be negative. Thus, an ESS is possible only if the transmission rate $\beta(v)$ increases at a diminishing rate as virulence increases (i.e., it must be increasing and concave down). Technically, it is not necessary that $\beta(v)$ be increasing and concave down for all values of v, but only at the specific value $v = v^*$. Therefore, a more precise statement is that, if an ESS exists, it can occur only for those values of v where the function $\beta(v)$ is increasing and concave down. Both of these requirements are true for our previous choice of $\beta(v) = \tau\, v/(k + v)$.

Let us assume that $\beta(v)$ satisfies the above two restrictions. We can now determine how the ESS level of virulence changes as background host mortality rate d increases. From condition (12.15c) of Recipe 12.2 we know that the ESS virulence must satisfy the equation

$$\left.\frac{\partial r(v_m,v,d)}{\partial v_m}\right|_{\substack{v_m=v^* \\ v=v^*}} = 0, \tag{12.31}$$

where we have now explicitly indicated that the mutant growth rate r depends on the parameter d as well as on v_m and v. Because both v_m and v are evaluated at v^*, equation (12.31) is a function of only v^* and d. Therefore, to clarify the notation, we define the function $F(v,d) = \partial r(v_m,v,d)/\partial v_m|_{v_m=v}$ and let $F(v^*,d) = F(v,d)|_{v=v^*}$. Equation (12.31) can then be written as $F(v^*,d) = 0$. This equation implicitly defines the ESS virulence as a function of the parameter d. In other words, any change in the parameter d must be met with a corresponding change in the ESS level of virulence v^*, such that the equation $F(v^*,d) = 0$ continues to hold.

Our goal is to determine how the ESS changes as the parameter d increases. Mathematically, this is given by the sign of the derivative of v^* with respect to d. But we do not have an expression for v^* so how can we calculate its derivative? The key insight lies in recognizing that we can calculate this derivative indirectly, by implicitly differentiating the equation $F(v^*,d) = 0$ with respect to d, treating v^* as a function of d (Appendix 2). This gives

$$\frac{\partial F(v^*,d)}{\partial v^*}\frac{\mathrm{d}v^*}{\mathrm{d}d} + \frac{\partial F(v^*,d)}{\partial d} = 0, \tag{12.32}$$

which can be rearranged as

$$\frac{\mathrm{d}v^*}{\mathrm{d}d} = -\frac{\left(\dfrac{\partial F(v^*,d)}{\partial d}\right)}{\left(\dfrac{\partial F(v^*,d)}{\partial v^*}\right)}. \tag{12.33}$$

We emphasize that the partial derivative $\partial F(v^*,d)/\partial d$ treats the first argument v^* as a constant and the second argument d as the variable, while the partial derivative $\partial F(v^*,d)/\partial v^*$ treats v^* as the variable and d as a constant. We can now rewrite (12.33) more explicitly, using our definition of $F(v,d)$:

$$\frac{\mathrm{d}v^*}{\mathrm{d}d} = -\frac{\left(\dfrac{\partial}{\partial d}\left(\left.\dfrac{\partial r}{\partial v_m}\right|_{v_m=v}\right)\right)_{v=v^*}}{\left(\dfrac{\partial}{\partial v}\left(\left.\dfrac{\partial r}{\partial v_m}\right|_{v_m=v}\right)\right)_{v=v^*}}. \tag{12.34}$$

The denominator of equation (12.34) must be negative for the ESS of interest to be convergence stable (see equation (12.30d)). Assuming that the ESS is convergence stable, we therefore have

$$\frac{\mathrm{d}v^*}{\mathrm{d}d} \propto \left(\frac{\partial}{\partial d}\left(\left.\frac{\partial r}{\partial v_m}\right|_{v_m=v}\right)\right)_{v=v^*}. \tag{12.35}$$

Equation (12.35) can be evaluated using equation (12.30a) to find

$$\frac{\mathrm{d}v^*}{\mathrm{d}d} \propto \left(\frac{\partial}{\partial d}\left(\left.\frac{\partial r}{\partial v_m}\right|_{v_m=v}\right)\right)_{v=v^*} = \frac{1}{\beta(v^*)}\left.\frac{\mathrm{d}\beta}{\mathrm{d}v_m}\right|_{v_m=v^*}. \tag{12.36}$$

Given that $\beta(v)$ must be positive (it is a transmission rate) and must be increasing at v^* for v^* to be an ESS, the right-hand side of equation (12.36) must be positive. Thus, for this model, as long as there is a convergence-stable ESS, the ESS level of virulence will always increase as the background host mortality rate increases, no matter what the exact relationship between the transmission rate $\beta(v)$ and virulence v.

Recipe 12.3
The Effect of Parameters on the ESS

Often we can determine how parameters of a model affect ESS trait values even when the ESS is not known. Suppose that p is the parameter of interest and that r is the invasion fitness of a mutant allele in a continuous-time model. (r can be replaced by λ for a discrete-time model.)

Step 1: Determine any restrictions that must be met in order for an intermediate, convergence stable ESS to exist. This is done by evaluating conditions (12.15b), (12.15c), and (12.15d) of Recipe 12.2. Attempt to simplify conditions (12.15b) and (12.15d) by using the relationship that must hold among the functions and parameters given by condition (12.15c) (as we did in the above example).

Step 2: The trait value at a convergence stable ESS will change with a small increase in the parameter p in a direction given by the sign of

$$\frac{\mathrm{d}x^*}{\mathrm{d}p} \propto \left(\frac{\partial}{\partial p}\left(\left.\frac{\partial r}{\partial x_m}\right|_{x_m=x}\right)\right)_{x=x^*}.$$

A similar recipe can be applied to models with more than one trait of interest (Supplementary Material 12.3).

12.5 Evolutionary Invasion Analyses in Class-Structured Populations

Often we are interested in constructing evolutionary models for class-structured populations. For example, when we examined virulence evolution, we assumed that all susceptible hosts were identical. Instead, we might want to consider some form of host heterogeneity (e.g., more resistant and less resistant hosts), which would require us to model the different possible types of hosts explicitly (Gandon 2004; Ganusov et al. 2002). Similarly, if we were interested in the evolution of a trait that has different consequences when expressed by males versus females, then we would need to keep track of these two kinds of individuals explicitly. Here we shall see how the recipes presented above can be applied to such class-structured models.

12.5.1 Introductory Example of Class-Structured Evolutionary Invasion Analyses

Example: Revisiting the Evolution of Reproductive Effort

In the model of the evolution of reproductive effort in section 12.2, all offspring reach maturity within a single time step. Therefore we tracked the entire population by counting the number of mature individuals. If the time required for maturation is longer than a single time step, however, then we need to keep track of both immature and mature individuals. Here we model a plant population in which maturation takes two time steps. For simplicity suppose that the plant reproduces asexually. We considered a model of this type in Chapter 10, equation (10.1).

We use $J(t)$ and $A(t)$ to denote the numbers of juvenile and adult individuals at time t. As with model (12.1), we assume that the rate of establishment of new plants depends on the density of adults according to the Ricker model: $b(E)\, e^{-\alpha A(t)}$. Therefore, the number of juveniles in the next time step is $J(t+1) = b(E)\, e^{-\alpha A(t)} A(t)$. We model the number of adults in the next time step as $A(t+1) = p_j\, J(t) + p_a(E)\, A(t)$ where p_j and $p_a(E)$ are the probabilities of survival for juveniles and adults, respectively. As in model (12.1), we assume that the adult survival rate is a decreasing function of reproductive effort E. Juvenile survival is assumed to be independent of the reproductive effort E, because juvenile plants are not reproductively mature. The dynamics of the entire plant population is

$$\begin{aligned} J(t+1) &= b\, e^{-\alpha A(t)} A(t), \\ A(t+1) &= p_j\, J(t) + p_a\, A(t). \end{aligned} \tag{12.37}$$

The next stage is to find the equilibria of model (12.37) using the techniques of Chapter 8. We find

$$\hat{J} = 0, \hat{A} = 0 \tag{12.38a}$$

and

$$\hat{J} = \frac{1 - p_a}{\alpha\, p_j} \ln\left(\frac{b\, p_j}{1 - p_a}\right), \quad \hat{A} = \frac{1}{\alpha} \ln\left(\frac{b\, p_j}{1 - p_a}\right). \tag{12.38b}$$

We focus on equilibrium (12.38b), where the plant species is present. A linear stability analysis (Chapter 8) reveals that this equilibrium is locally stable as long as

$$\frac{1 - p_a}{p_j} < b < \frac{1 - p_a}{p_j} \exp\left[\frac{2 - p_a}{1 - p_a}\right].$$

Following Recipe 12.1, Step 2, we now obtain an expression for the growth rate of a mutant allele. To do so, we need to augment model (12.37) to allow for the presence of a mutant allele coding for a different reproductive effort. As before, we assume that there is a tradeoff between the birth rate b and adult survival p_a. The size of the mutant plant population will change through time according to a pair of equations that are identical in form to (12.37) but with reproductive effort E_m. The dynamics of the mutant and resident plant populations will also be coupled to one another through the effects of density dependence. The four recursion equations are therefore

$$\begin{aligned} J(t+1) &= b(E)\, e^{-\alpha(A(t)+A_m(t))} A(t), \\ A(t+1) &= p_j\, J(t) + p_a(E)\, A(t), \\ J_m(t+1) &= b(E_m)\, e^{-\alpha(A(t)+A_m(t))}\, A_m(t), \\ A_m(t+1) &= p_j\, J_m(t) + p_a(E_m)\, A_m(t). \end{aligned} \tag{12.39}$$

In the following we use the abbreviated notation $b_m = b(E_m)$ and $p_{a,m} = p_a(E_m)$ for mutant characteristics, and b and p_a for the corresponding resident values.

As expected, one equilibrium of model (12.39) is given by equilibrium (12.38b) along with $\hat{J}_m = 0$ and $\hat{A}_m = 0$, corresponding to the resident population at equilibrium without the mutant allele. It is the stability of this equilibrium that is of interest, because its stability will determine whether or not the mutant allele can increase when rare. Using the techniques of Chapter 8, stability of this equilibrium is determined by the eigenvalues of the local stability matrix

$$\begin{pmatrix} 0 & b\, e^{-\alpha\hat{A}}(1 - \alpha\hat{A}) & 0 & -b\,\alpha\,\hat{A}\, e^{-\alpha\hat{A}} \\ p_j & p_a & 0 & 0 \\ 0 & 0 & 0 & b_m\, e^{-\alpha\hat{A}} \\ 0 & 0 & p_j & p_{a,m} \end{pmatrix}. \tag{12.40a}$$

Matrix (12.40a) has the expected upper triangular form (see section 12.3.2). Specifically, it can be written as

$$\begin{pmatrix} \mathbf{J}_{\text{res}} & \mathbf{V} \\ \mathbf{0} & \mathbf{J}_{\text{mut}} \end{pmatrix} \tag{12.40b}$$

where $\mathbf{0}$ is a 2×2 submatrix of zeros, $\mathbf{J}_{\text{res}}$ is the 2×2 submatrix corresponding to the local stability matrix of model (12.37) (i.e., the resident population alone), **V** is the 2 × 2 submatrix containing the four elements in the upper right of matrix (12.40a), and $\mathbf{J}_{\text{mut}}$ is the 2 × 2 mutant submatrix given by

$$\mathbf{J}_{\text{mut}} = \begin{pmatrix} 0 & b_m e^{-\alpha \hat{A}} \\ p_j & p_{a,m} \end{pmatrix}. \tag{12.40c}$$

Assuming that the resident equilibrium is locally stable, the long-term reproductive factor of the mutant allele is given by the largest eigenvalue of submatrix $\mathbf{J}_{\text{mut}}$. Using equation (P2.7) from Primer 2, the two eigenvalues of $\mathbf{J}_{\text{mut}}$ are

$$\frac{1}{2}\left(p_{a,m} + \sqrt{p_{a,m}^2 + 4e^{-\alpha\hat{A}}\, b_m p_j}\right) \quad \text{and} \quad \frac{1}{2}\left(p_{a,m} - \sqrt{p_{a,m}^2 + 4e^{-\alpha\hat{A}}\, b_m p_j}\right). \tag{12.41}$$

Because $p_{a,m}$ and the term within the square root are positive, the first of these eigenvalues is always the largest in absolute value. Thus, we write the long-term reproductive factor of the mutant allele as

$$\lambda(E_m, E) = \frac{1}{2}\left(p_a(E_m) + \sqrt{p_a(E_m)^2 + 4\, e^{-\alpha\hat{A}}\, b(E_m)\, p_j}\right). \tag{12.42}$$

Moving to Step 3 of Recipe 12.1, the mutant allele increases when rare only if its reproductive factor is larger than one. This gives the invasion condition

$$\frac{1}{2}\left(p_a(E_m) + \sqrt{p_a(E_m)^2 + 4\, e^{-\alpha\hat{A}}\, b(E_m)\, p_j}\right) > 1. \tag{12.43a}$$

Condition (12.43a) can be simplified by moving all terms to the right except the square-root term:

$$\sqrt{p_a(E_m)^2 + 4\, e^{-\alpha\hat{A}}\, b(E_m)\, p_j} > 2 - p_a(E_m), \tag{12.43b}$$

and then squaring both sides:

$$p_a(E_m)^2 + 4\, e^{-\alpha\hat{A}}\, b(E_m)\, p_j > 4 - 4\, p_a(E_m) + p_a(E_m)^2. \tag{12.43c}$$

This condition can be simplified further to get

$$\frac{b(E_m)\, p_j}{1 - p_a(E_m)} > e^{\alpha \hat{A}}, \tag{12.43d}$$

which, using the equilibrium value for $\hat{A}$, becomes

$$\frac{b(E_m)\, p_j}{1 - p_a(E_m)} > \frac{b(E)\, p_j}{1 - p_a(E)}. \tag{12.44}$$

This completes Step 3. Equation (12.44) indicates that evolution maximizes the quantity $b(E)\, p_j/(1 - p_a(E))$. Thus, we can use the maximization techniques of Appendix 4 to determine the direction of evolution and any possible ESS trait values.

As with the example of reproductive effort in section 12.2, the quantity $b(E)p_j/(1 - p_a(E))$ can be interpreted as the lifetime reproductive output of an individual in the absence of density dependence. The only difference is that the lifetime reproductive output now includes the factor p_j, because an individual must first survive to adulthood before having any reproductive output.

12.5.2 Techniques for Simplifying Invasion Analyses of Class-Structured Populations

In an ideal world, Recipes 12.1 and 12.2 would lead to simple mathematical expressions that can be easily interpreted. Unfortunately, for class-structured models, this is often not the case. In particular, when there are n classes, the mutant submatrix will typically be an $n \times n$ matrix, and determining its leading eigenvalue can become difficult or even impossible. In such cases, there are two techniques that can be used to derive the fitness gradients required in Recipe 12.2, which do not require that we derive an expression for the leading eigenvalue.

We will illustrate the two techniques with model (12.39), even though we are able to derive an explicit expression (12.42) for the leading eigenvalue. The benefit of working with this example is that we can directly calculate the fitness gradient from (12.42) and compare it to the results of the new techniques. Specifically, we can calculate the derivative of λ with respect to E_m using (12.42) and then evaluate at $E_m = E$ to obtain

$$\left.\frac{\partial\lambda}{\partial E_m}\right|_{E_m=E} = \frac{1}{2}\left(\frac{\mathrm{d}p_a}{\mathrm{d}E} + \frac{p_a\frac{\mathrm{d}p_a}{\mathrm{d}E} + 2\,p_j e^{-\alpha\hat{A}}\frac{\mathrm{d}b}{\mathrm{d}E}}{\sqrt{p_a^2 + 4\,b\,e^{-\alpha\hat{A}}\,p_j}}\right)$$

$$= \frac{\left(\sqrt{p_a^2 + 4\,b\,e^{-\alpha\hat{A}}\,p_j} + p_a\right)\frac{\mathrm{d}p_a}{\mathrm{d}E} + 2\,p_j\,e^{-\alpha\hat{A}}\frac{\mathrm{d}b}{\mathrm{d}E}}{2\sqrt{p_a^2 + 4\,b\,e^{-\alpha\hat{A}}\,p_j}}. \tag{12.45a}$$

Because the denominator is positive, the sign of the fitness gradient is determined by the numerator

$$\left.\frac{\partial\lambda}{\partial E_m}\right|_{E_m=E} \propto \left(\sqrt{p_a^2 + 4\,b\,e^{-\alpha\hat{A}}\,p_j} + p_a\right)\frac{\mathrm{d}p_a}{\mathrm{d}E} + 2\,p_j\,e^{-\alpha\hat{A}}\frac{\mathrm{d}b}{\mathrm{d}E}. \tag{12.45b}$$

Plugging in the equilibrium value of $\hat{A}$ from (12.38b) and simplifying gives

$$\left.\frac{\partial\lambda}{\partial E_m}\right|_{E_m=E} \propto \frac{1}{1 - p_a}\frac{\mathrm{d}p_a}{\mathrm{d}E} + \frac{1}{b}\frac{\mathrm{d}b}{\mathrm{d}E}. \tag{12.45c}$$

Expression (12.45c) reveals that the fitness gradient for reproductive effort is composed of two terms. The first involves the derivative $\mathrm{d}p_a/\mathrm{d}E$, which is the marginal decrease in adult survival that comes from a small increase in reproductive effort. The second involves the derivative $\mathrm{d}b/\mathrm{d}E$, which is the marginal increase in reproduction that comes from a small increase in reproductive effort. Next we will derive (12.45c) using two techniques that do not require explicit knowledge of the eigenvalue.

The first technique for deriving an expression for the fitness gradient without explicitly calculating the leading eigenvalue works directly with the characteristic polynomial (see Definition P2.7) of the mutant submatrix $\mathbf{J}_{\text{mut}}$. The characteristic polynomial of matrix (12.40c) is

$$\lambda(E_m,E)^2 - \lambda(E_m,E)\, p_a(E_m) - b(E_m)\, e^{-\alpha\hat{A}} p_j = 0. \tag{12.46}$$

All eigenvalues of $\mathbf{J}_{\text{mut}}$ must satisfy (12.46), including the leading eigenvalue $\lambda(E_m,E)$.

To derive an expression for the fitness gradient, we can now differentiate (12.46) implicitly with respect to E_m and evaluate at $E_m = E$ (see Appendix 2):

$$2\,\lambda(E,E)\left.\frac{\partial\lambda}{\partial E_m}\right|_{E_m=E} - \left.\frac{\partial\lambda}{\partial E_m}\right|_{E_m=E} p_a - \lambda(E,E)\left.\frac{\mathrm{d}p_a}{\mathrm{d}E_m}\right|_{E_m=E} - \left.\frac{\mathrm{d}b}{\mathrm{d}E_m}\right|_{E_m=E} e^{-\alpha\hat{A}} p_j = 0. \tag{12.47a}$$

Equation (12.47a) can be simplified further by noting that $\lambda(E,E) = 1$ for any value of E, and that $(\mathrm{d}x/\mathrm{d}E_m)|_{E_m=E} = \mathrm{d}x/\mathrm{d}E$ where $x = p_a$ or b. Therefore,

$$\left.\frac{\partial\lambda}{\partial E_m}\right|_{E_m=E}(2 - p_a) - \frac{\mathrm{d}p_a}{\mathrm{d}E} - \frac{\mathrm{d}b}{\mathrm{d}E}e^{-\alpha\hat{A}} p_j = 0. \tag{12.47b}$$

We can then solve for $(\partial\lambda/\partial E_m)|_{E_m=E}$ to obtain

$$\left.\frac{\partial\lambda}{\partial E_m}\right|_{E_m=E} = \frac{\dfrac{\mathrm{d}p_a}{\mathrm{d}E} + \dfrac{\mathrm{d}b}{\mathrm{d}E}e^{-\alpha\hat{A}} p_j}{2 - p_a}. \tag{12.47c}$$

As the denominator is positive, equation (12.47c) shows that the fitness gradient has the sign of

$$\left.\frac{\partial\lambda}{\partial E_m}\right|_{E_m=E} \propto \frac{\mathrm{d}p_a}{\mathrm{d}E} + \frac{\mathrm{d}b}{\mathrm{d}E}e^{-\alpha\hat{A}} p_j. \tag{12.47d}$$

Finally, plugging in the equilibrium value of $\hat{A}$ from (12.38b) and rearranging gives (12.45c).

Recipe 12.4

Deriving Fitness Gradients Directly from the Characteristic Polynomial

To derive the fitness gradient required in Recipes 12.1 and 12.2 for an invasion analysis, begin by identifying the mutant submatrix $\mathbf{J}_{\text{mut}}$ in Step 2 (Recipe 12.1).

Step 1: Derive the characteristic polynomial of $\mathbf{J}_{\text{mut}}$, specifying the dependence of each parameter, as well as the leading eigenvalue, on the mutant and resident trait values x_m and x.

Step 2: Implicitly differentiate the characteristic polynomial with respect to x_m and evaluate the result at $x_m = x$.

Step 3: As mentioned in Recipe 12.2, we know that $\lambda(x,x) = 1$ for discrete-time models and $r(x,x) = 0$ for continuous-time models. Make this substitution in the result of Step 2.

Step 4: Solve the resulting equation in Step 3 for $(\partial\lambda/\partial x_m)|_{x_m=x}$ (or equivalently $(\partial r/\partial x_m)|_{x_m=x}$ in a continuous-time model) and simplify the result.

The second technique for deriving an expression for the fitness gradient involves using the reproductive value of each mutant class, much as we did in Chapter 10 to explore the effect of a trait on the long-term growth of a structured population (see Box 10.2). Here, we wish to know the effect of changing the trait value on the long-term growth of the mutant allele, as measured by the leading eigenvalue of the mutant submatrix $\mathbf{J}_{\text{mut}}$.

The matrix $\mathbf{J}_{\text{mut}}$ has elements that are functions of the resident reproductive effort E and the mutant reproductive effort E_m. Let us emphasize this fact by using the notation $\mathbf{J}_{\text{mut}}(E_m,E)$. We can then follow through the calculations of Box 10.2 to obtain the analogue of equation (10.2.2b):

$$\left.\frac{\partial\lambda}{\partial E_m}\right|_{E_m=E} = \frac{\vec{v}^{\mathrm{T}}\left(\left.\frac{\partial \mathbf{J}_{\text{mut}}}{\partial E_m}\right|_{E_m=E}\right)\vec{u}}{\vec{v}^{\mathrm{T}}\vec{u}}, \tag{12.48a}$$

where $\vec{u}$ and $\vec{v}$ are the right and left eigenvectors of the matrix $\mathbf{J}_{\text{mut}}(E,E)$ associated with the eigenvalue of one. The vector $\vec{u}$ corresponds to the relative proportions of mutants in each class, and $\vec{v}$ corresponds to the long-term reproductive value of mutants in each class. Assuming that the eigenvectors are chosen such that $\vec{v}^{\mathrm{T}}\vec{u}$ is positive, we then have

$$\left.\frac{\partial\lambda}{\partial E_m}\right|_{E_m=E} \propto \left.\vec{v}^{\mathrm{T}}\frac{\partial \mathbf{J}_{\text{mut}}}{\partial E_m}\vec{u}\right|_{E_m=E}. \tag{12.48b}$$

We now apply this method to model (12.39). Expression (12.48b) becomes

$$\left.\frac{\partial\lambda}{\partial E_m}\right|_{E_m=E} \propto \begin{pmatrix} v_j & v_a \end{pmatrix}\begin{pmatrix} 0 & e^{-\alpha\hat{A}}\, db/dE \\ 0 & dp_a/dE \end{pmatrix}\begin{pmatrix} u_j \\ u_a \end{pmatrix}. \tag{12.49a}$$

Equation (12.49a) can be expanded as

$$\left.\frac{\partial\lambda}{\partial E_m}\right|_{E_m=E} \propto u_a\left(\frac{\mathrm{d}p_a}{\mathrm{d}E}\,v_a + e^{-\alpha\hat{A}}\,\frac{\mathrm{d}b}{\mathrm{d}E}\,v_j\right), \tag{12.49b}$$

whose sign is proportional to

$$\left.\frac{\partial\lambda}{\partial E_m}\right|_{E_m=E} \propto \frac{\mathrm{d}p_a}{\mathrm{d}E}\,v_a + e^{-\alpha\hat{A}}\,\frac{\mathrm{d}b}{\mathrm{d}E}\,v_j. \tag{12.49c}$$

Finally, plugging in the equilibrium value of $\hat{A}$ from (12.38b) and simplifying gives

$$\left.\frac{\partial\lambda}{\partial E_m}\right|_{E_m=E} \propto \frac{1}{1-p_a}\,\frac{\mathrm{d}p_a}{\mathrm{d}E}\,v_a + \frac{1}{b\,p_j}\,\frac{\mathrm{d}b}{\mathrm{d}E}\,v_j. \tag{12.49d}$$

Although seemingly different, equations (12.45c) and (12.49d) are equivalent. To see this, we must calculate the reproductive values in (12.49d). These are obtained from the left eigenvector of $\mathbf{J}_{\mathrm{mut}}(E,E)$; i.e., the mutant submatrix when the mutant trait value equals that of the resident:

$$(v_j \quad v_a)\begin{pmatrix} 0 & b\,e^{-\alpha\hat{A}} \\ p_j & p_a \end{pmatrix} = \lambda(E,E)\;(v_j \quad v_a), \tag{12.50}$$

where $\lambda(E,E) = 1$. Carrying out the matrix multiplication, the first column gives the relationship $v_a\,p_j = v_j$. In other words, the reproductive value of a juvenile is a fraction p_j of the reproductive value of an adult, because juveniles might die before they first reproduce. Plugging this relationship into (12.49d), we end up with exactly expression (12.45c).

Recipe 12.5

Deriving Fitness Gradients in Terms of Reproductive Value

To derive the fitness gradient required in Recipe 12.2 for an invasion analysis, begin by identifying the mutant submatrix $\mathbf{J}_{\mathrm{mut}}$ in Step 2 of Recipe 12.1.

Step 1: Calculate $(\partial\mathbf{J}_{\mathrm{mut}}/\partial x_m)|_{x_m=x}$, where every element in $\mathbf{J}_{\mathrm{mut}}$ is differentiated with respect to x_m and the result evaluated at $x_m = x$.

Step 2: Letting the vectors $\vec{u}$ and $\vec{v}^{\mathrm{T}}$ represent the relative proportions and reproductive values of mutants in each class when the mutant trait value is identical to the resident trait value, the fitness gradient has the same sign as

$$\left.\frac{\partial\lambda}{\partial x_m}\right|_{x_m=x} \propto \vec{v}^{\mathrm{T}}\left(\left.\frac{\partial\mathbf{J}_{\mathrm{mut}}}{\partial x_m}\right|_{x_m=x}\right)\vec{u}. \tag{12.51}$$

(continued)

Recipe 12.5 ***(continued)***

Step 3: To obtain an explicit expression for the fitness gradient, determine the right eigenvector $\vec{u}$ and the left eigenvector $\vec{v}^{\mathrm{T}}$ associated with an eigenvalue of one from the matrix $\mathbf{J}_{\mathrm{mut}}\big|_{x_m=x}$ (chosen such that $\vec{v}^{\mathrm{T}}\vec{u} > 0$).

Recipes 12.4 and 12.5 are two alternative routes to calculating the fitness gradient without requiring explicit knowledge of the leading eigenvalue describing the growth of the mutant allele. In this particular example, we need not bother using these techniques because we can directly derive the required eigenvalue and its derivative. For models where calculating the eigenvalue is difficult or impossible, however, these two techniques become invaluable. Which of the two recipes should be used depends on the model at hand. One of the main benefits of Recipe 12.5 over 12.4 is that it provides an expression for the fitness gradient in terms of the reproductive values. These expressions are typically easier to interpret. For instance, expression (12.49d) reveals that the fitness gradient is composed of the marginal benefit in fecundity that comes from increased reproductive effort, $(1/(b\,p_j))\,\mathrm{d}b/\mathrm{d}E$, and the marginal cost in terms of adult mortality, $(1/(1-p_a))\,\mathrm{d}p_a/\mathrm{d}E$. Because the benefits of fecundity are realized as the production of surviving juveniles whereas the costs of mortality are realized as the loss of an adult, these terms are weighted by the relevant reproductive value: the fecundity benefit is weighted by the reproductive value of a juvenile, v_j, whereas the mortality cost is weighted by the reproductive value of an adult, v_a. The reproductive values convert the fecundity benefits and the mortality costs into a common currency so that they can be combined.

The main drawback of Recipe 12.5 over 12.4 is that the fitness gradient obtained is still an implicit function of the model parameters unless we carry out Step 3 of Recipe 12.5. This requires deriving expressions for the eigenvectors of the mutant submatrix when the mutant trait is the same as that of the resident. It might be easier to calculate the characteristic polynomial needed in Recipe 12.4 than the eigenvectors needed in Recipe 12.5, or harder, depending on the model.

Finally, for mutant alleles with a small effect, we can use the fitness gradient calculated from either Recipe 12.4 or 12.5 to approximate the invasion fitness. This is because the invasion fitness can be rewritten using a Taylor series as

$$\lambda(x_m,x) = 1 + \left.\frac{\partial \lambda}{\partial x_m}\right|_{x_m=x}(x_m - x) + O(x_m - x)^2 \quad \text{in a discrete-time model,}$$

$$r(x_m,x) = \left.\frac{\partial r}{\partial x_m}\right|_{x_m=x}(x_m - x) + O(x_m - x)^2 \quad \text{in a continuous-time model}$$

(see equation (12.10)). Furthermore, we can use the fitness gradient in Recipe 12.2 to identify trait values that are convergence stable and to identify any ESS.

12.5.3 Further Examples of Evolutionary Invasion Analyses with Population Structure

We conclude with two more examples that illustrate the breadth of the above techniques and that involve two other commonly encountered types of class structure: (i) spatial population structure, and (ii) genetic population structure.

Example: The Genetic Architecture of Adaptation near Range Boundaries

Most species have relatively well-defined geographic ranges. Often the edge of a species' range coincides with an abrupt habitat change such as the boundary between land and water. In other cases there is no obvious qualitative change in habitat. Instead there is a continuous change across the species' range in various habitat parameters such as temperature, rainfall, amount of sunlight, etc. Recent models have demonstrated that the range of a species can nevertheless end abruptly in such situations if there is a tradeoff between individual performance in different habitats (García Ramos and Kirkpatrick 1997; Holt 2003; Kirkpatrick and Barton 1997). Under these conditions, a habitat in which the species happens to be well adapted (which we will term the "central habitat") will maintain a higher density of individuals than a habitat in which it is poorly adapted (which we will term the "boundary habitat"). As a result, most of the population tends to experience the central habitat, thereby causing natural selection to reinforce the degree of adaptation to this habitat at the expense of the boundary habitat (Gomulkiewicz et al. 1999; Holt 1996a; Holt 1996b; Holt 2003; Holt and Gomulkiewicz 1997).

One way in which a diploid species might partially escape the tradeoff in performance between the two habitats and adapt to the boundary habitat is through the invasion of recessive alleles conferring higher performance in the boundary regions. Because population sizes tend to be small and sparse in the boundary habitat, inbreeding is more likely to occur in these habitats. This inbreeding will increase the frequency with which newly introduced recessive alleles are found in the homozygous state. The large population sizes of the central habitat, however, will experience less inbreeding. This means that recessive alleles will typically be found in heterozygotes in the central population and will thereby be "shielded" from selection. We will conduct an invasion analysis to determine when adaptation to boundary habitats is possible. We will then explore how an allele's ability to invade depends on how recessive it is, using Recipe 12.5.

Consider a plant species with an annual life cycle. In this example we take the trait of interest to be flowering time, and we suppose that the optimal flowering time in the central habitat is earlier than that of the boundary habitat. Thus, flowering time mediates the performance tradeoff between habitats. Plant genotypes that flower early in the season are assumed to have a high reproductive success if they are located in the central habitat but a low reproductive success if they are located in the boundary habitat. The opposite pattern is assumed to hold for genotypes that flower late in the season. We use $W_c(T)$ to denote the number of gametes (50% of which are ovules and 50% of which are pollen grains) produced by a plant with flowering time T in the central habitat. Analogously, we use $W_b(T)$ to denote this quantity in the boundary habitat. The tradeoff is incorporated by assuming $W_b(T)$ increases with flowering time T, while $W_c(T)$ decreases with T.

The life cycle in each habitat is illustrated in Figure 12.6 and is composed of the following steps: (i) gametes fuse to form new seedlings, (ii) seedlings undergo selection as they mature and become adults, (iii) the adults produce gametes, and (iv) gamete dispersal occurs. The cycle then repeats. Biologically, the only gametes that disperse in a plant population are pollen, not ovules, but we simplify the model by ignoring this fact (you might wish to consider how you could expand the model to avoid this simplification). Furthermore, we suppose that the gene controlling flowering time is on an autosome, and thus all individuals carry two alleles for flowering time. If an individual is heterozygous, carrying both a resident allele coding for flowering time T and a mutant allele coding for flowering time T_m, then its flowering time is $T_{het} = h\,T_m + (1 - h)\,T$, where h is the degree of dominance of the mutant allele.

Rather than tracking the *number* of mutant alleles as we have done in previous examples, here we track the *frequency* of mutant alleles. We start with a population fixed for a resident allele, and we then imagine introducing a mutant allele and determining the conditions under which it will spread. Unlike in previous models, however, we no longer need to specify a "resident" system in Step 2 of Recipe 12.1 because we already know that the frequency of the resident before the mutant appears is one. Consequently, this step of the invasion analysis involves less work. Ignoring the population dynamics comes with a price, however, in that we can no longer allow the fitness of different genotypes to depend on population density (as in Problems 3.13 and 3.21).

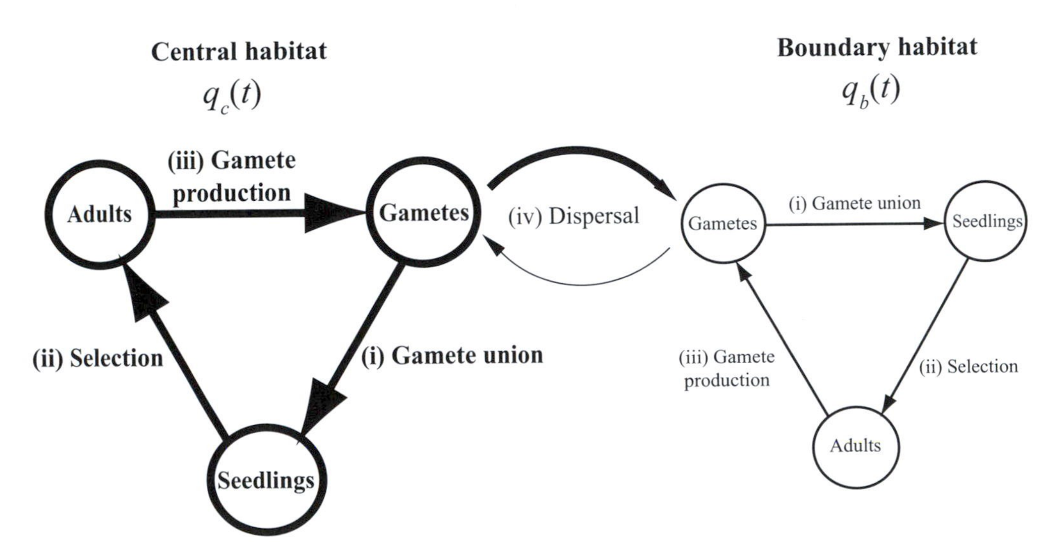

Figure 12.6: Life-cycle diagram of an annual plant species. The boundary habitat is assumed to be less productive overall, which is why its life cycle is depicted as being smaller. The life cycle in the central habitat is depicted as moving counterclockwise solely to make the illustration of dispersal more transparent. The population in each habitat is censused at the gamete stage. Roman numerals correspond to the steps in the life cycle outlined in the text.

We use $q_c(t)$ and $q_b(t)$ to denote the frequency of a mutant allele in the gamete pool in generation t for the central and boundary habitats, respectively (Figure 12.6). We now need to derive equations for the dynamics of a rare mutant allele in these two classes. We do so by first deriving a full model for the joint mutant-resident dynamics using the life-cycle diagram in Figure 12.6. (i) In both habitats, gametes first combine to form diploid seedlings. If there is inbreeding within the habitat, gametes carrying the same allele are more likely to unite than expected by random chance. Denoting the degree of inbreeding in each habitat by f_c and f_b, the frequencies of zygotes formed are (see Problem 3.18)

$$\begin{aligned} \text{frequency}(T_m,T_m) &= (1 - f_i)\, q_i^2 + f_i\, q_i, \\ \text{frequency}(T_m,T) &= (1 - f_i)\, 2q_i\, (1 - q_i), \\ \text{frequency}(T,T) &= (1 - f_i)(1 - q_i)^2 + f_i\, (1 - q_i), \end{aligned} \tag{12.52}$$

where the index i stands for either the central ($i = c$) or the boundary habitat ($i = b$). (ii) After selection, the frequencies of these seedlings become

$$\begin{aligned} \text{frequency}(T_m,T_m) &= \frac{((1 - f_i)\, q_i^2 + f_i\, q_i) W_i(T_m)}{\overline{W}_i}, \\ \text{frequency}(T_m,T) &= \frac{((1 - f_i)\, 2\, q_i\, (1 - q_i)) W_i(T_{het})}{\overline{W}_i}, \\ \text{frequency}(T,T) &= \frac{((1 - f_i)(1 - q_i)^2 + f_i\, (1 - q_i)) W_i(T)}{\overline{W}_I}, \end{aligned} \tag{12.53}$$

where the mean fitness in habitat i, $\overline{W}_i$, is given by the sum of the numerators of (12.53). (iii) Gamete production then occurs. The frequency of the mutant allele in each habitat after gamete production is

$$q_i\, G_i, \tag{12.54a}$$

where

$$G_i = \frac{((1 - f_i)\, q_i + f_i)\, W_i(T_m) + ((1 - f_i)(1 - q_i))\, W_i(T_{het})}{\overline{W}_i}. \tag{12.54b}$$

(iv) Finally, after dispersal, a fraction ρ_i of the gamete pool of habitat i is made up of immigrant gametes. Therefore, we have

$$\begin{pmatrix} q_c(t + 1) \\ q_b(t + 1) \end{pmatrix} = \begin{pmatrix} G_c\, (1 - \rho_c) & G_b\, \rho_c \\ G_c\, \rho_b & G_b\, (1 - \rho_b) \end{pmatrix} \begin{pmatrix} q_c(t) \\ q_b(t) \end{pmatrix}. \tag{12.55}$$

We are now ready to conduct a local stability analysis of model (12.55) for the equilibrium $\hat{q}_c = \hat{q}_b = 0$. Doing so, and supposing that there is no inbreeding in the central habitat ($f_c = 0$), gives the local stability matrix

$$\mathbf{J}_{\text{mut}}(T_m,T) = \begin{pmatrix} G_c(T_m,T)\, (1 - \rho_c) & G_b(T_m,T)\, \rho_c \\ G_c(T_m,T)\, \rho_b & G_b(T_m,T)\, (1 - \rho_b) \end{pmatrix}, \tag{12.56a}$$

where

$$G_c(T_m,T) = \frac{W_c(T_{het})}{W_c(T)}, \tag{12.56b}$$

$$G_b(T_m,T) = \frac{f_b\, W_b(T_m) + (1 - f_b)\, W_b(T_{het})}{W_b(T)}. \tag{12.56c}$$

The reproductive factor of the mutant allele when rare (in terms of its frequency) is given by the leading eigenvalue $\lambda(T_m,T)$ of matrix (12.56a).

It is possible to obtain an explicit expression for the leading eigenvalue and then use it in Step 3 of Recipe 12.1 to determine when mutants can spread that increase adaptation to the boundary habitat. But here we will show how Recipe 12.5 can be used instead. Specifically, we suppose that the mutant allele codes for a flowering time that is similar to that of the resident allele. Plugging the definition $T_{het} = h\, T_m + (1 - h)\, T$ into (12.56), we can use Recipe 12.5 to derive the fitness gradient. Skipping over the calculations, we get

$$\left.\frac{\partial\lambda}{\partial T_m}\right|_{T_m=T} \propto \rho_c \frac{dW_b}{dT_m}\frac{1}{W_b}(f_b + (1 - f_b)h) + \rho_b \frac{dW_c}{dT_m}\frac{1}{W_c} h. \tag{12.57}$$

The sign of expression (12.57) determines whether or not the mutant allele can invade. The factor $(dW_b/dT_m)\ 1/W_b$ represents the proportional increase in reproductive success in the boundary habitat that comes from a unit increase in flowering time, and the factor $(dW_c/dT_m)\ 1/W_c$ represents the proportional decrease in reproductive success in the central habitat that comes from such an increase. Let us suppose that there is a tradeoff such that the magnitude of these fitness effects is the same in the two habitats,

$$\frac{dW_b}{dT_m}\frac{1}{W_b} = -\frac{dW_c}{dT_m}\frac{1}{W_c}.$$

In this case, the mutant allele will invade provided that

$$\rho_c\,(f_b + (1 - f_b)\,h) - \rho_b\,h > 0. \tag{12.58a}$$

By assumption, the productivity of the central habitat is greater than that of the boundary habitat, so that immigrant alleles will typically comprise a larger fraction of the population in the boundary habitat than in the central habitat ($\rho_b > \rho_c$). In this case, (12.58a) will be satisfied only when the level of dominance is sufficiently low:

$$h < \frac{f_b}{\frac{\rho_b}{\rho_c} - (1 - f_b)}. \tag{12.58b}$$

We can draw several interesting conclusions from condition (12.58b). First, if there is no inbreeding in the boundary habitat (i.e., $f_b = 0$) then condition

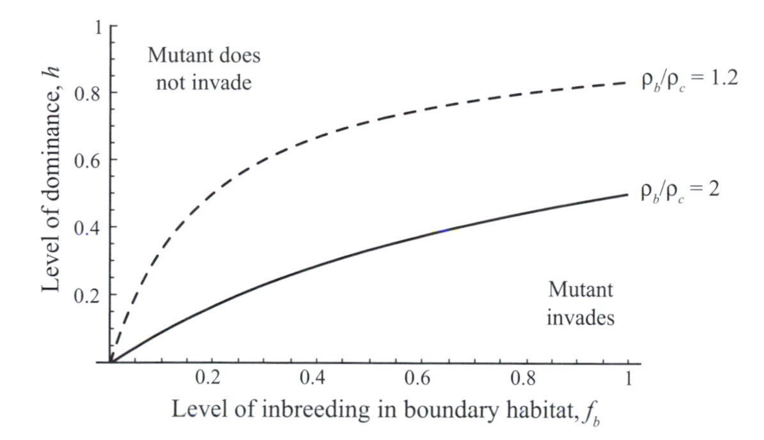

Figure 12.7: Adaptation to a boundary habitat. Mutants adapting a population to a boundary habitat can invade if their level of dominance is below the curves. Solid curve corresponds to the case where twice as many alleles are immigrant in the boundary relative to the central habitat ($\rho_b/\rho_c = 2$), while the dashed curve corresponds to a ratio of only $\rho_b/\rho_c = 1.2$. Inbreeding in the boundary habitat (*x*-axis) facilitates adaptation to that environment.

(12.58b) is never satisfied; mutant alleles conferring adaptation to the marginal habitat will never invade. This confirms the results of previous studies demonstrating that natural selection reinforces adaptation to habitats with the highest productivity (García Ramos and Kirkpatrick 1997; Holt 2003; Kirkpatrick and Barton 1997). Second, as we anticipated, if the mutant is recessive then the condition for invasion is more easily met (Figure 12.7). Finally, adaptation to the boundary habitat is more likely when the fraction of immigrants in the boundary habitat is not that much greater than in the central habitat (ρ_b/ρ_c near one). This confirms the idea that restricted gene flow aids the process of adaptation to boundary habitats.

Interestingly, condition (12.58b) also reveals that mutant alleles conferring adaptation to the boundary habitat need not be completely recessive to invade. Provided that the degree of inbreeding in the boundary habitat is large enough and ρ_b/ρ_c is small enough, additive or even partially dominant alleles can invade. Inbreeding increases the degree to which the mutant allele is expressed in the boundary habitat when it is rare. Assuming higher inbreeding levels in the boundary habitat than in the central habitat increases the efficacy of selection on the mutant allele and promotes adaptation to the boundary habitat.

Example: The Evolution of Dominance

Another important form of population structure is genetic structure, where a mutant allele can occur on different genetic backgrounds. In population genetics, models that track the spread of a mutant allele altering some trait of interest in a genetically structured model are often referred to as *modifier models* (Box 12.5). In particular, the mutant is assumed to modify the form of selection experienced by other genes, either directly (e.g., by altering gene expression or dominance levels) or indirectly by altering the array of offspring produced at the other genes (e.g., by altering the mutation rate, recombination rate, or migration rate).

A *modifier model* is a population-genetic model that tracks changes in allele frequency at a gene that alters the way that other genes experience selection.

We have already explored a modifier model in Chapter 8, when we considered the evolution of haploid and diploid life cycles. In that model, one gene was subject to mutation and selection, creating genetic structure with individuals carrying zero, one, or two copies of a deleterious allele. In addition, we considered a second gene, called a *modifier gene*, that altered the life cycle of the

Box 12.5: Optimization, Game Theory, Adaptive Dynamics, and Modifier Models

Although we refer to the techniques of this chapter as *evolutionary invasion analyses*, the same concepts have historically been used in several different contexts. We briefly mention connections among these.

Evolutionary ecologists have long used *optimality* arguments when attempting to predict the long-term outcome of evolution, arguing that traits evolve in a manner that maximizes a particular fitness measure. Unfortunately, many optimality models are analyzed without first performing an explicit evolutionary invasion analysis to prove that this fitness measure is maximized. It can be difficult to know ahead of time what function (if any) we expect evolution to maximize. Under certain assumptions we can use an evolutionary invasion analysis to prove that a quantity is maximized (Mylius and Diekmann 1995). In many contexts, however, no optimization criterion exists. One should always begin with a full invasion analysis and only proceed to using optimality techniques once it has been proven that evolution does maximize some quantity.

Game-theoretic models are based on the very same principles of evolutionary invasion analysis presented in this chapter, although many such models do not have an explicit genetic model underlying them or even an explicit population-dynamic model. The vast majority of game-theoretic models have been developed to understand the evolution of social interactions and have consequently ignored many of the genetic and ecological details of the species in question. Instead, these models focus on other quantities besides the number of individuals or genes (e.g., the amount of some desirable quantity, such as money or spare time). As this chapter has demonstrated, however, we can consider the spread of alternative strategies across a broad array of genetic and ecological contexts.

Recently, there has been interest in a type of evolutionary invasion analysis referred to as *adaptive dynamics* (Dieckmann and Law 1996). One of the main features that sets adaptive dynamics apart from the other invasion analyses discussed in the text is that it provides an explicit mathematical framework for determining the long-term course of evolution, by following a series of mutational steps. An underlying assumption of all the invasion analyses of the text is that evolution is a mutation-limited process. The population reaches an equilibrium while it contains only a single allele, and then a new mutation arises and either replaces the resident or dies out. If it replaces the resident, a new equilibrium is attained. At this stage, another mutation arises and the process repeats. We have used this stepwise process implicitly when inferring the direction of evolution. Adaptive dynamics is based on the same process, but it goes a step further by providing an explicit stochastic model of this stepwise, mutation-limited evolutionary process. In sections 15.2.3 and 15.3.2 of Chapter 15, we illustrate how to derive a stochastic model for the long-term dynamics of mutation-limited evolution.

Finally, population geneticists have used evolutionary invasion analyses stretching back to some of the early work of R. A. Fisher and Sewall Wright (see section 12.5.3), often referring to these analyses as modifier models. As with optimization, game theory, and adaptive dynamics, a modifier model is also a special case of the general evolutionary invasion analyses presented in this chapter. With modifier models, the mutant allele in question increases or decreases in frequency as a result of the way that it modifies selection acting on other genes (or modifies the rules governing the reproduction and transmission of other genes). As a result, one must typically model the dynamics of two or more genes. The life-cycle example of Chapter 8 and the dominance example of section 12.5.3 are two examples of modifier models.

organism. We imagined introducing a mutant allele that modified the probability that the organism would undergo meiosis early in life (and so have a predominantly haploid existence) or late in life (and so be predominantly diploid). Here, we consider another example with our new invasion analysis perspective in mind, one that explores the evolution of dominance.

In 1930, Fisher noted that the "pronounced tendency of the mutant gene to be recessive, to the gene of wild type from which it arises, calls for explanation." Fisher proposed an explanation in which the degree of dominance is an evolved characteristic. He supposed that there is a second gene that modifies the degree of dominance of alleles at the gene of interest and argued that modifier alleles increasing the dominance of the wild-type allele at the first gene would be favored. Here we construct a mathematical model to explore Fisher's hypothesis in more detail.

In modeling the evolution of dominance, we assume that there are two genes, one of which is at mutation-selection balance, the other of which modifies the degree of dominance of the wild-type allele at the first gene (e.g., a regulatory gene controlling expression levels). A mutant allele is then introduced at the modifier gene, and the conditions for its invasion are determined.

We have already completed Step 1 of Recipe 12.1—the decision variable is the degree of dominance of the wild-type allele. Step 2 requires us to determine the reproductive factor of a mutant allele coding for a different level of dominance. Fortunately, we have already developed the recursion equations (8.43) for a two-gene model in Chapter 8. Here, let A_1 denote the wild-type allele and A_2 the deleterious mutant at the gene under selection, **A**. We suppose that the fitness of an A_1A_1 individual is standardized to one, while the fitness of an A_2A_2 individual is $1 - s$. The fitness of an A_1A_2 heterozygote is assumed to depend on the genotype at the modifier gene **B**:

$$\begin{aligned} \text{fitness of } A_1A_2 \text{ in } B_1B_1 \text{ individuals} &= 1 - h_{11}\, s, \\ \text{fitness of } A_1A_2 \text{ in } B_1B_2 \text{ individuals} &= 1 - h_{12}\, s, \\ \text{fitness of } A_1A_2 \text{ in } B_2B_2 \text{ individuals} &= 1 - h_{22}\, s, \end{aligned}$$

where $h_{ij} = 0$ if the wild type allele is dominant. There are thus three genetically determined classes of individuals (A_1A_1, A_1A_2, and A_2A_2), and the form of selection experienced by the heterozygous class depends on the **B** gene.

Using x_1, x_2, x_3, and x_4 to denote the frequency of the gametes A_1B_1, A_1B_2, A_2B_1, and A_2B_2, respectively, then equations (8.43) describe the frequencies of each genotype after selection and reproduction. The fitnesses w_{ij} are specified by the above definitions of selection and dominance (see Table 12.1). To incorporate the effects of mutation from the wild-type allele A_1 to the deleterious allele A_2, we assume that mutation occurs during gamete formation such that a proportion μ of the gametes A_1B_1 and A_1B_2 mutate to A_2B_1 and A_2B_2, respectively, while a proportion $1 - \mu$ remain unaltered. We ignore back mutation, as the deleterious allele is assumed to be rare. Incorporating mutation, equations (8.43) become

$$x_1(t+1) = (1-\mu)\left(x_1(t)\sum_{j=1}^{4} x_j(t)\left(\frac{w_{1j} + w_{j1}}{2\overline{w}}\right) - rD^*\right)$$

TABLE 12.1:
Genotypic fitnesses in a model exploring the evolution of dominance.

Chromosome From Father

Chromosome From Mother	A_1B_1 (freq. x_1)	A_1B_2 (freq. x_2)	A_2B_1 (freq. x_3)	A_2B_2 (freq. x_4)
A_1B_1 (freq. x_1)	$w_{11} = 1$	$w_{12} = 1$	$w_{13} = 1 - h_{11}s$	$w_{14} = 1 - h_{12}s$
A_1B_2 (freq. x_2)	$w_{21} = 1$	$w_{22} = 1$	$w_{23} = 1 - h_{12}s$	$w_{24} = 1 - h_{22}s$
A_2B_1 (freq. x_3)	$w_{31} = 1 - h_{11}s$	$w_{32} = 1 - h_{12}s$	$w_{33} = 1 - s$	$w_{34} = 1 - s$
A_2B_2 (freq. x_4)	$w_{41} = 1 - h_{12}s$	$w_{42} = 1 - h_{22}s$	$w_{43} = 1 - s$	$w_{44} = 1 - s$

$$x_2(t+1) = (1-\mu)\left(x_2(t)\sum_{j=1}^{4} x_j(t)\left(\frac{w_{2j}+w_{j2}}{2\overline{w}}\right) + rD^*\right)$$

$$x_3(t+1) = x_3(t)\sum_{j=1}^{4} x_j(t)\left(\frac{w_{3j}+w_{j3}}{2\overline{w}}\right) + rD^* + \mu\left(x_1(t)\sum_{j=1}^{4} x_j(t)\left(\frac{w_{1j}+w_{j1}}{2\overline{w}}\right) - rD^*\right)$$

$$x_4(t+1) = x_4(t)\sum_{j=1}^{4} x_j(t)\left(\frac{w_{4j}+w_{j4}}{2\overline{w}}\right) - rD^* + \mu\left(x_2(t)\sum_{j=1}^{4} x_j(t)\left(\frac{w_{2j}+w_{j2}}{2\overline{w}}\right) + rD^*\right) \tag{12.59}$$

where D* is the linkage disequilibrium measured after selection,

$$D^* = x_1x_4\left(\frac{w_{14}+w_{41}}{2\overline{w}}\right) - x_2x_3\left(\frac{w_{23}+w_{32}}{2\overline{w}}\right),$$

and $\overline{w}$ is the average fitness,

$$\overline{w} = \sum_{i=1}^{4}\sum_{j=1}^{4} x_i(t)\,x_j(t)\left(\frac{w_{ij}+w_{ji}}{2}\right).$$

Model (12.59) gives the general dynamics of this two-gene model regardless of the allele frequencies, but we are primarily interested in the conditions under which a mutant allele at the modifier gene **B** can invade. To address this question, we must first analyze the stability of the resident equilibrium where the **B** gene is fixed for allele B_1 ($x_2 = x_4 = 0$). Here, we focus on a resident population that lacks dominance (i.e., $h_{11} = 1/2$). This assumption allows us to calculate the mutation-selection balance equilibrium explicitly:

$$\hat{x}_1 = 1 - \frac{2\mu}{s\,(1+\mu)}, \quad \hat{x}_3 = \frac{2\mu}{s\,(1+\mu)}. \tag{12.60}$$

In an accompanying on-line lab, we describe the steps needed to analyze a more general model where the resident modifier allele codes for any level of dominance h_{11}. The lab guides you through multiple approaches, including a direct calculation of the eigenvalue, a perturbation analysis as in section 8.5, and the method described in Recipe 12.5, giving you first-hand experience with the strengths and weaknesses of each approach.

If you calculated the local stability matrix from (12.59), you would find that it does not have the expected block-triangular form. This is because the recursions for the mutant modifier allele (x_2 and x_4) are not grouped together at the end (see section 12.3.2). Thus, we must first reorder equations (12.59) to place the dynamics for x_1 and x_3 first, followed by the dynamics for the mutant types, x_2 and x_4. Reorganizing equations (12.59) in this way, as well as changing the order in which the recursions are differentiated (with respect to x_1 in the first column, then x_3, then x_2, and finally x_4 in the last column), the stability matrix becomes

$$\begin{pmatrix} \mathbf{J}_{\text{res}} & \mathbf{V} \\ \mathbf{0} & \mathbf{J}_{\text{mut}} \end{pmatrix}$$

where $\mathbf{J}_{\text{res}}$ is the resident submatrix and $\mathbf{J}_{\text{mut}}$ is the mutant submatrix:

$$\mathbf{J}_{\text{mut}} = \begin{pmatrix} 1 + \mu(1 - 2h_{12}) - \dfrac{2\mu r\,(1 - h_{12}s)}{s} & \dfrac{r(1 - h_{12}s)(s - 2\mu + s\mu)}{s} \\ \dfrac{\mu(1 + \mu - 2h_{12}\mu)}{1 - \mu} + \dfrac{2\mu r\,(1 - h_{12}s)}{s} & 1 - \dfrac{h_{12}(s - 2\mu + s\mu)}{1 - \mu} - \dfrac{r(1 - h_{12}s)(s - 2\mu + s\mu)}{s} \end{pmatrix}. \tag{12.61}$$

With a bit of algebra (or with the help of *Mathematica*), you can check that the mutant submatrix has a leading eigenvalue of one when the modifier has no effect (i.e., when $h_{12} = 1/2$).

We will assume that modifier alleles have small effects, in which case the fitness gradient determines the direction of evolution. For higher dominance levels to evolve, $\partial\lambda/\partial h_{12}|_{h_{12}=1/2}$ must be negative because smaller values of h_{12} correspond to higher levels of dominance by the wild-type allele. We can calculate this fitness gradient directly by first deriving the leading eigenvalue of (12.61), but for the sake of variety we use Recipe 12.4.

The characteristic polynomial of matrix (12.61) is a bit cumbersome but has the form

$$\lambda^2 + a(h_{12}, s, \mu)\,\lambda + b(h_{12}, s, \mu) = 0 \tag{12.62a}$$

where a and b are functions of the three parameters h_{12}, s, and μ. We proceed to differentiate equation (12.62a) implicitly (Appendix 2) and evaluate the result at $h_{12} = 1/2$ and $\lambda = 1$. Carrying out these calculations for matrix (12.61), we obtain

$$\left(2\frac{\partial\lambda}{\partial h_{12}} + \frac{\partial a}{\partial h_{12}} + a\frac{\partial\lambda}{\partial h_{12}} + \frac{\partial b}{\partial h_{12}}\right)\bigg|_{h_{12}=1/2} = 0 \tag{12.62b}$$

Equation (12.62b) can then be solved for $\partial\lambda/\partial h_{12}$:

$$\left.\frac{\partial\lambda}{\partial h_{12}}\right|_{h_{12}=1/2} = -\frac{2(4\,r + s\,(1-2r))(s\,(1+\mu)-2\mu)\mu}{s\,(s\,(1+\mu)-2\mu+r\,(2-s)(1-\mu^2))}. \quad (12.62c)$$

Because 2μ must be less than $s\,(1+\mu)$ for the equilibrium frequencies (12.60) to lie between zero and one, the second parenthetical term in the numerator as well as the denominator of (12.62c) must be positive. Therefore, we have

$$\left.\frac{\partial\lambda}{\partial h_{12}}\right|_{h_{12}=1/2} \propto -(4\,r + s\,(1-2\,r)). \quad (12.62d)$$

Expression (12.62d) is always negative, and we conclude that modifier alleles that increase the level of dominance of the wild type ($h_{12} < 1/2$) should always invade when rare (although technically we have proven this only for modifiers of weak effect).

Interestingly, Wright (1929) used the same modeling framework to turn Fisher's explanation on its head. Although the mutant reproductive factor will be greater than one when $h_{12} < h_{11}$ (shown above for $h_{11} = 1/2$ and for general values of h in the on-line lab), the reproductive factor is greater than one by only a very small amount. For modifier alleles with a weak effect on dominance, this amount is $(\partial\lambda/\partial h_{12})\,(h_{12} - h_{11})$, where $\partial\lambda/\partial h_{12}$ is proportional to the mutation rate, which is typically very small (see equation (12.62c)). Even for modifier alleles with a large effect on dominance, the leading eigenvalue differs from one by an amount proportional to μ (see online lab). Therefore, the rate of increase of the B_2 allele predicted by the model is exceedingly slow and is easily overwhelmed by other forces (such as selection or mutation acting directly on the modifier gene or random genetic drift). This led Wright to conclude that dominance is unlikely to have evolved to mask the deleterious effects of mutations, and he favored metabolic explanations for this phenomenon (Wright 1929, 1934). This example serves to emphasize that the magnitude of the fitness gradient, as well as its sign, is biologically important. Small fitness gradients imply that the invasion of the mutant allele will occur very slowly, making it especially likely that other processes not included in the model could affect the direction of evolution.

12.6 Concluding Message

In this chapter, we have presented an overview of evolutionary invasion analysis (see also Box 12.5). From the examples included, it can be seen that there are a great many different types of questions that can be addressed using this technique. Despite the variety of models that we have examined, all of the analyses were based on analyzing the growth of mutants within a population. This growth is described by the leading eigenvalue involving the mutant in a local stability analysis. This method allowed us to predict the direction of trait evolution and to identify trait values that cannot be invaded by alternative alleles (*evolutionarily stable strategies*).

In the first examples considered, the spread of the mutant type could be described by a single equation (sections 12.2–12.4). That is, there was a single mutant type within a population of residents. In the latter examples, we demonstrated how the same type of invasion analysis can be conducted for a variety of different types of population structure, including age, stage, and size structure, as well as spatial and genetic structure (section 12.5).

Although the mathematical methods used in evolutionary invasion analyses are identical to methods that we have seen in previous chapters, evolutionary invasion analyses often involve quite complicated models (e.g., with resident and mutant alleles and, sometimes, with multiple types of individuals). Thus, this chapter has emphasized a number of helpful approximations (Recipes 12.3–12.5) that can allow progress to be made (see also the perturbation method described in section 8.4 and in the on-line lab on dominance). The techniques that work depend on the model, however, and therefore gaining experience with several examples is especially helpful. We provide more examples of evolutionary invasion analyses in the Supplementary Material 12.1.

Problems

Problem 12.1: Here you will derive condition (12.5) for a diploid plant that reproduces sexually. Label the resident allele A and the mutant allele a, and use N_{AA}, N_{Aa}, and N_{aa} to denote the number of each of the three genotypes in the population. For each genotype ij, a fraction p_{ij} survives from one time step to the next, which means that the recursions have the form $N_{ij}(t+1) = N_{ij}(t)\, p_{ij}$ + new ij offspring. (a) Suppose that $b_{ij}\, e^{-\alpha\, (N_{AA}(t)+N_{Aa}(t)+N_{aa}(t))}$ is the per capita number of surviving eggs produced by a plant of genotype ij. Derive an expression for the fraction, k, of all eggs produced that contain the A allele. (b) Assume that each plant produces sperm in proportion to its egg production. Thus, the frequency of sperm carrying allele A will also be k from part (a). Show that, if mating occurs at random among gametes, and if B denotes the total number of eggs produced, we have

$$N_{AA}(t+1) = N_{AA}(t)\, p_{AA} + k^2 B,$$
$$N_{Aa}(t+1) = N_{Aa}(t)\, p_{Aa} + 2k(1-k)\, B,$$
$$N_{aa}(t+1) = N_{aa}(t)\, p_{aa} + (1-k)^2 B,$$

where $B = (N_{AA}(t)\, b_{AA} + N_{Aa}(t)\, b_{Aa} + N_{aa}(t)\, b_{aa})e^{-\alpha(N_{AA}(t)+N_{Aa}(t)+N_{aa}(t))}$. (c) Show that if the mutant allele is absent, the population reaches the equilibrium $\hat{N}_{aa} = 0$, $\hat{N}_{Aa} = 0$, $\hat{N}_{AA} = (1/\alpha)\ln(b_{AA}/(1-p_{AA}))$. (d) Derive the local stability matrix for the above model at this equilibrium and demonstrate that, assuming $p_{aa} < 1$, the reproductive factor is again given by condition (12.4c).

Problem 12.2: Here we add inbreeding to the model for the evolution of reproductive effort in a diploid sexual species presented in Problem 12.1. In Problem 3.18 of

Chapter 3 we introduced a quantity f which we called the inbreeding coefficient, and used this to account for nonrandom union of gametes. Specifically, if k is the frequency of the A allele, the zygote frequencies with inbreeding are freq(AA) = $f\,k + (1 - f)\,k^2$, freq(Aa) = $(1 - f)\,2\,k\,(1 - k)$, freq(aa) = $f\,(1 - k) + (1 - f)(1 - k)^2$. (a) Use these altered genotypic frequencies in the model of reproductive effort from Problem 12.1 to derive the local stability matrix. (b) Identify the submatrices $\mathbf{J}_{\text{res}}$ and $\mathbf{J}_{\text{mut}}$ in your answer to (a). Show that, unless $f = 0$, the leading eigenvalue of $\mathbf{J}_{\text{mut}}$ depends on the characteristics of the homozygous mutant, which occur at an appreciable frequency because of inbreeding.

Problem 12.3: Use $K(s) = \kappa\, e^{-(s-s_0)^2/a_K}$ and $\alpha(s_m, s) = e^{-(s_m - s)^2/a_\alpha}$ to evaluate condition (12.19) explicitly, and show that it can be written as $-(s_m - s^*)^2/a_\alpha - (s^* - s_0)^2/a_K + (s_m - s_0)^2/a_K > 0$. As with the model of dispersal, we cannot write this in the form $f(s^*) > f(s_m)$, indicating that there is no function that is maximized by evolution.

Problem 12.4: (a) For the model of dispersal evolution, use equation (12.7) to demonstrate that the second derivative condition (12.13b) for an ESS is satisfied by $d^* = 1/(1 + c)$. (b) Demonstrate that the convergence stability condition (12.15b) in Recipe 12.2 is also satisfied.

Problem 12.5: Draw pairwise invasibility plots for the model of dispersal (see Box 12.4). (a) Use (12.7) to solve $\lambda(d_m, d) = 1$ for the mutant strategy, d_m. (b) Plot d_m on the vertical axis versus d on the horizontal axis in three pairwise invasibility plots, one for $c = 0$, $1/2$, and 1, respectively. (c) Shade in those regions in which mutant strategies can invade by determining whether $\lambda(d_m, d) > 1$ at various points. (d) In a few sentences, interpret what these plots say about the long-term evolution of dispersal.

Problem 12.6: Consider the model of resource competition given by equations (12.17), where the carrying capacity function K and the competition coefficient α are left unspecified. By considering (12.18), demonstrate that, if the population is currently at a resource utilization strategy s, then any mutant strategy s_m that increases the carrying capacity K can invade regardless of the functional form of K because of the assumption that $\alpha(s_m, s)$ is less than one. Because mutants increasing the carrying capacity can always invade, we expect that values of s will converge upon a point that maximizes K. On the other hand, because $\alpha(s_m, s)$ is less than one, it remains possible for mutants to spread that slightly reduce the carrying capacity. These considerations suggest that convergence to a point that may or may not be evolutionarily stable is likely to be a general feature of this resource competition model.

Problem 12.7: One hypothesis for the evolution of elaborate male traits such as the large tails of birds of paradise proposes that large tails in males serve as honest indicators of a male's genetic quality, causing the male to be more attractive to females. This hypothesis was met with considerable skepticism because some felt that there was nothing preventing a male from "cheating" and producing a large tail that falsely advertises its quality. In this problem, you will derive the condition that must be met for natural selection to favor males that honestly advertise their quality.

Let q denote a male's quality and z denote his tail size. Suppose that the growth rate of a mutant allele coding for an altered tail length z_m is given by $\lambda(z_m, z) = P(z_m, q)\, M(z_m, z)$. Here P is his probability of survival to mating (which is a decreasing function of tail size z_m and an increasing function of quality q), and M is his mating success (which is an increasing function of his tail size z_m and a decreasing function of the tail size of resident males z). (a) Using condition (12.15c) in Recipe 12.2, derive a condition that must be satisfied by an ESS tail size, z^*. (b) Suppose the functions P and M are such that any value of z^* that satisfies this condition also satisfies the second derivative condition (12.15d) and the convergence stability condition (12.15b) in Recipe 12.2. Use the techniques of Recipe 12.3 to show that, if tail length is to evolve in such a way that larger tails are associated with higher quality (i.e., $dz^*/dq > 0$), then the following condition must be met:

$$\frac{\partial}{\partial q}\left(\frac{\partial P/\partial z_m}{P}\right)_{\substack{z_m = z^* \\ z = z^*}} > 0.$$

[Hint: You will need to use the fact that neither M nor its derivatives depend on q, and you will need to use your result from (a) to substitute in for $(\partial M(z_m, z)/\partial z_m)_{z_m = z^*, z = z^*}$.] (c) Provide a biological interpretation of this result.

Problem 12.8: In the text you saw how evolutionary invasion analyses can be used to make predictions about the evolution of feeding strategies in organisms such as daphnia. To do so we derived expression (12.18) for the growth factor $\lambda(s_m, s)$ of a rare mutant daphnia with feeding strategy s_m in a population in which most individuals had feeding strategy s:

$$\lambda(s_m,s) = 1 + r - r\frac{\alpha(s_m, s)\, K(s)}{K(s_m)}.$$

(a) Using the functional forms, $K(s) = \kappa\, e^{-s^2/a}$ and $\alpha(s_m,s) = \exp[\beta^2]\exp[-(s_m - s + \beta)^2]$, derive an equation that an ESS feeding strategy must satisfy using condition (12.15c), Recipe 12.2. (b) This competition coefficient is asymmetric whereas that used in the text is symmetric. Provide a biological explanation for why the condition obtained in (a) differs from that of the text. (c) Derive an inequality that must be met for the result obtained in (a) to represent a maximum rather than a minimum of the growth factor $\lambda(s_m,s)$, and thus for the value of s^* to be an ESS.

Problem 12.9: Here we model the evolution of pathogen replication rate under the assumption that infected hosts transmit more pathogens when they have an intermediate parasite load. Suppose that the pathogen of interest has a life cycle that can be modeled in discrete generations. There is a global pool of pathogen dispersal stages, and at the beginning of each generation these pathogens colonize all available hosts. The pathogens then replicate within the hosts, and dispersal stages are released into a global pool. The life cycle then starts again. Suppose that the number of available hosts in each generation is very large and that all infections begin with n individual pathogens. Each of these pathogens then replicate within the host. To simplify matters, we assume that each pathogen produces r descendants. If all colonizing pathogens have replication strategy r, then the host will harbour a total of $P = n\, r$ pathogens after replication. We would like to determine the ESS value of r.

Suppose that a mutant parasite with replication strategy r_m appears in a host with $n - 1$ other parasites having replication strategy r. The growth factor of the mutant parasite in one generation can be written as

$$\lambda(r_m, r) = \frac{r_m}{r_m + (n - 1)r}\, T(r_m + (n - 1)r).$$

Here, $r_m/(r_m + (n-1)\,r)$ is the fraction of the pathogens within a host that are mutant among those hosts that harbor a mutant pathogen. The function $T(x)$ describes the total number of pathogen dispersal stages that are released by an infected host that carries x parasites. Assume that this function has an intermediate maximum, and let $T(x) = k\,x\,(1-x)$. (a) What is the ESS replication strategy? (b) How does the ESS replication strategy change as the number of competing pathogens, n, increases? Provide a biological explanation. (c) What is the total pathogen output of the host at the ESS? (d) How does the total pathogen output of the host at the ESS change as the number of competing pathogens, n, increases? Provide a biological explanation.

Problem 12.10: The dynamics of a pathogen that can be transmitted both vertically to offspring and horizontally can be described by the following system of differential equations:

$$\frac{dS}{dt} = b\,S - u\,S + \varepsilon\,b\,(1-v)\,I - \beta\,S\,I,$$

$$\frac{dI}{dt} = \varepsilon\,b\,v\,I + \beta\,S\,I - u\,I - \alpha\,I.$$

Here S and I are the numbers of susceptible and infected hosts, b is the per capita birth rate, u is the per capita death rate, α is the per capita disease-induced death rate, ε is the factor by which the birth rate is reduced for an infected host, v is the fraction of an infected host's offspring that are also infected, and β is the disease's horizontal transmission rate. Suppose that $b > u$. (a) What are the equilibria of the model? (b) What are the conditions under which the endemic equilibrium (with nonzero S and I) is locally stable? (c) Suppose that the condition obtained in (b) holds, and imagine introducing another pathogen strain that has a different value of β, and a different value of α. Derive an expression (involving the parameters β and α) that is maximized by an ESS parasite strain that is able to exclude all other parasite strains. (d) Suppose that there is a tradeoff between β and α such that β is given by the function $\beta = \alpha^n$ where $0 < n < 1$. What is the level of virulence α expressed by the strain that can exclude all other strains?

Problem 12.11: Consider the evolution of reproductive effort in a class-structured model where the survival probabilities as well as the birth rates decrease exponentially with the adult population size:

$$J(t+1) = b\,e^{-\alpha A(t)} A(t),$$
$$A(t+1) = p_j\,e^{-\alpha A(t)} J(t) + p_a\,e^{-\alpha A(t)} A(t).$$

(a) Show that one equilibrium of the resident population is given by

$$\hat{J} = \frac{2b \ln\left(\dfrac{p_a + \sqrt{p_a^2 + 4\,b\,p_j}}{2}\right)}{p_a + \sqrt{p_a^2 + 4bp_j}},$$

$$\hat{A} = \frac{1}{\alpha} \ln\left(\frac{p_a + \sqrt{p_a^2 + 4\,b\,p_j}}{2}\right).$$

(b) Extend the recursions to include the dynamics of a rare mutant allele, and show that the mutant submatrix is

$$\mathbf{J}_{\text{mut}} = \begin{pmatrix} 0 & b_m\, e^{-\alpha\hat{A}} \\ p_j\, e^{-\alpha\hat{A}} & p_{a,m}\, e^{-\alpha\hat{A}} \end{pmatrix}.$$

(c) Show that the leading eigenvalue of $\mathbf{J}_{\text{mut}}$, giving the long-term reproductive factor of the mutant allele, is $\lambda(E_m, E) = e^{-\alpha\hat{A}}\left(p_a(E_m) + \sqrt{p_a^2(E_m) + 4\, b(E_m)\, p_j}\right)/2$. (d) Demonstrate that evolution maximizes $\left(p_a(E) + \sqrt{p_a^2(E) + 4\, b(E)\, p_j}\right)/2$ in this model.

Problem 12.12: Consider again the model reproductive effort examined in Problem 12.11. (a) Demonstrate that the fitness gradient for reproductive effort can be written

$$\left.\frac{\partial\lambda}{\partial E_m}\right|_{E_m=E} \propto \left(\sqrt{p_a^2 + 4\, b\, p_j} + p_a\right)\frac{\mathrm{d}p_a}{\mathrm{d}E} + 2p_j\frac{\mathrm{d}b}{\mathrm{d}E}$$

using $\lambda(E_m,E)$ given in Problem 12.11c. (b) Derive the characteristic polynomial for the mutant submatrix in Problem 12.11b and then use Recipe 12.4 to obtain the fitness gradient. (c) Use the mutant submatrix in Problem 12.11b in Recipe 12.5 to show that the fitness gradient can also be written

$$\left.\frac{\partial\lambda}{\partial E_m}\right|_{E_m=E} \propto \frac{\mathrm{d}b}{\mathrm{d}E}v_j + \frac{\mathrm{d}p_a}{\mathrm{d}E}v_a.$$

Provide a biological interpretation for this expression. (d) Demonstrate that the reproductive values satisfy the relationship $v_j = \left(2p_j/\left(p_a + \sqrt{p_a^2 + 4\, b\, p_j}\right)\right)v_a$ (Step 3 of Recipe 12.5). (e) Confirm that the fitness gradients in (a)–(c) are the same.

Problem 12.13: In model (12.55) and (12.56) for adaptation near range boundaries we assumed that the level of inbreeding in the central patch was zero. (a) Derive the mutant submatrix for this model under the assumption that the level of inbreeding in the central patch is f_c. (b) How is expression (12.57) altered in this case? (c) Derive the level of dominance, h, below which a mutant allele can invade equivalent to (12.58b). As in the text, assume that the trade-off in fitness of the mutant allele is given by $(\mathrm{d}W_b/\mathrm{d}z)\ 1/W_b = -(\mathrm{d}W_c/\mathrm{d}z)\ 1/W_c$.

Problem 12.14: Consider model (12.39) for the evolution of reproductive effort in a population of juveniles and mature individuals. In the text we assumed that fecundity is an increasing function of reproductive effort, whereas adult survival was a decreasing function of reproductive effort. Now suppose that juvenile survival also decreases as the level of reproductive effort increases, perhaps because energy is shunted toward the development of larger reproductive organs as a juvenile. Use Recipe 12.5 to demonstrate that a mutant allele coding for a slightly larger reproductive effort than that currently being used will invade if

$$u_a\left(\frac{\mathrm{d}b}{\mathrm{d}E}v_j\, e^{-\alpha\hat{A}} + \frac{\mathrm{d}p_a}{\mathrm{d}E}v_a\right) + u_j\frac{\mathrm{d}p_j}{\mathrm{d}E}v_a > 0.$$

Provide a biological interpretation of this inequality.

Problem 12.15: Although we have treated mutation rates as if they were a fixed parameter, mutation rates can themselves evolve through genetic changes at genes involved in DNA replication and repair. Here we consider a haploid model with two genes: one subject to deleterious mutations (gene **A**) and a second that alters mutation rates (gene **M**). Let each gene have two alleles and let the frequency of the four types of chromosomes be freq(M_1 A_1) = x_1, freq(M_1 A_2) = x_2, freq(M_2 A_1) = x_3, freq(M_2 A_2) = x_4. Let the life cycle consist of selection in the haploid phase (with selection coefficient s against A_2), followed by mating and meiosis (with recombination rate r between the two genes), and finally mutation. We ignore back mutation and assume that mutations from A_1 to A_2 occur at rate μ_1 among haploids that carry M_1 but at rate μ_2 among haploids that carry M_2. The recursions for this model are then

$$x_1(t+1) = (1-\mu_1)\left(\frac{x_1}{\overline{w}} - r\,D_s\right),$$

$$x_2(t+1) = \mu_1\left(\frac{x_1}{\overline{w}} - r\,D_s\right) + \left(\frac{x_2(1-s)}{\overline{w}} + r\,D_s\right),$$

$$x_3(t+1) = (1-\mu_2)\left(\frac{x_3}{\overline{w}} + r\,D_s\right),$$

$$x_4(t+1) = \mu_2\left(\frac{x_3}{\overline{w}} + r\,D_s\right) + \left(\frac{x_4(1-s)}{\overline{w}} - r\,D_s\right),$$

where the mean fitness among haploids is $\overline{w} = 1 - s(x_2 + x_4)$ and the disequilibrium after selection is $D_s = (x_1x_4 - x_2x_3)(1-s)/\overline{w}^2$. (a) Before the appearance of the mutant modifier allele M_2, confirm that the mutation-selection balance equilibrium is $\hat{x}_2 = \mu_1/s$ and $\hat{x}_1 = 1 - \hat{x}_2$. (b) Derive the 2×2 mutant submatrix $\mathbf{J}_{\text{mut}}$ assuming that allele M_2 is rare. (c) Confirm that the eigenvalues of this matrix are 1 and $(1-r)(1-s)$ when mutation is absent ($\mu_1 = \mu_2 = 0$). (d) Take the Taylor series of the characteristic polynomial Det($\mathbf{J}_{\text{mut}} - \lambda\mathbf{I}$) = 0 with respect to the small parameter ζ, assuming that mutations are rare such that $\mu_1 = \tilde{\mu}_1\zeta$ and $\mu_2 = \tilde{\mu}_2\zeta$ and that the leading eigenvalue is near one, $\lambda = 1 + \tilde{\lambda}_1\zeta$. (e) Solve the resulting equation for $\tilde{\lambda}_1$ and factor to get the leading eigenvalue to linear order in ζ. (f) Use this leading eigenvalue to prove that a rare modifier allele spreads only if it decreases the mutation rate. [A computational software package such as *Mathematica* would help.]

Problem 12.16: Here, we use Recipe 12.5 to demonstrate that selection favors lower mutation rates in the model of Problem 12.15. (a) and (b) See Problem 12.15. (c) Find the left and right eigenvectors associated with the eigenvalue equal to one of the mutant submatrix evaluated at $\mu_1 = \mu_2$. (d) Calculate the matrix $(\partial\mathbf{J}_{\text{mut}}/\partial\mu_2)|_{\mu_2=\mu_1}$. (e) Evaluate and simplify

$$\left.\frac{\partial\lambda}{\partial\mu_2}\right|_{\mu_2=\mu_1} = \left.\frac{\vec{v}^{\mathrm{T}}\dfrac{\partial\mathbf{J}_{\text{mut}}}{\partial\mu_2}\vec{u}}{\vec{v}^{\mathrm{T}}\vec{u}}\right|_{\mu_2=\mu_1}$$

(f) Using the fact that $\lambda \cong 1 + (\partial\lambda/\partial\mu_2)|_{\mu_2=\mu_1} (\mu_2 - \mu_1)$, use your results from (e) to show that the rare modifier allele M_2 spreads ($\lambda > 1$) if it reduces the mutation rate. This analysis assumes that the new allele M_2 causes only a small difference in the trait ($\mu_2 - \mu_1$ small), while the analysis in Problem 12.15 assumes that the mutation rates are small.

Problem 12.17: Consider the epidemiological model (12.23). This model assumed that there was a single type of host in the population, but often hosts differ in their response to infection. Suppose that there are two host types that, when infected, differ in their transmission rates β

$$\frac{dS_1}{dt} = \theta - d\,S_1 - \beta_1(v)\,I_1\,S_1 - \beta_2(v)\,I_2\,S_1,$$

$$\frac{dS_2}{dt} = \theta - d\,S_2 - \beta_1(v)\,I_1\,S_2 - \beta_2(v)\,I_2\,S_2,$$

$$\frac{dI_1}{dt} = \beta_1(v)\,I_1\,S_1 + \beta_2(v)\,I_2\,S_1 - (d + v)\,I_1,$$

$$\frac{dI_2}{dt} = \beta_1(v)\,I_1\,S_2 + \beta_2(v)\,I_2\,S_2 - (d + v)\,I_2,$$

where S_i and I_i are the numbers of susceptible and infected hosts of type i, and $\beta_i(v)$ are the transmission rates, written to make it clear that transmission rates depend on virulence as in model (12.23). (a) Provide a biological explanation for each of the terms in the above model. (b) What are the equilibria of the above model? (c) Suppose that the resident strain has reached a stable equilibrium. What is the 4×4 Jacobian matrix $\mathbf{J}_{\text{res}}$ that determines the stability of the equilibrium where the resident strain is present and the mutant strain is absent? (d) If a mutant pathogen strain appears with a different level of virulence, what is the 2×2 mutant submatrix $\mathbf{J}_{\text{mut}}$ that governs the ability of the mutant strain to invade? (e) Use the trace and determinant conditions (Rule P2.11) to obtain an inequality that must be satisfied for the mutant strain to invade. (f) Rewrite this inequality in the form of $f(v_m) > f(v)$ to identify a function that is maximized over evolutionary time. Compare the result to equation (12.28b).

Problem 12.18: The majority of populations that harbor pathogens are also subject to predation. Here we will incorporate a predator into model (12.25) for the evolution of virulence. Assume that the predator consumes prey at a rate that is proportional to the numbers of predators and prey (a mass-action assumption; see Chapter 3). Our model is then

$$\frac{dS}{dt} = \theta - d\,S - \beta(v)\,S\,I - \beta(v_m)\,S\,I_m - a\,S\,P,$$

$$\frac{dI}{dt} = \beta(v)\,S\,I - (d + v)\,I - a\,I\,P,$$

$$\frac{dP}{dt} = a\,S\,P + a\,I\,P + a\,I_m\,P - \delta\,P,$$

$$\frac{dI_m}{dt} = \beta(v_m)\,S\,I_m - (d + v_m)\,I_m - a\,I_m\,P,$$

where a is the attack rate of predator on the host population (assumed to be the same for both susceptible and infected hosts), and δ is the per capita mortality rate of the predator. (a) Before the mutant allele appears ($I_m = 0$), what is the equilibrium where both pathogen and predator are present? (b) Derive the local stability matrix for the augmented model and evaluate it at the equilibrium in part (a). It should be block triangular. (c) Show that the growth rate of the mutant pathogen with the altered level of virulence is

$$r(v_m, v) = v - v_m + \frac{a\,\theta\,(\beta(v_m) - \beta(v))}{\beta(v)\,\delta - a\,v}.$$

Unlike model (12.25), there is no longer a quantity that is maximized by evolution (i.e., there is no function, f, allowing us to cannot rewrite $r(v_m,v) > 0$ as $f(v_m) > f(v)$).

Problem 12.19: Here you will apply Recipe 12.3 to the model of virulence evolution with predation in Problem 12.18. The goal is to determine how background mortality rate affects the ESS level of virulence when predation is occurring. Again we leave the function $\beta(v)$ unspecified. (a) Use $r(v_m,v)$ given in Problem 12.18c to show that, if v^* is an ESS level of virulence, then $(\mathrm{d}\beta(v_m)/\mathrm{d}v_m)_{v_m = v^*} = (\beta(v^*)\,\delta - a\,v^*)/(a\,\theta)$. Use the fact that $\hat{S} = a\,\theta/(\beta(v)\,\delta - a\,v)$ at the endemic equilibrium to infer what the sign of this derivative must be if an ESS is to exist. (b) Show that $(\mathrm{d}^2\beta(v)/\mathrm{d}v^2)_{v = v^*}\, a\,\theta/(\beta(v^*)\,\delta - a\,v^*) < 0$ must hold as well for v^* to be an ESS. Use the fact that $\hat{S} = a\,\theta/(\beta(v)\,\delta - a\,v)$ at the endemic equilibrium to infer a second necessary property of $\beta(v)$ if there is to exist an ESS. (c) Show that if v^* is also to satisfy the convergence stability condition (12.15b) then the following must hold:

$$\frac{a\,\theta}{\beta(v^*)\,\delta - a\,v^*}\left(\frac{\mathrm{d}^2\beta(v)}{\mathrm{d}v^2}\right)_{v=v^*} - \frac{a\,\theta\left(\delta\left(\frac{\mathrm{d}\beta(v)}{\mathrm{d}v}\right)_{v=v^*} - a\right)}{(\beta(v^*)\,\delta - a\,v^*)^2}\left(\frac{\mathrm{d}\beta(v)}{\mathrm{d}v}\right)_{v=v^*} < 0.$$

(d) Simplify the above expression for the convergence stability condition as suggested in Step 1 of Recipe 12.3 to get

$$\left(\frac{\mathrm{d}^2\beta(v)}{\mathrm{d}v^2}\right)_{v=v^*}\frac{a\,\theta}{\beta(v^*)\,\delta - a\,v^*} < \frac{\delta}{a\,\theta} - \frac{a}{\beta(v^*)\,\delta - a\,v^*}.$$

(e) Provide a logical argument that if v^* satisfies the condition in part (b), then it will also satisfy this convergence stability condition. (f) Use Step 2 in Recipe 12.3 to show that the ESS value of virulence is no longer affected by host background mortality rate, unlike the model (12.25) without predation.

Further Reading

For further information on evolutionary invasion analysis, see

- Dugatkin, L. A. and H. K. Reeve (eds.). 1998. *Game Theory and Animal Behavior*. Oxford University Press, Oxford.
- Earn, D. J. D. and R. A. Johnstone. 2008. *Game Theory for Biologists*. Princeton University Press, Princeton, N.J.

- Maynard Smith, J. 1982. *Evolution and the Theory of Games*. Cambridge University Press, Cambridge.
- Metz, J. A. J., R. M. Nisbet, and S.A.H. Geritz. 1992. How should we define "fitness" for general ecological scenarios? *Trends Ecol. Evol.* 7:198–202.
- Metz, J. A. J., S. D. Mylius, and O. Diekmann. 1996. When does evolution optimise? On the relation between types of density dependence and evolutionarily stable life history parameters. Working Paper, IIASA http://www.iiasa.ac.at/Publications/ Documents/ WP-96-004.pdf

References

Abrams, P. A., H. Matsuda, and Y. Harada. 1993. Evolutionarily unstable fitness maxima and stable fitness minima of continuous traits. *Evol. Ecol.* 7:465–487.

Anderson, R. M., and R. M. May. 1982. Coevolution of hosts and parasites. *Parasitology* 85:411–426.

Christiansen, F. B. 1991. On conditions for evolutionary stability for a continuously varying character. *Am. Nat.* 138:37–50.

Dieckmann, U., and R. Law. 1996. The dynamical theory of coevolution: a derivation from stochastic ecological processes. *J. Math. Biol.* 34:579–612.

Dieckmann, U., and M. Doebeli. 1999. On the origin of species by sympatric speciation. *Nature* 400:354–357.

Eshel, I. 1983. Evolutionary and continuous stability. *J. Theor. Biol.* 103:99–111.

Eshel, I., and U. Motro. 1981. Kin selection and strong evolutionary stability of mutual help. *Theor. Popul. Biol.* 19:420–433.

Fisher, R. A. 1930. *The Genetical Theory of Natural Selection*. Clarendon Press, Oxford.

Gandon, S. 2004. Evolution of multihost parasites. *Evolution* 58:455–469.

Ganusov, V. V., C. T. Bergstrom, and R. Antia. 2002. Within-host population dynamics and the evolution of microparasites in a heterogeneous host population. *Evolution* 56:213–223.

García Ramos, G., and M. Kirkpatrick. 1997. Genetic models of adaptation and gene flow in peripheral populations. *Evolution* 51:21–28.

Geritz, S.A.H., É. Kisdi, G. Meszéna, and J. A. J. Metz. 1998. Evolutionarily singular strategies and the adaptive growth and branching of the evolutionary tree. *Evol. Ecol.* 12:35–57.

Gomulkiewicz, R., R. D. Holt, and M. Barfield. 1999. The effects of density dependence and immigration on local adaptation and niche evolution in a black-hole sink environment. *Theor. Popul. Biol.* 55:283–296.

Hamilton, W. D., and R. M. May. 1977. Dispersal in stable habitats. *Nature* 269:578–581.

Holt, R. D. 1996a. Adaptive evolution in source-sink environments: direct and indirect effects of density-dependence on niche evolution. *Oikos* 75:182–192.

Holt, R. D. 1996b. Demographic constraints in evolution: towards unifying the evolutionary theories of senescence and niche conservatism. *Evol. Ecol.* 10:1–11.

Holt, R. D. 2003. On the evolutionary ecology of species' ranges. *Evol. Ecol. Research* 5:159–178.

Holt, R. D., and R. Gomulkiewicz. 1997. How does immigration influence local adaptation? A reexamination of a familiar paradigm. *Am. Nat.* 149:563–572.

Kirkpatrick, M., and N. H. Barton. 1997. Evolution of a species' range. *Am. Nat.* 150:1–23.

Mylius, S. D., and O. Diekmann. 1995. On evolutionarily stable life histories, optimization and the need to be specific about density dependence. *Oikos* 74:218–224.

Nowak, M., and K. Sigmund. 1989. Oscillations in the evolution of reciprocity. *J. Theor. Biol.* 137:21–26.

Taylor, P. D. 1989. Evolutionary stability in one-parameter models under weak selection. *Theoretical Population Biology* 36:125–143.

Wright, S. 1929. Fisher's theory of dominance. *Am. Nat.* 274–279.

Wright, S. 1934. Physiological and evolutionary theories of dominance. *Am. Nat.* 24–53.

PRIMER 3
Probability Theory

Probability theory, or the mathematical description of chance events, is arguably one of the areas of mathematics that has provided the most insight into biology. This primer serves as a basic introduction to probability theory, providing the background material necessary for Chapters 13–15.

P3.1 An Introduction to Probability

Before introducing the concept of probability, we must introduce the concept of a *trial*. A trial can be any sort of occurrence, like the birth of a child or the flowering of a plant. We are interested in trials that can have more than one possible outcome. For example, a baby might be a boy or a girl. A plant might produce any number of seeds from zero to thousands. Because more than one outcome is possible, we can consider the outcome to be a variable, specifically a *random variable*, which we denote by a capital letter (e.g., X). Once a trial has happened, the random variable takes on a specific value. For example, if a boy is born we could write X = "boy" if we wanted to describe the outcome in words, or we could write $X = 0$ if we were counting the number of girls.

Before the trial actually occurs, we can consider quantifying the chance or *probability* that a particular outcome will be realized. We denote a particular outcome of a random variable by a lower-case letter (e.g., x). Some outcomes will have a high chance of occurring and others will be extremely unlikely. We can write the probability that the random variable X takes on the value x as $P(X = x)$ or just $P(x)$. There are two ways to think about probabilities:

- Frequency interpretation: A probability is understood as the frequency of a particular outcome across the course of many trials.
- Subjective interpretation: A probability is understood as a subjective belief or opinion of the chance that a particular outcome will be realized.

For example, when a baby rhinoceros is born, you might think that there is roughly a 50% probability that the baby is male, $P(X = \text{"male"}) = 0.5$. This opinion might be based on previous observations that about half of all mammals are male; this would be a frequentist's perspective. Alternatively, your opinion might be based on the idea that sex chromosomes should segregate 50:50 (i.e., half of all sperm should bear an X chromosome and half a Y chromosome); this would be a subjectivist's perspective. Most of us think about probabilities in both ways, depending on the situation.

Venn Diagram	Set Language	Set Notation
(a)	 The set	Ω
(b)	 Subset A (or event A)	A
(c)	 Complement of A	A^C
(d)	 Intersection of A and B	$A \cap B$ or AB
(e)	 Union of A and B	$A \cup B$
(f)	 A and B are disjoint (mutually exclusive)	$AB = \emptyset$
(g)	 A is a subset of B	$A \subset B$

Figure P3.1: Venn diagrams. The probability of an outcome can be represented as an area (shaded grey) within a Venn diagram, which is a square of area one. (a) If an outcome is certain to occur (with probability one), the entire square is shaded. (b) If outcome A has a probability less than one of occurring (say, p), a fraction p of the square is shaded. (c) The complement of A represents any outcome other than A, so that a fraction $1 - p$ of the square is shaded. (d) The probability that two outcomes, A and B, both occur is given by the area of their intersection. (e) The probability that A or B or both occurs is represented by the total area inside the shaded regions (their intersection must be counted only once). (f) If two outcomes A and B have no intersection, then they are mutually exclusive (they cannot both occur). (g) An outcome A is a subset of B if its area is entirely encompassed within the area of B, in this case B will always be observed when A occurs.

Because a probability represents the chance that a trial has a particular outcome, any probability must lie between 0 and 1 (or, equivalently, between 0% and 100%). An aid to visualizing probabilities is a "Venn diagram" (Figure P3.1), which is a square whose area is one. The area of the whole square (one) represents the probability that the trial has any outcome (including, potentially, that nothing happens). We can subdivide the square into subsets, where each subset represents a potential outcome and the area of the subset represents the probability of observing that outcome. For example, if your sister is pregnant, you might think that there is a 1/7 chance that the baby will be born on any particular day of the week, e.g., $P(X =$ "Monday"$) = 0.14$. In this example, the potential outcomes (days of the week) are "disjoint" or "mutually exclusive," meaning that only one of the alternative outcomes is possible—the baby cannot be born both on a Monday and on a Tuesday. For trials with mutually exclusive outcomes, the Venn diagram can be partitioned into nonoverlapping subsets, and the following rule applies:

Rule P3.1: Probabilities of Mutually Exclusive Outcomes
If a trial can result in only one of a set of possible outcomes, the outcomes are said to be "mutually exclusive." The probabilities of mutually exclusive outcomes sum to one:

$$\sum_{i=1}^{\#\text{ of outcomes}} P(X = x_i) = 1.$$

As a special case of Rule P3.1, one can always partition the outcomes of a trial into one outcome of interest, A, and its complement, A^C. The complement represents "not A," and the probability of observing the complement is the probability of not observing A. For example, the baby might be born on a Monday (A) with probability 1/7 or on any other day of the week (A^C) with probability 6/7.

Rule P3.2: Complement Rule
The probability of an outcome plus the probability of its complement sum to one:

$$P(X = A) + P(X = A^C) = 1.$$

This rule is easy to visualize using a Venn diagram (Figure P3.1c). Rule P3.2 is extremely handy, because it is sometimes easier to calculate the probability of the complement of an outcome of interest. For example, if you are monitoring the populations of lizards on five islands and you want to know the probability that one, two, three, four, or five of the populations goes extinct over the course of a year, then the easiest way to calculate $P(X =$ "one or more extinctions") is to calculate the complement $P(X =$ "no extinctions"). Rule P3.2 then tells us that $P(X =$ "one or more extinctions") equals one minus $P(X =$ "no extinctions").

The outcomes of a trial need not be mutually exclusive. For example, if you are observing interactions between two fish in a five-minute interval, you might observe no contact, aggressive contact, mating, or avoidance behavior, but you might very well see more than one of these outcomes in the same period (e.g., aggression and mating). If two outcomes are not mutually exclusive, then there is some probability that both will be observed. On a Venn diagram, the intersection of the two outcomes represents this probability. We can write the probability that both A and B occur using $P(X = A \cap B)$ where $\cap$ is called the intersection and represents "and" (see Figure P3.1d). Following convention, we can drop the "$X =$ " and the "$\cap$" in such probability statements and write $P(X = A \cap B)$ as $P(A\ B)$. $P(A\ B)$ is read as "the probability that the random variable X has both the outcome A *and* the outcome B."

The outcomes A and B are said to be *independent* if the probability of observing both, $P(A\ B)$, equals the product of each outcome's probability, $P(A)\ P(B)$. When outcomes A and B are independent, observing A provides no information about whether or not B will be observed. For example, imagine throwing two dice—as long as you don't have any tricks up your sleeve, the number showing on the first die will have no influence on the number showing on the second die; they will be independent events. In biology, independence is often assumed for trials involving different individuals who are separated in time and space and who have had no contact. For example, the day of the month in which a woman in Vancouver and a woman in New York start menstruating might reasonably be independent of one another, but this is not true for women living in close proximity (Preti et al. 1986). Mutually exclusive outcomes are never independent because their intersection $P(A\ B)$ is zero and not $P(A)\ P(B)$; for example, a single die thrown cannot show both a "two" and a "five" as these are mutually exclusive.

Often we are interested in knowing the probability of outcome A or B or both (that is, A and/or B). In a Venn diagram, this probability is represented by the total area of the subsets A and B. We write this probability as $P(A \cup B)$, where $\cup$ is called the "union" and represents "and/or" (see Figure P3.1e). For example, we might be interested in the probability that a forest patch is decimated by fire or disease or both. To calculate the union of two subsets, we could add together the two subsets, but then we would be counting their intersection twice. Thus, to find the union, we must subtract the intersection from the sum of the subsets.

Rule P3.3: Inclusion-Exclusion Rule

The probability of outcome A or B or both is the sum of each outcome's probability minus the probability that both occur:

$$P(A \cup B) = P(A) + P(B) - P(AB).$$

CAUTION: It can be tempting to interpret $\cup$ as "and" whereas $\cup$ really represents "and/or." It can help to remember that $\cup$ represents the total area ("union") in a Venn diagram. Instead, it is $\cap$ that represents "and," where $\cap$ specifies the area in a Venn diagram within which both A and B occur (their intersection).

Because the intersection is known for two independent outcomes ($P(A)\ P(B)$) and for two mutually exclusive outcomes (zero), we can calculate the probability of A and B as well as the probability of A or B or both:

Rule P3.4: Independent Outcomes

(a) If outcomes A and B are independent, the probability of observing both outcomes is the product of observing each separately:

$$P(A \cap B) = P(A\ B) = P(A)\ P(B).$$

(b) Using Rule P3.3, the probability of A or B (or both) is then

$$P(A \cup B) = P(A) + P(B) - P(A)\ P(B).$$

Rule P3.5: Mutually Exclusive Outcomes

(a) If outcomes A and B are mutually exclusive, the probability of observing both outcomes is zero:

$$P(A \cap B) = P(A\ B) = 0.$$

(b) Using Rule P3.3, the probability of A or B (or both) is then

$$P(A \cup B) = P(A) + P(B).$$

These rules are fairly intuitive, in part because we have experience with games involving these probability calculations. For example, if you take a randomly shuffled deck of 52 cards, the probability that the first card you turn over is the queen of spades is 1/52. Because 1/52 is the product of the probability of observing a queen, P("queen") = (1/13), and the probability of observing a spade, P("spade") = (1/4), these two outcomes are independent. Thus, we can calculate the probability that a queen or a spade (or both) shows up from Rule P3.4 as P("queen" ∪ "spades") = P("queen") + P("spades") − P("queen" ∩ "spades") = (1/13) + (1/4) − (1/52), which equals 16/52. We can get the same answer by counting the number of queens (4) and the number of spades that are not queens (12) out of 52 cards (=16/52), where this calculation avoids counting the intersection (the queen of spades) twice. As another example, the probability of getting a red card is 1/2, which equals the probability of getting a heart (1/4) plus the probability of getting a diamond (1/4). In this case, we don't have to subtract off the intersection, because "heart" and "diamond" are mutually exclusive outcomes (Rule P3.5).

Exercise P3.1: For each question, write the answer as P(insert appropriate description) = solution, and state any assumptions that you make.

(a) In a forest, imagine that 1% of trees are infected by fungal rot and 0.1% have owl nests. What is the probability that a tree has both fungal rot and an owl nest if the two are independent? If the two are mutually exclusive?

(b) Individuals of blood type O that are Rhesus negative are universal donors. If 46% of individuals have blood type O, if 16% of individuals are Rhesus negative, and if the two blood types are independent of one another, what is the probability that a randomly chosen individual is a universal donor (O−)?

(c) In a population, 46% of individuals have blood type O, 40% have blood type A, 10% have blood type B, and 4% have blood type AB. An individual with blood type A can receive transfusions from people with blood type O or A. What is the probability that a donor has the appropriate blood type for a patient of blood type A?

(d) Two independent studies are performed to test the same null hypothesis. What is the probability that one or both of the studies obtains a

(continued)

Exercise P3.1 (*continued*)

significant result and rejects the null hypothesis even if the null hypothesis is true? Assume that, in each study, there is a 0.05 probability of rejecting the null hypothesis.

Answers to the exercises are provided at the end of the primer.

P3.2 Conditional Probabilities and Bayes' Theorem

Unless two outcomes are independent, the probability of observing one outcome depends on whether the other outcome is observed. *Conditional probabilities* describe the relationship between outcomes.

Rule P3.6: Conditional Probability

Given that outcome B has occurred, we write the probability of observing outcome A as $P(A \mid B)$. The "|" can be read as "given that" or "conditional upon." By definition, $P(A \mid B)$ equals

$$P(A \mid B) = \frac{P(A\,B)}{P(B)}.$$

That is, the probability of observing A given that B has occurred, $P(A \mid B)$, is the fraction of cases in which B occurs, $P(B)$, that A also occurs, $P(A\,B)$.

For independent outcomes, $P(A \mid B) = P(A)$, because observing B provides no information about whether or not A has occurred.

For mutually exclusive outcomes, $P(A \mid B) = 0$, because observing B implies that A has not occurred.

Conditional probabilities can make it easier to determine the probability that two outcomes are both observed. The probability of both A and B occurring, $P(A\,B)$, is the probability of observing B times the probability that, among those cases in which B occurs, A occurs:

$$P(A\,B) = P(B)\,P(A \mid B). \qquad \text{(P3.1a)}$$

Rearranging (P3.1a) we get the definition for $P(A \mid B)$ given in Rule P3.6. Of course, the same reasoning allows us to write this joint probability as

$$P(A\,B) = P(A)\,P(B \mid A), \qquad \text{(P3.1b)}$$

which is the probability of observing A times the probability of observing B given that A has occurred.

These formulae look simple enough but they are extremely powerful. They immediately lead to one of the most important theorems in probability:

Rule P3.7: Bayes' Theorem

Because the joint probability of observing two outcomes, $P(A\ B)$, equals both $P(B)\ P(A \mid B)$ and $P(A)\ P(B \mid A)$, we can determine one conditional probability from the other using Bayes' theorem:

$$P(B \mid A) = \frac{P(A \mid B)\ P(B)}{P(A)}.$$

As an example, suppose we want to calculate the probability that a person will die of lung cancer given that they smoke. We could study a cohort of individuals, determining which ones smoke and which ones don't and tracking them until they died. At that point we could calculate the fraction of smokers who died of lung cancer. In this example, we are trying to calculate the conditional probability $P(\text{death due to lung cancer} \mid \text{smoker})$. Using Bayes' rule, however, there is an alternative way to calculate this probability:

$$P(\text{death due to lung cancer} \mid \text{smoker}) = \frac{P(\text{smoker} \mid \text{death due to lung cancer})\ P\ (\text{death due to lung cancer})}{P(\text{smoker})}.$$

The probabilities on the right-hand side have already been estimated (Shopland 1995), allowing us to estimate the risk that a smoker dies of lung cancer without the above-mentioned study. $P(\text{smoker} \mid \text{death due to lung cancer})$ is estimated as the fraction of people that have died of lung cancer who are smokers. $P(\text{death due to lung cancer})$ is estimated from death records, and $P(\text{smoker})$ is estimated by polling an appropriate control population (a population similar in age drawn from similar environments). Using the data in Shopland (1995), $P(\text{smoker} \mid \text{death due to lung cancer}) = 0.9$, $P(\text{death due to lung cancer}) = 0.3$, and $P(\text{smoker}) = 0.5$, the probability that a smoker will die of lung cancer is estimated as $(0.9)(0.3)/(0.5) = 0.54$. Similar calculations for nonsmokers give a probability of death of only 0.06 (Exercise P3.2c). Thus, smokers have a nearly tenfold higher risk of dying of lung cancer compared to nonsmokers.

Bayes' theorem is widely used in scientific inference, using a methodology known as Bayesian analysis (see Hilborn and Mangel 1997). As described in Supplementary Material P3.1, Bayesian analysis allows scientists to infer aspects of the biological world that are hard to measure directly.

Exercise P3.2:

(a) If the probability of having green eyes is 10%, the probability of having brown hair is 75%, and the probability of having both green eyes and brown hair is 9%, what is the probability of having brown hair given that you have green eyes?

(*continued*)

Exercise P3.2 *(continued)*

(b) Ability to taste phenylthiocarbamide (PTC) is thought to be determined by a single dominant gene with incomplete penetrance. Among North American Caucasians, there is a 70% chance of being able to taste PTC [P(taster) = 0.7]. If everybody who tastes PTC is a carrier [P(carrier | taster) = 1] and if 80% of the population carries the gene [P(carrier) = 0.8], what is the penetrance of the gene? That is, what is the probability of tasting PTC if you are a carrier, P(taster | carrier)?

(c) Write a formula for the risk of dying of lung cancer given that a person does not smoke in terms of P(smoker | death due to lung cancer), P(death due to lung cancer), and P(smoker). Estimate the risk of death due to lung cancer among nonsmokers using P(smoker | death due to lung cancer) = 0.9, P(death due to lung cancer) = 0.3, and P(smoker) = 0.5.

We can also use conditional probabilities to calculate the overall probability of outcome A, $P(A)$, when A occurs in the context of a set of mutually exclusive outcomes, B_i, of a second random variable:

Rule P3.8: Law of Total Probability

Suppose that the outcomes, B_i, consist of n mutually exclusive events whose probabilities sum to one (i.e., $\sum_{i=1}^{n} P(B_i) = 1$). Then the probability of A is equal to the sum of the probabilities of A given each outcome B_i, weighted by the probability of each outcome B_i occurring:

$$P(A) = P(A \mid B_1)\, P(B_1) + P(A \mid B_2)\, P(B_2) + \cdots + P(A \mid B_n)\, P(B_n).$$

For example, Rule P3.8 can be used to calculate the overall probability of a randomly chosen individual contracting the flu, $P(A)$, when some individuals have had a flu shot (B_1) and others have not (B_2). If vaccinated individuals have a probability of infection of $P(A \mid B_1) = 0.01$ and nonvaccinated individuals have a probability of infection of $P(A \mid B_2) = 0.2$, then the overall probability of contracting the flu is $P(A) = 0.01\, P(B_1) + 0.2\, P(B_2)$ according to Rule P3.8. Thus, if 90% of the population were vaccinated ($P(B_1) = 0.9$, $P(B_2) = 0.1$), the probability of a randomly chosen individual contracting the flu would be 0.029.

An important concept that we will see repeatedly in this primer is the "expected value" or mean value of a random variable X. The expected value, denoted $E[X]$, can be thought of as the average outcome that would be observed if the trial were repeated infinitely many times (see more precise Definitions P3.2 and P3.9 below). There is a useful formula for calculating the expected value of a random variable that we present here because of its analogy to the law of total probability.

Rule P3.9: Law of Total Expectation

Suppose that the outcomes, B_i, of a second random variable consist of n mutually exclusive events whose probabilities sum to one (i.e., $\sum_{i=1}^{n} P(B_i) = 1$). Then the expectation of a random variable X is equal to the sum of the expectation of X given each outcome B_i, weighted by the probability of each outcome B_i occurring:

$$E[X] = E[X \mid B_1]\, P(B_1) + E[X \mid B_2]\, P(B_2) + \cdots + E[X \mid B_n]\, P(B_n).$$

The law of total expectation is analogous to the law of total probability. Either the expected value of the random variable or the probability of a particular outcome can be calculated by summing the conditional values over all possible outcomes, B_i, of another random variable. For example, Rule P3.9 can be used to calculate the expected fitness (the "mean fitness") of a population consisting of three genotypes: AA, Aa, and aa. Here, the genotype is a second random variable with three mutually exclusive outcomes. The expected fitness is then $E[W] = E[W \mid AA]\, P(AA) + E[W \mid Aa]\, P(Aa) + E[W \mid aa]\, P(aa)$. In Chapter 3, we used subscripts to write the expected fitness conditional on being AA as $E[W \mid AA] = W_{AA}$. If we also assume that the genotype frequencies at time t are at Hardy-Weinberg proportions: $P(AA) = p(t)^2$, $P(Aa) = 2\, p(t)\, q(t)$, $P(aa) = q(t)^2$, the expected fitness becomes $E[W] = W_{AA}\, p(t)^2 + W_{Aa}\, 2\, p(t)\, q(t) + W_{aa}\, q(t)^2$, which equals the mean fitness in equation (3.12). The law of total expectation is particularly helpful when it is easier to describe the distribution of X conditional on the state of another factor.

P3.3 Discrete Probability Distributions

The first step in incorporating stochasticity into a model is to determine what process (or processes) has chance outcomes and then to describe the outcome of this process by a random variable. The next step is to describe the "probability distribution" for that random variable, which specifies how likely it is for the random variable to take on various values. In this section, we consider *discrete probability distributions*, where the random variable has a discrete set of mutually exclusive outcomes (e.g., 0, 1, 2). In section P3.4, we describe random variables whose outcomes can be any point along a continuum (e.g., any real number between 0 and 1). In both cases, we show how important quantities like the mean and the variance can be derived. Key attributes of all of the distributions are summarized in tables at the end of the Primer.

We start with the simplest discrete probability distribution describing the outcome of a single Bernoulli trial:

Definition P3.1:
A Bernoulli trial has two possible outcomes, say "zero" and "one," where the probability of the outcome "one" equals $P(X = 1) = p$.

We will often refer to an outcome of one as a "success," despite the fact that outcome one is not always desirable (e.g., if "one" represents "death"). Because there are only two outcomes, observing zero is the complement of observing one. Thus, by Rule P3.2, $P(X = 0) + P(X = 1) = 1$, and the probability of observing zero is $P(X = 0) = 1 - p$ (Figure P3.1c). For example, p might be the probability of having a successful crop (outcome 1), and $1 - p$ would be the probability of having a crop failure (outcome 0).

In describing a probability distribution, we assume that the outcomes form a mutually exclusive set (e.g., success versus failure) and that we have described all possible outcomes. As a consequence, Rule P3.1 tells us:

Rule P3.10: The Sum of a Discrete Probability Distribution
The sum of $P(X = x_i)$ over all outcomes, x_i, equals one:

$$\sum_{x_i} P(X = x_i) = 1.$$

The notation Σ_{x_i} in Rule P3.10 means the sum over all outcomes, x_i. For a Bernoulli trial, $P(X = 0) + P(X = 1) = (1 - p) + p$, which does equal one. For the distributions described in this primer, Rule P3.10 has been checked. If you want to develop a new probability distribution, however, you must confirm that your distribution obeys Rule P3.10.

Once a probability distribution has been specified, the distribution can be plotted. Typically, histograms are used, with the area of each bar representing the probability of observing the outcome labeled on the horizontal axis (Figure P3.2).

Besides plotting a probability distribution, the two most important quantities that we might wish to know about a distribution are its mean and its variance. We write the *mean* (or *average*, or *expectation*) of a random variable X as μ or $E[X]$, calculated as follows:

Definition P3.2: The Mean of a Discrete Random Variable
The mean (or average) of a discrete random variable is the sum of the value of each outcome weighted by the probability of that outcome:

$$\mu = E[X] = \sum_{x_i} x_i \, P(X = x_i).$$

(a)

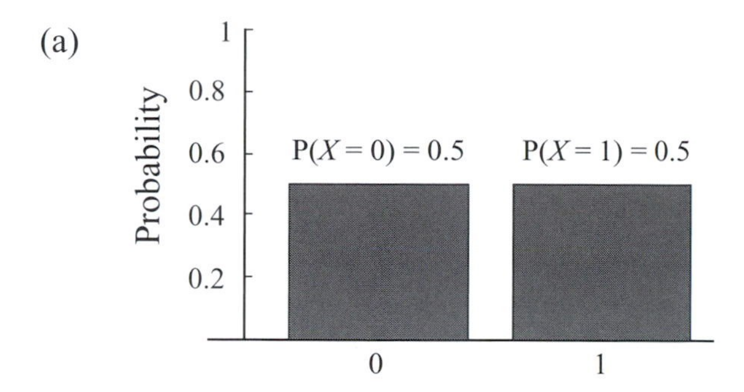

(b)

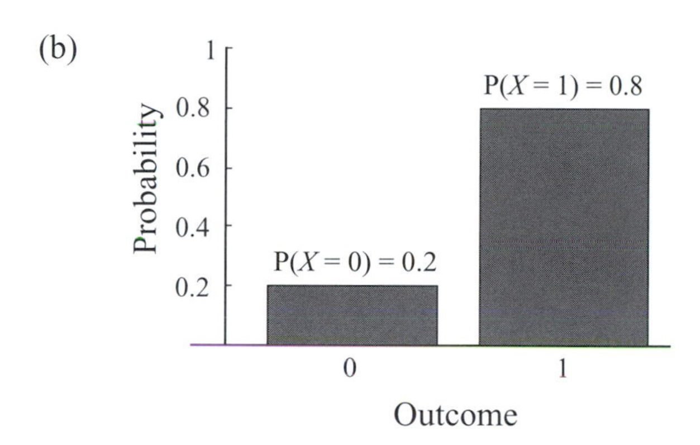

Figure P3.2: Histograms. The area of each bar represents the probability of observing outcome 0 or outcome 1 in a Bernoulli trial. Throughout, we will draw bars whose widths are equal and arbitrarily set to one, so that the height of the bar gives the probability of observing the outcome. The heights will then always sum to one. (a) Outcomes 0 and 1 are equally probable (as in a coin toss). (b) Outcome 1 is four times more likely than outcome 0.

One way to visualize the mean is to imagine balancing the histogram depicting the probability distribution on your finger, assuming that the weight of each bar is proportional to its height. The histogram will balance perfectly when you place your finger exactly at the mean. Thus, the mean is the "center of mass" of the probability distribution. For a Bernoulli trial, the mean equals p. This follows from Definition P3.2, which tells us that $E[X]$ is $0 \times P(X = 0) + 1 \times P(X = 1)$, which equals $0 \times (1 - p) + 1 \times (p) = p$. For example, if we let 0 represent crop failure and 1 represent crop success and if there is a 90% chance of a successful crop ($P(X = 1) = p = 0.9$), then the mean outcome will be 0.9. In any one year, the crop will either fail (outcome 0) or be successful (outcome 1), but we can think of the expected value as the average outcome that we would see after an indefinitely large number of years.

Often, we want to know how dispersed the random variable is around its mean. One measure of dispersion is the *variance*, which is often written as σ^2 or as Var[X]:

Definition P3.3: The Variance of a Discrete Random Variable

The variance of a discrete random variable is the expected value of $(X - \mu)^2$ over the probability distribution. It is calculated as the sum of the squared distance of each outcome from the mean, weighted by the probability of that outcome:

$$\text{Var}[X] = E[(X - \mu)^2] = \sum_{x_i} (x_i - \mu)^2 \, P(X = x_i).$$

For example, the variance of the distribution describing a Bernoulli trial is equal to $p\,(1 - p)$ because $E[(X - \mu)^2] = (0 - p)^2\ P(X = 0) + (1 - p)^2\ P(X = 1) = p^2(1 - p) + (1 - p)^2\,(p)$, which factors into $p\,(1 - p)\,\{(1 - p) + p\} = p\,(1 - p)$.

There are several alternative ways to measure dispersion around the mean, the two most important being the standard deviation and the coefficient of variation. The *standard deviation* is the square root of the variance, represented by σ. The standard deviation has the same units as the random variable and the mean. In contrast, the variance is in terms of these units squared. The *coefficient of variation*, CV, equals the standard deviation divided by the mean, σ/μ, and is sometimes expressed as a percentage, $\sigma/\mu \times 100\%$. The CV is a dimensionless measure of the variability around the mean. It has the advantage of being the same regardless of the measurement scale used (e.g., centimeters or kilometers).

Table P3.1 lists several useful facts that can simplify matters when calculating expectations and variances. For example, we can use the rules of Table P3.1 to derive a second formula for the variance. We can always expand the square in $E[(X - \mu)^2]$ as $E[X^2 - 2X\mu + \mu^2]$. According to Table P3.1, the expectation of a sum equals the sum of the expectations, yielding $E[X^2] + E[-2X\mu] + E[\mu^2]$. The mean μ is a constant parameter; all such constants can be factored out of expectations, leaving $E[X^2] - 2\,\mu\,E[X] + \mu^2\,E[1]$. Finally, because $E[X] = \mu$ and $E[1] = 1$, we can rewrite the variance as

$$\mathrm{Var}[X] = E[X^2] - \mu^2. \tag{P3.2}$$

The expectation and the variance are two descriptors of a probability distribution and are sometimes referred to as the first and second *central moments* of the distribution. This terminology reflects the fact that they are expectations of the first and second powers of the random variable, after subtracting the mean so that the distributions are "centered" around the mean. In some cases, you might be interested in knowing the skew or kurtosis (peakedness) of a distribution, which are quantities related to higher moments (the third and fourth moments, respectively). After becoming familiar with the material in this Primer, consult Appendix 5 for a general method for finding moments of a distribution using "moment generating functions."

In the following sections, we describe a number of probability distributions that commonly arise in biology. In each case, we provide an overview of the distribution, specify its mean and variance, and describe the contexts in which the distribution is likely to arise. Having a good intuitive sense for the context of each probability distribution is extremely useful. It makes it easier to solve many probabilistic problems that arise in biology by allowing you to make connections between the problem and known facts about probability distributions. Furthermore, in order to incorporate stochasticity into any biological model, you must first choose the most appropriate probability distribution, which is easier to do if you have a good sense of the different possibilities.

TABLE P3.1
Some useful rules of expectations and variances. The rules involving summations assume a discrete probability distribution, but analogous formulas involving integrals exist for continuous probability distributions.

Rule	Notes
$E[c] = c$	If c is a constant
$E[c\,X] = c\,E[X]$	If c is a constant
$E[g(X)] = \sum_{x_i} g(x_i)P(X = x_i)$	The expectation of the function $g(X)$ of a random variable
Geometric mean $X = \prod_{x_i} x_i^{P(X=x_i)}$	The geometric mean of a random variable
Harmonic mean $X = \dfrac{1}{\sum_{x_i} \frac{1}{x_i} P(X = x_i)}$	The harmonic mean of a random variable
$E[f(X,Y)] = \sum_{x_i}\sum_{y_j} f(x_i,y_j)P(X = x_i \cap Y = y_j)$	The expectation of a function $f(X, Y)$ involving two random variables
$E[X + Y] = E[X] + E[Y]$	The expectation of a sum is the sum of the expectations
$E[X\,Y] = E[X]\,E[Y]$	If X and Y are independent random variables, the expectation of a product is the product of the expectations
$\text{Var}[c] = 0$	If c is a constant
$\text{Var}[c\,X] = c^2\,\text{Var}[X]$	If c is a constant
$\text{Var}[X + Y] = \text{Var}[X] + \text{Var}[Y]$	If X and Y are independent random variables
$\text{Var}[X + Y] = \text{Var}[X] + \text{Var}[Y] + 2\,\text{Cov}[X\,Y]$	If X and Y are not independent
$\text{Cov}[X\,Y] = E[X\,Y] - E[X]\,E[Y]$	$\text{Cov}[X\,Y]$ describes the "covariance" between X and Y. It equals zero if X and Y are independent.
$\rho = \dfrac{\text{Cov}[X,Y]}{\sigma_x \sigma_y}$	The "correlation" coefficient standardizes the covariance by the standard deviation of X and Y
$\text{Cov}[X\,Y] = E[\text{Cov}_i[X_i\,Y_i]] - \text{Cov}[E_i[X_i]\,E_i[Y_i]]$	The *covariance decomposition theorem* calculates the covariance over a set of mutually exclusive classes, i. On the right, Cov[ ] and E[ ] are calculated across classes, weighted by the proportion of the population in each class, p_i, while Cov_i[ ] and E_i[ ] are the covariances and expectations within a class.

Exercise P3.3:

(a) Calculate the variance for the distribution describing a Bernoulli trial using the definition $\text{Var}[X] = E[X^2] - \mu^2$, and show that the variance equals $p\,(1 - p)$.

(continued)

Exercise P3.3 (*continued*)

(b) Imagine doing an experiment involving two independent Bernoulli trials. The total number of successes could be 0, 1, or 2. Determine the probability of each outcome. Confirm that these probabilities sum to one (Rule P3.10). Determine the mean outcome using Definition P3.2. Determine the variance in the outcome using Definition P3.3.

(c) Show that you can obtain your answers more easily for the mean and variance of two Bernoulli trials using the following facts from Table P3.1: $E[X + Y] = E[X] + E[Y]$ and $\text{Var}[X + Y] = \text{Var}[X] + \text{Var}[Y]$, where X represents the outcome from one Bernoulli trial and Y represent the outcome from the second Bernoulli trial.

P3.3.1 Binomial Distribution

The binomial distribution generalizes a single Bernoulli trial to n independent trials. In each trial, there are two possible outcomes (say "zero" and "one"), where the probability of outcome "one" is p in every trial. The random variable in a binomial distribution is then the total number of ones observed in n trials, which takes on integer values from 0 to n.

Definition P3.4:
The binomial distribution describes the probability of observing a total of k "ones" in n independent Bernoulli trials:

$$P(X = k) = \binom{n}{k} p^k (1 - p)^{n-k}.$$

$\binom{n}{k}$ is read as "n choose k"; it equals $n!/(k!\,(n - k)!)$ and represents the number of different ways in which k "ones" can occur over the course of n trials (see Box P3.1 on page 559 at the end of this primer for more details). For example, $\binom{2}{1} = 2!/(1!\,1!) = 2$, which reflects the fact that there are two ways to get a single "one" in two trials—the "one" can occur on the first trial or on the second trial. By definition, 0! equals one, so that $\binom{2}{0} = 2!/(0!\,2!) = 1$, which reflects the fact that there is only one way to get zero "ones" in two trials—a "one" must not occur in the first trial or in the second trial.

For $p = 1/2$, the binomial distribution is symmetric and bell-shaped (Definition P1.6), while for p values near 0 or 1, the distribution becomes quite skewed (Figure P3.3).

The mean of a binomial random variable is

$$E[X] = n\,p. \tag{P3.3}$$

This follows from the fact that the binomial represents the sum of n random variables, each of which corresponds to a single Bernoulli trial (see Exercise

(a)

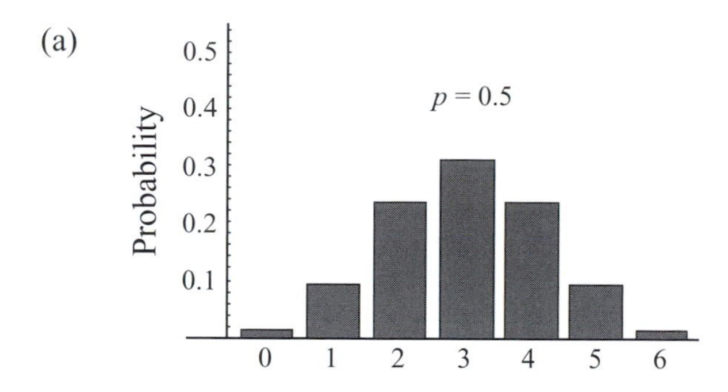

(b)

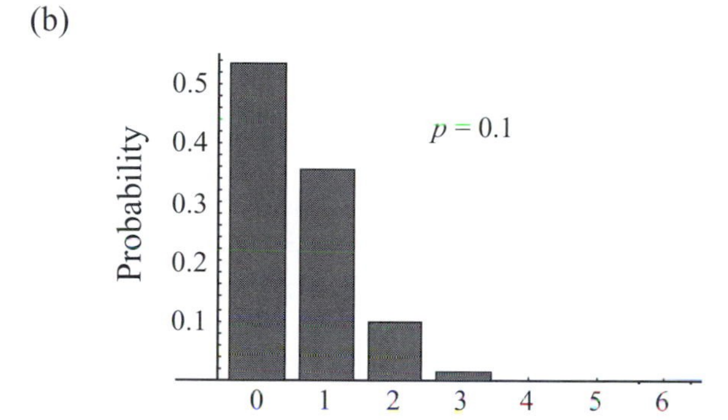

(c)

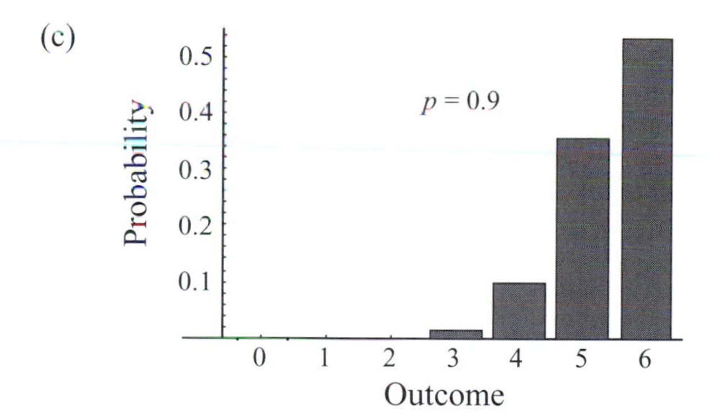

Figure P3.3: Binomial distribution (Definition P3.4). Each bar represents the probability of observing a particular number of successes (from zero to six) among six trials ($n = 6$). The probability of success is (a) $p = 0.5$, (b) $p = 0.1$, (c) $p = 0.9$.

P3.3). Because the expected value of a sum of random variables equals the sum of the expected values of each random variable (Table P3.1), and because $E[X] = p$ for each Bernoulli trial, the sum of n such trials has an expected value of $n\,p$.

Similarly, the variance of a binomial random variable equals

$$\text{Var}[X] = np\,(1 - p). \tag{P3.4}$$

This follows from the fact that the variance of a sum of independent random variables is the sum of the variance of each random variable (Table P3.1) and from the fact that $\text{Var}[X] = p\,(1 - p)$ for a single Bernoulli trial.

Examples

The binomial distribution arises when there are a number of independent trials and each trial results in one of two possible outcomes. For example, if there is a 50% probability of having a daughter ($p = 0.5$), the binomial distribution would describe the probability of observing a certain number of daughters in a family, e.g., three daughters among five children: $P(X = 3) = \binom{5}{3} (0.5)^3(0.5)^{5-3} = 5/16$. The binomial distribution also arises when a binary attribute (e.g., diseased or healthy, flowering or not flowering) is measured in a *sample* of size n taken from a population. This assumes that the sample is only a small fraction of the total population or that sampling occurs with replacement so that p does not change as we take our sample.

The binomial distribution is very helpful when interpreting data and simulations. For example, you might collect data on the number of frogs with and without limb defects near a nuclear reactor. Often, you will be collecting such data to estimate an unknown parameter p (e.g., the probability of limb defects). We can estimate p using equation (P3.3) by dividing both sides by n and replacing the expected number of ones, $E[X]$, with the observed number, x, giving an estimate of $\tilde{p} = x/n$, where we have written a tilde over the p to indicate that it is an estimate. We can also estimate $\text{Var}[X/n]$, which describes the variance in this estimate of p. Given that $\text{Var}[c\,X] = c^2\,\text{Var}[X]$ (Table P3.1), the variance of X/n is $\text{Var}[X]/n^2 = n\,p\,(1-p)/n^2 = p\,(1-p)/n$. Replacing p with its estimated value $\tilde{p}$, the variance of the estimate for p becomes $\tilde{p}\,(1-\tilde{p})/n$. Taking the square root of the variance, we get the standard deviation of the estimate for p (referred to as the "standard error of the proportion," *SE*): $SE = \sqrt{\tilde{p}(1-\tilde{p})/n}$. As a rule of thumb, the true value p has roughly a 95% chance of lying within two standard errors of the estimate $\tilde{p}$. (Note: Replacing p with $\tilde{p}$ in the variance introduces a bias, especially when $\tilde{p}$ is near 0 and 1. More exact treatments correct for this bias; see Zar (1998).)

P3.3.2 Multinomial Distribution

For some problems, each trial might have more than two possible outcomes. For example, you might want to classify the offspring of a cross as homozygous *AA*, heterozygous *Aa*, or homozygous *aa*. The multinomial provides such an extension to the binomial distribution.

Definition P3.5:

The multinomial distribution describes the probability of observing $\{k_1, k_2, \ldots, k_c\}$ individuals in each of c discrete categories, where the probability of observing an outcome in category i is p_i.

$$P(X = \{k_1, k_2, \ldots, k_c\}) = \frac{n!}{(k_1!)(k_2!)\cdots(k_c!)} p_1^{k_1}\, p_2^{k_2} \cdots p_c^{k_c}.$$

The expected number in category i and the variance in this number are

$$E[X_i] = n\,p_i, \tag{P3.5a}$$

$$\text{Var}[X_i] = n\,p_i\,(1 - p_i). \tag{P3.5b}$$

Example

If you were to survey plants within a tropical forest, the probability of observing a certain number of each species would be described by a multinomial distribution, with n equal to the total number of individuals that you sample and p_i equal to the proportion of plants of species i.

P3.3.3 Hypergeometric Distribution

In section P3.3.1, we mentioned that the binomial distribution arises when sampling n individuals from a population that has a proportion p of individuals of a certain type. Technically, this claim is true only if each individual sampled is replaced before the next individual is sampled, otherwise p will change as the sample is gathered, causing the outcome of each trial to depend on the outcomes of previous trials. If sampling occurs without replacement, the hypergeometric distribution describes the distribution of possible samples.

Definition P3.6:

The hypergeometric distribution describes the probability of observing k "ones" in a sample of size n, which is randomly drawn without replacement from a population of size N:

$$P(X = k) = \frac{\binom{N_1}{k}\binom{N_2}{n-k}}{\binom{N}{n}},$$

where $N_1 = N\,p$ is the number of "ones" and $N_2 = N\,(1 - p)$ is the number of "zeros" in the total population before sampling.

The probability distribution for a hypergeometric distribution looks complicated but it can be derived by counting up all of the types of samples that could occur (Box P3.1). The denominator represents the number of different ways (i.e., the number of combinations; Box P3.1) in which n individuals can be chosen without replacement from a population of size N, regardless of whether they are successes or failures. For example, there are three ways to chose two individuals ($n = 2$) from a population of size three ($N = 3$): either the first, the second, or the third individual can be left out. Out of all of these possibilities, we then need to count up all of those instances in which there were exactly k successes and $n - k$ failures. Moving to the numerator, the quantity $\binom{N_1}{k}$ is the number of ways (i.e., the number of combinations) in which k successes can be drawn (without replacement) from the subpopulation of N_1 successes without caring about the order in which they occur. For each of these, there are then $\binom{N_2}{n-k}$ different ways (i.e., combinations) in which the desired $n - k$ failures can be drawn (without replacement) from the subpopulation of N_2 failures. Thus the total number of ways in which we can obtain exactly k successes and $n - k$

failures is $\binom{N_1}{k}\binom{N_2}{n-k}$. Consequently, of the $\binom{N}{n}$ ways that we could sample n individuals from a population, only $\binom{N_1}{k}\binom{N_2}{n-k}$ of these will contain k successes and $n - k$ failures. The fraction of samples with k successes is thus given by Definition P3.6.

Using Definitions P3.2 and P3.3, the mean and variance of a hypergeometric random variable are

$$E[X] = n\,p, \tag{P3.6}$$

$$\text{Var}[X] = n\,p\,(1 - p)\frac{N - n}{N - 1}. \tag{P3.7}$$

The mean is the same as a binomial random variable (P3.3). But, the variance is a factor $(N - n)/(N - 1)$ smaller than the variance of a binomial random variable. The variance decreases toward zero as the sample size approaches the population size ($n \rightarrow N$), because the composition of the sample becomes nearly the same as the composition of the whole population. Conversely, if the sample size is very small relative to the population size ($n << N$), then $(N - n)/(N - 1)$ approaches one, and the hypergeometric distribution converges upon the binomial distribution.

Example

Imagine that you are studying the nesting behavior of puffins on an island, which contains $N = 100$ suitable nesting cavities. Of these nesting cavities, 30 are on a cliff face that is inaccessible to mammalian predators, while the remainder are on a grassy slope. You watch as the first $n = 20$ puffins choose cavities and begin nesting, and you observe that $k = 11$ choose cliff sites. Thus, among the first nesters, you observe a higher proportion (11/20 = 55%) using cliff sites than expected on the basis of the proportion of cliff sites (30/100 = 30%). The hypergeometric distribution can be used to determine the probability of observing exactly $k = 11$ nesting on the cliff:

$$P(X = 11) = \frac{\binom{30}{11}\binom{70}{9}}{\binom{100}{20}} = 0.0066.$$

Of greater interest is the probability that 11 or more early nesters choose cliff sites, which again can be calculated from the hypergeometric distribution: $P(X \geq 11) = \sum_{k=11}^{20} P(X = k) = 0.0085$. This probability is so low that you can conclude it is unlikely that early nesters are randomly choosing their nest sites and that they appear to prefer cliff sites.

P3.3.4 Geometric Distribution

All of the above distributions (binomial, multinomial, and hypergeometric) describe the number of outcomes that fall into different categories (e.g., cliff nesters vs noncliff nesters). Each outcome falls into some category, and these distributions predict "counts" in each category. In contrast, the distributions discussed next (geometric, negative binomial) were derived to describe how much time passes before a particular outcome (or set of outcomes) is observed. Despite this fundamental difference, the geometric distribution is again based on a series of Bernoulli trials.

> **Definition P3.7:**
> **The geometric distribution** describes the probability that, in a series of Bernoulli trials, the first success is observed on the kth trial:
>
> $$P(X = k) = p\,(1 - p)^{k-1}.$$

The geometric distribution is derived as follows. For the first success to occur on the kth trial requires that the previous $k - 1$ trials were unsuccessful. Assuming that each Bernoulli trial is independent of previous ones, the probability of $k - 1$ unsuccessful trials is the product of the probability that each trial is unsuccessful, which is $(1 - p)^{k-1}$ (Rule P3.4a). Following this series of failures, the kth trial will be successful with probability p. Because each trial is independent, we can multiply these terms together to get the geometric distribution.

Because at least one trial must occur to observe a success, k can be any integer greater than or equal to one. $P(X = k)$ always declines with increasing k because every trial that passes unsuccessfully decreases the probability by a factor $(1 - p)$. Thus, the event that there are no failures prior to the first success (i.e., $k = 1$) always has the highest probability (Figure P3.4).

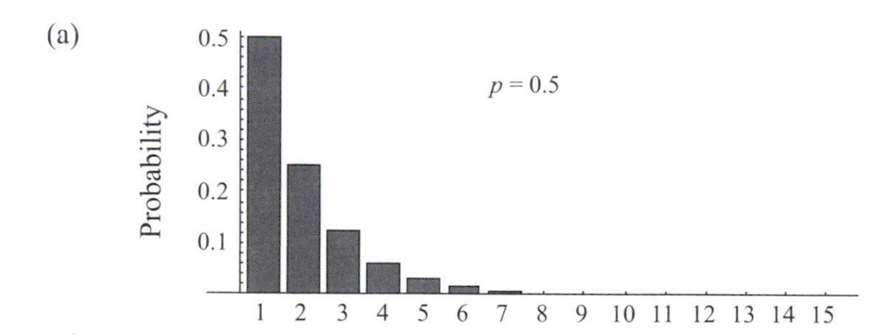

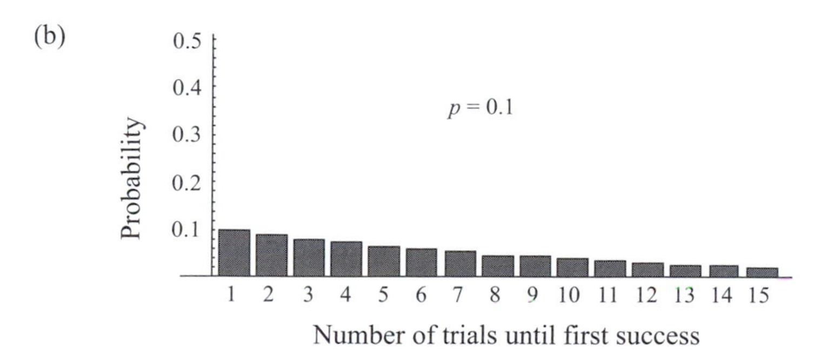

Figure P3.4: Geometric distribution (Definition P3.7). Each bar represents the probability that the first successful event occurs after a particular number of trials (from one to infinity). The probability of success in any one trial is (a) $p = 0.5$, (b) $p = 0.1$.

The mean of a geometric random variable is given by $E[X] = 1/p$. To get more comfortable working with sums, it is worth deriving this fact. We start with Definition P3.2 giving the mean:

$$E[X] = \sum_{k=1}^{\infty} k\, p\, (1 - p)^{k-1}. \tag{P3.8}$$

This sum is not one of those listed in Appendix A1, but consider taking the derivative of both sides of A1.20 with respect to a, giving us $\Sigma_{i=1}^{\infty} i\, a^{i-1} = 1/(1 - a)^2$ (To see this, it might help to think about writing out the summation as $a + a^2 + a^3 + \cdots$). If we factor out p from (P3.8) and let $a = 1 - p$, the sum in (P3.8) can be written as $p\Sigma_{k=1}^{\infty} k\, a^{k-1}$, which equals $p/(1 - a)^2$. Plugging in $a = 1 - p$, the mean equals

$$E[X] = \frac{1}{p}. \tag{P3.9}$$

In a similar fashion, the variance of the geometric random variable is

$$\text{Var}[X] = \frac{1 - p}{p^2}. \tag{P3.10}$$

Examples

The number of courtship displays made by a male before he successfully mates might be described by a geometric random variable. Here, each time a male displays is a Bernoulli trial resulting in a mating ("success") or not ("failure"). The key assumption for this process to be described by a geometric distribution is that the probability that a mating attempt succeeds remains constant over time and is not influenced by ("is independent of") the outcome of previous mating attempts. The geometric distribution might also describe the time until extinction of an endangered population that is censused yearly, if the probability of extinction is constant. Thus, with an annual extinction risk of 10% ($p = 0.1$), the expected time until extinction is ten years (i.e., $1/p = 1/0.1 = 10$). The variance in this case is pretty large (90 years squared). This means that the actual year in which the population goes extinct is very hard to predict, as suggested by Figure P3.4.

P3.3.5 Negative Binomial Distribution

The negative binomial distribution generalizes the geometric distribution and describes the waiting time until r "successes" have occurred:

Definition P3.8:

The negative binomial distribution describes the probability that, in a series of Bernoulli trials, the rth success is observed on the kth trial:

$$P(X = k) = \binom{k - 1}{r - 1} p^r (1 - p)^{k-r}.$$

For the rth success to occur on the kth trial, there must have been $r - 1$ successes in the previous $k - 1$ trials. We have already described the probability of observing a certain number of successes out of a total number of trials: it is given by the binomial distribution. Thus, we can write the probability distribution for the negative binomial as the product of the binomial probability of observing $r - 1$ successes out of $k - 1$ trials, $\binom{k-1}{r-1} p^{r-1} (1 - p)^{k-r}$, multiplied by p, the probability that the kth trial is a success.

Because at least r trials must occur to observe r successes, k can be any integer greater than or equal to r. Now, $P(X = k)$ does not always decline with an increasing numbers of trials (k). In fact, if we were waiting for a large number of successful outcomes, the negative binomial distribution has a bell shape (Figure P3.5).

We can think of the negative binomial distribution as describing the sum of r independent random variables: the sum of the waiting times before each of the r successful trials. Each of these waiting times follows a geometric distribution with mean $1/p$ and variance $(1 - p)/p^2$. Using the fact that the expectation of a sum is the sum of the expectations, the number of trials until the rth success is expected to equal

$$E[X] = \frac{r}{p}. \tag{P3.11}$$

Similarly, because the variance of a sum of independent random variables is the sum of the variance of each random variable (Table P3.1), the variance of a negative binomial random variable is:

$$\mathrm{Var}[X] = \frac{r(1 - p)}{p^2} \tag{P3.12}$$

(a)

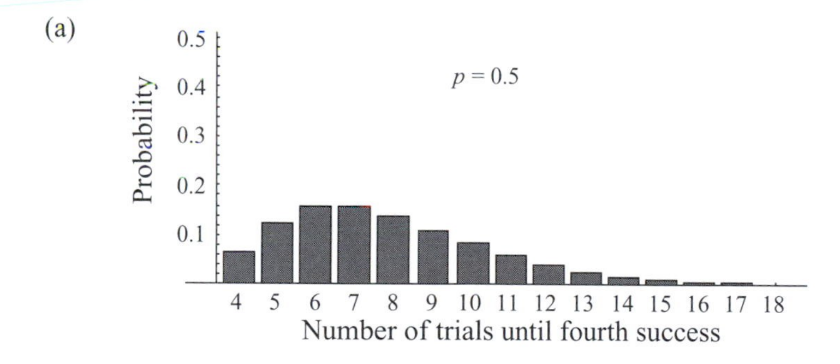

(b)

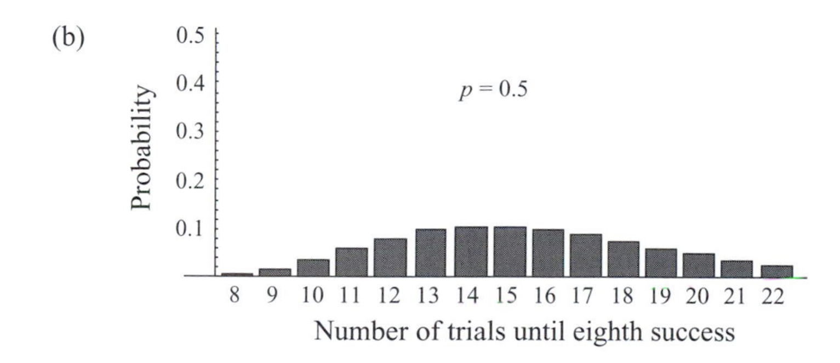

Figure P3.5: Negative binomial distribution (Definition P3.8). Each bar represents the probability that (a) the fourth successful event occurs after a particular number of trials (from four to infinity) and (b) the eighth successful event occurs after a particular number of trials (from eight to infinity). The probability of success in any one trial is $p = 0.5$ (Figure P3.4a describes the comparable probability distribution for the first successful event).

Example

If a predator must capture $r = 10$ prey before it can grow sufficiently large to reproduce, and if it has a 10% success rate per hunt ($p = 0.1$), the age of onset of reproduction would be described by the negative binomial distribution. On average, the predator must go on 100 hunts before it can reproduce, where the variance in this number is 900 hunts^2 ($SD = 30$ hunts).

P3.3.6 Poisson Distribution

The last of the discrete distributions that we will visit is the Poisson distribution. It differs fundamentally from the above distributions because it describes neither the numbers that fall into various categories (binomial, multinomial, and hypergeometric) nor the waiting time until a certain number of events have occurred (geometric, negative binomial). Rather, the Poisson distribution describes the number of events that occur in a given time period (or within a given area) when events occur randomly and independently over time (or space). The Poisson distribution naturally arises when counting the number of events witnessed during an observation period, such as the number of birds that stray onto an island within a year or the number of seedlings that germinate on a plot within a week.

Definition P3.9:

The Poisson distribution describes the probability of observing k events in a given space or time period when the expected number of events is μ and when each event occurs independently:

$$P(X = k) = \frac{e^{-\mu}\mu^k}{k!}$$

If you know the rate at which events occur per unit time (λ), then the expected number of events is $\mu = \lambda t$, where t is the time period of observation. Similarly, if you know the density of events per unit area (δ), then the expected number of events is $\mu = \delta A$, where A is the area under observation. The actual number of events that occurs, k, can be any integer from 0 to infinity. When μ is small, the Poisson distribution is skewed, and the probability of observing no events or only one is high. Alternatively, when μ is large, the Poisson distribution becomes bell shaped, with the most likely number of observations centered on μ (Figure P3.6).

The Poisson distribution has an unusual attribute in that its mean equals its variance:

$$\begin{aligned} E[X] &= \mu \\ &= \text{Var}[X]. \end{aligned} \tag{P3.13}$$

Another important attribute of the Poisson distribution is that the sum of a number of Poisson random variables is itself Poisson distributed (Supplementary

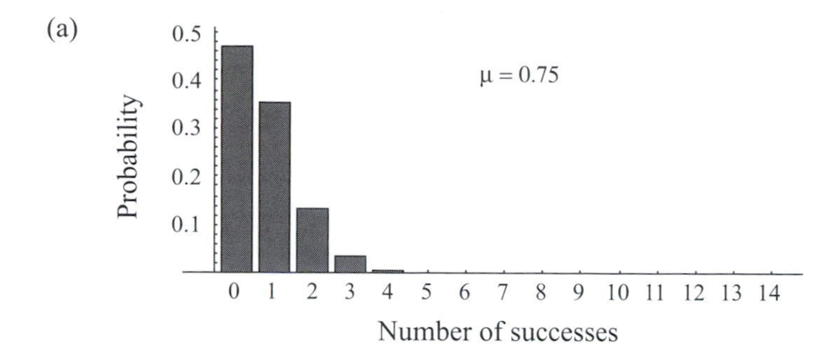

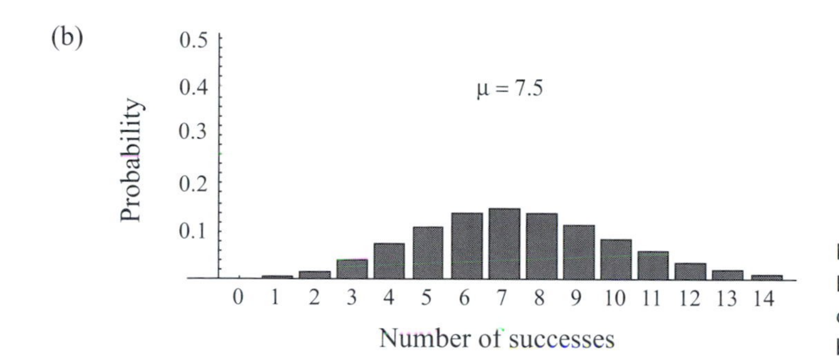

Figure P3.6: Poisson distribution (Definition P3.9). Each bar represents the probability of observing a certain number of events when the expected number is (a) $\mu = 0.75$ and (b) $\mu = 7.5$.

Material P3.2). For example, if the number of hemlock seedlings, the number of cedar seedlings, and the number of fir seedlings that emerge within a square meter is Poisson distributed, then their sum, describing the total number of tree seedlings that emerge, will also be Poisson distributed.

Examples

If hummingbirds arrive at a flower at a rate $\lambda = 0.2$ per minute, the expected number of visits in $t = 20$ minutes of observation would be $\mu = 4$. Assuming that hummingbirds arrive independently and randomly over time, we would expect the actual number of visits to be Poisson distributed with a mean and variance of 4. If the observed variance is significantly lower, this would call into question the assumption that hummingbird visits occur independently over time and indicates that birds tend to space out their visits, causing the visits to be more evenly distributed than expected under the Poisson distribution.

As another example, the Poisson distribution describes the number of new mutations that an individual is expected to carry. In the diploid human genome, there are about $A = 6.4 \times 10^9$ basepairs and the mutation rate per generation per basepair is approximately $\delta = 1.8 \times 10^{-8}$ (Kondrashov 2003). In this case, we are monitoring a particular area (A, here the stretch of DNA) for events that occur at a particular density (δ). The expected number of events is then $A \times \delta = 115.2$. According to a Poisson distribution, the variance should also equal 115.2, and the standard deviation should be $\sqrt{115.2} = 10.7$. Furthermore, if we plot the Poisson distribution with mean 115.2, we can predict that about 95% of us carry between 96 and 136 new mutations that were not present within our parents (Figure P3.7).

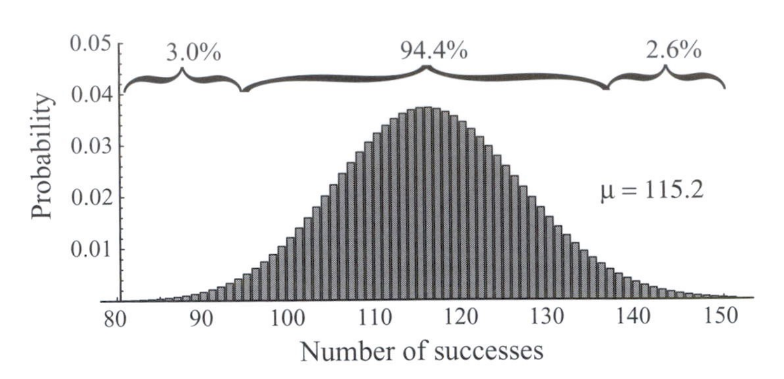

Figure P3.7: The distribution of the number of mutations according to a Poisson distribution with mean $\mu = 115.2$ (Definition P3.9), the number of mutations is expected to fall between 96 and 136 in 94.4% of cases. (From 0 to 95 accounts for 3.0% of the distribution; from 137 to infinity accounts for 2.6% of the distribution. These ranges were chosen to be the smallest possible ranges that accounted for >2.5% of the distribution each.)

Although we have used a Poisson distribution to describe the number of mutations, technically, this assumes that mutations can occur at any real-valued position and ignores the discrete nature of the nucleotides that make up chromosomes. In reality, a mutation alters the first nucleotide, the second nucleotide, . . . or the nth nucleotide in the sequence. We can capture the discrete nature of mutations using a binomial distribution to describe the probability of observing k mutations out of $n = 6.4 \times 10^9$ nucleotides, where the probability of a mutation (a "success") is $p = 1.8 \times 10^{-8}$ per generation. According to the binomial distribution, the mean number of mutations is $np = 115.2$ and the variance is $np\,(1 - p) \approx 115.2$. Interestingly, these are the same mean and variance predicted by the Poisson distribution. In fact, the binomial distribution converges upon a Poisson distribution whenever the probability of success, p, is small and the number of trials, n, is large. In this case, the variance of the binomial, $np\,(1 - p)$, is nearly equal to the mean, np, which is a property of the Poisson distribution. The higher moments also converge, as can be shown using moment generating functions (Appendix 5).

Exercise P3.4: Unlike the Poisson distribution, the sum of two independent random variables, each following a binomial distribution, is not generally binomial. Let X represent the outcome from n_1 Bernoulli trials each of which has a probability of success of p_1, and let Y represent the outcome from n_2 trials with probability of success p_2. What is the variance of the sum of these two random variables, $X + Y$? Show that $\text{Var}[X + Y]$ cannot be factored into the form $n\,p\,(1 - p)$ and so does not equal the variance expected if the sum were binomial, unless $p_1 = p_2$. As an example, consider $n_1 = 100$ trials with $p_1 = 0$ and $n_2 = 100$ trials with $p_2 = 1$. What is $\text{Var}[X + Y]$? For comparison, what variance would you expect from the binomial distribution with $n = 200$ trials and an average proportion of successes, $p = 1/2$?

P3.4 Continuous Probability Distributions

In the previous distributions, the possible outcomes were discrete (e.g., integers from 0 to n). What if you were interested in a random variable that could take on any real value (e.g., any point in time)? Random variables that can take on a continuum of possible values are known as *continuous random variables*. The procedures described above to calculate the total probability, mean, and variance for discrete random variables are similar to the procedures for continuous random variables, but there is one crucial difference. Imagine calculating the sum in Rule P3.10, $\sum_{x_i} P(X = x_i) = 1$, for a continuous random variable. Because there is a continuum of possible outcomes (e.g., all points in time), this sum would be infinitely large if $P(X = x_i)$ were finite for every possible value of x_i. Even for a continuous random variable, however, the total probability that the random variable takes on some value must be one, not infinity.

How is this discrepancy resolved? It is resolved by recognizing that, with a continuum of possible outcomes, the probability of any one particular outcome is not a finite number but is, instead, infinitesimally small. For example, if a continuous random variable lies between 0 and 1, the probability of it taking on the exact value of, say, 1/8 (i.e., 0.12500000 . . .) is essentially zero. The same is true for any other particular value. Because we cannot talk about the probability of any one outcome, we instead describe the probability that the random variable X falls within a small interval dx of x:

$$P(x < X < x + \mathrm{d}x) = f(x)\,\mathrm{d}x, \tag{P3.14}$$

where $f(x)$ is known as the "*probability density function*" describing the probability distribution for a continuous random variable. Typically, $f(x)$ is drawn as a curve over the region of possible outcomes x (Figure P3.8). Equation (P3.14) can be interpreted as the area of a histogram with a height of $f(x)$ and a very small width of dx, which gives the probability that the random variable falls within a region from x to x + dx. More generally, the area under the curve between any two points, a and b, equals the probability that the random variable falls between a and b (Figure P3.8):

$$P(a < X < b) = \int_a^b f(x)\mathrm{d}x. \tag{P3.15}$$

Thus, the probability density function tells us the regions in which the random variable is likely to fall (high $f(x)$) or unlikely to fall (low $f(x)$).

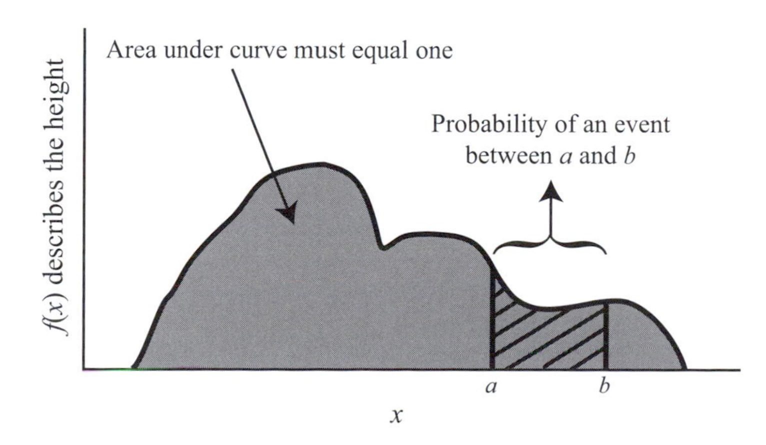

Figure P3.8: A probability density function with height $f(x)$. To represent a probability distribution, the area under the curve must equal one, and its height must never be negative. The probability of falling within any particular interval (e.g., between points a and b) is the area under the curve within this interval (hatched). Regions in which the curve is high ($f(x)$ is large) are more probable than regions in which the curve is low.

Replacing $P(X = x_i)$ with $f(x)\, dx$ and summations with integrals, we can proceed to analyze continuous random variables as before. In particular, because the random variable must take on some outcome, we have the following rule:

Rule P3.11: The Integral of a Continuous Probability Distribution

The integral of $f(x)$ over the range of possible outcomes must equal one:

$$\int_{\min}^{\max} f(x)\, dx = 1.$$

As with a discrete probability distribution, the mean of a continuous probability distribution equals the average value that would be obtained from an infinite number of draws from the distribution. To calculate this average, we use a definition analogous to Definition P3.2:

Definition P3.10a: The Mean of a Continuous Random Variable

The mean of a continuous random variable is given by integrating the value of each outcome x weighted by the probability density function $f(x)$ over the range of possible outcomes:

$$\mu = E[X] = \int_{\min}^{\max} x\, f(x)\, dx.$$

Again, the mean can be visualized as the "center of mass" of the probability distribution represented by the curve $f(x)$.

The variance of a continuous random variable is calculated in a similar fashion:

Definition P3.10b: The Variance of a Continuous Random Variable

The variance of a continuous random variable is the integral of the squared distance of each outcome x from the mean, weighted by the probability density function $f(x)$:

$$\text{Var}[X] = E[(X - \mu)^2] = \int_{-\infty}^{\infty} (x - \mu)^2 f(x)\, dx \ .$$

Equivalently, the variance equals

$$\text{Var}[X] = E[X^2] - \mu^2 = \left(\int_{-\infty}^{\infty} x^2 f(x)\, dx \right) - \mu^2.$$

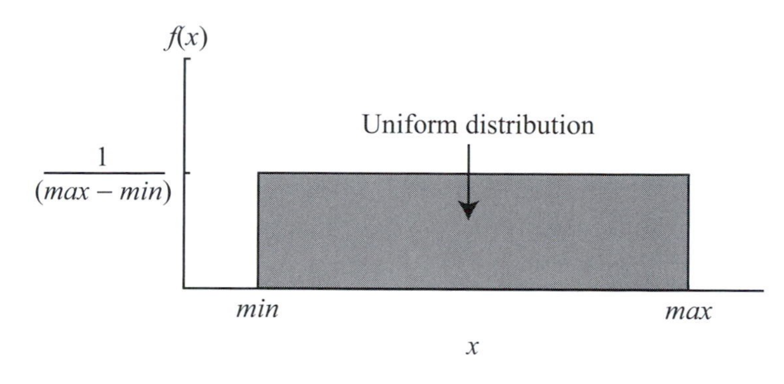

Figure P3.9: Uniform distribution (Definition P3.11). All outcomes between $x = min$ and $x = max$ are equally likely, and no outcome outside this region can occur.

We turn next to a description of some of the most important continuous probability distributions.

P3.4.1 Uniform Distribution

The uniform distribution is the simplest continuous probability distribution. It describes a random variable that is equally likely to fall at any point within a range from *min* to *max* (Figure P3.9). Within this range, the probability density function has a constant height $f(x) = h$, which is calculated from the fact that the integral over the range of possible values must equal one (Rule P3.12):

$$\int_{\min}^{\max} h \, \mathrm{d}x = h \max - h \min = 1, \qquad \text{(P3.16)}$$

Solving for h, we get $h = 1/(\max - \min)$.

Definition P3.11:

The uniform distribution describes the probability density at x for a random variable when all outcomes between min and max are equally likely:

$$f(x) = \frac{1}{\max - \min} \qquad \text{for } \min \leqslant x \leqslant \max$$

and $f(x) = 0$ for x outside of min and max.

Intuitively, the mean of a uniform probability density function occurs halfway between min and max. Indeed, applying Definition P3.9, we get

$$E[X] = \int_{\min}^{\max} x \frac{1}{\max - \min} \, \mathrm{d}x$$

$$= \left(\frac{x^2}{2} \frac{1}{\max - \min} \right) \Bigg|_{\min}^{\max}$$

$$= \frac{(\max^2 - \min^2)}{2(\max - \min)} \tag{P3.17}$$

$$= \frac{\max + \min}{2}.$$

The variance is not so easy to intuit, but it can be determined using Definition P3.10:

$$\begin{aligned}
\text{Var}[X] &= \int_{\min}^{\max} (x - \mu)^2 f(x)\, dx \\
&= \int_{\min}^{\max} \left(x - \frac{\max + \min}{2}\right)^2 \frac{1}{\max - \min}\, dx \\
&= \left(\frac{1}{3}\left(x - \frac{\max + \min}{2}\right)^3 \frac{1}{\max - \min}\right)\Bigg|_{\min}^{\max} \\
&= \frac{(\max - \min)^2}{12}.
\end{aligned} \tag{P3.18}$$

Examples

Imagine that you are studying mating behavior in *Drosophila*. The flies are in a cage and reproducing continuously. For your study, you watch the flies in ten-minute intervals (600 seconds) and record each time a mating takes place. You notice that out of 100 matings, none occur within the first 20 seconds. This makes you concerned that the initial handling might affect the behavior of the flies. To test this, you determine the probability that a mating occurs any time after 20 seconds:

$$\int_{20}^{600} \frac{1}{600}\, dx = \frac{29}{30}. \tag{P3.19}$$

This is the probability that one mating is observed after 20 seconds, assuming that the probability of mating is uniformly distributed over the 600 seconds of observation. The probability that all 100 matings occur after 20 seconds would then be $(29/30)^{100} = 0.034$ (Rule P3.4a). As this is a small probability, you conclude that handling might well have an influence on the flies, making them initially less likely to mate.

As another example, imagine a chromosome that is 2 Morgans in length. (A Morgan gives the distance along a chromosome within which one recombination event, or *crossover*, occurs, on average.) Among those chromosomes containing a single crossover, the mean observed position of the crossover is at 1 Morgan with a variance of 1/2. If crossovers occurred uniformly across the chromosome (min = 0 and max = 2 Morgans), we would expect the mean position to be at 1 Morgan with a variance of $(2 - 0)^2/12 = 1/3$. Thus, there is more variance than expected based on a uniform distribution of crossover positions. If this increase in variance were significant, it would suggest that crossovers are less likely to occur near the middle of the chromosome.

P3.4.2 Exponential Distribution

The exponential distribution arises when measuring the time until an event first occurs in continuous time.

> **Definition P3.12:**
> **The exponential distribution** describes the probability density of the waiting time x until an event first occurs under the assumption that events occur at a constant rate α per unit time:
>
> $$f(x) = \alpha\, e^{-\alpha x} \qquad \text{for } 0 \leq x \leq \infty$$

Here we write the waiting time as x rather than t to be consistent with the other probability distributions. We also use the rate parameter α to be consistent with the gamma distribution described next. In the biological literature, λ is often used as the rate parameter in place of α.

To derive the exponential distribution, consider calculating the probability $P(x)$ that no events occur before time x. From the results of previous chapters, we can write a recursion equation for P over a small time step dx as $P(x + \mathrm{d}x) = P(x)\,(1 - \alpha \mathrm{d}x)$. In other words, the probability that the event has still not occurred at time $x + \mathrm{d}x$ is just the probability that it had not occurred at time x, multiplied by the probability it does not occur in the time interval dx. As we saw in Box 2.6 of Chapter 2, we can rearrange this as $(P(x + \mathrm{d}x) - P(x))/\mathrm{d}x = -\alpha\, P(x)$ and then take that limit as dx gets small to obtain the differential equation $\mathrm{d}P/\mathrm{d}x = -\alpha\, P(x)$. This differential equation has the form of the exponential growth model and can be solved to get $P(x) = e^{-\alpha x}$ (see Chapter 6). For the event to occur for the first time near time x (i.e., between time x and $x + \mathrm{d}x$), we multiply the probability that the event does not happen before time x, $P(x) = e^{-\alpha x}$, by the probability that the event does occur in the short interval of time dx, which is $\alpha \mathrm{d}x$. This gives us $\alpha e^{-\alpha x}\, \mathrm{d}x$, which equals the exponential probability distribution $f(x)$ times the time interval dx.

The exponential distribution starts at height, α, when $x = 0$ and declines exponentially with x at rate α (Figure P3.10). The total area under the curve correctly integrates to one (Rule P3.11):

$$\int_0^\infty \alpha\, e^{-\alpha x}\, \mathrm{d}x = -\frac{\alpha\, e^{-\alpha x}}{\alpha}\Bigg|_0^\infty = 1. \tag{P3.20}$$

Using Definitions P3.10a and P3.10b, the mean and variance of an exponential random variable are:

$$E[X] = \frac{1}{\alpha}, \tag{P3.21}$$

$$\mathrm{Var}[X] = \frac{1}{\alpha^2} \tag{P3.22}$$

(see Exercise P3.6).

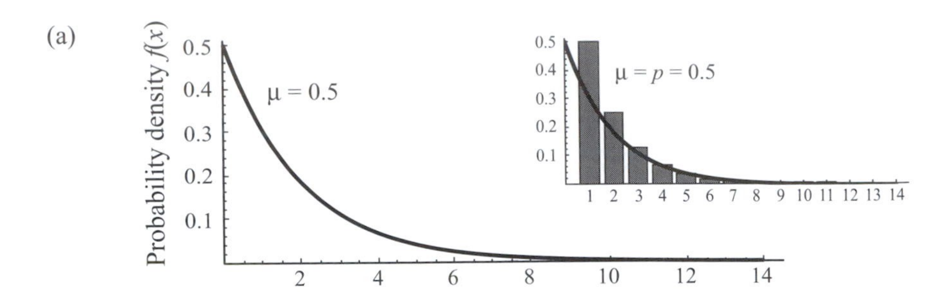

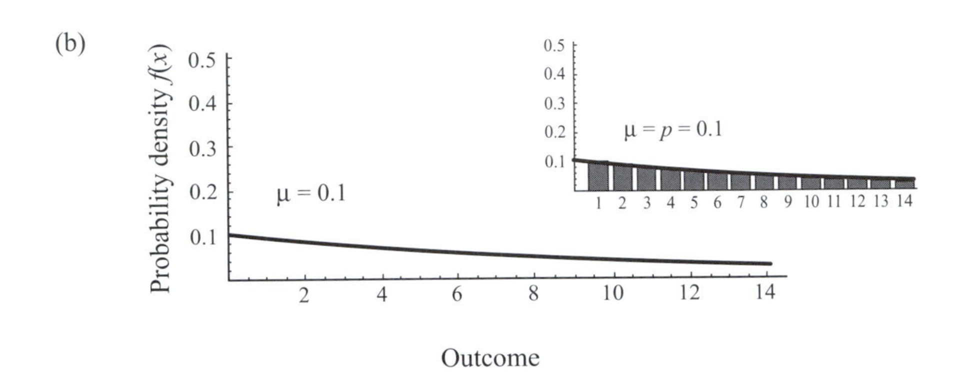

Figure P3.10: Exponential distribution (Definition P3.12). The probability density function, $f(x)$, declines exponentially with x. Although the expected outcomes are (a) $\mu = 0.5$ and (b) $\mu = 0.1$, the most likely outcomes are always near zero ($f(x)$ is highest at zero). The inset figures compare the exponential distribution (continuous curves) to the geometric distribution (discrete bars). The two distributions are similar when μ and p are similar and small.

The exponential distribution applies only if the rate of events per unit time is constant. For the rate to be constant, the probability that the event occurs in a short interval of time dx must always be the same, αdx. This is not the same as events occurring at regular intervals, which would exhibit a rate of zero except at these regular points in time. It is not necessary, however, for all events to be identical in kind; the exponential distribution continues to apply when events can be broken down into subcategories, just as we saw with the Poisson distribution. For example, the death rate from cancer might be α_1, the death rate from heart attacks might be α_2, and the death rate from other causes might be α_3. Yet if we are only interested in the time until death, its distribution would be exponential with a rate parameter equal to the total death rate $\alpha = \alpha_1 + \alpha_2 + \alpha_3$, as long as α is constant over time.

Examples

If individuals die at a constant rate α per unit time, the lifespan of the individual is described by an exponential distribution. Here, the lifespan is measured in continuous time so that an individual could have, for example, a lifespan of 70.23853 years. This differs from the geometric distribution, which can also be used to describe the age at which an individual dies but which measures lifespan in discrete age classes (e.g., 70 or 71 years). Which distribution is most appropriate depends on the precision desired as well as the information provided about the chance of death. If the chance of death is given as an instantaneous death rate, α, then the exponential distribution should be used. If the chance of death is given as the probability of death within a year, p, then the geometric distribution should be used. That said, we can always convert a death rate α into the mortality risk in one year using the integral

$$p = P(\text{death per year}) = \int_{x=0}^{1} \alpha\, e^{-\alpha x}\, dx = 1 - e^{-\alpha}. \tag{P3.23}$$

Equation (P3.23) provides a way of translating between an exponential distribution in continuous time and a geometric distribution in discrete time (see Exercise P3.5). As a numerical example, if the death rate is $\alpha = 0.1$ per year, then the probability of dying within a year is $p = 0.095$. These numbers predict similar mean life spans ($E[X] = 1/\alpha = 10$ years according to the exponential distribution versus $E[X] = 1/p = 10.5$ years according to the geometric distribution). In fact, whenever the rate of events is small, the exponential and geometric distributions are very similar in shape with $p \approx \alpha$ (Figure P3.10).

The exponential distribution also arises when measuring the distance traveled until a certain event occurs, assuming that the event occurs at a constant rate per distance. For example, if a bee is foraging and stops at flowers at a constant rate, α, per meter, then the distance until it stops at a flower would be exponentially distributed. If α were 0.05 per meter, then the mean distance traveled between flowers would be $E[X] = 1/\alpha = 20$ meters with a standard deviation of $SD[X] = 1/\alpha = 20$ meters. Essentially, we are measuring a waiting time in this example, but in terms of meters traveled rather than chronological time.

Exercise P3.5: Based on the exponential distribution, calculate the probability that an event occurs for the first time between the interval of time $k - 1$ and k. Rewrite your answer in terms of the annual mortality risk by replacing $e^{-\alpha}$ with $1 - p$ (see equation (P3.23)). Show that the result, which describes the probability that the event first occurs within the time interval between $k - 1$ and k, is the same as the geometric probability distribution.

Exercise P3.6: [Advanced]

(a) Calculate the mean for an exponential distribution using Definition P3.10a. Remember to restrict the range of x from 0 to positive infinity.
(b) Calculate the variance for an exponential distribution using Definition P3.10b. Remember to restrict the range of x from 0 to positive infinity.
(c) Calculate the mean and the variance for an exponential distribution using the fact that its moment generating function is $MGF(z) = \alpha/(\alpha - z)$ (see Appendix 5).

P3.4.3 Gamma Distribution

The gamma distribution generalizes the exponential distribution by describing the waiting time until β events occur:

Definition P3.13:

The gamma distribution describes the probability density that x amount of time passes before β events occur when each event occurs at a constant rate, α, per unit time:

$$f(x) = \frac{\alpha^{\beta}}{\Gamma(\beta)} x^{\beta-1} e^{-\alpha x} \qquad \text{for } 0 \leq x \leq \infty,$$

where $\Gamma(\beta) = \int_{y=0}^{\infty} e^{-y} y^{\beta-1} \, dy$ is known as the gamma function.

When $\beta = 1$, $\Gamma(1) = 1$ and the gamma distribution reduces to the exponential distribution.

The gamma distribution can be derived by drawing a connection to the Poisson distribution. For x to be the first time that β events have occurred, there must have been $\beta - 1$ events within the time interval from 0 to x. The probability of observing $\beta - 1$ events in a fixed time interval is described by the Poisson distribution, with an expected number of events equal to the rate of events times the time interval: $\mu = \alpha\, x$. This Poisson probability must then be multiplied by the probability that the βth and final event occurs between time x and $x + dx$ (i.e., α dx). The probability density function for observing β events for the first time near time x is therefore

$$f(x) = \underbrace{\left(\frac{e^{-\alpha x}(\alpha x)^{\beta-1}}{(\beta - 1)!}\right)}_{\substack{\text{Poisson distribution of observing} \\ \beta \,-\, 1 \text{ events given a mean of } \alpha\, x}} \alpha. \tag{P3.24}$$

This derivation assumes that β is an integer, but the gamma distribution is typically written in a more general fashion that allows for any positive value of β (see Definition P3.13). To generalize (P3.24), we replace the factorial $(\beta-1)!$, which is defined for integers only, by a new constant that is chosen to ensure that $f(x)$ integrates to one. Using equation (P3.24) in Rule P3.11, this constant must equal $\int_{x=0}^{\infty} e^{-\alpha x}\alpha^{\beta}x^{\beta-1}dx$. This integral can be simplified by rewriting it in terms of $y = \alpha\, x$ (and hence $dy = \alpha\, dx$), giving $\int_{y=0}^{\infty} e^{-y}\alpha^{\beta}(y/\alpha)^{\beta-1}(dy/\alpha)$. The α terms cancel out of this integral, leaving us with $\int_{y=0}^{\infty} e^{-y}y^{\beta-1}\, dy$, which is known as the gamma function, $\Gamma(\beta)$. The gamma function generalizes factorials to any real number; when β is an integer, $\Gamma(\beta) = (\beta-1)!$. For more facts involving gamma functions, see Abramowitz and Stegun (1972).

The gamma distribution in Definition P3.13 can be thought of as the continuous-time version of the negative binomial distribution. The main difference is that the probability density function is positive for the gamma distribution regardless of how little time has passed, because there is always some small chance that all β events occur in rapid succession. In contrast, we must wait until at least r trials have passed in discrete time before the probability of observing r events is positive with the negative binomial distribution.

The mean and variance of a gamma distribution can be calculated using Definition P3.10. It is easier, however, to use the fact that the gamma distribution represents the sum of β waiting times, each of which is exponentially distributed. Because the expectation of a sum of independent random variables is the sum of the expectations and the same is true for the variance (Table P3.1), we can multiply the mean and variance of the exponential distribution by β to get the mean and variance of the gamma distribution:

$$E[X] = \frac{\beta}{\alpha}, \tag{P3.25}$$

$$\mathrm{Var}[X] = \frac{\beta}{\alpha^2} \tag{P3.26}$$

(see Exercise P3.7).

Examples

Consider an experiment in which you wish to study $\beta = 100$ grooming events in a baboon colony. If the rate of grooming events is two per hour (α = 2/hour), then you would expect the study to take 50 hours (β/α = 100/2 hours) with a standard deviation of 5 hours ($\sqrt{\beta/\alpha^2}$). Furthermore, you can use the gamma distribution to tell you what the chances are that you complete the study by any given time. For example, if you were only able to have 60 hours of observation time, the probability that you will successfully observe 100 grooming events would be 97.2%, as calculated from the integral

$$\int_{x=0}^{60} \frac{\alpha^\beta}{\Gamma(\beta)} x^{\beta-1} e^{-\alpha x}\, dx = 0.972.$$

As another example, imagine that you are collecting truffles in a forest and you find one truffle every 200 meters. The rate at which you encounter truffles is thus α = 1/(200 meters). If you wish to collect 30 truffles (β = 30), you can expect to walk $E[X] = \beta/\alpha$ = 6000 meters. Again, this is fundamentally similar to a waiting time problem, where we are measuring the waiting time in terms of meters traveled.

The shape of the gamma distribution varies from L shaped when β is small to bell shaped when β is large (Figure P3.11, Exercise P3.7). Thus, β is often called the "shape" parameter for the gamma distribution. In contrast, α is called the "scale" parameter. Increasing or decreasing α while holding β constant does not change the shape of the distribution.

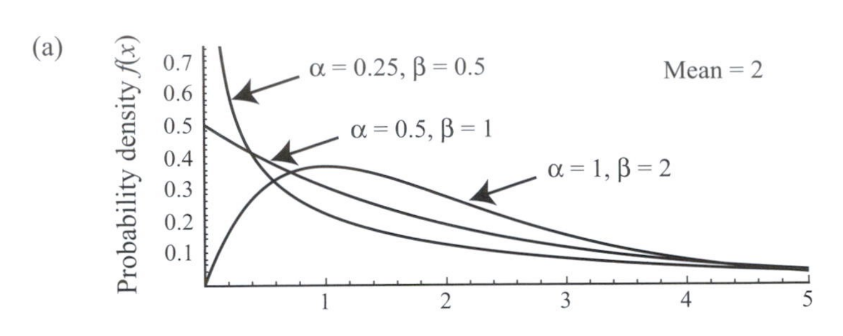

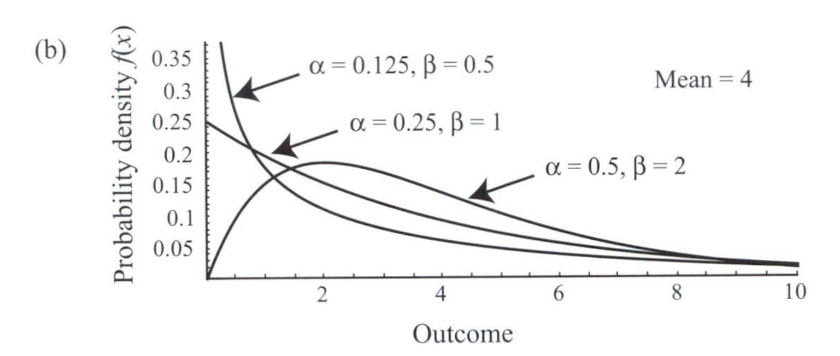

Figure P3.11: Gamma distribution (Definition P3.13). The probability density function, $f(x)$, is plotted for various values of α and β, holding the mean value constant at (a) $E[X] = 2$ and (b) $E[X] = 4$. When $\beta = 1$, the shape of the gamma distribution is the same as the exponential distribution. When $\beta < 1$, the distribution is more L-shaped, with substantial probability density near zero. When $\beta > 1$, the distribution is more bell-shaped. Because the same set of β values is used in (a) and (b), the shapes of the curves are the same, but the horizontal axis has expanded and the vertical axis has shrunk in (b) because the mean has doubled.

Because the gamma distribution is so flexible in shape, it is often used to describe the distribution of an unknown parameter. For example, the gamma distribution is used in analyzing DNA sequences to describe the variation in mutation rates among sites. The use of the gamma distribution is not rigorously justified in this case. This application does not involve the sum of waiting times, for example. Instead, the gamma distribution is used as a heuristic description of what the distribution of substitution rates might look like. Sequence data can then be used to estimate α and β, providing us with information about whether there is a large (β small) or small (β large) degree of variation among sites in mutation rate (Felsenstein 2004; Keightley 1994).

Exercise P3.7:

(a) Calculate the coefficient of variation for the gamma distribution.

(b) Rewrite the probability density function for the gamma distribution replacing α and β in terms of the mean and coefficient of variation.

(c) What must the coefficient of variation be for the gamma distribution to reduce to the exponential distribution? Would smaller values of the coefficient of variation correspond to more L-shaped or more bell-shaped distributions?

P3.4.4 Normal (Gaussian) Distribution

Arguably the most important distribution in biology is the normal or Gaussian distribution:

Definition P3.14a:

The normal distribution is bell-shaped, with a probability density function that falls off exponentially with the squared distance to the mean:

$$f(x) = \frac{e^{-(x-\mu)^2/(2\sigma^2)}}{\sqrt{2\pi\sigma^2}} \qquad \text{for } -\infty \leq x \leq \infty.$$

The denominator ensures that the distribution integrates to one. The mean and variance of the normal distribution are

$$E[X] = \mu, \tag{P3.27}$$

$$\text{Var}[X] = \sigma^2. \tag{P3.28}$$

When σ^2 is small (low variance), the probability density function is very narrow and drops off rapidly in height away from the mean, so that most observations are expected to lie near the mean. Conversely, when σ^2 is large (high variance), the probability density function is very broad (Figure P3.12a).

Historically, the normal distribution has appeared in many different contexts. The normal distribution was first described by Abraham de Moivre (1667–1754), who used it to approximate the binomial distribution and to provide gambling advice to rich patrons. Others, including Pierre Simon de Laplace (1749–1827) and Carl Friedrich Gauss (1777–1855), noticed that measurement errors tend to be normally distributed. In the nineteenth century, Adolphe Quetelet (1796–1874) and Francis Galton (1822–1911) observed that the heights and weights of human and animal populations, along with many other characteristics, roughly follow a normal distribution.

Why does the normal distribution play such a ubiquitous role? The reason lies in one of the most important theorems in statistics first developed by Laplace and known as the central limit theorem.

Rule P3.12: Central Limit Theorem

The sum (or the average) of n independent and identically distributed random variables tends toward a normal distribution as n goes to infinity.

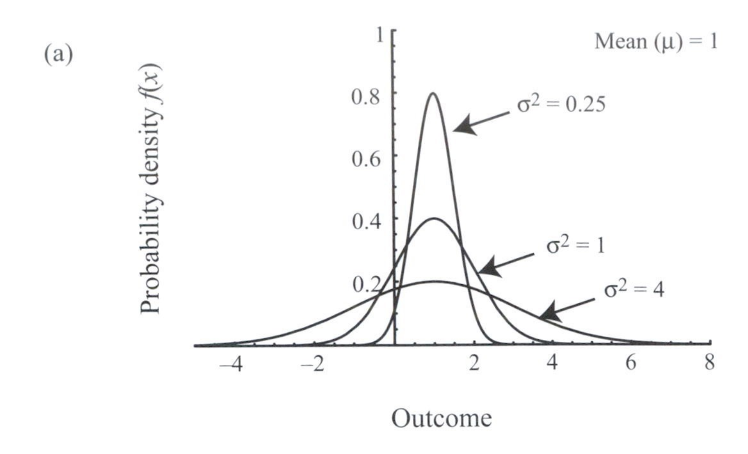

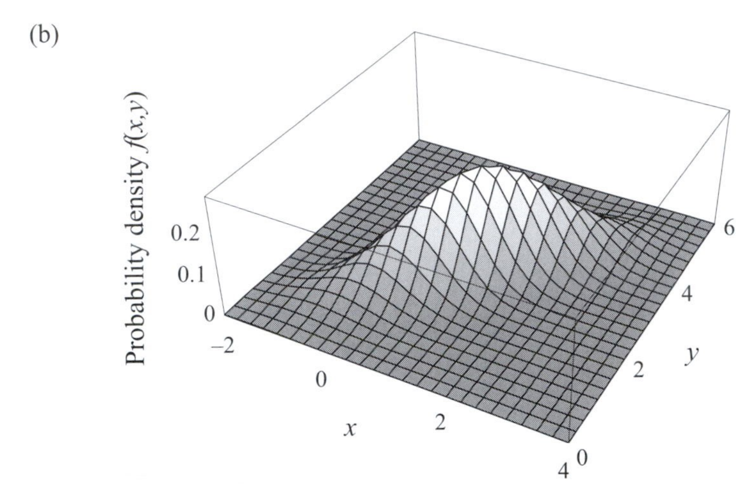

Figure P3.12: Normal distribution. (a) The probability density function, $f(x)$, of a normal distribution (Definition P3.14a) is plotted for various values of the variance, holding the mean value constant at $\mu = 1$. (b) The probability density function, $f(x,y)$, of a bivariate normal distribution (Definition P3.14b) is plotted assuming that the means are $\mu_x = 1$, $\mu_y = 3$, that the correlation between X and Y is $\rho = 0.9$, and that the variances are $\sigma_x^2 = \sigma_y^2 = 1$.

Technically, the central limit theorem requires that each random variable follow a distribution with finite mean and variance, as is the case for the distributions that we have considered. Variants of the central limit theorem have also been proven relaxing the requirement that each random variable is drawn from the same distribution and that the random variables are entirely independent of one another. Basically, as long as enough random variables are combined and these variables are nearly independent, then the combined effect of the random variables looks nearly normal in shape.

The central limit theorem explains why many of the distributions described in this Primer are, under certain circumstances, bell shaped. First, the binomial distribution involves summing the outcome of n independent Bernoulli trials. Thus, the normal distribution provides an excellent approximation for a binomial distribution, as long as n is sufficiently large that multiple successes ($n\,p$) and multiple failures ($n\,(1-p)$) are expected (Figure P3.13). Similarly, the negative binomial distribution involves summing the waiting times needed for r events to occur and is nearly normal in shape when r is large (see Figure P3.5). The same holds for the gamma distribution if we wait for a sum total of β events to occur in continuous time (see Figure P3.11). Even the Poisson

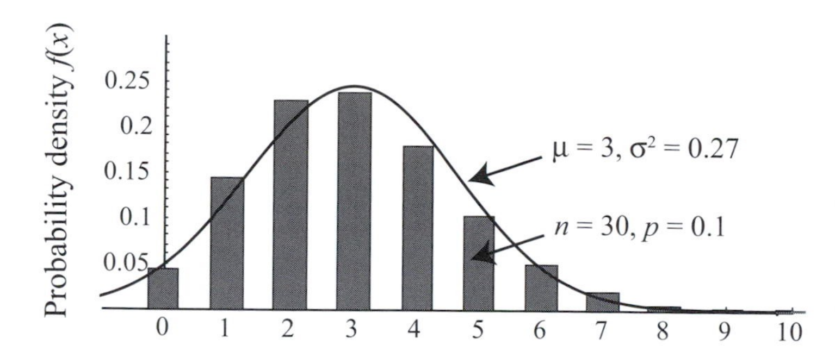

Figure P3.13: Approximating a binomial distribution with a normal distribution. When the number of trials, *n*, is large in a binomial distribution, its shape is approximately normal. Here we compare the normal distribution with mean $\mu = n\,p$ and variance $\sigma^2 = n\,p\,(1 - p)$ to a binomial distribution with parameters *n* and *p*. The fit is good in this example even though *n* is not very large.

distribution is approximately normal in shape when the total number of events is expected to be large (see Figure P3.6). As we increase the number of events being summed, every one of these distributions becomes more bell shaped, that is, more closely approximated by a normal distribution. These distributions are never exactly normal. For example, a negative outcome is not possible with these distributions, whereas negative outcomes are always possible with the normal distribution. Nevertheless, the discrepancy between the true distribution and the normal distribution becomes smaller as more random variables are summed.

The central limit theorem also helps explain why many traits follow a normal distribution, because such traits are typically influenced by a large number of factors (genetic and/or environmental). In this case, the random variable that is being summed (or averaged) is the contribution of each factor to the trait.

An important generalization of the normal distribution is the multivariate normal distribution. Here we present the two-variable (bivariate) case:

Definition P3.14b:

The bivariate normal distribution is a probability distribution for two random variables. It is bell shaped for both random variables and again the probability density function falls off exponentially with the squared distance to the mean of either variable:

$$f(x,y) = \frac{\exp\left[-\left(\left(\frac{x-\mu_x}{\sigma_x}\right)^2 + \left(\frac{y-\mu_y}{\sigma_y}\right)^2 - 2\,\rho\left(\frac{x-\mu_x}{\sigma_x}\right)\left(\frac{y-\mu_y}{\sigma_y}\right)\right)\Big/ 2(1-\rho^2)\right]}{2\,\pi\,\sigma_x\sigma_y\sqrt{1-\rho^2}}$$

for $-\infty \leq x \leq \infty$ and $-\infty \leq y \leq \infty$.

Again the denominator ensures that the integral of the distribution over both variables equals one. The mean and variance of the X variable is calculated just as it was for a one-variable probability density (i.e., from Definitions P3.9 and P3.10), only now we must integrate over both x and y. The same is true for the Y variable. For the bivariate normal, the mean and variance of the two random variables X and Y are

$$E[X] = \mu_x, E[Y] = \mu_y, \tag{P3.29}$$

$$\mathrm{Var}[X] = \sigma_x^2, \mathrm{Var}[Y] = \sigma_y^2. \tag{P3.30}$$

Now, however, there is an additional parameter, $-1 \leq \rho \leq 1$, which is known as the *correlation* coefficient. Positive values of ρ mean that larger than average values of X tend to be associated with larger than average values of Y and vice versa. Negative values of ρ mean that larger than average values of X tend to be associated with smaller than average values of Y and vice versa (Figure P3.12b). The correlation coefficient is related to the covariance of two random variables by the equation $\rho = \mathrm{Cov}[X,Y]/(\sigma_x \sigma_y)$ (Table P3.1).

P3.4.5 Log-Normal Distribution

The log-normal distribution arises when describing the *product* of a large number of independent and identically distributed random variables. If $Y = Y_1 Y_2 \cdots Y_n$, then we expect $X = \ln(Y)$ to become normally distributed as the number of variables increases. This follows from the fact that $\ln(Y) = \ln(Y_1) + \ln(Y_2) + \cdots + \ln(Y_n)$ is the sum of a number of random variables. Thus, the central limit theorem applies to $X = \ln(Y)$ and says that X tends toward a normal distribution. We then say that $Y = e^x$ has a log-normal distribution:

Definition P3.15:
The log-normal distribution describes the distribution of a random variable Y whose natural logarithm is normally distributed:

$$f(y) = \frac{e^{-(\ln(y)-m)^2/(2s^2)}}{y\sqrt{2\pi s^2}} \qquad \text{for } 0 \leq y \leq \infty.$$

The log-normal distribution can be derived from a normal distribution with mean m and variance s^2 (Definition P3.14a) by replacing x with $\ln(y)$ and then using the fact that $dx = d(\ln(y)) = (1/y)\, dy$.

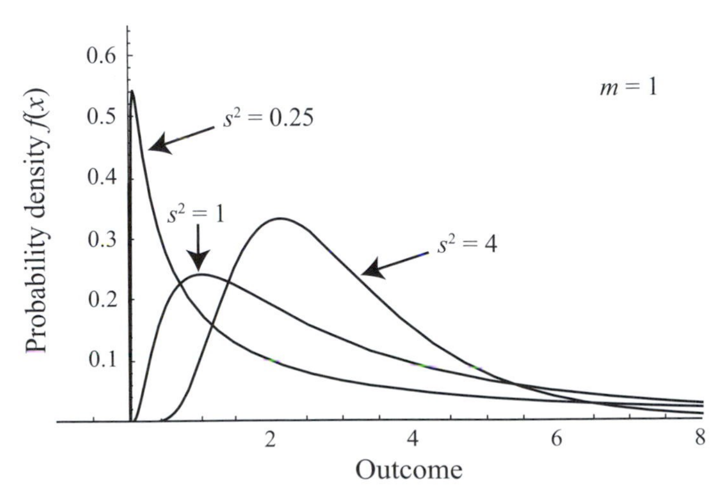

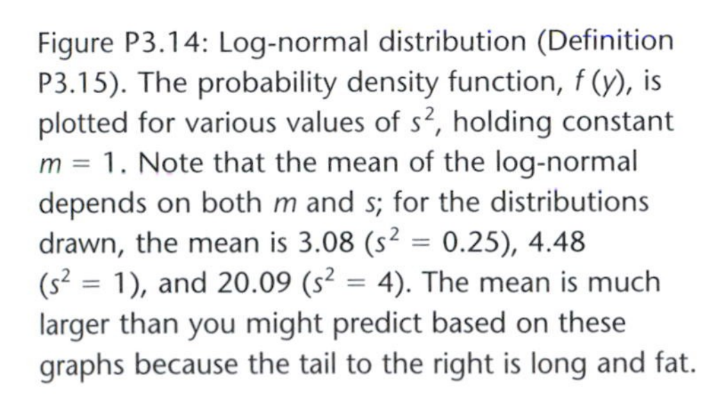

Figure P3.14: Log-normal distribution (Definition P3.15). The probability density function, $f(y)$, is plotted for various values of s^2, holding constant $m = 1$. Note that the mean of the log-normal depends on both m and s; for the distributions drawn, the mean is 3.08 ($s^2 = 0.25$), 4.48 ($s^2 = 1$), and 20.09 ($s^2 = 4$). The mean is much larger than you might predict based on these graphs because the tail to the right is long and fat.

The mean and variance of the log-normal distribution equal

$$E[Y] = e^{m+(s^2/2)}, \quad \text{(P3.31)}$$

$$\text{Var}[Y] = e^{2m+2s^2} - e^{2m+s^2} \quad \text{(P3.32)}$$

The log-normal distribution is asymmetrical with a long tail extending to the right, but it becomes more bell shaped as s^2 decreases (Figure P3.14).

Example

We would expect the abundance of a population to follow a log-normal distribution if growth rates (i.e., r in the exponential growth model) vary over time in an additive fashion, because the population size n is proportional to e^r. Indeed, population size surveys often reveal log-normal distributions in a variety of species, from diatoms to birds (Limpert et al. 2001). We would also expect survival times to be log-normally distributed if several factors have a multiplicative impact on survival (e.g., increasing or decreasing survival by a percentage). In fact, survival times after diagnosis with cancer have been shown to follow a log-normal distribution (Limpert et al. 2001).

P3.4.6 Beta Distribution

For the binomial distribution, we focused on a random variable describing the number of successes, k, out of n trials, where p and n were the parameters of the distribution. What if, instead, we wanted a distribution for the probability of success, p, given that we have observed k successes out of n trials? The appropriate distribution for the random variable, p, is the beta distribution with $a = k + 1$ and $b = n - k + 1$:

Definition P3.16:
The beta distribution describes a probability density for a proportion p:

$$f(p) = \frac{\Gamma(a+b)}{\Gamma(a)\,\Gamma(b)}\, p^{a-1}(1-p)^{b-1} \qquad \text{for } 0 \le p \le 1$$

where a and b are real and positive parameters and $\Gamma(a) = \int_{y=0}^{\infty} e^{-y}\, y^{a-1} dy$ is the gamma function.

To derive the beta distribution, we start by assuming that the probability density function for p is proportional to the binomial distribution, $n!/(k!(n-k)!)\, p^k\,(1-p)^{n-k}$. In other words, values of the random variable p that are more likely to yield k successes out of n trials are given greater probability density. Again, we can generalize this distribution to noninteger parameter values by replacing the binomial coefficient, $n!/(k!(n-k)!)$ (considered here to be a constant given the data), by a new constant chosen to ensure that the probability density function integrates to one. Using Rule P3.11, the constant by which we must divide $p^k(1-p)^{n-k}$ is $\int_{p=0}^{1} p^k\,(1-p)^{n-k}\, dp$. *Mathematica* comes in handy for this integration and gives

$$\int_{p=0}^{1} p^k(1-p)^{n-k}\, dp = \frac{\Gamma(k+1)\,\Gamma(n-k+1)}{\Gamma(n+2)}.$$

Finally, to obtain Definition P3.16, we rewrite the distribution in terms of the parameters, a and b, where $a = k + 1$ and $b = n - k + 1$, as it is more typically written.

The beta distribution has mean and variance

$$E[X] = \frac{a}{a + b}, \tag{P3.33}$$

$$\text{Var}[X] = \frac{a\,b}{(a + b)^2(1 + a + b)} \tag{P3.34}$$

(see Exercise P3.8).

The shape of the beta distribution is extremely flexible (Figure P3.15). It is bell shaped when a and b are similar in magnitude and large. It is a flat line when a and b are one. It is U shaped when a and b are similar in magnitude and smaller than one. The beta distribution can even be L shaped when b is much larger than a or J shaped when a is much larger than b.

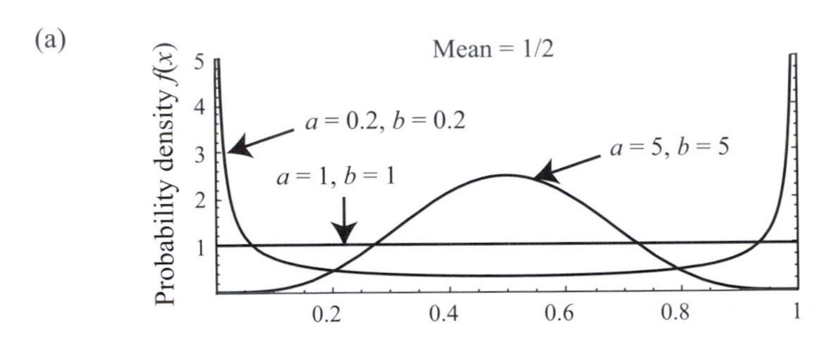

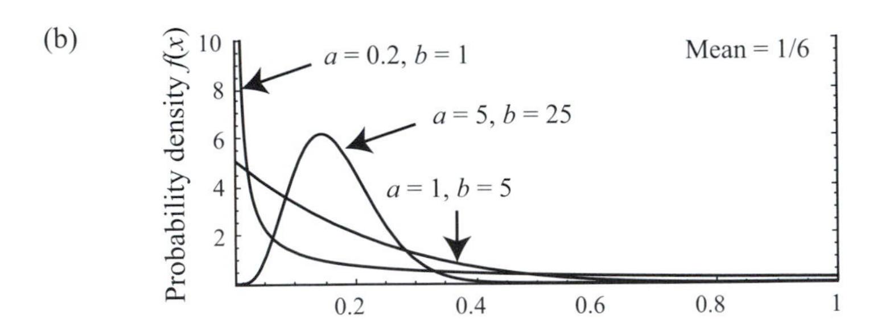

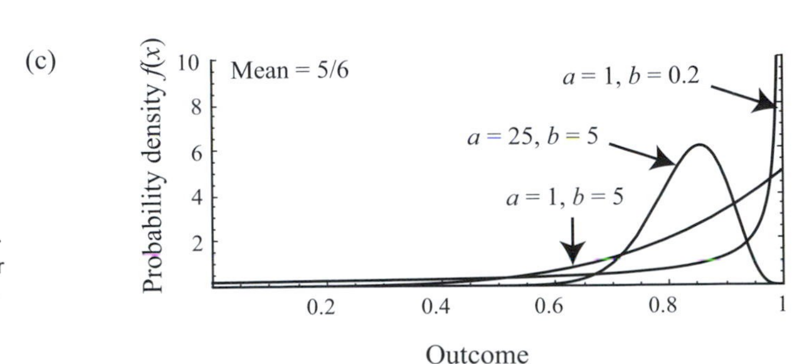

Figure P3.15: Beta distribution (Definition P3.16). The probability density function, $f(x)$, is plotted for various values of a and b, holding the mean value constant at (a) $E[X] = 1/2$, (b) $E[X] = 1/6$, and (c) $E[X] = 5/6$.

Example

As we shall see in Chapter 15, the beta distribution arises in genetic models describing the probability distribution of allele frequencies within a population (see equation (15.25); Crow and Kimura 1970; Turelli 1981). The most frequent context in which the beta distribution arises, however, is in Bayesian analysis where the beta distribution is often used as a prior probability distribution for parameters that lie between 0 and 1 (see Supplementary Material P3.1).

Exercise P3.8: Use the fact that $\int_{x=0}^{1} p^{a-1}(1 - p)^{b-1}dp = \Gamma(a)\Gamma(b)/\Gamma(a + b)$ and $\Gamma(a + 1) = a\Gamma(a)$ to find the mean of the beta distribution.

P3.4.7 Dirichlet Distribution

Just as the multinomial distribution generalizes the binomial distribution for outcomes involving more than two possible states, the Dirichlet distribution generalizes the beta distribution:

Definition P3.17:

The Dirichlet distribution describes the probability density function for the proportions p_i in each of c discrete categories:

$$f(p_1, p_2, \ldots, p_c) = \frac{\Gamma(a_1 + a_2 + \cdots + a_c)}{\Gamma(a_1)\Gamma(a_2)\cdots\Gamma(a_c)} p_1^{a_1-1} p_2^{a_2-1} \cdots p_c^{a_c-1}$$

for $0 \leq p_i \leq 1$

Here the a_i are real and positive parameters (akin to $k+1$ in the multinomial distribution, see Definition P3.5), and the proportions in each state must sum to one: $\sum_{i=1}^{c} p_i = 1$. For example, the Dirichlet distribution arises when describing the frequency distribution of multiple alleles at a locus.

P3.5 The (Insert Your Name Here) Distribution

While we have discussed a number of classic distributions that arise often, it is important to recognize that there are an unlimited number of probability distributions. For many problems, the appropriate distribution might be one of those discussed in this Primer. It is definitely possible, however, that the appropriate distribution is a new one. At that point you should take the plunge and describe your very own distribution. The rules and definitions described in this Primer allow you to check that your distribution correctly sums to one and to determine such things as the mean and variance of the distribution.

For example, you might want to model a population in which there are two types of males, and where females have a preference for one type over the other. To begin, you choose a simple procedure by which females decide on their mates. First, a female randomly encounters a male from the population. If she encounters a type-1 male, she mates with him. If, however, she encounters a type-2 male, she mates with him with probability ϕ. If she remains unmated, she tries again, and the same rules apply. However, after two attempts, the female mates with any male she encounters. Say that you are particularly interested in knowing how often females mate in one, two, or three attempts. You could simulate the above process a number of times to answer this question, but it is much easier to develop the probability distribution.

If the proportion of males that are type 1 is p, then the probability that a female mates in the first attempt is $P(X = 1) = p + (1 - p)\,\phi$, which we will define as F. Of the remaining $1 - F$ females, a similar proportion mates in the second attempt. Thus, $P(X = 2) = (1 - F)\,F$. Any females that remain unmated then mate at the third attempt, so that $P(X = 3) = (1 - F)\,(1 - F)$. This completes our derivation of a new probability distribution:

$$\begin{aligned} P(X = 1) &= F, \\ P(X = 2) &= (1 - F)\,F, \\ P(X = 3) &= (1 - F)^2. \end{aligned} \tag{P3.35}$$

These probabilities sum to one and so obey Rule P3.10. The mean can be calculated using Definition P3.2 as

$$\begin{aligned} E[X] &= P(X = 1) + 2\,P(X = 2) + 3\,P(X = 3) \\ &= 3 - 3F + F^2. \end{aligned} \tag{P3.36}$$

If males of the preferred type are common (p high) or if females are inclined to mate even with males of the second type (ϕ high), then $F = p + (1 - p)\phi$ will be near one and the mean will be near one (Figure P3.16). After a little bit of algebra it is also possible to show that the variance of this distribution equals $\mathrm{Var}[X] = F(1 - F)\,(5 - 5F + F^2)$.

Equation (P3.35) describes the probability that a particular female mates at the first, second, or third attempt. We can use these probabilities within the multinomial distribution (Definition P3.2), however, to describe the number of females mating at the first, second, or third attempt within a population of n females. This would be a lot faster than simulating each female as she chooses her mate, especially if there were thousands of females!

The main distributions described in this Primer are summarized in Tables P3.2 (discrete probability distributions) and P3.3 (continuous probability distributions). These tables provide a quick reference describing the probability distribution, as well as its mean and variance. In addition, moment generating functions are given where they exist and are simple enough to be useful. As described in Appendix 5, moment generating functions provide a quick and relatively painless method for finding higher moments of a distribution. Finally, Tables P3.2 and P3.3 provide one-sentence descriptions of when we might expect each distribution to apply. Remember, however, that probability distributions are not written in stone. If you are interested in a process that is not well described by any probability distribution that you know, forge ahead and develop the appropriate distribution on your own. Who knows, you might go down in history as the person for whom a probability distribution is named!

Exercise P3.9: In equation (P3.35), we described the probability distribution for the number of mating attempts made by a female before she mates. Calculate the overall probability that she mates with a male of type 1. Relate this result to the mean number of trials (P3.36). Use the complement rule (Rule P3.2) to determine the probability that she mates with a male of type 2.

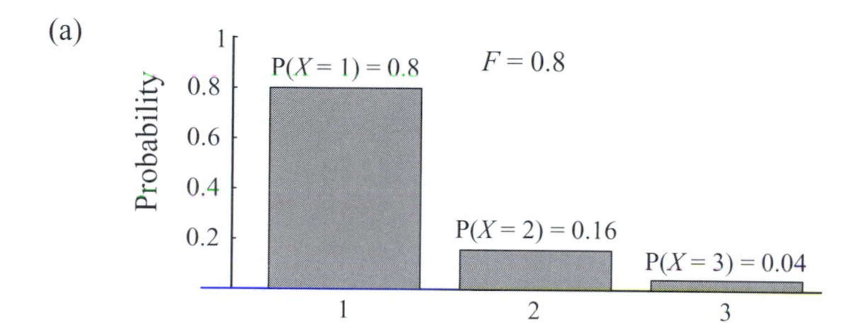

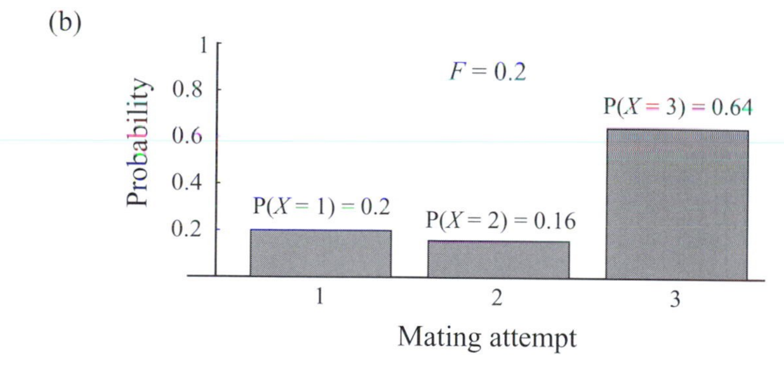

Figure P3.16: Mating probabilities. The probability that a female mates on her nth encounter with a male, according to the probability distribution defined by equation (P3.35). In (a), females are likely to mate with a randomly encountered male ($F = 0.8$), while in (b), females are more choosy ($F = 0.2$).

TABLE P3.2
Discrete probability distributions. Moment generating functions $MGF(z)$ are given where they exist in a useful form (see Appendix 5).

Binomial Distribution Parameters: n, p

Definition P3.4 $P(X = k) = \binom{n}{k} p^k (1 - p)^{n-k}$ for $k = 0,1,2, \ldots, n$

$$E[X] = np$$
$$\text{Var}[X] = np(1-p)$$
$$MGF[z] = (1 - p + e^z p)^n$$

Circumstances: The binomial arises when there are n independent events, each of which can have two outcomes ("success" or "failure"). The probability of observing a total of k successes is $P(X = k)$.

Hypergeometric Distribution. Parameters: N, p, n, where $N_1 = p\,N$ and $N_2 = (1 - p)\,N$

Definition P3.6 $P(X = k) = \dfrac{\binom{N_1}{k}\binom{N_2}{n-k}}{\binom{N}{n}}$ for $k = 0,1,2, \ldots, \min(n, N_1)$

$$E[X] = n\,p$$
$$\text{Var}[X] = n\,p(1 - p)\frac{N - n}{N - 1}$$

Circumstances: The hypergeometric describes the probability of observing k successes when n objects are sampled without replacement from a total pool of N objects, of which a fraction, p, represent a successful outcome.

Geometric Distribution. Parameter: p

Definition P3.7 $P(X = k) = p\,(1-p)^{k-1}$ for $k = 1, 2, 3, \ldots$

$$E[X] = \frac{1}{p}$$
$$\text{Var}[X] = \frac{(1 - p)}{p^2}$$
$$MGF[z] = \frac{e^z p}{1 - (1 - p)e^z}$$

Circumstances: The geometric arises when measuring the number of independent trials, k, until the first success.

Negative Binomial Distribution. Parameters: r, p.

Definition P3.8 $P(X = k) = \binom{k-1}{r-1} p^r (1 - p)^{k-r}$ for $k = r, r + 1, r + 2, \ldots$

$$E[X] = \frac{r}{p}$$
$$\text{Var}[X] = \frac{r\,(1 - p)}{p^2}$$
$$MGF[z] = \left(\frac{e^z p}{1 - (1 - p)e^z}\right)^r$$

Circumstances: The negative binomial arises when measuring the number of independent trials, k, until the rth success.

Poisson Distribution. Parameter: μ

Definition P3.9

$$P(X = k) = \frac{e^{-\mu}\mu^k}{k!} \qquad \text{for } k = 0,1,2,\ldots$$

$$E[X] = \mu$$

$$\mathrm{Var}[X] = \mu$$

$$MGF[z] = e^{\mu(e^z-1)}$$

Circumstances: The Poisson arises when measuring the number of independent events, k, that occur in a certain period (or area) of observation. The Poisson also arises as an approximation to the binomial when n is large and p is small, in which case the mean and variance are well approximated by $\mu = np$.

TABLE P3.3

Continuous probability distributions. Moment generating functions $MGF(z)$ are given where they exist in a useful form (see Appendix 5).

Uniform Distribution. Parameters: min, max

Definition P3.11

$$f(x) = \frac{1}{\max - \min} \qquad \text{for } \min \le x \le \max$$

$$E[X] = \frac{\max + \min}{2}$$

$$\mathrm{Var}[X] = \frac{(\max - \min)^2}{12}$$

$$MGF[z] = \frac{e^{\max z} - e^{\min z}}{(\max - \min)\, z}$$

Circumstances: The uniform distribution arises whenever you are interested in describing where an event occurs for events that have the same chance of occurring anywhere between two points (min and max).

Exponential Distribution. Parameters: α

Definition P3.12

$$f(x) = \alpha e^{-\alpha x} \qquad \text{for } 0 \le x \le \infty$$

$$E[X] = \frac{1}{\alpha}$$

$$\mathrm{Var}[X] = \frac{1}{\alpha^2}$$

$$MGF[z] = \frac{\alpha}{\alpha - z} \qquad \text{for } z < \alpha$$

Circumstances: The exponential distribution arises when measuring the amount of time that passes, x, until an event occurs, measured in continuous time.

Gamma Distribution. Parameters: α, β

Definition P3.13

$$f(x) = \frac{\alpha^\beta}{\Gamma(\beta)} x^{\beta-1} e^{-\alpha x} \qquad \text{for } 0 \le x \le \infty$$

$$E[X] = \frac{\beta}{\alpha}$$

$$\mathrm{Var}[X] = \frac{\beta}{\alpha^2}$$

$$MGF[z] = \left(\frac{\alpha}{\alpha - z}\right)^\beta \qquad \text{for } z < \alpha$$

(continued)

TABLE P3.3 *continued*

Circumstances: The gamma distribution arises when measuring the amount of time that passes, x, until β independent events occur, measured in continuous time.

Normal Distribution. Parameters: μ, σ^2

Definition P3.14a $$f(x) = \frac{e^{-(x-\mu)^2/(2\sigma^2)}}{\sqrt{2\pi\sigma^2}} \qquad \text{for } -\infty \leq x \leq \infty$$

$$E[X] = \mu$$

$$\text{Var}[X] = \sigma^2$$

$$MGF[z] = e^{z\,\mu + z^2\sigma^2/2}$$

Circumstances: The normal distribution arises when several factors sum (or average) together to influence an outcome.

Bivariate Normal Distribution. Parameters: $\mu_x, \mu_y, \sigma_x, \sigma_y, \rho$

Definition P3.14b: $$f(x,y) = \frac{\exp\left[-\dfrac{z}{2(1-\rho^2)}\right]}{2\,\pi\,\sigma_x\sigma_y\sqrt{1-\rho^2}}$$

$$\text{where } z = \left(\frac{x-\mu_x}{\sigma_x}\right)^2 + \left(\frac{y-\mu_y}{\sigma_y}\right)^2 - 2\,\rho\left(\frac{x-\mu_x}{\sigma_x}\right)\left(\frac{y-\mu_y}{\sigma_y}\right) \quad \text{for } -\infty \leq x \leq \infty \text{ and } -\infty \leq y \leq \infty$$

$$E[X] = \mu_x, \quad E[Y] = \mu_y$$

$$\text{Var}[X] = \sigma_x^2, \quad \text{Var}[Y] = \sigma_y^2, \quad \text{Correlation}[X,Y] = \rho$$

Log-normal Distribution. Parameters: m, s^2

Definition P3.15 $$f(y) = \frac{e^{-(\ln(y)-m)^2/(2s^2)}}{y\sqrt{2\pi\,s^2}} \qquad \text{for } 0 \leq y \leq \infty$$

$$E[Y] = \ln\left(\frac{m^2}{\sqrt{m^2+s^2}}\right)$$

$$\text{Var}[Y] = \sqrt{\ln\left(1+\frac{s^2}{m^2}\right)}$$

Circumstances: The log-normal distribution arises when several factors have multiplicative effects on an outcome.

Beta Distribution. Parameters: a, b

Definition P3.16 $$f(p) = \frac{\Gamma(a+b)}{\Gamma(a)\,\Gamma(b)}\,p^{a-1}(1-p)^{b-1}, \qquad \text{for } 0 \leq p \leq 1$$

$$E[X] = \frac{a}{a+b}$$

$$\text{Var}[X] = \frac{ab}{(a-b)^2\,(1+a+b)}$$

Circumstances: The beta distribution arises when estimating an unknown probability or proportion, p.

Box P3.1: Counting and combinatorics

In the description of the binomial distribution, we encountered a quantity $\binom{n}{k}$, read as "n choose k." This quantity, often referred to as the *binomial coefficient*, is an integer that counts up the number of different ways in which k "ones" can occur over the course of n trials. There are other probability distributions that require us to count up the various ways in which things can happen, including the hypergeometric distribution (Definition P3.6) and the negative binomial distribution (Definition P3.8). Being able to enumerate the possible outcomes of a process is invaluable in many other areas of mathematical modeling, as well. The area of mathematics devoted to understanding arrangements of sets of items is *combinatorics*. Here we describe the basics of combinatorics and derive the binomial coefficient.

As an example, consider counting up the number of different genetic strains of a DNA virus that are possible if the virus's genome is 1000 base pairs long. The easiest way to evaluate the total count is to begin with the first position and consider all of the possible nucleotides that might be present (4 for A, C, T, or G). For each one of these, we then move on to the second position and count up the number of possibilities (4 again). Thus, considering only the first two positions, there are $4 \times 4 = 16$ possibilities (AA, AC, AT, AG, CA, CC, CT, CG, TA, TC, TT, TG, GA, GC, GT, and GG). Proceeding onward using similar reasoning, there are $4 \times 4 \times 4 = 4^3$ possibilities for the first three positions (AAA, AAC, AAT, etc.), and each additional position gives 4 times as many possibilities. Over the entire genome, there will thus be 4^{1000} possible sequences. The different possibilities are referred to as *permutations*, and the above process enumerates all of the possible permutations of the genome. In particular, these are the permutations possible with *replacement*; after we assign a particular basepair (say, G) to the first site, we *replace* G back in the list of possibilities (A, C, T, or G) for the next site. More generally:

Rule P3.1.1: Enumerating Permutations with Replacement
Suppose that there are k positions, each of which can be occupied by any one of n different items. The total number of permutations possible is n^k. Because there are always n choices, this is the total number of permutations with replacement.

A related question focuses on enumerating all possible permutations without replacement. As an example, imagine you are studying a migratory bird species and that you have marked the entire population with leg bands so that you can identify each individual. Each season you record the order in which the birds arrive back on the breeding ground. If there are 25 birds in the population, let's count how many different ways the population of birds could return.

Again we can count the number of the possible choices for the first bird to return, the second bird to return, and so on. There are 25 possible choices for the first bird (bird 1, 2, 3, etc.). For each of these, however, there are only 24 possible choices for the second bird because the second bird to return cannot also be the first bird. That is, we do not *replace* the first bird in the list of possibilities for subsequent trials. Thus, considering only the first two birds, there are $25 \times 24 = 600$ possible orderings. Considering the 23 possible birds that can return next, there are

(continued)

Box P3.1 *(continued)*

$25 \times 24 \times 23 = 13{,}800$ possible orderings of the first three birds. Repeating this process and assuming that all 25 birds return, the total number of possible permutations without replacement is $25 \times 24 \times 23 \times \ldots \times 2 \times 1$, which we can write as 25! ("25 factorial").

What if you were only interested in the order of the first k birds to arrive at the breeding ground? By the time you get to the kth bird, $25 - (k - 1)$ birds remain, so that there will be $25 \times 24 \times \ldots \times (25 - k + 1)$ possible orderings. This result can be written more compactly as

$$\frac{25!}{(25-k)!} = \frac{25 \times 24 \times \cdots \times (25-k+1) \times (25-k) \times \cdots \times 1}{(25-k) \times \cdots \times 1}$$

$$= 25 \times 24 \times \cdots \times (25 - k + 1).$$

If we watch all of the birds return ($k = 25$), this equals 25!/0!. By definition, 0! equals one, so we regain the result from the previous paragraph that there are 25! possible permutations without replacement of the entire set of birds. More generally:

Rule P3.1.2: Enumerating Permutations without Replacement
Suppose that there are n different items, each of which is placed in one of k positions, and once placed it is removed from the list of items. The total number of permutations without replacement is then $n \times (n-1) \times (n-2) \times \ldots \times (n-k+1)$ or, equivalently, $n!/(n-k)!$. This result assumes that $n \geq k$ and that $0! = 1$, so that the number of permutations is $n!$ when $k = n$.

The number of permutations without replacement counts the various possibilities when the order in which the various items occur is of interest. For instance, in the bird example, the number of different *orderings* in which the first k birds can arrive is $25!/(25-k)!$. If, instead, we are interested in the identity of the first k birds, but we do not care about the order in which they appear, then we must group together all equivalent possibilities and count only the number of *unordered* possibilities. For example, if we label each bird from A to Y and observe two birds, one possible combination is bird M and bird G, where we do not care whether bird M or G arrived first. The total number of unordered possibilities is referred to as the number of *combinations*.

Returning to the example of migratory birds, let us count the number of possible combinations of birds that might be found in the first three arrivals of the season. Our calculations above reveal that there are $25 \times 24 \times 23 = 13{,}800$ ordered permutations, but some of these contain the same three birds in different orders. For example, suppose that the first three to arrive are birds D, S, and G (in that order). This is one of the 13,800 permutations just mentioned, but S,

(continued)

Box P3.1 *(continued)*

D, and G (in that order) is another of these. From the standpoint of enumerating the possible combinations, however, these two outcomes are the same (because the order does not matter). In fact, there are $3 \times 2 \times 1 = 3!$ different ordering of birds D, G, S that might occur and that would all be considered as the same combination ({D, G, S}, {D, S, G}, {G, D, S}, {G, S, D}, {S, D, G}, and {S, G, D}). Furthermore, this is true for any three specific birds that might be the first to arrive. Therefore, the number of combinations possible for the first three birds is given by the number of permutations of the arrivals (i.e., $25 \times 24 \times 23 = 13{,}800$) divided by the number of equivalent orderings, 3!. The result, $(25 \times 24 \times 23)/3!$, equals 2300 and can be rewritten using Rule P3.1.2 as $25!/((25 - 3)!\,3!)$, which, by definition, equals the binomial coefficient $\binom{25}{3}$. And, in general,

Rule P3.1.3: Enumerating Combinations without Replacement

Suppose that there are n different items, each of which is placed in one of k positions ($n \geq k$) and once placed it is removed from the list of items. The total number of combinations of the k items that are possible equals the binomial coefficient

$$\binom{n}{k} = \frac{n!}{(n-k)!\,k!}.$$

At this point, we are ready to connect the number of combinations to the binomial distribution. Say that we want to know how many ways in which k "ones" can occur in n trials. Let us label the trials as A, B, C, etc., just as we labeled the birds above. Given that k ones occur, we want to keep track of the trial in which they occur. This amounts to assigning one of the n trial names to each of the k ones that have occurred. There is nothing to distinguish the k ones from one another, however, so we only wish to count the number of distinct combinations of k trial numbers. For example, if $k = 3$ ones occur in $n = 25$ trials, one possible outcome is that there was a "one" at trial D, trial G, and trial S. Again, we must group together the 3! different permutations of trial numbers ({D, G, S}, {D, S, G}, etc.) just as we did in the bird example. As a result, there are $n!/((n-k)!\,k!) = \binom{n}{k}$ unordered ways in which the n trial names could be matched to the k ones that have occurred (Rule P3.1.3).

Further Reading

For an introductory text on probability theory, consult

- Pitman, J. 1997. *Probability*. Springer-Verlag, Berlin.
- Larsen, R. J. and M. L. Marx. 2001. *An Introduction to Probability and Its Applications*, 3rd ed. Prentice-Hall, Englewood Cliffs, NJ.

- Taylor, H. M. and S. Karlin. 1998. *An Introduction to Stochastic Modeling*, 3rd ed. Academic Press, New York.

For a more advanced text on probability theory, consult

- Rice, J. A. 1995. *Mathematical Statistics and Data Analysis*, 2nd ed. Duxbury Press, Belmont, Calif.

References

Abramowitz, M., and I. A. Stegun. 1972. *Handbook of Mathematical Functions*. Dover Publications, New York.

Crow, J. F., and M. Kimura. 1970. *An Introduction to Population Genetics Theory*. Harper & Row, New York.

Felsenstein, J. 2004. *Inferring Phylogenies*. Sinauer Associates, Sunderland, Mass.

Hilborn, R., and M. Mangel. 1997. *The Ecological Detective*. Princeton University Press, Princeton, N.J.

Keightley, P. D. 1994. The distribution of mutation effects in *Drosophila melanogaster*. *Genetics* 138:1315–1322.

Kondrashov, A. S. 2003. Direct estimates of human per nucleotide mutation rates at 20 loci causing Mendelian diseases. *Hum. Mutat.* 21:12–27.

Limpert, E., W. A. Stahel, and M. Abbt. 2001. Log-normal distributions across the sciences: Keys and clues. *Bioscience* 51:341–352.

Preti, G., W. B. Cutler, C. R. Garcia, G. R. Huggins, and H. J. Lawley. 1986. Human axillary secretions influence women's menstrual cycles: The role of donor extract of females. *Horm. Behav.* 20:474–482.

Shopland, D. R. 1995. Tobacco use and its contribution to early cancer mortality with a special emphasis on cigarette smoking. *Environ. Health Perspect. Suppl.* 103(S8):131–142.

Turelli, M. 1981. Temporally varying selection on multiple alleles: a diffusion analysis. J. Math. Biol. 13:115–129.

Zar, J. H. 1998. *Biostatistical Analysis*. Prentice-Hall, Upper Saddle River, N.J.

Answers to Exercises

Exercise P3.1

(a) According to Rule P3.4, if the two events are independent, P("tree has fungal rot" ∩ "tree has an owl nest") = (1/100)(1/1000) = 10^{-5}. According to Rule P3.5, if the two events are mutually exclusive, P("tree has fungal rot" ∩ "tree has an owl nest") = 0.

(b) According to Rule P3.4 for independent events, P("blood type O" ∩ "Rhesus negative") = (0.46)(0.16) = 0.0736. That is, roughly 7% of the population is expected to be O−.

(c) According to Rule P3.5 for mutually exclusive events, P("blood type O" ∩ "blood type A") = 0.46 + 0.40 = 0.86. That is, the probability that the donor is acceptable is 86%.

(d) According to Rule P3. 4 for independent events, P("first study is significant" ∪ "second study is significant") = (0.05) + (0.05) − $(0.05)^2$ = 0.0975. Thus, when two studies are performed, there is nearly a 10% chance that at least one of the studies will conclude, incorrectly, that a true hypothesis is false. An alternative way to reach the same answer is by using the complement rule P3.2. The probability that at least one of the two studies obtains a significant result is equal to 1 − P("neither study obtains a significant result"). Because the two studies are independent, we can use Rule P3.4 to calculate P("neither study obtains a significant result") = $(1 - 0.05)^2$. This second method gives the same probability, $1 - (1 - 0.05)^2 = 0.0975$, that one or both of the results is significant.

Exercise P3.2

(a) We start by rewriting the question in terms of probability statements. We want to know P(brown hair | green eyes) given that P(green eyes) = 0.10, P(brown hair) = 0.75, P(brown hair ∩ green eyes) = 0.09. Equation (P3.1a) then can be used to write P(brown hair ∩ green eyes) = P(brown hair | green eyes) P(green eyes). Thus, P(brown hair | green eyes) = 0.09/0.10 and there is a 90% chance of having brown hair given that you have green eyes.

(b) Using Bayes' Rule P3.7, P(taster | carrier) = P(carrier | taster) P(taster)/ P(carrier), which equals (1) (0.7)/(0.8) = 0.875.

(c) Using Bayes' Rule P3.7, P(death due to lung cancer | not a smoker) = P(not a smoker | death due to lung cancer) P(death due to lung cancer) / P(not a smoker). Because P(not a smoker) is the complement of P(smoker), we can use the complement Rule P3.2 to write P(not a smoker) = 1 − P(smoker). The complement rule also applies to conditional statements, so that P(not a smoker | death due to lung cancer) = 1 − P(smoker | death due to lung cancer). Altogether, we get the formula: P(death due to lung cancer | not a smoker) = (1 − P(smoker | death due to lung cancer)) P(death due to lung cancer) /(1 − P(smoker)). Using the data, the risk of death due to lung cancer among non-smokers is P(death due to lung cancer | not a smoker) = (1 − 0.9) (0.3) /(1 − 0.5) = 0.06.

Exercise P3.3

(a) The expected value of X^2 equals $E[X^2] = 0^2 \times P(X = 0) + 1^2 \times P(X = 1) = 0^2 \times (1 - p) + 1^2 \times (p) = p$. Subtracting off the square of the mean of a Bernoulli trial, $\mu^2 = p^2$, gives the variance, $\text{Var}[X] = E[X^2] - \mu^2 = p - p^2$, which again equals $p\,(1 - p)$.

(b) With two Bernoulli trials, the probability of no successes is $P(X = 0) = (1 - p)^2$ (getting a failure on the first trial and then independently getting a failure on the second trial), the probability of getting a single success is $P(X = 1) = p\,(1 - p) + (1 - p)\,p = 2\,p\,(1 - p)$ (having a success followed by a failure or vice versa), and the probability of getting two successes, is $P(X = 2) = p^2$. The sum of these probabilities is $(1 - p)^2 + 2\,p\,(1 - p) + p^2$, which factors to one, as it should. The expected outcome is given by the formula, $E[X] = 0 \times P(X = 0) + 1 \times P(X = 1) + 2 \times P(X = 2)$, which evaluates to $0 \times (1 - p)^2 + 1 \times 2p\,(1 - p) + 2 \times p^2$, which equals $2p$. The variance of the outcome is slightly easier to calculate using $\text{Var}[X] = E[X^2]-\mu^2$. First, we calculate the expected value of X^2: $E[X^2] = 0^2 \times P(X = 0) + 1^2 \times P(X = 1) + 2^2 \times P(X = 2)$, which equals $2p(1-p) + 4p^2$. We then subtract off μ^2, where μ is the mean of two Bernoulli trials, which we have already calculated as $\mu = E[X] = 2p$, leaving us with $\text{Var}[X] = 2\,p\,(1-p)$.

(c) Because $E[X] = E[Y] = p$ for a single Bernoulli trial, the expected outcome from two independent Bernoulli trials is $E[X + Y] = E[X] + E[Y] = 2p$. Because $\text{Var}[X] = \text{Var}[Y] = p(1-p)$ for a single Bernoulli trial, $\text{Var}[X + Y] = \text{Var}[X] + \text{Var}[Y] = 2\,p\,(1-p)$. These results are identical to those obtained in (b).

Exercise P3.4

Using the fact that the variance of a sum is the sum of the variance for independent random variables, $\text{Var}[X + Y] = n_1\,p_1\,(1-p_1) + n_2\,p_2\,(1-p_2)$. For general values of p_1 and p_2, $\text{Var}[X + Y]$ cannot be factored and so cannot be written in the form $np\,(1-p)$. Only if the probability of success is the same for each trial, does the variance factor into the form of the variance of a binomial distribution, $n\,p\,(1-p)$, where $n = (n_1 + n_2)$ and $p = p_1 = p_2$. When $n_1 = 100$ and $p_1 = 1$, $\text{Var}[X] = 0$. Similarly, when $n_2 = 100$ and $p_2 = 1$, $\text{Var}[Y] = 0$. In this example, we would always observe 100 failures and 100 successes, and $\text{Var}[X + Y] = 0$, which is much lower than the expected variance of a binomial, $n\,p\,(1-p) = 50$, for the same total number of events ($n = 200$) and average probability of success ($p = 1/2$).

Exercise P3.5

According to the exponential distribution, the probability, $P(k-1 < X < k)$, that an event occurs between $k-1$ and k is given by $\int_{x=k-1}^{k} \alpha\,e^{-\alpha x} = -e^{-\alpha k} + e^{-\alpha(k-1)}$. Replacing $e^{-\alpha}$ with $1-p$ gives $-(1-p)^k + (1-p)^{k-1}$, which factors to $p(1-p)^{k-1}$. This is the same formula as the probability that an event is first observed at time step k in a geometric distribution (Definition P3.7).

Exercise P3.6

(a) The mean of the exponential distribution is given by (Definition P3.10a)

$$\mu = E[X] = \int_0^\infty x\,\alpha\,e^{-\alpha x}\,dx.$$

Integrating by parts (Rule A2.29 from Appendix 2) with $u = x$ and $v = -e^{-\alpha x}$, $\int x\,\alpha e^{-\alpha x}\,dx = -x\,e^{-\alpha x} - e^{-\alpha x}/\alpha$. In the limit as x goes to positive infinity the indefinite integral goes to zero (both $x\,e^{-\alpha x}$ and $e^{-\alpha x}$ approach 0 as x increases), while at $x = 0$ the indefinite integral becomes $-1/\alpha$. Thus the definite integral from $x = 0$ to infinity is $\mu = 1/\alpha$.

(b) The variance of the exponential distribution is given by (Definition P3.10b)

$$\mathrm{Var}[X] = E[X^2] - \mu^2 = \left(\int_0^\infty x^2\, \alpha\, e^{-\alpha x}\, \mathrm{d}x \right) - \frac{1}{\alpha^2}$$

Integrating by parts (Rule A2.29), starting with $u = x^2$ and $v = -e^{-\alpha x}$, $\int x^2\, \alpha\, e^{-\alpha x}\, \mathrm{d}x = -x^2\, e^{-\alpha x} - \int 2\, x\, e^{-\alpha x}\, \mathrm{d}x$. From part (a), we know that $(2/\alpha) \int x\, \alpha\, e^{-\alpha x}\, \mathrm{d}x = (2/\alpha)\, (-x\, e^{-\alpha x} - e^{-\alpha x}/\alpha)$. Evaluating the definite integral then gives $E[X^2] = 2/\alpha^2$, so that $\mathrm{Var}[X] = 2/\alpha^2 - 1/\alpha^2 = 1/\alpha^2$.

(c) The mean is given by $(\mathrm{d}(MGF(z))/\mathrm{d}z)|_{z=0}$, which equals $\alpha/(\alpha - z)^2|_{z=0} = 1/\alpha$. The variance can be calculated using $\mathrm{Var}[X] = E[X^2] - \mu^2$ (Definition P3.10b). $E[X^2]$ is given by $(\mathrm{d}^2(MGF(z))/\mathrm{d}z^2)|_{z=0}$, which equals $2\alpha/(\alpha - z)^3|_{z=0} = 2/\alpha^2$. Thus the variance equals $2/\alpha^2 - (1/\alpha)^2 = 1/\alpha^2$. Alternatively, the variance can be calculated directly from the central moment generating function $CMGF(z) = e^{-z/\alpha}\, \alpha/(\alpha - z)$ as $(\mathrm{d}^2(CMGF(z))/\mathrm{d}z^2)|_{z=0}$, which also equals $1/\alpha^2$.

Exercise P3.7

(a) $CV = \sqrt{\dfrac{\mathrm{Var}[X]}{E[X]^2}} = \dfrac{1}{\sqrt{\beta}}$.

(b) Rearranging (a), $\beta = 1/CV^2$ and thus $\alpha = 1/(\mu CV^2)$. This allows us to rewrite the probability density function for the gamma distribution as

$$f(x) = \frac{(\mu e CV^2)^{-(1/CV^2)}}{\Gamma\left(\dfrac{1}{CV^2}\right)} x^{(1/CV^2)-1} e^{-x/(\mu\, CV^2)}.$$

(c) The coefficient of variation for an exponential distribution equals 1. Smaller values of CV (i.e., larger values of β) correspond to more bell-shaped distributions. As the coefficient of variation goes to zero, the probability density function narrows, and most outcomes are observed near the mean, μ.

Exercise P3.8

Because $\int_{p=0}^{1} p^{a-1}(1-p)^{b-1}\, \mathrm{d}p = \Gamma(a)\Gamma(b)/\Gamma(a+b)$, it must be the case that $\int_{p=0}^{1} p\, p^{a-1}\, (1-p)^{b-1}\, \mathrm{d}p = \Gamma(a+1)\Gamma(b)/\Gamma(a+b+1)$. Thus, the expectation of the beta distribution is $E[X] = \int_{p=0}^{1} p\, f(p)\, \mathrm{d}p = \Gamma(a+1)\, \Gamma(b)\, \Gamma(a+b)/\Gamma(a+b+1)\, \Gamma(a)\, \Gamma(b)$. Because $\Gamma(a+1) = a\, \Gamma(a)$ and $\Gamma(a+b+1) = (a+b)\Gamma(a+b)$, this reduces to $E[X] = a/(a+b)$.

Exercise P3.9

In each trial a female has a chance p of encountering a male of type 1. The chance that she has not mated before the kth trial is $(1 - F)^{k-1}$. Multiplying these two together and summing over all possible numbers of trials, we find that the probability that a female mates with a male of type 1 is $p + p\,(1 - F) + p\,(1 - F)^2$, which simplifies to $p\,(3 - 3\,F + F^2)$. This equals p times the mean number of trials, which in hindsight, makes sense. Because there are only two mutually exclusive outcomes (she mates with a male of type 1 or with a male of type 2), the probability that she mates with a male of type 2 is $1 - p\,(3 - 3\,F + F^2)$. Note that if we ignore the mechanics behind how a choice was made, whether a female mates with a male of type 1 or type 2 is described by a Bernoulli trial.

CHAPTER 13

Probabilistic Models

Chapter Goals:

- To give examples of dynamical models involving chance events
- To discuss how to incorporate chance events into simulations

Chapter Concepts:

- Demographic stochasticity
- Environmental stochasticity
- Birth-death process
- Wright-Fisher model
- Random genetic drift
- Individual-based model
- Moran model
- Coalescent theory

13.1 Introduction

All of the models considered so far have been deterministic; that is, the models predict that the system will be at one specific state at any given time. If a deterministic model forecasts that $n(t) = 50$, then the implication is that there will be exactly 50 individuals at time t. The real world is never so certain. Individuals may fail to reproduce or produce a bonanza crop of offspring simply by chance. Even if 50 is the most likely number of individuals, we might find 49 or 51 individuals, and there might even be some chance that the population is extinct or that it numbers in the millions. To account for such uncertainty, we must broaden our models. We need models that describe the realm of possible states; such models are known as "stochastic" or "probabilistic" models.

> **Definition 13.1: Stochastic Model**
> A model describing how the probability of a system being in different states changes over time.

Before embarking on the material in this chapter, first familiarize yourself with the principles of probability theory introduced in Primer 3. The core of this chapter focuses on developing stochastic models and simulating them, much as we did in Chapter 4 for deterministic models. Then, in Chapters 14 and 15, we introduce various methods that can be used to analyze stochastic models.

This chapter frequently relies on drawing random numbers from a probability distribution. Although computers cannot generate truly random numbers (everything they do is specified deterministically by computer code), there are many programs that generate "pseudo-random" numbers (see Press 2002). Pseudo-random numbers are determined by an algorithm in such a way that it is difficult to detect a pattern between successive numbers. For example, it is difficult to detect a pattern in the series: 1, 5, 9, 2, 6, 5, 3, 5, 8, 9, 7, 9, 3, 2, 3, . . . , but in fact these numbers are the digits in π (3.14159265358979323 . . .) that follow 3.14. Thus, we could use an algorithm that calculates π to get a series of pseudo-random integers between zero and nine. More sophisticated algorithms are described in Press (2002), which also discusses how random numbers can be drawn from different probability distributions (e.g., Poisson, binomial, normal, etc.). In this chapter, we use the random number generators of *Mathematica* to simulate stochastic models, and we provide the code for generating each figure in the on-line supplementary material.

In the next four sections, we introduce the most fundamental stochastic models in ecology and evolution. Sections 13.2 and 13.3 describe stochastic models of population growth in discrete and continuous time, respectively. Similarly, sections 13.4 and 13.5 describe stochastic models of allele frequency change. To give a flavor for the breadth of stochastic models, we then explore three other models. Section 13.6 develops a stochastic model of cancer to illustrate how new models are explored. Section 13.7 introduces the concept of a spatially explicit stochastic model, which tracks the number and location of individuals within a population. Finally, section 13.8 is slightly more advanced and describes a relatively new and important branch of evolutionary theory, known as *coalescent theory*, which traces the ancestry of a sample.

13.2 Models of Population Growth

We begin by developing a stochastic model of population growth. A general deterministic model of population growth in discrete time is $n(t+1) = R\, n(t)$, where R might be a constant as in the exponential model or a function of the current density as in the logistic model. The equivalent stochastic model describes the *probability* of observing $n(t+1)$ individuals at time $t+1$, given that there are $n(t)$ individuals at time t. Again, a stochastic model might consider the passage of time to occur in discrete time steps or continuously.

Consider a species that reproduces once per season, at which point all of the parents die (i.e., nonoverlapping generations). To determine the number of individuals in the next generation, we must know the probability distribution describing the number of offspring per parent (Figure 13.1). That is, we must specify the probability that each reproducing parent is replaced in the next time unit by 0, 1, 2, etc. offspring. For a species with separate sexes, this exponential model counts only females and assumes that there are always enough males to fertilize these females. For a species that is hermaphroditic, each individual within the population is considered to be a reproducing parent. The distribution of offspring number will vary from species to species, but a simple (albeit arbitrary) choice is that the number of surviving offspring per parent follows a Poisson distribution (Figure P3.6). A Poisson distribution has only one parameter, μ, which gives both its mean and its variance. If the population size were initially $n(t) = 10$ and the mean number of offspring per parent, R, were 1.2, then there would be $n(t+1) = 1.2 \times 10 = 12$ offspring in a deterministic model. Even though 1.2 is the expected number of offspring per parent, however, any one parent will have a random number of offspring, which we draw from a Poisson distribution with mean $R = 1.2$. For example, the number of offspring per parent for each of the ten parents might be

$$4, 2, 0, 4, 1, 1, 1, 0, 0, 0$$

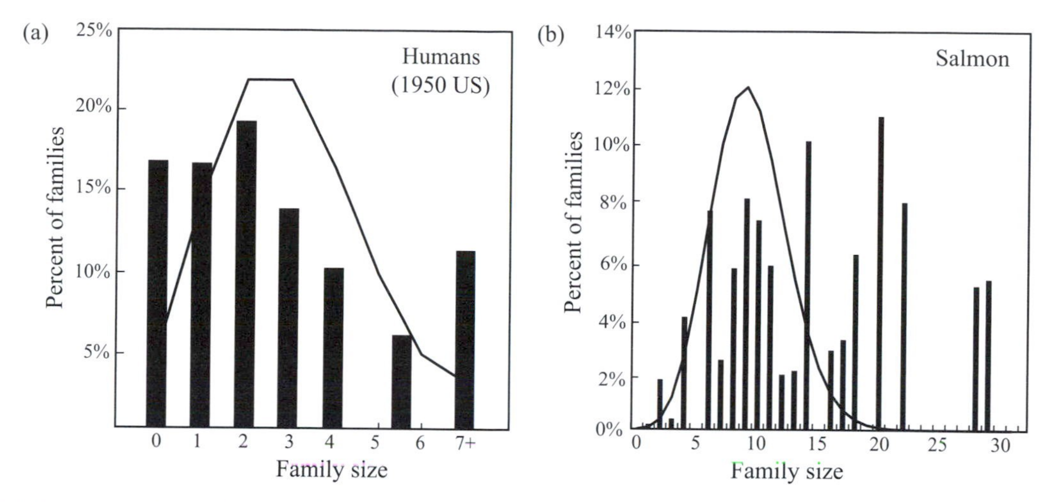

Figure 13.1: Family size distributions. (a) Distribution of family sizes for humans based on Kojima and Kelleher (1962). (b) Distribution of the number of offspring that survive and return to spawn per female in pink salmon, based on Figure 3b in Geiger *et al.* (1997). For these species, family size is more variable than predicted by the Poisson distribution (solid curves) with the same mean as the empirical distribution (histograms).

for a total of 13 surviving offspring. (We used *Mathematica* to draw these random numbers from a Poisson distribution.)

In this example, we expected the population to increase in size (from 10 to 12), but it actually increased even more (to 13). By chance, two of the parents left a surprisingly large number of offspring (four). Retracing our steps and drawing another random set of ten numbers from a Poisson distribution with mean $R = 1.2$ gives an entirely different outcome:

$$0, 1, 0, 1, 1, 1, 1, 3, 0, 1$$

for a total of 9 surviving offspring. In this case, the population size decreased.

To simulate population growth using a stochastic model, we could use random numbers to specify the number of offspring per parent in each subsequent generation. Given $n(t)$ parents at time t, the numbers of offspring per parent could be randomly drawn and the total set to $n(t+1)$. Repeating the process to determine how many offspring are born to each of these parents would give us $n(t+2)$. We could repeat this procedure for as many generations as desired. The simulation, however, would get slower and slower as the population size increased, because we must draw $n(t)$ random numbers, each one specifying the number of offspring per parent.

Fortunately, knowledge of probability theory can help us. We only care about the total number of offspring, and therefore we need only draw a single random number from a distribution that represents the sum of $n(t)$ draws from a Poisson distribution with mean R. The sum of $n(t)$ numbers drawn from a Poisson distribution with mean R is known to follow a Poisson distribution with mean $\mu = R\, n(t)$ (Supplementary Material P3.2). Thus, we can simulate a population in which $R = 1.2$ and $n(0) = 10$ by drawing a single random number from a Poisson with mean $\mu = 1.2 \times 10 = 12$. Using *Mathematica*, we obtained a random number of offspring equal to $n(1) = 21$. To get $n(2)$, we then

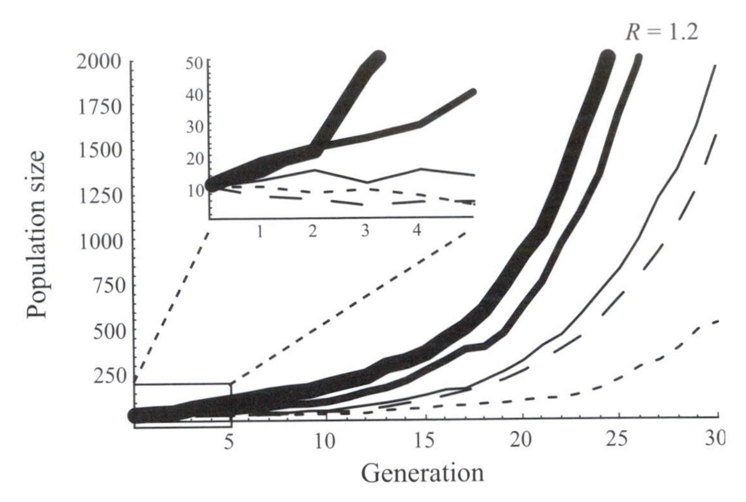

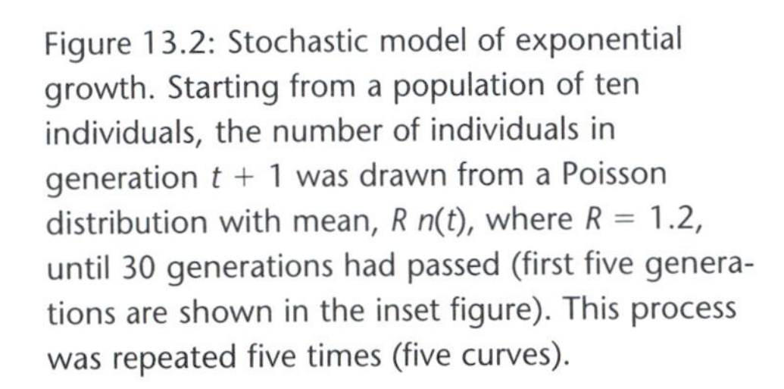
Figure 13.2: Stochastic model of exponential growth. Starting from a population of ten individuals, the number of individuals in generation $t + 1$ was drawn from a Poisson distribution with mean, $R\, n(t)$, where $R = 1.2$, until 30 generations had passed (first five generations are shown in the inset figure). This process was repeated five times (five curves).

drew a random number from a Poisson with mean $1.2 \times 21 = 25.2$, generating 26 offspring, by chance. In Figure 13.2, we show the resulting trajectory of population growth over 30 generations. Finally, we started the whole process over again from $n(0) = 10$ to generate the different curves (replicates).

The different curves in Figure 13.2 look as if they were drawn using different reproductive ratios R, but they weren't. In each case, $R = 1.2$. In the case of the top curve, the parents just happened to have more offspring early on in the simulation than in the case of the bottom curve. As in many stochastic models, there is a lot of variability in the outcome. Consequently, it is important to run several replicates of a stochastic simulation, starting with the same initial conditions and parameters, but drawing new random numbers each time step. We can then summarize the outcomes to draw conclusions. For example, we ran 100 replicate simulations with $n(0) = 10$ and $R = 1.2$. On average, 2470 offspring were alive after 30 generations. The standard deviation was 1739 offspring, indicating that the replicates varied substantially from one another. Indeed, the population had gone extinct in 3 of the 100 replicates. This variability in outcome is referred to as *demographic stochasticity*.

Variability in population size caused by chance differences in the number of surviving offspring per parent is known as *demographic stochasticity*.

The above simulations modeled exponential growth, where the mean number of offspring per parent, R, was the same regardless of population size. It is easy to incorporate density dependence by specifying how the mean of the Poisson distribution, $\mu = R(n)\, n(t)$, depends on the current population size. For example, we can run a stochastic simulation of the logistic model (3.5a) using $R(n) = 1 + r\,(1 - n(t)/K)$. If we let $r = 0.2$, R would again be 1.2 at low population sizes ($n(t) << K$). As the population size gets larger, however, the mean number of offspring per parent drops. With $n(0) = 10$, $r = 0.2$, and $K = 100$, the total number of offspring is Poisson distributed with mean $R(n(0))\, n(0) = (1 + 0.2\,(1 - 10/100))\, 10 = 11.8$. When we drew such a random number, we got $n(1) = 12$. In the next generation, the sum total number of offspring would follow a Poisson distribution with mean $\mu\, n(1) = (1 + 0.2\,(1 - 12/100))\, 12 = 14.1$, from which we drew a random number of $n(2) = 16$. Following this

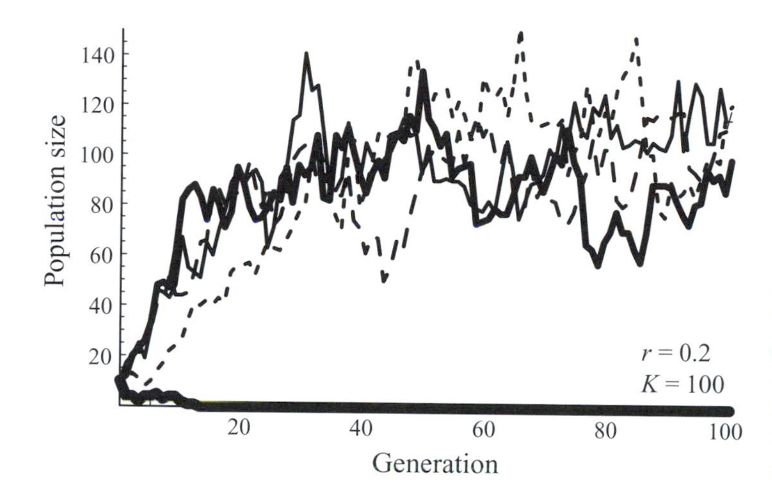

Figure 13.3: Stochastic model of logistic growth. Starting from a population of ten individuals, the number of individuals in generation $t + 1$ was drawn from a Poisson distribution with mean, $(1 + r\,(1 - n(t)/K))\,n(t)$, where $r = 0.2$ and $K = 100$, until 100 generations had passed. This process was repeated five times (five curves).

procedure for 100 generations and repeating the entire process five times gave us the data for Figure 13.3.

Although one replicate population out of five went extinct (again due to demographic stochasticity), the other four hovered around the carrying capacity of 100 and exhibited much less variability than Figure 13.2. Density dependence dampened the amount of demographic stochasticity by reducing the subsequent growth in those populations that happened to grow rapidly early on.

In Figures 13.2 and 13.3, we held the parameters R, r, and K constant, but environmental fluctuations can cause the parameters of a model to vary as well. This is referred to as *environmental stochasticity*. We can incorporate environmental stochasticity in the exponential growth model of Figure 13.2 by drawing the mean number of offspring per parent, R, from a probability distribution. For simplicity, assume that there are good years and bad years, with reproductive ratios R_g and R_b. If the chance that a year is good is p, the type of year will represent a Bernoulli random variable (Primer 3). We model environmental stochasticity by drawing a random number to determine the type of year. Specifically, each year, we draw a random number between 0 and 1 (uniformly); if the random number is less than p, the year is good; otherwise it is bad (Figure 13.4).

Variability in population size caused by chance fluctuations in the environment is known as *environmental stochasticity*.

The results in Figure 13.4 are dramatically different from Figure 13.2. The population size plummets during bad years, causing the trajectories to fluctuate wildly. Consequently, the risk of extinction is much higher. Indeed, out of 100 replicates with an average R of 1.2 and $n(0) = 10$, extinction occurred for 37 of the populations within 30 generations compared to only 3 with demographic stochasticity alone. Furthermore, the population size at generation 30 was smaller, on average (1775 versus 2470), with a much greater standard deviation (11,689 versus 1739).

These stochastic models of population growth exhibit fluctuations in population size regardless of the growth rate r. We also saw fluctuations in population size in the entirely deterministic model of logistic growth in discrete time when growth rates were high (Figure 4.2 and Box 4.1). Given data on changes over time in the size of a population, it can be difficult to determine the source of fluctuations (demographic stochasticity, environmental changes, or chaos).

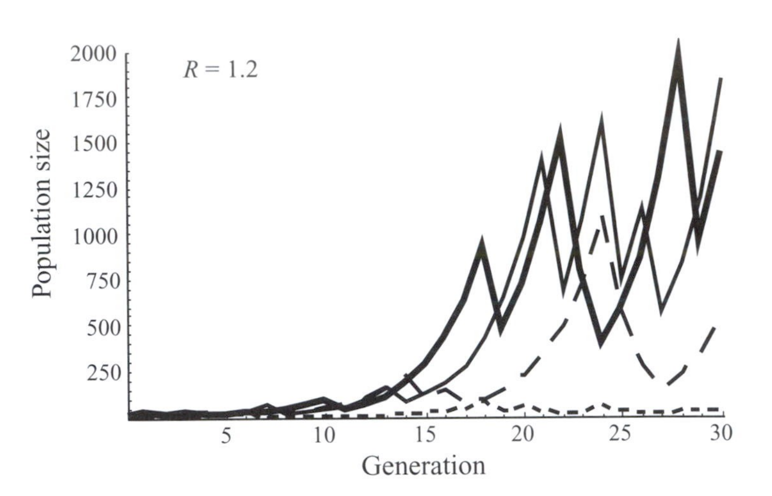

Figure 13.4: Exponential growth with demographic and environmental stochasticity. The probability of a good environment was $p = 0.7$. Each year, a random number, X, was drawn uniformly between 0 and 1 to determine if the current environment was good (if $X < 0.7$) or bad (if $X > 0.7$), where the reproductive factors in good and bad environments were $R_g = 1.5$ and $R_b = 0.5$, respectively. The average growth factor, $p\,R_g + (1 - p)\,R_b = 1.2$, is the same as Figure 13.2. One replicate went extinct after eight generations.

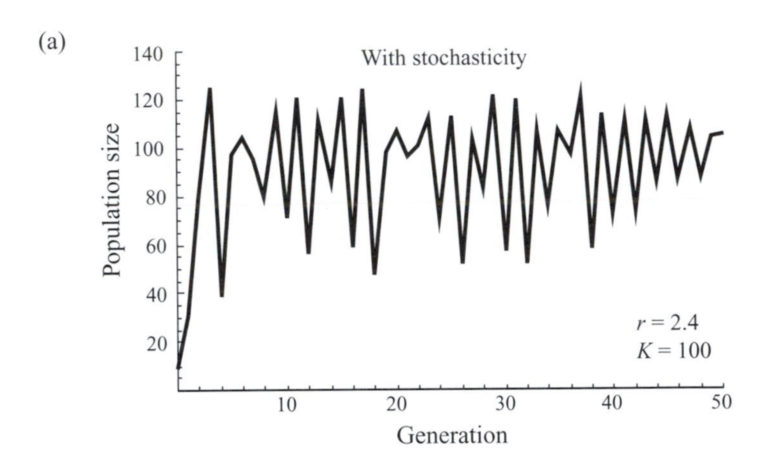

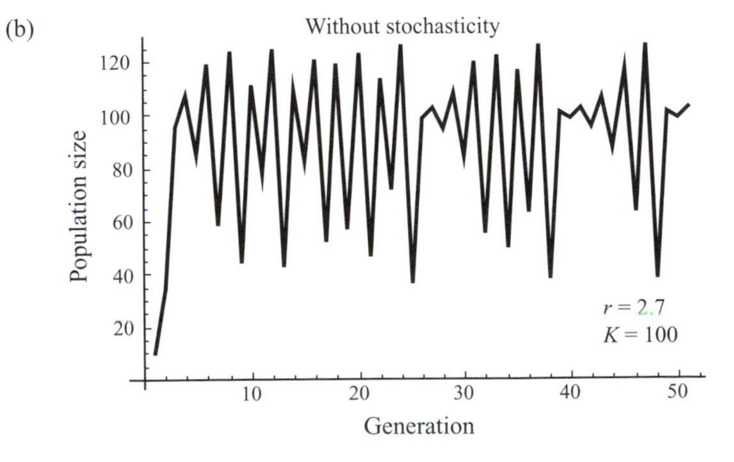

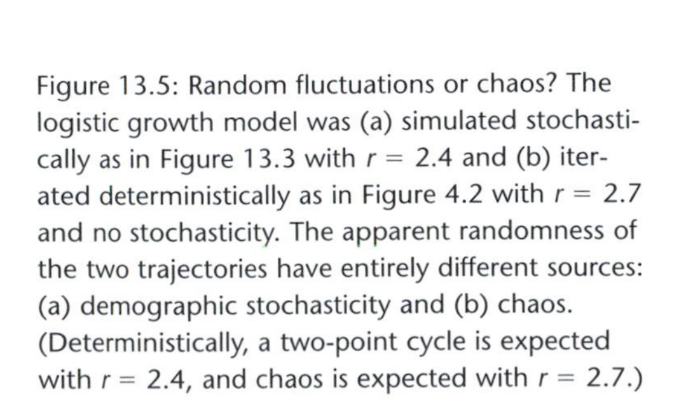
Figure 13.5: Random fluctuations or chaos? The logistic growth model was (a) simulated stochastically as in Figure 13.3 with $r = 2.4$ and (b) iterated deterministically as in Figure 4.2 with $r = 2.7$ and no stochasticity. The apparent randomness of the two trajectories have entirely different sources: (a) demographic stochasticity and (b) chaos. (Deterministically, a two-point cycle is expected with $r = 2.4$, and chaos is expected with $r = 2.7$.)

This point is illustrated in Figure 13.5, where panel (a) is a simulation of the stochastic logistic model with Poisson variation in the number of offspring per parent with $r = 2.4$ and panel (b) is a simulation of the deterministic logistic model (3.5a) with no variation in offspring number per parent and $r = 2.7$. These graphs look very similar, but they differ fundamentally in that the second graph is not random at all—each population size is exactly determined by the population size in the previous generation according to equation (3.5a).

More generally, several mechanisms can be acting simultaneously to affect population dynamics. The statistical field known as "time series analysis" was born to interpret data measured over time and to identify underlying dynamic forces. For example, spectral analysis determines whether there are cycles of particular frequencies within time series and can be used to ascribe these cycles to abiotic (e.g., climatic) and biotic (e.g., predator-prey) fluctuations (e.g., Loeuille and Ghil 2004). The interested reader is referred to Bjørnstad and Grenfell (2001), who review the literature on time series analysis applied to animal population dynamics, and to Kaplan and Glass (1995) for an introduction to time series analysis.

13.3 Birth-Death Models

In the previous section, generations were discrete and the entire population reproduced simultaneously. For populations in which reproduction is not synchronized, we need a different class of models. Imagine a vial of yeast. Yeast replicate by binary fission, but not every cell divides at the same time. If we were to track the population, we might see one cell divide and then another. Starting from only a few cells in the vial, we would initially observe few events per minute because there are so few cells replicating. As the population of cells expands, more and more new cells would be created each minute, causing cell "births" to occur in rapid succession.

How might we simulate this scenario? Let us start with a single cell. The chance that the cell replicates in any small unit of time Δt is $b\,\Delta t$, where b stands for the birth rate. As long as b is constant, the waiting time until the cell divides is exponentially distributed (see Definition P3.12) with mean $1/b$. For example, under nutrient-rich conditions, the mean time to cell division is approximately 90 minutes ($b = 0.011$ divisions per minute). In a simulation, we could draw a random number from the exponential distribution with parameter $\alpha = b$ to simulate the waiting time until cell division. Using *Mathematica*, we drew a waiting time of 83 minutes. Now we have two cells. As long as we don't care which cell divides, the total rate of cell division is twice what it was before, $\alpha = 2b$, and the distribution of waiting times is still exponential (Primer 3). Again using *Mathematica*, we drew a waiting time for the next cell division of 44 minutes from an exponential distribution with $\alpha = 0.022$. We could thus illustrate population growth as a series of steps rising from one cell to two cells at 83 minutes, to three cells after another 44 minutes, etc. To calculate the length of each step, we would draw a random number from an exponential distribution with mean $\alpha = b\,n(t)$ where $n(t)$ is the number of cells at time t (Figure 13.6).

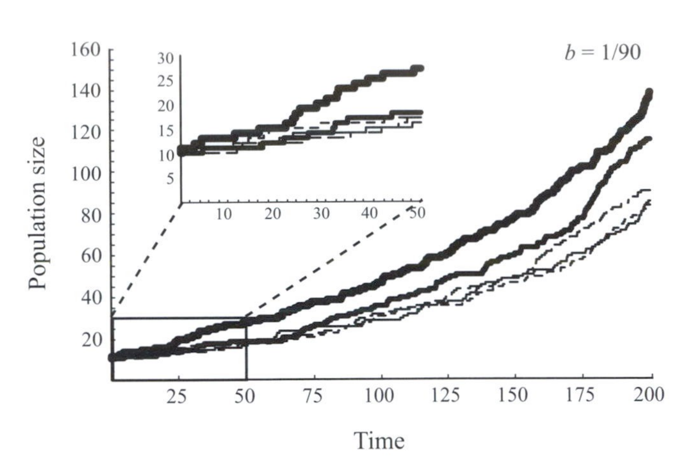

Figure 13.6: Birth process. A cell is chosen divide at a time randomly drawn from an exponential distribution with mean $b\,n(t)$, where $n(t)$ is the population size after the previous cell division. The division rate per cell was $b = 1/90$ per unit of time, and the initial population size was $n(0) = 10$. Five replicates are shown, and the inset shows the first 50 minutes.

This stochastic model is known as a *pure-birth process* or a *Yule process*, in honor of George Udny Yule (1924), who used this model to fit data on the number of species per genus assuming that speciation was akin to a birth. There is quite a bit of variation generated by a birth process, especially early on when few individuals are replicating (Figure 13.6 inset). The variation is not, however, as dramatic as in the stochastic model with discrete generations illustrated in Figure 13.2. In particular, the steps always rise upward because we allow only births within the population, but no deaths.

A *birth-death process* tracks changes to a population through births and deaths, assuming that only one event happens at a time.

We can extend the birth process to account for deaths by allowing individuals to die at rate d per individual per unit time as well as replicate at rate b. Such a model is known as a *birth-death process*. With $n(t)$ individuals, the waiting time until the next event happens, regardless of whether it is a birth or a death, depends on the total rate of events $\alpha = (b + d)\,n(t)$. When the event occurs, however, we must classify it as a birth or a death in order to track the resulting change in the population size.

In general, the chance that an event is a birth is given by $b/(b+d)$. For example, there is a 50% chance that the event is a birth when the birth and death rates are equal ($b = d$). This expression is fairly intuitive, but we can derive it formally using Rule P3.6. We wish to know the probability that a birth occurs in a time interval, Δt, given that either a birth or a death occurs in this interval. Using Rule P3.6, $P(\text{birth} \mid \text{birth or death}) = P(\text{birth} \cap \text{birth or death})/P(\text{birth or death})$. The event "birth ∩ birth or death" is read "birth and a birth or a death," and it can occur only if a birth occurs, which happens with probability $b\,\Delta t$; so $P(\text{birth} \cap \text{birth or death}) = b\,\Delta t$. Also, the probability of a birth or death is just $P(\text{birth or death}) = (b + d)\,\Delta t$. Therefore, we have $P(\text{birth} \mid \text{birth or death}) = b/(b + d)$.

We will analyze a birth-death process in Chapter 14, but to prepare for this analysis, let us summarize the behavior of the model in terms of the transitions possible in a small amount of time, Δt. Using an upper-case N to denote the random variable "population size", the probability that the population size at time $t + \Delta t$ is j, given that the population size at time t was i, is

$$p_{ji}(\Delta t) = P(N(t + \Delta t) = j \mid N(t) = i), \tag{13.1}$$

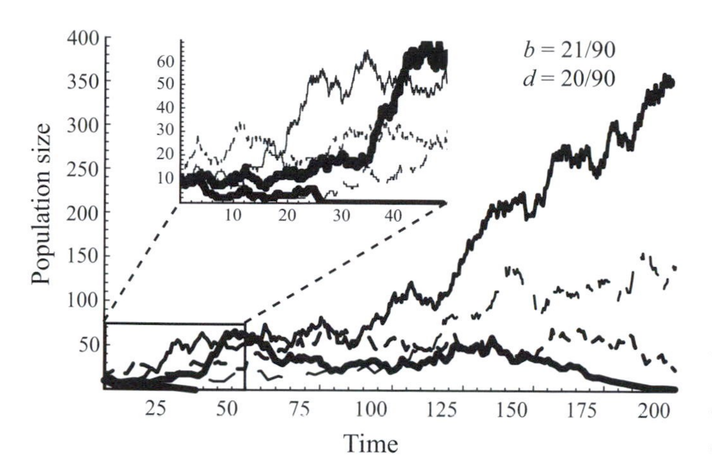

Figure 13.7: Birth-death process. The simulations of Figure 13.6 were repeated but with a birth rate of $b = 21/90$ and a death rate of $d = 20/90$, so that the net growth rate was $b - d = 1/90$ per unit time as in Figure 13.6. The inset figure shows the first 50 minutes. One population went extinct after 26 minutes.

where $p_{ji}(\Delta t)$ denotes the "transition probability" within a time period Δt. In a very short amount of time (so short that at most one event can occur) the transition probabilities $p_{ji}(\Delta t)$ are approximately

$$p_{ji}(\Delta t) = \begin{cases} b\, i\, \Delta t & \text{for } j = i + 1 \quad \text{(a birth)}, \\ d\, i\, \Delta t & \text{for } j = i - 1 \quad \text{(a death)}, \\ 1 - (b + d)\, i\, \Delta t & \text{for } j = i \quad \text{(no change)}, \\ 0 & \text{for } j \neq i - 1, i, i + 1 \quad \text{(other changes)}. \end{cases} \tag{13.2}$$

Figure 13.7 illustrates how adding deaths to the birth-process changes the dynamics (compare to Figure 13.6). Although the net growth rate $(b - d)$ is the same (1/90), the inclusion of deaths causes the population to grow more erratically. In fact, one of the five replicates went extinct at $t = 26$.

So far, we have assumed that the per capita birth and death rates are constant, regardless of population size. It is easy to generalize this birth-death model to incorporate density dependence, by making either the birth or death rate a function of the number of individuals. Although it is possible to incorporate density dependence in a number of different ways, it is often assumed that competition among individuals acts to reduce the replication rate, and that the death rate remains constant (Renshaw 1991). For example, the per capita birth rate might decrease linearly with population size, as in the logistic model, giving the transition probabilities

$$p_{ji}(\Delta t) = \begin{cases} b\, i\left(1 - \frac{i}{K}\right)\Delta t & \text{for } j = i + 1 \quad \text{(a birth)}, \\ d\, i\, \Delta t & \text{for } j = i - 1 \quad \text{(a death)}, \\ 1 - \left(b\left(1 - \frac{i}{K}\right) + d\right) i\, \Delta t & \text{for } j = i \quad \text{(no change)}, \\ 0 & \text{for } j \neq i - 1, i, i + 1 \quad \text{(other changes)}. \end{cases} \tag{13.3}$$

Here, the probability of a birth is zero at K, which represents a limit to the population size. To revise the simulations, all we have to do is update the birth and death rates each time the population size changes. It is also possible to incorporate temporal variation in the birth and death rates due to environmental fluctuations; such models are known as "nonhomogeneous birth-death processes."

Birth-death models have been applied to many other biological problems. For example, birth-death models have been used to describe changes in the number of repeats at microsatellites, which are stretches of DNA containing several copies in a row of a short motif (e.g., GAGAGAGA . . .) (Edwards et al. 1992; Ohta and Kimura 1973; Valdes et al. 1993). The birth-death model has also been used to describe the process of speciation (akin to birth) and extinction (akin to death), providing an interesting null model to describe the generation of biodiversity (Harvey et al. 1994; Nee et al. 1994a; Nee et al. 1995; Purvis et al. 1995). We will return to birth-death models in Chapter 14, where we describe analytical techniques that can be used to determine such things as the probability that the system is at any particular size, the probability of extinction, and the expected time until extinction.

13.4 Wright-Fisher Model of Allele Frequency Change

Next, we turn to a class of stochastic models that have played an important role in evolutionary biology. In the previous sections, the stochastic models focused on the total number of individuals within a population. Stochastic models can also be used to track the frequency of various types. We will again consider two different types of models. In this section, as in section 13.2, we assume that the entire population reproduces simultaneously, so that the generations are discrete and nonoverlapping. In the next section, we assume that generations are overlapping and that individuals are born and die at random points in time, as in the birth-death model of section 13.3. Again, the focus here will be on the development of these models and their simulation, laying the groundwork for the analytical techniques presented in subsequent chapters.

Consider a population that has a constant size, N, and only two types of individuals (A and a), as in the one-locus, two-allele haploid model (see extension to diploids in Problem 13.4). The deterministic model of this process, equation (3.8c), predicts that the frequency of type A at time $t+1$ will be exactly $p(t+1) = W_A\, p(t)/(W_A\, p(t) + W_a\,(1-p(t)))$, where W_i represents the relative fitness of each type. By chance, however, individuals of type A might happen to leave more or fewer offspring in any given generation, so that $p(t+1)$ will have a probability distribution centered around this deterministic prediction.

In the *Wright-Fisher model*, N offspring are sampled with replacement from the parental generation, which then dies. This sampling process causes random fluctuations in allele frequencies.

We first tackle the so-called "neutral" case where individuals are equally fit ($W_A = W_a = 1$). If the population size remains constant at N, and if the initial frequency of type A is $p(0)$, we can imagine individuals producing an infinite number of propagules (seeds, spores, etc.) from which a total of N surviving offspring are sampled. This thought experiment implies that the number of copies of allele A among the offspring should be binomially distributed with a mean of $N\,p(0)$ (see Primer 3). Thus, to simulate the Wright-Fisher model, we draw a random number from the binomial distribution with parameters N and $p(0)$.

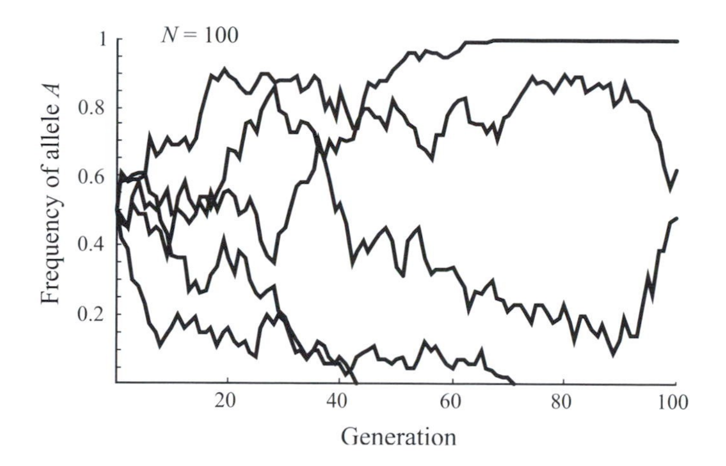

Figure 13.8: The Wright-Fisher model without selection. Each generation, offspring were chosen by randomly drawing from the alleles (*A* and *a*) carried by the parents with replacement (equivalent to binomial sampling). The population was assumed to be haploid and of constant size, $N = 100$. The frequency of allele *A* is plotted over time, starting with $p(0) = 0.5$.

The result j is the number of copies of the allele in the next generation, and $p(1) = j/N$. When we did this for an initial allele frequency of $p(0) = 1/2$ in a population of size $N = 100$, we drew 42 copies of allele *A*, so $p(1) = 0.42$. We can then find $p(2)$ by drawing a random number from a binomial distribution with parameters N and $p(1)$. Extending this process over 100 generations and repeating it five times, gave us the data for Figure 13.8.

The results are completely different from what we would expect based on the deterministic model of Chapter 3. According to equation (3.8c), when relative fitnesses are equal ($W_A = W_a = 1$), the allele frequency should stay constant ($p(t + 1) = p(t)$). In Figure 13.8, however, the allele frequencies rise and fall by chance over time. This process, whereby random sampling of offspring causes allele frequencies to vary from their deterministic expectation, is known as *random genetic drift*. These chance events led to the loss of the *A* allele at generation 43 in one replicate and at generation 71 in another. Conversely, the *A* allele became fixed within the population at generation 67 in a third replicate. A polymorphism remained in two of the replicates at generation 100, but eventually the *A* allele would have been lost or fixed had we continued to run the simulations.

The process whereby random sampling of offspring causes allele frequencies to vary from their deterministic expectation is known as *genetic drift*.

Here, we have been using simulations to determine the probability that, at some future point in time, the population will be composed of a certain proportion $p(t)$ of type *A*. There is a faster way to calculate this probability distribution in small populations, which will provide us with a good background for the analysis in Chapter 14. First, because sampling N surviving offspring randomly and independently from all possible offspring is described by a binomial distribution, we can use Definition P3.4 to write down the probability that there are j individuals of type *A* at time $t + 1$ given that there were i individuals of type *A* at time t. Using an upper-case X to denote the random variable "number of type *A* individuals", the transition probabilities for the Wright-Fisher model are

$$\begin{aligned} p_{ji} &= P(X(t+1) = j \mid X(t) = i) \\ &= \binom{N}{j}\left(\frac{i}{N}\right)^{j}\left(1 - \frac{i}{N}\right)^{N-j}, \end{aligned} \tag{13.4}$$

where p_{ji} denotes the transition probability within one generation. Equation (13.4) is the formula for the binomial distribution (Definition P3.4), but with p written as i/N.

We can use (13.4) to describe a "transition probability matrix" for the Wright-Fisher model, which gives the probability of going from any state i to any state j in one generation. Because we could have anywhere from 0, 1, 2, to N copies of type A, this matrix has $N+1$ rows and columns. For example, in a population of size four, the transition probability matrix is

$$\mathbf{M} = \begin{pmatrix} p_{00} & p_{01} & p_{02} & p_{03} & p_{04} \\ p_{10} & p_{11} & p_{12} & p_{13} & p_{14} \\ p_{20} & p_{21} & p_{22} & p_{23} & p_{24} \\ p_{30} & p_{31} & p_{32} & p_{33} & p_{34} \\ p_{40} & p_{41} & p_{42} & p_{43} & p_{44} \end{pmatrix}$$

$$= \begin{pmatrix} 1 & \frac{81}{256} & \frac{1}{16} & \frac{1}{256} & 0 \\ 0 & \frac{108}{256} & \frac{4}{16} & \frac{12}{256} & 0 \\ 0 & \frac{54}{256} & \frac{6}{16} & \frac{54}{256} & 0 \\ 0 & \frac{12}{256} & \frac{4}{16} & \frac{108}{256} & 0 \\ 0 & \frac{1}{256} & \frac{1}{16} & \frac{81}{256} & 1 \end{pmatrix}. \tag{13.5}$$

Each column sums to one because a population that starts with i copies of the allele must have some number between 0 and N copies in the next generation: $\sum_{j=0}^{N} p_{ji} = 1$. The first and last columns are particularly simple because there is no mutation; if nobody is type A ($i = 0$; first column) or if everybody is type A ($i = N$; last column), then no further changes are possible.

The helpful part about writing (13.5) in matrix form is that it can be iterated using the rules of matrix multiplication (Primer 2). $\mathbf{M}^2$ tells us the probability that there are j copies at time $t+2$ given that there were i copies at time t. In general, $\mathbf{M}^t$ tells us the probability that there are j copies at time t given that there were i copies at time 0. For example, calculating $\mathbf{M}^{1000}$ using equation (13.5) (using a mathematical software package) gives

$$\mathbf{M}^{1000} = \begin{pmatrix} 1 & 0.75 & 0.5 & 0.25 & 0 \\ 0 & 0 & 0 & 0 & 0 \\ 0 & 0 & 0 & 0 & 0 \\ 0 & 0 & 0 & 0 & 0 \\ 0 & 0.25 & 0.5 & 0.75 & 1 \end{pmatrix}. \tag{13.6}$$

(The zeros in the middle of this matrix aren't exactly zero, but they are less than 10^{-126}.)

We can also represent the initial state of the system using a vector

$$\begin{pmatrix} P(X(0)=0) \\ P(X(0)=1) \\ P(X(0)=2) \\ P(X(0)=3) \\ P(X(0)=4) \end{pmatrix}. \tag{13.7}$$

For example, if the population initially had two copies of the allele, then $P(X(0)=2)=1$ and all other entries in this vector are zero. Multiplying $\mathbf{M}^{1000}$ on the right by this initial vector, we find

$$\begin{pmatrix} 1 & 0.75 & 0.5 & 0.25 & 0 \\ 0 & 0 & 0 & 0 & 0 \\ 0 & 0 & 0 & 0 & 0 \\ 0 & 0 & 0 & 0 & 0 \\ 0 & 0.25 & 0.5 & 0.75 & 1 \end{pmatrix} \begin{pmatrix} 0 \\ 0 \\ 1 \\ 0 \\ 0 \end{pmatrix} = \begin{pmatrix} 0.5 \\ 0 \\ 0 \\ 0 \\ 0.5 \end{pmatrix}.$$

The vector on the right indicates that there is a 50% chance that type A will be lost ($j=0$) after 1000 generations and a 50% chance that type A will be fixed ($j=4$). If instead, the system initially had one copy of the allele, then $P(X(0)=1)=1$ and the remaining terms in vector (13.7) are zero. Now when we multiply $\mathbf{M}^{1000}$ on the right by this initial vector, we find that there is a 75% chance that type A will be lost and a 25% chance that it will be fixed after 1000 generations. These results suggest that if we start with i copies of type A, then type A will eventually be lost with probability $1-i/N$ and fixed with probability i/N.

Writing this stochastic model in terms of a transition probability matrix suggests that we could apply the matrix techniques used in Primer 2 and Chapters 7–9 to understand stochastic models. This is exactly right, and we shall do so in the next chapter. Once again, eigenvalues and eigenvectors play a key role in analyzing stochastic models. At least for small population sizes, however, we can get an exact numerical solution just by calculating $\mathbf{M}^t$, against which we can check any theoretical prediction.

The Wright-Fisher model can be extended to incorporate fitness differences, mutation, multiple loci, etc. In reality, many of these processes are themselves stochastic, but a shortcut is often taken by assuming that these processes affect the number of propagules and their allele frequencies. If the number of propagules is very large, then these processes can be described by a deterministic recursion (e.g., using equation (3.8c) for selection or $(1-\mu)\,p(t)+\nu\,(1-p(t))$ for the allele frequency after mutation). As a consequence, sampling occurs only once, when the N adult individuals are chosen from the propagules.

As an example, Figure 13.9 illustrates simulations of the Wright-Fisher model with selection. In this figure, the A allele is 10% more fit than the a type and begins at a frequency of $p(0)=0.05$. The simulations are run for populations of size (a) $N=100$ and (b) $N=10{,}000$. In both cases, the alleles rise in

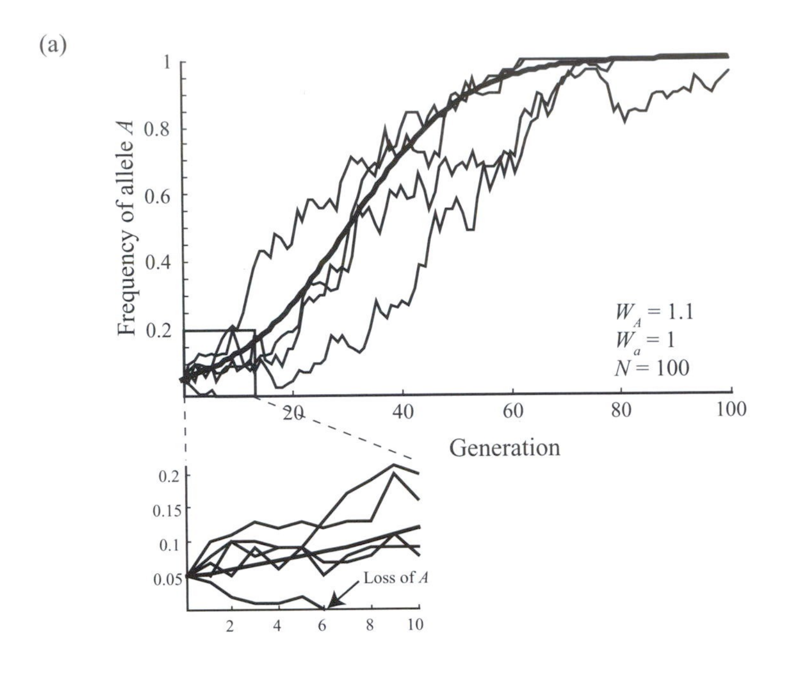

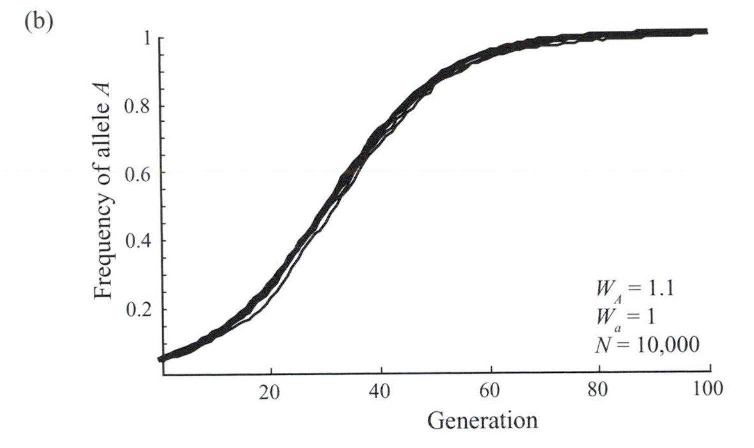

Figure 13.9: The Wright-Fisher model with selection. Simulations were carried out as in Figure 13.8 except that offspring alleles were more likely to be chosen from parents carrying the more fit *A* allele ($W_A = 1.1$, $W_a = 1$). For the thin curves, population size was set to (a) $N = 100$ and (b) $N = 10{,}000$ (five replicates each). For comparison, the thick curves illustrate the deterministic trajectory ($N = \infty$), obtained by iterating equation (3.8c). The frequency of allele *A* is plotted over time, starting with $p(0) = 0.05$. The favorable allele *A* was lost at generation 6 in one replicate with $N = 100$ (inset).

frequency towards fixation within 100 generations, roughly following the S-shaped trajectory seen in deterministic models (bold curve). When the population size is small, the Wright-Fisher model exhibits more variability around the deterministic trajectory than when the population size is large. This is consistent with the fact that the variance in the frequency of allele *A* due to sampling should be $p\,(1 - p)/N$ under the binomial distribution (see section P3.3.1). In fact, when N is only 100, we observe extinction of the beneficial allele in one of the five replicates (see inset figure). When N is 10,000, however, none of the replicates go extinct, and there is little variability in the trajectory.

These figures illustrate an important point: adding stochasticity to a model need not cause major changes to the results. In populations of small size ($N = 100$), we have seen allele frequency change when there should have been none (the neutral case, Figure 13.8), and we have witnessed the loss of a beneficial allele, which we would expect to fix (Figure 13.9a). In populations of large size, however, there is less random genetic drift. Consequently, when the amount of chance (here represented by variation in samples from the binomial distribution) is small relative to other forces like selection, stochastic models can behave very much like deterministic models.

To simulate more than two types within a population (e.g., more than two alleles at a locus, or multiple genotypes at two loci), the above method must be modified by drawing from a multinomial distribution, with parameters N and p_i where the p_i are the frequencies of the various types, so that $\sum_{i=1}^{c} p_i = 1$ when there are c types (see Definition P3.5). This works well unless the number of types becomes very large. For example, with two alleles at each of 100 loci, there are 2^{100} possible haploid genotypes (Box P3.1). This number is greater than 10^{30}, which is much larger than any population. When there are too many types, drawing random numbers from a multinomial distribution grinds to a halt. What else can you do? The alternative is to develop an *individual-based model* where you mimic the production of each offspring within the population, one at a time (Deangelis and Gross 1992).

An *individual-based model* is a simulation where each individual is tracked explicitly, along with its properties (e.g., genotype, location, age, etc.).

Typically, in an individual-based model for the above process, you randomly draw a gamete from a mother and a gamete from a father within the parental population, and unite these to form a diploid offspring (followed by meiosis if you wish to produce a haploid offspring). If the fitness of the offspring is W and the maximum fitness is W_{max}, you can then test to see if your offspring survives selection by drawing a random number uniformly between 0 and 1. If that random number is less than W/W_{max} the offspring becomes one of the N surviving individuals in the next generation, otherwise you start again by choosing new parents at random. This procedure works well unless the population size is very large.

13.5 Moran Model of Allele Frequency Change

In the Wright-Fisher model, we assumed that the entire population reproduced simultaneously. Intuitively, one might think that random genetic drift would be exaggerated by having the entire population replicate at once. To check this intuition, we explore a model where only one individual reproduces at a time. One way to do this would be to expand the birth-death process to allow multiple types of individuals (e.g., types of alleles) and to track the numbers of each type. In this case, you would observe both changes to the population size and to the frequencies of each type. But what if you wanted to hold the population size constant, to compare the results to those of the Wright-Fisher model?

The easiest way to adapt the birth-death process, holding the population size constant, is to couple each birth event with a death event. Whenever an individual is chosen to give birth, another individual is randomly chosen to die. Typically, the individual chosen to die can be any individual in the population, including the parent of the new offspring, but not the new offspring itself. It is also typical to track the population only at those discrete points in time where a birth-death event occurs, measuring time in terms of the number

In the *Moran model*, a randomly chosen individual reproduces, followed by the death of a randomly chosen individual. This sampling process also causes genetic drift.

of events that have happened rather than in chronological time. This evolutionary model is known as the *Moran model* (Moran 1962).

We focus on a population of size N with only two types A and a, where the number of copies of A is i and the frequency of A is $p = i/N$. If all individuals are equally fit, then the chance that an A-type parent is chosen to replicate is p. Thus, after one birth-death event, the number of copies of A goes up by one if the individual chosen to replicate carries the A allele (with probability p) and the individual chosen to die carries the a allele (with probability $1 - p$), giving an overall probability of $p\,(1 - p)$. Similarly, the number of copies of A goes down by one (if a replicates and A dies), with probability $(1 - p)\,p$. Finally, the number of copies stays the same if the individual chosen to replicate is the same type as the individual chosen to die, which happens with probability $p^2 + (1 - p)^2$. These calculations allow us to write down the probability of going from i copies of type A to j copies:

$$p_{ji} = P(X(t+1) = j \mid X(t) = i), \tag{13.8}$$

where p_{ji} denotes the transition probability after one birth-death event, and $X(t)$ is a random variable representing the number of copies of type A at time t. For the Moran model, the transition probabilities p_{ji} are

$$p_{ji} = \begin{cases} p\,(1-p) & \text{for } j = i+1 & \text{(increase by one),} \\ (1-p)\,p & \text{for } j = i-1 & \text{(decrease by one),} \\ p^2 + (1-p)^2 & \text{for } j = i & \text{(no change),} \\ 0 & \text{for } j \neq i-1, i, i+1 & \text{(other changes),} \end{cases} \tag{13.9}$$

where $p = i/N$. The key assumption of the Moran model is that the transition probability is zero for transitions that differ from the current state by more than one A allele.

Figure 13.10 illustrates the outcome of five replicate simulations of the Moran model starting with $i = 50$ copies of type A in a population of size $N = 100$. The simulations look similar to those from the Wright-Fisher model without selection (Figure 13.8). There are differences, however, as the inset figure shows. The allele frequency only jumps by $+/- \; 1/N$ in the Moran model, whereas much larger jumps can occur in the Wright-Fisher model. The main qualitative difference, however, is the scale along the x axis. There are only 100 generations represented in Figure 13.8 of the Wright-Fisher model, but 10,000 birth-death events represented in Figure 13.10 of the Moran model. You might be tempted to conclude that the Moran model exhibits less drift, but this is not a fair comparison. One time step in the Wright-Fisher model involves N births followed by the death of all N parents and so is more equivalent to N birth-death events in the Moran model. Thus, Figures 13.8 and 13.10 both represent the same total number of generations (100) with $N = 100$. Over this time period, and with only five replicates each, it is unclear which model exhibits more drift.

Given that no clear conclusions emerge from a few replicate simulations, we must run many more replicate simulations to compare the Wright-Fisher and Moran models. Starting with $p(0) = 0.5$ in a population of size 100, we ran 500

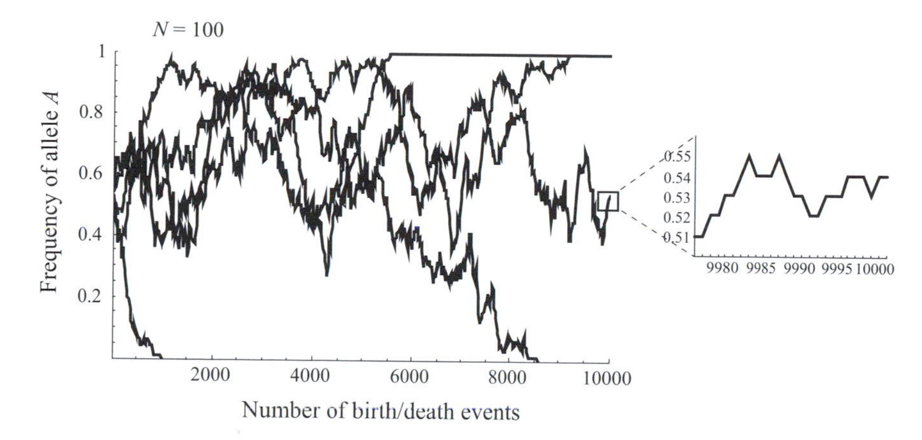

Figure 13.10: The Moran model without selection. At each time step, an individual was randomly chosen to give birth, after which an individual other than the new offspring was randomly chosen to die. The population was assumed to be haploid and of constant size, $N = 100$. The frequency of allele A is plotted over time, starting with $p(0) = 0.5$.

replicate simulations until fixation or loss of type A. By coincidence, the A type was lost 48.4% of the time using both the Moran and the Wright-Fisher model. In the Moran model, however, it took only 66.3 generations ($SE = 2.2$), on average, which was approximately half the time until loss or fixation in the Wright-Fisher model (133.3 generations with $SE = 4.6$).

The above results show that polymorphism is lost significantly faster in the Moran model than in the Wright-Fisher model. This result seems counterintuitive, because the Moran model makes only little jumps in frequency, whereas the Wright-Fisher model can make large jumps. A clue that can help us to understand this result is provided by the variance in reproductive success in the two models. When reproductive success is more variable, stochasticity (here, random genetic drift) plays a stronger role, and polymorphism will be lost by chance more rapidly.

In the Wright-Fisher model, the variance in reproductive success of single individuals, σ_r^2, is given by the binomial variance $N\,p\,(1-p)$ from equation (P3.4), when there is a single individual (i.e., with $p = 1/N$). Thus, $\sigma_r^2 = 1 - 1/N$. To calculate the variance in reproductive success over a single birth-death event in the Moran model, we use the formula for calculating variance (Definition P3.3), summing the squared change in number of copies over all possible transitions using (13.9):

$$\begin{aligned} p(1-p)(+1)^2 + (1-p)p(-1)^2 + (1-2p(1-p))(0)^2 \\ = 2p(1-p). \end{aligned}$$

Because the variance of a sum of independent random variables is the sum of the variances (Table P3.1), the total variance in reproductive success over N such birth-death events is $2Np(1-p)$ per generation. Again, because we are interested in the variance in reproductive success of a focal individual, we set $p = 1/N$, demonstrating that $\sigma_r^2 = 2 - 2/N$. Thus, the Moran model exhibits twice the variance in reproductive success, and consequently more random genetic drift, than the Wright-Fisher model (Ewens 1979). At an intuitive level, the Moran model is more variable because sampling occurs twice, when choosing which individual replicates and when choosing which individual dies.

The Moran model can be extended to incorporate processes such as selection and mutation by modifying the transition probabilities (Problem 13.6). For example, selection can be incorporated by altering the chance that an individual is chosen to reproduce. With selection, type A is chosen to give birth with a probability, p', equal to the frequency of type A weighted by its fitness (W_A) divided by the mean fitness: $p' = W_A\, p/\overline{W}$, where $\overline{W} = W_A\, p + W_a\,(1 - p)$. That is, p' is the same as the frequency change due to one generation of selection in the standard deterministic model of haploid selection (see equation (3.8c)). Making the key assumption that only one birth-death event occurs per time step, the transition probabilities are

$$p_{ji} = \begin{cases} p'(1-p) & \text{for } j = i+1 \quad \text{(increase by one)}, \\ (1-p')\,p & \text{for } j = i-1 \quad \text{(decrease by one)}, \\ p'p + (1-p')(1-p) & \text{for } j = i \quad \text{(no change)}, \\ 0 & \text{for } j \neq i-1, i, i+1 \quad \text{(other changes)}. \end{cases} \tag{13.10}$$

Here we have assumed that individuals are chosen at random to die, because we did not want to impose two bouts of selection on the population per generation. Other choices are equally plausible, however. You could impose viability selection on the death probabilities instead of (or in addition to) fertility selection on the birth probabilities.

In Chapter 14, we shall derive several important results using the Moran model, including the probability of fixation (or loss) and the time until fixation (or loss). These analytical results assume that there are only two types of individuals, so that we can count the number of one type and infer the number of the other. You can explore the Moran with more than two types by running simulations akin to Figure 13.10 by developing appropriate rules for who gives birth and who dies.

13.6 Cancer Development

The above examples are well-known and provide good background for how stochastic models can be constructed. In this section, we develop another example and model the occurrence of retinoblastoma, a cancer of the eye. This example will help illustrate how stochasticity can be incorporated into models investigating a wide variety of problems in biology.

Retinoblastoma is the most common eye cancer among children, with a worldwide incidence of about 5 in 100,000 children (Knudson 1971, 1993). The genetics of retinoblastoma are highly unusual. The mutation responsible for heritable cases of retinoblastoma occurs at the RB-1 gene on the long arm of chromosome 13 (Lohmann 1999). RB-1 is a tumour suppressor gene, and mutations in this gene disrupt control of the cell cycle. At a cellular level, the RB-1 mutation is recessive; the cell cycle is normal as long as there is one wild-type allele in the cell. At an individual level, however, the mutation is dominant with a *penetrance* ρ of about 95%, meaning that about 95% individuals born with one mutant and one wild-type allele develop eye cancer (Knudson 1971, 1993).

How can a heterozygous individual get cancer when heterozygous cells are normal? The resolution of this paradox lies in the fact that somatic mutations occur sporadically during development, causing some cells in the eye to lose heterozygosity. It is those few mutant cells that lose their one copy of the wild-type allele that are responsible for retinoblastoma. Loss of heterozygosity (LOH) can occur by several mechanisms during mitosis (Lohmann 1999), including gene deletion, chromosome loss, mitotic recombination, and point mutations. Understanding the development of retinoblastoma requires a stochastic model, because the chance timing of mutational events determines whether cancer develops, as well as its severity.

Figure 13.11 illustrates the development of the vertebrate retina. The single-celled zygote undergoes five binary cell divisions to reach the 32-celled blastula stage. Experiments performed at this stage in *Xenopus* indicate that only nine of these blastomere cells (a through i) contribute to the retina of each eye (Huang and Moody 1993). These cells then undergo a series of n cell divisions. Averaged over the 32 blastomere cells, n must be $\sim$41 to account for the approximately 10^{14} cells in the human body (Moffett et al. 1993). The retina is composed of $\sim 1.5 \times 10^8$ cells (Bron et al. 1997; Dreher et al. 1992), but only three of the seven major retinal cell types (horizontal, amacrine, and Müller cells) appear to have the potential to proliferate into retinoblastoma in RB-1 homozygous mutant cells (Chen et al. 2004). Based on counts of these three cell types (Dreher et al. 1992; Van Driel et al. 1990), the total number of retinal cells that have the potential to cause retinoblastoma in one fully formed eye, C, is $\sim 2 \times 10^7$.

The experiments of Huang and Moody (1993) also indicate that different fractions of retinal cells descend from each blastomere cell (see inset table in Figure 13.11). We will call these fractions f_a through f_i. For example, the cell D1.1.1 (marked as "a") contributes 49.7% of the cells in the left retina. The exact cell fate is determined later in development, so each blastomere contributes to the different cell types in the retina (Huang and Moody 1993). We incorporate these observations by letting $f_y\,C$ equal the number of susceptible cells contributed by the blastomere cell y to the left retina.

To model stochastic mutation, we assume that mutations occur during DNA replication (i.e., at discrete points in time). Whether a daughter cell produced by a heterozygous parent cell is mutant represents a random variable with two possible outcomes (a Bernoulli trial): with probability μ it is mutant, and with probability $1 - \mu$ it remains heterozygous. Unfortunately, we do not know exactly when each progenitor cell divides in the development of the retina. As a preliminary map of development, we considered Figure 13.12. Phase 1 consists of the five cell divisions leading to the blastula. In phase 2, cell divisions produce all of the cell types in the body, and we assume that only one daughter cell per division remains in the lineage leading to the retina. In phase 3, the stem cells of the retina proliferate, with all daughter cells contributing to the retina. The number of divisions in phase 3, m_y, is chosen to ensure that

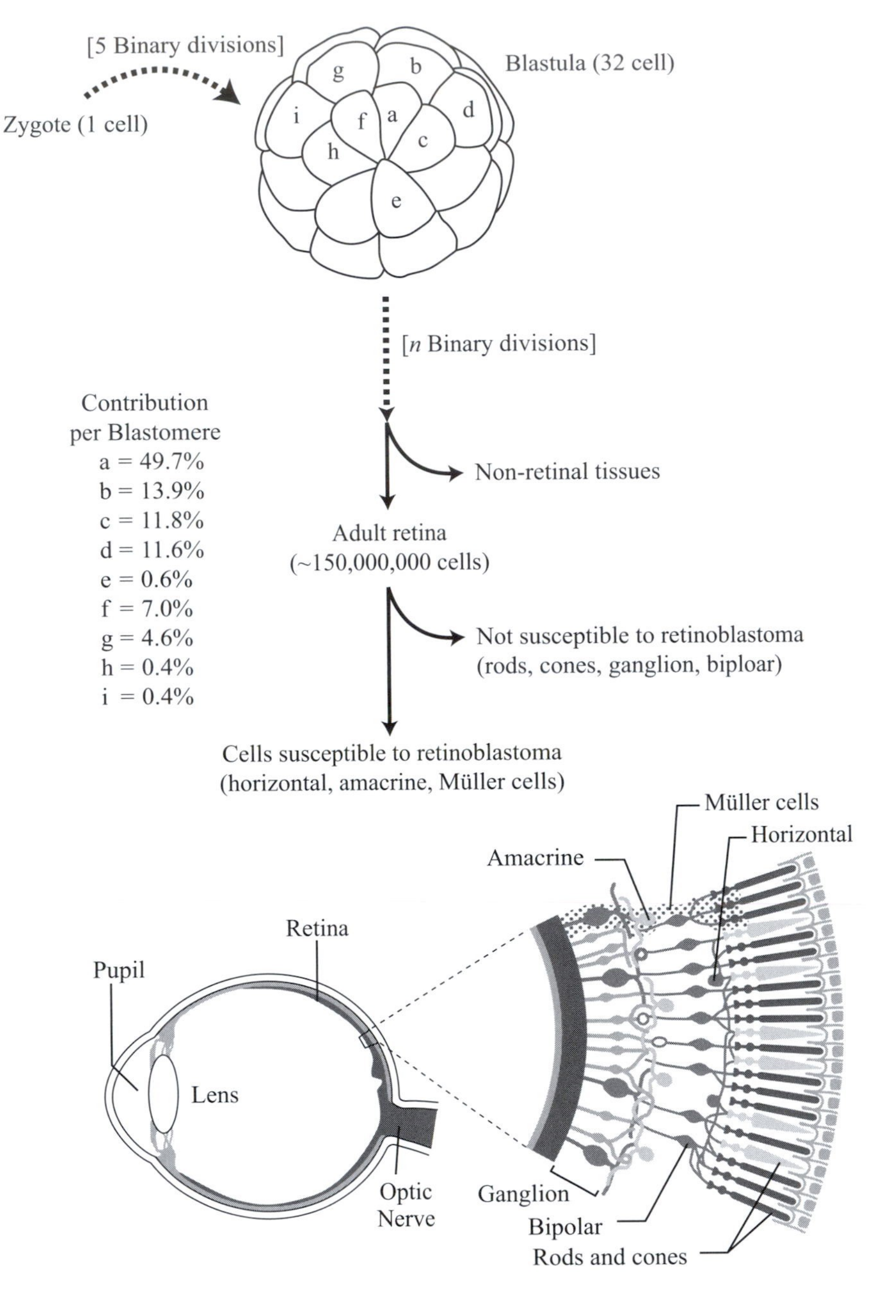

Figure 13.11: Development of the retina. Development from the zygote (top left), through the blastula stage (top center), to the eye (bottom center) is illustrated (http://webvision.med.utah.edu). Percentages indicate the fraction of retinal cells of the left eye derived from each of the blastomeres marked *a* through *i* (Huang and Moody 1993).

blastomere y contributes the appropriate number of susceptible cells to the fully developed retina, $f_y C$. (For a more precise calculation, we allow a fraction p_y of the cells to undergo an additional cell division to get exactly $f_y C$ cells.)

The bulk of mutations causing a loss of heterozygosity are likely to happen when there are many cells (i.e., many Bernoulli trials), which occurs when the

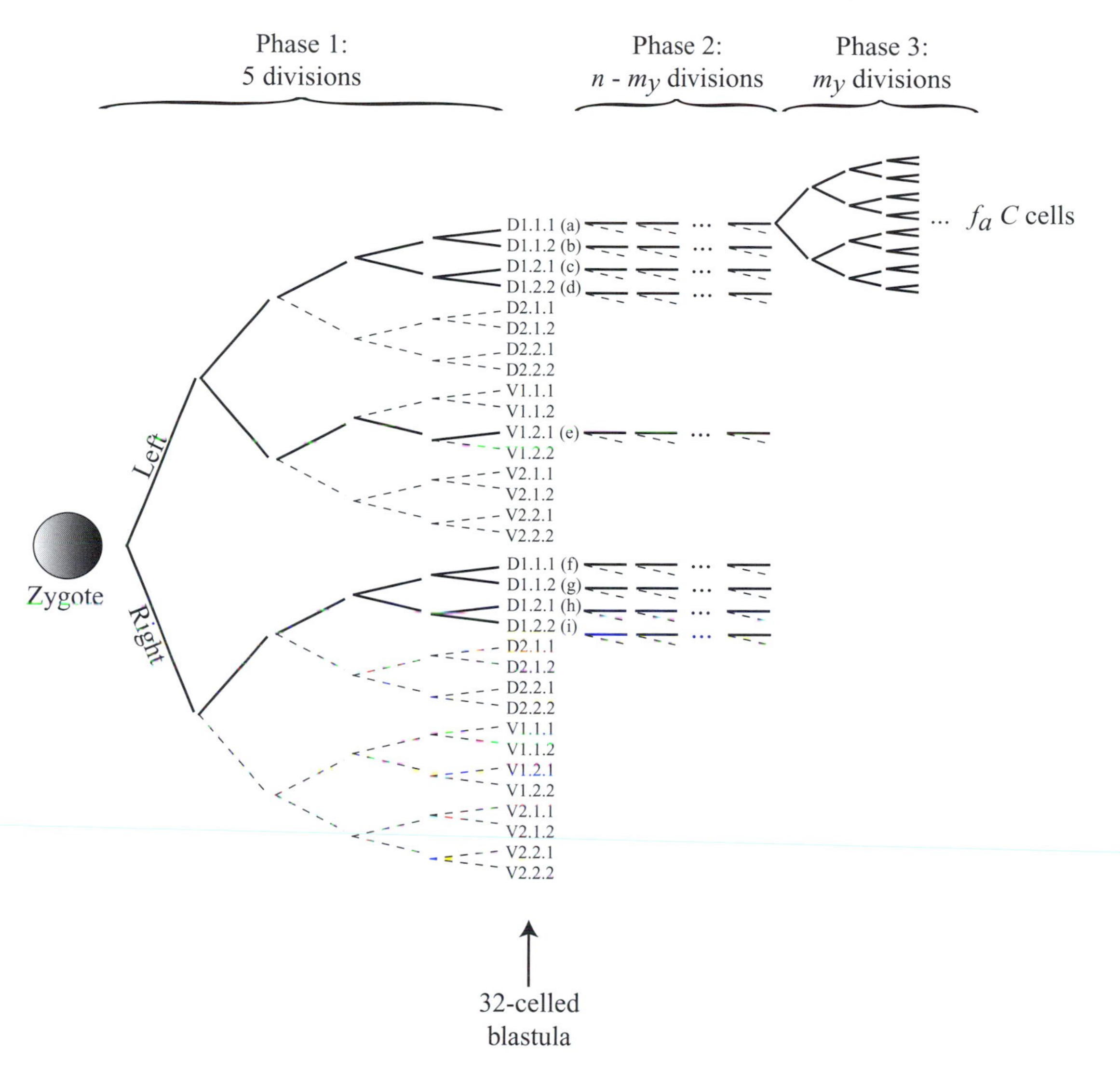

Figure 13.12: A cell-lineage map leading to the eye. With time proceeding from left to right, lines connect parent cells to daughter cells; solid lines indicate lineages that contribute to the pool of susceptible retina cells, while dashed lines indicate lineages that do not contribute to the retina. Phase 1 consists of the five cell divisions from the zygote to the blastula. Phase 2 consists of the cell divisions between the blastula and the stem cells that generate the retina. Phase 3 consists of the proliferation stage during which the retina develops from a series of binary divisions. The exact details in phases 2 and 3 are not known.

eye is nearly fully developed (phase 3 of Figure 13.12). Thus, we might expect that the exact number of cell divisions during phase 3, m_y, would be much more critical than the number in phase 2, $n - m_y$.

Our goal is to characterize the probability that retinoblastoma occurs and in what form: in one eye or both, and with multiple tumors per eye or only one. If we carried out a Bernoulli trial for every daughter cell illustrated in Figure 13.12, however, simulating development would be quite slow. We can speed up the process by simulating mutations among the $x(t)$ daughter cells produced at cell

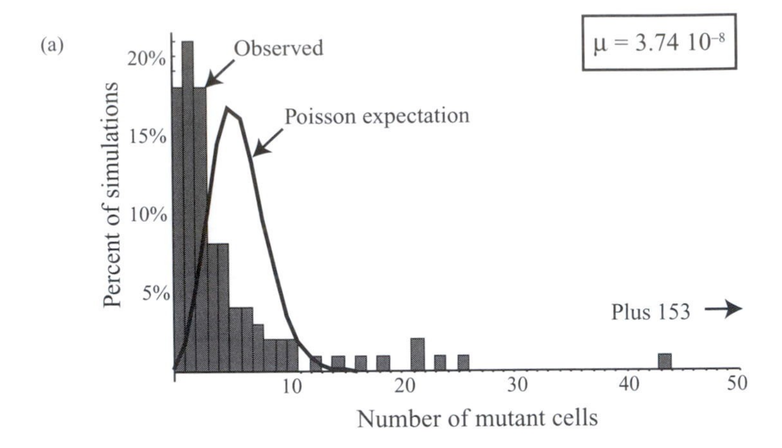

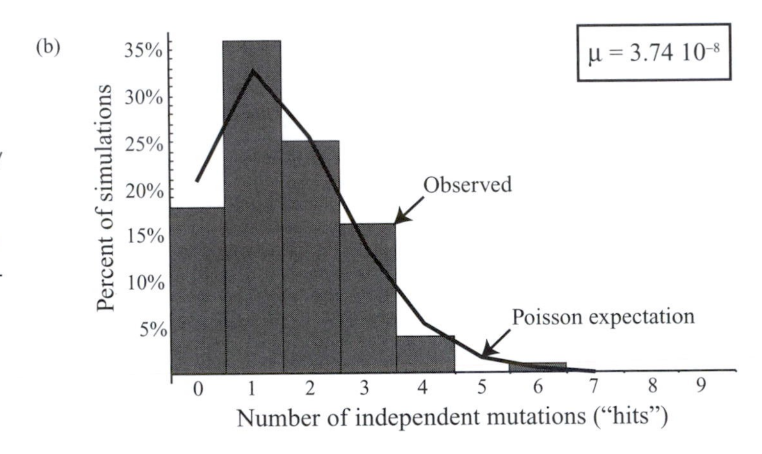

Figure 13.13: A stochastic model of mutation leading to retinoblastoma. Simulations were based on the exact sequence of cell replication described in Figure 13.12 and replicated 100 times. Starting with a heterozygous zygote, the number of cells that lose the wild-type allele in the tth round of cell division was drawn randomly from a binomial distribution with parameters $x(t)$ (the number of daughter cells) and μ (the mutation rate). (a) A histogram of the total number of mutant cells per eye. (b) A histogram of the number of distinct mutational events leading to the cancerous cells in an eye. Curves illustrate a Poisson distribution with the same mean as the observed distribution. $\mu = 3.74 \times 10^{-8}$, $C = 2 \times 10^7$, $n = 41$.

division t by the parent cells that remain heterozygous. The number of these daughter cells that lose the wild-type allele is a random variable drawn from a Binomial distribution with probability μ and a number of trials equal to $x(t)$. The daughter cells that remain heterozygous then produce $x(t + 1)$ daughter cells, and the process continues. All mutant cells and their descendents are kept track of separately, as these are assumed to remain mutant. We used this method to generate the histograms in Figure 13.13, replicating the process of development 100 times and using a mutation rate of $\mu = 3.74 \times 10^{-8}$ per daughter cell (as estimated below). Figure 13.13a illustrates the total number of homozygous mutant cells that developed within the left eye of each simulated "individual." Many of these mutant cells descended from the same mutation. Figure 13.13b illustrates the number of independent mutations that led to the observed number of mutant cells in the left eye.

These two histograms tell an interesting story. In the second histogram, the number of mutational events closely follows a Poisson distribution, as expected if mutations occur independently at a small rate in a large number of Bernoulli trials (recall that the Poisson distribution is an excellent approximation to the binomial distribution in this case). Technically, the LOH (loss of heterozygosity) mutations do not occur independently, because the descendants of a LOH mutation cannot have a further LOH mutation. Nevertheless, because most mutations happen late in development when there are many cells, there is little opportunity for further mutation. The first histogram is decidedly not Poisson and has "fat tails" (leptokurtosis). That is, there is a much higher probability of observing many LOH cells, or none, than expected based on the mean number of LOH cells.

The great variability in outcomes observed in our model is typical of a *jackpot* distribution. A jackpot distribution is one where there is a small chance of getting a very large outcome (akin to the small chance of winning a lottery). Such a distribution arises naturally when modeling mutation in a growing population of cells, because there is a small chance that a mutation happens early and is carried by many descendent cells. Although our model incorporates more developmental details, the results are fundamentally similar to a model developed by Luria and Delbruck (1943). These authors carried out a series of experiments growing bacteria in liquid culture and afterwards exposing the cells to a novel environment (a bacteriophage). They then counted up the number of resistant cells and observed a jackpot distribution—some cultures contained many resistant cells while most had few. Luria and Delbruck then used a mathematical model of mutation to demonstrate that mutations must have occurred during the growth of the population, before exposure to the novel environment, and not in response to the novel environment—only then is a jackpot distribution expected. This result became a cornerstone of modern genetics. The Luria-Delbruck model also forms the basis for an important method used to calculate mutation rates, known as the *fluctuation test*.

The results of Figure 13.13b can be used to predict the form of retinoblastoma. The probability that an eye is not affected is estimated by the height of the bar at 0: $p_0 = 0.18$. Using this estimate, we can calculate the probability of observing no retinoblastoma, retinoblastoma in one eye (unilateral), and retinoblastoma in both eyes (bilateral) among individuals that inherit the RB-1 mutant allele. Assuming that the two eyes represent independent sampling events, each with a probability p_0 of being unaffected, these probabilities are given by the binomial distribution:

$$P(\text{no retinoblastoma}) = p_0^2 = 0.032,$$
$$P(\text{unilateral retinoblastoma}) = 2p_0(1 - p_0) = 0.295,$$
$$P(\text{bilateral retinoblastoma}) = (1 - p_0)^2 = 0.672.$$

Furthermore, there is a pretty high probability, 46%, that an eye contains multiple tumors (summing the bars from 2 onward in Figure 13.13b).

Even with the fairly complicated model of development illustrated in Figure 13.12, we can make some general predictions using the probability theory introduced in Primer 3. To do so, we need to derive formulas for the values of p_0, p_1, and p_{2+}, rather than estimating them from simulations. Calculating p_0 is the most straightforward, so we focus only on p_0 and on the questions that can be answered with this quantity.

If S is the total number of daughter cells produced throughout the development of one eye, then the probability that none of these are mutant is

$$p_0 = (1 - \mu)^S \tag{13.11}$$

(Definition P3.4 with $k = 0$), where μ is the mutation rate per daughter cell. Equation (13.11) provides us with a way to relate the mutation rate to the penetrance of the mutation (i.e., to the probability that an individual is not affected). To calculate S, we count the number of daughter cells ever produced that contribute to the susceptible population of retinal cells in one eye. Using Figure 13.12, S is very nearly 4×10^7, almost all of which arise in Phase 3.

Using equation (13.11), the probability of being free of symptoms in both eyes is given by $p_0^2 = (1 - \mu)^{2S}$. One minus this quantity gives the probability of getting a tumor in at least one eye (the penetrance): $\rho = 1 - (1 - \mu)^{2S}$. We can rearrange this equation to solve for the mutation rate: $\mu = 1 - (1 - \rho)^{1/(2S)}$. Given the observed penetrance, $\rho = 0.95$, and $S = 4 \times 10^7$, the estimated mutation rate is $\mu = 3.74 \times 10^{-8}$ per daughter cell produced, as used above.

Is this estimated mutation rate per daughter cell reasonable? The observed mutation rate at RB-1 is 8×10^{-6} per individual generation (Knudson 1993). The number of cell divisions within humans has been estimated as ~179 divisions from zygote to zygote (averaged across sexes and assuming a generation time of 25 years; Vogel and Rathenberg 1975). Thus, the observed mutation rate corresponds to a mutation rate of 4.47×10^{-8} per cell division, which is reasonably close to our estimated mutation rate of 3.74×10^{-8}.

We can also use equation (13.11) to predict the form of retinoblastoma:

$$\begin{aligned} P(\text{no retinoblastoma}) &= p_0^2 = (1 - \mu)^{2S}, \\ P(\text{unilateral retinoblastoma}) &= 2p_0(1 - p_0) = 2(1 - \mu)^S (1 - (1 - \mu)^S), \\ P(\text{bilateral retinoblastoma}) &= (1 - p_0)^2 = (1 - (1 - \mu)^S)^2. \end{aligned}$$

Using $\mu = 3.74 \times 10^{-8}$, these calculations predict that, of individuals initially carrying the RB-1 mutation, 5% should be symptom free, 35% should develop unilateral retinoblastoma, and 60% should develop bilateral retinoblastoma. These predictions are consistent with observations (Knudson 1971) and with the simulation results presented above. Interestingly, these results depend only on μ and S.

Our model of retinoblastoma could be improved by taking into account a more sophisticated version of development than illustrated in Figure 13.12. Yet our model provides insight into which details matter most. As mentioned earlier, the exact number of cell divisions during phases 1 and 2 has a negligible influence on the number of mutations that arise. In fact, our results were nearly unchanged when we replaced Figure 13.12 with a simple series of binary cell divisions. While our results are not sensitive to events during phases 1 and 2, they would be sensitive to events late in development, including the exact number of cells in the retina (C) and the extent of cell births and deaths in phase 3.

13.7 Cellular Automata—A Model of Extinction and Recolonization

In previous models, we ignored the spatial location of individuals. Space often matters, however, because individuals tend to interact and breed locally and might not migrate over long distances relative to the range of the species. For example, HIV is highly spatially structured in different tissues within an infected individual (Frost et al. 2001). Only by accounting for this structure do models generate reasonable predictions for the level of genetic variability observed in HIV and the ability of HIV to respond to antiretroviral drugs. Although some models of spatially structured populations are analytically tractable (see Chapter 15 as well as examples in Nisbet and Gurney (1982) and Renshaw (1991)), many are not. Numerical analysis of spatial models has thus played an important role in biology.

A commonly used type of spatial model is a *cellular automaton*. An *automaton* is a machine or robot that carries out a series of instructions. A cellular automaton is an array of automata arranged in a lattice or grid, where each automaton is assigned its own position or *cell*. Typically, the grid lies in one or two dimensions, and cell shapes are uniform (as in the square grid illustrated in Figure 13.14). But the exact size and shape of a cellular automaton is flexible. One of the more famous cellular automata is the game of life, invented by John H. Conway to mimic births and deaths in a spatially arranged population (Gardner 1983).

To simulate a biological process on a cellular automaton, you must first specify the initial states of each cell and the instructions that each automaton

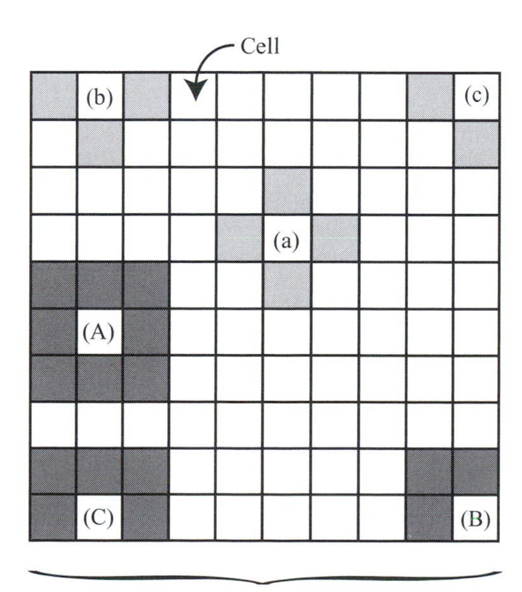

Figure 13.14: A cellular automaton. Each cell in this 10 × 10 grid is either inhabited or empty and can receive migrants from n nearest neighbor cells. Allowing migration from only the vertical and horizontal nearest neighbors, the light grey cells are potential sources of migrants to the focal cells: (a) a center cell (n = 4), (b) an edge cell (n = 3), (c) a corner cell (n = 2). Allowing migration from the vertical, horizontal, and diagonal nearest neighbors, the dark gray cells are potential sources of migrants to the focal cells: (A) a center cell (n = 8), (B) an edge cell (n = 5), (C) a corner cell (n = 3).

must use to determine their state in the next time step. The instructions to be carried out typically depend on the states of the surrounding cells. For example, the original game of life was played on a square grid, with each cell being dead (0) or alive (1). The number of live cells in the eight cells surrounding a given focal cell was then counted (n). If n was two, the state of the focal cell (alive or dead) remained unchanged. If n was three, the focal cell was set to 1 (alive) regardless of what it was before. In all other cases, the focal cell was set to 0. These cases roughly describe survival, birth, and death in the presence of local reproduction and competition. The exact rules were not chosen to portray growth in any particular species, per se, but to generate interesting spatial patterns, without exploding or imploding too rapidly.

In the game of life, the rules for updating the cells are deterministic, but stochastic rules are commonly used in cellular automaton models. As an example, we develop a cellular automaton model of extinction and recolonization (see Problems 5.12 and 5.13). In the nonspatial model, a fraction of patches, $p(t)$, is occupied at time step t. Of the $1 - p(t)$ unoccupied sites, a fraction $m\, p(t)$ are recolonized from occupied patches. Subsequently, each occupied site suffers a risk of extinction e through catastrophic events such as fire or disease. The resulting recursion equation for the deterministic model is (see Problem 5.12)

$$p(t + 1) = (1 - e)(p(t) + m\, p(t)\, (1 - p(t))). \tag{13.12}$$

To be more realistic, we model a spatial version of this model, where each cell in a 10×10 square grid represents a patch (empty or occupied) and where extinction and recolonization are stochastic events. Recolonization of an empty patch at position $\{i,j\}$ occurs with probability $m\, f_{i,j}$, where m is the recolonization rate per patch and $f_{i,j}$ is the number of neighboring patches that are occupied. This process is repeated for each unoccupied cell in the grid. In our simulations, we considered the neighborhood size to consist of the eight nearest cells (Figure 13.14A). We run into a problem, however, when we consider cells on the edge of the grid, which don't have eight neighbors. There are two approaches for handling the edges of a grid. First, the grid can be "wrapped around" to make a torus (a donut shape), so that, for example, a cell on the left edge can receive migrants from cells on the right edge. This procedure ensures that the edge cells and the central cells follow the same rules and is thought to represent populations larger than the grid size more accurately. The second approach assumes that the environment outside of the habitat is inhospitable, so that edge and corner cells really have fewer neighbors (Figure 13.14). We used this second approach in our simulations.

Next, we consider extinction. For each occupied site on the grid, we choose a random number uniformly between 0 and 1. If the random number is less than e, the population goes extinct, otherwise the site remains occupied.

Simulations of this extinction-recolonization model are illustrated in Figure 13.15 (the *Mathematica* code used to generate the figure is available on the book website). Colonization causes the spread of populations to adjacent cells and generates clusters of occupied cells. Extinction, however, causes sites

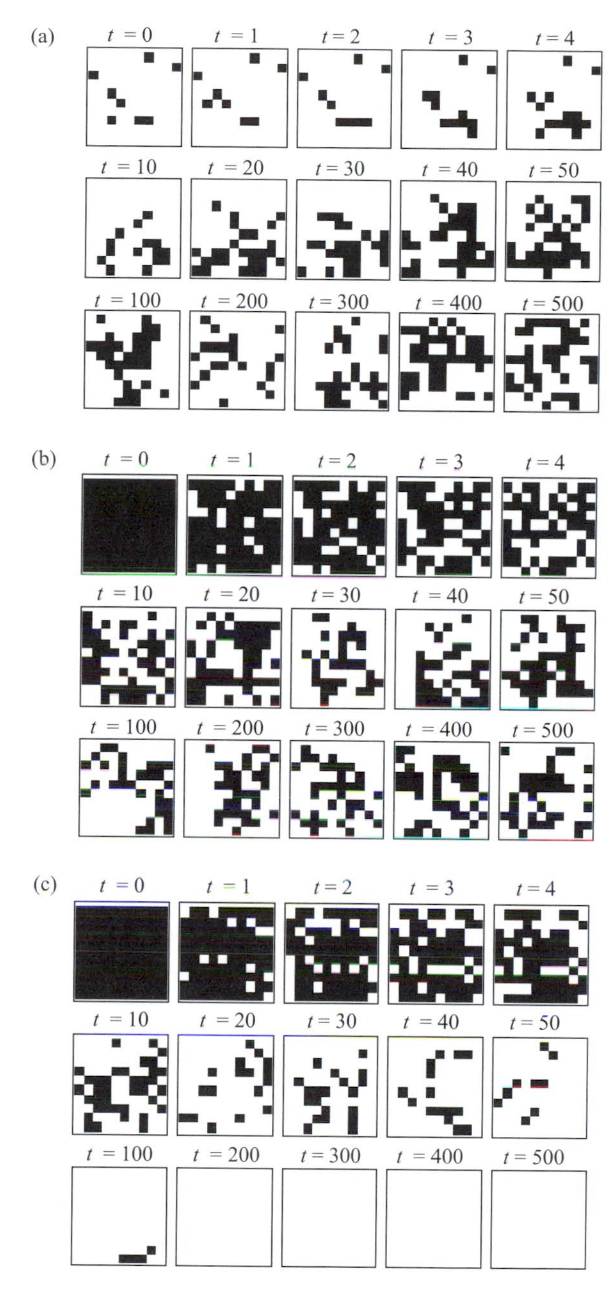

Figure 13.15: Simulations of the extinction-recolonization model. Occupied (black) and empty (clear) patches are shown on a 10×10 grid of sites. The simulations are run for 500 generations, and a snapshot of the metapopulation is shown at several intermediate time points. (a) $m = 0.05$, $e = 0.16$, initial fraction of filled sites = 10%, final fraction of filled sites = 41%, (b) $m = 0.05$, $e = 0.16$, initial fraction of filled sites = 100%, final fraction of filled sites = 36%, (c) $m = 0.03$, $e = 0.16$, initial fraction of filled sites = 100%, final fraction of filled sites = 0% (extinction).

that were previously occupied (black) to become empty (clear). Over time, the grid approaches a balance between filled and empty sites that is roughly the same whether 10% of sites were initially filled (Figure 13.15a) or 100% (Figure

13.15b). Eventually, however, the ensemble of populations goes extinct, but this takes many generations (t = 23,343 and 31,282 generations in the simulations of Figures 13.15a and 13.15b, respectively). In contrast, if we reduce the migration rate or increase the extinction rate, extinction happens much more rapidly, on average (e.g., t = 144 in the simulations of Figure 13.15c).

The main advantage of cellular automaton models is that they allow us to explore the dynamics of a population arranged over space, so that we can determine how summary statistics such as the mean extinction time of a species depend on the spatial arrangement and connectedness of populations.

13.8 Looking Backward in Time—Coalescent Theory

In all of the stochastic models considered so far, we imagine time running forward. While this is natural, there are some problems for which it is faster and easier to imagine time running backwards. This isn't as crazy as it seems. For example, when you draw your family tree, you start with yourself in the present and go backward in time through your parents, grandparents, etc. This example provides a good explanation for why you might want to run time backwards. To draw your family tree forward in time, you would have to start with every individual alive in, say, 1700 A.D. Then you would figure out who gave birth to whom, and draw every generation to the present day. Having traced every family lineage to the present, you would then throw out almost all of this information, keeping only those lineages that led to you. Ridiculous. Working backward in time thus makes sense when you are interested in a focal individual living in the present and in the historical processes leading up to that individual.

Here we explore a model based on the assumptions of the Wright-Fisher model, but that is run backward in time. The analysis of this model has led to an important new branch of mathematical biology known as *coalescent theory*. We begin in the present ($t = 0$), focusing on a certain number of alleles (n) sampled from a population of N individuals. To simplify the situation, we assume that the population is haploid, but all of the following results apply to a randomly mating diploid population as long as we replace N with $2N$, the number of alleles in a diploid population of N individuals.

In the current generation ($t = 0$), there is a different individual alive for each of the n alleles. For now, we do not keep track of whether the alleles encode the same DNA sequences, but rather only whether the alleles are carried by different individuals. In the previous generation ($t = 1$), there is some chance that two of the alleles descended from the exact same parent allele, meaning that there were only $n - 1$ different parent alleles that gave birth to the n alleles sampled today (Figure 13.16). This event, whereby n offspring alleles descended from only $n - 1$ parent alleles, is known as a *coalescent event*, and represents two lineages coming together ("coalescing") into one. If we predict that the alleles in our sample are likely to have coalesced in the recent past, then these alleles should be closely related and similar to one another. Conversely, alleles that are predicted to coalesce in the distant past should be less similar. Coalescent theory has had such a great impact because it predicts the relatedness among

Coalescent theory describes the probability that alleles in a sample descend from the same ancestral allele at time t in the past.

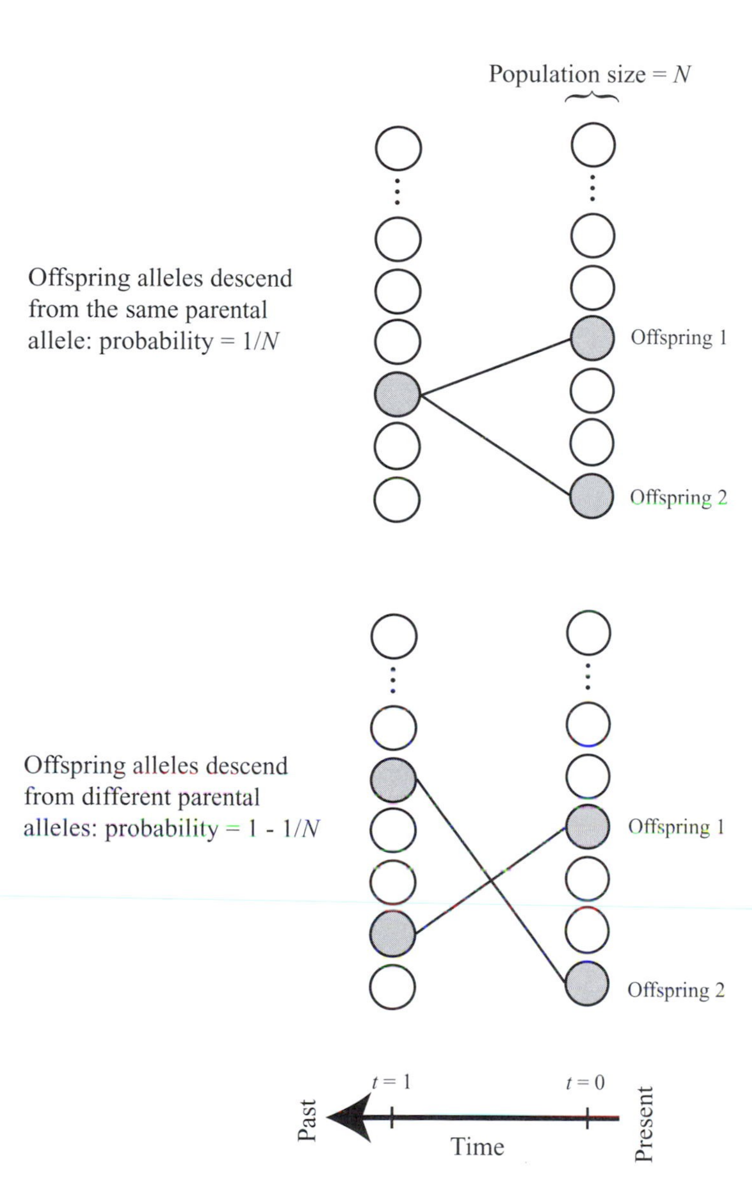

Figure 13.16: Descent of alleles from parents to offspring. The probability that two focal alleles (shaded circles) were born from the same parent allele is $1/N$ (top); this is called a coalescent event. Otherwise, the two focal alleles were born from different parent alleles (bottom). In Figures 13.16–13.18, time runs from the present on the right to the past on the left.

samples of individuals, predictions that can be tested using the DNA sequences carried by these individuals (Felsenstein 2004; Hudson and Kaplan 1995; Rosenberg and Nordborg 2002).

Our first aim is to describe the chance that a coalescent event happens t generations in the past, starting with a sample of only two alleles. The time in the past at which these two alleles coalesce, T_2, is the random variable of interest, and we seek the probability distribution for T_2. Given that all individuals in the population reproduce simultaneously (as in the Wright-Fisher model), what is the probability that two alleles descend from the same parent allele? The first sampled allele must have had some parent (with probability equal to one), but the second sampled allele could have had the same parent (with probability $1/N$) or a different parent (with probability $1 - 1/N$). Thus, the probability that there was a coalescent event in the previous generation ($t = 1$) is $1/N$. If the

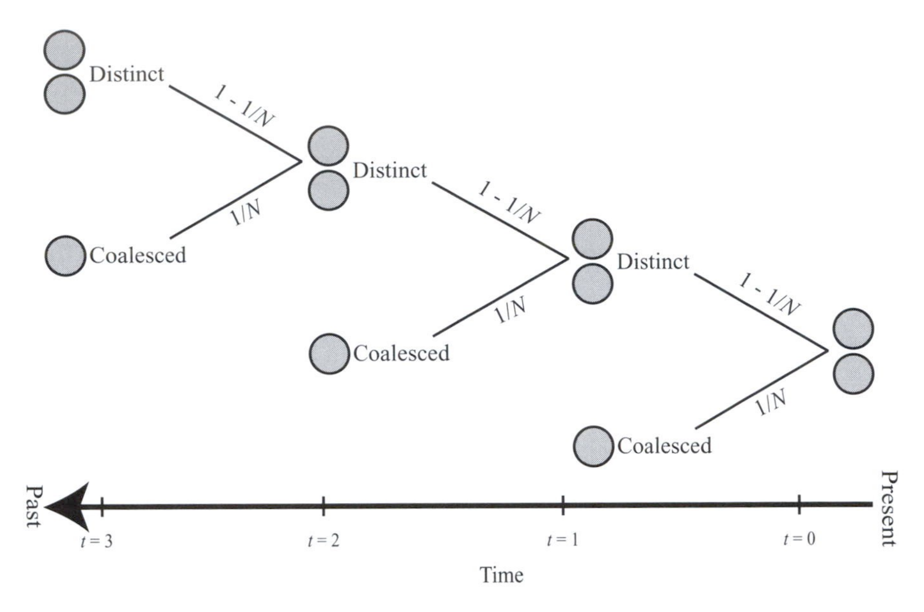

Figure 13.17: A decision tree for the coalescent model. Given two sampled alleles (shaded circles), there are two possibilities: the alleles share the same parent in the previous generation (coalesce) or they remain distinct. Once the two alleles coalesce, the coalescent process is over, and we no longer trace their history. The probability of one particular outcome is calculated as the product of the probabilities along the path to that outcome. For example, the probability of a coalescent at time $t = 3$ is given by $(1 - 1/N)$ $(1 - 1/N)$ $(1/N)$.

alleles coalesced, we know how related the sampled alleles are: they are siblings. If the alleles did not coalesce, then we are right back to where we started: with two alleles (now at $t = 1$), which may or may not have descended from the same ancestral allele at $t = 2$. And, again, the probability that they are descended from the same ancestral allele at $t = 2$ is $1/N$, in which case the alleles represent cousins. We can write all of these possibilities in the form of a decision tree (Figure 13.17).

From this decision tree, we can calculate the probability that the sampled alleles coalesce at generation t, counted backward in time. The probability of coalescence in the current generation is zero, $P(T_2 = 0) = 0$, because we know we sampled two different alleles. We have already figured out that the probability of coalescence at time $t = 1$ is $P(T_2 = 1) = 1/N$. The probability that the coalescent event occurs at time $t = 2$ is $P(T_2 = 2) = (1 - 1/N)\, 1/N$, which equals the probability that the alleles had not coalesced at time $t = 1$ (contributing the $1 - 1/N$ term) but did coalesce at time $t = 2$. In general, the probability that the two alleles coalesce t generations in the past is given by

$$P(T_2 = t) = \left(1 - \frac{1}{N}\right)^{t-1} \frac{1}{N}. \tag{13.13}$$

Expression (13.13) is what we are after—it is the probability distribution for the time to coalescence for two alleles, T_2. A comparison of equation (13.13) with Definition (P3.7) reveals that the waiting time for the coalescence of two alleles has a geometric probability distribution with parameter $p = 1/N$. As a result, we can immediately use the properties of the geometric distribution to infer that the mean time until coalescence of two alleles in a haploid population of size N is $E[T_2] = 1/p$ or N generations. Thus, the larger the population, the less likely it is that our two sampled individuals are close relatives. We also know that a geometric random variable has a variance $(1 - p)/p^2$ (see Table P3.2), from which we can calculate the variance in coalescent times as $N(N - 1)$. Thus, the exact time of coalescence is extremely variable.

If it takes N generations, on average, for a sample of two alleles to coalesce, you might think that it would take an incredibly long time for a sample of n alleles to coalesce into one allele. Our next aim is to find the probability distribution for the time, M_n, until the most recent common ancestor (MRCA) of n alleles sampled from a population of size N. The random variable M_n can be viewed as the sum of several independent random variables representing the time until each coalescent event: the time until the n sampled alleles coalesce into $n - 1$ alleles, plus the time that the $n - 1$ alleles coalesce into $n - 2$ alleles, etc., until only one allele remains. Defining T_i as the time until i alleles coalesce into $i - 1$ alleles, $M_n = T_n + T_{n-1} + \cdots + T_2$. We have already calculated the probability distribution for T_2; now we need to find the probability distribution for T_i when there are more than two alleles.

When there are i sampled alleles, the probability that none of them coalesce in the previous generation is given by the probability that each allele descends from different parents. The first allele must descend from some parent allele (probability = 1), the second allele must descend from a different parent allele from the first (probability = $1 - 1/N$), the third allele must descend from a different parent allele than either the first or the second (probability = $1 - 2/N$), etc. Writing $p(i)$ as the probability that there is at least one coalescent event in the preceding generation, the probability that there is *not* a coalescent event, $1 - p(i)$, is the probability that all i alleles descended from different parent alleles:

$$1 - p(i) = \left(1 - \frac{1}{N}\right)\left(1 - \frac{2}{N}\right)\cdots\left(1 - \frac{i-1}{N}\right). \qquad (13.14)$$

Standard coalescent theory makes the assumption that the population size N is large relative to the sample size i, and then approximates (13.14) using a Taylor series (Recipe P1.2, Primer 1). Assuming $1/N$ to be small and ignoring terms that are $O(1/N^2)$, equation (13.14) becomes

$$\begin{aligned} 1 - p(i) &\approx 1 - \frac{1}{N} - \frac{2}{N} \cdots - \frac{i-1}{N} \\ &= 1 - \frac{1}{N}\sum_{j=1}^{i-1} j \\ &= 1 - \frac{i(i-1)}{2N}, \end{aligned} \qquad (13.15)$$

where Rule A1.18 is used to evaluate the sum. Therefore, the probability of at least one coalescent event is $p(i) = i(i - 1)/(2N)$, which is sometimes written using the binomial coefficient as $p(i) \approx \binom{i}{2}/N$ (see Box P3.1). Technically, $p(i)$ describes the probability of *one or more* coalescent events in the preceding generation, but it is very unlikely that more than one coalescent event occurs when the population size is much larger than the sample size ($N >> i$). In this case, the probability that i alleles are descended from $i - 1$ alleles is very nearly equal to $p(i)$.

Following the same logic leading up to equation (13.13), the probability that it takes t generations for i alleles to coalesce into $i - 1$ alleles (represented by the random variable T_i) is

$$P(T_i = t) = (1 - p(i))^{t-1}\, p(i). \tag{13.16}$$

Again, this is a geometric distribution, now with parameter $p(i)$. As a result, the mean time until i alleles coalesce to $i - 1$ alleles is $1/p(i)$, or $2N/(i\ (i - 1))$ generations.

Given the above results, the time until the most recent common ancestor of n alleles, M_n, is given by the sum of several geometrically distributed random variables, T_i, each with their own parameter $p(i)$. Unfortunately, the probability distribution for a random variable given by the sum of different geometric random variables is not known in any simple form. Nevertheless, we can derive the expected (or mean) time until the MRCA for n alleles: $E[M_n] = E[T_n + T_{n-1} + \cdots + T_2]$ as $E[M_n] = E[T_n] + E[T_{n-1}] + \cdots + E[T_2]$, because the expectation of a sum equals the sum of the expectations (Table P3.1). Therefore, the expected time until the MRCA of n alleles is

$$\begin{aligned} E[M_n] &= \frac{2N}{n(n-1)} + \frac{2N}{(n-1)(n-2)} + \cdots + N \\ &= 2N \sum_{i=2}^{n} \frac{1}{i\,(i-1)} = 2N\,\frac{n-1}{n}. \end{aligned} \tag{13.17}$$

The last sum in (13.17) can be evaluated by induction (see Problem 13.8).

Result (13.17) is pretty amazing. The average time until the MRCA of all n alleles in a sample is less than twice the average time until the ancestor of only two alleles, no matter how large the sample. When there are lots of alleles, not much time passes before a coalescent event takes places because there are many pairs of alleles that could potentially coalesce.

Now that we have described the probability distribution for coalescent times, we can simulate these coalescent events to give us a better feeling for the ways in which a sample of alleles are likely to be related. To carry out these simulations starting with n alleles, we draw a random number from a geometric distribution with parameter $p(n)$ to get the time frame over which there remain n distinct alleles within the population. At this randomly drawn time, a coalescent event occurs, and we join together the branches for two alleles. We then repeat the process for the remaining $n - 1$ alleles, until we reach the MRCA. We carried out this coalescent simulation starting with a sample of ten alleles in Figure 13.18, repeating the process four times to illustrate the variability in tree length and tree shape. Notice that coalescent events occur more rapidly near the present ($t = 0$) because there are more pairs of alleles that can potentially

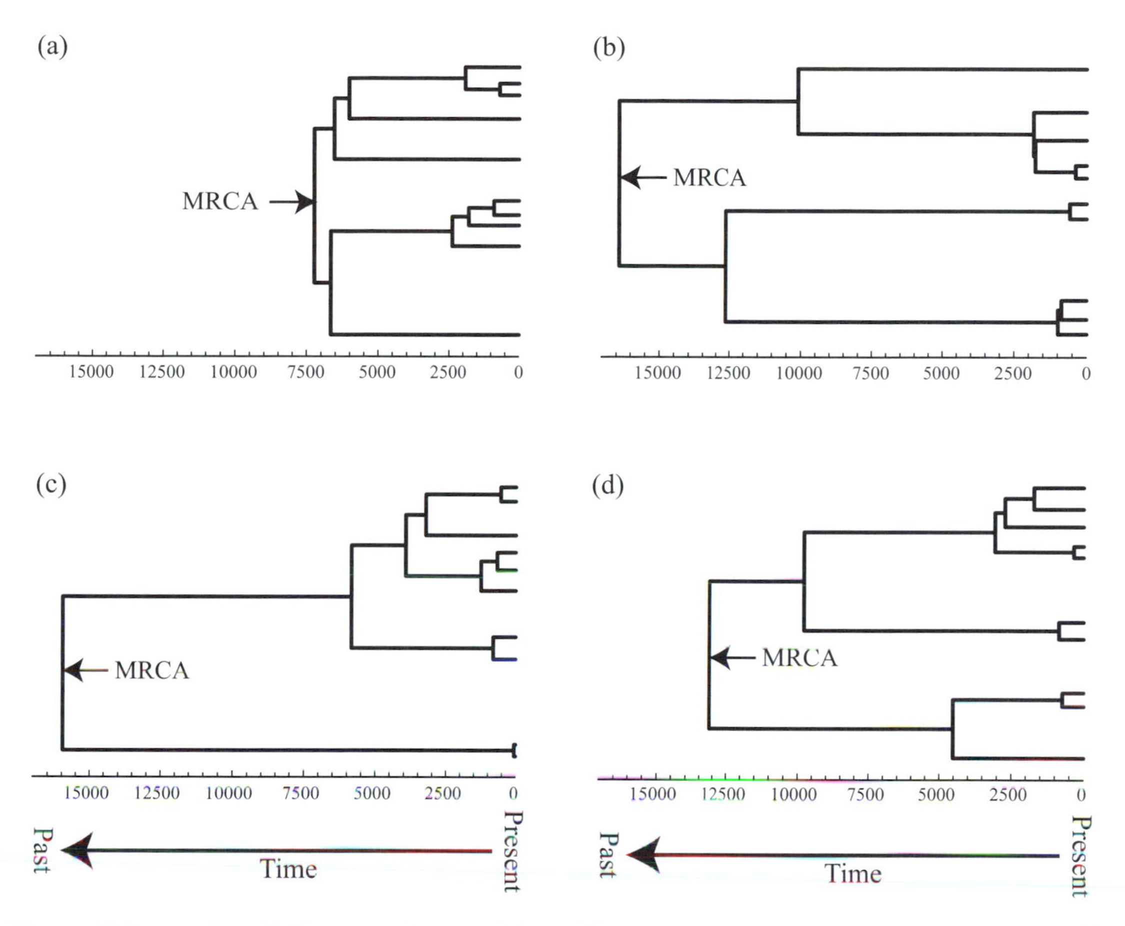

Figure 13.18: Coalescent simulations. Independent simulations of the coalescent process are illustrated in each panel. Starting at the present ($t = 0$, far right) with a sample of $n = 10$ alleles in a population of size $N = 10{,}000$, the time until a coalescent event was determined by randomly drawing times from the geometric distribution given by equation (13.16). At that point, two branches were randomly merged. The process was repeated until only one lineage remained (the most recent common ancestor, MRCA).

coalesce and that the final coalescent event between two alleles is, on average, the longest one.

The coalescent thus provides us with a description of the types of phylogenetic trees expected under the null model of a population of constant size in the absence of selection. While this description is interesting in and of itself, the real value of coalescent theory is that mutation can be overlaid on top of the phylogenies to describe patterns expected within a sample of DNA sequences. Each time a mutation occurs, it alters the DNA sequence in that individual and all of its descendants.

If mutations occur continuously over time at a constant rate, mutations can be imagined as raining down on the phylogenies drawn in Figure 13.18. The total number of mutations would then follow a Poisson distribution with a mean equal to the mutation rate times the total amount of time represented by all of the branches on the tree. For shorter trees (e.g., Figure 13.18a), we would thus expect fewer mutations and less genetic variability among the sampled sequences than in longer trees (e.g., Figure 13.18b).

Alternatively, if mutations occur only at discrete points in time (e.g., at meiosis), then the total number of mutations that occur in the history of a sample would follow a binomial distribution with parameters equal to the mutation rate and the total number of events (meioses) throughout all of the branches in the tree. Fortunately, because the binomial distribution converges upon a Poisson distribution when events are rare and when there are a large numbers of trials (see section P3.3.6 and Appendix 5), it makes little difference whether we model mutations as arising continually over time or at specific points in the life cycle (e.g., meiosis). Here, we assume that mutation is a continuous process, as is more common in coalescent theory.

To give you a flavor of the sorts of results that can be generated using coalescent theory, we will illustrate how to calculate two important quantities: (i) the probability that two sampled alleles are genetically identical and (ii) the number of segregating sites in a sample of n alleles. Throughout, we assume that mutations occur at rate μ per generation per sequence and that each mutation is unique (i.e., changes a different base pair within the sequence).

Let us consider the first question—what is the probability that two sampled alleles are identical? This question is impossible to answer without knowing how related the alleles are. Fortunately, the probability distribution (13.13) tells us the probability that the two alleles last shared a common ancestor at time t in the past, P(coalescence at time t). But how do we rewrite the probability that the alleles are identical using this information? The answer is to use the law of total probability (Rule P3.8). Because the coalescent times represent a set of mutually exclusive events, we can rewrite the probability that the two alleles are identical by summing the conditional probability that the alleles are identical given their coalescent time, over all coalescent times:

$$P(\text{alleles identical}) = \sum_{t=1}^{\infty} P(\text{alleles identical}|\text{coalescence at time } t) P(\text{coalescence at time } t).$$

The benefit of this expression is that we have now broken down the probability into pieces that are easier to calculate. The probability of coalescence at time t is given by (13.13): $P(\text{coalescence at time } t) = (1 - 1/N)^{t-1}(1/N)$. And $P(\text{alleles identical} \mid \text{coalescence at time } t)$ equals the probability that no mutations occur along either of the two branches leading from the common ancestor to the two sequences, amounting to a total branch length of $2t$ if they coalesced at time t. The probability of no mutations in this time period is given by the probability of drawing no events ($k = 0$) from a Poisson distribution with an expected number of mutations of $2\mu t$ (see Definition P3. 6). This probability is $P(\text{alleles identical} \mid \text{coalescence at time } t) = e^{-2\mu t}$. Summing over all coalescent times, the probability that the two sequences are identical is

$$P(\text{alleles identical}) = \sum_{t=1}^{\infty} e^{-2\mu t}\left(1 - \frac{1}{N}\right)^{t-1}\frac{1}{N}$$
$$= \frac{1}{1 + (e^{2\mu} - 1)N}. \tag{13.18a}$$

The sum in the first line of (13.18a) can be interpreted as a constant (here $e^{-2\mu}$) raised to the power of the random variable (here t), and averaged over its probability distribution (here the geometric distribution). This sum defines the moment generating function of a distribution (see Appendix 5). Thus, rather than having to simplify this summation from scratch, we can use the moment generating function of the geometric distribution (see Table P3.2 with $z = -2\mu$) to obtain the second line in (13.18a).

Equation (13.18a) can be simplified further by assuming that the mutation rate is small but that $N\mu$ is not small. First, (13.18a) is rewritten in terms of $\theta = 2N\mu$, the expected number of differences between two sequences in a haploid population (see Problem 13.9), by replacing N with $\theta/(2\mu)$. Next, the limit of (13.18a) is taken as the mutation rate goes to zero but θ is held constant (see Appendix 2). The probability of identity given by (13.18a) is then very nearly equal to

$$P(\text{alleles identical}) = \frac{1}{1 + 2N\mu}. \tag{13.18b}$$

This result makes qualitative sense. Two alleles will be more similar when P(alleles identical) is near one, which requires a low mutation rate and/or a small population size, so that the average coalescent time is short.

Next let us consider the second question—how many nucleotide sites in the DNA are likely to vary within a sample of n alleles? Any mutation that occurs in the history of the sample since the most recent common ancestor will cause a nucleotide difference between some individuals in the sample. We refer to this polymorphic nucleotide as a *segregating site*. Thus the number of segregating sites in the sample is equal to the number of mutations that occur over all of the branches of the tree. The number of segregating sites, S, is a random variable because mutations occur randomly over the tree and the length of the tree is determined by the random coalescence times. In general we might attempt to derive the probability distribution for S, but here we focus only on its expected value (i.e., the mean number of segregating sites, $E[S]$). Again, it would be impossible to calculate the number of segregating sites without knowing how much time has passed along the lineages leading to the present day sample from their most recent common ancestor. To proceed, we condition on the total length of a tree, L, summed over all branches and use the law of total expectation (Rule P3.9) to rewrite $E[S]$ as

$$E[S] = \sum_{l} E[S|L = l]\, P(L = l). \tag{13.19}$$

Equation (13.19) decomposes our problem into smaller pieces that can be evaluated.

The term $E[S \mid L = l]$ is the expected number of segregating sites given that the total tree length is l generations, which will be a Poisson random variable with mean $\mu\, l$. Plugging this result into equation (13.19), the expected number of segregating sites becomes $E[S] = \mu \sum_{l} l\, P(L = l)$, or just $E[S] = \mu\, E[L]$ (see Definition P3.2).

Finally, we need to calculate $E[L]$. We already know that the coalescent time when there are i distinct alleles is geometrically distributed with mean $2N/(i\,(i-1))$ generations. The sum of the branch lengths within a phylogenetic tree while there are i alleles is thus, on average, $2N/(i\,(i-1))$ multiplied by i, the number of branches in the tree during this period. Summing over all possible numbers of branches, i, the mean total length of a tree, $E[L]$, is

$$E[L] = \sum_{i=2}^{n} i\,\frac{2N}{i\,(i-1)} = 2N\sum_{i=2}^{n}\frac{1}{(i-1)} = 2N\sum_{i=1}^{n-1}\frac{1}{i}. \tag{13.20}$$

The expected number of segregating sites within a sample is therefore given by

$$E[S] = \mu\,E[L] = 2N\mu\sum_{i=1}^{n-1}\frac{1}{i} = \theta\sum_{i=1}^{n-1}\frac{1}{i}.$$

Interestingly, both the probability of identity and the expected number of segregating sites depend on the same quantity $\theta = 2N\mu$, which measures the expected difference between two sequences under the Wright-Fisher model in a haploid population (see Problem 13.9). (In a diploid population, $\theta = 4N\mu$.) The fact that these quantities should be related to one another was used by Tajima (1983) to test the null hypothesis that genes have been evolving neutrally, without selection (Tajima's D statistic).

The power of coalescent theory is that one can obtain the expected values of various properties (like the number of segregating sites) and compare them against data. You might wonder, however, whether it is reasonable to assume that the Wright-Fisher model is correct. Indeed, one way to think about coalescent theory is that it provides us with a null hypothesis that should hold if the history of the sample involved nothing other than neutral sampling in a population of constant size. If the patterns within the sampled sequences do not match these expectations, then we infer that some other process is happening. This other process might have been selection, but it might also have been changes to the population size and/or migration. Although it is difficult to extend coalescent theory to describe selection (see Krone and Neuhauser 1997; Neuhauser and Krone 1997), the theory has been extended in various ways to account for changing population size, migration, founder events, etc. (Felsenstein 2004; Hudson and Kaplan 1995; Rosenberg and Nordborg 2002).

13.9 Concluding Message

In this chapter we have described how many of the classic deterministic models introduced in Chapter 3 can be extended to incorporate stochasticity. Discrete-time models of exponential and logistic growth were extended in section 13.2 to allow for variation in family size. Continuous-time models of population growth were extended in section 13.3, using a birth-death model that allows for replication and death at random points in time. Population-genetic models of allele frequency change were extended in section 13.4 using the classic Wright-Fisher model (where reproduction is simultaneous) and in section 13.5 using the Moran model (where only one individual reproduces at a time).

In addition to these classic models, we developed three other stochastic models, chosen to illustrate different ways in which chance events can be modeled. The model of retinoblastoma, a cancer of the eye, describes the stochastic way in which mutations arise during development. The cellular automaton model of extinction-recolonization describes the stochastic way in which populations spread over space, as well as the stochastic nature of extinction. Finally, the coalescent model describes the stochastic way in which individuals are related to one another. In each of these models, chance plays a key role in determining the outcome, as we explored through the use of probability theory (Primer 3) and simulation. In the next two chapters, we describe methods of analyzing such stochastic models that allow us to draw more general conclusions than are possible from simulations alone.

Problems

Problem 13.1: In the text, we considered an environment that varies over time between a good state and a bad state, where the probability of being in either state at time $t + 1$ is given by

$$\begin{pmatrix} P_g(t+1) \\ P_b(t+1) \end{pmatrix} = \begin{pmatrix} p_{gg} & p_{gb} \\ p_{bg} & p_{bb} \end{pmatrix} \begin{pmatrix} P_g(t) \\ P_b(t) \end{pmatrix}$$

Generalize this model by allowing three states: good (g), bad (b), and recovering (r). Assume that if the environment is bad, it either remains bad or enters the recovering state, that if the environment is recovering, it either continues to recover or enters the good state, and that if the environment is good, it either remains good or turns bad. No other transitions are possible. Write down the matrix equation describing these transition probabilities.

Problem 13.2: Here you will simulate the model of exponential growth with environmental and demographic stochasticity when the environment is correlated from year to year. As in Figure 13.4, assume that there are good ($R_g = 1.5$) and bad ($R_b = 0.5$) environments. Now, if the current environment is good, assume that it remains good with probability 0.85, while if the current environment is bad it remains bad with probability 0.65:

$$\begin{pmatrix} P_g(t+1) \\ P_b(t+1) \end{pmatrix} = \begin{pmatrix} 0.85 & 0.35 \\ 0.15 & 0.65 \end{pmatrix} \begin{pmatrix} P_g(t) \\ P_b(t) \end{pmatrix}$$

These numbers were chosen so that, over the long run, the environment is good 70% of the time as in Figure 13.4, but now the environment has a higher probability of remaining in the same state for several years in a row (given by the diagonal elements). (a) Generate figures for these simulations as in Figure 13.4. (b) Simulate the model 100 times and count how often the population goes extinct within 30 generations. [Hint: Use the state of the environment in the previous generation and the transition matrix to obtain the probability that the environment will be good in the next time step, p. Use this p value as we did in Figure 13.4.]

Problem 13.3: Alter the birth-death model of population growth to allow death rates to depend on the current population size, i, rather than birth rates. (a) Determine the transition probabilities (see equation (13.3) for density-dependent birth rates). (b) According to your answer to (a), at what population size is there the lowest probability of dying per time? Does this make sense?

Problem 13.4: The Wright-Fisher model can be extended to describe a population of N diploid individuals rather than N haploid individuals if it is assumed that alleles can come from any parent and are united at random in the offspring. (a) Write down the equivalent to equation (13.4) for a diploid population with N individuals. (b) Write down the equivalent to equation (13.13) for a diploid population with N individuals. (c) Infer the average time to coalescence for a pair of alleles drawn from a diploid population. (d) Explain why your answer for the diploid model differs from the expected coalescence time of N generations for a haploid population.

Problem 13.5: (a) For the Moran model (13.8), write p_{ji} in the form of a transition probability matrix for a population of size $N = 4$. Confirm that the columns sum to one. (b) Repeat for a population of any size N by filling in the "—" entries in the matrix

$$\begin{pmatrix} - & - & - & \cdots & - \\ - & - & - & \cdots & - \\ - & - & - & \cdots & - \\ \cdots & \cdots & \cdots & \cdots & \cdots \\ - & - & - & \cdots & - \end{pmatrix}.$$

The transition probability matrix for the Moran model is tridiagonal, with zeros everywhere except the entries on, immediately above, or immediately below the diagonal.

Problem 13.6: Modify equations (13.9) for the Moran model of allele frequency change to take into account mutation. Assume that mutations occur only during reproduction (births) and that there is a probability μ that allele A mutates to a and a probability ν that allele a mutates to A.

Problem 13.7: In Figure 13.19, we show an empty cell (A) surrounded by cells in a cellular automaton (dark cells are occupied, white cells are empty). A monsoon causes the top left cell to go extinct in the current generation. What is the probability that the central cell is recolonized at the next time step if (a) the eight nearest neighbors serve as a migrant source and extinction occurs before migration, (b) the eight nearest neighbors serve as a migrant source and migration occurs before extinction, and (c) the four nearest neighbors serve as a migrant source and extinction occurs before migration. [Define m and e as in the derivation of (13.12) in the text.]

Problem 13.8: Prove that $\sum_{i=2}^{n} 1/(i\,(i-1)) = (n-1)/n$, which was needed to derive the average time until the most recent common ancestor of a sample of n alleles,

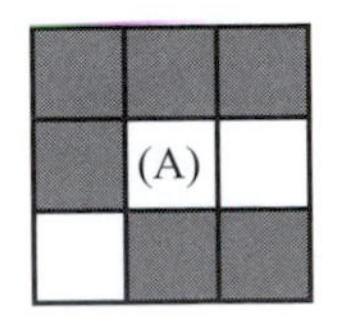

Figure 13.19: Recolonization from neighboring sites

equation (13.17). (a) Calculate and simplify the sum $\sum_{i=2}^{n} 1/(i\,(i-1))$ for $n = 2$, 3, and 4. The resulting pattern suggests that $\sum_{i=2}^{n} 1/(i\,(i-1)) = (n-1)/n$. (b) Assume that this equation holds true for $n-1$, so that $\sum_{i=2}^{n-1} 1/(i(i-1)) = (n-2)/(n-1)$. Show that you can add $1/(n\,(n-1))$ to both sides of the equation to prove that the equation also holds true for n. Because the equation is correct for $n = 2$ and remains true every time n is increased by one, you have proven by induction that $\sum_{i=2}^{n} 1/(i\,(i-1)) = (n-1)/n$.

Problem 13.9: Calculate the average number of differences between two sequences, $E[D]$, using coalescent theory when mutations occur continuously over time at a rate μ per generation per sequence length. (a) Rewrite $E[D]$ in terms of $E[D \mid$ coalescence at time $t]$ using the law of total expectation (Rule P3.9). (b) Calculate $E[D \mid$ coalescence at time $t]$. (c) Use the probability distribution for coalescent times, P(coalescence at time t) $= (1 - 1/N)^{t-1}\, 1/N$, as well as your answers to parts (a) and (b) to determine the average number of differences between two sequences, $E[D]$.

Further Reading

For more information on the mathematical underpinnings of stochastic models, see

- Taylor, H. M., and S. Karlin. 1998. *An Introduction to Stochastic Modeling*. Academic Press, San Diego.
- Allen, L.J.S. 2003. *An Introduction to Stochastic Processes with Applications to Biology*. Pearson/prentice Hall, Upper Saddle River, N.J.

For more examples of stochastic models in ecology, see

- Renshaw, E. 1991. *Modelling Biological Populations in Space and Time*. Cambridge University Press, Cambridge.
- Nisbet, R. M., and W.S.C. Gurney. 1982. *Modelling Fluctuating Populations*. Wiley, Chichester.
- Hubbell, S. P. 2001. *The Unified Neutral theory of Biodiversity and Biogeography*. Princeton University Press, Princeton, N.J.

For more examples of stochastic models in evolution, see

- Ewens, W. J. 1979. *Mathematical Population Genetics*. Springer-Verlag, Berlin.
- Crow, J. F., and M. Kimura. 1970. *An Introduction to Population Genetics Theory*. Harper & Row, New York.

References

Bjørnstad, O. N., and B. T. Grenfell. 2001. Noisy clockwork: Time series analysis of population fluctuations in animals. *Science* 293:638–43.

Bron, A. J., R. C. Tripathi, B. J. Tripathi, and E. Wolff. 1997. *Wolff's Anatomy of the Eye and Orbit*. Chapman & Hall Medical, London.

Chen, D., I. Livne-bar, J. Vanderluit, R. Slack, M. Agochiya, and R. Bremner. 2004. Cell-specific effects of RB or RB/p107 loss on retinal development implicate an intrinsically death resistant cell-of-origin in retinoblastoma. *Cancer Cell* 5:539–551.

Deangelis, D. L., and L. J. Gross (eds.). 1992. *Individual-Based Models and Approaches in Ecology: Populations, Communities, and Ecosystems*. Chapman and Hall, New York.

Dreher, Z., S. R. Robinson, and C. Distler. 1992. Müller cells in vascular and avascular retinae: A survey of seven mammals. *J. Comp. Neurol.* 323:59–80.

Edwards, A., H. A. Hammond, L. Jin, C. T. Caskey, and R. Chakraborty. 1992. Genetic variation at five trimeric and tetrameric tandem repeat loci in four human population groups. *Genomics* 12:241–253.

Ewens, W. J. 1979. *Mathematical Population Genetics*. Springer-Verlag, Berlin.

Felsenstein, J. 2004. *Inferring Phylogenies*. Sinauer Associates, Sunderland, Mass.

Frost, S.D.W., M.-J. Dumaurier, S. Wain-Hobson, and A.J.L. Brown. 2001. Genetic drift and within-host metapopulation dynamics of HIV-1 infection. *Proc. Natl. Acad. Sci. U.S.A.* 98:6975–6980.

Gardner, M. 1983. *Wheels, Life, and other Mathematical Amusements*. W. H. Freeman, New York.

Geiger, H. J., W. W. Smoker, L. A. Zhivotovsky, and A. J. Gharrett. 1997. Variability of family size and marine survival in pink salmon (Oncorhynchus gorbuscha) has implications for conservation biology and human use. *Can. J. Fish. Aquat. Sci.* 54:2684–2690.

Harvey, P. H., E. C. Holmes, A. Ø. Mooers, and S. Nee. 1994. Inferring evolutionary processes from molecular phylogenies. Pp. 313–333 in R. W. Scotland, D. J. Siebert and D. M. Williams, eds. *Models in Phylogeny Reconstruction*. Clarendon Press, Oxford.

Huang, S., and S. A. Moody. 1993. The retinal fate of Xenopus cleavage stage progenitors is dependent upon blastomere position and competence: Studies of normal and regulated clones. *J. Neurosci.* 13:3193–3210.

Hudson, R. R., and N. L. Kaplan. 1995. The coalescent process and background selection. *Philos Trans. R. Soc. London, Ser. B* 349:19–23.

Kaplan, D., and D. Glass. 1995. *Understanding Nonlinear Dynamics*. Springer-Verlag, New York.

Knudson, A. G. 1971. Mutation and cancer: statistical study of retinoblastoma. *Proc. Natl. Acad. Sci. U.S.A.* 68:820–823.

Knudson, A. G. 1993. Antioncogenes and human cancer. *Proc. Natl. Acad. Sci. U.S.A.* 90:10914–10921.

Kojima, K., and T. M. Kelleher. 1962. The survival of mutant genes. *Am. Nat.* 96:329–343.

Krone, S. M., and C. Neuhauser. 1997. Ancestral Processes with Selection. *Theor. Popul. Biol.* 51:210–37.

Loeuille, N., and M. Ghil. 2004. Intrinsic and climatic factors in North-American animal population dynamics. *BMC Ecol.* 4:6.

Lohmann, D. R. 1999. RB1 gene mutations in retinoblastoma. *Hum. Mutat.* 14:283–8.

Luria, S. E., and M. Delbruck. 1943. Mutations of bacteria from virus sensitivity to virus resistance. *Genetics* 28:491–511.

Moffett, D. F., S. B. Moffett, and C. L. Schauf. 1993. *Human Physiology—Foundations and Frontiers*. Mosby, St. Louis.

Moran, P. 1962. *The Statistical Processes of Evolutionary Theory*. Clarendon Press, Oxford.

Nee, S., E. C. Holmes, R. M. May, and P. H. Harvey. 1994a. Extinction rates can be estimated from molecular phylogenies. *Philos. Trans. R. Soc. London Ser. B* 344:77–82.

Nee, S., E. C. Holmes, R. M. May, and P. H. Harvey. 1995. Estimating extinction from molecular phylogenies. Pp. 164–182 in J. L. Lawton and R. M. May, eds. *Extinction Rates*. Oxford University Press, Oxford.

Nee, S., R. M. May, and P. H. Harvey. 1994b. The reconstructed evolutionary process. *Philos. Trans. R. Soc. London, Ser. B* 344:305–311.

Neuhauser, C., and S. M. Krone. 1997. The genealogy of samples in models with selection. *Genetics* 145:519–534.

Nisbet, R. M., and W. S. C. Gurney. 1982. *Modelling Fluctuating Populations*. Wiley, New York.

Ohta, T., and M. Kimura. 1973. A model of mutation appropriate to estimate the number of electrophoretically detectable alleles in a finite population. *Genet. Res.* 22:201–204.

Press, W. H. 2002. *Numerical Recipes in C++ : The Art of Scientific Computing*. Cambridge University Press, Cambridge.

Press, W. H., B. P. Flannery, S. A. Teukolsky, and W. T. Vetterling. 1992. *Numerical recipes in C: The Art of Scientific Computing*. Cambridge University Press, Cambridge.

Purvis, A., S. Nee, and P. H. Harvey. 1995. Macroevolutionary inferences from primate phylogeny. *Proc. R. Soc. London, Ser. B* 260:329–333.

Renshaw, E. 1991. *Modelling Biological Populations in Space and Time*. Cambridge University Press, Cambridge.

Rosenberg, N. A., and M. Nordborg. 2002. Genealogical trees, coalescent theory and the analysis of genetic polymorphisms. *Nat. Rev. Genet.* 3:380–390.

Tajima, F. 1983. Evolutionary relationship of DNA sequences in finite populations. *Genetics* 105:437–460.

Valdes, A. M., M. Slatkin, and N. B. Freimer. 1993. Allele frequencies at microsatellite loci: the stepwise mutation model revisited. *Genetics* 133:737–749.

Van Driel, D., J. M. Provis, and F. A. Billson. 1990. Early differentiation of ganglion, amacrine, bipolar, and Muller cells in the developing fovea of human retina. *J. Comp. Neurol.* 291:203–219.

Vogel, F., and R. Rathenberg. 1975. Spontaneous mutation in man. *Adv. Hum. Genet.* 5:223–318.

Yule, G. U. 1924. A mathematical theory of evolution, based on the conclusions of Dr. J. C. Willis. *Philos. Trans. R. Soc. London, Ser. B* 213:21–87.

CHAPTER 14

Analyzing Discrete Stochastic Models

Chapter Goals:

- To identify the stationary distribution of discrete stochastic models
- To calculate the probability of absorption in a particular state
- To calculate the time until absorption in a particular state

Chapter Concepts:

- Markov model
- Transition probability matrix/stochastic matrix
- Absorption probability
- Time to absorption

14.1 Introduction

In the next two chapters, we describe important methods for analyzing stochastic models. Many of the concepts are similar to those introduced earlier for deterministic models. In a deterministic model, we write down a recursion or differential equation to describe how the state of a system changes over time. In a stochastic model, we write down transition probabilities to describe how the probability that a system is in any particular state changes over time. In a deterministic model, we find equilibrium values by determining when the state of the system remains unchanged over time. Analogously, in a stochastic model, we find the *stationary distribution*, which describes the probability of being in any state, and is found by determining when the probability distribution remains unchanged over time. In a deterministic model, we might search for a general solution, telling us the state of the system at any future point in time. Similarly, in a stochastic model, we might solve for the probability of being in any particular state at any future point in time.

There are many methods for analyzing stochastic models, and we can only describe a fraction of them in two chapters. We therefore focus on methods that are commonly encountered, with a bias towards methods that are straightforward to apply but that yield insightful and interesting results. In this chapter we consider models with random variables that take on discrete values (e.g., the number of individuals). Most of the models examined are also discrete-time models, although we illustrate one example of a continuous-time discrete-state model. Models for continuous random variables are considered in Chapter 15.

In section 14.2, we introduce two-state Markov chains. Section 14.3 then generalizes these results to multiple states, and demonstrates how to calculate the stationary distribution, probability of absorption, and expected time until absorption for such models. Section 14.4 introduces a discrete-state, continuous-time stochastic model referred to as a birth-death process. Finally, section 14.5 presents a third important type of stochastic model referred to as a branching process.

A *Markov model* is a stochastic model in which transitions depend only on the most recent state and not on any previous states.

14.2 Two-State Markov Models

All of the models that we consider, including those introduced in Chapter 13, are known as *Markov models* or *Markov chains*, which are characterized by the following property:

Definition 14.1: Markov Property

If the probability that a system is in any particular state at time t depends only on the state of the system $X(t-1)$ at time $t-1$ and not on any previous states, the system is said to satisfy the Markov property.

In other words, the future transitions that occur in a Markov model depend only on the current state and not on the history of past transitions. Markov models are thus said to be "memoryless."

If we represent the state of the system at time t by the random variable, $X(t)$, then the Markov property can be written as:

$$\begin{aligned} P(X(t) = x(t) \mid X(0) = x(0), X(1) = x(1), \ldots, X(t-1) = x(t-1)), \\ = P(X(t) = x(t) \mid X(t-1) = x(t-1)). \end{aligned} \tag{14.1}$$

Reading equation (14.1) in words, the probability that the system is in state $x(t)$ at time t given ("|") that it is in state $x(0)$ at time 0, $x(1)$ at time 1, . . . and $x(t-1)$ at time $t-1$ equals the probability that the system is in state $x(t)$ at time t given only the information that it is in state $x(t-1)$ at time $t-1$.

From the current state, the probability that the system is in any particular state in the next time step is determined by the transition probability:

Definition 14.2: Transition Probability

The probability that a system is in state $x(t)$ at time t, given that it is in state $x(t-1)$ at time $t-1$, is referred to as the *transition probability*, $P(X(t) = x(t) \mid X(t-1) = x(t-1))$.

A Markov model is fully described by the transition probabilities from each state to all other states, along with the initial state of the system. Here, we assume that these transition probabilities are constant over time. Advanced texts also treat transition probabilities that vary over time ("nonstationary" probabilities).

The simplest Markov model describes a system consisting of only two states. An example is an environment that is either good or bad for a particular species, which we write as g and b, respectively. A two-state Markov model would also describe the probability that a nest hole is or is not occupied by a bird, or the probability that a specific nucleotide is a purine or a pyrimidine in a DNA sequence. The transition probabilities for this model are

$$\begin{aligned} P(X(t) = g \mid X(t-1) = g) = P_{gg}, \\ P(X(t) = b \mid X(t-1) = g) = P_{bg}, \\ P(X(t) = g \mid X(t-1) = b) = P_{gb}, \\ P(X(t) = b \mid X(t-1) = b) = P_{bb}. \end{aligned} \tag{14.2}$$

Using these probabilities, we can write the possible transitions in matrix form:

$$\begin{pmatrix} P(X(t) = g) \\ P(X(t) = b) \end{pmatrix} = \begin{pmatrix} p_{gg} & p_{gb} \\ p_{bg} & p_{bb} \end{pmatrix} \begin{pmatrix} P(X(t-1) = g) \\ P(X(t-1) = b) \end{pmatrix} \tag{14.3a}$$

Equation (14.3a) is a standard matrix equation for a linear model involving two variables. In this case, however, the two variables are related. At any time t the system must be in either the good environment or the bad environment. Thus, the probabilities of being in these two states must sum to one: $P(X(t) = g) + P(X(t) = b) = 1$. Consequently, we can eliminate one variable from the model by writing $P(X(t) = b) = 1 - P(X(t) = g)$. Similarly, the system will either change or stay the same after one time step; there are no other possible outcomes. Because these are mutually exclusive events, the probability of these two outcomes must sum to one: $p_{gg} + p_{bg} = 1$ and $p_{gb} + p_{bb} = 1$, which implies that the columns of the matrix in (14.3a) sum to one. Consequently, we can eliminate two parameters from the model by writing $p_{gg} = 1 - p_{bg}$ and $p_{bb} = 1 - p_{gb}$.

Multiplying out (14.3a), we get a recursion equation describing changes over time in the probability that the environment is good:

$$\begin{aligned} P(X(t) = g) &= p_{gg} P(X(t-1) = g) + p_{gb} P(X(t-1) = b) \\ &= (1 - p_{bg}) P(X(t-1) = g) + p_{gb}(1 - P(X(t-1) = g)) \end{aligned} \tag{14.3b}$$

where we have used both $P(X(t) = b) = 1 - P(X(t) = g)$ and $p_{gg} = 1 - p_{bg}$ in the second step. Equation (14.3b) says that the environment will be good at time t if the environment was good in the previous time step and remained good (first term on the right), or if the environment was bad but then became good (second term on the right).

Equation (14.3b) is a linear, affine recursion equation involving one variable. This is easier to recognize if we rewrite the variable $P(X(t) = g)$ as $x(t)$, giving

$$\begin{aligned} x(t) &= (1 - p_{bg})\, x(t-1) + p_{gb}(1 - x(t-1)) \\ &= (1 - p_{bg} - p_{gb})\, x(t-1) + p_{gb}. \end{aligned} \tag{14.4}$$

We already know how to analyze recursion equations of this form (Recipe 6.1). First, the equilibrium of (14.4) is found by setting $x(t)$ and $x(t-1)$ to $\hat{x}$ and solving for $\hat{x}$, giving $\hat{x} = p_{gb}/(p_{bg} + p_{gb})$. But what does this equilibrium represent? Remember that $x(t)$ is the probability that the environment is good at time t, and this probability changes stochastically through time according to equation (14.4). The quantity $\hat{x}$ is the equilibrium value of this probability. If the probability that the environment is good at time $t - 1$ is $p_{gb}/(p_{bg} + p_{gb})$, then the same will be true at time t, and for all future points in time. Similarly, if the probability that the environment is bad starts at $1 - (p_{gb}/(p_{bg} + p_{gb})) = p_{bg}/(p_{bg} + p_{gb})$, the probability that the environment is bad remains unchanged over time. The *realized* state of the environment will change back and forth randomly over time, but the *probability* that it is in the good state will remain at $\hat{x}$. This equilibrium is referred to as the *stationary distribution* of the model. Just as with equilibria in deterministic models, there can be more than one stationary distribution in stochastic models.

Definition 14.3: Stationary Distribution

If the distribution describing the probability that a system is in any particular state remains unchanged from one time point to the next, then the probability distribution is said to be a *stationary distribution*. The probability that the state of the system is x at the stationary distribution will be represented by $\pi(x)$.

Using the above notation, we can write the stationary distribution as

$$\pi(g) = \frac{p_{gb}}{p_{bg} + p_{gb}}, \quad \pi(b) = \frac{p_{bg}}{p_{bg} + p_{gb}}. \tag{14.5}$$

This stationary distribution makes sense. If it is twice as likely that a bad environment turns good than a good environment turns bad (say, $p_{gb} = 0.02$ and $p_{bg} = 0.01$), then we are twice as likely to encounter a good environment at the stationary distribution: $\pi(g) = 2/3$ and $\pi(b) = 1/3$.

If we do not start out at the stationary distribution, however, then the probability that the environment is good (or bad) in the next time step will *not* be $\pi(g)$ (or $\pi(b)$). For example, if we know that the environment was good last year, and if $p_{bg} = 0.01$, then the environment will likely remain good for a period of time.

Given that the system starts out in a particular state, we can describe the probability that the system will be in any particular state at any time in the future using the general solution for an affine model. Following Recipe 6.1, we find that

$$x(t) = (1 - p_{bg} - p_{gb})^t\, x(0) + (1 - (1 - p_{bg} - p_{gb})^t)\frac{p_{gb}}{p_{bg} + p_{gb}}. \tag{14.6}$$

This is the general solution for the probability that the environment is good at time t, $x(t) = P(X(0) = g)$, given the probability that the environment was good at time 0, $x(0) = P(X(0) = g)$. As time passes, $(1 - p_{bg} - p_{gb})^t$ in the general solution (14.6) gets smaller (assuming that p_{bg} and p_{gb} are not both zero or one). Thus, after a large amount of time, the probability that the environment is good approaches $p_{gb}/(p_{bg} + p_{gb})$, as expected from the stationary distribution.

Many stochastic models have states that, once reached, are never left. Such states are referred to as *absorbing states*. Absorbing states are a special kind of stationary distribution represented by a vector, $\vec{\pi}$, whose elements are all zero except for the absorbing state in question (whose element is one). For example, if we were modeling the number of individuals of a species, we could set up a Markov model to describe the probability that there are a certain number of individuals at time t. In such a model, the state 0 would be an absorbing state because, if the population size ever reached zero, it would remain there for all future time. The vector, $\vec{\pi} = (1,0,0,\dots)^T$, would be the stationary distribution representing this absorbing state.

Definition 14.4: Absorbing State

An absorbing state is a state (or set of states) that a Markov chain can reach but never leave. If state i is an absorbing state, then $p_{ji} = 0$ for any j other than i and $p_{ii} = 1$. An absorbing state necessarily represents a stationary distribution.

Throughout this chapter we assume that each absorbing state corresponds to a single state only. Advanced texts also treat cases where absorption occurs to a set of states, among which transitions are still possible (see "Further Reading").

When a stochastic model has at least one absorbing state, we often want to know the probability that a particular absorbing state is reached. Additionally, we want to know how long it takes before an absorbing state is reached (the "waiting time"). We can address both of these questions by applying the mathematical techniques presented in earlier chapters.

We first modify the simple two-state model so that one of the states is an absorbing state. For example, imagine that the two-state model describes the probability that a species is present ("good") or extinct ("bad"). In this case, p_{bg} would represent the probability of extinction per time step, and p_{gb} would be 0, because the species is irretrievably lost once it has gone extinct. Thus, in the following, we suppose that the bad state (extinction) is absorbing.

It is clear that the system will eventually reach the bad state as long as there is some probability that the good state turns bad ($p_{bg} > 0$). This fact is confirmed by the general solution (14.6) with p_{gb} set to 0: $x(t) = (1 - p_{bg})^t x(0)$, which indicates that the probability of being in the good state, $x(t)$, goes to zero as time goes to infinity as long as $p_{bg} > 0$. Having started in state j, we can write the probability of eventually being absorbed in state i as

$$\rho_{j,i} = P(X(\infty) = j \mid X(t) = i). \tag{14.7}$$

For this two-state model, $\rho_{b,b} = 1$ because the bad state is absorbing and so the system will remain there forevermore. Similarly, $\rho_{b,g} = 1$ because the system will eventually become absorbed in the bad state as long as $p_{bg} > 0$. Thus, regardless of the initial state of the system, the system will eventually enter the bad state if we wait long enough.

A Markov model that has one or more absorbing states and a finite number of nonabsorbing states will eventually become "absorbed." This assumes only that the absorbing states have some nonzero probability of being reached (e.g., $p_{bg} > 0$ in the previous example). This fact is enormously insightful. For example, given that the number of individuals of any species must have some finite maximum value, that it is always possible that every individual fails to reproduce, and that the extinction of the species represents an absorbing state, the theory of stochastic processes tells us that all species (including our own) must eventually go extinct. It could, however, take a very long time for this to happen. We turn next to calculating such waiting times.

Given that the state of the system is initially good, how long must we wait before it turns bad? The waiting time before state b is reached represents a random variable, which we denote by T. This waiting time is characterized by some probability distribution, describing the probability that the bad state is first reached at time 1, time 2, etc. We will calculate this probability distribution and then determine the mean waiting time, given that the initial state is good.

Suppose that the system is in the good state at time 0, and let the probability that the system turns bad for the first time at time t be denoted by $P(T = t \mid X(0) = g)$. The probability that the system turns bad after only one time step (at $t = 1$) is given by the probability of a transition from a good to a bad state, thus, $P(T = 1 \mid X(0) = g) = p_{bg}$. For all future time points, we can calculate the waiting time using the following logic. For the system to reach the bad state for the first time at $t > 1$, the system must make a transition from a good state at time 0 to a good state at time 1 and then wait until time t for the first transition to a bad state:

$$P(T = t \mid X(0) = g) = p_{gg}\, P(T = t \mid X(1) = g). \tag{14.8}$$

Under the assumption that the transition probabilities do not change over time, the probability of waiting from time 1 to time t before first reaching the bad state, $P(T = t \mid X(1) = g)$, equals the probability of waiting from time 0 to time $t - 1$ before first reaching the bad state, $P(T = t - 1 \mid X(0) = g)$; the same amount of time passes in either case. Thus, we can write (14.8) as

$$P(T = t \mid X(0) = g) = p_{gg}\, P(T = t - 1 \mid X(0) = g). \tag{14.9}$$

If we let $y(t) = P(T = t \mid X(0) = g)$ represent the probability that the waiting time is t time steps, given that the system starts in the good state at time zero, then equation (14.9) can be written as $y(t) = p_{gg}\, y(t - 1)$. This is a linear recursion equation and has the same form as the exponential growth model (3.1b). As with the exponential growth model, we can solve (14.9) by iteration to get

$$P(T = t \mid X(0) = g) = p_{gg}^{t-1} P(T = 1 \mid X(0) = g). \tag{14.10}$$

Plugging in the probability that the system hits the bad state in one time step, $P(T = 1 \mid X(0) = g) = p_{bg}$, we get the distribution of waiting times until the bad state is reached:

$$P(T = t \mid X(0) = g) = p_{gg}^{t-1}\, p_{bg}. \tag{14.11}$$

This waiting time distribution is the geometric distribution with $p_{bg} = p$ and $p_{gg} = 1 - p$ (Definition P3.7).

Although we are able to determine the full distribution of waiting times for this model, the distribution of waiting times becomes very difficult if not impossible to derive for Markov models involving more than two states. Nevertheless, it is usually possible to derive the mean waiting time, as we shall see in section 14.2.4. In the two-state model, calculating the expected waiting time is easy as it equals the mean of the geometric distribution, $1/p$, from equation (P3.9b). Writing the expected waiting time until the system enters the bad state starting in the good state at time 0 as

$$\overline{T}_g = E[T \mid X(0) = g], \tag{14.12}$$

we have $\overline{T}_g = 1/p_{bg}$. Intuitively, the lower the probability that the system moves from the good state to the bad state (the lower p_{bg}), the longer it will take, on average, before the system turns bad.

14.3 Multistate Markov Models

In the two-state model considered in the previous section, the probability that the system is in one state equals one minus the probability that the system is in the other state. Thus, we could express the probability that the system is in any particular state at time t as a linear equation involving a single variable (14.4). We then solved this model for several quantities of interest, including the stationary distribution, the probability of absorption, and the waiting time until absorption. These methods can be readily generalized to stochastic models with multiple states. Instead of a single linear equation, we will now have a set of linear equations, which we write in matrix form. Consequently, we can use the general methods from Primer 2 and Chapters 7 and 9 to analyze stochastic models involving more than two states.

14.3.1 Constructing a Multistate Markov Model

To illustrate the general approach for analyzing multistate Markov models, let's consider a five-state Markov chain describing, in simplified form, the interaction between two species. We assume that resources are sufficient to sustain at least one of the species and that the system can be divided roughly into five categories:

State 1—species 1 is common, species 2 is extinct.
State 2—species 1 is common, species 2 is at risk.
State 3—species 1 and species 2 are both common.
State 4—species 1 is at risk, species 2 is common.
State 5—species 1 is extinct, species 2 is common.

We will use a Markov model to determine the likely state of the system, to determine which species is more likely to go extinct, and to determine the expected time to extinction.

To construct the model, we must specify the probabilities of the system moving between the different possible states. Our model has a total of five states and, therefore, $5 \times 5 = 25$ transition probabilities. In the present model,

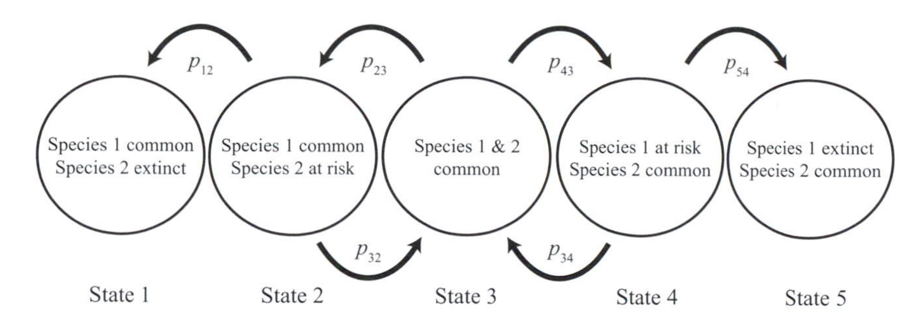

Figure 14.1: The five states for the model of interspecific interactions. Arrows indicate the probability of a transition among states.

we assume that only adjacent states can be reached from one another, as illustrated schematically in Figure 14.1. The transition matrix for the corresponding Markov chain is:

$$\mathbf{M} = \begin{pmatrix} 1 & p_{12} & 0 & 0 & 0 \\ 0 & 1 - p_{12} - p_{32} & p_{23} & 0 & 0 \\ 0 & p_{32} & 1 - p_{23} - p_{43} & p_{34} & 0 \\ 0 & 0 & p_{43} & 1 - p_{34} - p_{54} & 0 \\ 0 & 0 & 0 & p_{54} & 1 \end{pmatrix}. \quad (14.13)$$

The elements $P(X(t) = j \mid X(t-1)) = p_{ji}$ in matrix (14.13) represent the probabilities of moving from state i to state j in a single time step and are therefore non-negative. The elements in each column of matrix (14.13) sum to one, reflecting the fact that, if we start out at state i, then we must end up somewhere in the next time step. Matrices such as (14.13) whose elements are non-negative and whose columns sum to one are referred to as *transition probability matrices* or *stochastic matrices*. The stochastic matrix (14.13) contains many zero elements, reflecting the fact that transitions can occur only between neighboring states. Furthermore, there are two columns in (14.13) that have a 1 on the diagonal and zeroes everywhere else. Such columns represent absorbing states because, once reached, the probability of going to any other state is zero. In this model, the two absorbing states (1 and 5) correspond to the extinction of species 2 and species 1, respectively.

A *transition probability matrix* or a *stochastic matrix* describes the probability of moving between any two states of the system.

Figure 14.2 illustrates simulations of the system over time. In these simulations, a number y was randomly drawn each generation from a uniform distribution between 0 and 1. This number was then compared to the transition probabilities to determine the state of the system at the next time step given the current state. For example, if the system is currently in state 3, then the system moves to state 2 if $0 < y < p_{23}$, moves to state 4 if $p_{23} < y < p_{23} + p_{43}$, and otherwise stays in state 3. In this way, we ensure that the system behaves in a manner consistent with the transition probability matrix (14.13).

Using the law of total probability, we can calculate the state of the system at time t by summing over all possible states of the system in the previous time

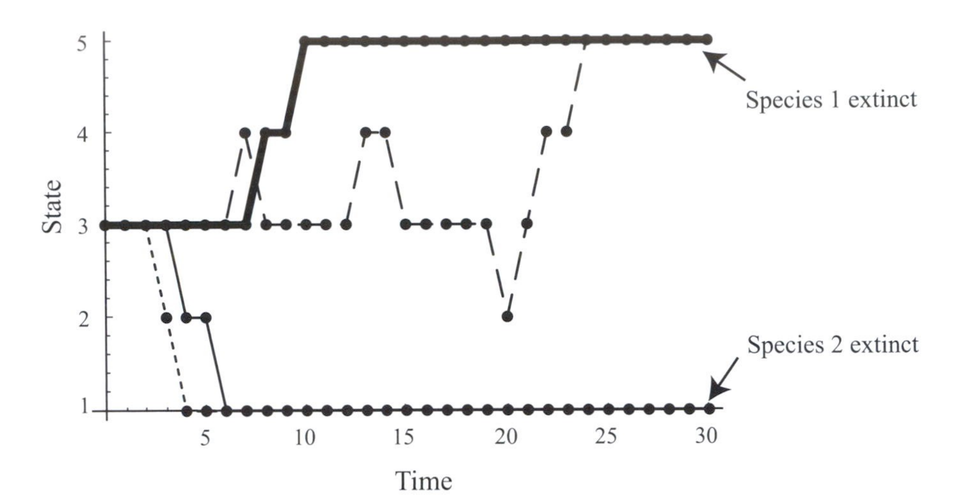

Figure 14.2: Simulations in the two-species model. The system begins in state 3 (both species common) and has a probability of moving from state i to state j of p_{ji} at each time point, according to (14.13) with $\alpha = p_{12} = p_{54} = 0.2$, $\beta = p_{32} = p_{34} = 0.35$, and $\lambda = p_{23} = p_{43} = 0.12$. Four simulations are shown (dots connected by different line styles). At each point in time, the system is in one and only one state. Once an absorbing state (1 or 5) is reached, the system remains at this state.

step. For example, the probability that the system is in state 2 at time t, $P(X(t) = 2)$, can be calculated as

$$\begin{aligned} P(X(t) = 2) = p_{21}\, P(X(t-1) = 1) &+ p_{22}\, P(X(t-1) = 2) \\ + p_{23}\, P(X(t-1) = 3) &+ p_{24}\, P(X(t-1) = 4) \\ + p_{25}\, P(X(t-1) = 5). & \end{aligned} \tag{14.14}$$

Equation (14.14) is a recursion equation for $P(X(t) = 2)$, in terms of the probability that the system was in any of the five possible states at time $t - 1$. Similar calculations can be done for all five states, resulting in five recursion equations of the form of (14.14).

It is easier to write these recursion equations in matrix form, however:

$$\vec{x}(t) = \mathbf{M}\, \vec{x}(t-1), \tag{14.15}$$

where $\mathbf{M}$ is given by (14.13) and

$$\vec{x}(t) = \begin{pmatrix} P(X(t) = 1) \\ P(X(t) = 2) \\ P(X(t) = 3) \\ P(X(t) = 4) \\ P(X(t) = 5) \end{pmatrix}.$$

You should take some time to verify that (14.15) is correct, i.e., that it generates recursions equations of the form of (14.14). Figure 14.3 illustrates how the probability of being in each state changes over time using (14.15).

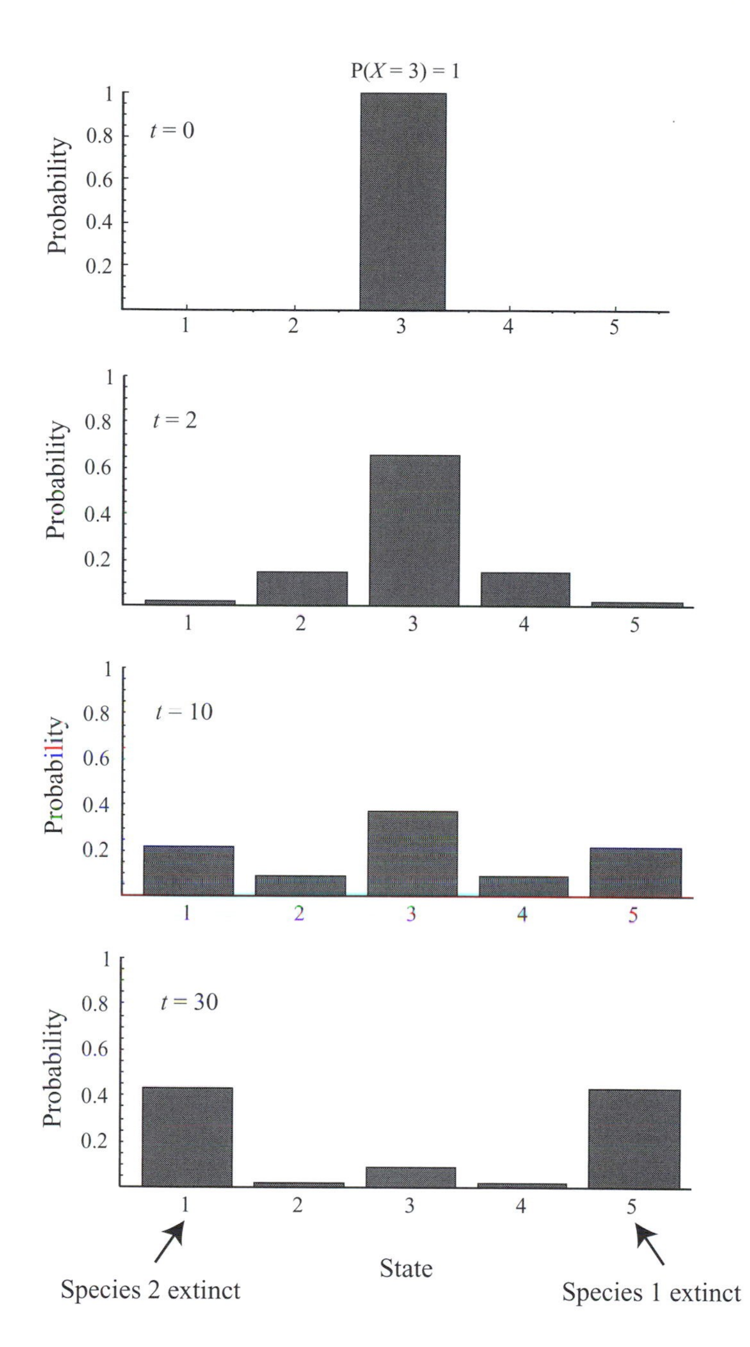

Figure 14.3: Probability distribution for the two-species model of interactions. Even though the system is only in one state at any point in time (Figure 14.2), we can describe the *probability* that the system is in a particular state with a probability distribution. Here, we show how this probability distribution changes over time, using equation (14.15). The system begins in state 3 and has the same transition probabilities as in Figure 14.2.

For stochastic matrices, it is possible to show that there is always at least one eigenvalue equal to +1 (Problem 14.2) and that no other eigenvalue is ever larger than one in absolute value. As a result, unlike the equilibria of the deterministic, discrete-time linear models of Chapter 7, we never have to worry about a stationary distribution in such stochastic models being unstable. There might, however, be more than one eigenvalue with an absolute value of one. In particular, if there is more than one absorbing state then there will be more than one eigenvalue equal to +1. For example, our model of interspecific interactions (14.13) has two absorbing states (extinction of either of the two species), and therefore we expect to find more than one eigenvalue equal to one. There is also the possibility of negative or complex eigenvalues with absolute values of one, but in these cases the Markov chain visits states in a stereotypical sequence (e.g., the environment always alternates between being good then bad then good . . .) and is said to be *periodic* (see Problem 14.3). Here, we focus only on non-periodic Markov chains (see "Further Reading").

The above results can be generalized to Markov chains involving any number of states:

Rule 14.1: A Markov Chain for a System with m States

For a Markov chain with m states, the $m \times m$ transition probability matrix is

$$\mathbf{M} = \begin{pmatrix} p_{11} & p_{12} & \cdots & p_{1m} \\ p_{21} & p_{22} & \cdots & p_{2m} \\ \cdots & \cdots & \cdots & \cdots \\ p_{m1} & p_{m2} & \cdots & p_{mm} \end{pmatrix}, \tag{14.16}$$

with non-negative elements whose columns sum to one; i.e.,

$$\sum_{j=1}^{m} p_{ji} = 1. \tag{14.17}$$

Using the law of total probability, the probability that the system is in state j at time t, $P(X(t) = j)$, can be written as a function of the state of the system at time $t - 1$:

$$P(X(t) = j) = \sum_{i=1}^{m} p_{ji}\, P(X(t-1) = i). \tag{14.18}$$

This results in m recursion equations that can be written as $\vec{x}(t) = \mathbf{M}\,\vec{x}(t-1)$, where $\vec{x}(t)$ is a vector giving the probability of being in each of the m states at time t.

The transition probability matrix $\mathbf{M}$ will have one eigenvalue equal to one if there are no absorbing states and k eigenvalues equal to one if there are k distinct absorbing states (i.e., k states such that $p_{ii} = 1$). Assuming that the model does not exhibit periodic behavior, all other eigenvalues are less than or equal to one in absolute value.

14.3.2 Finding the Stationary Distribution

The first step in analyzing our Markov chain is to look for a stationary distribution, just as we did in the two-state model. This is the probabilistic equilibrium of the model. The stationary distribution of any Markov chain can be represented by a vector $\vec{\pi}$ whose elements give the probability of being in each state:

Recipe 14.1
The Stationary Distribution

By definition, the stationary distribution remains constant from one time step to the next, and therefore it must satisfy

$$\vec{\pi} = \mathbf{M}\,\vec{\pi}, \tag{14.19}$$

where the elements of $\vec{\pi}$ are normalized to sum to one. According to Definition (P2.4a), the stationary distribution $\vec{\pi}$ is the right eigenvector of the transition matrix $\mathbf{M}$ associated with the leading eigenvalue of one.

Suppose that the model does not exhibit periodic behavior. Then,

(1) If there is zero or one absorbing states, then the transition matrix $\mathbf{M}$ will have a single eigenvalue equal to $+1$. There is then a single, unique eigenvector $\vec{\pi}$ describing the stationary distribution. The elements of $\vec{\pi}$ give the probability of residing in any one of the m states of the system after enough time has passed.

(2) If there are $k > 1$ absorbing states, then the transition matrix $\mathbf{M}$ will have k eigenvalues equal to $+1$. Although there are an infinite number of possible choices of eigenvectors associated with an eigenvalue of $+1$, a natural choice is to use the k stationary distributions associated with each of the k absorbing states as eigenvectors (i.e., if state j is an absorbing state, then the vector with a one in the jth position and zeros elsewhere is one of the eigenvectors).

This recipe should seem familiar. Our study of the dynamics of class-structured populations in Chapter 10 also revealed that the long-term proportion of a population in each class (the so-called *stable class distribution*) is given by the leading right eigenvector of the transition matrix.

The transition probability matrix (14.13) for the model of interspecific interactions has two absorbing states, state 1 and state 5. Therefore, Recipe 14.1 tells us to expect two eigenvalues equal to one, which we write as λ_1 and λ_5 to reflect their association with the two absorbing states, 1 and 5. Recipe 14.1 also suggests that we choose leading eigenvectors that correspond to the two absorbing states $\vec{\pi}_1 = (1 \quad 0 \quad 0 \quad 0 \quad 0)^{\mathrm{T}}$ and $\vec{\pi}_5 = (0 \quad 0 \quad 0 \quad 0 \quad 1)^{\mathrm{T}}$ (recall, a roman T denotes transpose). The system eventually gets "absorbed" into one of these two stationary distributions.

For models like (14.13), where there is more than one stationary distribution, we need to determine which of these is eventually reached after a sufficiently long period of time. At any time t, we can calculate the probability distribution by solving equation (14.15) for $x(t)$ just as we did with linear matrix equations in Chapter 9:

$$\vec{x}(t) = \mathbf{M}^t\, \vec{x}(0), \tag{14.20}$$

where $\vec{x}(0)$ is the vector whose elements give the probability that the system is in each of the various states at time zero. As discussed in Chapter 9, $\mathbf{M}^t$ can be rewritten in terms of the eigenvalues and eigenvectors of $\mathbf{M}$. In particular, if we write the eigenvalues of $\mathbf{M}$ along the diagonal of a diagonal matrix $\mathbf{D}$, and the corresponding right eigenvectors in the columns of a matrix $\mathbf{A}$, then

$$\vec{x}(t) = \mathbf{A}\,\mathbf{D}^t\,\mathbf{A}^{-1}\,\vec{x}(0) \tag{14.21}$$

(Recipe 9.1). Recall that the matrix, $\mathbf{A}^{-1}$ is the inverse of matrix $\mathbf{A}$, and its rows contain the "left" eigenvectors, which we write as $\vec{\rho}$.

As time passes, $\mathbf{A}\,\mathbf{D}^t\mathbf{A}^{-1}$ becomes dominated by the eigenvalue(s) with the largest absolute value. In the two-species model (14.13), the leading eigenvalues are $\lambda_1 = \lambda_5 = 1$. Following the analysis presented in Box 9.1 of Chapter 9, we can approximate the general solution (14.20) with

$$\vec{x}(t) \approx \lambda_1^t\, \vec{\pi}_1\, \vec{\rho}_1^{\mathrm{T}}\, \vec{x}(0) + \lambda_5^t\, \vec{\pi}_5\, \vec{\rho}_5^{\mathrm{T}}\, \vec{x}(0) \tag{14.22a}$$

(where $\vec{\pi}_i$ is a column vector and $\vec{\rho}_i^{\mathrm{T}}$ is a row vector). Writing $\vec{x}(t)$ as $\vec{x}(\infty)$ to emphasize that the above approximation applies only after enough time has passed and plugging in $\lambda_1 = 1$ and $\lambda_5 = 1$, we have

$$\vec{x}(\infty) \approx \vec{\pi}_1\, \vec{\rho}_1^{\mathrm{T}}\, \vec{x}(0) + \vec{\pi}_5\, \vec{\rho}_5^{\mathrm{T}}\, \vec{x}(0). \tag{14.22b}$$

The row vectors $\vec{\rho}_1^{\mathrm{T}}$ and $\vec{\rho}_5^{\mathrm{T}}$ are the left eigenvectors associated with $\lambda_1 = 1$ and $\lambda_5 = 1$. More generally, an m-state Markov model will have a term in equation (14.22b) corresponding to each absorbing state.

Once again, the leading left eigenvectors play a fundamental role in the general solution of Markov models, just as they did in the class-structured models of Chapter 9. There, the leading left eigenvector represented the ultimate contribution of individuals in the different classes to the long-term growth of the population (the "reproductive value" of the class). As we shall demonstrate in the next section, with the stationary distributions $\vec{\pi}_i$ chosen to correspond to the absorbing states, the ith element of the left eigenvectors $\vec{\rho}_j^{\mathrm{T}}$ represents the probability of eventual absorption from state i into the absorbing state, j. In particular, $\vec{\rho}_1^{\mathrm{T}}\, \vec{x}(0)$ in (14.22b) can be expanded out as $\vec{\rho}_1^{\mathrm{T}}\, \vec{x}(0) = \rho_{11}\, x_1(0) + \rho_{12}\, x_2(0) + \rho_{13}\, x_3(0) + \rho_{14}\, x_4(0) + \rho_{15}\, x_5(0)$, where $x_i(0)$ is the probability that the system was in state i at time zero. An analogous equation holds for $\vec{\rho}_5^{\mathrm{T}}\, \vec{x}(0)$. Therefore, $\vec{\rho}_1^{\mathrm{T}}\, \vec{x}(0)$and $\vec{\rho}_5^{\mathrm{T}}\, \vec{x}(0)$ represent the total probability of eventual absorption into states 1 and 5, respectively, given the starting state $\vec{x}(0)$. Writing these total probabilities as $U_1 = \vec{\rho}_1^{\mathrm{T}}\, \vec{x}(0)$ and $U_5 = \vec{\rho}_5^{\mathrm{T}}\, \vec{x}(0)$, equation (14.22b) becomes

$$\vec{x}(\infty) \approx \vec{\pi}_1 U_1 + \vec{\pi}_5 U_5. \tag{14.22c}$$

Equation (14.22c) reveals that the system eventually ends up in one of the two absorbing states, with the probability of ending up in absorbing state i given by U_i, which depends on the initial state. For example, in Figure 14.3 where the system starts in state 3 with symmetrical transition probabilities, U_1 and U_5 both equal 1/2, and by generation 30, the probability distribution is very nearly given by $\vec{x}(\infty) \approx (1/2 \quad 0 \quad 0 \quad 0 \quad 1/2)^{\mathrm{T}}$.

14.3.3 Calculating the Probability of Absorption

Let us now demonstrate that the probability of absorption is, indeed, given by a leading left eigenvector of the transition matrix. Taking model (14.13) describing interactions between two species as an example, consider calculating the probability that species 2 is eventually driven to extinction (i.e., the system gets absorbed into state 1) given that it is currently in state 3. Mathematically, this is written as $P(X(\infty) = 1 \mid X(0) = 3)$. Let's first break down this absorption probability using the law of total probability (Rule P3.8) by conditioning on the mutually exclusive sets of states to which the system could move at time $t = 1$:

The *absorption probability* describes the probability of reaching and permanently remaining in a particular state of the system.

$$\begin{aligned} P(X(\infty) = 1 \mid X(0) = 3) = {} & P(X(\infty) = 1 \mid X(1) = 1, X(0) = 3)\, p_{13} \\ & + P(X(\infty) = 1 \mid X(1) = 2, X(0) = 3)\, p_{23} \\ & + P(X(\infty) = 1 \mid X(1) = 3, X(0) = 3)\, p_{33} \\ & + P(X(\infty) = 1 \mid X(1) = 4, X(0) = 3)\, p_{43} \\ & + P(X(\infty) = 1 \mid X(1) = 5, X(0) = 3)\, p_{53}. \end{aligned} \tag{14.23}$$

Although we have considered the possibility that the system moves to any of the five states, not all of these transitions are possible from state 3 (e.g., $p_{13} = 0$). But tracking all terms at this stage is helpful to see how the calculations work more generally.

Equation (14.23) can be simplified using the Markov property (14.1). The probability of ultimately being absorbed in state 1, $P(X(\infty) = 1 \mid X(1) = 2, X(0) = 3)$, depends only on the state of the system at time $t = 1$, $P(X(\infty) = 1 \mid X(1) = 2)$ and is not influenced by any previous states of the system. Moreover, under the assumption that the transition probabilities are constant over time, there should be no difference between the absorption probabilities starting time at $t = 1$ or at $t = 0$. For example, the probability of ultimately being absorbed in state 1 should be the same if we last measured the system in state 2 at time $t = 1$ or if we last measured the system in state 2 at time $t = 0$: $P(X(\infty) = 1 \mid X(1) = 2) = P(X(\infty) = 1 \mid X(0) = 2)$. Making these simplifications and defining the probability of absorption into state j starting from state i as $\rho_{j,i} = P(X(\infty) = j \mid X(0) = i)$, equation (14.23) can be rewritten as

$$\rho_{1,3} = \rho_{1,1}\, p_{13} + \rho_{1,2}\, p_{23} + \rho_{1,3}\, p_{33} + \rho_{1,4}\, p_{43} + \rho_{1,5}\, p_{53}. \tag{14.24}$$

An expression analogous to (14.24) can be derived for all five possible starting states. Writing the results in matrix form, we get

$$(\rho_{1,1} \quad \rho_{1,2} \quad \rho_{1,3} \quad \rho_{1,4} \quad \rho_{1,5}) = (\rho_{1,1} \quad \rho_{1,2} \quad \rho_{1,3} \quad \rho_{1,4} \quad \rho_{1,5})\,\mathbf{M} \tag{14.25a}$$

or

$$\vec{\rho}_1^{\mathrm{T}} = \vec{\rho}_1^{\mathrm{T}} \mathbf{M}, \tag{14.25b}$$

where $\vec{\rho}_1^{\mathrm{T}}$ is a row vector whose elements are the probabilities of eventual absorption into state 1 from each of the five possible starting states. By the definition of a left eigenvector (P2.12), equation (14.25b) reveals that the vector, $\vec{\rho}_1^{\mathrm{T}}$, is a left eigenvector of the transition matrix $\mathbf{M}$ associated with an eigenvalue of one.

Repeating this procedure for absorbing state 5, the probabilities of eventual absorption into state 5 from each of the five possible starting states are given by the elements of the vector $\vec{\rho}_5^{\mathrm{T}}$, which satisfies the equation: $\vec{\rho}_5^{\mathrm{T}} = \vec{\rho}_5^{\mathrm{T}} \mathbf{M}$. Note, however, that when calculating the left eigenvectors, we must chose the length and direction of each eigenvector such that $\rho_{1,1} = 1$ and $\rho_{5,1} = 0$ for the vector $\vec{\rho}_1^{\mathrm{T}}$ and that $\rho_{5,1} = 0$ and $\rho_{5,5} = 1$ for the vector $\vec{\rho}_5^{\mathrm{T}}$. These choices ensure that the elements of $\vec{\rho}_1^{\mathrm{T}}$ represent the probability of eventual absorption into absorbing state 1 and that the elements of $\vec{\rho}_5^{\mathrm{T}}$ represent the probability of eventual absorption into absorbing state 5.

Taking our example further, we can simplify things by supposing that the two species have identical effects on one another. This means that $p_{12} = p_{54} \equiv \alpha$, which is the probability of extinction of one of the two species given that we are in a state where one of them is at risk. We also set $p_{32} = p_{34} \equiv \beta$, which is the probability of recovering to state 3 in which both species are common given that one species is at risk, and $p_{23} = p_{43} = \gamma$, which is the probability that one species becomes at risk given that the two are initially common. In this case our model is symmetric, and we obtain the following transition probability matrix:

$$\mathbf{M} = \begin{pmatrix} 1 & \alpha & 0 & 0 & 0 \\ 0 & 1-\alpha-\beta & \gamma & 0 & 0 \\ 0 & \beta & 1-2\gamma & \beta & 0 \\ 0 & 0 & \gamma & 1-\alpha-\beta & 0 \\ 0 & 0 & 0 & \alpha & 1 \end{pmatrix} \tag{14.26}$$

The left eigenvectors of $\mathbf{M}$ are

$$\vec{\rho}_1^{\mathrm{T}} = \begin{pmatrix} 1 & \frac{2\alpha+\beta}{2(\alpha+\beta)} & \frac{1}{2} & \frac{\beta}{2(\alpha+\beta)} & 0 \end{pmatrix} \quad \text{satisfying } \rho_{1,1} = 1 \text{ and } \rho_{1,5} = 0, \tag{14.27a}$$

$$\vec{\rho}_5^{\mathrm{T}} = \begin{pmatrix} 0 & \frac{\beta}{2(\alpha+\beta)} & \frac{1}{2} & \frac{2\alpha+\beta}{2(\alpha+\beta)} & 1 \end{pmatrix} \quad \text{satisfying } \rho_{5,1} = 0 \text{ and } \rho_{5,5} = 1. \tag{14.27b}$$

The elements of vector (14.27a) give the probability of species 2 going extinct (i.e., absorption in state 1), given that the system starts in one of the five different possible states. Analogously, the elements of vector (14.27b) give the probability of species 1 going extinct (i.e., absorption in state 5). Observe that the ith elements in these two vectors sum to one, $\rho_{1,i} + \rho_{5,i} = 1$, reflecting the

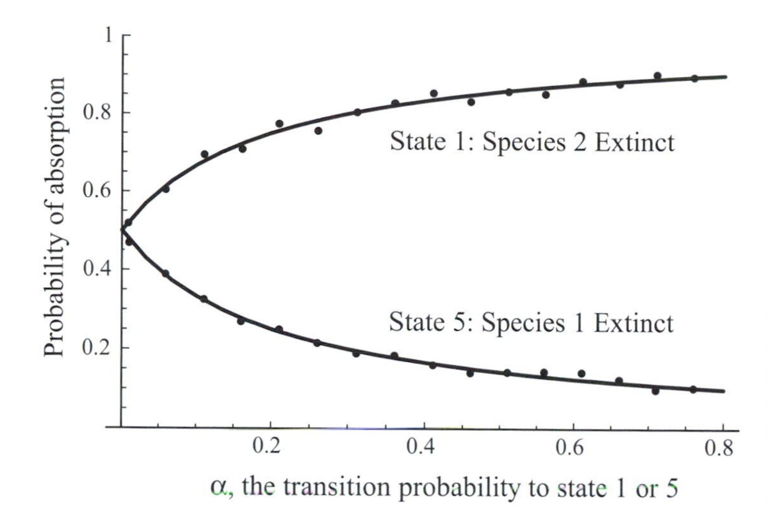

Figure 14.4: Absorption probabilities for the two-species model. Probability of eventual absorption in state 5 (bottom) and in state 1 (top), given that the system starts in state 2, as a function of the parameter $\alpha = p_{12} = p_{54}$. Curves are analytical results from equations (14.27) and dots depict the average of 1000 replicates in a simulation of the model. Parameter values: $\beta = p_{32} = p_{34} = 0.2$, $\gamma = p_{23} = p_{43} = 0.25$.

fact that the system eventually gets absorbed into one of these two absorbing states regardless of the initial state, i.

To interpret vectors (14.27), first consider starting in state $i = 1$. From the first element of vector $\vec{\rho}_1^{\,T}$, we find that absorption in state 1 is guaranteed ($\rho_{11} = 1$), because the system is already there. Conversely, from the first element of $\vec{\rho}_5^{\,T}$, we find that absorption in state 5 is impossible ($\rho_{5,1} = 0$), because the system is already absorbed in state 1. Next, consider the starting state $i = 2$, where species 2 is at risk. From the second element of $\vec{\rho}_1^{\,T}$, the probability of absorption into state 1 is $(2\alpha + \beta)/(2(\alpha + \beta))$, which is greater than the probability of absorption in state 5 given by $\beta/(2(\alpha + \beta))$ because extinction of species 2 (state 1) is more easily reached when species 2 is initially at risk (state 2), as illustrated in Figure 14.4. Finally, consider starting in state $i = 3$. The third element of both vectors is 1/2 because of our assumption of symmetry. Both species are equally likely to go extinct first when both species are initially common.

The left eigenvectors (14.27), along with the right eigenvectors, $\vec{\pi}_1 = (1 \quad 0 \quad 0 \quad 0 \quad 0)^{\mathrm{T}}$ and $\vec{\pi}_5 = (0 \quad 0 \quad 0 \quad 0 \quad 1)^{\mathrm{T}}$, and the eigenvalues $\lambda_1 = \lambda_5 = 1$, can then be used in the approximate general solution (14.22b) to predict the eventual state of the system:

$$\begin{pmatrix} x_1(\infty) \\ x_2(\infty) \\ x_3(\infty) \\ x_4(\infty) \\ x_5(\infty) \end{pmatrix} \simeq \begin{pmatrix} x_1(0) + \dfrac{(2\alpha + \beta)\, x_2(0)}{2(\alpha + \beta)} + \dfrac{x_3(0)}{2} + \dfrac{\beta\, x_4(0)}{2(\alpha + \beta)} \\ 0 \\ 0 \\ 0 \\ \dfrac{\beta\, x_2(0)}{2(\alpha + \beta)} + \dfrac{x_3(0)}{2} + \dfrac{(2\alpha + \beta)\, x_4(0)}{2(\alpha + \beta)} + x_5(0) \end{pmatrix} \tag{14.28}$$

(Problem 14.13). After enough time has passed, the probability that the system is in each of the five states approaches (14.28). Altogether, the stationary distributions (i.e., the right eigenvectors) and the probability of absorption (i.e., the left eigenvectors) combine with the starting conditions to determine the long-term probability that the system resides in each of the two absorbing states.

Recipe 14.2
The Probability of Absorption into State j
With one or more absorbing states, the probability of eventual absorption into absorbing state j is given by $\vec{\rho}_j^{\mathrm{T}}$, the leading left eigenvector of $\mathbf{M}$, that satisfies

$$\vec{\rho}_j^{\mathrm{T}} = \vec{\rho}_j^{\mathrm{T}}\mathbf{M} \quad \text{subject to } \rho_{j,j} = 1 \text{ and } \rho_{j,l} = 0, \tag{14.29}$$

where l is any absorbing state different from j, and the roman T denotes transpose. Specifically, the ith element of $\vec{\rho}_j^{\mathrm{T}}$ describes the probability of being absorbed in state j when initially in state i.

14.3.4 Calculating the Expected Time until Absorption

We turn now to the waiting time until absorption occurs. We start by focusing on the waiting time until the system reaches any of its absorbing states. Afterwards, we consider the waiting time until the system reaches a particular absorbing state, conditioned upon the fact that it eventually does so. This section involves a great deal of derivation, but the results are summarized at the end in Recipe 14.3.

The *time to absorption* describes the time taken for the system to first reaching an absorbing state.

Let T_i be a random variable representing the waiting time until absorption in any one of the absorbing states, given that the system is initially in state i. The probability that T_i takes on any particular value will be described by a probability distribution. We only calculate the mean of this distribution, $\overline{T}_i = E[T_i | X(0) = i]$ (see Ewens 1979, p. 71, for the variance). We emphasize that T_i represents the waiting time, *measured from when we start waiting*. Given that the system is in state i when we start waiting and that the transition probabilities are constant over time, the expected waiting time is the same if we start waiting at time 0 or start waiting at time 1 (or any other arbitrary time). That is, $E[T_i | X(0) = i] = E[T_i | X(1) = i]$. On the other hand, if we start waiting at time 0 and watch as the system moves from a nonabsorbed state i to state i' at time 1, the expected waiting time, denoted by $E[T_i | X(1) = i', X(0) = i]$, equals $E[T_{i'} | X(1) = i']$ (the waiting time had we started waiting at time $t = 1$ with the system in state i') plus the one time step that we waited between $t = 0$ and $t = 1$. Putting these facts together, the expected waiting time is

$$\begin{aligned} \overline{T}_i = E[T_i | X(0) = i] &= E[T_i | X(1) = i', X(0) = i] \\ &= E[T_{i'} | X(1) = i'] + 1 \\ &= E[T_{i'} | X(0) = i'] + 1 \\ &= \overline{T}_{i'} + 1, \end{aligned} \tag{14.30}$$

which will come in handy below.

By necessity, the expected waiting time until absorption is zero if the system starts in one of the absorbing states. Thus, for model (14.13), $\overline{T}_1 = 0$ and $\overline{T}_5 = 0$. If the system starts in a nonabsorbing state i, then we calculate $\overline{T}_i = E[T_i|X(0) = i]$ by first rewriting this expectation in terms of the mutually exclusive set of states where the system might go in one time step, using the law of total expectation (Rule P3.9). Starting model (14.13) in the nonabsorbing state i, we get

$$\begin{aligned} E[T_i|X(0) = i] = {} & E[T_1|X(1) = 1, X(0) = i]\, p_{1i} \\ & + E[T_2|X(1) = 2, X(0) = i]\, p_{2i} \\ & + E[T_3|X(1) = 3, X(0) = i]\, p_{3i} \\ & + E[T_4|X(1) = 4, X(0) = i]\, p_{4i} \\ & + E[T_5|X(1) = 5, X(0) = i]\, p_{5i}. \end{aligned} \tag{14.31a}$$

Substituting in from (14.30) then gives

$$\begin{aligned} \overline{T}_i &= (\overline{T}_1 + 1)\, p_{1i} + (\overline{T}_2 + 1)\, p_{2i} + (\overline{T}_3 + 1)\, p_{3i} \\ &\quad + (\overline{T}_4 + 1)\, p_{4i} + (\overline{T}_5 + 1)\, p_{5i} \\ &= \overline{T}_1\, p_{1i} + \overline{T}_2\, p_{2i} + \overline{T}_3\, p_{3i} + \overline{T}_4\, p_{4i} + \overline{T}_5\, p_{5i} + 1, \end{aligned} \tag{14.31b}$$

where we have used (14.17) to simplify the last line. Equation (14.31b) makes sense. The total waiting time starting at time 0 is the average waiting time across all possible states of the system at time 1, plus one for the intervening time step.

We now have five waiting times: the waiting times from the two absorbing states, $\overline{T}_1 = 0$ and $\overline{T}_5 = 0$, and the waiting times from the three nonabsorbing states, $\overline{T}_2$, $\overline{T}_3$, and $\overline{T}_4$, given by (14.31b). We can write these five waiting times in a row vector $\vec{\overline{T}}$ that satisfies

$$(\overline{T}_1 \quad \overline{T}_2 \quad \overline{T}_3 \quad \overline{T}_4 \quad \overline{T}_5) = (\overline{T}_1 \quad \overline{T}_2 \quad \overline{T}_3 \quad \overline{T}_4 \quad \overline{T}_5)\, \mathbf{M} + (0 \quad 1 \quad 1 \quad 1 \quad 0) \tag{14.32a}$$

or

$$\vec{\overline{T}}^{\mathrm{T}} = \vec{\overline{T}}^{\mathrm{T}} \mathbf{M} + \vec{b}^{\mathrm{T}}, \tag{14.32b}$$

where $\vec{b}^{\mathrm{T}}$ is a row vector containing a one for each nonabsorbing state and a zero for each absorbing state (again, the roman T denotes transpose).

Equation (14.32) represents a set of linear equations, which can be solved, subject to the condition that $\overline{T}_1 = 0$ and $\overline{T}_5 = 0$, using Recipe P2.1 in Primer 2. Alternatively, we might try to solve (14.32b) by multiplying the left-hand side by the 5×5 identity matrix and factoring to get $\vec{\overline{T}}^{\mathrm{T}}(\mathbf{I} - \mathbf{M}) = \vec{b}^{\mathrm{T}}$. At this point, we could try to multiply both sides on the right by $(\mathbf{I} - \mathbf{M})^{-1}$, except that we run into a problem. Because the first and last columns of $\mathbf{M}$ (corresponding to the absorbing states) contain ones on the diagonal and zeros elsewhere, $(\mathbf{I} - \mathbf{M})$ has two columns of zeros and cannot be inverted. That is, $(\mathbf{I} - \mathbf{M})$ is a *singular* matrix (Primer 2). Indeed, equation (14.32) provides no information about the values of $\overline{T}_1$ and $\overline{T}_5$, which we defined as zero on logical grounds.

Given that $\overline{T}_1$ and $\overline{T}_5$ are zero, we can eliminate the first and last elements of the vectors in (14.32), both of which have the form $0 = 0 + 0$, leaving us with:

$$\vec{\overline{T}}_u^{\,\mathrm{T}} = \vec{\overline{T}}_u^{\,\mathrm{T}}\, \mathbf{M}_u + \vec{b}_u^{\,\mathrm{T}}, \tag{14.33}$$

where the subscript u refers to the subset of the vectors and matrices involving unabsorbed states, only. Specifically, $\vec{\overline{T}}_u^{\,\mathrm{T}} = (\overline{T}_2 \quad \overline{T}_3 \quad \overline{T}_4)$,

$$\mathbf{M}_u = \begin{pmatrix} 1 - p_{12} - p_{32} & p_{23} & 0 \\ p_{32} & 1 - p_{23} - p_{43} & p_{34} \\ 0 & p_{43} & 1 - p_{34} - p_{54} \end{pmatrix}, \tag{14.34}$$

and $\vec{b}_u^{\,\mathrm{T}} = (1 \quad 1 \quad 1)$. We can now multiply the left-hand side of (14.33) by the 3×3 identity matrix and factor to get $\vec{\overline{T}}_u^{\,\mathrm{T}}(\mathbf{I} - \mathbf{M}_u) = \vec{b}_u^{\,\mathrm{T}}$, where $(\mathbf{I} - \mathbf{M}_u)$ can be inverted to give an explicit solution for the average times until absorption:

$$\vec{\overline{T}}_u^{\,\mathrm{T}} = \vec{b}_u^{\,\mathrm{T}}(\mathbf{I} - \mathbf{M}_u)^{-1}, \tag{14.35}$$

along with $\overline{T}_1 = 0$ and $\overline{T}_5 = 0$.

Returning to the symmetric matrix (14.26) for the two-species model, equation (14.35) can be solved to give

$$\vec{\overline{T}} = \left(0 \quad \frac{\beta + 2\gamma}{2\alpha\gamma} \quad \frac{\alpha + \beta + 2\gamma}{2\alpha\gamma} \quad \frac{\beta + 2\gamma}{2\alpha\gamma} \quad 0\right). \tag{14.36}$$

Equation (14.36) reveals that as γ (the probability that the system moves away from state 3 where both species are common) increases, the expected time until one of the species goes extinct decreases (see Figure 14.5). Intuitively, if γ is large, then it will rarely be the case that both species are common, which decreases the amount of time that must pass before one of the two species goes extinct.

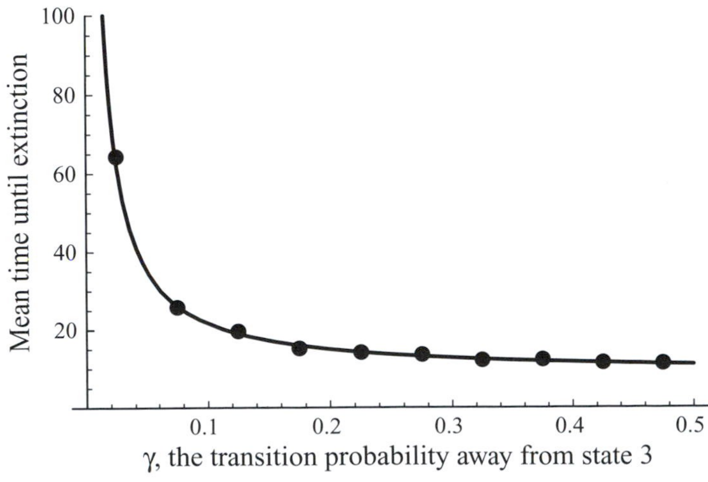

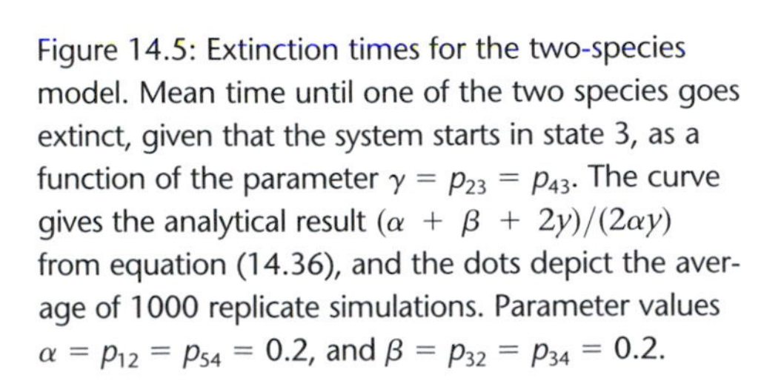
Figure 14.5: Extinction times for the two-species model. Mean time until one of the two species goes extinct, given that the system starts in state 3, as a function of the parameter $\gamma = p_{23} = p_{43}$. The curve gives the analytical result $(\alpha + \beta + 2y)/(2\alpha y)$ from equation (14.36), and the dots depict the average of 1000 replicate simulations. Parameter values $\alpha = p_{12} = p_{54} = 0.2$, and $\beta = p_{32} = p_{34} = 0.2$.

Recipe 14.3
The Mean Time until Absorption in Any Absorbing State

The mean time until absorption into any one of the k absorbing states, given that the system starts in state i, is given by the element, $\overline{T}_i$, of the row vector $\vec{\overline{T}}^{\mathrm{T}}$ that satisfies

$$\vec{\overline{T}}^{\mathrm{T}} = \vec{\overline{T}}^{\mathrm{T}}\mathbf{M} + \vec{b}^{\mathrm{T}} \quad \text{subject to } \overline{T}_a = 0 \text{ for all absorbing states } a. \tag{14.37}$$

The row vector $\vec{b}^{\mathrm{T}}$ contains a one in every position corresponding to a nonabsorbing state and a zero in every position corresponding to an absorbing state. The solution to equation (14.37) equals $\overline{T}_a = 0$ for all absorbing states a, and

$$\vec{\overline{T}}_u^{\mathrm{T}} = \vec{b}_u^{\mathrm{T}}(\mathbf{I} - \mathbf{M}_u)^{-1}, \tag{14.38}$$

for all unabsorbed states, where the subscript u refers to the vectors and matrices that remain after removing all rows and columns involving the absorbing states (e.g., $\vec{b}_u^{\mathrm{T}}$ is a vector of length $m - k$ consisting of ones only).

The above assumes that we are interested in the time until absorption in any absorbing state. For many problems, it is worth knowing the time until absorption in a particular state, conditioned on that state being reached. For example, we might wish to know the time until a particular allele becomes fixed within a population, among those cases where fixation occurs. The above calculations can be easily extended to calculate such *conditional times to absorption.*

Let us return to our two competing species and suppose that we are interested in the time until species 2 goes extinct, $T_{(1),i}$, starting in state i. We use the (1) subscript to emphasize that we are only interested in those cases where the system eventually becomes absorbed in state 1. Again, we can use the law of total expectation to calculate the expected time until absorption in state 1 starting from any particular starting state:

$$\begin{aligned} E[T_{(1),i}|X(0) = i] = {} & E[T_{(1),1}|X(1) = 1, X(0) = i]\,\tilde{p}_{1i} \\ & + E[T_{(1),2}|X(1) = 2, X(0) = i]\,\tilde{p}_{2i} \\ & + E[T_{(1),3}|X(1) = 3, X(0) = i]\,\tilde{p}_{3i} \\ & + E[T_{(1),4}|X(1) = 4, X(0) = i]\,\tilde{p}_{4i} \\ & + E[T_{(1),5}|X(1) = 5, X(0) = i]\,\tilde{p}_{5i}. \end{aligned} \tag{14.39}$$

The key difference between equation (14.39) and (14.31a) is that the transitions that occur must remain consistent with ultimate absorption in state 1. Thus, we define $\tilde{p}_{ji}$ as the probability of a transition from state i to state j, conditional on ultimately fixing on state 1: $\tilde{p}_{ji} = P(X(1) = j|X(0) = i$ and absorption in state 1).

Writing A_1 for the event ultimate absorption in state 1 and using Bayes' theorem (Rule P3.7, Primer 3), this conditional transition probability equals

$$\begin{aligned}\widetilde{p}_{ji} &= P(X(1) = j \mid X(0) = i, A_1) \\ &= \frac{P(A_1 \mid X(1) = j, X(0) = i)\, P(X(1) = j \mid X(0) = i)}{P(A_1 \mid X(0) = i)}.\end{aligned} \tag{14.40a}$$

We have already calculated all of the elements of this ratio. $P(A_1|X(0) = i)$ is the probability of absorption in state 1 with the system starting in state i, which equals $\rho_{1,i}$. Similarly, $P(A_1 \mid X(1) = j, X(0) = i)$ is the probability of absorption in state 1 starting in state j at time 1, which equals $\rho_{1,j}$ (further conditioning on the state at time 0 does not matter because of the Markov property). Finally, $P(X(1) = j \mid X(0) = i)$ is the original transition probability p_{ji}. Thus, the conditional transition probabilities needed in (14.39) are given by

$$\widetilde{p}_{ji} = \frac{\rho_{1,j}}{\rho_{1,i}}\, p_{ji}. \tag{14.40b}$$

Equation (14.40b) indicates that transitions toward an absorbing state of interest become more likely, while transitions away from that absorbing state become less likely, when we consider only those cases that ultimately reach the absorbing state of interest. For example, if the system is currently in state 4 and we condition on the system ultimately reaching state 1, a transition to state 3 is more likely (because $\rho_{1,3}/\rho_{1,4} > 1$) and a transition to state 5 is impossible (because $\rho_{1,5}/\rho_{1,4} = 0$).

Plugging in (14.40b) along with (14.30) into (14.39), the expected waiting time conditioned on reaching state 1 and starting in a nonabsorbed state i becomes

$$\begin{aligned}\overline{T}_{(1),i} &= (\overline{T}_{(1),1} + 1)\,\widetilde{p}_{1i} + (\overline{T}_{(1),2} + 1)\,\widetilde{p}_{2i} \\ &\quad + (\overline{T}_{(1),3} + 1)\,\widetilde{p}_{3i} + (\overline{T}_{(1),4} + 1)\,\widetilde{p}_{4i} + (\overline{T}_{(1),5} + 1)\,\widetilde{p}_{5i} \\ &= \overline{T}_{(1),1}\widetilde{p}_{1i} + \overline{T}_{(1),2}\,\widetilde{p}_{2i} + \overline{T}_{(1),3}\,\widetilde{p}_{3i} + \overline{T}_{(1),4}\,\widetilde{p}_{4i} + \overline{T}_{(1),5}\,\widetilde{p}_{5i} + 1.\end{aligned} \tag{14.41}$$

In addition, $\overline{T}_{(1),1} = 0$ because the system is already at the absorbing state of interest, and $\overline{T}_{(1),5} = \infty$ because the system can never reach state 1 from the other absorbing state.

Altogether, the conditional waiting times satisfy

$$\begin{aligned}&(\overline{T}_{(1),1} \quad \overline{T}_{(1),2} \quad \overline{T}_{(1),3} \quad \overline{T}_{(1),4} \quad \overline{T}_{(1),5}) \\ &\quad = (\overline{T}_{(1),1} \quad \overline{T}_{(1),2} \quad \overline{T}_{(1),3} \quad \overline{T}_{(1),4} \quad \overline{T}_{(1),5})\, \mathbf{U} \circ \mathbf{M} + (0 \quad 1 \quad 1 \quad 1 \quad 1)\end{aligned} \tag{14.42a}$$

or

$$\vec{\overline{T}}_{(1)}^{\mathrm{T}} = \vec{\overline{T}}_{(1)}^{\mathrm{T}}(\mathbf{U} \circ \mathbf{M}) + \vec{b}^{\mathrm{T}}, \tag{14.42b}$$

where $\mathbf{U}$ is a matrix containing elements $\rho_{1,j}/\rho_{1,i}$ in the jth row and ith column, $\mathbf{M}$ is the original transition probability matrix (14.13), and the matrix operator

"$\circ$" represents the direct product of these two matrices (simply taking the *ij*th element of the first matrix and multiplying it by the *ij*th element of the second matrix). One other difference in (14.42b) is that the last element in vector $\vec{b}^{\mathrm{T}}$, corresponding to the other non-absorbing state, now equals one, so that the last element in the vector (14.42a) gives $\overline{T}_{(1),5} = \overline{T}_{(1),5} + 1$, which requires that $\overline{T}_{(1),5} = \infty$.

Again, equation (14.42) can be solved by hand using Recipe P2.1 in Primer 2, or we can eliminate the absorbing states from (14.42), whose solutions we already know, to get

$$\vec{\overline{T}}_{u,(1)}^{\mathrm{T}} = \vec{\overline{T}}_{u,(1)}^{\mathrm{T}} (\mathbf{U}_u \circ \mathbf{M}_u) + \vec{b}_u^{\mathrm{T}}, \tag{14.43}$$

where $\vec{\overline{T}}_{u,(1)}^{\mathrm{T}} = (\overline{T}_{(1),2} \quad \overline{T}_{(1),3} \quad \overline{T}_{(1),4})$,

$$\mathbf{U}_u = \begin{pmatrix} \frac{\rho_{1,2}}{\rho_{1,2}} & \frac{\rho_{1,2}}{\rho_{1,3}} & \frac{\rho_{1,2}}{\rho_{1,4}} \\ \frac{\rho_{1,3}}{\rho_{1,2}} & \frac{\rho_{1,3}}{\rho_{1,3}} & \frac{\rho_{1,3}}{\rho_{1,4}} \\ \frac{\rho_{1,4}}{\rho_{1,2}} & \frac{\rho_{1,4}}{\rho_{1,3}} & \frac{\rho_{1,4}}{\rho_{1,4}} \end{pmatrix}, \tag{14.44}$$

$\mathbf{M}_u$ is again (14.34), and $\vec{b}_u^{\mathrm{T}} = (1 \quad 1 \quad 1)$. Equation (14.43) can be solved to give an explicit solution for the average time until absorption among those cases that become absorbed in state 1:

$$\vec{\overline{T}}_{u,(1)}^{\mathrm{T}} = \vec{b}_u^{\mathrm{T}} (\mathbf{I} - \mathbf{U}_u \circ \mathbf{M}_u)^{-1}, \tag{14.45}$$

along with $\overline{T}_{(1),1} = 0$ and $\overline{T}_{(1),5} = \infty$.

For the symmetric matrix (14.26) where the two species are identical, (14.45) gives

$$\vec{\overline{T}}_{(1)}^{\mathrm{T}} = \vec{\overline{T}}^{\mathrm{T}} + \left(0 \quad -\frac{\beta(\alpha + \beta + 2\gamma)}{2(\alpha + \beta)(2\alpha + \beta)\gamma} \quad 0 \quad \frac{(\alpha + \beta + 2\gamma)}{2(\alpha + \beta)\gamma} \quad \infty\right), \tag{14.46a}$$

$$\vec{\overline{T}}_{(5)}^{\mathrm{T}} = \vec{\overline{T}}^{\mathrm{T}} + \left(\infty \quad \frac{(\alpha + \beta + 2\gamma)}{2(\alpha + \beta)\gamma} \quad 0 \quad -\frac{\beta(\alpha + \beta + 2\gamma)}{2(\alpha + \beta)(2\alpha + \beta)\gamma} \quad 0\right). \tag{14.46b}$$

We have written (14.46) as the difference from the unconditional time to absorption, $\vec{\overline{T}}^{\mathrm{T}}$, to emphasize the impact on the expected waiting time of conditioning on reaching a particular absorbing state. As we would expect, the time to absorption is decreased from starting states near an absorbing state of interest but is increased from distant starting states. Figure 14.6 illustrates this effect when species 1 is initially rare and species 2 common (state 4); the conditional time to extinction of species 1 (state 5) is then shorter, on average, than the conditional time to extinction of species 2 (state 1).

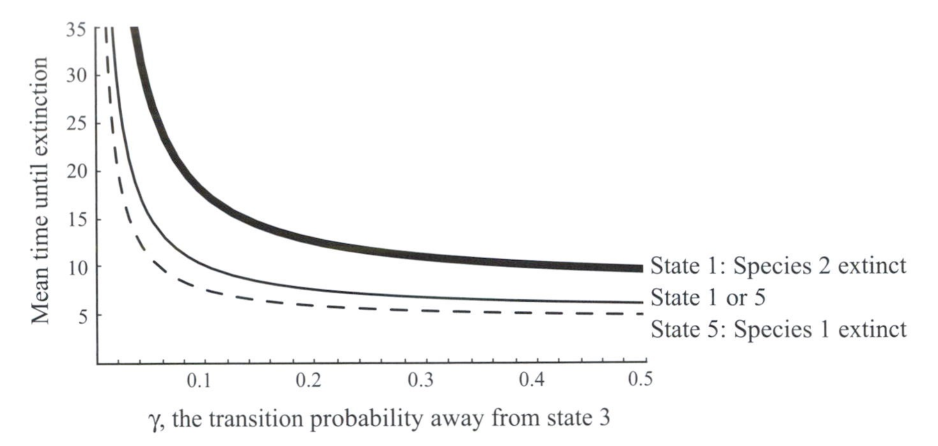

Figure 14.6: Extinction times for the two-species model of interactions. Starting from state 4, the mean time until one of the two species goes extinct (thin solid curve) is compared to the time to extinction of species 1 given that state 5 is ultimately reached (dashed curve) and to the time to extinction of species 2 given that state 1 is ultimately reached (thick solid curve). Parameter values are $\alpha = p_{12} = p_{54} = 0.2$, and $\beta = p_{32} = p_{34} = 0.2$.

Recipe 14.4
The Mean Time until Absorption in a Particular Absorbing State

The mean time until absorption into a particular absorbing state, a, given that the system starts in state i, is given by the element, $\overline{T}_{(a),i}$, of the row vector $\vec{\overline{T}}_{(a)}^{\mathrm{T}}$ that satisfies

$$\vec{\overline{T}}_{(a)}^{\mathrm{T}} = \vec{\overline{T}}_{(a)}^{\mathrm{T}} (\mathbf{U} \circ \mathbf{M}) + \vec{b}^{\mathrm{T}} \quad \text{subject to } \overline{T}_{(a),a} = 0 \text{ for absorbing state } a, \text{ and } \overline{T}_{(a),l} = \infty \text{ for all other absorbing states } l. \tag{14.47}$$

The row vector $\vec{b}^{\mathrm{T}}$ now contains ones in every position except the ath position, which contains a zero. The matrix $\mathbf{U}$ contains elements $\rho_{a,j}/\rho_{a,i}$ in the jth row and ith column, and the matrix operator $\circ$ represents a direct product of matrices (multiplying the ijth element of the first matrix by the ijth element of the second matrix). The solution to equation (14.47) equals $\overline{T}_{(a),a} = 0$ for absorbing state a, $\overline{T}_{(a),l} = \infty$ for all other absorbing states l, and

$$\vec{\overline{T}}_{u,(a)}^{\mathrm{T}} = \vec{b}_u^{\mathrm{T}} (\mathbf{I} - \mathbf{U}_u \circ \mathbf{M}_u)^{-1}, \tag{14.48}$$

for all unabsorbed states, where the subscript u refers to the vectors and matrices that remain after removing all rows and columns involving the absorbing states (e.g., $\vec{b}_u^{\mathrm{T}}$ is again a vector of length $m - k$ consisting of ones only).

When there is only one absorbing state a, the system is eventually absorbed in that state regardless of its current state, so that the

(continued)

Recipe 14.4 ***(continued)***

elements of matrix **U** equal one. In this case, the conditional and unconditional expected waiting times starting from state i are the same, $\overline{T}_{(a),i} = \overline{T}_i$. When there are k absorbing states, the conditional and unconditional expected waiting times are related to one another by

$$\overline{T}_i = \sum_{a=1}^{k} (\rho_{a,i}\,\overline{T}_{(a),i}) \tag{14.49}$$

(Rule P3.9), which states that the expected waiting time to absorption in any absorbing state equals the expected waiting time to absorption in each absorbing state times the probability of absorption in that state, summed over all absorbing states.

14.4 Birth-Death Models

The stochastic methods described above work well whenever there is a small and finite number of states (as in sections 14.2 and 14.3) or when the transition probability matrix contains actual numbers rather than parameters. For other types of problems, however, it can be impractical to apply the matrix operations to calculate the general solution (14.21), the probability of reaching a particular absorbing state (14.29), or the waiting times (14.38) and (14.48). For most complicated Markov models, the only workable approach is to approximate the transitions, e.g., using a diffusion approach (see Chapter 15). For certain matrices, however, exact results can be calculated even for very large matrices. An important example is the transition probability matrix that results from the birth-death process introduced in Chapter 13.

14.4.1 A General Birth-Death Process

Any matrix that has zeros everywhere except on the diagonal, right below the diagonal, and right above the diagonal is known as a *continuant* matrix. That is, if the transition probabilities obey the following rules:

$$p_{ji} = \left\{\begin{array}{lll} b_i & \text{for } j = i+1 & \text{(increase by one),} \\ d_i & \text{for } j = i-1 & \text{(decrease by one),} \\ 1 - b_i - d_i & \text{for } j = i & \text{(no change),} \\ 0 & \text{for } j \neq i-1, i, i+1 & \text{(other changes),} \end{array}\right\} \tag{14.50}$$

then the resulting matrix is continuant:

$$\mathbf{M} = \begin{pmatrix} 1 - b_0 & d_1 & 0 & \cdots & 0 \\ b_0 & 1 - b_1 - d_1 & d_2 & \cdots & 0 \\ 0 & b_1 & 1 - b_2 - d_2 & \cdots & 0 \\ \cdots & \cdots & \cdots & \cdots & \cdots \\ 0 & 0 & 0 & \cdots & 1 - d_m \end{pmatrix}. \tag{14.51}$$

Here, we have started indexing the states with 0 rather than 1. Birth-death models typically track the number of some quantity (e.g., the number of individuals, the number of microsatellite repeats, or the number of species), and therefore 0 is the lowest possible state. As a result, the transition matrix (14.51) is an $(m + 1) \times (m + 1)$ matrix.

The birth-death model describing the number of individuals within a population (13.2) is described by a continuant matrix as long as we keep track of time in small discrete units (small enough so that only one birth or death can occur per unit time, regardless of the population size) and if we assume that there is some maximum attainable population size, m, with $b_m = 0$. Many of the results can be generalized to the case of an infinite number of states, which we shall mention as appropriate. The Moran model of allele frequency change is also described by a continuant matrix, either without selection (13.9) or with selection (13.10).

Because transitions occur only among neighboring states, it is possible to derive many quantities of interest for any model whose transition probability matrix is continuant.

14.4.2 The Stationary Distribution for a Birth-Death Process

We start by calculating the stationary distribution $\vec{\pi}$ under the assumption that all b_i and d_i are positive (in this case, there are no absorbing states). The stationary distribution must satisfy $\vec{\pi} = \mathbf{M}\,\vec{\pi}$ (equation (14.19)), which can be rewritten as $(\mathbf{I} - \mathbf{M})\,\vec{\pi} = \vec{0}$, where $\vec{0}$ is a vector of zeros. Using the matrix (14.51)

$$(\mathbf{I} - \mathbf{M})\,\vec{\pi} = \begin{pmatrix} b_0\pi_0 - d_1\pi_1 \\ -b_0\pi_0 + b_1\pi_1 + d_1\pi_1 - d_2\pi_2 \\ -b_1\pi_1 + b_2\pi_2 + d_2\pi_2 - d_3\pi_3 \\ \cdots \\ -b_{m-2}\pi_{m-2} + b_{m-1}\pi_{m-1} + d_{m-1}\pi_{m-1} - d_m\pi_m \\ -b_{m-1}\pi_{m-1} + d_m\pi_m \end{pmatrix} = \vec{0}. \quad (14.53)$$

Even easier, we can add the first row of (14.52) to the second row to cancel out two of the terms in the second row. Because each row equals zero at the stationary distribution, this summation does not alter the fact that the second row equals zero. Then we can add the result to the third row and simplify, etc., to get

$$\begin{pmatrix} b_0\pi_0 - d_1\pi_1 \\ b_1\pi_1 - d_2\pi_2 \\ b_2\pi_2 - d_3\pi_3 \\ \cdots \\ b_{m-1}\pi_{m-1} - d_m\pi_m \\ -b_{m-1}\pi_{m-1} + d_m\pi_m \end{pmatrix} = \vec{0}. \quad (14.53)$$

According to (14.53), $b_{i-1}\pi_{i-1} - d_i\,\pi_i = 0$, which can be rewritten as a recursion equation, $\pi_i = (b_{i-1}/d_i)\pi_{i-1}$. We can solve this by brute-force iteration just as we did with the exponential growth model: $\pi_i = (b_{i-1}/d_i) \ldots (b_1/d_2)\,(b_0/d_1)\,\pi_0$. To simplify the presentation, let us define

$$\begin{cases} \kappa_0 = 1 \\ \kappa_i = \dfrac{b_{i-1}}{d_i} \cdots \dfrac{b_1}{d_2}\dfrac{b_0}{d_1} \quad \text{for } i > 0, \end{cases} \tag{14.54}$$

so that $\pi_i = \kappa_i\,\pi_0$. Dividing each π_i by their sum, we get the stationary distribution

$$\pi_i = \frac{\kappa_i}{\sum_{j=0}^{m} \kappa_j}. \tag{14.55}$$

Similar calculations can be carried out even if there is an infinite number of states (i.e., $m = \infty$), but a stationary distribution exists only if the sum $\sum_{j=0}^{\infty} \kappa_j$ is finite.

Equation (14.55) continues to apply if there is one absorbing state. For example, extinction represents an absorbing state in the birth-death model of population growth with $b_0 = 0$. In this case the stationary distribution (14.55) simplifies to $\pi_0 = 1$, and there is eventually zero probability of being in any other state. This confirms the fact mentioned earlier that all populations ultimately go extinct. Analogous calculations can be conducted when there are multiple absorbing states (Problem 14.14).

14.4.3 The Probability of Absorption for a Birth-Death Process

In some birth-death models, there are two absorbing states. In the Moran model of allele frequency change without mutation (13.9) and (13.10), the two absorbing states represent loss of the allele and fixation of the allele. Similarly, if we are concerned about a population at risk of extinction, there might be some size above which we declare the population "viable" and stop careful monitoring of the population. We could then consider this large population size as an absorbing state along with an absorbing state at size 0. A birth-death model can then be used to ask whether the population will go extinct before it reaches the minimum viable population size, as well as to ask how long we should expect to have to monitor the population. In such models, $b_0 = 0$ and $d_m = 0$, so that the system never leaves an absorbing state.

For a birth-death model with two absorbing states, we derive the probability that a particular absorbing state (say 0) is ultimately reached in Box 14.1. The calculations are a bit involved but require only algebra. Here we focus on three examples of special interest: the probability of extinction of a population undergoing a birth-death process, the probability of fixation of a neutral allele, and the probability of fixation of a favorable allele.

In section 13.3, we described a model of population growth where the probability of a birth or a death in a very small time step, Δt, was equal to $b\,i\,\Delta t$ and $d\,i\,\Delta t$. Simulations of this process (Figure 13.7) indicated that populations could

Box 14.1: The Probability of Loss in a Birth-Death Model

Here we derive the probability that a birth-death process is ultimately absorbed in state 0 assuming that there are two absorbing states: 0 and m. The probability of absorption in any particular state a can be calculated from Recipe 14.2. Evaluating (14.29) gives the series of equations:

$$\begin{aligned}\rho_{a,1} &= d_1\rho_{a,0} + (1 - b_1 - d_1)\rho_{a,1} + b_1\rho_{a,2},\\ \rho_{a,2} &= d_1\rho_{a,1} + (1 - b_2 - d_2)\rho_{a,3} + b_2\rho_{a,3},\\ &\vdots\\ \rho_{a,m-1} &= d_{m-1}\rho_{a,m-2} + (1 - b_{m-1} - d_{m-1})\rho_{a,m-1} + b_{m-1}\rho_{a,m},\end{aligned} \tag{14.1.1}$$

together with $\rho_{0,0} = 1$ and $\rho_{0,m} = 0$ if we are interested in absorption at state $a = 0$, or $\rho_{m,0} = 0$ and $\rho_{m,m} = 1$ if we are interested in absorption at state m. Solving the above equations is easier if we rearrange each one into a more symmetrical form:

$$\begin{aligned}(\rho_{a,2} - \rho_{a,1}) &= \frac{d_1}{b_1}(\rho_{a,1} - \rho_{a,0}),\\ (\rho_{a,3} - \rho_{a,2}) &= \frac{d_2}{b_2}(\rho_{a,2} - \rho_{a,1}),\\ &\vdots\\ (\rho_{a,m} - \rho_{a,m-1}) &= \frac{d_{m-1}}{b_{m-1}}(\rho_{a,m-1} - \rho_{a,m-2}).\end{aligned} \tag{14.1.2}$$

By iteration, the above tells us that $(\rho_{a,i+1} - \rho_{a,i}) = (d_i/b_i)\ldots(d_2/b_2)(d_1/b_1)(\rho_{a,1} - \rho_{a,0})$. To simplify the presentation, let us define

$$\begin{cases}\gamma_0 = 1\\ \gamma_i = \dfrac{d_i}{b_i}\cdots\dfrac{d_2}{b_2}\dfrac{d_1}{b_1} & \text{for } i > 0\end{cases} \tag{14.1.3}$$

so that

$$(\rho_{a,i+1} - \rho_{a,i}) = \gamma_i(\rho_{a,1} - \rho_{a,0}). \tag{14.1.4}$$

Equation (14.1.4) can be rewritten as a recursion equation for $\rho_{a,i}$, to give $\rho_{a,i+1} = \gamma_i(\rho_{a,1} - \rho_{a,0}) + \rho_{a,i}$. We can solve this recursion by brute force iteration. For $i = 0$ we have $\rho_{a,1} = \gamma_0(\rho_{a,1} - \rho_{a,0}) + \rho_{a,0}$. For $i = 1$ we have $\rho_{a,2} = \gamma_1(\rho_{a,1} - \rho_{a,0}) + \rho_{a,1}$. Now substituting in the previous result for $\rho_{a,1}$ gives $\rho_{a,2} = \gamma_1(\rho_{a,1} - \rho_{a,0}) + \gamma_0(\rho_{a,1} - \rho_{a,0}) + \rho_{a,0}$ or $\rho_{a,2} = (\gamma_1 + \gamma_0)(\rho_{a,1} - \rho_{a,0}) + \rho_{a,0}$. Carrying on in this fashion, we obtain the solution:

$$\rho_{a,i} = \left(\sum_{k=0}^{i-1}\gamma_k\right)(\rho_{a,1} - \rho_{a,0}) + \rho_{a,0}. \tag{14.1.5}$$

(continued)

Box 14.1 *(continued)*

To complete the calculation, we must specify $\rho_{a,0}$ and $\rho_{a,m}$. Focusing on the probability of loss (i.e., absorption in state '0'), we can substitute $\rho_{0,0} = 1$ and $\rho_{0,m} = 0$ into (14.1.5) to get $\rho_{0,m} = 0 = \left(\sum_{k=0}^{m-1} \gamma_k\right)(\rho_{0,1} - 1) + 1$, which can be rearranged to find $(\rho_{0,1} - 1) = -1/\sum_{k=0}^{m-1} \gamma_k$. This can then be substituted back into (14.1.5) for $(\rho_1 - \rho_0)$ to obtain the final result for the probability of loss given $n_0 = i$ initial copies:

$$\rho_{0,i} = 1 - \frac{\sum_{k=0}^{i-1} \gamma_k}{\sum_{k=0}^{m-1} \gamma_k}. \tag{14.1.6}$$

Because there are only two absorbing states, the probability of ultimately being absorbed in state m is just $\rho_{m,i} = 1 - \rho_{0,i}$. This provides the probability that the allele becomes fixed in the Moran model.

Although we have assumed a fixed number of states, we can take the limit of (14.1.6) as m tends to infinity to obtain some useful results about birth-death processes over an infinite domain

$$\rho_{0,i} = 1 - \frac{\sum_{k=0}^{i=1} \gamma_k}{\sum_{k=0}^{m-1} \gamma_k} \xrightarrow{m \to \infty} 1 - \frac{\sum_{k=0}^{i=1} \gamma_k}{\sum_{k=0}^{\infty} \gamma_k}. \tag{14.1.7}$$

There are two possibilities. If $\sum_{k=0}^{\infty} \gamma_k = \infty$, then $\rho_{0,i}$ will be one for any finite starting state, and the system will always tend towards extinction. If $\sum_{k=0}^{\infty} \gamma_k$ is finite, then there will be some chance of being absorbed in state 0 given by (14.1.7), and otherwise the system will march off to infinity.

go extinct even with a positive net growth rate $(b - d)$. We can use the results of Box 14.1 to calculate the probability of extinction, ρ_{0,n_0}, from an initial population size of n_0. Because the probability of death divided by the probability of birth is d/b regardless of the population size, the constants (14.1.3) needed to calculate the fixation probability simplify to $\gamma_i = (d/b)^i$.

From (14.1.6) and using rule (A1.20), we find that the probability that the population goes extinct before reaching a minimum viable population size m equals

$$\rho_{0,n_0} = \frac{\left(\frac{d}{b}\right)^{n_0} - \left(\frac{d}{b}\right)^{m}}{1 - \left(\frac{d}{b}\right)^{m}}. \tag{14.56}$$

Conversely, the probability that the population survives to reach the viable population size before going extinct equals $\rho_{m,n_0} = 1 - \rho_{0,n_0}$ (Problem 14.8).

Next, let's suppose that the possible population size is unbounded ($m \to \infty$). According to Box 14.1, the population is certain to go extinct at some point if $\sum_{i=0}^{\infty}\gamma_i = \infty$. This condition is necessarily true if the death rate is greater than the birth rate. If the birth rate is greater than the death rate, however, then there is always some finite probability of extinction, even though the population typically grows to infinite size. Using equation (14.1.7), the probability of extinction of a population of initial size n_0 is

$$\rho_{0,n_0} = \left(\frac{d}{b}\right)^{n_0} \tag{14.57}$$

(see Problem 14.9). Equation (14.57) confirms that a population might go extinct even if its birth rate exceeds its death rate, and even though there is no upper limit to the population size. To illustrate this analytical result, we ran 100 simulations using the parameters of Figure 13.7. In those simulations with $b = 21/90$, $d = 20/90$, and an initial population size of $n_0 = 10$, the observed probability of extinction was 0.58 with a 95% confidence interval of $(0.48 - 0.68)$, which includes the predicted probability of extinction, $\rho_{0,n_0} = 0.614$.

In section 13.5, we described the Moran model of allele frequency change. In the Moran model, the population size is held constant ($m = N$), but at each event the number of copies of a particular allele increases by one ("birth"), decreases by one ("death"), or remains the same, as described by equation (13.9). When all individuals are equally fit (the neutral case), the probability of death and the probability of birth are the same; $p\,(1 - p)$, where $p = i/N$. Thus, $d_i/b_i = 1$ for all i, and from (14.1.3) $\gamma_i = 1$. Using equation (14.1.6), we can determine the probability that the allele is ultimately lost (i.e., the probability that the birth-death process hits the absorbing state 0 before it hits N) when starting with n_0 initial copies: $\rho_{0,n_0} = 1 - n_0/N$ (see Problem 14.10). Conversely, the probability that an allele ultimately fixes is $\rho_{N,n_0} = 1 - \rho_{0,n_0} = n_0/N$. This is a classic result of evolutionary theory: in the absence of selection, mutation, and other evolutionary forces, the probability that a neutral allele eventually fixes within a population is given by its initial frequency.

With selection, the fixation probability of an allele is no longer equal to its initial frequency. In equation (13.10), we described how the transition probabilities are modified to account for selection in the Moran model. In this case, the probability of death divided by the probability of birth becomes $d_i/b_i = W_a/W_A$ (Problem 14.11). Thus, from (14.1.3), $\gamma_i = (W_a/W_A)^i$. Using (14.1.6), the probability that the A allele is ultimately lost equals

$$\rho_{0,n_0} = 1 - \frac{1 - \left(\dfrac{W_a}{W_A}\right)^{n_0}}{1 - \left(\dfrac{W_a}{W_A}\right)^{N}}, \tag{14.58a}$$

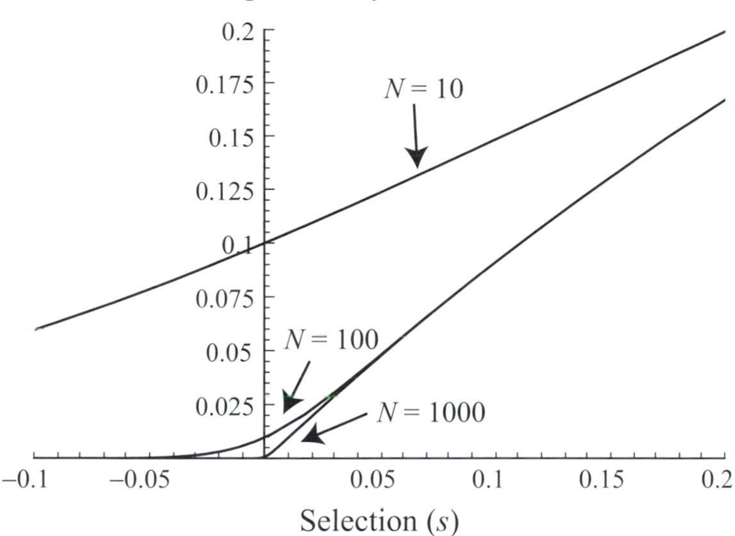

Figure 14.7: The probability of fixation in the Moran model. Starting from a single allele ($p_0 = 1/N$), the exact probability of fixation of allele A given by (14.58b) is evaluated as a function of the selection coefficient, s, where $W_A = 1 + s$ and $W_a = 1$. A haploid population of size N is assumed.

while the probability that it ultimately fixes is

$$\rho_{N,n_0} = \frac{1 - \left(\dfrac{W_a}{W_A}\right)^{n_0}}{1 - \left(\dfrac{W_a}{W_A}\right)^{N}} \tag{14.58b}$$

(Problem 14.11). If there is initially a single copy of the A allele, its fixation probability varies as a function of the population size and the selection coefficient, s, defined as $W_A/W_a = 1 + s$ (Figure 14.7). As expected, when selection is absent, the fixation probability is $1/N$, which is where the curves intersect the vertical axis. When the population is large, the fixation probability rises nearly linearly with the selection coefficient (see Problem 14.11). Figure 14.7 also shows that the A allele can fix even if it decreases fitness ($s < 0$), although the fixation probability is vanishingly small unless the selective disadvantage and the population size are small.

14.4.4 The Expected Time to Absorption for a Birth-Death Process

The time that it takes for a birth-death model to reach an absorbing state is derived in Box 14.2. These formulas were used in Figure 14.8 to answer a question posed earlier: what is the time frame over which we must monitor a population at risk of extinction, given that we stop monitoring the population if it goes extinct or reaches a threshold population size, such as a minimum viable population size? The waiting time is short for small populations, then rises, and falls again for populations near the threshold population size, reflecting the fact that populations near absorbing states are likely to be absorbed

Box 14.2: The Time until Absorption in a Birth-Death Model

The time until absorption in a birth-death process can be calculated for the birth-death model (14.51) using the general equation (14.37):

$$\begin{aligned}
\overline{T}_1 &= (d_1\overline{T}_0 + (1 - b_1 - d_1)\overline{T}_1 + b_1\overline{T}_2) + 1,\\
\overline{T}_2 &= (d_2\overline{T}_1 + (1 - b_2 - d_2)\overline{T}_2 + b_2\overline{T}_3) + 1,\\
\overline{T}_3 &= (d_3\overline{T}_2 + (1 - b_3 - d_3)\overline{T}_3 + b_3\overline{T}_4) + 1,\\
&\vdots\\
\overline{T}_{m-1} &= (d_{m-1}\overline{T}_{m-2} + (1 - b_{m-1} - d_{m-1})\overline{T}_{m-1} + b_{m-1}\overline{T}_m) + 1,
\end{aligned} \tag{14.2.1}$$

subject to $\overline{T}_0 = 0$ and $\overline{T}_m = 0$, as the process is already fixed in these states. We can rearrange (14.2.1) into a more symmetrical form, just as we did in Box 14.1:

$$\begin{aligned}
(\overline{T}_2 - \overline{T}_1) &= \frac{d_1}{b_1}(\overline{T}_1 - \overline{T}_0) - \frac{1}{b_1},\\
(\overline{T}_3 - \overline{T}_2) &= \frac{d_2}{b_2}(\overline{T}_2 - \overline{T}_1) - \frac{1}{b_2},\\
&\vdots\\
(\overline{T}_m - \overline{T}_{m-1}) &= \frac{d_{m-1}}{b_{m-1}}(\overline{T}_{m-1} - \overline{T}_{m-2}) - \frac{1}{b_{m-1}}.
\end{aligned} \tag{14.2.2}$$

By plugging the first equation into the second, simplifying, and repeating this process for each subsequent term, we obtain

$$(\overline{T}_{i+1} - \overline{T}_i) = \gamma_i(\overline{T}_1 - \overline{T}_0) - \sum_{j=1}^{i} \frac{\gamma_i}{b_j\gamma_j}, \tag{14.2.3}$$

where γ_i is given by (14.1.3).

Equation (14.2.3) can be rewritten as a general recursion equation $(\overline{T}_{i+1} = \gamma_i(\overline{T}_1 - \overline{T}_0) - \sum_{j=1}^{i}\gamma_i/(b_j\gamma_j) + \overline{T}_i$, which can be solved by direct iteration. Starting with $i = 1$ we have $\overline{T}_2 = \gamma_1(\overline{T}_1 - \overline{T}_0) - \sum_{j=1}^{1}\gamma_1/(b_j\gamma_j) + \overline{T}_1$. The general recursion tells us that we must add $\gamma_i(\overline{T}_1 - \overline{T}_0) - \sum_{j=1}^{i}\gamma_i/(b_j\gamma_j)$ to get $\overline{T}_{i+1}$ from $\overline{T}_i$. Thus, for $i = 2$, we add $\gamma_2(\overline{T}_1 - \overline{T}_0) - \sum_{j=1}^{2}\gamma_2/(b_j\gamma_j)$ to $\overline{T}_2$ to get

$$\overline{T}_3 = (\gamma_1 + \gamma_2)(\overline{T}_1 - \overline{T}_0) - \left(\sum_{j=1}^{1}\frac{\gamma_1}{b_j\gamma_j} + \sum_{j=1}^{2}\frac{\gamma_2}{b_j\gamma_j}\right) + \overline{T}_1. \tag{14.2.4}$$

Repeating this procedure, the solution is

$$\overline{T}_{i+1} = \left(\sum_{k=1}^{i}\gamma_k\right)(\overline{T}_1 - \overline{T}_0) - \sum_{k=1}^{i}\sum_{j=1}^{k}\frac{\gamma_k}{b_j\gamma_j} + \overline{T}_1. \tag{14.2.5}$$

Because $\overline{T}_0 = 0$, equation (14.2.5) can be further simplified as

(continued)

Box 14.2 *(continued)*

$$\begin{aligned}\overline{T}_{i+1} &= \left(\sum_{k=1}^{i}\gamma_k\right)\overline{T}_1 - \sum_{k=1}^{i}\sum_{j=1}^{k}\frac{\gamma_k}{b_j\gamma_j} + \overline{T}_1 \\ &= \left(\sum_{k=0}^{i}\gamma_k\right)\overline{T}_1 - \sum_{k=1}^{i}\sum_{j=1}^{k}\frac{\gamma_k}{b_j\gamma_j}.\end{aligned} \tag{14.2.6}$$

Because we know that $\overline{T}_m = 0$, we can set $i = m - 1$ and solve for the waiting time $\overline{T}_1$ if starting with one copy:

$$\overline{T}_1 = \frac{\sum_{k=1}^{m-1}\sum_{j=1}^{k}\frac{\gamma_k}{b_j\gamma_j}}{\sum_{k=0}^{m-1}\gamma_k}. \tag{14.2.7}$$

Substituting (14.2.7) back into (14.2.7) gives the mean time until absorption starting from any state i:

$$\overline{T}_i = \frac{\sum_{k=0}^{i-1}\gamma_k}{\sum_{k=0}^{m-1}\gamma_k}\sum_{k=1}^{m-1}\sum_{j=1}^{k}\frac{\gamma_k}{b_j\gamma_j} - \sum_{k=1}^{i-1}\sum_{j=1}^{k}\frac{\gamma_k}{b_j\gamma_j}. \tag{14.2.8}$$

Equation (14.2.8) describes the average waiting time until absorption in any of the absorbing states. Ewens (1979, p. 73) describes similar results for the waiting time until absorption in a particular state and for the waiting times when there is only one absorbing state.

soon. Interestingly, the total waiting time is more sensitive to the per capita birth rate relative to the death rate (b/d) than to the absolute rates of births and deaths ($b + d$). If $b/d > 1$, the population tends to grow, if $b/d < 1$, the population tends to shrink, and in either case it tends to move more rapidly to absorption. Here, we have assumed that the time step taken is so small (e.g., an hour) that the probability of a birth or a death in a time step is always less than one regardless of the population size ($b\,m + d\,m < 1$).

14.5 Branching Processes

In the previous section, we showed how exact results can be obtained from a special type of Markov process, the birth-death process, where a limited number of transitions are possible. Analytical progress is also possible when you

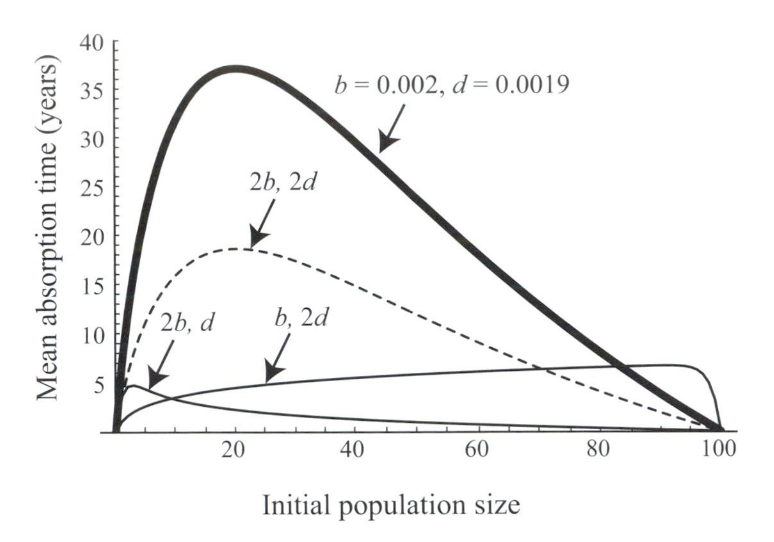

Figure 14.8: Waiting time until extinction or recovery. Using the birth-death model of population growth (13.2), the average time frame over which an at-risk population must be monitored is given by (14.2.8), assuming that monitoring stops if the population goes extinct or reaches size $m = 100$. The per capita birth and death probabilities were set to $b = 0.002$ and $d = 0.0019$ per day (thick curve). Using (14.2.8), it is possible to show that doubling both the birth and death rates halves the waiting time until absorption because events happen twice as fast (dashed curve). Doubling only the birth rate or only the death rate (thin curves), however, causes a much more dramatic change in the waiting time, because such changes affect whether the population grows or shrinks.

focus not on a particular model but on a particular type of question. The theory of branching processes addresses the following question: What ultimately happens when a small number of a new type arises? Will the type become established or disappear? Even if the type ultimately disappears, does it do so slowly or rapidly? And how many descendents does it have before it disappears? A branching process is an example of a Markov process where the types being studied replicate ("branch") or die. In a branching process, it is assumed that there is an absorbing boundary at 0 (extinction or loss) and that there is no upper bound to the possible number of a given type.

Historically, branching processes were developed to study the distribution of last names within human populations (see review in Mode 1971). The theory of branching processes was used to describe the fate of a single new surname that arises within a population. From this narrow initial focus, branching processes have found numerous applications. For example, branching processes have been used to determine whether a newly introduced species becomes established or goes extinct. They have also been used to determine whether a novel allele introduced by mutation fixes or is lost from a population and to determine the severity of disease outbreaks. The theory of branching processes has been extended to account for the fact that not all individuals carrying the novel type are identical (Mode 1971), for example, individuals may differ in age (Charlesworth 1994), in spatial location (Barton 1993), or in genetic background (Barton 1995).

14.5.1 Probability of Establishment

We first focus on the probability of extinction versus establishment. The probability that a novel type (a surname, a species, or an allele) that arises in a single copy eventually goes extinct is $\rho_{0,1}$, which we will refer to as P_{loss}. Conversely, the probability of establishment (or fixation) in a population of unlimited size is $\rho_{\infty,1}$, which we will call P_{fix}. Ultimately, one of these two outcomes must occur, $P_{loss} + P_{fix} = 1$. Haldane (1927) determined the probability of establishment assuming that the number of copies of the novel type in the next generation follows a Poisson distribution with a mean of $1 + s$. His reasoning was that, on average, most populations in the world must have a long-term average reproductive value near one to persist without growing out of control, but that occasionally new types might arise with a slight reproductive advantage, measured by s.

To analyze this model, we could try to proceed by setting up the transition matrix as we did earlier. Doing this is problematic for two reasons. First, because there is no upper bound on the population size, the transition matrix would be infinitely large. We could try to handle this as we did with the birth-death model in Box 14.1 by starting with a finite transition matrix and then taking the limit as the size of the matrix gets larger and larger. But then, we could not use the Poisson distribution for the number of offspring per parent, because the Poisson distribution does not have a finite maximum. The second problem is that the matrix would not contain any zeros, because it is possible to move between any two nonzero states depending on how many offspring each current individual has. This means that the matrix would not have a simple form and would be extremely cumbersome for reasonably large population sizes.

The key insight made by Haldane (1927) was that the probability that a type ultimately leaves no descendants must equal the probability that each offspring produced by this type leaves no descendants. This is yet another example where the law of total probability (Rule P3.8) comes in handy. Mathematically, we have

$$\begin{aligned} P_{loss} = & \sum_{j=0}^{\infty} P(\text{ultimate extinction} \mid j \text{ offspring produced}) \\ & \times P(j \text{ offspring produced}). \end{aligned} \tag{14.59}$$

If the fate of each offspring is independent of the others, then the probability that all j offspring fail to leave descendants is the product of the probability that each one fails to leave descendants. Assuming that the environment is not changing, then each of these offspring face exactly the same situation faced by the original individual, and therefore this probability is P_{loss}^{j}. Plugging this into equation (14.59) along with the probability of having j offspring under the Poisson distribution gives (see Definition P3.9 of Primer 3):

$$\begin{aligned} P_{loss} &= \sum_{j=0}^{\infty} P_{loss}^{j}\, e^{-(1+s)} \frac{(1+s)^j}{j!} \\ &= e^{-(1+s)} \sum_{j=0}^{\infty} \frac{(P_{loss}(1+s))^j}{j!}. \end{aligned} \tag{14.60}$$

Although the sum in (14.60) looks complicated, it can be simplified using the Taylor series for the exponential function from equation (P1.14), $e^x = 1/0! + x/1! + x^2/2! + \ldots$, which for $x = P_{loss}(1 + s)$ implies that

$$e^{P_{loss}(1+s)} = \sum_{j=0}^{\infty} \frac{(P_{loss}(1 + s))^j}{j!}.$$

Consequently, the probability that the new type is ultimately lost is

$$\begin{aligned} P_{loss} &= e^{-(1+s)}\, e^{P_{loss}(1+s)} \\ &= e^{-(1-P_{loss})(1+s)}. \end{aligned} \tag{14.61}$$

If the probability of loss satisfies (14.61), then the probability of establishment must solve $1 - P_{fix} = e^{-(1+s)P_{fix}}$. When selection is absent ($s = 0$), this becomes $1 - P_{fix} = e^{-P_{fix}}$, whose only solution is $P_{fix} = 0$, indicating that the allele will never become established by drift in an infinitely large population. Assuming that both P_{fix} and s are proportional to some small quantity, ε (i.e., $P_{fix} = \widetilde{P}_{fix}\,\varepsilon$ and $s = \tilde{s}\,\varepsilon$), we can take the Taylor series of $1 - P_{fix} = e^{-(1+s)P_{fix}}$ to second order in ε using Recipe P1.2:

$$\begin{aligned} 1 - \widetilde{P}_{fix}\,\varepsilon &= e^{-(1+\tilde{s}\,\varepsilon)\widetilde{P}_{fix}\,\varepsilon} \\ &= 1 - \widetilde{P}_{fix}\,\varepsilon + \frac{(\widetilde{P}_{fix}^2 - 2\,\tilde{s}\,\widetilde{P}_{fix})}{2}\varepsilon^2 + O(\varepsilon^3) \end{aligned} \tag{14.62}$$

(this is another example of a perturbation analysis; see Box 5.1). Solving for P_{fix} (via $\widetilde{P}_{fix}$), we get a key result in evolutionary biology: the probability of establishment of a favorable allele is approximately $P_{fix} = 2s$.

One of the critical assumptions in the above derivation was that the number of offspring follows a Poisson distribution. This is approximately true in the Wright-Fisher model for large populations (Feldman 1966; Karlin and McGregor 1964), but it is not generally true in nature (Figure 13.1). But we can repeat the above procedure for any probability distribution describing the probability of having j offspring, $P(Y = j)$ to get

$$P_{loss} = \sum_{j=0}^{\infty} P(Y = j) P_{loss}^j. \tag{14.63}$$

If we define $z = \ln(P_{loss})$, so that $P_{loss} = e^z$, equation (14.63) turns into the moment generating function for the offspring distribution (see Appendix 5 for a discussion of moment generating functions):

$$\begin{aligned} P_{loss} &= \sum_{j=0}^{\infty} P(Y = j)\, e^{zj} \\ &= MGF[z] \\ &= MGF[\ln(P_{loss})]. \end{aligned} \tag{14.64}$$

The probability of establishment is therefore given by

$$1 - P_{fix} = MGF[\ln(1 - P_{fix})]. \tag{14.65}$$

Assuming that P_{fix} is small, the Taylor series of this result is

$$\begin{aligned} 1 - P_{fix} &= MGF[0] + P_{fix}\left(- \frac{dMGF[0]}{dP_{fix}}\bigg|_{P_{fix}=0} \right) \\ &+ \frac{P_{fix}^2}{2}\left(\frac{d^2MGF[0]}{dP_{fix}^2}\bigg|_{P_{fix}=0} - \frac{dMGF[0]}{dP_{fix}}\bigg|_{P_{fix}=0} \right) + O(P_{fix}^3). \end{aligned} \tag{14.66}$$

The whole power of moment generating functions is that $MGF[0] = 1$ and their derivatives are related to known quantities such as the mean and variance (Appendix 5). Using equations (A5.4) and (A5.5), $(dMGF[z]/dz)|_{z=0}$ equals the expected number of offspring, $E[Y] = (1 + s)$, and $(d^2MGF[z]/dz^2)|_{z=0}$ equals $E[Y^2]$. Using equation (P3.2) for the variance in number of offspring per parent: $\sigma^2 = E[Y^2] - E[Y]^2$, equation (14.66) becomes

$$1 - P_{fix} = 1 - P_{fix}(-(1 + s)) + \frac{1}{2}P_{fix}^2(\sigma^2 + (1 + s)^2 - (1 + s)). \tag{14.67}$$

Again assuming that both $P_{fix} = \tilde{P}_{fix}\varepsilon$ and $s = \tilde{s}\varepsilon$ are proportional to some small quantity ε, the Taylor series of (14.67) is

$$1 - \tilde{P}_{fix}\varepsilon = 1 - \tilde{P}_{fix}\,\varepsilon - \left(\tilde{P}_{fix}\,\tilde{s} - \frac{1}{2}\tilde{P}_{fix}^2\sigma^2 \right)\varepsilon^2 + O(\varepsilon^3). \tag{14.68}$$

Solving for P_{fix}, the probability of establishment of a new type is approximately

$$P_{fix} = \frac{2s}{\sigma^2} + O(s^2), \tag{14.69}$$

regardless of the distribution describing the number of offspring per parent. For a Poisson distribution, $E[Y] = \sigma^2 = 1 + s$, so that (14.69) is again $2s$ to leading order in selection, as calculated by Haldane (1927).

In Chapter 13, we noted that the variance in offspring number per parent is approximately one in the Wright-Fisher model but two in the Moran model of allele frequency change. Thus, (14.69) predicts that the fixation probability of an allele with selective advantage s is approximately $2s$ in the Wright-Fisher model but only s in the Moran model (see Problem 14.11). The greater the role of chance in the dynamics of a species, the less likely it is for favorable types to spread.

Equation (14.69) is an extremely general result. It applies for any newly introduced type, whether it be a new species on an uninhabited island, a new mutation, a new surname, or a new cultural innovation. The probability that the novel type ultimately becomes established is proportional to the expected number of offspring in excess of the number needed to replace the parent, s, and is inversely proportional to the variability in the number of offspring per parent. Although the above discussion assumes that a novel type invades a resident population, the method works even if there is no resident population as long as the novel type has $1 + s$ offspring per parent and can potentially grow to a large population size. The key assumptions are that s is small (otherwise (14.65) must be used), that the potential population size is large, that density-dependence does not impact the spread of the novel type, and that the novel types do not directly compete or interfere with each others' spread.

14.5.2 Mean and Variance of a Branching Process

The above section illustrates how you can determine the probability of establishment of a novel type using a branching process. Yet we often would like to know something about the probability distribution describing where the system is at different points in time. For example, how many individuals do we expect to bear a new allele, a new disease, or a new surname after a given amount of time? Similarly, what is the cumulative number of cases expected before the extinction of a new type (e.g., a new disease)? Such questions can also be addressed using branching processes, as we describe in Supplementary Material 14.1. Once again moment generating functions provide an incredibly powerful tool, allowing us to calculate the mean and variance of the number of a new type at any given time.

14.6 Concluding Message

In this chapter, we have introduced various methods that can be used to analyze stochastic models as well as some of their most important applications. For some problems, the exact methods described in section 14.3 are tractable, allowing analytical solutions to such things as the stationary distribution, the probability of ultimately reaching a particular state, and the waiting time until a state is reached. Even when the exact solutions are not analytically tractable, the methods can often be used to get numerical results without having to carry out thousands of replicate simulations. For birth-death models, explicit formulas can be obtained for quantities of interest because of the fact that transitions only occur among neighboring states. And, for questions relating to the fate of a rare novel type, branching processes can be used to obtain results even in cases where it is difficult to write down a transition probability matrix.

In the next chapter, we explore how the results for stochastic processes can be extended to processes occurring over continuous state spaces. An important example of such stochastic processes is known as a *diffusion process*. Aside from being of interest as a stochastic model for continuous variables, the diffusion process also provides a powerful way to approximate stochastic models with a discrete state space, particularly when the range of discrete states is large and the methods of this chapter become unwieldy.

Problems

Problem 14.1: Find the stationary distribution and the general solution for the probability distribution of a two-state Markov model in which the transition probability to a good state is the same whether the system is currently in a good or bad state (i.e., $p_{gg} = p_{gb} = p$). Provide an intuitive explanation of your results, which should depend only on the single parameter p.

Problem 14.2: The left eigenvector associated with the eigenvalue λ of a matrix $\mathbf{M}$ is defined by $\vec{v}\,\mathbf{M} = \lambda \vec{v}$. (a) Explain why $\vec{v} = (1,1,\ldots,1)$ is always a left eigenvector of the transition probability matrix $\mathbf{M}$ defined in equation (14.16). (b) What is the eigenvalue associated with this eigenvector?

Problem 14.3: Consider the transition probability matrices

$$\mathbf{A} = \begin{pmatrix} 0 & 1 \\ 1 & 0 \end{pmatrix} \quad \text{and} \quad \mathbf{B} = \begin{pmatrix} 0 & 0 & 1 \\ 1 & 0 & 0 \\ 0 & 1 & 0 \end{pmatrix}.$$

(a) Show that the Markov chains corresponding to $\mathbf{A}$ and $\mathbf{B}$ are periodic by considering what happens over time if the system starts out in state 0, $P(X_0 = 0) = 1$. (b) Find the eigenvalues of $\mathbf{A}$ and $\mathbf{B}$ and show that they both have more than one eigenvalue whose absolute value is one. (c) Using Appendix 3, show that $\mathbf{A}$ and $\mathbf{B}$ are irreducible (every state can be reached from every other state) and imprimitive (given a particular starting point, there is never a point in time with a nonzero probability of being in all possible states). Even without any absorbing states, such irreducible and imprimitive matrices can have more than one eigenvalue whose magnitude is one (result (2) of the Perron-Frobenius theorem; Appendix 3).

Problem 14.4: Consider a forest patch, which can exist in one of three states: unoccupied (state 1), occupied by an early successional forest (state 2), or occupied by a late successional forest (state 3). Assume that a patch has an annual probability of forest loss (due to fire, disease, clear-cutting, etc.) of e_2 and e_3 for early and late successional forests, respectively. Subsequently, assume that empty patches are converted into early successional forests with probability b_1 and that early successional forests that have not been deforested are converted into late successional forests with probability b_2 per year. The transition probability matrix for the forest patch is then

$$\mathbf{M} = \begin{pmatrix} 1 - b_1 & e_2 & e_3 \\ b_1 & (1 - e_2)(1 - b_2) & 0 \\ 0 & (1 - e_2)\, b_2 & 1 - e_3 \end{pmatrix}, \tag{14.32}$$

which has one eigenvalue equal to one. (a) Find the right eigenvector $\vec{\pi}$ associated with $\lambda = 1$ and normalize this vector so that its elements sum to one. Describe what each element in the eigenvector tells us, and describe the effect of having relatively high ($e > b$) or low ($e < b$) extinction rates. (b) Consider what happens following some environmental change (global warming or the presence of grazing animals) that prevents the first successional event ($b_1 = 0$). As a consequence, the system now has an absorbing state (loss of the forest patch), as can be verified by the fact that $\mathbf{M}$ now has a column with zeros everywhere other than on the diagonal. Calculate the mean time until forest loss, starting with a first successional forest or from a second successional forest. (c) What are your answers to part (b) when the extinction rate is the same for both early and late successional forests ($e_1 = e_2 = e$)? Does this answer make sense based on your knowledge of probability theory (see Table P3.2)?

Problem 14.5: Consider a patch within a forest subject to slash-and-burn agriculture. Assume that the vegetation that grows best on the patch depends on whether the patch was occupied by an early or late successional forest before being cleared. For example, corn (state 1) might grow better when an early successional forest (state 2) is cleared, but beans (state 4) might grow better after a late successional forest (state 3) is cleared. Assume that the annual probability that a patch is cleared is e_2 and e_3 for early and late successional forests, that early successional forests become late successional forests with probability b_2, and that the patch remains agricultural land once cleared. The transition probability matrix then becomes:

$$\mathbf{M} = \begin{pmatrix} 1 & e_2 & 0 & 0 \\ 0 & (1-e_2)(1-b_2) & 0 & 0 \\ 0 & (1-e_2)b_2 & 1-e_3 & 0 \\ 0 & 0 & e_3 & 1 \end{pmatrix}.$$

(a) What are the four eigenvalues of this matrix? (Use Rule P2.28.) (b) What are the absorbing states of $\mathbf{M}$? (c) Calculate the probability that the patch ultimately becomes a corn field (state 1) starting from either states 2 or 3 (early or late successional forests). (d) Calculate the probability that the patch ultimately becomes a bean field (state 4) starting from either states 2 or 3. (e) What should the sum of your answers to (c) and (d) equal? (f) Calculate the mean time until clearing (i.e., reaching any absorbing state) starting from either states 2 or 3.

Problem 14.6: Calculate the characteristic polynomial, Det($\mathbf{M}-\lambda\mathbf{I}$), for matrix (14.26) to show that $\mathbf{M}$ has two eigenvalues equal to one. [Hint: Use Rule P2.22.]

Problem 14.7: (a) Calculate the probability of absorption into state 5 for the transition probability matrix (14.13) using equation (14.29) and the restriction that $\rho_{5,1} = 0$ and $\rho_{5,5} = 1$. (b) Confirm that your answer gives (14.27b) when $p_{12} = p_{54} \equiv \alpha$, $p_{32} = p_{34} \equiv \beta$, and $p_{23} = p_{43} = \gamma$.

Problem 14.8: Using the birth-death model of population growth (13.2) and the results of Box 14.1, calculate the probability ρ_{m,n_0} that a population at risk of extinction reaches a minimum viable population size, m (at which point the population is no longer monitored) rather than going extinct. Assume that the probability of a birth or a death is given by $b\,i$ or $d\,i$, respectively, where i is the current number of individuals within the population. Check that your answer equals $1 - \rho_{0,n_0}$, where ρ_{0,n_0} is given by (14.56).

Problem 14.9: Using the transition probabilities given by the birth-death model (13.2) and equation (14.1.7), show that a population will go extinct with probability $\rho_{0,\,n_0} = (d/b)^{n_0}$ even though the birth rate exceeds the death rate, when the initial population size is n_0 and the population has no upper limit in size. [Hint: Use rules A1.19 and A1.20.]

Problem 14.10: For the Moran model of allele frequency change, use (14.1.6) and the transition probabilities given in (13.9) to show that the probability that a neutral allele is lost is $\rho_{0,n_0} = 1 - n_0/N$, where n_0 is the initial number of copies.

Problem 14.11: For the Moran model of allele frequency change, use (14.1.6) and the transition probabilities given by (13.10) to (a) confirm that $d_i/b_i = W_a/W_A$, (b) derive equation (14.58a) for the probability that allele A is eventually lost from a population, (c) find the limit toward which the fixation probability (14.58b) tends as the population size increases, assuming that A is more fit than a, (d) take the Taylor series of your answer to part (c) to linear order in the selection coefficient, s, where $W_A = (1 + s)W_a$. Does your answer to (d) agree with Figure 14.7 for populations of large size? [Hint: Use Rule A1.19.]

Problem 14.12: Suppose that a new strain of avian influenza has been found in the human population. You are asked to make predictions about a potential pandemic, and so you decide to model the spread of this new strain using a birth-death model in discrete time. You thus choose the time steps to be small enough so that, in each time step, the number of infections can either increase by one, decrease by one, or remain the same. The probability that the number of infections increases by one is $\beta\, S\, I$, where I is the current number of infections, S is the number of susceptible people (assumed to be constant), and β is the transmission rate. The probability that the number of infections decreases by one is $(c + d)\, I$, where c and d are the rates of recovery and death, respectively. There are two outcomes of interest: either the avian strain dies out, or the number of infections reaches m, at which point the public health infrastructure is overwhelmed and a global pandemic is inevitable. (a) Write down the transition probabilities for the model that correspond to equations (14.50). (b) Suppose that $\beta\, S > c + d$ so that, in a deterministic model, the number of infections would increase. What is the probability that a global pandemic will occur, given that we start with one initial infection? (c) On average, how long will we have to wait before the fate of the outbreak is certain (i.e., before the disease either dies out or m individuals are infected)?

Problem 14.13: For a model with k absorbing states, Recipe 14.1 suggests choosing right eigenvectors that correspond to the stationary distributions for each absorbing state. That is, for the jth absorbing state, we choose the eigenvector, $\vec{\pi}_j$, whose elements are all zero except for the jth element, which equals one. Furthermore, Recipe 14.2 suggests choosing left eigenvectors such that the jth element of $\overline{\rho}_j$ is one and the elements whose index corresponds to any absorbing state other than j are zero (the remaining elements fall between 0 and 1). Show that, if we place the right eigenvectors in the first k columns of $\mathbf{A}$ and the left eigenvectors in the first k rows of $\mathbf{A}^{-1}$, that multiplying these rows and columns together is consistent with $\mathbf{A}^{-1}\mathbf{A} = \mathbf{I}$. Consequently, these choices of eigenvectors can be used in the general solution (14.21).

Problem 14.14: Consider the birth-death model (14.50) with an absorbing state at zero and at m. In this case, $b_0 = 0$ and $d_m = 0$. (a) Confirm that $\vec{\pi}_0$, where the first element is one and all remaining elements are zero, represents a stationary distribution of the model. (b) Confirm that $\vec{\pi}_m$, where the last element is one and all remaining elements are zero, represents a stationary distribution. (c) More generally, show that any vector whose first and last elements sum to one and whose remaining elements are zero also represents a stationary distribution. While the vectors in (c) represent an infinite number of possible choices for the right eigenvectors, the vectors in (a) and (b) are easier to interpret (Recipe 14.1).

Further Reading

For more information on the mathematical underpinnings of stochastic models, see:

- Karlin, S., and H. M. Taylor. 1981. *A Second Course in Stochastic Processes*. Academic Press, New York.
- Allen, L.J.S. 2003. *An Introduction to Stochastic Processes with Applications to Biology*. Pearson/Prentice-Hall, Upper Saddle River, N.J.

For more information on stochastic models in evolution and ecology, see:

- Ewens, W. J. 1979. *Mathematical Population Genetics*. Springer-Verlag, Berlin.
- Nisbet, R. M., and W. S. C. Gurney. 1982. *Modelling Fluctuating Populations*. Wiley, New York.
- Lande, R., and B.-E. Sæther. 2003. *Stochastic Population Dynamics in Ecology and Conservation*. Oxford University Press, Oxford.

References

Barton, N. H. 1993. The probability of fixation of a favoured allele in a subdivided population. *Genet. Res. Cambridge* 62:149–158.

Barton, N. H. 1995. Linkage and the limits to natural selection. *Genetics* 140:821–841.

Charlesworth, B. 1994. *Evolution in Age-Structured Populations*. Cambridge University Press, Cambridge.

Ewens, W. J. 1979. *Mathematical Population Genetics*. Springer-Verlag, Berlin.

Feldman, M. W. 1966. On the offspring number distribution in a genetic population. *J. Appl. Probab.* 3:129–141.

Haldane, J. B. S. 1927. A mathematical theory of natural and artificial selection, Part V: Selection and mutation. *Proc. Cambridge Philos. Soc.* 23:838–844.

Karlin, S., and J. McGregor. 1964. Direct product branching processes and related Markov chains. *Proc. Natl. Acad. Sci. U. S. A.* 51:598–602.

Mode, C. J. 1971. *Multitype Branching Processes Theory and Applications*. Elsevier, New York.

CHAPTER 15

Analyzing Continuous Stochastic Models—Diffusion in Time and Space

Chapter Goals:

- To develop stochastic models with continuous state spaces based on diffusion
- To obtain general solutions, stationary distributions, probabilities of absorption, and waiting times for diffusion models

Chapter Concepts:

- Diffusion model
- Drift coefficient
- Diffusion coefficient
- Transition probability density function
- Brownian motion

15.1 Introduction

In this chapter we present techniques for constructing and analyzing stochastic models with continuous state spaces. All of the models presented in Chapters 13 and 14 assume that the state of the system takes on discrete values. Some processes are more naturally modeled by allowing the state space to be continuous. For example, suppose we want to model the movement of an individual within its home range. There will likely be a great deal of stochasticity in movement, and at each point in time, the individual might be in any of a continuum of possible locations. Consequently, it would be appropriate to use a stochastic model with a continuous state space, representing the different possible spatial locations.

Aside from the fact that some processes are more naturally described by a continuous state space, some discrete models that are difficult to analyze can be well approximated using a stochastic model with a continuous state space. For example, the Wright-Fisher model for allele frequency change in a finite population (section 13.4) is difficult to analyze exactly, but it can be well approximated using a stochastic model with a continuous state space representing the allele frequency. Thus, techniques for analyzing models with a continuous state space often provide an alternative route for the approximate analysis of discrete-state models.

In section 15.2, we begin by describing different ways in which stochastic models with a continuous state space can be constructed. We focus on models referred to as *diffusion models*, which are based on the idea that variables disperse (or *diffuse*) away from their original location or state. In section 15.3, we then develop techniques allowing us to obtain general solutions, stationary distributions, probabilities of absorption, and waiting times for many diffusion models. Finally, in section 15.4 we illustrate how these mathematical techniques can be used to model the deterministic dynamics of populations in spatial settings.

15.2 Constructing Diffusion Models

In this section we use examples to illustrate different ways in which stochastic models with a continuous state space can be constructed. Interestingly, these different examples all generate the same type of equation for modeling diffusion.

The first example begins with a model having a discrete state space and then "takes the limit" as the state space becomes continuous. The second example proceeds directly to the construction of a model assuming a continuous state space. The third example considers how a continuous trait (e.g., height, growth rate) might evolve over time, using a diffusion model to track the long-term dynamics of a trait when mutations are limiting. Finally, the fourth example illustrates how a model having a discrete state space can be approximated using a continuous-state-space model. As with stochastic models having a discrete state space, models with a continuous state space can be formulated in either discrete or continuous time. It is often more natural to work with continuous time, however, and we do this in all of the examples below.

15.2.1 Modeling Individual Movement in a Continuous Spatial Habitat (Part 1)

Let us suppose that we are studying the movement of a species of stream-dwelling fish. Our ultimate goal is to understand how decisions that individuals make in terms of their movement either up- or downstream translate into patterns of population abundance at different locations. To begin, we will construct a stochastic model for the location $x(t)$ of a single fish in discrete time (Figure 15.1) using the techniques from Chapter 14. We call $X(t)$ the *state variable*, as it describes the state of the system at time t. Suppose that the stream can be divided into discrete locations that are equally spaced at a distance of Δx and that time proceeds in steps of length Δt.

As in section 13.3, we first specify the transition probabilities between the different spatial locations within a single time step. Let us suppose that the time step Δt is short enough that the fish can only move between two neighboring locations within a single time step (Figure 15.1). By analogy with the birth-death model (14.50), we can write the transition probabilities p_{yx} from spatial location x to y in one time step as

$$p_{yx} = \begin{cases} u & \text{for } y = x + \Delta x & \text{(upstream by one step)}, \\ d & \text{for } y = x - \Delta x & \text{(downstream by one step)}, \\ 1 - u - d & \text{for } y = x & \text{(no movement)}, \\ 0 & \text{for } y \neq x - \Delta x, x, x + \Delta x & \text{(other movements)}, \end{cases} \tag{15.1}$$

where u and d are the probabilities of moving up- and downstream, respectively. Let us label the site that is farthest downstream as 0, and successive sites by

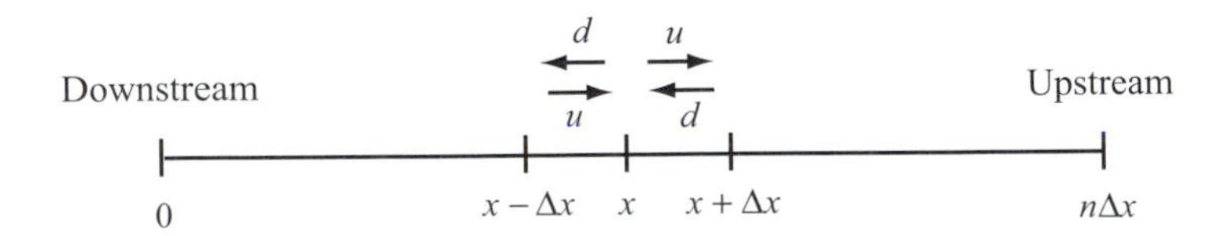

Figure 15.1: A schematic diagram of fish movement within a stream. The diffusion model is obtained by shrinking the distance, Δx, between sites to zero. The probabilities of moving upstream or downstream in a time step, Δt, are given by u and d, respectively.

their distance from this initial site as Δx, $2\,\Delta x, \ldots, n\,\Delta x$, where n is the total number of sites in addition to the first one. The probability that the fish is in each of the $n + 1$ sites at time t is given by the vector

$$\vec{x}(t) = \begin{pmatrix} P(X(t) = 0) \\ P(X(t) = \Delta x) \\ P(X(t) = 2\,\Delta x) \\ \vdots \\ P(X(t) = n\,\Delta x) \end{pmatrix}. \tag{15.2}$$

Changes over time are then described by

$$\vec{x}(t + \Delta t) = \mathbf{M}\,\vec{x}(t) \tag{15.3}$$

(see equation (14.15)), where $\mathbf{M}$ is the transition probability matrix whose elements are given by (15.1).

We can proceed to analyze model (15.3) using the techniques for birth-death processes in Chapter 14. For example, if we stop our observations once the fish leaves the region defined by the $n + 1$ sites (i.e., $d = 0$ when $x = 0$ and $u = 0$ when $x = n$), then we can use Box 14.1 to calculate the probability that the fish leaves the region upstream rather than downstream (Problem 15.1). Similarly, we could use Box 14.2 to calculate the expected time until the fish leaves the region.

Instead, we will use this model as a starting place for deriving the analogous model with a continuous state space. Carrying out the matrix multiplication in (15.3), the probability of being in a particular site x at time $t + \Delta t$ is given by

$$\begin{aligned} P(X(t + \Delta t) = x) = P(X(t) = x - \Delta x)\,u + P(X(t) = x)\,(1 - u - d) \\ + P(X(t) = x + \Delta x)\,d. \end{aligned} \tag{15.4a}$$

To simplify the notation, define $f(x,t)$ as the probability that the fish is at location x at time t; that is, $f(x,t) \equiv P(X(t) = x)$. The above equation can then be written as

$$f(x, t + \Delta t) = f(x - \Delta x,t)\,u + f(x,t)\,(1 - u - d) + f(x + \Delta x,t)\,d. \tag{15.4b}$$

The key to deriving the continuous state space model is to consider what happens as the distance between spatial locations, Δx, gets smaller. As we do this, we must also shrink the time step, Δt. Otherwise the assumption that the fish can move only between neighboring locations in one time step would become unrealistic. Thus we want to consider the limit as both Δx and Δt go to zero. We start by using the Taylor series (Recipe P1.2) to rewrite (15.4b) as a polynomial function of Δx:

$$\begin{aligned} f(x, t + \Delta t) = {} & \left(f(x,t) - \frac{\partial f(x,t)}{\partial x}\Delta x + \frac{\partial^2 f(x,t)}{\partial x^2}\frac{1}{2}\Delta x^2 + O(\Delta x^3)\right)u \\ & + f(x,t)(1 - u - d) \\ & + \left(f(x, t) + \frac{\partial f(x,t)}{\partial x}\Delta x + \frac{\partial^2 f(x,t)}{\partial x^2}\frac{1}{2}\Delta x^2 + O(\Delta x^3)\right)d. \end{aligned} \tag{15.5a}$$

Gathering together similar terms involving Δx, we get

$$f(x,t+\Delta t)-f(x,t) = -\frac{\partial f(x,t)}{\partial x}(u-d)\Delta x+\frac{\partial^2 f(x,t)}{\partial x^2}\frac{1}{2}(u+d)\Delta x^2+O(\Delta x^3). \tag{15.5b}$$

Equation (15.5b) gives the change in location within a time unit, Δt. If we divide by Δt, we get the rate of change in location:

$$\frac{f(x,t+\Delta t)-f(x,t)}{\Delta t} = -\frac{\partial f(x,t)}{\partial x}(u-d)\frac{\Delta x}{\Delta t}+\frac{\partial^2 f(x,t)}{\partial x^2}\frac{1}{2}(u+d)\frac{\Delta x^2}{\Delta t}+\frac{O(\Delta x^3)}{\Delta t}. \tag{15.5c}$$

We can now take the limit of (15.5c) as both Δx and Δt go to zero. By the definition of a derivative (Appendix 2), the limit of the left-hand side equals $\partial f(x,t)/\partial t$, which is a partial derivative because f is a function that depends on both location x and time t. To take the limit of the right-hand side, however, we must make some assumption about the relative rates at which Δx and Δt go to zero. By assumption, diffusion models choose these relative rates such that changes to the system depend on the first two leading terms in (15.5c), $\partial f(x,t)/\partial x$ and $\partial^2 f(x,t)/\partial x^2$, but not on higher-order terms. Specifically, diffusion models assume that

A *diffusion model* describes changes over time in the state of a system when the state can take on a continuum of values and can change by only a very small amount in a small amount of time.

$$\begin{aligned}\lim_{\Delta x,\Delta t\to 0}(u-d)\frac{\Delta x}{\Delta t} &= \mu,\\ \lim_{\Delta x,\Delta t\to 0}(u+d)\frac{\Delta x^2}{\Delta t} &= \sigma^2,\\ \lim_{\Delta x,\Delta t\to 0}\frac{O(\Delta x^3)}{\Delta t} &= 0,\end{aligned} \tag{15.6}$$

where μ and σ^2 are finite numbers (with $\sigma^2 > 0$). Using these assumptions, the probability density function $f(x,t)$ that the fish is in location x at time t satisfies

$$\frac{\partial f(x,t)}{\partial t} = -\mu\frac{\partial f(x,t)}{\partial x}+\frac{\sigma^2}{2}\frac{\partial^2 f(x,t)}{\partial x^2}. \tag{15.7}$$

The *drift coefficient* in a diffusion model describes the expected (or mean) amount of movement per unit time.

The *diffusion coefficient* describes variance in the movement of a system per unit time.

The function $f(x,t)$ now represents a continuous probability distribution specifying the probability that the fish is in an infinitesimal region of the continuous state space located at point x (see section P3.4). How this spatial probability distribution changes over time is described by the *partial differential equation* (15.7). Equation (15.7) can be analyzed to make predictions about the location of the fish using the techniques that we will introduce in section 15.3.

Before proceeding to the next example, it is worth making a couple of general remarks about the terms in (15.7). The quantities μ and σ^2 are referred to as the infinitesimal *mean* and *variance* (or the *drift* and *diffusion* coefficients) of the stochastic process. To appreciate where this terminology comes from, we calculate the mean or expected distance moved, $E[\Delta X]$, by the fish in a single time step of the discrete model (15.1). The fish moves a distance Δx upstream with probability u, and distance $-\Delta x$ downstream with probability d. It remains in the same location with probability $1 - u - d$. Therefore, from Definition P3.2, the expected distance moved is $E[\Delta X] = \Delta x\, u - \Delta x\, d + 0\,(1 - u - d) = (u - d)\Delta x$. Consequently $(u - d)\, \Delta x/\Delta t$ is the expected movement *per unit time* in the discrete model. In the limit of small Δx and Δt, this value becomes μ according to (15.6), which gives the expected movement per unit time in the continuous model.

Next, let's calculate the variance in the distance moved in a single time step of the discrete model (15.1): $\text{Var}[\Delta X] = E[(\Delta X)^2] - E[\Delta X]^2$ (equation (P3.2)). The average displacement squared is $E[(\Delta x)^2] = (\Delta x)^2\, u + (-\Delta x)^2\, d + 0\,(1 - u - d) = (u + d)\,\Delta x^2$, and therefore the variance in the movement is $(u + d)\,\Delta x^2 - ((u - d)\Delta x)^2$. If we now divide this by Δt, we obtain the variance in movement per unit time: $(u + d)\,(\Delta x^2)/(\Delta t) - (u - d)\,(\Delta x/\Delta t)\Delta x$. Taking the limit as both Δx and Δt go to zero and using conditions (15.6) then gives $\lim_{\Delta x, \Delta t \to 0} ((u + d)\,(\Delta x^2)/(\Delta t) - (u - d)\,(\Delta x/\Delta t)\Delta x) = \sigma^2 - \mu\, 0$ or simply σ^2. This reveals that σ^2 represents the variance in movement per unit time in the continuous model (15.7).

Equation (15.7) is known as a *diffusion equation with drift* and also as the *forward Kolmogorov equation* (or, sometimes, an *advection-diffusion equation*). The term diffusion refers specifically to the random, undirected movement described by the variance in movement per unit time, σ^2. If σ^2 were zero, then the system would always move by an amount equal to the expected rate of change, μ, and the model would be deterministic. Random, undirected movement is also called *Brownian motion*, named after the Scottish botanist Robert Brown (1773–1858), who studied fertilization in the flower *Clarkia pulchella*. He noticed that pollen grains suspended in water exhibited erratic movements over time. Similar movements are exhibited by inorganic particles and were described mathematically by Albert Einstein in 1905 as a result of collisions between the observed particles and unobserved liquid (or gas) molecules.

Later in this chapter we will use equations such as (15.7) to model allele frequency change in finite populations. This sometimes causes confusion due to an unfortunate conflict in terminology. In genetics, the random fluctuations in allele frequency due to chance events are described as "random genetic drift." In mathematics, however, "drift" refers to the tendency μ to move in a particular direction, in the sense of drifting down a river. As we shall see, in neutral models of random genetic drift, there is no (mathematical) drift, i.e., the allele frequencies do not systematically rise or systematically fall over time and $\mu = 0$. Thus, "random genetic drift" would be better called "random genetic diffusion," but the term is now too widely used to change.

15.2.2 Modeling Individual Movement in a Continuous Spatial Habitat (Part 2)

The example above illustrates an important conceptual connection between a discrete-time stochastic model in discrete state space (the birth-death model) and a continuous-time stochastic model in continuous state space (diffusion with drift). It is often the case, however, that stochastic models in continuous state space are constructed directly rather than deriving them from models with a discrete state space. The next example reexamines the model of fish movement using this approach.

Let us work directly with the continuous probability distribution that describes the probability of the fish being in an infinitesimally small spatial region at location x at time t. Over any period of time, we are interested in knowing the *transition probability density function*, $\phi(x,t \mid x_0,t_0)$, that specifies the probability density of moving to location x at time t, given that the fish is at location x_0 at time t_0.

The *transition probability density function* specifies the probability density of the system moving between any two values in any given amount of time.

The next step is to describe how this probability density function changes over time. For generality we allow the movement of the fish to depend on its current location (e.g., perhaps it moves downstream most often when it is very far upstream and vice versa), but we do not allow the movement behavior to depend on time (that is, we focus on *time-homogeneous* models). Although the transition probability density function could, in principle, allow any amount of change in a short period of time, it is difficult to make progress with this level of generality. Consequently, we focus on models for which only small amounts of change can occur in small amounts of time. Under these conditions, the function $\phi(x,t \mid x_0,t_0)$ can be shown to satisfy the *forward Kolmogorov equation*:

$$\frac{\partial \phi(x,t \mid x_0,t_0)}{\partial t} = -\frac{\partial(\mu(x)\,\phi(x,t \mid x_0,t_0))}{\partial x} + \frac{1}{2}\frac{\partial^2(\sigma^2(x)\,\phi(x,t \mid x_0,t_0))}{\partial x^2}. \quad (15.8)$$

A derivation of equation (15.8) is provided by Allen (2003, pp. 376–377) and by Ewens (1979, pp. 116–117).

Equation (15.8) is a partial differential equation having a similar form to that of equation (15.7). In fact, equation (15.8) is again a model of diffusion with drift, except that we now allow the drift and diffusion coefficients to depend on x (spatial location in this example). In particular, the drift coefficient $\mu(x)$ is the expected distance moved by the fish per unit time for the continuous stochastic process, given that it is currently at location x, defined as

$$\mu(x) = \lim_{\Delta t \to 0} \frac{E[X(t+\Delta t) - X(t) \mid X(t) = x]}{\Delta t} \quad (15.9a)$$

Similarly, the diffusion coefficient $\sigma^2(x)$ is the variance in the distance moved per unit time, given that it is currently at location x, defined as

$$\sigma^2(x) = \lim_{\Delta t \to 0} \frac{E[(X(t+\Delta t) - X(t))^2 \mid X(t) = x]}{\Delta t}. \quad (15.9b)$$

Technically, to get a variance we must subtract the mean squared from (15.9b), but in the limit this term is zero:

$$\lim_{\Delta t \to 0} \frac{E[X(t+\Delta t) - X(t) \mid X(t)]^2}{\Delta t} = \lim_{\Delta t \to 0} \frac{(\mu(x)\Delta t)^2}{\Delta t} = 0. \quad (15.10)$$

There is, however, a difference in notation between (15.7) and (15.8), which stems from the fact that $\phi(x,t \mid x_0,t_0)$ specifies explicitly how the probability density at time t depends on the state of the system at the previous time t_0. In contrast, we used $f(x,t)$ in (15.7) to describe the probability density at time t without specifying the initial conditions. Besides being more explicit, the notation $\phi(x,t|x_0,t_0)$ allows us to derive a second partial differential equation that is extremely useful for determining the behavior of the stochastic process (Box 15.1). This equation is referred to as the *backward Kolmogorov equation*:

$$\frac{\partial\phi(x,t \mid x_0,t_0)}{\partial t_0} = -\mu(x_0)\frac{\partial\phi(x,t \mid x_0,t_0)}{\partial x_0} - \frac{1}{2}\sigma^2(x_0)\frac{\partial^2\phi(x,t \mid x_0,t_0)}{\partial x_0^2}. \quad (15.11)$$

Comparing (15.8) and (15.11), we see that the derivatives of ϕ are now taken with respect to the initial variables, x_0 and t_0, rather than with respect to x and t as in the forward equation. In section 15.3, we will use both the forward and backward Kolmogorov equations to derive a number of key results from diffusion models, but first we present two more examples illustrating the breadth of problems that can be described by diffusion models.

15.2.3 Modeling the Long-Term Evolutionary Dynamics of a Trait when Mutations Are Limiting

In Chapter 12 we introduced evolutionary invasion analysis. At the heart of this technique is an assumption that evolution is mutation limited. Specifically, it is assumed that there is a continuum of different possible alleles, coding for a continuum of different possible trait values that an individual might express. The key assumption is that mutations among these alleles are rare, so that the population reaches an equilibrium while it contains only a single allele. Then a new mutation arises and either replaces the former resident allele or dies out. If it replaces the resident allele, then a new trait value is attained in the population.

In Chapter 12 we inferred the direction of evolution under this process by determining which alleles can invade which populations. But to make predictions about the trait values that are likely to be observed over time, we must construct an explicit model for the stochastic appearance and fixation of mutations altering the trait of interest. A modeling framework referred to as adaptive dynamics has been developed to do exactly this (Dieckmann and Law 1996; Box 12.5), and here we develop an analogous approach based on the diffusion equation.

We begin by defining $\phi(x,t|x_0,t_0)$ as the continuous probability density function describing the probability that the population is in state x at time t (i.e., the allele coding for trait value x is resident in the population at time t) given the initial trait value x_0 at time t_0. A diffusion model can be used to describe the dynamics of $\phi(x,t \mid x_0,t_0)$ if we assume that the resident trait value does not change very much in a small amount of time. This is analogous to the assumption in sections 15.2.1 and 15.2.2 that a fish makes only small movements in small amounts of time. Assuming that large evolutionary jumps do not occur, the probability distribution $\phi(x,t \mid x_0,t_0)$ will satisfy the diffusion equations (15.8) and (15.11). The diffusion coefficient $\sigma^2(x)$ now represents the random, undirected evolutionary change in trait space (owing to mutation),

Box 15.1: Deriving the Backward Kolmogorov Equation

The Kolmogorov equations (15.8) and (15.11) are differential equations that are satisfied by the probability distribution $\phi(x,t \mid x_0,t_0)$ describing the probability density that the system is at position x at time t, given that it was at position x_0 at time t_0. These equations have been derived in a number of ways, but here we focus only on the backward equation (15.11). Throughout this box, we assume without proof that the probability distribution ϕ is a continuous function whose derivates (to arbitrary order) are finite. (Within this seemingly innocuous statement is hidden a great deal of complexity. Mathematically rigorous derivations of a diffusion equation must demonstrate that this assumption holds.)

Before we begin, we first establish an important equation that $\phi(x,t|x_0,t_0)$ must satisfy. Consider our example of fish movement, and imagine a fish that starts in location x_0 at time t_0 and ends up at location x at time t. We can obtain an expression for $\phi(x,t \mid x_0,t_0)$ by considering all of the potential locations, x_1, of the fish at some intermediate point in time, t_1. In particular, $\phi(x,t \mid x_0,t_0)$ can be expressed as the probability of moving from x_0 to x_1 between times t_0 and t_1, and then moving from x_1 to x between times t_1 and t, evaluated over all possible intermediate states, x_1. We can write this logical statement mathematically as

$$\phi(x,t \mid x_0,t_0) = \int \phi(x,t \mid x_1,t_1)\, \phi(x_1,t_1 \mid x_0,t_0)\, \mathrm{d}x_1, \tag{15.1.1}$$

where this integral (and those that follow) is evaluated over the range of possible values of the random variable. Equation (15.1.1) is known as the Chapman-Kolmogorov equation, and it will come in handy below.

Let us now derive the backward equation (15.11). Consider an intermediate point in time, t_1, between the present time t and the initial time point t_0 but one that is very close to t_0. In this case we can write $t_1 = t_0 + \Delta t$, where Δt is very small. The crux of the derivation revolves around the assumption that the change in the random variable $X(t)$ over the short time period Δt is small enough that we can use a Taylor series with respect to this change (Primer 1). In particular, after a small amount of time Δt elapses, the value of X is assumed to change by a small amount Δx. Thus we can write the value of X at time t_1 as $x_1 = x_0 + \Delta x$.

Our goal is to derive an expression for the derivative, $\partial\phi/\partial t_0$. We start by using the definition for the derivative from Appendix A2:

$$\frac{\partial\phi(x,t \mid x_0,t_0)}{\partial t_0} \equiv \lim_{\Delta t\to 0}\frac{\phi(x,t \mid x_0,t_0+\Delta t) - \phi(x,t \mid x_0,t_0)}{\Delta t} \tag{15.1.2}$$

To obtain the desired expression, we must obtain an expression for the ratio $(\phi(x,t \mid x_0,t_0 + \Delta t) - \phi(x,t \mid x_0,t_0))/\Delta t$. First, we can replace $\phi(x,t \mid x_0,t_0)$ in this ratio with the Chapman-Kolmogorov equation (15.1.1). In addition, we know that $\int\phi(x_1,t_1 \mid x_0,t_0)\, \mathrm{d}x_1 = 1$ because the fish must be located somewhere at time t_1. Consequently, we are free to replace $\phi(x,t \mid x_0,t_0 + \Delta t)$ in (15.1.2) with $\phi(x,t \mid x_0,t_0 + \Delta t) \int\phi(x_1,t_1 \mid x_0,t_0)\, \mathrm{d}x_1$. Making these replacements we have

$$\frac{\phi(x,t \mid x_0,t_0+\Delta t)\displaystyle\int \phi(x_1,t_1 \mid x_0,t_0)\,\mathrm{d}x_1 - \displaystyle\int \phi(x,t \mid x_1,t_1)\phi(x_1,t_1 \mid x_0,t_0)\,\mathrm{d}x_1}{\Delta t}. \tag{15.1.3a}$$

(continued)

Box 15.1 *(continued)*

By factoring and using $t_1 = t_0 + \Delta t$, we get

$$\frac{\displaystyle\int \{\phi(x,t|x_0,t_1) - \phi(x,t|x_1,t_1)\}\,\phi(x_1,t_1|x_0,t_0)\,dx_1}{\Delta t}. \tag{15.1.3b}$$

Next, we use the fact that $x_1 = x_0 + \Delta x$ and take the Taylor series (Recipe P1.2) of the term within the curly braces with respect to x, near the point $x_1 = x_0$ (i.e., $\Delta x = 0$):

$$\frac{\displaystyle\int \left\{ -\Delta x \frac{\partial \phi(x,t|x_0,t_1)}{\partial x_0} - \frac{\Delta x^2}{2}\frac{\partial^2 \phi(x,t|x_0,t_1)}{\partial x_0^2} - O(\Delta x^3) \right\} \phi(x_1,t_1|x_0,t_0)\,dx_1}{\Delta t} \tag{15.1.4a}$$

At this point, we drop the higher-order terms $O(\Delta x^3)$. Doing so is equivalent to assuming that process does not make large "jumps" in small time intervals (see p. 327 in Allen 2003). Replacing Δx, with $x_1 - x_0$ and factoring out terms that do not depend on x_1 leaves

$$-\frac{\dfrac{\partial \phi(x,t|x_0,t_1)}{\partial x_0}\displaystyle\int (x_1 - x_0)\,\phi(x_1,t_1|x_0,t_0)\,dx_1}{\Delta t} - \frac{1}{2}\frac{\dfrac{\partial^2 \phi(x,t|x_0,t_1)}{\partial x_0^2}\displaystyle\int (x_1 - x_0)^2\,\phi(x_1,t_1|x_0,t_0)\,dx_1}{\Delta t}. \tag{15.1.4b}$$

Finally, we replace t_1 with $t_0 + \Delta t$ and take the limit as $\Delta t \to 0$, allowing us to write (15.1.2) as

$$\frac{\partial \phi(x,t\mid x_0,t_0)}{\partial t_0} = -\mu(x_0)\frac{\partial(x,t\mid x_0,t_0)}{\partial x_0} - \frac{1}{2}\sigma^2(x_0)\frac{\partial^2\phi(x,t\mid x_0,t_0)}{\partial x_0^2}, \tag{15.1.5}$$

where

$$\mu(x_0) = \lim_{\Delta t\to 0}\frac{\displaystyle\int (x_1 - x_0)\,\phi(x_1,t_0 + \Delta t\mid x_0,t_0)\,dx_1}{\Delta t},$$

$$\sigma^2(x_0) = \lim_{\Delta t\to 0}\frac{\displaystyle\int (x_1 - x_0)^2\,\phi(x_1,t_0 + \Delta t\mid x_0,t_0)\,dx_1}{\Delta t}. \tag{15.1.6}$$

Equations (15.1.6) can be more easily interpreted in terms of the expected rate of change. By definition, the expected value $E[g(X)]$ of a function $g(X)$ is given by multiplying $g(X)$ by the probability density function for the random variable X and integrating over all possible values of the random variable (see Table P3.1). Thus, the expected change in the random variable over a time step Δt, raised to the power k, is defined as

$$E[(X(t_0 + \Delta t) - X(t_0))^k \mid X(t_0) = x_0] = \int (x_1 - x_0)^k\,\phi(x_1,t_0 + \Delta t\mid x_0,t_0)dx_1. \tag{15.1.7}$$

(continued)

Box 15.1 *(continued)*

Consequently, the functions $\mu(x_0)$ and $\sigma^2(x_0)$ in (15.1.6) are the same drift and diffusion coefficients used in equations (15.9). Equation (15.1.5), or the equivalent (15.11), is referred to as the backward Kolmogorov equation because its derivatives depend on the original position and time, not on the current position and time as in the forward equation.

while the drift coefficient $\mu(x)$ represents directional evolutionary change in trait space (owing to natural selection).

We can derive expressions for the drift and diffusion coefficients of evolutionary change in trait space, $\mu(x)$ and $\sigma^2(x)$, based on the underlying evolutionary processes. From equations (15.1.6) of Box 15.1, these two quantities are

$$\mu(x) = \lim_{\Delta t \to 0} \frac{\int (x_m - x)\, \phi\, (x_m, t_0 + \Delta t \mid x, t_0)\, dx_m}{\Delta t},$$
$$\sigma^2(x) = \lim_{\Delta t \to 0} \frac{\int (x_m - x)^2 \phi(x_m, t_0 + \Delta t \mid x, t_0)\, dx_m}{\Delta t}, \tag{15.12}$$

where $\phi(x_m, t_0 + \Delta t \mid x, t_0)$ is the probability that the population moves from trait x to trait x_m during the time interval Δt. Next, we specify how the transitions $\phi(x_m, t_0 + \Delta t \mid x, t_0)$ are related to the underlying processes of mutation and natural selection.

The probability that a population moves to state x_m from state x in the time interval Δt equals the probability that a mutation of type x_m occurs in that time interval, which we denote by $M(x_m, x; \Delta t)$, times the probability that this new mutation ultimately replaces the resident allele (i.e., the probability of fixation), which we denote by $U(x_m, x)$ (the time scale is assumed to be so long that the fixation or loss of the new allele is nearly instantaneous). The above drift and diffusion coefficients can then be written as

$$\mu(x) = \lim_{\Delta t \to 0} \frac{\int (x_m - x)\, M\, (x_m, x; \Delta t)\, U\, (x_m, x)\, dx_m}{\Delta t},$$
$$\sigma^2(x) = \lim_{\Delta t \to 0} \frac{\int (x_m - x)^2 M\, (x_m, x; \Delta t)\, U\, (x_m, x)\, dx_m}{\Delta t}. \tag{15.13}$$

The diffusion equations (15.8) and (15.11), along with the drift and diffusion parameters (15.13), describe how the probability distribution for the resident trait value changes through time as a result of mutation and selection. This gives us information about the probable states of a population over time, which we shall explore further in section 15.3.2. The expected trait value over time can also be derived for this diffusion model using the forward Kolmogorov equation, as described in Supplementary Material 15.1.

15.2.4 Diffusion Models as Approximations to Discrete-State-Space Models

In some models, a biological phenomenon of interest will inherently have a discrete state space (e.g., the number of copies of a particular allele in a finite population, or the number of islands occupied by a plant species). Even when a discrete state space is more natural, many stochastic models can be analyzed by approximating the process using a continuous state space model such as the diffusion model with drift. In essence, the approximation treats both the state variable and time as continuous, distilling the stochastic process down to two quantities: the expected rate of change in the variable and the amount of variability around this expectation.

Let us take a relatively simple example to illustrate. The Moran model for allele frequency change in a finite haploid population was introduced in section 13.5 of Chapter 13 and analyzed in section 14.4 of Chapter 14. This is a stochastic model in which the state space is inherently discrete: the random variable $X(t)$ can take on any of the discrete values from zero to N. Suppose that there is no natural selection acting, in which case the expected change in $X(t)$ is zero regardless of its current value. Although the expected change is zero, the value of X will nevertheless change by chance. In section 13.5, we calculated the variance of the change in X per time unit as $2\,p\,(1-p)$, where $p = X/N$ is the frequency of the allele.

As a first step toward using a diffusion model to approximate the Moran process, we model the dynamics of the variable p, assuming that the drift and diffusion coefficients are given by $\mu(p) = 0$ and $\sigma^2(p) = \sigma^2(X/N) = 2\,p\,(1-p)/N^2$, where we have used the fact from Table P3.1 that $\sigma^2(c\,Y) = c^2\,\sigma^2(Y)$. (In Box 15.2, we describe how to derive these coefficients more formally; see Problem 15.2.) In this case, the forward diffusion equation (15.8) simplifies to

$$\frac{\partial \phi(p,t \mid p_0,t_0)}{\partial t} = \frac{1}{2}\,\frac{\partial^2\left(\dfrac{2p(1-p)}{N^2}\,\phi(p,t \mid p_0,t_0)\right)}{\partial p^2} \tag{15.14}$$

In Figure 15.2, we use *Mathematica* to solve the partial differential equation (15.14) numerically and then compare the results to exact iterations of the Moran model. Even though the population size consists of only ten haploid individuals, the diffusion approximation provides an amazingly accurate description of the change in allele frequency over time.

To some extent, the fit of the diffusion in this case is a bit lucky, as we have not formally derived the drift and diffusion coefficients nor have we shown that the diffusion approximation is reasonable for this model. Fortunately, there is a procedure for determining whether a diffusion approximation is valid for a discrete-state-space model (Box 15.2). Not all discrete processes can be

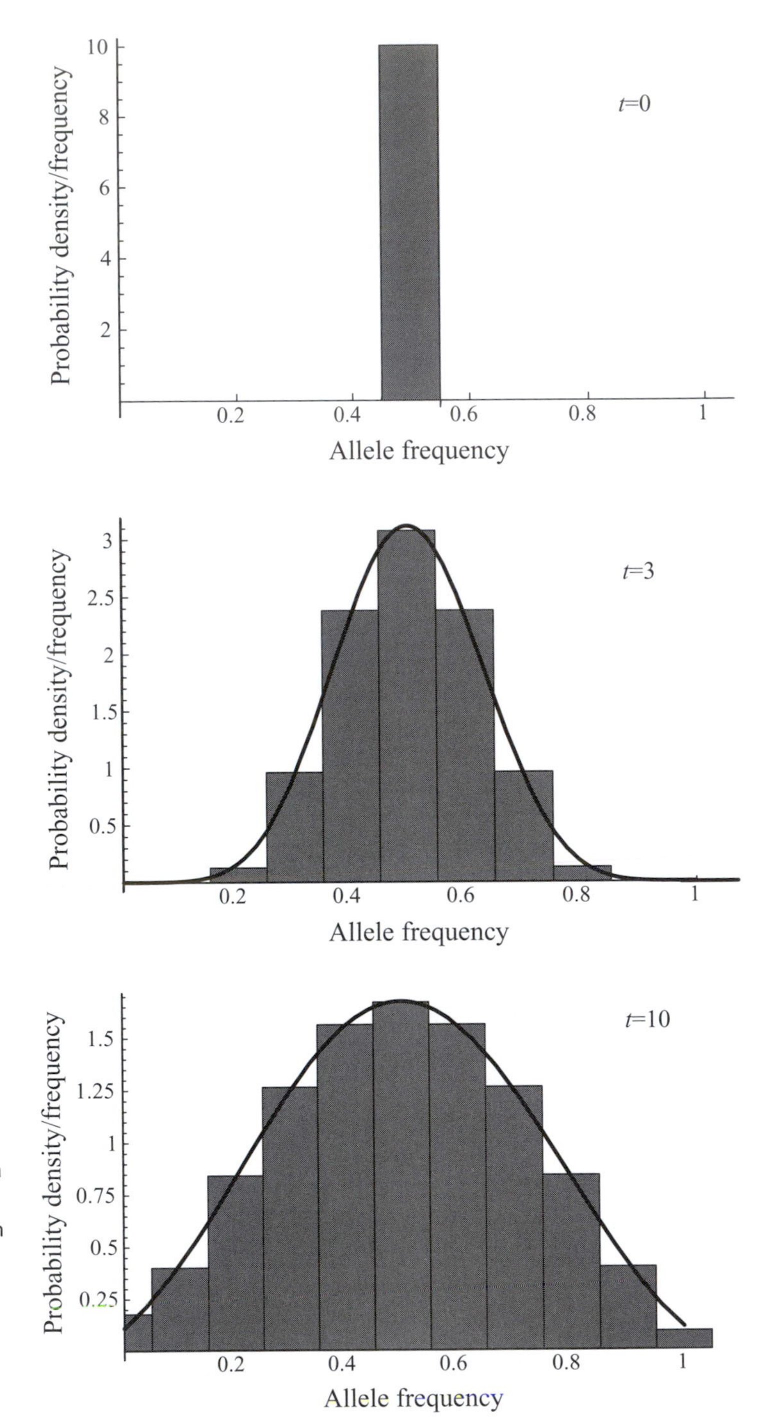

Figure 15.2: Allele frequency distributions in the Moran model. The distribution of allele frequencies is shown at three different points in time for a neutral allele in a finite haploid population. Bars illustrate the predicted frequency distribution from iterations of the exact matrix model using equation (13.9), and the curve illustrates the probability density obtained by numerically solving the partial differential equation (15.14) using *Mathematica*. Initial allele frequency is $p = 0.5$, and population size is $N = 10$.

adequately approximated in this way, but when they can, the techniques of Box 15.2 provide a way of choosing the drift and diffusion coefficients as well as the appropriate scales along which to measure the state variable and time. These techniques determine whether or not a diffusion model is likely to yield a reasonable approximation for a discrete-state-space model.

Box 15.2: Deriving Diffusion Approximations

In a diffusion analysis, the random variable and time are both treated as continuous variables, yet diffusion models are often used to approximate stochastic models with discrete variables and discrete time steps. In this box we illustrate how to go about deriving and justifying such an approximation. The key lies in choosing an appropriate transformation of the original model. Even if the original discrete-time, discrete-state-space model might not be well approximated by a diffusion model, perhaps some transformation of the model will be. If so, the diffusion approximation is not an approximation for the original model *per se* but rather, an approximation for the transformed model. Thus, to make predictions in terms of the original model's variables using a diffusion approximation, you must then transform the results of the diffusion model back into the original variables.

Consider a discrete-time discrete-variable stochastic model with transition probabilities p_{ji} determining the probability of moving from i to j in a single time step. We denote the random variable governed by this process as $Y(t)$. By summing up over all possible transitions that might occur in one time step, we can calculate the moments of the change in the variable Y (describing the mean change, the variance in this change, etc.) as

$$\begin{aligned} &\text{First moment} && E[Y(t+1) - Y(t) \mid Y(t) = i] = \sum_{j=0}^{m} (j-i) p_{ji}, \\ &\text{Second moment} && E[(Y(t+1) - Y(t))^2 \mid Y(t) = i] = \sum_{j=0}^{m} (j-i)^2 p_{ji}, \\ &\text{Third moment} && E[(Y(t+1) - Y(t))^3 \mid Y(t) = i] = \sum_{j=0}^{m} (j-i)^3 p_{ji}, \end{aligned} \tag{15.2.1}$$

where m is the maximum possible value of the random variable.

The next step is to transform the original state variable Y and the original time variable t into new state and time variables X and T, chosen such that, for large enough m, the third moment is essentially zero whereas the first moment remains finite, and the second moment remains finite and positive. A transformation of the state variable that often works is

$$X = \frac{Y}{m^{\beta}}, \tag{15.2.2}$$

Thus, while the original variable Y takes on discrete values between zero and m, the transformed variable X takes on discrete values between zero and $m^{1-\beta}$, where β is a constant that will be chosen shortly. Similarly, a transformation of the time variable that often proves useful is

$$T = \frac{t}{m^{\alpha}}, \tag{15.2.3}$$

where α is a constant that will also be chosen shortly. Consequently, one unit of time in the transformed model corresponds to m^{α} units of time in the original model.

(continued)

Box 15.2 *(continued)*

Let us take a step back for a moment to digest what we have done. The original random variable Y changes from one time step to the next, where the length of this time step might potentially be quite large depending upon what we are modeling. Furthermore, the size of the changes in Y that occur over these time steps might also be quite large. The transformed model, however, has different properties. It, too, is a discrete–time, discrete-state-space model, but each discrete step occupies a much smaller fraction of a unit in time. In particular, each time step of the original model corresponds to only $1/m^{\alpha}$ units of time in the transformed model. Therefore, as long as the size of the state space, m, is reasonably large, the time steps of the transformed model will appear very small. Additionally, in each of these small time steps, the transformed variable X changes only by the small amount $\Delta Y/m^{\beta}$. Therefore, we might expect this transformed model to be better approximated by a diffusion model because its variables change in a more continuous fashion.

We now determine the moments of change in the transformed model, given the moments of change (15.2.1) in the original model. We begin by transforming the state space to get the moments of change in the transformed variable X but keeping the original time variable t. Across one of the original time steps, the first moment of change in the transformed variable X starting from position X (corresponding to i in the original state space) is, from (15.2.2),

$$E[X(t+1) - X(t) \mid X(t) = x] = E\left[\frac{Y(t+1)}{m^{\beta}} - \frac{Y(t)}{m^{\beta}}\,\middle|\,\frac{Y(t)}{m^{\beta}} = \frac{i}{m^{\beta}}\right]. \tag{15.2.4a}$$

Using (15.2.1), this equals

$$\begin{aligned} E[X(t+1) - X(t) \mid X(t) = x] &= \frac{1}{m^{\beta}} E[Y(t+1) - Y(t) \mid Y(t) = i] \\ &= \frac{1}{m^{\beta}} \sum_{j=0}^{m} (j - i)\, p_{ji}. \end{aligned} \tag{15.2.4b}$$

Equation (15.2.4b) describes the amount of change in one time step in the original time scale. To obtain the *rate of change* in the original time scale we must divide by the length of the time step. Because we have defined the time steps to be of length one in the original model, this means that we must divide by one to get the rate of change in the original model. To complete the transformation we must now calculate the rate of change measured in the new time scale. Starting at time point T and using (15.2.3), the change in the transformed time variable that occurs over one time step in the original time scale is $\Delta T = (t+1)/m^{\alpha} - (t)/m^{\alpha} = 1/m^{\alpha}$. Therefore, the rate of change of the first moment in the fully transformed model equals equation (15.2.4b) divided by ΔT. Setting $\Delta T = 1/m^{\alpha}$ on the right, we get

$$\text{First moment} \quad \frac{E[X(T+\Delta T) - X(T) \mid X(T) = x]}{\Delta T} = \frac{m^{\alpha}}{m^{\beta}} \sum_{j=0}^{m} (j - i)\, p_{ji}. \tag{15.2.5a}$$

(continued)

Box 15.2 *(continued)*

Similar calculations give the rate of change of the second and third moments as

Second moment
$$\frac{E[X(T + \Delta T) - X(T))^2 | X(T) = x]}{\Delta T} = \frac{m^\alpha}{m^{2\beta}} \sum_{j=0}^{m} (j - i)^2 p_{ji}, \quad (15.2.5b)$$

Third moment
$$\frac{E[X(T + \Delta T) - X(T))^3 | X(T) = x]}{\Delta T} = \frac{m^\alpha}{m^{3\beta}} \sum_{j=0}^{m} (j - i)^3 p_{ji}. \quad (15.2.5c)$$

We now have two descriptions of the biological process of interest: (i) our original model, whose moments are given by (15.2.1), and (ii) our transformed model, whose moments are given by (15.2.5). To approximate a model using a diffusion equation, we must demonstrate that: (a) the rate of change in the mean is a finite number (including, potentially, zero); (b) the rate of change in the variance is a positive and finite number; and (c) the rate of change of any one of the higher-order moments is zero (see pp. 157–165 in Karlin and Taylor 1981 and pp. 327–328 in Allen 2003), although, typically, this is proven for the third or fourth moment (Karlin and Taylor 1981).

Usually the original description of the biological process will not satisfy these three requirements exactly. For example, changes in the original random variable Y might be large, causing nonzero rates of change of the higher-order moments. The same might be true with the transformed model, except that we are now free to choose α and β. The goal is to choose them in such a way that rules (a)–(c) are approximately satisfied. By "approximately" we mean that, as the size of the original state space, m, gets larger and larger (so that the discrete time steps in the transformed model, along with the changes in X, get smaller and smaller), conditions (a)–(c) are better met. Mathematically, in the limit as m goes to infinity, conditions (a)–(c) should hold exactly. Under these conditions, a diffusion model should yield a reasonable approximation for the transformed model so long as the actual value of m is not too small.

To summarize, as long as the state space, m, is not too small, a diffusion approximation will provide a reasonable description of the dynamics provided that we can choose values of α and β such that:

First moment
$$\mu(x) = \lim_{m\to\infty} \frac{m^\alpha}{m^\beta} \sum_{j=0}^{m} (j - i)\, p_{ji},$$

Second moment
$$\sigma^2(x) = \lim_{m\to\infty} \frac{m^\alpha}{m^{2\beta}} \sum_{j=0}^{m} (j - i)^2 p_{ji}, \quad (15.2.6)$$

Third moment
$$0 = \lim_{m\to\infty} \frac{m^\alpha}{m^{3\beta}} \sum_{j=0}^{m} (j - i)^3 p_{ji},$$

where $\mu(x)$ is finite and $\sigma^2(x)$ is finite and positive. These coefficients may then be used in the Kolmogorov equations (15.8) and (15.11) as a continuous-time, continuous-state-space approximation to the transformed model.

Given a discrete stochastic model, the first step in deriving a diffusion approximation is to obtain the moments (15.2.1) based on the transition probabilities for the original model. These

(continued)

Box 15.2 *(continued)*

moments are then plugged in for the sums in (15.2.6) and appropriate values of α and β are chosen. If the restrictions in (15.2.6) cannot be met, consider whether there is a scaling of the parameters (in addition to the transformation of the variables) that will satisfy the restrictions (e.g., multiplying the parameters by m). If the restrictions are still not met, consider whether there is some nondecreasing function $g(m)$ other than m^{α} that can be used to rescale the variables. If it is not possible to transform the model to satisfy (15.2.6), then the biological process might not be accurately described by a diffusion approximation. Finally, it is important to remember that you should reverse the transformations when you wish to interpret the results of a diffusion analysis in terms of the original variables and parameters.

Before leaving the subject, it should be mentioned that the diffusion approximation is often used without performing the transformation described in this box and ensuring that the approximation is valid. That is, the first and second moments for the original model (15.2.1) are naively taken to be the drift and diffusion coefficients (as we did in section 15.2.4). This approach sometimes works, yielding results that match simulations. But only by carrying out the above procedure can you be assured that the diffusion approximation is valid (see Problem 15.8).

15.3 Analyzing the Diffusion Equation with Drift

As with the other dynamical models that we have considered in this book, it would be ideal to derive a general solution for the diffusion equation with drift. Occasionally this is possible, but typically it isn't. Nevertheless, when a general solution is not available, we can still obtain many useful results analytically, just as we have done in previous chapters. In this section we illustrate some of the more important results for diffusion models. As it is often easier to appreciate a result when you know where it comes from, we provide derivations of the stationary distribution, the probability of absorption, and expected time to absorption in sections 15.3.2, 15.3.3, and Supplementary Material 15.2, respectively. Having derived these results, Recipes 15.1–15.3 summarize the solutions and provide a quick way to obtain results from any diffusion model by plugging in the appropriate drift and diffusion coefficients. These recipes involve integrals that can be calculated analytically or, when this is not possible, numerically. Thus, the main advantage of diffusion theory is that it can be used to reduce the analysis of a complex stochastic model to the evaluation of a handful of integrals.

Brownian motion describes random movement through space, with constant drift and diffusion coefficients.

15.3.1 General Solutions to the Diffusion Model

Although it is not possible to derive a general solution to the diffusion model for arbitrary drift and diffusion coefficients, $\mu(x)$ and $\sigma^2(x)$, it is possible to do so for certain special cases. One such example is constant Brownian motion, which is characterized by $\mu(x) = \mu$ and $\sigma^2(x) = \sigma^2$ for all x. In the classic

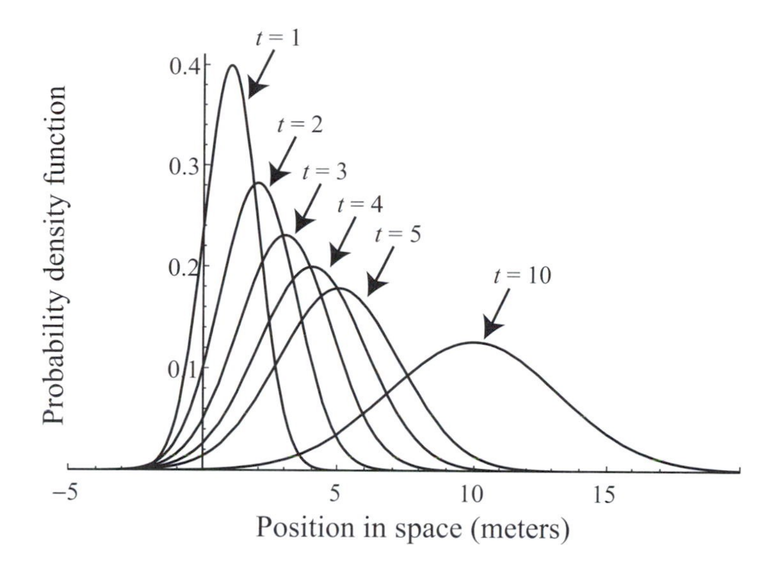

Figure 15.3: Diffusion over space with Brownian motion. Assuming that particles are initially located at $x_0 = 0$ at time $t_0 = 0$, equation (15.15) describes how the spatial distribution of particles spreads over time under Brownian motion. Parameters were set to $\mu = 1$ meter per unit time and $\sigma^2 = 1$ meter^2 per unit time. The probability density of particles becomes normally distributed with a breadth that increases as a square-root function of time.

Brownian motion model, x represents location in one-dimensional space, and the particles are assumed to start at a particular position x_0 at time t_0. This is exactly the case for our model of fish movement in equation (15.7). With constant drift and diffusion parameters, a solution for the probability density function for the location of the fish at time t (using either the backward or forward Kolmogorov equation) is

$$\phi(x,t \mid x_0,t_0) = \frac{e^{-(x-(x_0+(t-t_0)\,\mu))^2/(2\,(t-t_0)\,\sigma^2)}}{\sqrt{2\,\pi\,(t-t_0)\,\sigma^2}} \tag{15.15}$$

(see Problem 15.3).

The solution (15.15) is the probability density function for a normal distribution (see Definition P3.14a) with mean of $x_0 + (t - t_0)\,\mu$ and variance $(t - t_0)\,\sigma^2$. From this we conclude that the expected location of the fish moves away from its starting location, x_0, at a rate that depends on the drift parameter μ. According to equations (15.6), the drift parameter is proportional to the difference between upstream movement, u, and downstream movement, d. When prevailing currents cause $d > u$, the expected location of the fish is further and further downstream as time passes. From Equation (15.15) we can also see that the variance in the location of the fish increases linearly with time, at a rate proportional to the diffusion coefficient, σ^2. Consequently, the standard deviation, which measures the breadth of the normal curve, rises with the square root of time (Figure 15.3).

15.3.2 The Stationary Distribution of a Diffusion Model

While it is fantastic to obtain a general solution from a diffusion model, it is often not possible to solve the partial differential equations (15.8) or (15.11). Nevertheless, we can seek the stationary distribution of a diffusion model, just as we did for the discrete state space models in Chapter 14. By definition, a stationary distribution, $\pi(x)$, does not change over time: $\partial\pi(x)/\partial t = 0$. If the system is at a stationary distribution at time t_0, then the probability density at any future time t will also be at the stationary distribution and must satisfy $\pi(x) = \int \pi(x_0)\, \phi(x,t \mid x_0,t_0)\, dx_0$, when integrated over all possible initial states, x_0. Using this fact, we can multiply both sides of the forward Kolmogorov equation (15.8) by $\pi(x_0)$ and integrate over initial states x_0 to get

$$\int \pi(x_0) \frac{\partial\phi(x,t \mid x_0,t_0)}{\partial t}\, dx_0 = -\int \pi(x_0) \frac{\partial(\mu(x)\, \phi(x,t \mid x_0,t_0))}{\partial x}\, dx_0 + \frac{1}{2}\int \pi(x_0) \frac{\partial^2(\sigma^2(x)\, \phi(x,t \mid x_0,t_0))}{\partial x^2}\, dx_0.$$

Because $\pi(x_0)$ does not depend on time, we can interchange the order of integration and differentiation to get

$$\frac{\partial \int \pi(x_0)\, \phi(x,t \mid x_0,t_0)\, dx_0}{\partial t} = -\frac{\partial \int \pi(x_0)\, \mu(x)\, \phi(x,t \mid x_0,t_0)\, dx_0}{\partial x} + \frac{1}{2}\frac{\partial^2 \int \pi(x_0)\, \sigma^2(x)\, \phi(x,t \mid x_0,t_0)\, dx_0}{\partial x^2}$$

Plugging $\pi(x) = \int \pi(x_0)\, \phi(x,t \mid x_0,t_0)\, dx_0$ into each term and recalling that $\partial\pi(x)/\partial t = 0$ leaves

$$0 = -\frac{d(\mu(x)\, \pi(x))}{dx} + \frac{1}{2}\frac{d^2(\sigma^2(x)\, \pi(x))}{dx^2}. \tag{15.16}$$

Because the state of the system at the stationary distribution does not depend on time, equation (15.16) is now an ordinary differential equation involving derivatives with respect to x only, rather than a partial differential equation. Equation (15.16) must be satisfied by any potential stationary distribution, $\pi(x)$.

To solve equation (15.16) for the stationary distribution, $\pi(x)$, first integrate both sides with respect to x:

$$c_0 = -\mu(x)\, \pi(x) + \frac{1}{2}\frac{d(\sigma^2(x)\, \pi(x))}{dx}, \tag{15.17a}$$

where c_0 is a constant of integration. At this point, a change of notation makes things clearer. If we define the function $n(x) = \sigma^2(x)\, \pi(x)$ and rearrange, (15.17a) becomes

$$\frac{dn(x)}{dx} = \frac{2\mu(x)}{\sigma^2(x)} n(x) + c_1, \tag{15.17b}$$

where $c_1 = 2\, c_0$. Equation (15.17b) is a more familiar ordinary differential equation, of the sort described in Box 6.2. Specifically, (15.17b) is an example of a linear differential equation (6.2.1), and therefore its solution is

$$n(x) = e^{A(x)}\left(c_1 \int e^{-A(x)} dx + c_2\right) \tag{15.17c}$$

where

$$A(x) = \int \frac{2\mu(x)}{\sigma^2(x)} dx.$$

Using the fact that $n(x) = \sigma^2(x)\, \pi(x)$, we now have an explicit equation for the stationary distribution:

Recipe 15.1

The Stationary Distribution of the Diffusion Equation The stationary distribution of the diffusion equation with drift parameter $\mu(x)$ and diffusion parameter $\sigma^2(x)$ is

$$\pi(x) = \frac{e^{A(x)}\left(c_1 \int e^{-A(x)} dx + c_2\right)}{\sigma^2(x)} \tag{15.18}$$

where $A(x) = \int 2\mu(x)/\sigma^2(x)\, dx$. The constants of integration, c_1 and c_2, must be chosen such that $\pi(x)$ integrates to one, is never negative, and satisfies any other required conditions at the endpoints of the allowable values of x. Only if there are such choices of constants does the stationary distribution exist.

It is always possible to check whether or not the stationary distribution that you derive is correct by substituting it into the forward Kolmogorov equation. If you have not made any errors, then carrying out the differentiations on the right-hand side of equation (15.8) should result in zero (by the definition of a stationary distribution).

Example: The Stationary Distribution for a Trait when Mutations Are Limiting

Here we apply Recipe 15.1 to the model of mutation-limited evolution with drift and diffusion coefficients given by (15.13). To make things more concrete, let's implement this recipe for the model (12.16) from Chapter 12 involving the evolution of daphnia feeding strategies, where the population size of daphnia was described by a deterministic recursion equation. The growth factor of a mutant daphnia strain with feeding strategy s_m in a population with resident feeding strategy s was given by equation (12.18):

$$\lambda = 1 + r - r\,\frac{\alpha(s_m, s)\, K(s)}{K(s_m)}.$$

where $\alpha(s_m,s)$ is the competition coefficient and $K(s)$ is the carrying capacity.

To construct a stochastic model for this question we make slightly different assumptions than those of model (12.16). In particular, we assume that the population of daphnia is fixed at size N and undergoes a birth-death process described by the Moran model. These assumptions allow us to use the results of Chapter 14 to describe the probability that a mutant allele reaches fixation in the population.

First, let us specify the mutant distribution $M(s_m,s;\Delta t)$, which describes the probability distribution of mutants with strategy s_m arising within a population whose feeding strategy is s over a time interval Δt. We suppose that the probability that one of the N individuals in the population gives birth to a mutant individual in time interval Δt is given by $N\,\eta\,\Delta t$, where η is the per capita rate at which mutant offspring are produced. Given that a mutant offspring is produced, we then suppose that its phenotype is normally distributed with a mean equal to the current trait value s and a variance of ν (see Definition P3.14a). Therefore,

$$M(s_m,s;\Delta t) = N\eta\Delta t\frac{e^{-(s_m-s)^2/(2\nu)}}{\sqrt{2\pi\nu}}, \tag{15.19}$$

Next, we use the birth-death model to describe the fixation probability of a mutant allele arising within the population of daphnia, $U(s_m,s)$. We assume that the per capita birth rate is constant at $b = r = 1$, and that the per capita death rate is $d(s_m,s) = \alpha(s_m,s)\,K(s)\,/\,K(s_m)$. Consequently, when the mutant type has the same feeding strategy as the resident type, the birth rate equals the death rate, and the mutation is completely neutral. Furthermore, we assume that the amount of resources follows a bell-shaped distribution (Definition P1.6) with a peak at $s = 0$ using the function $K(s) = Ke^{-s^2/a_K}$, and that the competition coefficient is $\alpha(s_m,s) = e^{-(s_m-s)^2/a_\alpha}$ as in Chapter 12.

The above birth-death model implicitly assumes that the death rate does not change as the number of mutant individuals in the population increases. In reality this is not the case because eventually the mutant becomes so abundant that it will begin to experience interactions with other mutant daphnia. This can be remedied by making the death rate function $d(s_m,s)$ depend on the frequency of mutant individuals. We ignore this complication here, under the assumption that the fixation probability of the mutant is determined while the mutant is still rare. The probability of fixation of the mutant type in a population of N daphnia equals one minus the probability of loss, given by equation (14.56) for a birth-death model:

$$U(s_m,s) = \frac{1 - d(s_m,s)}{1 - d(s_m,s)^N}. \tag{15.20}$$

To obtain tractable solutions for the drift and diffusion coefficients in (15.13), we use a linear approximation to (15.20), under the assumption that the difference between the mutant strategy and the resident strategy is not too large.

(Identical results are obtained if we use a more accurate quadratic approximation.) Using Recipe P1.2 to calculate the first two terms of the Taylor series of (15.20) with respect to s_m, we get

$$U(s_m, s) = \frac{1}{N} - \frac{s\,(N-1)}{N\,a_K}\,(s_m - s). \tag{15.21}$$

(Evaluating the terms in the Taylor series in the limit as s_m goes to s requires L'Hôpital's rule, as described in Appendix 2.)

Plugging (15.19) and (15.21) into expressions (15.13) gives

$$\begin{aligned}\mu(s) &= \eta \int (s_m - s)\frac{e^{-(s_m-s)^2/(2\nu)}}{\sqrt{2\pi\nu}}\left(1 - \frac{s\,(N-1)}{a_K}\,(s_m - s)\right) ds_m \\ &= -\eta\,\frac{s(N-1)}{a_K}\,\nu, \end{aligned} \tag{15.22a}$$

$$\begin{aligned}\sigma^2(s) &= \eta \int (s_m - s)^2\,\frac{e^{-(s_m-s)^2/(2\nu)}}{\sqrt{2\pi\,\nu}}\left(1 - \frac{s\,(N-1)}{a_K}\,(s_m - s)\right) ds_m \\ &= \eta\,\nu \end{aligned} \tag{15.22b}$$

where we have used the fact that $\int (s_m - s)^i\,(e^{-(s_m-s)^2/(2\nu)}/\sqrt{2\pi\nu})\,ds_m$ is the ith central moment of the normal distribution and equals 0 for $i = 1$ (the first central moment), ν for $i = 2$ (the second central moment), and 0 for $i = 3$ (the third central moment) (Appendix 5, equations (A5.10)).

Equations (15.22) give the drift and diffusion coefficients, but we must also ensure that the rate of change of any one of the higher order moments is zero before we can validly use a diffusion model (see pp. 157–165 in Karlin and Taylor (1981) and pp. 327–328 in Allen (2003)). In Problem 15.4, you are asked to demonstrate that, by rescaling time, and measuring it in units of the mutational variance, ν, the higher order moments vanish as ν goes to zero. Furthermore, the infinitesimal mean and variance then become $\mu(s) = -\eta\,(s\,(N-1)/a_K)$ and $\sigma^2(s) = \eta$. Therefore, on this new time scale, the diffusion equation with these rescaled moments should provide a suitable model for mutation-limited evolution provided that the mutational variance, ν, is not too large. This is analogous to the principle we employed in Box 15.2, where we saw that we could use a diffusion equation to approximate a discrete stochastic model so long as the state space was large enough.

We are now ready to follow Recipe 15.1 to find the stationary distribution. From Recipe 15.1 we have $A(s) = \int 2\,\mu(s)/\sigma^2(s)\,ds = \int -2\,s\,(N-1)/a_K\,ds$, giving $A(s) = -2\,s^2\,(N-1)/a_K$ (the constant of integration is already included in (15.18)). Substituting into (15.18), we obtain

$$\pi(s) = \frac{e^{-(2\,(N-1)\,s^2/a_K)}\left(c_1 \int e^{(2\,(N-1)\,s^2/a_K)}\,ds + c_2\right)}{\eta} \tag{15.23}$$

The final step is to choose the constants of integration in (15.23). This must be done such that the probability density is never negative and integrates to one. Because the potential values of s are anywhere from $-\infty$ to $+\infty$, we must ensure that the probability density decays to zero as s gets very large or very small. The term inside the integral gets very large as s goes to $-\infty$ or to $+\infty$, and therefore we must set $c_1 = 0$ to ensure that (15.23) is finite. The remaining terms in equation (15.23), $e^{-2\,s^2\,(N-1)/a_K} c_2/\eta$, integrate to $\sqrt{\pi\, a_K/(2\,(N-1))}\; c_2/\eta$, which we set equal to one and solve for the constant, c_2. The stationary distribution thus equals

$$\pi(s) = \frac{e^{-s^2/2V}}{\sqrt{2\pi V}} \tag{15.24}$$

where $V = a_K/(4\,(N-1))$. Interestingly, the probability density (15.24) is a normal distribution with a mean centered at zero and a variance of $a_K/(4\,(N-1))$.

Once the stationary distribution is reached, the population continues to change stochastically from one resident trait value to another. But if we observe the population at some point in time, then (15.24) describes the probability density that the observed state will be s. From the form of (15.24), we can see that the population is most likely to have a feeding strategy yielding the greatest carrying capacity (i.e., $s = 0$). The feeding strategies will vary widely around this optimum when the resource distribution is wide (large a_K) and when the daphnia population size is small (more genetic drift). Conversely, the feeding strategies will cluster tightly around the optimum in very large populations.

In Chapter 12, we found that evolutionary branching occurred when the resource distribution was wider than the competition function ($a_K > a_\alpha$), in which case a polymorphism was expected. Yet no such diversification occurs in the above stochastic model of mutation-limited evolution. What causes this discrepancy? We constructed the diffusion model above to describe the state of the population under the assumption that the population is always monomorphic. By assumption, when mutations arise, they either disappear or fix and immediately replace the parental strain. The possibility that the population becomes polymorphic is therefore not allowed. Thus, this diffusion approach is limited to describing the probability distribution of populations as long as they remain, for most periods of time, dominated by single alleles.

Example: The Wright-Fisher model

Next, we consider the Wright-Fisher model of allele frequency change with mutation and selection in a haploid population. This model was introduced in section 13.4 as an example of a discrete-time, discrete-state stochastic model. The discrete states of the model represent the number of copies of an allele within a population, ranging from zero up to a maximum of N. Unless the population size is small, this model is difficult to analyze using the techniques of Chapter 14, so let us approximate the model with a diffusion equation using Box 15.2. In Box 15.3, we show that the Wright-Fisher model can be approximated by a diffusion if we measure time in units of N generations and keep

Box 15.3: The Drift and Diffusion Parameters in the Wright-Fisher Model

Here we derive a diffusion approximation for the Wright-Fisher model of allele frequency change in a haploid population using the methods of Box 15.2. Our goal is to show that there is a valid diffusion approximation for the Wright-Fisher model as the population size gets large. First we consider the case of no mutation or selection.

In Chapter 13, we introduced the Wright-Fisher model and pointed out that its transition probabilities are binomially distributed. Specifically, using $Y(t)$ to denote the number of copies of allele A at time t, the probability of a transition from i copies to j copies is

$$p_{ji} = P(Y(t+1) = j \mid Y(t) = i)$$

$$= \binom{N}{j}(p)^j(1-p)^{N-j},$$

where N is the size of the population and p is the allele frequency ($p = i/N$) when there is no selection or mutation. Using the mean of the binomial (Table 3.2), equation (15.2.1) becomes

$$\text{First moment} \quad E[Y(t+1) - Y(t) \mid Y(t) = i] = E[Y(t+1) \mid Y(t) = i] - E[Y(t) \mid Y(t) = i]$$

$$= Np - i,$$

which equals $N(i/N) - i = 0$. This reveals that there is no directional tendency in allele frequency, as we would expect in the absence of selection and mutation.

Because $E[Y(t+1) \mid Y(t) = i] = i$, we can rewrite the second moment in (15.2.1), $E[(Y(t+1) - Y(t))^2 \mid Y(t) = i]$ as $E[(Y(t+1) - E[Y(t+1)])^2 \mid Y(t) = i]$, which is the formula for the variance of $Y(t+1)$. For a binomial distribution, this variance is given by equation (Table 3.2)

$$\text{Second moment} \quad E[(Y(t+1) - E[Y(t+1)])^2 \mid Y(t) = i] = N\left(\frac{i}{N}\right)\left(1 - \frac{i}{N}\right).$$

Similarly, the third moment in (15.2.1) for the Wright-Fisher model is given by the third central moment of the binomial distribution. From the calculations of Appendix 5, this moment equals

$$\text{Third moment} \quad E[(Y(t+1) - E[Y(t+1)])^3 \mid Y(t) = i] = N\left(\frac{i}{N}\right)\left(1 - \frac{i}{N}\right)\left(1 - 2\frac{i}{N}\right).$$

In terms of these original variables, the 3rd moment will not go to zero in the limit as the population size gets larger. Can we transform the model using equations (15.2.2) and (15.2.3) so that the diffusion approximation holds? To find out, we must find the moments in terms of the transformed variables by plugging in the above expressions into equation (15.2.6):

$$\text{First moment} \qquad \mu(x) = 0,$$

$$\text{Second moment} \qquad \sigma^2(x) = \lim_{m\to\infty} \frac{m^\alpha}{m^{2\beta}} N\left(\frac{i}{N}\right)\left(1 - \frac{i}{N}\right),$$

$$\text{Third moment} \qquad 0 = \lim_{m\to\infty} \frac{m^\alpha}{m^{3\beta}} N\left(\frac{i}{N}\right)\left(1 - \frac{i}{N}\right)\left(1 - 2\frac{i}{N}\right).$$

(continued)

Box 15.3 ***(continued)***

In the Wright-Fisher model, the system size is $N + 1$, because the original variable $Y(t)$ can take on any value from 0 to N. Although we could set $m = N + 1$, it is inconvenient to keep track of the +1 term. Because any transformation that satisfies the above moment equations will suffice, we set $m = N$. Next, we rewrite the i's (which are particular values of the random variable Y) as x's (which are particular values of the transformed variable X) using equation (15.2.2), giving $i = x N^{\beta}$. Making these substitutions, the new moments become

$$\text{First moment} \qquad \mu(x) = 0,$$

$$\text{Second moment} \qquad \sigma^2(x) = \lim_{N\to\infty} \frac{N^{\alpha}}{N^{2\beta}} N \left(\frac{x\,N^{\beta}}{N}\right)\left(1 - \frac{x\,N^{\beta}}{N}\right),$$

$$\text{Third moment} \quad 0 = \lim_{N\to\infty} \frac{N^{\alpha}}{N^{3\beta}} N \left(\frac{x\,N^{\beta}}{N}\right)\left(1 - \frac{x\,N^{\beta}}{N}\right)\left(1 - 2\frac{x\,N^{\beta}}{N}\right).$$

All that remains is to simplify these expressions and choose the values of α and β so that the appropriate restrictions are met. The second moment expands to become

$$\sigma^2(x) = \lim_{N\to\infty}\left(\frac{N^{\alpha}x}{N^{\beta}} - \frac{N^{\alpha}x^2}{N}\right), \tag{15.3.1}$$

and the third moment expands to become

$$0 = \lim_{N\to\infty}\left(\frac{N^{\alpha}x}{N^{2\beta}} - 3\frac{N^{\alpha}x^2}{N^{\beta+1}} + 2\frac{N^{\alpha}x^3}{N^2}\right), \tag{15.3.2}$$

For the variance to be positive and finite as N becomes infinitely large, α must equal β. If we let $\alpha = \beta = 1$, the transformed random variable, $x = i/N$, measures the allele frequency, p, which is a very natural scale. With this choice, the third moment does approach 0 as N becomes large. Having met the requirements for the moments, the neutral Wright-Fisher model can be approximated by a diffusion model in the transformed state space where we keep track of the allele frequency $p = i/N$, and measure time in units of N generations. With these transformed variables, the drift and diffusion coefficients are $\mu(p) = 0$, $\sigma^2(p) = p\,(1 - p)$.

(Although $\alpha = \beta = 1$ is a natural choice, other choices can obey the requirements of a diffusion approximation. For example, $\alpha = \beta = 1/2$ can be used to provide a diffusion approximation that more accurately describes the fine fluctuations in allele frequencies while the allele is rare; see Karlin and Taylor 1981, pp. 180–182.)

Selection and mutation can be incorporated into the model by assuming that these processes act deterministically on the large number of propagules produced by the population. For example, the allele frequency after selection followed by mutation in a haploid population would be

$$p' = (1 - v_1)\frac{(1 + s)\,p}{(1 + s)\,p + (1 - p)} + v_2\frac{(1 - p)}{(1 + s)\,p + (1 - p)} \tag{15.3.3}$$

(continued)

Box 15.3 *(continued)*

where s is the selective advantage of allele A, ν_1 is the mutation rate from A to a, and ν_2 is the mutation rate from a to A. Because we want the diffusion approximation to be consistent with the neutral model when selection and mutation become weak, we will continue to use the transformation with $m = N$ and $\alpha = \beta = 1$ (we will show in a moment that the variance remains positive and finite under this choice, as required). Now, the number of copies of allele A changes by $E[Y(t+1) - Y(t) \mid Y(t=i)] = Np' - i = N(p' - p)$. Plugging this result into (15.2.6), we get

$$\mu(p) = \lim_{N\to\infty} N(p' - p). \tag{15.3.4}$$

This drift parameter measures the expected change in the transformed variable (the allele frequency) per unit of transformed time (corresponding to N generations). For the drift parameter to be finite in the limit, the forces causing a change in allele frequency, $(p' - p)$, must be weak, on the order of $1/N$. To ensure that this restriction is met, let us define $\psi = s\,N$, $\theta_1 = \nu_1 N$, and $\theta_2 = \nu_2 N$. In this case, (15.3.3) becomes

$$p' = \left(1 - \frac{\theta_1}{N}\right)\frac{\left(1 + \frac{\psi}{N}\right)p}{\left(1 + \frac{\psi}{N}\right)p + (1-p)} + \frac{\theta_2}{N}\frac{(1-p)}{\left(1 + \frac{\psi}{N}\right)p + (1-p)}. \tag{15.3.5}$$

Factoring $N(p' - p)$ and taking the limit as N goes to infinity leaves us with the drift parameter

$$\mu(p) = \lim_{N\to\infty} N(p' - p) = \psi\,p\,(1-p) - \theta_1\,p + \theta_2(1-p). \tag{15.3.6}$$

The drift parameter is finite only if the selection and mutation rates are small, such that $\psi = s\,N$, $\theta_1 = \nu_1 N$, and $\theta_2 = \nu_2 N$ remain finite even in the limit as the population size grows to infinity.

Under the above assumptions, the diffusion parameter for the Wright-Fisher model remains $\sigma^2(p) = p\,(1-p)$ even with selection and mutation. To demonstrate this fact, let us find the relationship between $E[(Y(t+1) - Y(t))^2 \mid Y(t) = i]$ and the variance of the binomial, $E[(Y(t+1) - E[Y(t+1)])^2 \mid Y(t) = i]$, which we know equals $N\,p'(1-p')$. Expanding both expectations using the rules given in Table P3.1, we find that

$$\begin{aligned} E[(Y(t+1) - Y(t))^2 \mid Y(t) = i] &= E[(Y(t+1) - E[Y(t+1)])^2 \mid Y(t) = i] \\ &\quad + E[Y(t+1) \mid Y(t) = i]^2 - 2\,E[Y(t+1) \mid Y(t) = i]\,i + i^2. \end{aligned} \tag{15.3.7}$$

Because sampling follows a binomial distribution, the expected number of A alleles after sampling is $E[Y(t+1) \mid Y(t) = i] = N\,p'$, allowing us to rewrite (15.3.7) as

$$E[(Y(t+1) - Y(t))^2 \mid Y(t) = i] = N\,p'(1-p') + (N\,p')^2 - 2\,N\,p'i + i^2. \tag{15.3.8}$$

Plugging (15.3.5) into (15.3.8) and the result into $\sigma^2(p) = \lim_{N\to\infty} (E[(Y(t+1) - Y(t))^2 \mid Y(t) = i]/N)$, we again get $\sigma^2(p) = p\,(1-p)$. Following a similar procedure, the third moment can be shown to equal zero in the limit.

track of the frequency p of allele A. With these variables, we find that the drift coefficient is $\mu(p) = -N\,\nu_1\,p + N\,\nu_2\,(1-p) + N\,s\,p\,(1-p)$ and the diffusion coefficient is $\sigma^2(p) = p\,(1-p)$, where s is the selective advantage of allele A, ν_1 is the mutation rate from A to a, and ν_2 is the mutation rate from a to A.

With these drift and diffusion coefficients, we can follow Recipe 15.1 to find the stationary distribution. Evaluating $A(p)$ first, we have $A(p) = 2N\nu_1 \ln(1-p) + 2\,N\,\nu_2 \ln(p) + 2\,N\,s\,p$ and $e^{A(p)} = (1-p)^{2N\nu_1}(p)^{2N\nu_2}e^{2Nsp}$. Substituting these into expression (15.18), we must next choose the constants of integration, c_1 and c_2. It is always worth trying to identify a stationary distribution by first setting $c_1 = 0$, as this eliminates the integral in (15.18). If this choice fails to identify the stationary distribution of interest, other choices of c_1 can then be made. (The choice of $c_1 = 0$ in the Wright-Fisher model can also be formally motivated, see p. 222 of Karlin and Taylor 1981.) Setting c_1 to zero, we immediately find the stationary distribution from (15.18):

$$\pi(p) = c_2(1-p)^{2N\nu_1-1}(p)^{2N\nu_2-1}e^{2Nsp}$$

where

$$c_2 = 1/\left(\int_{p=0}^{1}(1-p)^{2N\nu_1-1}(p)^{2N\nu_2-1}e^{2Nsp}dp\right). \qquad (15.25)$$

The stationary distribution (15.25) describing allele frequencies in the presence of selection and mutation is known as Wright's distribution in honor of its discoverer, the evolutionary biologist Sewall Wright (see Crow and Kimura 1970 for the equivalent distribution in a diploid population). It is illustrated for various parameters in Figure 15.4. In the absence of selection ($s = 0$), equation (15.25) represents the probability density function for a beta distribution (see definition P3.16).

15.3.3 Probability of Absorption in a Diffusion Model

Let us now derive the probability that the stochastic processes reaches a particular point, b. We assume that the point of interest, b, is the maximum (or minimum) state that can be reached by the stochastic process and that once this state is reached, the stochastic process remains there (i.e., $x = b$ is an "absorbing boundary"). While the forward equation is particularly helpful in deriving the stationary distribution, the backward equation is more useful for deriving the absorption probability.

We define $u(b,t \mid x_0,t_0)$ as the probability of reaching b by time t from initial position x_0 at t_0. Because we have assumed that b is an absorbing boundary, $u(b,t \mid x_0,t_0)$ equals the probability that the process makes a transition to b over this time period, i.e., $\phi(b,t \mid x_0,t_0)$. Thus, we can replace ϕ with u in the backward Kolmogorov equation (15.11):

$$\frac{\partial u(b,t \mid x_0,t_0)}{\partial t_0} = -\mu(x_0)\frac{\partial u(b,t \mid x_0,t_0)}{\partial x_0} - \frac{1}{2}\sigma^2(x_0)\frac{\partial^2 u(b,t \mid x_0,t_0)}{\partial x_0^2}, \qquad (15.26)$$

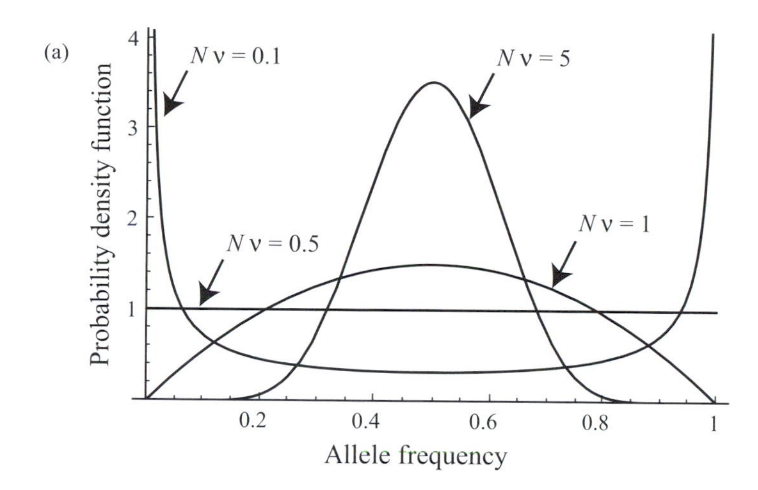

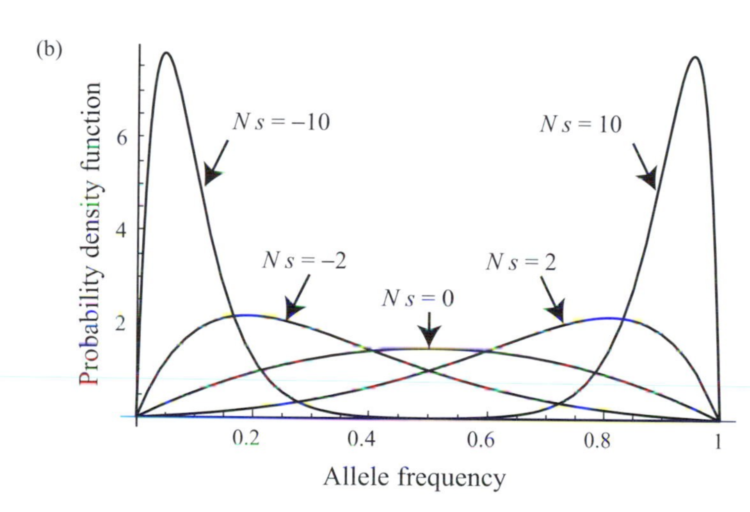

Figure 15.4: Wright's distribution. Wright's distribution describes the frequency of allele A based on a diffusion approximation of the Wright-Fisher model. The stationary probability density function (15.25) is plotted as a function of the allele frequency, assuming equal forward and backward mutation rates ($\nu_1 = \nu_2 = \nu$). (a) In the absence of selection ($N\,s = 0$), the lower the mutation rate, the more likely the allele is nearly lost or fixed within the population. (b) In the presence of selection with a mutation rate of $N\,\nu = 1$, the more strongly selection favors allele A, the more frequent the A allele is likely to be. *Mathematica* was used for numerical integration. Notice that an increase in the population size is equivalent to an increase in both mutation and selection.

which considers the target point b to be held constant and the initial points to be varied, $u(b,t \mid x_0,t_0)$.

After a long amount of time has passed, $u(b,t \mid x_0,t_0)$ typically approaches a constant value that depends only on its initial position. Let's define $u(b \mid x_0) = \lim_{t\to\infty} u(b,t \mid x_0,t_0)$ as the probability of ultimately reaching b from initial position x_0. Because $u(b \mid x_0)$ does not depend on time, $\partial u(b \mid x_0)/\partial t_0 = 0$, and the backward equation becomes an ordinary differential equation:

$$0 = -\mu(x_0)\frac{\mathrm{d}u(b \mid x_0)}{\mathrm{d}x_0} - \frac{1}{2}\sigma^2(x_0)\frac{\mathrm{d}^2u(b \mid x_0)}{\mathrm{d}x_0^2}. \qquad (15.27\text{a})$$

To solve (15.27a), it is again helpful to change notation to make the calculations more transparent. In particular, if we define $y \equiv \mathrm{d}u(b \mid x_0)/\mathrm{d}x_0$ then (15.27a) can be written as

$$0 = -\mu(x_0)\,y - \frac{1}{2}\sigma^2(x_0)\frac{\mathrm{d}y}{\mathrm{d}x_0}. \qquad (15.27\text{b})$$

Equation (15.27b) is now an ordinary differential equation that can be solved using a separation of variables (Recipe 6.2). Placing terms involving y on the left and those involving x_0 on the right, we can integrate both sides to get

$$\begin{aligned} \int \frac{1}{y} \mathrm{d}y &= -\int \frac{2\mu(x_0)}{\sigma^2(x_0)} \mathrm{d}x_0 \\ &= -A(x_0) \end{aligned} \tag{15.28}$$

where again $A(x_0) = \int 2\mu(x_0)/\sigma^2(x_0)\, \mathrm{d}x_0$. Performing the integration on the left-hand side, we have

$$\ln(y) = -A(x_0) + c_0, \tag{15.29}$$

where c_0 is a constant of integration. We can then exponentiate both sides, replace y with $\mathrm{d}u(b \mid x_0)/\mathrm{d}x_0$, and rename the constant e^{c_0} as c_1 to obtain the differential equation

$$\frac{\mathrm{d}u(b \mid x_0)}{\mathrm{d}x_0} = e^{-A(x_0)}\, c_1. \tag{15.30}$$

Performing another separation of variables, the solution is

$$\begin{aligned} u(b \mid x_0) &= \int e^{-A(x_0)}\, c_1 \, \mathrm{d}x_0 + c_2 \\ &= c_1 S(x_0) + c_2, \end{aligned} \tag{15.31}$$

where c_2 is another constant of integration and

$$S(x) = \int e^{-A(x)}\, dx. \tag{15.32}$$

To solve for the constants of integration, we assume that the random variable x_0 ranges from a to b and that both a and b are absorbing boundaries. If we start the system at the absorbing boundary a, it will never reach state b. This implies that $u(b \mid a) = 0$ and, from equation (15.31), that $c_2 = -c_1\, S(a)$. Substituting this requirement into equation (15.31) gives

$$u(b \mid x_0) = c_1\, (S(x_0) - S(a)). \tag{15.33}$$

Conversely, if we start in state b then the probability of hitting state b is one; i.e., $u(b \mid b) = 1$. Using this fact in equation (15.33), we can solve for $c_1 = 1/(S(b) - S(a))$. Thus the probability that the system eventually reaches the point b is

$$u(b \mid x_0) = \frac{S(x_0) - S(a)}{S(b) - S(a)}. \tag{15.34}$$

Because the difference between two indefinite integrals, $S(x) - S(y)$, equals the definite integral evaluated from y to x, we can write equation (15.34) in two equivalent forms:

Recipe 15.2

The Probability of Absorption in a Diffusion Model

The probability that a diffusion process that begins in state x_0 ultimately reaches the absorbing state b (the upper limit to the range) rather than the absorbing state a (the lower limit) is

$$u(b \mid x_0) = \frac{S(x_0) - S(a)}{S(b) - S(a)}, \tag{15.35a}$$

$$= \frac{\int_{x=a}^{x_0} e^{-A(x)}\,\mathrm{d}x}{\int_{x=a}^{b} e^{-A(x)}\,\mathrm{d}x}, \tag{15.35b}$$

where $A(x) = \int 2\mu(x)/\sigma^2(x)\,\mathrm{d}x$ and $S(x) = \int e^{-A(x)}\,\mathrm{d}x$. By the complement rule (Rule P3.2), the probability that the system ultimately hits a instead is

$$u(a \mid x_0) = 1 - u(b \mid x_0) = \frac{S(b) - S(x_0)}{S(b) - S(a)} \tag{15.36a}$$

$$= \frac{\int_{x=x_0}^{b} e^{-A(x)}\,\mathrm{d}x}{\int_{x=a}^{b} e^{-A(x)}\,\mathrm{d}x}. \tag{15.36b}$$

Recipe 15.2 is extremely useful and can be used for any diffusion model with absorbing states. Even when the integrals cannot be evaluated explicitly, equations (15.35) and (15.36) can be numerically integrated to obtain the absorption probabilities.

Example: Modeling Individual Movement in a Continuous Spatial Habitat

Let us return to our fish example and suppose that the drift and diffusion coefficients are the constants μ and σ^2, where the drift term is positive (reflecting the fact that the fish tends to move upstream more often than downstream). Suppose you are monitoring only a portion of the stream, from a location labeled 0 downstream to a location labeled b upstream. Once the fish leaves this part of the stream, you stop monitoring. What is the probability that the fish leaves the region at the downstream end (point 0) instead of the upstream end?

We can answer this question using Recipe 15.2. First we must calculate $A(x) = \int (2\mu/\sigma^2)\, dx$. Because the drift and diffusion coefficients are constants, $A(x) = 2\mu x/\sigma^2$. We can save some time if we evaluate $S(x) = \int e^{-A(x)}\, dx$ next, because we can then use equation (15.36a) without having to perform any further integrations. Carrying out this integration, we have $S(x) = -(\sigma^2/2\mu)\ e^{-2\mu x/\sigma^2}$. Consequently, the probability that the fish leaves at the downstream end of the monitored region is

$$u(0 \mid x_0) = \frac{S(b) - S(x_0)}{S(b) - S(0)} = \frac{-\dfrac{\sigma^2}{2\mu} e^{-2\mu b/\sigma^2} + \dfrac{\sigma^2}{2\mu} e^{-2\mu x_0/\sigma^2}}{-\dfrac{\sigma^2}{2\mu} e^{-2\mu b/\sigma^2} + \dfrac{\sigma^2}{2\mu}} = \frac{-e^{-2\mu b/\sigma^2} + e^{-2\mu x_0/\sigma^2}}{-e^{-2\mu b/\sigma^2} + 1}. \tag{15.37}$$

Equation (15.37) answers our question, but we can simplify it even further if we suppose that the fish's rate of movement is slow relative to length of the stream. In this case we might expect the drift coefficient to be small, and we can expand (15.37) as a Taylor series in μ:

$$u(0 \mid x_0) = \frac{b - x_0}{b} - \frac{x_0(b - x_0)}{b\,\sigma^2}\mu. \tag{15.38}$$

Equation (15.38) reveals that, in the absence of any directional movement ($\mu = 0$) the probability of leaving the region at the downstream end is given by where the fish starts, measured as a proportion of the distance from the original position to the downstream end: $(b - x_0)/b$. If the fish has a small tendency to move upstream, however, this exit probability is decreased by an amount $x_0\,(b - x_0)/(b\,\sigma^2)\mu$. The larger the variance in movement, the smaller this directional effect is, because the fish is likely to exit sooner as a result of random diffusive movement, before substantial directional movement has taken place.

While framed in terms of the movement of the fish, equations (15.37) and (15.38) apply to any model of diffusion with constant drift and diffusion coefficients (i.e., to any Brownian motion model) along a single axis. Generalizations to movement in more than one dimension are given by Karlin and Taylor (1981).

Example: Fixation in the Wright-Fisher Model

As a second example, let us calculate the probability that an allele becomes fixed within a haploid population using the Wright-Fisher model with selection but no mutation ($\nu_1 = 0$, $\nu_2 = 0$). From Box 15.3, the drift and diffusion parameters are $\mu(p) = \psi\, p\,(1 - p)$ and $\sigma^2(p) = p\,(1 - p)$, where p represents the allele frequency, N is the size of the haploid population, and $\psi = N\,s$ (see Problem 15.7 for the equivalent result in diploids). With these drift and diffusion parameters, $A(p) = 2\,\psi\,p$ and $S(p) = \int e^{-2\,\psi\,p}\,dp = -1/(2\,\psi)\,e^{-2\,\psi\,p}$. According to equation (15.35a), the probability of fixation of an allele whose initial frequency is p_0 is then

$$\begin{aligned} u(1|p_0) &= \frac{S(p_0) - S(0)}{S(1) - S(0)} \\ &= \frac{-e^{-2\,Ns\,p_0} + 1}{-e^{-2\,Ns} + 1} \end{aligned} \tag{15.39}$$

The diffusion approximation for the probability of fixation (15.39) is a classic result in evolutionary biology (Kimura 1957, 1962). In Figure 15.5, we compare the diffusion approximation (15.39) to exact numerical results for the probability of fixation, which can be obtained using (14.29) for small to moderately sized populations. Even though the diffusion approximation technically assumes that the population size is large and that selection is weak (see Box 15.3), the diffusion provides a remarkably good approximation to the fixation probability even for very small populations.

In the absence of selection and mutation, the fixation probability can be found from (15.39) by taking the limit as s goes to zero using L'Hôpital's rule (A2.30), in which case $u(1 \mid p_0) = p_0$. This shows that the fixation probability equals the initial allele frequency, as we might expect when all alleles are equally fit. It is instructive, however, to rederive the fixation probability by noting that $A(p)$ is zero whenever $\mu(p)$ is zero, so that $S(p)$ becomes p. The fixation probability is then easy to solve using equation (15.35a) and again gives $u(1 \mid p_0) = (p_0 - 0)/(1 - 0) = p_0$. This makes it clear that the fixation probability will equal p_0 whenever $\mu(p)$ is zero, regardless of the amount of variation caused by chance events during reproduction (as measured by the diffusion coefficient).

15.3.4 Time to Absorption in a Diffusion Model

While a stochastic model eventually becomes absorbed whenever absorbing states exist and can be reached, whether absorption occurs slowly or quickly

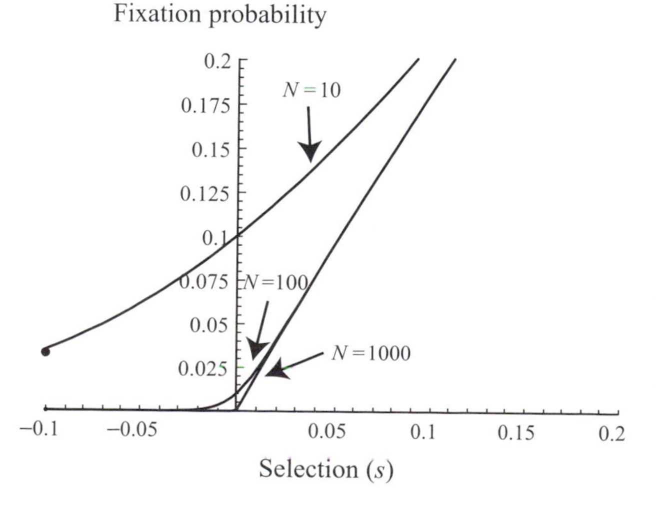

Figure 15.5: The probability of fixation in the Wright-Fisher model. Starting from a single allele ($p_0 = 1/N$), the probability that the allele ultimately fixes within a population is shown as a function of its selection coefficient, s. The diffusion approximation using (15.39) (curves) is compared to exact numerical results using (14.29) for $N = 10$ and $N = 100$.

cannot be determined from the absorption probability, $u(b|x_0)$. To determine the relevance of the absorbing states for any biological model, we need to know the time frame over which absorption occurs. Does the fixation of an allele or the extinction of a population tend to occur over the course of a few years or millions of years? We derive the expected waiting time to absorption in Supplementary Material 15.2 (it is similar in flavor to the derivation of the absorption probability) and summarize the results here.

Recipe 15.3
Waiting Time until Absorption in a Diffusion Model

Suppose that there are two absorbing state labeled a and b. If the system begins in state x_0, the expected waiting time until absorption in state b, conditioned upon reaching state b, is

$$\bar{t}_b(x_0) = 2(S(b) - S(a))\left(\int_{y=x_0}^{b} \frac{u(b|y)(1 - u(b|y))}{e^{-A(y)}\sigma^2(y)} dy + \frac{(1 - u(b|x_0))}{u(b|x_0)} \int_{y=a}^{x_0} \frac{u(b|y)^2}{e^{-A(y)}\sigma^2(y)} dy \right). \quad (15.40)$$

Similarly, the expected waiting time until absorption in state a, conditioned upon reaching state a, is

$$\bar{t}_a(x_0) = 2(S(b) - S(a))\left(\frac{(1 - u(a|x_0))}{u(a|x_0)} \int_{y=x_0}^{b} \frac{u(a|y)^2}{e^{-A(y)}\sigma^2(y)} dy + \int_{y=a}^{x_0} \frac{u(a|y)(1 - u(a|y))}{e^{-A(y)}\sigma^2(y)} dy \right). \quad (15.41)$$

The expected time until absorption in any state, $\bar{t}(x_0)$, is the average of the above two equations, each weighted by the probability of absorption in the corresponding state: $\bar{t}(x_0) = u(b|x_0)\,\bar{t}_b(x_0) + u(a|x_0)\,\bar{t}_a(x_0)$ (see Rule P3.9). Using the fact that the fixation probabilities sum to one, $u(a|x_0) + u(b|x_0) = 1$, the expected waiting time until absorption in either state is given by

$$\bar{t}(x_0) = 2(S(b) - S(a))\left(u(b|x_0) \int_{y=x_0}^{b} \frac{(1 - u(b|y))}{e^{-A(y)}\sigma^2(y)} dy + (1 - u(b|x_0)) \int_{y=a}^{x_0} \frac{u(b|y)}{e^{-A(y)}\sigma^2(y)} dy \right). \quad (15.42)$$

If there is only one absorbing state in the model, the expected waiting time until absorption is

$$\bar{t}(x_0) = 2 \int_{y=a}^{x_0} e^{-A(z)} \int_{y=z}^{b} \frac{e^{A(y)}}{\sigma^2(y)} dy\, dz, \quad (15.43)$$

where we assume that all processes are eventually absorbed in state a and that state b is a (potentially infinite) upper boundary that is either unattainable or rapidly left (see Karlin and Taylor 1981; Lande 1993).

Example: Expected Time to Absorption in the Wright-Fisher Model

Let us apply Recipe 15.3 to the Wright-Fisher model of allele frequency change. We focus on the case where there is no selection or mutation, in which case there are two absorbing states at $a = 0$ (allele loss) and $b = 1$ (allele fixation). The drift and diffusion coefficients are given by $\mu(p) = 0$ and $\sigma^2(p) = p(1 - p)$ (see Box 15.3). The first step is to calculate the various quantities needed in the waiting time; specifically, we have $A(p) = 0$, $S(p) = p$, and $u(1 \mid p_0) = p_0$. Plugging these results into (15.40) and integrating (see Problem 15.5), we find that the average time until fixation starting from an initial allele frequency p_0 is

$$\bar{t}_1(p_0) = -2\frac{1 - p_0}{p_0} \ln(1 - p_0). \tag{15.44}$$

The unit of time in (15.44) is measured in whatever time scale is required for the diffusion approximation to hold. In Box 15.3, the appropriate scale for the Wright-Fisher model measured time in units of N generations in a haploid population. Thus, we must multiply (15.44) by N to obtain the fixation time in generations. (This result also applies to a diploid population if we replace N with $2N$.) The result is plotted in Figure 15.6 as a function of the initial allele frequency. The fixation time is on the order of N generations and drops only when the initial allele frequency is very near one. This reflects the fact that allele frequencies vary so much under random genetic drift that the initial allele frequency is not very predictive of the exact time at which fixation occurs.

Of particular interest is the time until fixation for an allele that appears in a single copy ($p_0 = 1/N$). Taking the Taylor series of (15.44) with respect to p_0 using Recipe P1.3, we find that the mean time until fixation is, to leading order, $2N$ generations in a haploid population ($4N$ in a diploid population). Interestingly, this result is consistent with the time that it takes for a population to coalesce down to a single individual (see equation (13.17) with n large). This parallel makes sense; looking either forward in time or backward in time, the fixation time and the coalescence time measure how long it takes, on average, until a single ancestor gives rise to every member of a population. The waiting time

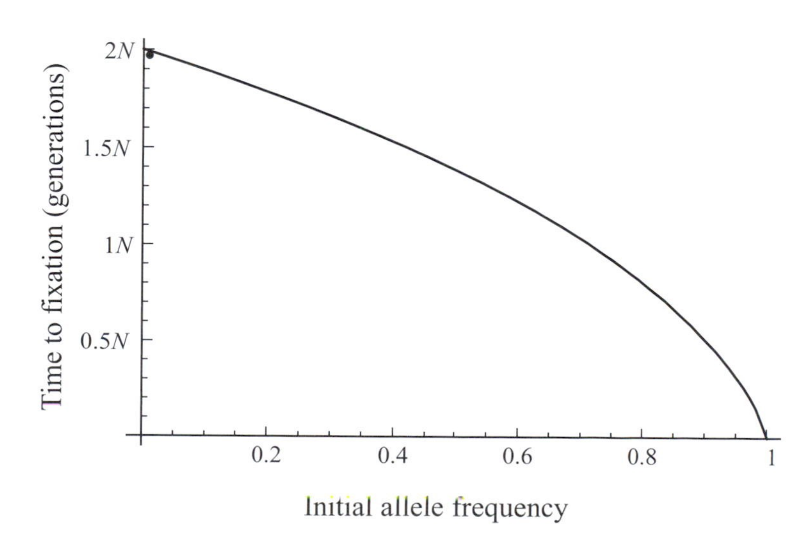

Figure 15.6: Fixation time for neutral alleles in the Wright-Fisher model. The expected time to fixation of a neutral allele is plotted as a function of the initial allele frequency. The waiting time is averaged across only those cases in which the neutral allele ultimately fixes. The solid curve is based on the diffusion approximation (15.44), while the dots give the exact waiting time for a population of size $N = 100$ using equation (14.48).

until fixation is slightly longer than the coalescent time because the fixation time includes any periods during which the stochastic process leaves and later returns to the state with only a single ancestor ($p = 1/N$) whereas the coalescent time excludes these periods.

In the presence of selection in the Wright-Fisher model, we previously found the probability of fixation is given by (15.39) and $A(p) = 2\,\psi\,p$, where $\psi = N\,s$. Plugging these values into (15.40) results in an integral that cannot be evaluated for general values of the parameters. Nevertheless, we can integrate the equation numerically using software such as *Mathematica* (Figure 15.7). One interesting finding is that the time until fixation is the same whether the selection coefficient s is positive or negative (Ewens 1979). At first, this result makes no sense. Clearly, if an allele is favorable, it should fix faster when selection is stronger. What is less obvious is that if an allele is disfavored, it also tends to fix faster when selection against it is stronger. The reason is that there must be large and rapid chance increases in the allele frequency for a disfavored allele to rise to fixation in the face of strong selection against it. Such chance increases are unlikely, and a disfavored allele has a low fixation probability. But given that a disfavored allele does fix, it does so over the same average time frame as a favored allele experiencing selection of the same magnitude.

A similar result occurs in the birth-death model of a population at risk of extinction. The average time until a population becomes established (i.e.,

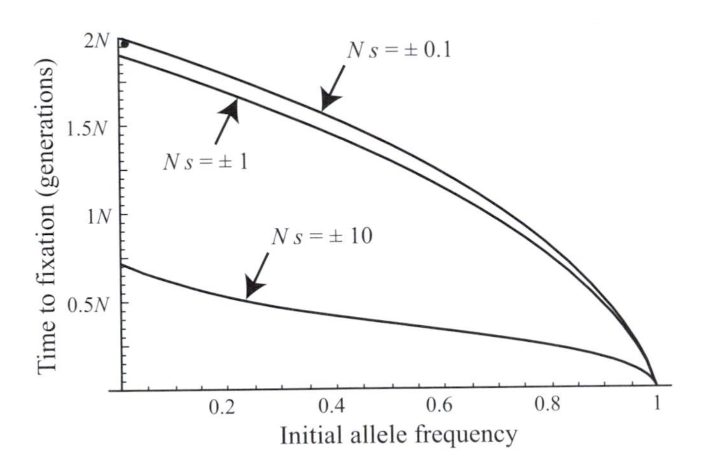

Figure 15.7: Fixation time for selected alleles in the Wright-Fisher model. The expected time to fixation of a selected allele is plotted as a function of the initial allele frequency of A (p_0). The waiting time is averaged across those cases in which the A allele ultimately fixes. The solid curve is based on numerical integration of the diffusion approximation (15.40), and the dots give the exact waiting time for a population of size $N = 100$ using equation (14.48). Parameters used were $N\,s = -10, -1, -0.1, 0.1, 1,$ and 10. Surprisingly, the waiting time for a deleterious allele A to fix cannot be distinguished from the waiting time for a beneficial allele, given the same magnitude of selection. (For the exact numerical results, the fitnesses were set to $W_a = 1$, $W_A = 1 + s$ when A was beneficial and to $W_a = 1 + s$, $W_A = 1$ when A was deleterious. This choice keeps the strength of selection equivalent in the two cases.)

reaches a threshold size m) is the same whether the birth rate is greater than the death rate or vice versa (Figure 15.8a). The same is true for the average time until extinction (Figure 15.8b). Of course, if deaths are more common than births, the most likely fate of the population is extinction (see Problem 15.6). But, assuming that the population has risen from some initial size to m, there is no way to know for sure whether the population truly had a higher birth rate than death rate or whether there happened, by some small chance, to be more births than deaths during the monitoring period despite the fact that the expected birth rate is less than the death rate.

The above results depend critically on the assumption that there are two absorbing boundaries. For example, compare Figure 15.8 to Figure 15.9, which is based on the same birth-death model, except that we now impose a "hard carrying capacity" on the population at K. Specifically, when $i = K$, the probability of a birth drops to zero. This model has only one absorbing boundary: extinction of the population. As shown in Figure 15.9, the average time until extinction based on (15.43) now rises dramatically when the birth rate exceeds the death rate.

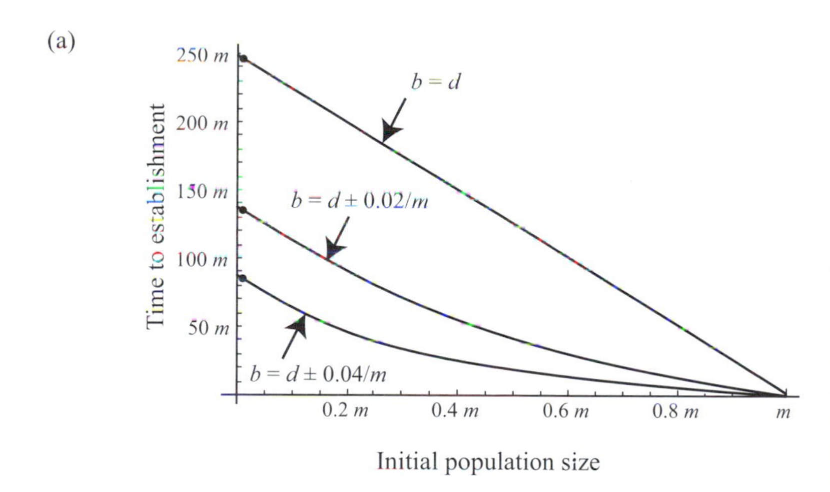

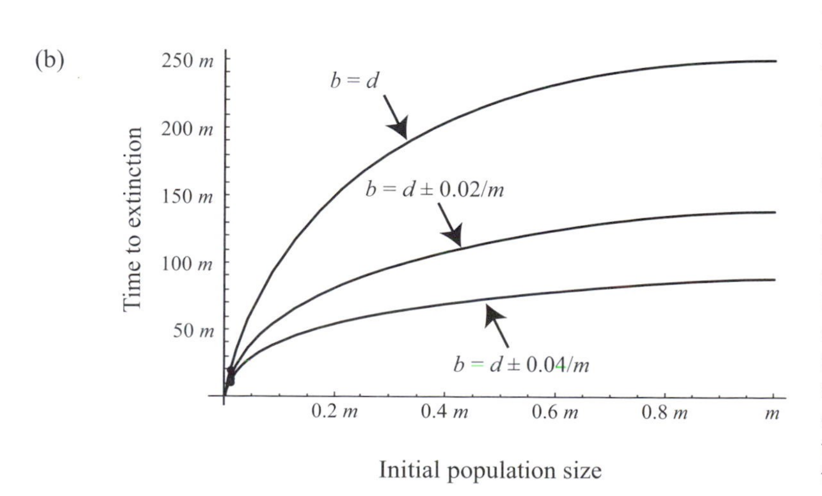

Figure 15.8: Expected waiting time until extinction or recovery. The expected waiting time until an at-risk population becomes (a) established or (b) extinct is plotted as a function of the initial population size. Population dynamics were assumed to follow a simple birth-death process (section 14.4). Using the drift and diffusion coefficients from Problem 15.6, the solid curves are based on the diffusion approximations (15.40) and (15.41), and the dots give the average waiting time using the exact equation (14.48) with $m = 100$. The horizontal axis represents the initial population size, measured as a fraction of the threshold size, m. The vertical axis represents the waiting time, scaled by m. The total probability of an event per individual per time step was held constant at $b + d = 0.004$, so that a single individual undergoes, on average, one birth or death every 250 time steps, with $b = d$ (highest curves), $b = d \pm 0.02/m$ (middle curves), and $b = d \pm 0.04/m$ (lowest curves). The waiting time depends on whether births and deaths differ in frequency but not on which is more frequent.

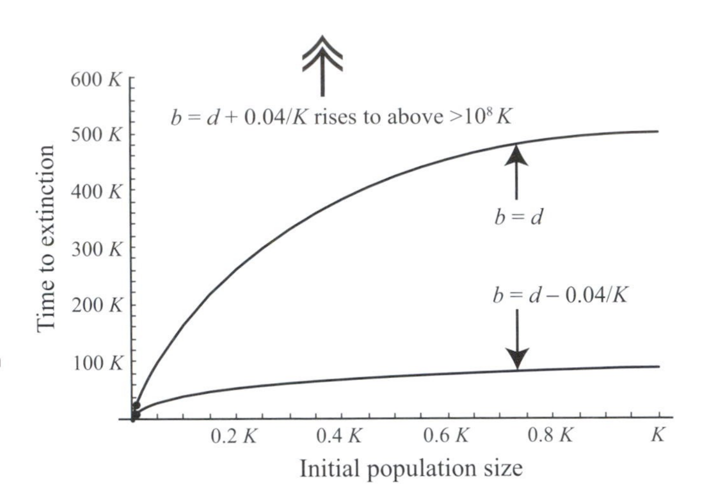

Figure 15.9: The average waiting time until extinction of a population with a hard carrying capacity of K. Population dynamics were assumed to follow the simple birth-death process (section 14.4) with a hard carrying capacity of K. The solid curves are based on the diffusion approximation (15.43) with drift and diffusion coefficients given in Problem 15.6, and the dots give the average waiting time for $K = 100$ using the exact equation (14.38). Again, $b + d = 0.004$, with $b = d + 0.04/K$ (highest curve rises to above 10^8 K time steps and is off the scale of the figure), $b = d$ (middle curve), and $b = d - 0.04/K$ (lowest curve).

15.4 Modeling Populations in Space Using the Diffusion Equation

In section 15.2.1, we introduced a stochastic model of fish movement and showed that a diffusion model could be used to describe the probability that an individual fish is found at position x at time t. What if we were tracking an entire population of fish, each of which moves stochastically over space? Provided that the movement of each individual is independent of all others and provided that there is a large number of individuals in the population, then $\phi(x,t \mid x_0,t_0)$ can be thought of as describing the *fraction* of a population currently found at position x (e.g., the distribution of a population of fish over space). The requirement that the population be large follows from the frequency interpretation of a probability (Primer 3); we can interpret the probability of an event as the fraction of times it occurs in a very large number of trials. Interestingly, under this interpretation, the stochastic equations describing individual movement provide a *deterministic* description of the entire population over space.

Example: Sperm Movement in a Reproductive Tract

The reproductive tract of females in many species can be viewed as a one-dimensional structure through which sperm travel to reach and fertilize an egg. Typically there are many millions of sperm transferred to a female during a mating event, and therefore it is probably not unreasonable to suppose that the number at any given location is always very large. Furthermore, it seems reasonable to model the movement of each spermatozoa as being stochastic, using a diffusion model with drift.

Let us suppose that the drift rate and the variance are constant throughout the reproductive tract, and denote them by μ and σ^2. Thus, μ represents the average direction of movement of the individual sperm at a particular location, and σ^2 represents the variance in the movement of an individual. We index each location along the reproductive tract by the distance to the egg and suppose that the egg is located at $x = b$. We will view both ends of the reproductive tract as being absorbing boundaries with the rationale that sperm that exit the tract at $x = 0$ are lost, and sperm that reach the egg ($x = b$) attempt to fertilize the egg. We can then use Recipe 15.2 to calculate the probability that any given sperm ends up at the egg (i.e., absorption at $x = b$). Alternatively, given the assumption that each of the millions of sperm move independently, this Recipe can also be interpreted as giving us the proportion of the sperm that end up at the egg.

In fact, the calculations of $A(x)$ and $S(x)$ are identical to those for the model of fish movement: $A(x) = (2\mu/\sigma^2) \int 1\, dx = 2\mu x/\sigma^2$ and $S(x) = -(\sigma^2/2\mu)\, e^{-2\mu x/\sigma^2}$. Therefore, from equation (15.35a), the proportion of the sperm that make it to the egg is

$$
\begin{aligned}
u(b \mid x_0) &= \frac{S(x_0) - S(0)}{S(b) - S(0)} \\
&= \frac{-e^{-2\mu x_0/\sigma^2} + 1}{-e^{-2\mu b/\sigma^2} + 1}
\end{aligned}
\tag{15.45}
$$

where x_0 is the initial location of all the sperm, which we assume to be nearer 0 than b. Equation (15.45) reveals that, unless there is a large amount of directionality in movement (i.e., large μ), a very small proportion of the sperm actually make it to the egg (Figure 15.10). This might be one reason why such large numbers of sperm are transferred during mating. Indeed, it has been hypothesized that the very long and intricate reproductive tracts of some animals have evolved to select vigorous sperm with the greatest motility (Parker 1970).

Example: The Drift Paradox

Let us return to our model of movement in a stream, but now we consider a population of aquatic invertebrates (e.g., the aquatic stage of mayflies). Because invertebrates are typically quite small, it is reasonable to suppose that the population size at any given location in the stream is quite large. Thus, if each individual moves independently and stochastically, we can interpret a model of diffusion with drift as giving the dynamics of the number of individuals in each location.

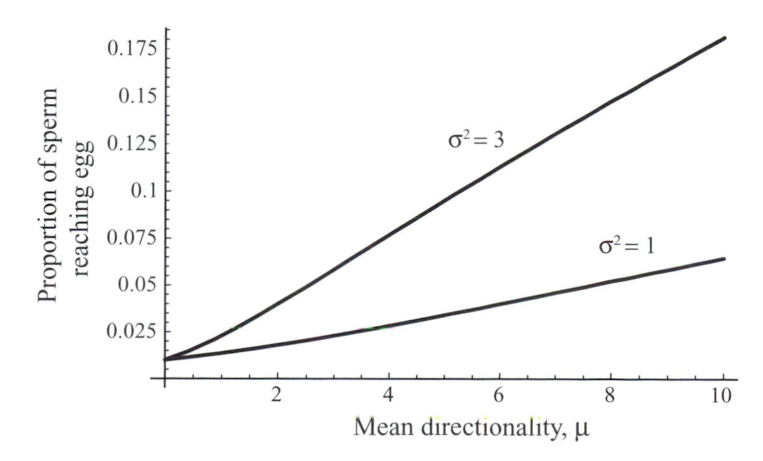

Figure 15.10: Proportion of sperm reaching an egg. It was assumed that sperm start out 1/100 of the way along the reproductive tract ($x_0 = 0.01$, $b = 1$) and travel according to a diffusion model with drift coefficient μ and diffusion coefficient σ.

Let us keep the model as general as possible and suppose that the average direction of movement of individuals at location x, as well as the variance in their movement, depends on the location in the stream. Assuming that movement occurs continuously over time and space, the population size at each location, denoted by $n(x)$, obeys the diffusion equation with drift (15.8); that is,

$$\frac{\partial n(x, t)}{\partial t} = -\frac{\partial(\mu(x)\, n(x, t))}{\partial x} + \frac{1}{2}\frac{\partial^2(\sigma^2(x)\, n(x, t))}{\partial x^2} \tag{15.46}$$

To complete the model we need to specify what happens to the population at the boundaries $x = b$ (taken to be the upper reaches of the stream) and $x = 0$ (taken to be where the stream empties into a lake or ocean). It is reasonable to suppose that there is a reflecting boundary at the top of the stream (i.e., individuals cannot go past this point but they are not "absorbed" there either) and that there is an absorbing boundary at $x = 0$ (the invertebrates exit the stream).

What does the above model predict for the dynamics of the invertebrate population? Our results from section 15.3.3 on the probability of absorption reveal that the population of invertebrates eventually go extinct in the stream because there is only one absorbing state. In other words, eventually 100% of the population of invertebrates will be "absorbed" at state $x = 0$. This finding is referred to as the "drift paradox" (Speirs and Gurney 2001). Clearly populations of invertebrates exist in streams, so how are they able to persist?

There are several resolutions to this paradox, but two possibilities stand out. The first is that, even if the invertebrate population eventually disappears, the time to extinction might be exceedingly long. The second is that we have neglected an important feature of the biology of invertebrate populations: reproduction! To remain faithful to the stochastic underpinnings of the original model, we must model not only how each individual moves stochastically over time, but also how they interact and reproduce with other individuals and how they die.

Constructing a complete stochastic model for the above processes is a difficult enterprise, and simplifications are often made. One of the most common is to interpret the stochastic model of movement as a deterministic description of a large population, to which we add further deterministic components to describe other processes of interest.

As an example, reproduction can be included in model (15.46) by supposing that the birth rate of new invertebrates at location x is given by some function, $f(n(x, t))$:

$$\frac{\partial n(x, t)}{\partial t} = f(n(x, t)) - \frac{\partial(\mu(x)\, n(x, t))}{\partial x} + \frac{1}{2}\frac{\partial^2(\sigma^2(x)\, n(x, t))}{\partial x^2}. \tag{15.47}$$

Ignoring the last two terms describing movement, $\partial n(x, t)/\partial t = f(n(x, t))$ is akin to the deterministic models of population growth considered in Chapter 3 (e.g., exponential or logistic growth). This seemingly small addition, however, brings with it major changes in behavior. In all stochastic models of movement considered so far, individuals were neither created nor destroyed (although they might "pile up" at a boundary). In contrast, equations such as (15.47) now allow the population to grow or shrink over time. The analysis of spatial diffusion equations with varying population sizes is considerably more difficult and beyond the scope of this book (interested readers should consult Britton 1986; Edelstein-Keshet 1988; Kot 2001; Okubo and Levin 2001).

Interestingly, analyses of the drift paradox have shown that reproduction does not guarantee the persistence of aquatic invertebrates unless the rate of reproduction is sufficiently high (Speirs and Gurney 2001). Alternative explanations for the persistence of stream dwelling populations have also been explored using diffusion models, including the importance of zones where drift and diffusion rates are minimal, such as along the edge of a river (Pachepsky et al. 2004).

Example: Hybrid Zones

In Supplementary Material 15.3, we illustrate how a diffusion approach can be extended to explore the dynamics of evolutionary change in a spatial setting. We consider a model of how the frequencies of two types vary across a "hybrid zone," which is a region in space where two types meet and produce less fit hybrids. The model again assumes that the population is large at every point in space, so that a diffusion equation can be used to describe the effects of dispersal on the population distribution over space. Selection is then incorporated as an additional deterministic process, much as population growth was added to equation (15.47). In the case of a hybrid zone, it is possible to make substantial analytical progress, and we describe results on the shape of the hybrid zone when a balance is reached between dispersal and selection.

15.5 Concluding Message

In this chapter, we have introduced various methods for constructing and analyzing stochastic models with continuous state spaces, focusing on a diffusion model with drift. Diffusion models have played an important role in many areas of science, including biology. For example, they have been used to describe the diffusion of retroviruses used in gene therapy (Chuck et al. 1996), the diffusion of water through normal and injured brain tissue (Melhem 2002), and changes in lizard morphology over evolutionary time (Schluter et al. 1997). They have also been used to estimate the time until extinction of populations in the presence of demographic and/or environmental stochasticity (Lande 1993) and to assess the minimum population size needed to reduce the risk of extinction during a time period of interest (e.g., less than 1% extinction risk over 1000 years). Such analyses have played an important role in scientific assessments of the level of risk facing endangered species (Clegg 1995). From this small selection of applications alone, it is clear that diffusion methods form an important group of mathematical tools in biology.

Problems

Problem 15.1: For the model of fish movement in section 15.2.1 derive the probability that the fish exits the monitored region at the downstream end, using the results for the birth-death model from Chapter 14. Compare your result to equation (15.37) for the continuous-space model.

Problem 15.2: Use the techniques of Box 15.2 to derive a diffusion approximation for the Moran model for a neutral allele. (a) Show that the choices $\alpha = 2$ and $\beta = 1$ are appropriate scaling parameters. This implies that one time unit in the diffusion approximation corresponds to N^2 time units in the original model. (b) Calculate the drift and diffusion coefficients from (15.2.6). (c) Use your results from (b) in the forward Kolmogorov equation (15.8) to derive a diffusion approximation for the Moran model. (d) Show that your result is consistent with (15.14) by rescaling time.

Problem 15.3: For the Brownian motion model, verify that the general solution (15.15) satisfies (a) the forward Kolmogorov equation (15.8) and (b) the backward Kolmogorov equation (15.11).

Problem 15.4: (a) Following analogous calculations to those in equation (15.22), show that the rate of change of the nth moment for the adaptive dynamics model specified by equations (15.19) and (15.21) is given by $\eta\, \mathcal{M}_n - (\eta\, (N-1)\, s/a_K)\, \mathcal{M}_{n+1}$, where $\mathcal{M}_n$ denotes the nth central moment of the mutational distribution, which is assumed to be normal: $e^{-(s_m - s)^2/(2v)}/\sqrt{2\pi v}$. (b) Use central moment generating functions (Appendix 5) to show that all odd moments of a normal distribution are zero. (c) Use central moment generating functions to show that the even moments of a normal distribution are proportional to the variance raised to a power that increases for higher and higher moments. (d) Use the results from (a) to (c) to argue that all higher moments of the adaptive dynamics model will therefore be negligible provided that the mutational variance v is very small.

Problem 15.5: In this problem, we examine the diffusion approximation to the Wright-Fisher model of allele frequency change in haploid population in the absence of selection and mutation. Use the drift and diffusion parameters $\mu(p) = 0$ and $\sigma^2(p) = p\,(1-p)$ from Box 15.3 to (a) derive the average time until fixation given that the allele is fixed, $\overline{t_1}(p_0)$, and confirm that it equals (15.44), (b) derive the average time until loss given that the allele is lost, $\overline{t_0}(p_0)$, and (c) derive the average time until loss or fixation, regardless of which occurs first, $\bar{t}(p_0)$. (d) Finally, take the Taylor Series with respect to p_0 of your results to parts (a) – (c) assuming that the allele is initially rare, then set $p_0 = 1/N$ to obtain the waiting times starting from a single mutation. In each case, specify the time scale. [Hint: You can check your answer to part (b) by thinking about the relationship between $\overline{t_1}(p_0)$ and $\overline{t_0}(1 - p_0)$ in this neutral model.]

Problem 15.6: Derive a diffusion approximation to the birth-death model describing the number of individuals in a population that is at risk of extinction. Assume that the birth and death probabilities are given by $b\, i$ and $d\, i$, where i is the population size. Also assume that the population is tracked until it reaches a minimum viable population size m. (a) Use the method outlined in Box 15.2 to derive the drift and diffusion parameters and to show that the third moment equals zero. [Hint: You can use the scale parameters $\alpha = 1$ and $\beta = 1$, but you will need to assume that $(b - d)$ is small in the same way that we assumed s was small in Box 15.3.] (b) From these drift and diffusion parameters, show that $A(x) = \int 2\mu(x)/\sigma^2(x)\, dx$ is a linear function of $x = i/m$. Use this fact and equation (15.35) to write down the probability that the population reaches the minimum viable population size before going extinct. (c) Plot your result from (b) and the exact result (14.29) for this birth-death model for your choice of parameters. How well does the diffusion approximation perform?

Problem 15.7: Derive a diffusion approximation for the diploid version of the Wright-Fisher model with selection in the absence of mutation, following the steps in Box 15.3. We assume that random mating produces N diploid individuals that experience viability and fertility selection such that the total frequency of the A allele among their gametes becomes

$$p' = \frac{(1 + s)\, p^2 + p\,(1 - p)(1 + h\,s)}{(1 + s)\, p^2 + 2\,p\,(1 - p)(1 + h\,s) + (1 - p)^2}.$$

$2N$ gametes are then sampled at random to produce the next generation of diploids, following a binomial distribution with parameters $2N$ and p'. (a) Show that the diffusion approximation is valid if we rescale selection using $\psi = 2Ns$ and the scaling parameters $\alpha = \beta = 1$. Specifically, show that the drift and diffusion coefficients are finite. [EXTRA CHALLENGE: Show that the third moment is zero.] (b) Using this scaling, what are the drift and diffusion coefficients and how is time measured? (c) Use equation (15.35b) to write the probability of fixation for an allele A that is initially at frequency p_0 (do not evaluate the integrals). (d) If allele A is beneficial, its fate is generally determined while it remains rare. In this case, show that we can approximate $A(p) = \int 2\mu(p)/\sigma^2(p)\, dp$ by $A(p) = \int 2\,h\,\psi\, dp$. Having done so, carry out the integrals in (15.35b) and derive the probability of fixation. Compare your result to the fixation probability in the haploid model (15.39). [Note: In general, the integrals in (c) cannot be evaluated analytically. For alleles whose fates are not decided while rare, such as recessive beneficial alleles and deleterious alleles, the integrals in (c) must be evaluated numerically to determine the probability of fixation.]

Problem 15.8: Consider a birth-death model of population growth incorporating density dependence. Specifically, in a time step, a birth occurs with a density-dependent probability $b\,i\,(1 - i/K)$, and a death occurs with a density-independent probability $d\,i$, where i is the current population size and K is the maximum population size (the birth probability is zero at $i = K$). (a) Calculate the first three moments describing the change in population size (15.2.1). (b) Assuming that a diffusion approximation is valid and using the first and second moments from (a) for the drift and diffusion coefficients, plot the average time to extinction for various values of d between 0.001 and 0.01, using $i = 10$, $K = 100$, and $b = 0.002$. Because there is only one absorbing boundary (at $i = 0$) in this model, use equation (15.43) and a mathematical software package to evaluate the integrals numerically. (c) Show that your results from (b) are similar to a numerical evaluation of the exact time to extinction using Recipe 14.3. (d) Show that the restrictions (15.2.6) required for the diffusion approximation to be valid can be met by choosing $\alpha = \beta = 1/2$ (with $m = K$), as long as the intrinsic birth rate minus the death rate is small, such that $(b - d)\,K^{1/2} = \psi$. (e) Repeat part (b) using the drift and diffusion coefficients from part (d). [Hint: The results should be the same if you rescale the variables appropriately.]

Further Reading

For more information on the mathematical underpinnings of diffusion models, see

- Karlin, S., and H. M. Taylor. 1981. *A Second Course in Stochastic Processes*. Academic Press, New York.
- Allen, L. J. S. 2003. *An Introduction to Stochastic Processes with Applications to Biology*. Pearson Prentice Hall, Upper Saddle River, N.J.

For more information on diffusion models in evolution and ecology, see

- Crow, J. F., and M. Kimura. 1970. *An Introduction to Population Genetics Theory*. Harper & Row, New York.
- Ewens, W. J. 1979. *Mathematical Population Genetics*. Springer-Verlag, Berlin.
- Lande, R., and B.-E. Sæther. 2003. *Stochastic Population Dynamics in Ecology and Conservation*. Oxford University Press, Oxford.
- Nisbet, R. M., and W.S.C. Gurney. 1982. *Modelling Fluctuating Populations*. Wiley, New York.
- Rice, S. 2004. *Evolutionary Theory: Mathematical and Conceptual Foundations*. Sinauer Associates. Sunderland, Mass.

References

Allen, L. J. S. 2003. *An Introduction to Stochastic Processes with Applications to Biology*. Pearson prentice Hall, Upper Saddle River, N.J.

Britton, N. F. 1986. *Reaction-Diffusion Equations and Their Applications to Biology*. Academic Press, New York.

Chuck, A. S., M. F. Clarke, and B. O. Palsson. 1996. Retroviral infection is limited by Brownian motion. *Hum. Gene Therapy* 7:1527–1534.

Clegg, M. T. 1995. *Science and the Endangered Species Act*. National Academy Press, Washington D.C.

Crow, J. F., and M. Kimura. 1970. *An Introduction to Population Genetics Theory*. Harper & Row, New York.

Dieckmann, U., and R. Law. 1996. The dynamical theory of coevolution: a derivation from stochastic ecological processes. *J. Math. Biol.* 34:579–612.

Edelstein-Keshet, L. 1988. *Mathematical Models in Biology*. McGraw-Hill, New York.

Ewens, W. J. 1979. *Mathematical Population Genetics*. Springer-Verlag, Berlin.

Karlin, S., and H. M. Taylor. 1981. *A Second Course in Stochastic Processes*. Academic Press, New York.

Kimura, M. 1957. Some problems of stochastic processes in genetics. *Ann. Math. Stat.* 28:882–901.

Kimura, M. 1962. On the probability of fixation of mutant genes in a population. *Genetics* 47:713–719.

Kot, M. 2001. *Elements of Mathematical Ecology*. Cambridge University Press, Cambridge.

Lande, R. 1993. Risks of population extinction from demographic and environmental stochasticity and random catastrophes. *Am. Nat.* 142:911–927.

Melhem, E. R. 2002. Time-course of apparent diffusion coefficient in neonatal brain injury: the first piece of the puzzle. *Neurology* 59:798–799.

Okubo, A., and S. A. Levin. 2001. *Diffusion and Ecological Problems*. Springer-Verlag, Berlin.

Pachepsky, E., F. Lutscher, R. M. Nisbet, and M. A. Lewis. 2004. Persistence, spread and the drift paradox. *Theor. Popul. Biol.* 67:61–73.

Parker, G. 1970. Sperm competition and its evolutionary consequences in the insects. *Biol. Rev.* 45:525–556.

Schluter, D., T. D. Price, A. Ø. Mooers, and D. Ludwig. 1997. Likelihood of ancestor states in adaptive radiation. *Evolution* 51:1699–1711.

Speirs, D. C., and W.S.C. Gurney. 2001. Population persistence in rivers and estuaries. *Ecology* 85:1219–1237.

Taylor, P., and T. Day. 1997. Evolutionary stability under the replicator and the gradient dynamics. *Evol. Ecol.* 11:579–590.

EPILOGUE
The Art of Mathematical Modeling in Biology

In this book, we have covered a lot of mathematical and biological terrain. We began in Chapter 2 by laying out a series of steps involved in modeling, emphasizing that mathematical biology is both a science and an art. Although we have illustrated the artistic aspect of modeling throughout the book, it is by far the most difficult part to learn. Only after gaining experience with models can we begin to appreciate the art of modeling and to be able to discern good models from poor ones. With the collection of examples presented in the preceding chapters, we are now in a better position to return to this artistic side of modeling. In this epilogue we do so by asking two main questions: What are models for? and What makes a model "good"?

What are models for? There is no simple answer to this question, but there are at least two broad goals that most models have: (i) shaping our understanding and intuition about biological phenomena, and/or (ii) making quantitative predictions about biological phenomena.

For the goal of shaping our understanding and intuition, the model itself is not really the ultimate object of interest. Rather, the model is used to uncover new aspects of the biology that can then be incorporated into our thinking without reference to the model itself. The models of HIV replication by Ho et al. (1995), Nowak et al. (1995), and Phillips (1996) introduced in Chapter 1 are good examples. Although all of these models were intended to make accurate quantitative predictions about the within-host dynamics of HIV replication, ultimately the goal of each was to revise our thinking about the processes occurring during HIV infection. The lasting impact of these studies is not so much in the models themselves, as in the way they have altered our understanding of HIV dynamics.

The model of Williams et al. (2001) introduced in Chapter 1 is more aligned with the second goal: making quantitative predictions. The goal of this model was to generate quantitative estimates of the risk of contracting HIV as a function of a woman's age in Hlabisa, South Africa. The model was not intended to change our intuition about HIV transmission, but rather to quantify who is most at risk, so that these age groups could be better served. This quantitative information is useful from the standpoint of developing public health policy, and it can be further updated, using the model, as more precise estimates of HIV prevalence become available.

More generally, however, most models usually accomplish a blend of both goals. For example, the models of Ho et al. (1995) and Nowak et al. (1995) not only reshaped the way we think about HIV infections (demonstrating that there is an active and continual turnover of HIV in the body), but they also provided quantitative predictions about the rate of HIV replication during an infection. Similarly, the model of Blower et al. (2000) that tracked the dynamics of HIV spread within the gay male population of San Francisco (introduced in Chapter 1), not only made important quantitative predictions about HIV spread in San Francisco, but it also provided the initially counterintuitive insight that disease incidence can actually go up as a result of the availability of anti-retroviral therapy.

What makes a model "good"? This question is even more difficult to answer. The most obvious answer is that a model is good if it correctly describes the phenomenon of interest. But this is giving a model too much credit—it can never correctly and fully describe a biological phenomenon, if for no other reason than we cannot perceive or measure all aspects of any phenomenon. We might be able to determine whether a model is patently false, but we can never say that it is correct.

What we can assess is whether a model is useful. Some models are better than others at accomplishing the two goals mentioned above: to provide understanding and to provide quantitative predictions. As an example, consider Newton's model of object movement under the force of gravity. This model is incredibly useful, as it allows physicists to predict the position of objects over time. This model turned out to be incorrect, as Einstein showed with his theory of relativity, but it remains a very useful approximation.

These same principles hold for modeling in biology. Models are never correct, but some models are better than others at providing understanding or reasonably accurate predictions. Importantly, the utility of a model must be judged relative to other available models. At any given time, we place most faith in models that best accomplish our goals. As such, modeling is always an iterative process, with current models continually being revised. We must then judge whether a revised model is better than the previous model. Although there is no single criterion for making this decision, factors that seem to play a role include the breadth and generality of phenomena that the model can explain, the extent to which the model can be tested empirically, and the elegance and simplicity of how the model explains nature.

The various versions of the crab-anemone model developed in section 7.4 provide a good example of these issues. The model was developed to address whether including mortality differences of crabs on anemones of different colors, coupled with measured preferences of crabs for these anemones, could explain observed patterns of anemone use by crabs in nature. The first version of the model made predictions that ran counter to what we know about nature; namely, exponential growth to an infinite number of occupied sea anemones or decay to zero. Consequently, we revised the model. The second version of the model altered the way that crab recruitment to the population was specified and fixed the problem noticed in the first version, but then we noticed another prediction that was clearly wrong: the equilibrium numbers of occupied anemones did not depend on the number of available anemones of each color. As a result, we revised the model a third time, adding the assumption that a crab cannot take over an already occupied anemone. Unlike the second model, our third model correctly predicted the observed patterns of anemone use in nature, given experimental data on the model parameters.

The above example illustrates how useful it can be to seek out secondary predictions made by a model; that is, predictions that stem from the analysis, but that were not originally the object of study (e.g., the nature of the population dynamics in the crab-anemone model). The more the secondary predictions of a model match up with observations or assumptions about the biological system, the more stock we can place in the model.

Finally, simplicity and elegance are also critically important in determining which models withstand the test of time. Although you might think that elegance plays no role in science, most people can appreciate elegant results when they see them—a proof that takes only a few lines of mathematics and that stuns you with a realization is always going to be more satisfying than a barrage of equations demonstrating the same fact. Furthermore, it is not uncommon that the most elegant solutions are also the most general. Take our model of virulence evolution in section 12.4. Although we were able to make predictions about how background host mortality should affect virulence evolution by specifying particular functional forms for all aspects of the model (see discussion following equation (12.28b)), we were able to obtain much more general predictions by leaving the exact form of these functions unspecified (see equation 12.36). The resulting predictions stemming from this more general analysis appear more elegant as well.

The same holds for the example of density-dependent natural selection in section 8.3. There we specified functional forms for all aspects of the model, and, after a somewhat messy analysis, we were able to deduce useful predictions about evolution under density-dependent population growth. In Problem 8.15, however, you were asked to generalize this model without specifying the particular functional forms for all aspects of the model. Doing so generates a more elegant and a more general interpretation of the results.

One final reason why simplicity matters in model construction is that we often do not have enough information about parameter values to use complex models appropriately. Suppose, for example, we wanted to develop a model to make predictions about a novel disease that is currently spreading in the human population. We could build a model with vast amounts of detail, tracking the dynamics of each infection, as well as the contacts made by each individual in the population, but we would typically have very little information available to specify these parameter values. Furthermore, many of these details probably don't matter, with the main result depending largely on the average transmission rate and the distribution of contacts per person. Consequently, it might be much better to use an over-simplified model for these processes, but one for which more accurate parameter estimates are available. Starting simple, we could then add complexity (e.g., does variation in transmission rate affect the results?) and determine whether the results of the model are dramatically changed. If the added element of complexity is potentially present in nature, and if its presence has a large impact, then its inclusion in our model is essential to either goal of modeling: providing understanding or accurate predictions. Paraphrasing Albert Einstein, we encourage you to build your own models to address biological questions that interest you, aiming for models that are as simple as possible, but no simpler.

Appendix 1

Commonly Used Mathematical Rules

A1.1 Rules for Algebraic Functions

The following rules help to simplify functions involving powers and fractions:

Rule A1.1: $a^x a^y = a^{x+y}$

Rule A1.2: $\dfrac{1}{a^y} = a^{-y}$

Rule A1.3: $\dfrac{a^x}{a^y} = a^{x-y}$

Rule A1.4: $(a^x)^y = a^{xy}$

Rule A1.5: $a^{1/x} = \sqrt[x]{a}$; in particular, $a^{1/2} = \sqrt{a}$

Rule A1.6: $(a^x)^{1/x} = a$

Rule A1.7: $(a\,c)^x = a^x c^x$

Rule A1.8: $\left(\dfrac{a}{c}\right)^x = \dfrac{a^x}{c^x}$

Rule A1.9: $\dfrac{1}{(a_1 + b_1 x)(a_2 + b_2 x)} = \dfrac{A}{(a_1 + b_1 x)} + \dfrac{B}{(a_2 + b_2 x)}$ (partial fractions)

where $A = -\dfrac{b_1}{(a_1 b_2 - a_2 b_1)}$ and $B = \dfrac{b_2}{(a_1 b_2 - a_2 b_1)}$

Rule A1.10: If $a\,x^2 + b\,x + c = 0$ then $x = \dfrac{-b \pm \sqrt{b^2 - 4ac}}{2a}$ (quadratic formula)

A1.2 Rules for Logarithmic and Exponential Functions

The following rules help to simplify functions involving logarithms. On the left are rules for logarithms in any base, b, defined by the fact that if $y = b^x$ then $\log_b(y) = x$. On the right are specific rules for natural logs in base $e = 2.71\ldots$, defined by the fact that if $y = e^x$ then $\ln(y) = \log_e(y) = x$.

Rule A1.11: $\log_b(a^t) = t\,\log_b(a)$ $\qquad$ $\ln(a^t) = t\,\ln(a)$

Rule A1.12: $\log_b(a\,c) = \log_b(a) + \log_b(c)$ $\qquad$ $\ln(a\,c) = \ln(a) + \ln(c)$

Rule A1.13: $\log_b\left(\dfrac{1}{c}\right) = -\log_b(c)$ $\qquad$ $\ln\left(\dfrac{1}{c}\right) = -\ln(c)$

Rule A1.14: $\log_b\left(\frac{a}{c}\right) = \log_b(a) - \log_b(c)$ $\quad$ $\ln\left(\frac{a}{c}\right) = \ln(a) - \ln(c)$

Rule A1.15: $\log_b(b^x) = x$ $\quad$ $\ln(e^x) = x$

Rule A1.16: $b^{\log_b(x)} = x$ $\quad$ $e^{\ln(x)} = x$

A1.3 Some Important Sums

The following rules describe how certain sums can be evaluated and written in simpler terms. To interpret a sum, read $\Sigma_{i=1}^{n} f(i)$ as "the sum of the values of $f(i)$ starting with i equals one, then two, then three, etc., up until $i = n$." That is, $\Sigma_{i=1}^{n} f(i) = f(1) + f(2) + f(3) + \cdots + f(n)$. Sums starting with $i = 1$ are given on the left and starting with $i = 0$ on the right.

Rule A1.17: $\sum_{i=1}^{n} a = n\,a$ $\quad$ $\sum_{i=0}^{n} a = (n + 1)\,a$

Rule A1.18: $\sum_{i=1}^{n} i = \frac{n\,(n + 1)}{2}$ $\quad$ $\sum_{i=0}^{n} i = \frac{n\,(n + 1)}{2}$

Rule A1.19: $\sum_{i=1}^{n} a^i = \frac{a - a^{n+1}}{1 - a}$ $\quad$ $\sum_{i=0}^{n} a^i = \frac{1 - a^{n+1}}{1 - a}$

Rule A1.20: $\sum_{i=1}^{\infty} a^i = \frac{a}{1 - a}$ if $|a| < 1$ $\quad$ $\sum_{i=0}^{\infty} a^i = \frac{1}{1 - a}$ if $|a| < 1$

Rule A1.21: $\frac{1}{n}\sum_{i=1}^{n} x_i = \bar{x}$ $\quad$ (arithmetic mean)

Rule A1.22: $\frac{1}{(n - 1)}\sum_{i=1}^{n} (x_i - \bar{x})^2 = \mathrm{Var}(x)$ $\quad$ (sample variance)

Note that, in Rule A1.22, if the variance is based on the true value of the mean (i.e., the mean is known without error), then the sum should be divided by (n) rather than $(n - 1)$. For example, if the x_i values are known for every member of a population rather than just a sample, then the variance is given by $(1/n)\Sigma_{i=1}^{n} (x_i - \bar{x})^2$. The order in which a sum is taken does not matter, so that $\Sigma_{i=1}^{n} (f(i) + g(i)) = \left(\Sigma_{i=1}^{n} f(i)\right) + \left(\Sigma_{i=1}^{n} g(i)\right)$, and constants can always be factored out of a sum, $\Sigma_{i=1}^{n} (a f(i)) = a\Sigma_{i=1}^{n} f(i)$.

A1.4 Some Important Products

The following rules describe how certain products can be evaluated. To interpret a product, read $\Pi_{i=1}^{n} f(i)$ as "the product of the values of $f(i)$ starting with i equals one, then two, then three, etc., up until $i = n$." That is, $\Pi_{i=1}^{n} f(i) = f(1)\,f(2)\,f(3)\cdots f(n)$.

Rule A1.23: $\prod_{i=1}^{n} a = a^n$

Rule A1.24: $\prod_{i=1}^{n} i = n!$ $\quad$ (n factorial)

Rule A1.25: $\prod_{i=1}^{n} a^i = a^{n(n+1)/2}$

Rule A1.26: $\left(\prod_{i=1}^{n} x_i\right)^{1/n} = \sqrt[n]{\prod_{i=1}^{n} x_i} = \bar{x}_h$ (geometric mean)

Note that the order in which a product is taken does not matter, so that $\prod_{i=1}^{n} (f(i)g(i)) = \left(\prod_{i=1}^{n} f(i)\right) \times \left(\prod_{i=1}^{n} g(i)\right)$, and that constants can be factored out of each term in a product, $\prod_{i=1}^{n} (a\, f(i)) = a^n \prod_{i=1}^{n} f(i)$.

A1.5 Inequalities

The following rules are used to simplify functions involving the inequalities "$<$" (less than) and "$>$" (greater than):

Rule A1.27: If $x + a > y + b$, then $x > y + b - a$. The direction of an inequality is unchanged by addition or subtraction.

Rule A1.28: If $x/a > y/b$, then $x > y\, a/b$ if a is positive, while $x < y\, a/b$ if a is negative. The direction of an inequality must be reversed when multiplying or dividing by a negative number.

Exercise A1.1: The following questions review algebraic techniques needed throughout the text.

(a) Solve $2x^2 - 7x + 3 = 0$ for x.
(b) Simplify $((a\,x)^2 - a^2)/(ax - a)$ as much as possible.
(c) Factor both sides of $x^3 - yx^2 = x - y$. What are the three possible values of x that ensure that this equation holds true?
(d) Solve $\ln(x^t) = 1/2$ for t.
(e) Write $\ln(a\,x) + \ln(b\,x) - \ln(c)$ as a single logarithmic function $\ln()$.
(f) Solve $x^{rt} = 100$ for t. [Hint: take the logarithm of both sides.]
(g) What does the sum $\sum_{i=1}^{n} 1$ equal?
(h) What does the product $\prod_{i=1}^{n} 1$ equal?
(i) Evaluate and simplify $\sum_{i=1}^{n} (2i - 1)$.
(j) If $x/(-3) + 5 > 15$, is x greater than some number or less than some number? What is that number?

Answers to Exercise

Exercise A1.1

(a) Using the quadratic formula (Rule A1.10) with $a = 2$, $b = -7$, and $c = 3$, the two solutions of $2x^2 - 7x + 3 = 0$ are $x = \left(7 \pm \sqrt{49 - 4(2)(3)}\right)/4 = (7 \pm 5)/4$. That is, $x = 3$ and $x = 1/2$. Alternatively, we could try to factor $2x^2 - 7x + 3$ in various combinations to show that it equals $(2x - 1)(x - 3)$, which gives the same answer.

(b) $$\frac{(a\,x)^2 - a^2}{a\,x - a} = \frac{a^2x^2 - a^2}{a\,x - a} = \frac{a^2(x^2 - 1)}{a\,(x - 1)} = \frac{a^2(x + 1)(x - 1)}{a\,(x - 1)}.$$
$$= a\,(x + 1)$$

It is worth remembering that $(x^2 - 1)$ can be factored as $(x + 1)(x - 1)$; alternatively, the quadratic formula (Rule A1.10) can be used to show that the two roots of $(x^2 - 1) = 0$ are $x = -1$ and $x = +1$, indicating that we can factor $(x^2 - 1)$ as $(x + 1)(x - 1)$.

(c) Factoring both sides gives $x^2\,(x - y) = (x - y)$. For this equation to hold true, either x^2 must equal one or $(x - y)$ must equal zero. Three possible solutions for x are thus $x = -1$, $x = +1$, and $x = y$.

(d) Using Rule A1.11, $\ln(x^t) = t\ln(x) = 1/2$. Thus, $t = 1/(2\ln(x))$, which we can also write as $1/\ln(x^2)$ (both are correct).

(e) Using Rules A1.12 and A1.14, $\ln(a\,x) + \ln(b\,x) - \ln(c) = \ln(a\,b\,x^2/c)$.

(f) Solving for terms in the exponent is made easier by taking the logarithm (in any base) of both sides: $\ln(x^{rt}) = r\,t\,\ln(x) = \ln(100)$. Thus, $t = \ln(100)/(r\ln(x))$ or, equally, $t = \ln(100)/\ln(x^r)$.

(g) Using Rule A1.17, $\Sigma_{i=1}^{n} 1 = n$.

(h) Using Rule A1.23, $\Pi_{i=1}^{n} 1 = 1$.

(i) We can rewrite $\Sigma_{i=1}^{n}(2i - 1)$ as $\left(2\Sigma_{i=1}^{n} i\right) - \left(\Sigma_{i=1}^{n} 1\right)$, which according to Rules A1.17 and A1.18 equals $(n\,(n + 1)) - (n) = n^2$.

(j) Adding (-5) to both sides, we get $x/(-3) > 10$. Multiplying both sides by (-3), we get $x < -30$ (x must be less than -30). Note that this last operation required that we reverse the inequality.

Appendix 2

Some Important Rules from Calculus

A2.1 Concepts

In this appendix, we review basic concepts and formulae from calculus that are used repeatedly in the text. We assume that you have learned this material in the past and provide exercises to help refresh your memory. See Neuhauser (2003) for additional review.

Let us consider a function $f(x)$ of an independent variable x. The function can be represented as a curve drawn on a two-dimensional plot (Figure A2.1). The rate at which the height of $f(x)$ changes as x is varied is described by the *derivative* of the function. By definition, the derivative of a function $f(x)$ with respect to x is

$$\frac{\mathrm{d}f(x)}{\mathrm{d}x} \equiv \lim_{\Delta x \to 0}\left[\frac{f(x + \Delta x) - f(x)}{\Delta x}\right].$$

In words, the derivative of f with respect to x is defined as ("$\equiv$") the change in f over an interval Δx (that is, $f(x + \Delta x) - f(x)$), divided by the length of the interval (Δx), as the interval is reduced to zero ("$\lim_{\Delta x \to 0}$"). A graphical way to think about the derivative is that it equals the slope of the line tangent to the function at point, x. For example, in Figure A2.1, we plot the function $f(x) = x^3 - 12x^2 + 36x - 20$, whose derivative is $\mathrm{d}f(x)/\mathrm{d}x = 3x^2 - 24x + 36$. At $x = 1$, the slope of the tangent line ("rise over run") would be 15 (thin line). Whenever the function $f(x)$ is flat, for example at a local maximum or minimum, the derivative is zero. In Figure A2.1, the function has a derivative of zero at both $x = 2$ and $x = 6$. Another way to think about derivatives is that they measure the sensitivity of the height $f(x)$ to changes in x. This mental picture helps explain why derivatives are so important in biology, because we often want to describe how sensitive a quantity of interest (e.g., the growth of a population) is to some other quantity (e.g., the current population size).

If the derivative of a function $f(x)$ is $g(x)$, then the *antiderivative* of $g(x)$ is $f(x)$. Thus, antidifferentiation ("*integration*") undoes the process of taking the derivative of a function. We can represent the antiderivative of $g(x)$ with respect to x using an indefinite integral:

$$\int g(x)\,\mathrm{d}x = f(x).$$

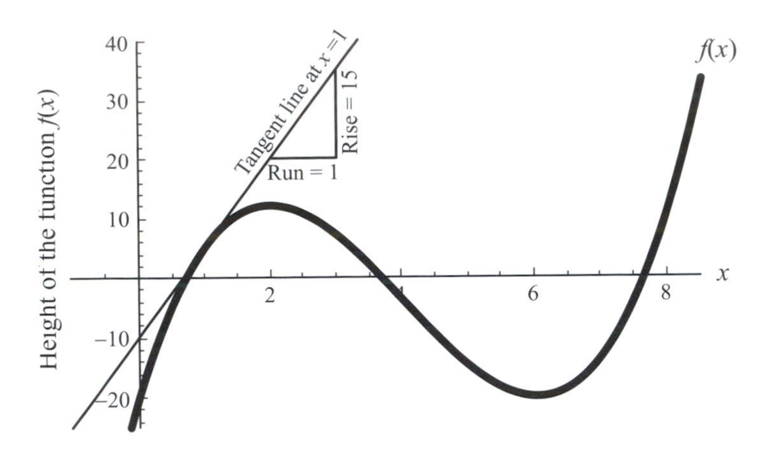

Figure A2.1: Slopes and derivatives. A plot of the function $f(x) = x^3 - 12x^2 + 36x - 20$ (thick curve) and its tangent line at $x = 1$ (thin line). The slope of the tangent line equals 15, which is described by the derivative of $f(x)$ at $x = 1$ and which can be confirmed by dividing the "rise" in the tangent (15) over a "run" in x of 1.

But not only is $f(x)$ the antiderivative of $g(x)$, so too is $f(x) + c$. That is, indefinite integrals are not unique. Thus, when taking an indefinite integral, we add a "constant of integration" (c) to the result to indicate that any possible value of c would work:

$$\int g(x)\ \mathrm{d}x = f(x) + c.$$

For example, if $g(x) = 2x$, then its antiderivative would be any function $x^2 + c$, including both $f(x) = x^2$ and $f(x) = x^2 - 20$ (as can be confirmed by taking the derivative of both possibilities for $f(x)$).

If derivatives are measures of rates of change ("slopes"), antiderivatives are measures of "areas." Indefinite integrals represent the area under the curve $g(x)$ without specifying the range of values of x that we want to consider. If we take an indefinite integral evaluated at $x = b$ and subtract off the indefinite integral evaluated at $x = a$, then we get the definite integral:

$$\int_{x=a}^{b} g(x)\ dx;$$

this definite integral is the area under the function $g(x)$ between points a and b. This result is known as "the fundamental theorem of calculus." Figure A2.2 illustrates the integration of the function $g(x) = -x^2 + 8x - 12$, whose indefinite integral is $\int g(x)\ \mathrm{d}x = -(1/3)x^3 + 4x^2 - 12x + c$. The area under the curve between $x = 5$ and $x = 6$ is $\int_{x=5}^{6} g(x)\ \mathrm{d}x = 5/3$, while the area under the curve between $x = 6$ and $x = 8$ is $\int_{x=6}^{8} g(x)\ \mathrm{d}x = -32/3$. The total area under the curve between $x = 5$ and $x = 8$ can be found either by evaluating the definite integral $\int_{x=5}^{8} g(x)\ \mathrm{d}x = -9$ or by adding together the two areas between $x = 5$ and 6 and between $x = 6$ and 8; either way the answer is -9. The fact that this area is a negative number indicates that the curve lies mainly below the horizontal axis.

It is helpful to think of an area as a sum of rectangles whose width is very small and whose height is given by $g(x)$ (inset in Figure A2.2). This mental

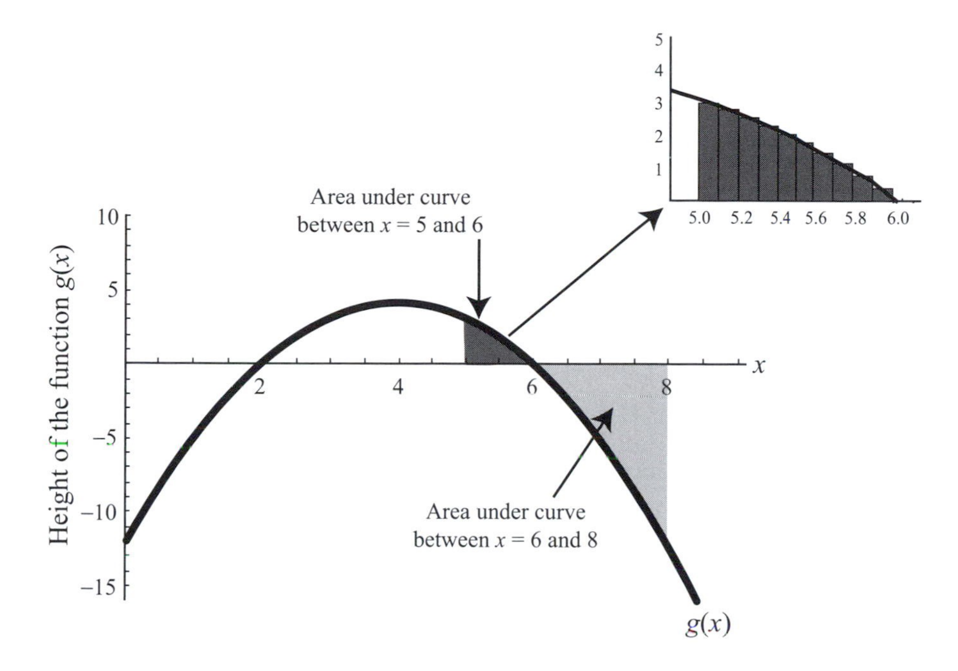

Figure A2.2: Areas and integration. A plot of the function $g(x) = -x^2 + 8x - 12$. The area under the curve can be found by integration and is 5/3 between $x = 5$ and $x = 6$ (dark shaded area) and $-32/3$ between $x = 6$ and $x = 8$ (light shaded area). The inset shows how the area under the curve can be approximated by filling in the area with a series of rectangles whose height is given by $g(x)$. The sum of these rectangles is an approximation to the area found by integration. This approximation improves when using more rectangles of smaller width.

image helps to explain the importance of integration in biology, because we often want to describe the sum total effect of a process. For example, we might want to determine the sum total effect of past selection or the sum total change to the size of a population over an interval of time.

A2.2 Derivatives

The following rules describe some of the more important derivatives:

Rule A2.1: $$\frac{\mathrm{d}a}{\mathrm{d}x} = 0$$ (derivative of a constant)

Rule A2.2: $$\frac{\mathrm{d}(a\,f(x))}{\mathrm{d}x} = a\frac{\mathrm{d}f(x)}{\mathrm{d}x}$$ (factoring out a constant)

Rule A2.3: $$\frac{\mathrm{d}(f(x) + g(x))}{\mathrm{d}x} = \frac{\mathrm{d}f(x)}{\mathrm{d}x} + \frac{\mathrm{d}g(x)}{\mathrm{d}x}$$ (linearity property)

Rule A2.4: $\frac{d(a\,x)}{dx} = a$ (linear functions)

Rule A2.5: $\frac{dx^a}{dx} = a\,x^{a-1}$ (polynomial functions)

Rule A2.6: $\frac{d(e^{f(x)})}{dx} = e^{f(x)}\frac{df(x)}{dx}$ (exponential functions)

Rule A2.7a: $\frac{d(a^{f(x)})}{dx} = a^{f(x)}\frac{d(f(x))}{dx}\ln(a)$ (power functions)

Rule A2.7b: $\frac{d(f(x)^a)}{dx} = a\,f(x)^{a-1}\frac{d(f(x))}{dx}$ (power functions)

Rule A2.8: $\frac{d(\ln(f(x)))}{dx} = \frac{1}{f(x)}\frac{d(f(x))}{dx}$ (natural log functions)

Rule A2.9: $\frac{d(\log_b(f(x)))}{dx} = \frac{1}{\ln(b)\,f(x)}\frac{d(f(x))}{dx}$ (log functions in base b)

Rule A2.10: $\frac{d(\sin(f(x)))}{dx} = \cos(f(x))\frac{d(f(x))}{dx}$ (sine functions)

Rule A2.11: $\frac{d(\cos(f(x)))}{dx} = -\sin(f(x))\frac{d(f(x))}{dx}$ (cosine functions)

Rule A2.12: $\frac{d(f(x)\,g(x))}{dx} = \frac{d(f(x))}{dx}g(x) + f(x)\frac{d(g(x))}{dx}$ (product rule)

Rule A2.13: $\frac{d\left(\frac{f(x)}{g(x)}\right)}{dx} = \frac{\frac{d(f(x))}{dx}g(x) - f(x)\frac{d(g(x))}{dx}}{g(x)^2}$ (quotient rule)

Rule A2.14: $\frac{d(f(x(t)))}{dt} = \frac{df}{dx}\frac{dx}{dt}$ (chain rule: 1 variable)

Rule A2.15: $\frac{d(f(x(t), y(t)))}{dt} = \frac{\partial f}{\partial x}\frac{dx}{dt} + \frac{\partial f}{\partial y}\frac{dy}{dt}$ (chain rule: 2 variables)

Implicit differentiation is an important method that builds upon the chain rule. An *explicit* function describes how a variable x depends on another variable t in terms of an equation $x = f(t)$. Thus, when t changes, we have an explicit expression that tells us how x changes. In contrast, an *implicit* function has the form $f(x, t) = c$. Again we can view this as an equation governing how x must change whenever t changes; when t is varied, x must change in such a way that the function $f(x, t)$ remains equal to c. For example, the equation $x^2 + t^2 = 5$ implicitly describes x as a function of t. When t varies, x must also vary so that the sum $x^2 + t^2$ equals 5. Taking the derivative of implicit functions is called implicit differentiation; it can be carried out using the rules of calculus described above, and it can be a useful method for finding the derivative, dx/dt. For example, the derivative of $x^2 + t^2 = 5$ with respect to t is $d(x^2)/dt + d(t^2)/dt =$

d(5)/dt (using Rule A2.3). According to the chain rule (Rule A2.14), $d(x^2)/dt = (d(x^2)/dx)\,(dx/dt)$, which equals $2x\,dx/dt$. Furthermore, $d(t^2)/dt = 2t$ (Rule A2.5) and $d(5)/dt = 0$ (Rule A2.1). Altogether, we have $2x\,dx/dt + 2t = 0$, which demonstrates that the derivative of x with respect to t must equal $dx/dt = -t/x$. In this case, we could obtain the same result by first finding the explicit solutions to this function, $x = \sqrt{5 - t^2}$ and $x = -\sqrt{5 - t^2}$, calculating dx/dt for each, and showing that they both equal $dx/dt = -t/x$. The real power of implicit differentiation comes from the fact that you can use it to find the derivative of a function without an explicit expression for the function itself (which can sometimes be impossible to obtain).

A2.3 Integrals

The following describes some of the more important indefinite integrals ("antiderivatives"), where c is a constant of integration. Because a definite integral evaluated over the range a to b can always be obtained from an indefinite integral by plugging in $x = b$ and subtracting off the indefinite integral at $x = a$, we provide rules for indefinite integrals only.

Rule A2.16: $$\int a\,dx = a\,x + c$$ (integral of a constant)

Rule A2.17: $$\int a\,f(x)\,dx = a\int f(x)\,dx$$ (factoring out a constant)

Rule A2.18: $$\int f(x) + g(x)\,dx = \int f(x)\,dx + \int g(x)\,dx$$ (linearity property)

Rule A2.19: $$\int a\,x\,dx = \frac{a\,x^2}{2} + c$$ (linear functions)

Rule A2.20: $$\int x^n\,dx = \frac{x^{n+1}}{n+1} + c$$ (polynomial functions)

Rule A2.21: $$\int \frac{a}{x}\,dx = a\ln\left(|x|\right) + c$$ (fractional functions)

Rule A2.22: $$\int \frac{1}{(a_1 + b_1 x)(a_2 + b_2 x)}\,dx = \frac{\ln\left(\left|\frac{a_1 + b_1 x}{a_2 + b_2 x}\right|\right)}{a_2 b_1 - a_1 b_2} + c$$ (fractional functions)

Rule A2.23: $$\int e^{ax}\,dx = \frac{e^{ax}}{a} + c$$ (exponential functions)

Rule A2.24: $$\int a^{bx}\,dx = \frac{a^{bx}}{b\ln(a)} + c \quad \text{for } a > 0$$ (power functions)

Rule A2.25: $$\int \ln(x)\,dx = x\ln(x) - x + c \quad \text{for } x > 0$$ (natural log functions)

Rule A2.26: $$\int \sin(x)\,dx = -\cos(x) + c$$ (sine functions)

Rule A2.27: $\int \cos(x)\,dx = \sin(x) + c$ (cosine functions)

Rule A2.28: $\int f(x)\,dx = \int h(g(x))\frac{dg(x)}{dx}\,dx = \int h(u)\,du$
where $u = g(x)$ (integration by substitution)

Integration by substitution can be useful when the original function, $f(x)$, can be factored into the product of two terms, $h(g(x))\,dg(x)/dx$, where the first term depends on x only through the function, $g(x)$, and the second is the derivative of $g(x)$ with respect to x.

Rule A2.29: $\int u\,dv = u\,v - \int v\,du$ (integration by parts)

or, alternatively,

$$\int f(x)\,dx = \int \frac{d(g(x))}{dx} h(x)\,dx = g(x)\,h(x) - \int g(x)\frac{d(h(x))}{dx}\,dx$$

Integration by parts can be useful if the original function, $f(x)$, can be factored into the product of two terms $(d(g(x))/dx)\,h(x)$, where the first term is the derivative of another function $g(x)$. This method helps integrate functions whenever $g(x)\,d(h(x))/dx$ is easier to integrate than the original function.

A2.4 Limits

The rules of calculus are also useful for determining the *limit* of a function as a variable approaches a specific value. To denote the limit of $f(x)$ as x goes to a, we write $\lim_{x\to a} f(x)$. For many functions, the limit is straightforward to determine. For example, the limit of $f(x) = 2 + x^2$ as x goes to one is three, and the limit of $f(x) = e^{rx}$ as x goes to zero is one. In some cases, the limit as x approaches a depends on whether x starts above a ($\lim_{x\to a+} f(x)$) or below a ($\lim_{x\to a-} f(x)$). For example, the limit of $f(x) = 1/x$ as x goes to zero is $+\infty$ if x is initially positive but $-\infty$ if x is initially negative (Figure A2.3).

In certain cases, however, the limit is not obvious. For example, what is the limit of $(e^{rx} - e^{sx})/x$ as x goes to zero? The answer is unclear because both the numerator and the denominator are zero at $x = 0$. In such cases, we can use L'Hôpital's rule to determine the limit:

Rule A2.30: $\lim_{x\to a}\frac{f(x)}{g(x)} = \lim_{x\to a}\frac{f'(x)}{g'(x)}$ (L'Hôpital's rule)

L'Hôpital's rule requires that both $f(x)$ and $g(x)$ are zero at $x = a$ (or both are infinite), that both $f'(x) = df/dx$ and $g'(x) = dg/dx$ exist at $x = a$, and that $g(a)$ is not zero.

L'Hôpital's rule allows us to calculate limits of quotients such as $\lim_{x\to 0}((e^{rx} - e^{sx})/x)$. Specifically, Rule A2.30 tells us that $(e^{rx} - e^{sx})/x$ has the same limit as $f'(x)/g'(x) = (re^{rx} - se^{sx})/1$, whose limit as x goes to zero is easy to calculate: $(r - s)$. Occasionally, you must rearrange a function to apply L'Hôpital's

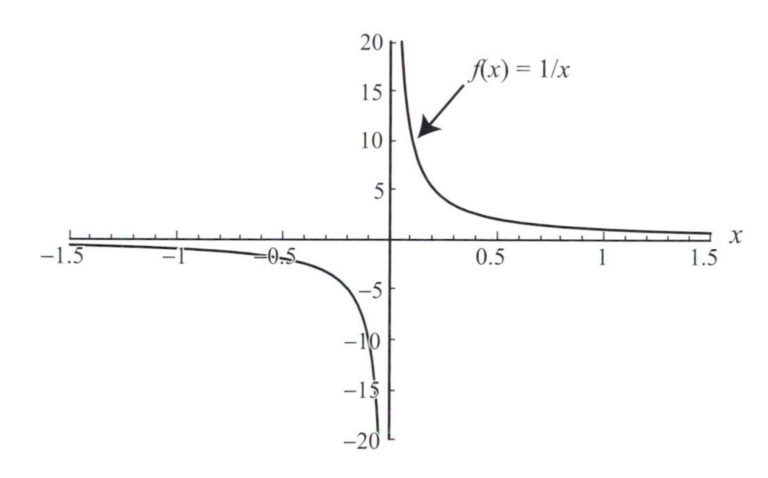

Figure A2.3: Limit of the inverse function. A plot of the function $f(x) = 1/x$, whose limit as x goes to zero depends on whether x is initially positive or negative.

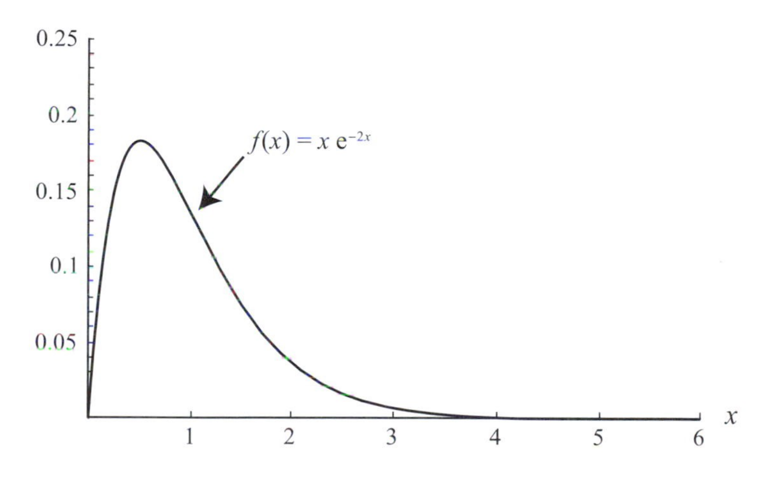

Figure A2.4: An application of L'Hôpital's rule. A plot of the function $f(x) = x\,e^{-rx}$ using $r = 2$, whose limit as x goes to infinity is zero, as can be shown using L'Hôpital's rule (A2.30).

rule. For example, to evaluate $\lim_{x\to\infty}(x\,e^{-rx})$ for positive r using L'Hôpital's rule, we must first write it as a quotient: $\lim_{x\to\infty}(x/e^{rx})$. In this example, both the numerator and denominator approach infinity, rather than zero, in the limit. L'Hôpital's rule can still be applied in such cases, allowing us to determine the limit of x/e^{rx} from the limit of $f'(x)/g'(x) = 1/(r\,e^{rx})$, which approaches zero as x goes to infinity (Figure A2.4).

Exercise A2.1: The following questions review calculus techniques needed throughout the text (treat everything but x as a constant).

(a) Find the derivative with respect to x of the following functions $f(x)$:

- $-x^2 + 8x - 12$
- $\dfrac{1}{x^2 + 1}$

Continued

Exercise A2.1 (*Continued*)

- e^{3x}
- $x^n + e^{a\,x}$
- $x^n\, e^{a\,x}$
- $\ln(x)$
- $\ln(a\, x^2)$
- $\cos(5x)$
- $\sin^2(b\, x^a)$

(b) Determine the following indefinite integrals (don't forget the constant of integration):

- $\int 3x^2\, \mathrm{d}x$
- $\int (4x + 5)\, \mathrm{d}x$
- $\int \frac{2}{x}\, \mathrm{d}x$
- $\int e^{2x}\, \mathrm{d}x$
- $\int x^n + e^{ax}\, \mathrm{d}x$
- $\int 2^x\, \mathrm{d}x$
- $\int \ln(x)\, \mathrm{d}x$ (use integration by parts)
- $\int e^x\, x\, \mathrm{d}x$ (use integration by parts)

(c) Calculate the following definite integrals:

- $\int_{x=1}^{2} 8\, x^3\, \mathrm{d}x$
- $\int_{x=0}^{3} (2x + 1)\, \mathrm{d}x$
- $\int_{x=0}^{1} e^{-r\,x}\, \mathrm{d}x$
- $\int_{x=0}^{\infty} e^{-r\,x}\, \mathrm{d}x$ (assume that r is positive)

(d) Calculate the following limits:

- $\lim_{x \to 0} \frac{\sin(x)}{1 + x}$
- $\lim_{x \to 0} \frac{\sin(x)}{x}$

References

Neuhauser, C. 2003. *Calculus for Biology and Medicine.* Prentice-Hall, Upper Saddle River, N.J.

Answer to Exercise

Exercise A.2.1

(a) $\dfrac{df}{dx}$ is

- $-2x + 8$ (using Rule A2.5)
- $\dfrac{-2x}{(x^2 + 1)^2}$ (using quotient Rule A2.13)
- $3\, e^{3x}$ (using Rule A2.6)
- $n\, x^{n-1} + a\, e^{a\,x}$ (using Rules A2.3, A2.5, and A2.6)
- $n\, x^{n-1}\, e^{a\,x} + a\, x^n\, e^{a\,x}$ (using product rule A2.12)
- $\dfrac{1}{x}$ (using Rule A2.8)
- $\dfrac{2}{x}$ (using Rule A2.8)
- $-5\,\sin(5x)$ (using Rule A2.11)
- $2\, b\, a\, x^{a-1}\,\sin(b\, x^a)\,\cos(b\, x^a)$ (using Rule A2.7b and A2.10)

(b) The indefinite integrals are

- $x^3 + c$ (using Rule A2.20)
- $2x^2 + 5x + c$ (using Rule A2.20)
- $2\,\ln(|x|) + c$ (using Rule A2.21)
- $\dfrac{e^{2x}}{2} + c$ (using Rule A2.23)
- $\dfrac{x^{n+1}}{n+1} + \dfrac{e^{ax}}{a} + c$ (using Rules A2.18, A2.20, A2.23)
- $\dfrac{2^x}{\ln(2)} + c$ (using Rule A2.24)
- Letting $u = \ln(x)$ and $dv = dx$ so that $du = 1/x\, dx$ and $v = x$,
  $\int \ln(x)\, dx = x\,\ln(x) - \int 1\, dx = x\,\ln(x) - x + c$ (using Rule A2.29)
- Letting $u = x$ and $dv = e^x\, dx$ so that $du = 1\, dx$ and $v = e^x$,
  $\int e^x\, x\, dx = e^x x - \int e^x\, dx = e^x x - e^x + c$ (using Rule A2.29)

(c) The definite integrals are

- $\int_{x=1}^{2} 8\, x^3\, dx = 2\, x^4\Big|_{x=1}^{2} = 32 - 2 = 30$ (using Rule A2.20)

- $$\int_{x=0}^{3} (2x + 1)\,dx = (x^2 + x)\Big|_{x=0}^{3} = 12 - 0 = 12 \qquad \text{(using Rule A2.20)}$$
- $$\int_{x=0}^{1} e^{-rx}\,dx = -\frac{e^{-rx}}{r}\Big|_{x=0}^{1} = -\frac{e^{-r}}{r} + \frac{1}{r} \qquad \text{(using Rule A2.23)}$$
- $$\int_{x=0}^{\infty} e^{-rx}\,dx = -\frac{e^{-rx}}{r}\Big|_{x=0}^{\infty} = -\frac{e^{-r\infty}}{r} + \frac{1}{r} = \frac{1}{r} \quad \text{for } r > 0;$$

  the exponential function e^{-rx} decreases with increasing x when r is positive, so that e^{-rx} evaluated at $x = \infty$ is zero.

(d) The limits are:

- $\lim_{x \to 0} \sin(x)/(1 + x) = 0$. We do not need to invoke L'Hôpital's rule, because the denominator equals one at $x = 0$.
- $\lim_{x \to 0} \sin(x)/(x) = \lim_{x \to 0} \cos(x)/1 = 1$ Here, we have used L'Hôpital's rule, because the numerator and denominator are both zero at $x = 0$.

Appendix 3

The Perron-Frobenius Theorem

Before going into the specifics of this very useful theorem we need to introduce some definitions. The Perron-Frobenius theorem is applicable only to non-negative matrices, and therefore we begin with this definition. We then define irreducible and primitive matrices. Non-negative matrices are classified into "reducible" and "irreducible" matrices, and irreducible matrices are further divided into "primitive" and "imprimitive" matrices.

A3.1 Definitions

Positive Matrix—A matrix $\mathbf{M}$ is positive if every element of $\mathbf{M}$ is positive.

Non-Negative Matrix—A matrix $\mathbf{M}$ is non-negative if every element of $\mathbf{M}$ is zero or positive.

Irreducible Matrix—A non-negative matrix $\mathbf{M}$ is irreducible (or ergodic) if any state can be reached from any other state in a finite number of time steps. In other words, irreducible transition matrices have the property that, if we begin with a population in only one state (i.e., all of the variables are zero except one), then there is some (finite) time in the future at which any of the variables takes on a positive value. This future time point might be the same or different for each variable.

Reducible Matrix—A non-negative matrix $\mathbf{M}$ is reducible if it is not possible to reach all classes of the model from all other classes in a finite number of time steps. Thus, one or more variable will remain zero for all time if the dynamics are started from certain initial states.

A positive matrix is automatically irreducible, because every variable will be positive after only one time step upon multiplication by $\mathbf{M}$. The same reasoning applies to any matrix that is positive when taken to some power (see "Primitive Matrix"). If it is not apparent whether a matrix is irreducible, the following Recipe can be applied:

Recipe A3.1

A matrix $\mathbf{M}$ is irreducible if and only if all elements of $(\mathbf{I} + \mathbf{M})^{n-1}$ are positive, where n is the number of classes (i.e., the dimension of $\mathbf{M}$) and $\mathbf{I}$ is the $n \times n$ identity matrix (Primer 2).

Primitive Matrix—A matrix **M** is primitive (or aperiodic) if all elements of the matrix are simultaneously positive when the matrix is raised to a high enough power. Primitive matrices are always irreducible.

Again, for relatively small matrices this can often be verified by multiplying the matrix by itself a few times. But how many times is enough? Fortunately, there is again a recipe to follow:

Recipe A3.2
A matrix **M** is primitive if and only if all elements of $\mathbf{M}^{n^2-2n+2}$ are positive, where n is the number of classes (i.e., the dimension of **M**). If **M** raised to a lower power than $n^2 - 2n + 2$ is positive, then $\mathbf{M}^{n^2-2n+2}$ will also be positive, and the matrix is primitive.

Imprimitive Matrix—A matrix **M** is imprimitive (or periodic) if it is irreducible but not primitive (i.e., it satisfies Recipe A3.1 but not A3.2). Such matrices allow the system to reach all classes of the model from all other classes in a finite number of time steps, but transitions between certain states occur only at periodic intervals.

The classification of a matrix can often be checked for specific examples using mathematical software such as *Mathematica*, Matlab, or Maple. When trying to determine which of these definitions applies to a particular matrix, you should first attempt to determine if the matrix is primitive (Recipe A3.2). If it is primitive, then it is also irreducible, and no further work is required. If not, then you should attempt Recipe A3.1 to determine if the matrix is irreducible, in which case it must be imprimitive. If both Recipe A3.2 and A3.1 fail, then the matrix is reducible. Once you have classified the matrix, you can apply the following theorem.

A3.2 The Perron-Frobenius Theorem

Suppose you have a non-negative matrix:

(1) If the matrix is irreducible and primitive, then one of its eigenvalues will be real, positive, and strictly larger in absolute value than all other eigenvalues. This eigenvalue is the leading eigenvalue. Furthermore, it is possible to choose right and left eigenvectors associated with the leading eigenvalue whose elements are strictly positive.

(2) If the matrix is irreducible and imprimitive, then one of its eigenvalues will be real and positive, but it need no longer be strictly larger in absolute value than the remaining eigenvalues. Rather, there might be other eigenvalues (real or complex) that have the same absolute value.

Nevertheless, the largest real-valued eigenvalue(s) is referred to as the leading eigenvalue. Again, the right and left eigenvectors associated with the leading eigenvalue can be chosen to have strictly positive elements.

(3) If the transition matrix is reducible, then one of its eigenvalues will be real, non-negative, and larger than (or equal to) all other eigenvalues in absolute value. If there is more than one eigenvalue with this largest magnitude (real or complex), then the one that has left and right eigenvectors containing non-negative real elements is referred to as the leading eigenvalue.

In Supplementary Material A3.1, we explore a model of seed dispersal among patches and use this model to illustrate the above classification and its implications.

Further Readings

The following books discuss the Perron-Frobenius theorem and related facts:

- Gantmacher, F. R. 1990. *Matrix Theory*, Vol. 1, Second Ed. Chelsea Publishing, London.
- Gantmacher, F. R. 2000. *Theory of Matrices*, Vol. 2, Second Ed. Chelsea Publishing, London.
- Seneta, E. 1973. *Non-Negative Matrices: An Introduction to Theory and Applications*. George Allen and Unwin, London.

Appendix 4

Finding Maxima and Minima of Functions

A4.1 Functions with One Variable

Suppose we have a function $W(z)$ where the variable z must lie within some interval $a \leq z \leq b$. Any value z^* that yields a local maximum of this function must satisfy one of the following three conditions:

(1) Intermediate local maximum:

$$\left.\frac{\mathrm{d}W}{\mathrm{d}z}\right|_{z=z^*} = 0 \quad \text{and} \quad \left.\frac{\mathrm{d}^2W}{\mathrm{d}z^2}\right|_{z=z^*} \leq 0. \tag{A4.1}$$

(2) Local maximum on the lower boundary:

$$\left.\frac{\mathrm{d}W}{\mathrm{d}z}\right|_{z=z^*=a} \leq 0 \quad \left(\text{and } \left.\frac{\mathrm{d}^2W}{\mathrm{d}z^2}\right|_{z=z^*=a} \leq 0 \text{ if equality holds}\right). \tag{A4.2}$$

(3) Local maximum on the upper boundary:

$$\left.\frac{\mathrm{d}W}{\mathrm{d}z}\right|_{z=z^*=b} \geq 0 \quad \left(\text{and } \left.\frac{\mathrm{d}^2W}{\mathrm{d}z^2}\right|_{z=z^*=b} \leq 0 \text{ if equality holds}\right). \tag{A4.3}$$

By "local" maximum we mean that z^* produces a larger value of the function W than any other "nearby" value of z. If z^* is to give a "global" maximum, then the corresponding value of W must be larger than that produced by any other possible value of z (nearby or otherwise). If there is only one value of z^* that satisfies one of the above three conditions, then this local maximum is also a global maximum by default. If there is more than one value of z^* that satisfies any of the above three conditions, then we must compare the value of W yielded by each to determine which results in a global maximum.

A4.1.1 Mathematical Derivation (Advanced)

It is useful to understand how the above results can be derived mathematically, as this will allow us to extend the results to the multiple-variable case, which is less obvious. We will only deal with the case of intermediate optima here.

Suppose z^* is a local maximum of $W(z)$. This means that $W(z) < W(z^*)$ for all z near $z = z^*$. We can expand the left-hand side of this inequality using the one-variable Taylor series (Recipe P1.2) near the value $z = z^*$ to obtain

$$W(z^*) + \left.\frac{dW}{dz}\right|_{z=z^*}(z - z^*) + \frac{1}{2!}\left.\frac{d^2W}{dz^2}\right|_{z=z^*}(z - z^*)^2 + O(z - z^*)^3 < W(z^*). \tag{A4.4a}$$

or

$$\left.\frac{dW}{dz}\right|_{z=z^*}(z - z^*) + \frac{1}{2!}\left.\frac{d^2W}{dz^2}\right|_{z=z^*}(z - z^*)^2 + O(z - z^*)^3 < 0, \tag{A4.4b}$$

for all z. Now because z is near z^*, powers of $(z - z^*)$ become progressively smaller, meaning that the first term in (A4.4b) will be the largest. Because $(z - z^*)$ might be either positive or negative depending on the value of z, for (A4.4b) to hold for all z near z^*, we require that $(dW/dz)|_{z=z^*} = 0$. This gives us the first condition in (A4.1). Given $(dW/dz)|_{z=z^*} = 0$, the largest term in (A4.4b) is now the second-derivative term, and the inequality then requires

$$\frac{1}{2!}\left.\frac{d^2W}{dz^2}\right|_{z=z^*}(z - z^*)^2 \leq 0. \tag{A4.4c}$$

Because $(z - z^*)^2$ is non-negative, requiring $(d^2W/dz^2)|_{z=z^*} \leq 0$ does the trick. If the first two terms of the Taylor series are zero, we then need to look at even higher-order terms to determine whether z^* is a local maximum or a local minimum.

A4.2 Functions with Multiple Variables

Given a function of two variables, $W(x,y)$, any pair of values (x^*,y^*) that yields an intermediate local maximum must satisfy

$$\left.\frac{\partial W}{\partial x}\right|_{\substack{x=x^*\\y=y^*}} = 0, \qquad \left.\frac{\partial W}{\partial y}\right|_{\substack{x=x^*\\y=y^*}} = 0. \tag{A4.5}$$

The second derivative condition in the multiple-variable case is not intuitively obvious. First we need to define the so-called "Hessian" matrix;

$$\mathbf{H} = \begin{pmatrix} \dfrac{\partial^2 W}{\partial x^2} & \dfrac{\partial^2 W}{\partial y\,\partial x} \\ \dfrac{\partial^2 W}{\partial x\,\partial y} & \dfrac{\partial^2 W}{\partial y^2} \end{pmatrix}_{\substack{x=x^*\\y=y^*}} \tag{A4.6}$$

The Hessian matrix (A4.6) is symmetric, and therefore its eigenvalues will be real (Rule P2.29). For (x^*,y^*) to represent a local maximum, the Hessian matrix (A4.6) must be negative semidefinite. Fortunately, to apply this criterion, you need not worry too much about what this means (if you are interested, then read the advanced section below). Rather, a nice result from linear algebra shows that a symmetric matrix is negative semidefinite if and only if all of its eigenvalues are nonpositive. Therefore, to demonstrate that z^* is a local maximum, we simply need to ensure that the Hessian matrix (A4.6) has no positive eigenvalues. The same requirement applies for functions of n variables.

A4.2.1 Mathematical Derivation (Advanced)

Suppose (x^*,y^*) is a local (intermediate) maximum of $W(x,y)$. This means that $W(x,y) < W(x^*,y^*)$ for all x near $x = x^*$ and y near $y = y^*$. We can expand the left-hand side of this inequality using the multiple-variable Taylor series (Box 8.1) near $x = x^*$, $y = y^*$:

$$W(x^*,y^*) + \left.\frac{\partial W}{\partial x}\right|_{\substack{x=x^*\\y=y^*}}(x - x^*) + \left.\frac{\partial W}{\partial y}\right|_{\substack{x=x^*\\y=y^*}}(y - y^*) + \frac{1}{2!}\left.\frac{\partial^2 W}{\partial x^2}\right|_{\substack{x=x^*\\y=y^*}}(x - x^*)^2$$

$$+ 2\frac{1}{2!}\left.\frac{\partial^2 W}{\partial y\,\partial x}\right|_{\substack{x=x^*\\y=y^*}}(x - x^*)(y - y^*) + \frac{1}{2!}\left.\frac{\partial^2 W}{\partial y^2}\right|_{\substack{x=x^*\\y=y^*}}(y - y^*)^2 \quad \text{(A4.7a)}$$

$$+ \cdots < W(x^*,y^*).$$

Dropping third-order and higher terms, we obtain the condition

$$\left.\frac{\partial W}{\partial x}\right|_{\substack{x=x^*\\y=y^*}}(x - x^*) + \left.\frac{\partial W}{\partial y}\right|_{\substack{x=x^*\\y=y^*}}(y - y^*) + \frac{1}{2!}\vec{v}^{\,\mathrm{T}}\,\mathbf{H}\,\vec{v} \leq 0, \quad \text{(A4.7b)}$$

where $\vec{v} = \begin{pmatrix} x - x^* \\ y - y^* \end{pmatrix}$. Because $(x - x^*)$ and $(y - y^*)$ might be either positive or negative depending on the value of x and y, inequality (A4.7) will hold for all x near $x = x^*$ and y near $y = y^*$ only if both $(\partial W/\partial x)|_{\substack{x=x^*\\y=y^*}} = 0$ and $(\partial W/\partial y)|_{\substack{x=x^*\\y=y^*}} = 0$. This gives us conditions (A4.5). Inequality (A4.7) then becomes

$$\frac{1}{2!}\vec{v}^{\,\mathrm{T}}\,\mathbf{H}\,\vec{v} \leq 0. \quad \text{(A4.7c)}$$

By definition, $\mathbf{H}$ is a *negative semidefinite matrix* if $\vec{v}^{\,\mathrm{T}}\,\mathbf{H}\,\vec{v} \leq 0$ and a *negative definite matrix* if $\vec{v}^{\,\mathrm{T}}\,\mathbf{H}\,\vec{v} < 0$ (with a strict inequality). A property of negative semidefinite matrices is that they have nonpositive eigenvalues, while negative definite matrices have strictly negative eigenvalues. Thus, (x^*,y^*) will be a local maximum if (A4.5) holds and if the eigenvalues of the Hessian matrix $\mathbf{H}$ are all negative. If the largest eigenvalue is zero, however, higher-order terms must be examined to ensure that $W(x,y) < W(x^*,y^*)$ for all points near (x^*,y^*).

Appendix 5

Moment-Generating Functions

Moment-generating functions are used to calculate the mean, variance, and higher moments of a probability distribution. By definition, the jth *moment* of a distribution is equal to the expected value of X^j. Thus, the first moment of a distribution is its mean, $E[X] = \mu$, and the second moment of a distribution, $E[X^2]$, is related to its variance (see equation P3.2 in Primer 3). Similarly, the jth *central moment* of a distribution is defined as the expected value of $(X - \mu)^j$, which subtracts off the distribution's center of mass (its mean) before calculating the expectation. The first central moment of a distribution is always zero, $E[X - \mu] = 0$, and the second central moment $E[(X - \mu)^2]$ is the variance of the distribution.

While we have focused in the text on the first two moments, it is sometimes important to know the higher moments of a distribution. In particular, the third central moment $E[(X - \mu)^3]$ provides a useful measure of the asymmetry of a distribution. This asymmetry is often measured by the *skewness* $E[(X - \mu)^3]/\sigma^3$, where σ is the standard deviation of the distribution (see Figure A5.1). Furthermore, the fourth central moment $E[(X - \mu)^4]$ provides a useful measure of the peakedness of a distribution. This peakedness is often measured by the *kurtosis* $E[(X - \mu)^4]/\sigma^4$ (see Figure A5.1).

For any discrete distribution of interest, we could calculate these higher moments by summing x_i^j over the probability distribution for X:

$$E[X^j] = \sum_{x_i} x_i^j P(X = x_i). \tag{A5.1}$$

Calculating moment after moment can be tedious, but fortunately there is a simpler way to calculate all of the moments of a distribution. To so this we need to introduce *moment-generating functions*. The moment-generating function of a distribution is defined as:

$$MGF[z] = \sum_{x_i} e^{z x_i} P(X = x_i). \tag{A5.2}$$

Equation (A5.2) sums over all values that the random variable can take (x_i) and involves a newly introduced "dummy" variable, z. For some distributions, the sum in (A5.2) cannot be evaluated, in which case there is no point in using moment generating functions. But for many distributions, the sum in (A5.2) can be evaluated. For example, the moment generating function for the binomial distribution is known to equal $MGF[z] = (1 - p + e^z p)^n$, and Table P3.2 provides

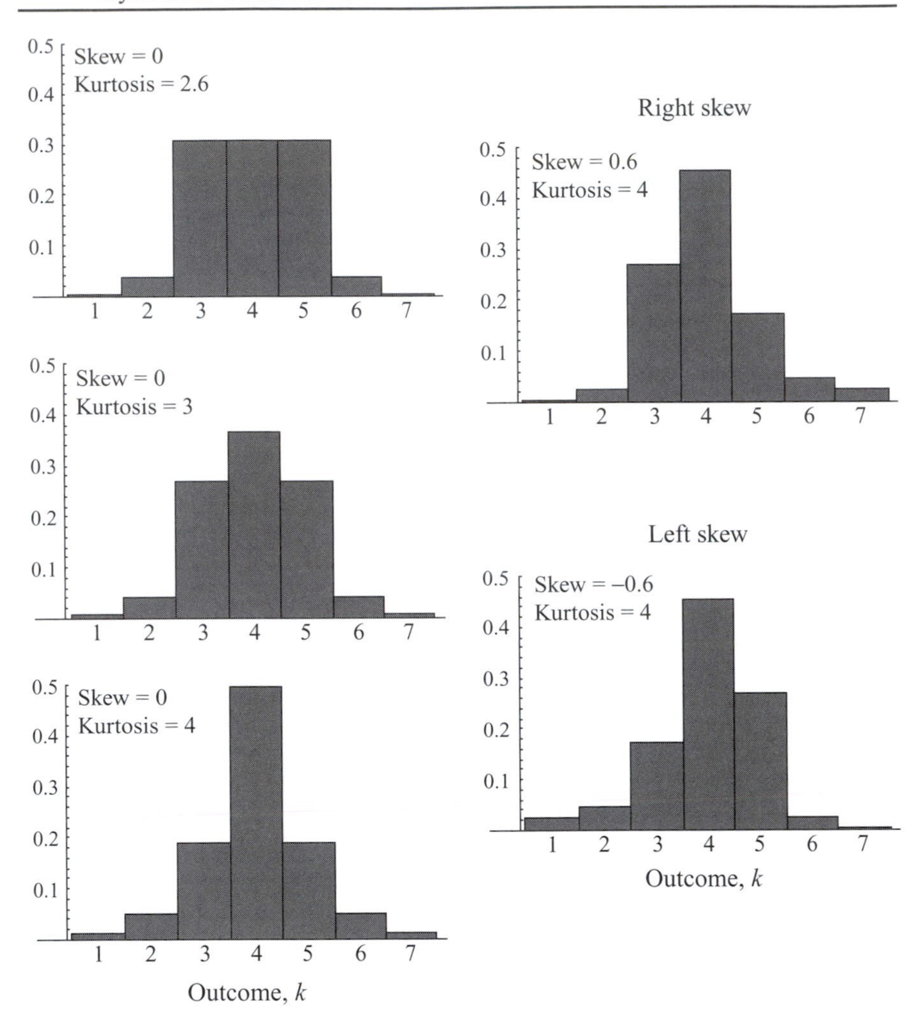

Figure A5.1: Skew and kurtosis of a distribution. A probability distribution with seven possible outcomes ($k = 1, \ldots, 7$) is used to illustrate skew and kurtosis. In each case, the mean was held at $\mu = 4$ and the variance at $\sigma^2 = 1$. The distributions on the left are symmetric (no skew), but the amount of kurtosis $E[(X - \mu)^4]/\sigma^4$ increases from top to bottom. Distributions with a high kurtosis are more "peaked" and have "fatter" tails (i.e., a higher probability of $k = 1$ or $k = 7$). The distributions on the right have skew $E[(X - \mu)^3]/\sigma^3$. The top distribution is "right skewed" (skew > 0), with a fatter tail on the right. The bottom distribution is "left skewed" (skew < 0), with a fatter tail on the left.

the moment generating functions for most of the discrete distributions introduced in Primer 3.

How do we use moment generating functions? Consider taking the derivative of (A5.2) with respect to the dummy variable z:

$$\frac{\mathrm{d}(MGF[z])}{\mathrm{d}z} = \sum_{x_i} x_i e^{z x_i} P(X = x_i). \tag{A5.3}$$

If you then set the dummy variable, z, to zero, we regain the formula for the mean (Definition P3.2):

$$\left.\frac{d(MGF[z])}{dz}\right|_{z=0} = \sum_{x_i} x_i\, P(X = x_i) = E[X]. \tag{A5.4}$$

Now consider taking the jth derivative of (A5.2) with respect to the dummy variable z and setting z to zero; doing so gives the jth moment of the distribution:

$$\left.\frac{d^j(MGF[z])}{dz^j}\right|_{z=0} = \sum_{x_i} x_i^j\, P(X = x_i) = E[X^j]. \tag{A5.5}$$

Furthermore, if we take the 0th derivative, we get $MGF[0] = \sum_{x_i} P(X = x_i) = 1$, which reflects the fact that the sum over a probability distribution must equal one.

To see how amazing these results are, let us work with the moment-generating function of the binomial distribution, $MGF[z] = (1 - p + e^z p)^n$. The derivative of this with respect to z is $n\, e^z p\, (1 - p + e^z p)^{n-1}$, which reduces to $n\, p$ when we set z to zero. Similarly, if we were interested in $E[X^3]$, we could calculate the third derivative with respect to z, set z to zero and find $E[X^3] = n\, p\, (1 - 3\, p + 3\, n\, p + 2\, p^2 - 3\, n\, p^2 + n^2\, p^2)$. As tedious as this is, it is much easier than having to figure out the sum in (A5.1).

Moment generating functions are extremely useful for showing how different distributions are related to one another. For example, we said in Primer 3 that the binomial distribution converges upon a Poisson distribution when p is small and n is large. In this case, the mean and variance of the binomial distribution are very nearly $\mu = n\, p$. This shows that the first two moments converge, but what about the higher moments? These also converge, as we can show by proving that the $MGF[z]$ of the binomial converges upon that of the Poisson distribution. The first step in the proof is to use the mean of the binomial, $\mu = n\, p$, to replace p with μ/n in the moment-generating function of the binomial:

$$MGF[z] = \left(1 - \frac{\mu}{n} + e^z \frac{\mu}{n}\right)^n.$$

In Box 7.4, we mentioned a remarkable relationship involving e, that $\lim_{n\to\infty} (1 + x/n)^n = e^x$. We can use this relationship to see how the $MGF[z]$ of the binomial changes as n gets large. By equating x to $\mu(e^z - 1)$, the $MGF[z]$ of the binomial converges upon $\lim_{n\to\infty}(1 + \mu(e^z - 1)/n)^n = e^{\mu\,(e^z-1)}$. But $e^{\mu\,(e^z-1)}$ is the moment-generating function for the Poisson distribution. This proof, which treats μ as a constant, demonstrates that all of the moments of the binomial converge upon the moments of the Poisson as n gets large if we hold the mean constant at μ.

Often, we are not so interested in the jth moment, $E[X^j]$, but rather in a related quantity, $E[(X - \mu)^j]$, known as the jth *central moment*. For example, the variance of a distribution is the second central moment (Definition P3.3).

To calculate central moments, we can just multiply the moment generating function by $e^{-z\mu}$. This gives us the *central moment generating function*, which can be written as

$$CMGF[z] = \sum_{x_i} e^{z\,(x_i-\mu)} P(X = x_i). \tag{A5.6}$$

Taking the *j*th derivative of (A5.6) with respect to the dummy variable z and setting z to zero, gives the *j*th central moment of the distribution:

$$\left.\frac{d^j(CMGF[z])}{dz^j}\right|_{z=0} = \sum_{x_i} (x_i - \mu)^j\, P(X = x_i) = E[(X - \mu)^j]. \tag{A5.7}$$

For example, multiplying the moment generating function for the binomial distribution by $e^{-z\mu}$ where μ is $n\,p$, we get the central moment-generating function $CMGF[z] = e^{-znp}(1 - p + e^z p)^n$. Taking the first derivative of this with respect to z gives $e^{-znp} n\, e^z p(1 - p + e^z p)^{n-1} - n\,p\, e^{-znp}(1 - p + e^z p)^n$, which reduces to 0 when we set z to zero. Of greater interest, we can apply this procedure to get the third central moment for the binomial distribution from the third derivative of $CMGF[z]$. Doing so, we find that $E[(X - \mu)^3] = n\,p(1 - p)\,(1 - 2p)$. This tells us that the binomial distribution is skewed unless $p = 1/2$.

The same principles apply to continuous distributions, but the above sums become integrals. Specifically, the moment generating function for a continuous distribution is defined as

$$MGF[z] = \int_a^b e^{zx} f(x)\, dx. \tag{A5.8}$$

Again, the integral in (A5.8) is evaluated over the entire range of values that the random variable can take (a,b). The central moment generating function is obtained by multiplying the moment generating function by $e^{-z\mu}$, if $MGF[z]$ is already known, or by direct calculation:

$$CMGF[z] = \int_a^b e^{z(x-\mu)} f(x)\, dx. \tag{A5.9}$$

Table P3.3 provides the moment generating functions for most of the continuous distributions discussed in Primer 3. As an example, the moment generating function for the normal distribution is defined as $MGF[z] = \int_{-\infty}^{\infty} e^{zx}\left(e^{-(x-\mu)^2/(2\sigma^2)}/\sqrt{2\pi\sigma^2}\right)dx$, which equals $MGF[z] = e^{z\mu + z^2\sigma^2/2}$. Because $CMGF[z] = e^{-z\mu}\, MGF[z]$, the central moment generating function for the normal distribution is even simpler: $CMGF[z] = e^{z^2\sigma^2/2}$. Using the central moment generating function, it becomes easy to calculate the first four central moments for the normal distribution:

$$E[(X - \mu)] = \left.\frac{d e^{z^2\sigma^2/2}}{dz}\right|_{z=0} = 0. \tag{A5.10a}$$

$$E[(X - \mu)^2] = \left.\frac{d^2 e^{z^2\sigma^2/2}}{dz^2}\right|_{z=0} = \sigma^2. \tag{A5.10b}$$

$$E[(X - \mu)^3] = \left. \frac{d^3 e^{z^2\sigma^2/2}}{dz^3} \right|_{z=0} = 0 \quad . \tag{A5.10c}$$

$$E[(X - \mu)^4] = \left. \frac{d^4 e^{z^2\sigma^2/2}}{dz^4} \right|_{z=0} = 3\sigma^4. \tag{A5.10d}$$

We can then conclude that the normal distribution has no skew, while it has a kurtosis equal to $E[(X - \mu)^4]/\sigma^4 = 3$. Only derivatives are needed in these calculations, allowing us to avoid any further integrals once the moment generating function has been calculated.

Before leaving the topic, one weakness and one strength of moment generating functions deserve mention. A weakness of moment generating functions is that they do not always exist, because e^{zx} can grow so fast when multiplied by $f(x)$ that the sum in (A5.2) (or the integral in (A5.8)) is infinite. This problem can be circumvented using the *characteristic function* of a distribution, which multiplies $f(x)$ by e^{izx} instead of e^{zx}. The characteristic function always converges and moments can be determined in a similar fashion, but the introduction of complex numbers is unnecessary for most problems. One powerful advantage of moment generating functions is that they can be combined and manipulated quickly to demonstrate important facts about probability distributions:

Rule A5.1: Properties of Moment-Generating Functions

(a) If the random variable X has moment generating function $MGF[z]$, then $Y = aX + b$ has the moment generating function $e^{zb}\, MGF[az]$.

(b) If the random variables X_i ($i = 1, \ldots, n$) have moment generating functions $MGF_i[z]$ and are independent, then their sum $\sum_{i=1}^{n} X_i$ has a moment generating function given by the product $\prod_{i=1}^{n} MGF_i[z]$.

(c) If the X_i are independent and identically distributed with moment generating function $MGF[z]$, the sum of a random number, N, of the X_i has moment generating function $MGF_N[\log(MGF[z])]$, where $MGF_N[z]$ is the moment generating function of N.

Rule A5.1 applies equally to central moment-generating functions.

Rule A5.1a can be used to calculate the effect of scaling and shifting a probability distribution on moments like the mean and the variance. For example, the binomial distribution is often used to describe how genetic drift affects the number of A alleles, in a population of N alleles, when p is the frequency of allele A before drift (see Chapters 14 and 15). Because the *number* of A alleles after genetic drift (X) follows a binomial distribution with $CMGF[z] = e^{-zNp}(1 - p + e^{z}p)^N$, Rule A5.1a tells us that the *frequency* of allele A after genetic drift ($Y = X/N$) follows a distribution with $CMGF[z] = e^{-zp}(1 - p + e^{z/N}p)^N$. By differentiating $CMGF[z]$ with respect to z and setting z to zero, the first three central moments of the frequency of allele A after drift are 0, $p\,(1 - p)/N$, and

$p(1 - p)(1 - 2p)/N^2$. Consequently, there is no expected change in the allele frequency (from the first moment), but the variance in allele frequency is proportional to $1/N$ (from the second moment) and is thus smaller in larger populations. Higher order moments, including the third central moment, are proportional to $1/N^2$, which is much smaller.

Rule A5.1b allows you to calculate moments for compound probability distributions, involving more than one random event. For example, if the trait of an individual reflects the sum effect of a genetic influence, summarized by a normal distribution with mean μ_1 and variance σ_1^2, and an environmental influence, summarized by a normal distribution with mean μ_2 and variance σ_2^2, the moment generating function for the trait will be the product of the moment generating functions for the genetic and environmental influences:

$$\begin{aligned} MGF[z] &= e^{z\,\mu_1 + z^2 \sigma_1{}^2/2}\, e^{z\,\mu_2 + z^2 \sigma_2{}^2/2} \\ &= e^{z(\mu_1 + \mu_2) + z^2(\sigma_1{}^2 + \sigma_2{}^2)/2}. \end{aligned} \tag{A5.11}$$

Because (A5.11) is the moment-generating function of a normal distribution with mean $(\mu_1 + \mu_2)$ and variance $(\sigma_1^2 + \sigma_2^2)$, the trait will remain normally distributed, even though it is influenced by multiple factors. This is a relatively painless way to prove that the sum of normally distributed random variables is itself normally distributed.

Exercise A5.1:

(a) Use equation (A5.8) to prove that the moment generating function of a uniform distribution is $MGF[z] = (e^{\max z} - e^{\min z})/((\max - \min)\, z)$.

(b) Calculate the mean of the uniform distribution from $E[X] = (d(MGF[z])/dz)|_{z=0}$. [Note: You will need to use L'Hôpital's Rule A2.30 to calculate the limit as z goes to zero.]

Exercise A5.2:

(a) Use the moment-generating function of the Poisson to show that the sum of two Poisson distribution with means μ_1 and μ_2 also follows a Poisson distribution.

(b) If the rate of offspring production is 0.1 per year during the first five years of reproductive life and then becomes 0.2 per year during the next five years, what are the mean and variance of the sum total number of offspring born per parent over the ten years?

References

Stuart, A., and J. K. Ord. 1987. *Kendall's Advanced Theory of Statistics, Vol. 1. Distribution Theory*. Oxford University Press, Oxford.

Answers to Exercises

Exercise A5.1

(a) Using the probability density function for a uniform distribution, $f(x) = 1/(\max - \min)$, equation (A5.8) becomes

$$MGF[z] = \int_{\min}^{\max} \frac{e^{zx}}{\max - \min} dx = \left. \frac{e^{zx}}{(\max - \min)\, z} \right|_{\min}^{\max},$$

which correctly evaluates to the moment-generating function of the uniform distribution, $MGF[z] = (e^{\max z} - e^{\min z})/((\max - \min)\, z)$.

(b) For a uniform distribution, $E[X] = (\mathrm{d}(MGF[z])/\mathrm{d}z)|_{z=0}$ equals

$$\frac{1}{(\max - \min)} \left. \frac{(\max e^{\max z} - \min e^{\min z})\, z - (e^{\max z} - e^{\min z})}{z^2} \right|_{z=0}.$$

Here, we run into a problem, because both the numerator and the denominator approach zero as z goes to zero. Fortunately, we can use L'Hôpital's rule (A2.30) to evaluate this function, allowing us to rewrite

$$\frac{1}{(\max - \min)} \lim_{z \to 0} \frac{(\max e^{\max z} - \min e^{\min z})\, z - (e^{\max z} - e^{\min z})}{z^2}$$

as

$$\frac{1}{(\max - \min)} \lim_{z \to 0} \frac{(\max^2 e^{\max z} - \min^2 e^{\min z})\, z}{2z}.$$

Canceling out the z in the numerator and denominator and taking the limit, we get the mean of the uniform distribution: $(\max^2 - \min^2)/(2\,(\max - \min)) = (\max + \min)/2$. While it is much easier to calculate the mean for the uniform distribution directly (see Primer 3), the same cannot be said for other distributions. Plus, this example illustrates what to do in cases where both the numerator and denominator equal zero in a moment generating function or its derivatives.

Exercise A5.2

(a) According to rule A5.1b, the moment-generating function for the sum of two Poisson distributions is given by $MGF[z] = e^{\mu_1 (e^z - 1)} e^{\mu_2 (e^z - 1)}$, which equals $MGF[z] = e^{(\mu_1 + \mu_2)(e^z - 1)}$. This moment-generating function is the moment generating function of a Poisson distribution with mean $(\mu_1 + \mu_2)$.

(b) The number of offspring during the first five years is Poisson distributed with mean 0.5 (= 0.1 × 5), and the number of offspring during the next five years is Poisson distributed with mean 1.0 (= 0.2 × 5). The sum total number of offspring is thus Poisson distributed with mean 1.5 offspring. Because the variance of the Poisson is equal to the mean, the variance is also 1.5. In this example, the rate of events varied over time, but in a known fashion. In this case, the total number of events remained Poisson distributed with mean given by the expected number of events during the whole time interval, $\left(0.1\frac{5}{10} + 0.2\frac{5}{10}\right) \times 10 = 1.5$. If, however, the expected number of events was itself a random variable (e.g., was gamma distributed with mean 1.5), then the process would no longer be Poisson (see Stuart and Ord 1987, p. 182).

Index of Definitions, Recipes, and Rules

Definitions

Recipes

Rules

General Index

When there are multiple page entries for a listing, and when one of these is the primary reference, it is listed first in italics.